A GENERAL RELATIVITY WORKBOOK

A GENERAL RELATIVITY WORKBOOK

Thomas A. Moore
Pomona College

UNIVERSITY SCIENCE BOOKS
MILL VALLEY, CALIFORNIA

University Science Books
www.uscibooks.com

Production Manager: Paul Anagnostopoulos
Text Design: Yvonne Tsang
Cover Design: Genette Itoko McGrew
Manuscript Editor: Lee Young
Proofreader: Rick Camp
Illustrators: Thomas Moore, Laurel Muller
Compositor: Cohographics
Printer & Binder: P.A. Hutchinson Company

Copyright © 2013 by University Science Books
Mill Valley, California

Print Book ISBN 978-1-891389-82-5
eBook ISBN 978-1-938787-32-4

This book is printed on acid-free paper.

Reproduction or translation of any part of this work beyond that permitted by Section 107 or 108 of the 1976 United States Copyright Act without the permission of the copyright owner is unlawful. Requests for permission or further information should be addressed to the Permissions Department, University Science Books.

Library of Congress Cataloging-in-Publication Data

Moore, Thomas A. (Thomas Andrew)
 A general relativity workbook / Thomas A. Moore, Pomona College.
    pages   cm
 Includes index.
 ISBN 978-1-891389-82-5 (alk. paper)
 1. General relativity (Physics)   I. Title.
 QC173.6.M66   2012
 530.11—dc23
                                               2012025909

Printed in North America
10 9 8 7 6 5 4 3

*For Joyce, whose miraculous love always supports me and
allows me to take risks with life that I could not face alone,*

*and for Edwin Taylor, whose book with Wheeler set me on this path decades ago,
and whose gracious support and friendship has kept me going.*

ONLINE STUDENT MANUAL WITH HINTS AND ANSWERS FOR SELECTED PROBLEMS

To provide extra help for readers (especially those studying this textbook outside of the context of a formal course), I am providing a special online student help manual. The goal of this manual is to provide users in any setting some extra feedback on exercises and problems that will help keep them on track and reassure them that they are doing things correctly, without providing so much information that students taking a university course for credit could use this manual to cheat and thereby short-cut the learning process. This is a bit of a balancing act, so you will not find in this manual complete solutions to any exercises or homework problems. Rather, you will find many hints and/or short (and often partial) answers to most of the problems in the book, a few complete solutions to problems not in the book, and other guidance that should give all readers useful feedback and direction without enabling short cuts around the hard personal work that is an essential part of the learning process. I have deliberately put this information online, as opposed to in the textbook, partly to encourage all readers to make an honest effort before resorting to the manual. You can find the manual at

www.uscibooks.com/GRW_answers.htm

I hope and intend that professors using the textbook in formal classes will find that this manual provides useful, not harmful, feedback to their students. In almost all of the exercises and problems, short answers are not even remotely equivalent to the comprehensive solutions called for, and I don't provide answers in the handful of cases where a short answer essentially is the solution.

If someone studying the book on his or her own really gets stuck on a problem, I invite that person to send me a question via email at tmoore@pomona.edu. This will help me update the hints to better address the difficulties that readers actually have.

Thomas A. Moore

CONTENTS

Preface *xv*

1. **INTRODUCTION** 1
   Concept Summary *2*
   Homework Problems *9*
   General Relativity in a Nutshell *11*

2. **REVIEW OF SPECIAL RELATIVITY** 13
   Concept Summary *14*
   Box 2.1 Overlapping IRFs Move with Constant Relative Velocities *19*
   Box 2.2 Unit Conversions Between SI and GR Units *20*
   Box 2.3 One Derivation of the Lorentz Transformation *21*
   Box 2.4 Lorentz Transformations and Rotations *25*
   Box 2.5 Frame-Independence of the Spacetime Interval *26*
   Box 2.6 Frame-Dependence of the Time Order of Events *26*
   Box 2.7 Proper Time Along a Path *27*
   Box 2.8 Length Contraction *27*
   Box 2.9 The Einstein Velocity Transformation *28*
   Homework Problems *29*

3. **FOUR-VECTORS** 31
   Concept Summary *32*
   Box 3.1 The Frame-Independence of the Scalar Product *36*
   Box 3.2 The Invariant Magnitude of the Four-Velocity *36*
   Box 3.3 The Low-Velocity Limit of u *37*
   Box 3.4 Conservation of Momentum or Four-momentum? *38*
   Box 3.5 Example: The GZK Cosmic-Ray Energy Cutoff *40*
   Homework Problems *42*

4. **INDEX NOTATION** 43
   Concept Summary *44*
   Box 4.1 Behavior of the Kronecker Delta *48*
   Box 4.2 EM Field Units in the GR Unit System *48*
   Box 4.3 Electromagnetic Equations in Index Notation *49*
   Box 4.4 Identifying Free and Bound Indices *50*
   Box 4.5 Rule Violations *50*
   Box 4.6 Example Derivations *51*
   Homework Problems *52*

5. ARBITRARY COORDINATES 53

Concept Summary *54*
Box 5.1 The Polar Coordinate Basis *58*
Box 5.2 Proof of the Metric Transformation Law *59*
Box 5.3 A 2D Example: Parabolic Coordinates *60*
Box 5.4 The LTEs as an Example General Transformation *62*
Box 5.5 The Metric Transformation Law in Flat Space *62*
Box 5.6 A Metric for a Sphere *63*
Homework Problems *63*

6. TENSOR EQUATIONS 65

Concept Summary *66*
Box 6.1 Example Gradient Covectors *70*
Box 6.2 Lowering Indices *71*
Box 6.3 The Inverse Metric *72*
Box 6.4 The Kronecker Delta Is a Tensor *73*
Box 6.5 Tensor Operations *73*
Homework Problems *75*

7. MAXWELL'S EQUATIONS 77

Concept Summary *78*
Box 7.1 Gauss's Law in Integral and Differential Form *82*
Box 7.2 The Derivative of m^2 *83*
Box 7.3 Raising and Lowering Indices in Cartesian Coordinates *83*
Box 7.4 The Tensor Equation for Conservation of Charge *84*
Box 7.5 The Antisymmetry of **F** Implies Charge Conservation *85*
Box 7.6 The Magnetic Potential *86*
Box 7.7 Proof of the Source-Free Maxwell Equations *87*
Homework Problems *88*

8. GEODESICS 89

Concept Summary *90*
Box 8.1 The Worldline of Longest Proper Time in Flat Spacetime *93*
Box 8.2 Derivation of the Euler-Lagrange Equation *94*
Box 8.3 Deriving the Second Form of the Geodesic Equation *95*
Box 8.4 Geodesics for Flat Space in Parabolic Coordinates *96*
Box 8.5 Geodesics for the Surface of a Sphere *98*
Box 8.6 The Geodesic Equation Does Not Determine the Scale of τ *100*
Box 8.7 Light Geodesics in Flat Spacetime *101*
Homework Problems *102*

9. THE SCHWARZSCHILD METRIC 105

Concept Summary *106*
Box 9.1 Radial Distance *110*
Box 9.2 Falling from Rest in Schwarzschild Spacetime *111*
Box 9.3 *GM* for the Earth and the Sun *112*
Box 9.4 The Gravitational Redshift for Weak Fields *112*
Homework Problems *114*

10. PARTICLE ORBITS   115

Concept Summary   *116*

Box 10.1   Schwarzschild Orbits Must Be Planar   *120*
Box 10.2   The Schwarzschild "Conservation of Energy" Equation   *121*
Box 10.3   Deriving Conservation of Newtonian Energy for Orbits   *122*
Box 10.4   The Radii of Circular Orbits   *122*
Box 10.5   Kepler's Third Law   *124*
Box 10.6   The Innermost Stable Circular Orbit (ISCO)   *125*
Box 10.7   The Energy Radiated by an Inspiraling Particle   *126*
Homework Problems   *127*

11. PRECESSION OF THE PERIHELION   129

Concept Summary   *130*

Box 11.1   Verifying the Orbital Equation for $u(\phi)$ *135*
Box 11.2   Verifying the Newtonian Orbital Equation   *135*
Box 11.3   Verifying the Equation for the Orbital "Wobble"   *136*
Box 11.4   Application to Mercury   *136*
Box 11.5   Constructing the Schwarzschild Embedding Diagram   *137*
Box 11.6   Calculating the Wedge Angle δ *138*
Box 11.7   A Computer Model for Schwarzschild Orbits   *138*
Homework Problems   *141*

12. PHOTON ORBITS   143

Concept Summary   *144*

Box 12.1   The Meaning of the Impact Parameter b *148*
Box 12.2   Derivation of the Equation of Motion for a Photon   *148*
Box 12.3   Features of the Effective Potential Energy Function for Light   *149*
Box 12.4   Photon Motion in Flat Space   *149*
Box 12.5   Evaluating 4-Vector Components in an Observer's Frame   *150*
Box 12.6   An Orthonormal Basis in Schwarzschild Coordinates   *150*
Box 12.7   Derivation of the Critical Angle for Photon Emission   *151*
Homework Problems   *152*

13. DEFLECTION OF LIGHT   153

Concept Summary   *154*

Box 13.1   Checking Equation 13.2   *159*
Box 13.2   The Differential Equation for the Shape of a Photon Orbit   *160*
Box 13.3   The Differential Equation for the Photon "Wobble"   *160*
Box 13.4   The Solution for $u(\phi)$ in the Large-r Limit   *161*
Box 13.5   The Maximum Angle of Light Deflection by the Sun   *161*
Box 13.6   The Lens Equation   *162*
Box 13.7   The Ratio of Image Brightness to the Source Brightness   *163*
Homework Problems   *164*

14. EVENT HORIZON   167

Concept Summary   *168*

Box 14.1   Finite Distance to $r = 2GM$ *172*
Box 14.2   Proper Time for Free Fall from $r = R$ to $r = 0$ *174*

Box 14.3   The Future Is Finite Inside the Event Horizon   *175*
Homework Problems   *176*

15. ALTERNATIVE COORDINATES   179

Concept Summary   *180*
Box 15.1   Calculating $\partial \mathring{t}/\partial r$ *184*
Box 15.2   The Global Rain Metric   *185*
Box 15.3   The Limits on $dr/d\mathring{t}$ Inside the Event Horizon   *185*
Box 15.4   Transforming to Kruskal-Szekeres Coordinates   *186*
Homework Problems   *188*

16. BLACK HOLE THERMODYNAMICS   189

Concept Summary   *190*
Box 16.1   Free-Fall Time to the Event Horizon from $r = 2GM + \varepsilon$ *194*
Box 16.2   Calculating E_∞ *195*
Box 16.3   Evaluating k_B, $\hbar$, and T for a Solar-Mass Black Hole   *196*
Box 16.4   Lifetime of a Black Hole   *197*
Homework Problems   *198*

17. THE ABSOLUTE GRADIENT   199

Concept Summary   *200*
Box 17.1   Absolute Gradient of a Vector   *204*
Box 17.2   Absolute Gradient of a Covector   *204*
Box 17.3   Symmetry of the Christoffel Symbols   *205*
Box 17.4   The Christoffel Symbols in Terms of the Metric   *205*
Box 17.5   Checking the Geodesic Equation   *206*
Box 17.6   A Trick for Calculating Christoffel Symbols   *206*
Box 17.7   The Local Flatness Theorem   *207*
Homework Problems   *210*

18. GEODESIC DEVIATION   211

Concept Summary   *212*
Box 18.1   Newtonian Tidal Deviation Near a Spherical Object   *216*
Box 18.2   Proving Equation 18.9   *217*
Box 18.3   The Absolute Derivative of **n**   *217*
Box 18.4   Proving Equation 18.14   *218*
Box 18.5   An Example of Calculating the Riemann Tensor   *218*
Homework Problems   *220*

19. THE RIEMANN TENSOR   221

Concept Summary   *222*
Box 19.1   The Riemann Tensor in a Locally Inertial Frame   *224*
Box 19.2   Symmetries of the Riemann Tensor   *225*
Box 19.3   Counting the Riemann Tensor's Independent Components   *226*
Box 19.4   The Bianchi Identity   *227*
Box 19.5   The Ricci Tensor Is Symmetric   *228*
Box 19.6   The Riemann and Ricci Tensors and R for a Sphere   *228*
Homework Problems   *230*

20. THE STRESS-ENERGY TENSOR 231

Concept Summary *232*

Box 20.1 Why the Source of Gravity Must Be Energy, Not Mass *236*
Box 20.2 Interpretation of T^{ij} in a Locally Inertial Frame *236*
Box 20.3 The Stress-Energy Tensor for a Perfect Fluid in Its Rest LIF *237*
Box 20.4 Equation 20.16 Reduces to Equation 20.15 *239*
Box 20.5 Fluid Dynamics from Conservation of Four-Momentum *239*
Homework Problems *241*

21. THE EINSTEIN EQUATION 243

Concept Summary *244*

Box 21.1 The Divergence of the Ricci Tensor *248*
Box 21.2 Finding the Value of b *249*
Box 21.3 Showing that $-R + 4\Lambda = \kappa T$ *250*
Homework Problems *251*

22. INTERPRETING THE EQUATION 253

Concept Summary *254*

Box 22.1 Conservation of Four-Momentum Implies $0 = \nabla_\nu(\rho_0 u^\nu)$ *258*
Box 22.2 The Inverse Metric in the Weak-Field Limit *258*
Box 22.3 The Riemann Tensor in the Weak-Field Limit *259*
Box 22.4 The Ricci Tensor in the Weak-Field Limit *260*
Box 22.5 The Stress-Energy Sources of the Metric Perturbation *261*
Box 22.6 The Geodesic Equation for a Slow Particle in a Weak Field *262*
Homework Problems *263*

23. THE SCHWARZSCHILD SOLUTION 265

Concept Summary *266*

Box 23.1 Diagonalizing the Spherically Symmetric Metric *270*
Box 23.2 The Components of the Ricci Tensor *271*
Box 23.3 Solving for B *274*
Box 23.4 Solving for $a(r)$ *275*
Box 23.5 The Christoffel Symbols with t-t as Subscripts *275*
Homework Problems *276*

24. THE UNIVERSE OBSERVED 279

Concept Summary *280*

Box 24.1 Measuring Astronomical Distances in the Solar System *284*
Box 24.2 Determining the Distance to Stellar Clusters *286*
Box 24.3 How the Doppler Shift Is Connected to Radial Speed *287*
Box 24.4 Values of the Hubble Constant *288*
Box 24.5 Every Point Is the Expansion's "Center" *288*
Box 24.6 The Evidence for Dark Matter *289*
Homework Problems *290*

25. A METRIC FOR THE COSMOS 293

Concept Summary *294*

Box 25.1 The Universal Ricci Tensor *298*

Box 25.2   Raising One Index of the Universal Ricci Tensor   *298*
Box 25.3   The Stress-Energy Tensor with One Index Lowered   *298*
Box 25.4   The Einstein Equation with One Index Lowered   *301*
Box 25.5   Verifying the Solutions for q *302*
Homework Problems   *303*

26. EVOLUTION OF THE UNIVERSE   305

Concept Summary   *306*
Box 26.1   The Other Components of the Einstein Equation   *310*
Box 26.2   Consequences of Local Energy/Momentum Conservation   *311*
Box 26.3   Deriving the Density/Scale Relationship for Radiation   *312*
Box 26.4   Deriving the Friedman Equation   *312*
Box 26.5   The Friedman Equation for the Present Time   *313*
Box 26.6   Deriving the Friedman Equation in Terms of the Omegas   *313*
Box 26.7   The Behavior of a Matter-Dominated Universe   *314*
Homework Problems   *315*

27. COSMIC IMPLICATIONS   317

Concept Summary   *318*
Box 27.1   Connecting the Redshift z to the Hubble Constant   *322*
Box 27.2   Deriving the Hubble Relation in Terms of Redshift z *322*
Box 27.3   The Luminosity Distance   *323*
Box 27.4   The Differential Equation for $a(\eta)$ *323*
Box 27.5   How to Generate a Numerical Solution for Equation 27.18   *324*
Homework Problems   *325*

28. THE EARLY UNIVERSE   327

Concept Summary   *328*
Box 28.1   Single-Component Universes   *332*
Box 28.2   The Transition to Matter Dominance   *333*
Box 28.3   The Time-Temperature Relation   *333*
Box 28.4   Neutrino Decoupling   *335*
Box 28.5   The Number Density of Photons   *337*
Homework Problems   *338*

29. CMB FLUCTUATIONS AND INFLATION   339

Concept Summary   *340*
Box 29.1   The Angular Width of the Largest CMB Fluctuations   *345*
Box 29.2   The Equation for $\Omega_k(t)$ *346*
Box 29.3   Cosmic Flatness at the End of Nucleosynthesis   *347*
Box 29.4   The Exponential Inflation Formula   *347*
Box 29.5   Inflation Calculations   *348*
Homework Problems   *349*

30. GAUGE FREEDOM   351

Concept Summary   *352*
Box 30.1   The Weak-Field Einstein Equation in Terms of $h_{\mu\nu}$ *355*
Box 30.2   The Trace-Reverse of $h_{\mu\nu}$ *356*

Box 30.3 The Weak-Field Einstein Equation in Terms of $H^{\mu\nu}$ *357*
Box 30.4 Gauge Transformations of the Metric Perturbations *358*
Box 30.5 A Gauge Transformation Does Not Change $R_{\alpha\beta\mu\nu}$ *359*
Box 30.6 Lorenz Gauge *360*
Box 30.7 Additional Gauge Freedom *361*
Homework Problems *361*

31. DETECTING GRAVITATIONAL WAVES 363

Concept Summary *364*
Box 31.1 Constraints on Our Trial Solution *368*
Box 31.2 The Transformation to Transverse-Traceless Gauge *369*
Box 31.3 A Particle at Rest Remains at Rest in TT Coordinates *371*
Box 31.4 The Effect of a Gravitational Wave on a Ring of Particles *372*
Homework Problems *373*

32. GRAVITATIONAL WAVE ENERGY 375

Concept Summary *376*
Box 32.1 The Ricci Tensor *379*
Box 32.2 The Averaged Curvature Scalar *379*
Box 32.3 The General Energy Density of a Gravitational Wave *379*
Homework Problems *382*

33. GENERATING GRAVITATIONAL WAVES 383

Concept Summary *384*
Box 33.1 $H^{t\mu}$ for a Compact Source Whose CM is at Rest *388*
Box 33.2 A Useful Identity *388*
Box 33.3 The Transverse-Traceless Components of $A^{\mu\nu}$ *390*
Box 33.4 How to Find $\ddot{\bar{H}}_{TT}^{jk}$ for Waves Moving in the $\vec{n}$ Direction *391*
Box 33.5 Flux in Terms of $\ddot{I}^{jk}$ *393*
Box 33.6 Evaluating the Integrals in the Power Calculation *394*
Homework Problems *395*

34. GRAVITATIONAL WAVE ASTRONOMY 397

Concept Summary *398*
Box 34.1 The Dumbbell $\ddot{I}^{jk}$ *402*
Box 34.2 The Power Radiated by a Rotating Dumbbell *403*
Box 34.3 The Total Energy of an Orbiting Binary Pair *404*
Box 34.4 The Time-Rate-of-Change of the Orbital Period *404*
Box 34.5 Characteristics of ι Boötis *405*
Homework Problems *406*

35. GRAVITOMAGNETISM 407

Concept Summary *408*
Box 35.1 The Lorenz Condition for the Potentials *412*
Box 35.2 The Maxwell Equations for the Gravitational Field *413*
Box 35.3 The Gravitational Lorentz Equation *414*
Box 35.4 The "Gravitomagnetic Moment" of a Spinning Object *414*
Box 35.5 Angular Speed of Gyroscope Precession *415*
Homework Problems *416*

36. THE KERR METRIC   417

Concept Summary   *418*
Box 36.1   Expanding $|\vec{R} - \vec{r}|^{-1}$ to First Order in r/R *421*
Box 36.2   The Integral for h^{tx} *422*
Box 36.3   Why the Other Terms in the Expansion Integrate to Zero   *423*
Box 36.4   Transforming the Weak-Field Solution to Polar Coordinates   *424*
Box 36.5   The Weak-Field Limit of the Kerr Metric   *425*
Homework Problems   *426*

37. PARTICLE ORBITS IN KERR SPACETIME   427

Concept Summary   *428*
Box 37.1   Calculating Expressions for $dt/d\tau$ and $d\phi/d\tau$ *431*
Box 37.2   Verify the Value of $[g_{t\phi}]^2 - g_{tt}g_{\phi\phi}$ *432*
Box 37.3   The "Energy-Conservation-Like" Equation of Motion   *433*
Box 37.4   Kepler's Third Law   *434*
Box 37.5   The Radii of ISCOs When $a = GM$ *435*
Homework Problems   *436*

38. ERGOREGION AND HORIZON   437

Concept Summary   *438*
Box 38.1   The Radii Where $g_{tt} = 0$ *441*
Box 38.2   The Angular Speed Range When dr and/or $d\theta \neq 0$ *442*
Box 38.3   Angular-Speed Limits in the Equatorial Plane   *443*
Box 38.4   The Metric of the Event Horizon's Surface   *444*
Box 38.5   The Area of the Outer Kerr Event Horizon   *445*
Box 38.6   Transformations Preserve the Metric Determinant's Sign   *445*
Homework Problems   *447*

39. NEGATIVE-ENERGY ORBITS   449

Concept Summary   *450*
Box 39.1   Quadratic Form for Conservation of Energy   *454*
Box 39.2   The Square Root Is Zero at the Event Horizon   *454*
Box 39.3   Negative e Is Possible Only in the Ergoregion   *456*
Box 39.4   The Fundamental Limit on δM in Terms of δS *457*
Box 39.5 $\delta M_{ir} \geq 0$ *458*
Box 39.6   The Spin Energy Contribution to a Black Hole's Mass   *459*
Homework Problems   *460*

Appendix: A Diagonal Metric Worksheet   *463*
Index   *467*

PREFACE

Introductory Comments. General relativity is one of the greatest triumphs of the human mind. Together with quantum field theory, general relativity lies at the foundation of contemporary physics and currently represents the most durable physical theory in existence, having survived nearly a century of development and increasingly rigorous testing without being contradicted or superseded. Long admired for its elegant beauty, general relativity has also (particularly in the past two decades) become an essential tool for working physicists. It provides the basis for understanding a huge variety of astrophysical phenomena ranging from active galactic nuclei, quasars, and pulsars to the formation, characteristics, and destiny of the universe itself. It has driven the development of new experimental tools for testing the theory and for the detection of gravitational waves that represent one of the most lively and challenging areas of contemporary physics. Even engineers are starting to have to pay attention to general relativity: making the Global Positioning System function correctly requires careful attention to general relativistic effects.

In some ways, general relativity was so far ahead of its time that it took a long time for instrumentation and applications to catch up sufficiently to make it more than than an intellectual adventure for the curious. However, as general relativity has now moved firmly into the mainstream of contemporary physics with a wide and growing variety of applications, teaching general relativity to undergraduate physics majors has become both relevant and important, and the need for appropriate and up-to-date undergraduate-level textbooks has become urgent.

Audience. This textbook seeks to support a one-semester introduction to general relativity for junior and/or senior undergraduates. It assumes only that students have taken multivariable calculus and some intermediate Newtonian mechanics beyond a standard treatment of mechanics and electricity and magnetism at the introductory level (though students who have also taken linear algebra, differential equations, some electrodynamics, and/or some special relativity will be able to move through the book more quickly and easily). This book has grown out of my experience teaching fourteen iterations of such an undergraduate course during my teaching career.

Those iterations have convinced me not only that undergraduates *can* develop a solid proficiency with the general relativity, but also that studying general relativity provides a superb introduction to the best practices of theoretical physics as well as a uniquely exciting and engaging introduction to ideas at the very frontier of physics, things that students rarely experience in other undergraduate courses.

Pedagogical Principles. Since students rarely see the tensor calculus used in general relativity in undergraduate mathematics courses, a course in general relativity must either teach this mathematics from scratch or seek to work around it (at some cost in coherence and depth of insight). In my experience, junior and senior undergraduates *can* master tensor calculus in an appropriately designed course, and that doing this is well worth the effort, as it provides the firm foundation needed for confidence and flexibility in confronting applications.

The pedagogical key for developing this mastery is for you (the student) to *personally own* the mathematics by working through most of the arguments and derivations *yourself.* Therefore, I have designed this textbook as a *workbook*. Each chapter opens with a concise core-concept presentation that helps you see the big picture without mathematical distraction. This presentation is keyed to subsequent "boxes" that I have designed to guide you in working through the supporting derivations as well as other details and applications whose direct presentation would obscure the core ideas. I have found this combination of overview and guided effort to be uniquely effective in building a practical understanding of the theory's core concepts and their mathematical foundations.

The overview-and-box design also helps keep you focused on the *physics* as opposed to the mathematics, underlining how the mathematics *supports* and *expresses* the physics. Other aspects of the textbook's design also support the principle that the physics should be foremost. I have ordered the topics so that the mathematics is presented not in one big lump but rather gradually and "as needed," thus allowing the physics to drive the presentation. For example, you will extensively practice using tensor notation by exploring real physical applications in flat space before learning about the geodesic equation that describes an object's motion in a curved spacetime. You will then spend a great deal of time exploring the physical implications of the geodesic equation in the particular curved spacetime surrounding a simple spherical object before learning the additional mathematical tools required to show *why* spacetime is curved in that particular way around a spherical object. Along the way, I use many "toy" examples in two-dimensional flat and curved spaces help develop your intuitive understanding of the physical meaning of the core ideas. The gradual development of the mathematics throughout the text also helps ensure that you have time to gain a firm footing for each step before continuing the climb.

The key to using this book successfully is working carefully through all the boxes in this book. Doing this will ultimately provide you with a range of experience and depth of understanding difficult to obtain any other way.

Chapter Dependencies. The chart that appears on each chapter's title page (and on the next page) shows how the major sections of the book depend on each other. For example, you can see from the chart that the **Introduction**, **Flat Space**, and **Tensors** sections (chapters 1 through 8) provide core material that every other section uses. After chapter 8, I strongly recommend going on to the **Schwarzschild Black Holes** section, because this will develop your understanding of how to work with curved spacetimes before having to wrestle with yet more math (and because black holes are fascinating applications of the theory). However, this is not essential; in a short course focused on cosmology, for example, one could go directly on to the **Calculus of Curvature**, **Einstein Equation**, and **Cosmology** sections. Note also that the final three sections (**Cosmology**, **Gravitational Waves**, and **Spinning Black Holes**) are completely independent of each other and can be explored in any order one might choose. However, all three of these sections require the **Calculus of Curvature** and **Einstein Equation** sections.

One also does not have to go all the way through the Schwarzschild section. The last three chapters (on black holes) are only necessary if you also plan to go through the last two chapters of the **Spinning Black Hole**s section (though it is hard to imagine why anyone would want to avoid learning about black holes!). One can easily omit the *Deflection of Light* chapter without loss of continuity. The *Precession of the Perihelion* chapter is necessary background for the *Deflection of Light* chapter, but you could omit both. The first two chapters are required for all of the other chapters in this section, and the fourth chapter on *Photon Orbits* presents a mathematical technique that is employed in certain homework problems throughout the rest of the book, but it is only absolutely required for the *Deflection of Light* and the *Black Hole Thermodynamics* chapters.

In the **Cosmology** section, the first four chapters provide core material and should all be included if this section is to be explored at all. The last two chapters, however, are completely optional; you can omit either both or the last, as desired.

INTRODUCTION

FLAT SPACETIME
- Review of Special Relativity
- Four-Vectors
- Index Notation

TENSORS
- Arbitrary Coordinates
- Tensor Equations
- Maxwell's Equations
- Geodesics

SCHWARZSCHILD BLACK HOLES
- The Schwarzschild Metric
- Particle Orbits
- Precession of the Perihelion*
- Photon Orbits
- Deflection of Light*
- Event Horizon*
- Alternative Coordinates*
- Black Hole Thermodynamics*

THE CALCULUS OF CURVATURE
- The Absolute Gradient
- Geodesic Deviation
- The Riemann Tensor

THE EINSTEIN EQUATION
- The Stress-Energy Tensor
- The Einstein Equation
- Interpreting the Equation
- The Schwarzschild Solution

COSMOLOGY
- The Universe Observed
- A Metric for the Cosmos
- Evolution of the Universe
- Cosmic Implications
- The Early Universe*
- CMB Fluctuations & Inflation*

GRAVITATIONAL WAVES
- Gauge Freedom
- Detecting Gravitational Waves
- Gravitational Wave Energy*
- Generating Gravitational Waves*
- Gravitational Wave Astronomy*

SPINNING BLACK HOLES
- Gravitomagnetism*
- The Kerr Metric*
- Kerr Particle Orbits*
- Ergoregion and Horizon*
- Negative-Energy Orbits*

this → depends on → this

A chart showing the chapters of the book grouped in their major sections and how those sections depend on each other. Chapters marked with a * are optional, though later optional chapters typically depend on earlier such chapters.

While in principle it is possible to stop after the first two chapters in the **Gravitational Wave** section, I think that a discussion of gravitational wave energy and generation is pretty important. I therefore recommend going through at least the first three chapters of this section if you want to explore gravitational waves at all.

One might reasonably elect to explore the *Gravitomagnetism* chapter alone in the **Spinning Black Holes** section, or stop after either the *Kerr Particle Orbits* chapter or the *Ergoregion and Horizon* chapter. However, the chapters in this section *do* need to be discussed in sequence; one cannot easily drop one from the middle.

The First Chapter. Please also note that the *first* chapter has a different structure than the others. After dealing with preliminaries, I usually end the first class session of the course I teach with a 40-minute interactive lecture. For the sake of completeness (and for later reference), I have provided in the first chapter what amounts to a polished transcript of that lecture. This chapter has no boxes because I don't expect my students to have read (or perhaps even own) the book before the first class. To help them track the lecture, I instead give them the two-sided handout that appears as the last two pages of the first chapter.

The Second Chapter. This chapter presents a very terse review of special relativity aimed primarily at students that have already encountered some relativity in a previous course. If you have not seen relativity before, you may find this chapter harder going. Even so, everything you need to know about special relativity for this book is presented there, and if you work through the chapter slowly, and do many of the homework problems, you should be fine. I have also included references to supplemental reading that you may find helpful.

Book Website. You can find a variety of other helpful information and supporting computer software on this textbook's website:

http://pages.pomona.edu/~tmoore/grw/

Please also feel free to email me suggestions, questions, and error notices: my email address is tmoore@pomona.edu.

Information for Instructors. So far in this preface, I have addressed issues of concern to all readers in language directed mostly to students. In the remainder, I want to specifically address issues of interest to instructors who are designing undergraduate courses around this book.

Course Pacing. I have designed the text so that (in my experience) *each chapter can generally be discussed in a single* (50-minute) *class session*, particularly if you use the format for class sessions I describe below. Your mileage may vary (for example, you may need to spend more time on chapter 2 if your students' background in special relativity is weak), but this general rule should help you appropriately pace the course.

You also have a lot flexibility in choosing which chapters to cover and which you might omit: there are at least twenty different chapter sequences that make sense. Be sure to examine thoroughly the section above on **Chapter Dependencies** before designing a syllabus that omits chapters. However, I find that I can usually get through the entire book in one semester.

Let me emphasize again that the last three sections (**Cosmology**, **Gravitational Waves**, and **Spinning Black Holes**) are independent; you can present them in any order. One of my colleagues likes to end the course with cosmology, which he thinks provides an exciting climax. I have made that section first of the three precisely because I *also* think it is the most important. If I am working through the book sequentially and run out of time, I'd rather do so in the Spinning Black Holes section than omit any of the cosmology material! I also find that students have many other pressures and concerns near the end of the semester, so I tend to schedule material that I consider *less* crucial toward the end. But you can certainly choose what works best for you and your students, and you have lots of flexibility to do so.

How to Spend Class Time. The workbook format will push students to gain mastery *only* if your course design somehow rewards students for filling out the boxes. The last time I taught the course, I asked several students chosen at random each class session to hand me their books, which I subsequently graded for thoughtful *effort* in filling out the boxes since the last time they submitted their book, with special emphasis on the chapter discussed in class that day. Each student's average grade for these random samples counted about 13% of their course grade. I arranged things so that each student was called on about five to six times a semester.

One of my colleagues at a another institution uses a different approach that may be even better. After determining which box exercises seemed easy enough to skip discussion, he then assigns each remaining box exercise to a student in a strict rotation (including himself in the rotation). The student must present the solution in front of the class. This strongly motivates the students to come to class prepared without having to assign a formal grade for preparation, and also makes class time a bit more active than the way I did it. I intend to use this approach myself the next time that I teach the course.

You might find some other approach better than either of these for your students, but I consider it very important when designing a course based on this book to find *some* way of rewarding students for doing work in the boxes before class.

In either of the approaches outlined above, we spend much of the class period discussing the challenges students encountered in going through the boxes. Because students have at least *tried* to work out the boxes before class, they typically bring good questions to the table, questions that directly address the difficulties they are experiencing personally. We are therefore able to spend class time efficiently addressing students' *actual needs*. If we have time (and we often do), I often work some example problems in class, targeted toward either some interesting physics and/or preparing them better to do the homework. In my experience, this approach to using class time is much more effective and efficient than lecturing would be.

I also recommend that you (the instructor) work through all the boxes in an assigned chapter *yourself* before class. (I myself do this every time I offer the course, even though I have worked through all the boxes several times now!) This will help refresh your memory, help isolate any issues that you might need to resolve for yourself before class, and (most importantly) help you anticipate and appreciate the difficulties that students will have with the boxes.

I intentionally designed most of the boxes so that they ask students to prove something, as the primary goal of the boxes is to help students gain ownership of the concepts and derivations discussed in the text. The homework problems are usually much more open-ended, providing opportunities for students to extend the ideas presented in the text, explore physical applications, and even think about new topics. Some of the problems are also designed to provide a basis for class discussion of topics not covered in the main text.

Homework. I typically assign about two homework problems per chapter: this is enough to keep students pretty busy. Homework problems for this class can be pretty challenging, and even the best students may not get them right the first time. Homework-grading schemes that focus *only* on the final results can therefore make students anxious. However, one can devise grading schemes that (1) allow students to engage difficult problems without anxiety, (2) provide them with an opportunity for further learning, and (3) make grading easier for you or your TAs. The "Course Design" section of the book's website provides a link to a page that discusses a scheme for grading homework that I strongly recommend that you consider: it not only encourages students to tackle tough problems without fearing failure, but I can also guarantee that it will save you time grading!

Resources for Instructors. Instructors adopting the text for classes should contact University Science Books at deskcopy@uscibooks.com for access to the instructor's manual.

I also welcome emails if you have questions, error notices, or other comments.

Appreciation. I am grateful to many people have helped bring this text to fruition. First, let me thank the students in my Physics 160 class (and particularly Nathan Reed and Ian Frank) for documenting errors in early versions and offering feedback. The overview/box idea grew out of conversations with Dayton Jones (more than thirty years ago). A fruitful correspondence with James Hartle helped me formulate my goals for this book, and his excellent text both taught and inspired me. I am grateful to Edwin Taylor, who has challenged me and broadened my perspective since I was a high school student, for personal support and insight. Thanks also to Tom Baumgarte and Ben Sugerman for trying initial versions of the book in their classes and offering thoughtful feedback. I am grateful to Tom Helliwell, Nandor Bokor, Nelson Christensen and his students (Tom Callister, Ross Cawthon, Andrew Chael, Micah Koller, Dustin Anderson, and David Miller, who all sent me individual reviews), Tom Baumgarte, Tom Carroll, Bryan van der Ende, and an unknown reviewer for reading a nearly final draft and offering a number of valuable suggestions and error corrections. Needless to say, any remaining errors are my own. I want to thank the developers of MathMagic (my equation-editing software) for extraordinary attention and help beyond the call of duty when I encountered various problems. Hilda Dinolfo and Christine Maynard were very helpful in printing copies for various early readers. I am very grateful to Sergio Picozzi and John Mallinckrodt for carefully reviewing the final draft and being willing to write such nice endorsements for the back cover. I want to thank the book's production team (Lee Young, Richard Camp, Yvonne Tsang, Genette Itako McGrew, and especially Laurel Muller and Paul Anagnostopoulos) for their excellent work, care, and extraordinary patience in dealing with a difficult book (and a sometimes difficult author). I also want to thank Jane Ellis, my managing editor at University Science Books, for her support, enthusiasm, and hard work in bringing this book to print, and for being willing to take a risk on a book that was a bit out of the ordinary.

Finally, let me thank my wife, Joyce, whose unfailing support for my writing habit is loving and gracious beyond the call of duty. I am very grateful to all!

Thomas A. Moore
Claremont, CA
July 11, 2012

1. INTRODUCTION

INTRODUCTION

FLAT SPACETIME
- Review of Special Relativity
- Four-Vectors
- Index Notation

TENSORS
- Arbitrary Coordinates
- Tensor Equations
- Maxwell's Equations
- Geodesics

SCHWARZSCHILD BLACK HOLES
- The Schwarzschild Metric
- Particle Orbits
- Precession of the Perihelion
- Photon Orbits
- Deflection of Light
- Event Horizon
- Alternative Coordinates
- Black Hole Thermodynamics

THE CALCULUS OF CURVATURE
- The Absolute Gradient
- Geodesic Deviation
- The Riemann Tensor

THE EINSTEIN EQUATION
- The Stress-Energy Tensor
- The Einstein Equation
- Interpreting the Equation
- The Schwarzschild Solution

COSMOLOGY
- The Universe Observed
- A Metric for the Cosmos
- Evolution of the Universe
- Cosmic Implications
- The Early Universe
- CMB Fluctuations & Inflation

GRAVITATIONAL WAVES
- Gauge Freedom
- Detecting Gravitational Waves
- Gravitational Wave Energy
- Generating Gravitational Waves
- Gravitational Wave Astronomy

SPINNING BLACK HOLES
- Gravitomagnetism
- The Kerr Metric
- Kerr Particle Orbits
- Ergoregion and Horizon
- Negative-Energy Orbits

this ⟶ depends on ⟶ this

1. INTRODUCTION

[**How to Use This Chapter.** This chapter has a different structure than the remaining chapters in this text (see the Preface). When I teach from this text, I do *not* have my students read this chapter; instead, I present the material appearing in this chapter in the form of a 40-minute lecture on the first day of class. To support that lecture, I give the students a handout that provides an outline of the ideas and also the main figures that appear in this chapter. (The handout appears at the end of this chapter and also is available on the text's website). I find this to be an efficient way to use time in the first session (since my students have done no reading in advance) and also an effective way to get them excited about the course.

I have therefore provided this chapter mostly for instructors (either to help them prepare for a similar lecture or to provide a reading assignment if they prefer to do other things with the first class session) and for those using this book for self-instruction. So that all users have a similar experience, I have structured this chapter more or less as a transcript of the opening lecture I give to my students, rather than using the workbook structure found in all of the remaining chapters.]

Introduction. General relativity, at its heart, is very simple. While it is true that its mathematics is at times challenging and its interpretation can be mind-bending, the theory's core concepts are straightforward, plausible, and easy to understand. This simplicity is the core of the theory's great elegance and beauty, and sets a high standard that other modern physical theories struggle to emulate.

In what follows, I will provide in a few pages a complete overview of the theory's conceptual structure. The entire rest of this book contains little more than details about and applications of these core ideas!

The Curious Equality of Gravitational and Inertial Mass. Consider first two particles with charges Q and q interacting electrostatically with each other but nothing else. Coulomb's law and Newton's second law then imply that

$$\frac{kQq}{r^2} = \left(\frac{kQ}{r^2}\right)q = F_e = m_I a \tag{1.1}$$

where k is Coulomb's constant, r is the particles' separation, F_e is the magnitude of the electrostatic force that Q exerts on q, a is the magnitude of q's acceleration, and m_I is that particle's inertial mass, expressing how strongly it resists being accelerated by a given force. We can (and typically do) interpret the quantity in parentheses as being the magnitude of the electric field $\vec{E}$ that the particle with charge Q creates at the other particle's position, and the quantity q as describing how strongly that other particle responds to or "couples" to that field.

Now consider two particles with masses M and m_G interacting gravitationally with each other but nothing else. Newton's law of universal gravitation and Newton's second law then imply that

$$\frac{GMm_G}{r^2} = \left(\frac{GM}{r^2}\right)m_G = F_g = m_I a \tag{1.2}$$

where G is the universal gravitational constant, F_g is the magnitude of the gravitational force that M exerts on m_G, and the other quantities are as before. Guided by equation 1.1, we interpret the quantity in parentheses as being the magnitude of the gravitational field $\vec{g}$ that the particle with mass M creates at the other particle's position and m_G as expressing how strongly that other particle couples with that field.

Now, I have put different subscripts on m_G and m_I to emphasize that, though they both describe properties of the same particle, conceptually these properties are quite different. The quantity m_I expresses the particle's **inertial mass**, that is, how strongly it resists being accelerated by a given force, while m_G expresses its **gravitational mass**, that is, how strongly it couples to a gravitational field. These are *completely distinct physical quantities expressing completely different concepts*. We should no more expect m_G to be linked with m_I than we would q to to be linked with m_I in equation 1.1.

So, canceling m_G and m_I on both sides of equation 1.2 (as we all have been trained to do) makes the *very* big assumption that *an object's inertial mass is the same as its gravitational mass*. Most physicists before Einstein simply *assumed* that this was true. But is it really? We now know that a gold nucleus, for example, has a smaller inertial mass (as measured in a mass spectrometer) than an equal number of protons and neutrons because of the gold nucleus's large binding energy. Might the gold nucleus's gravitational mass depend on simply the number of nucleons and not on the binding energy? What is the experimental evidence?

To see how we might answer this question, divide both sides of equation 1.2 by m_I but *don't* assume that m_G and m_I cancel. We get for the acceleration

$$\left(\frac{GM}{r^2}\right)\left(\frac{m_G}{m_I}\right) = a \quad (1.3)$$

If m_G and m_I are not the same, then the ratio m_G/m_I could be different for different objects, which would imply that they would experience different accelerations in the same gravitational field. This is something that we can investigate experimentally.

Galileo and Newton both provided some basic evidence for the equality of m_G and m_I (to about one part in a thousand) during the 1600s, and this satisfied the community for a long time. However, the question began to interest physicists again in the late 1800s. Starting with a famous experiment performed by Eötvös in 1890, physicists during the 20th century have designed increasingly sophisticated and accurate experiments using a number of different techniques. Current experiments have established that m_I and m_G are equal to within at least one part in a billion in a wide variety of circumstances, and the most precise experiments to date (which use a sensitive torsion balance to look for differences in the acceleration of different objects in the sun's gravitational field; see the website www.npl.washington.edu/eotwash/ for details) yield uncertainties of a few parts in 10^{13}.

Now, the fact that these two seemingly distinct quantities are the same to almost 13 significant digits begs for explanation. General relativity provides a simple and elegant explanation.

The Geodesic Hypothesis. The first step toward this explanation involves recognizing that if m_G were really equal to m_I, then equation 1.3 would imply that all objects in a given gravitational field experience the same acceleration, and thus that *all objects would follow the same trajectory in a given gravitational field* if launched from the same position with the same initial velocity, even if they differ in mass and/or other characteristics. Note that such a statement is *not* true in electrostatics: objects with different charges follow *different* trajectories in a given electric field, even if their initial positions and velocities are the same. But in the gravitational case, it is as if the trajectory were determined by the space through which the objects move rather than by anything about the objects.

But how can empty space determine a trajectory? In the two-dimensional space represented by a flat piece of paper, there is a unique path between any two points that has the shortest pathlength: that path is a straight line. In the two-dimensional space corresponding to the surface of a globe, the analogous paths are "great circles." Indeed, in the two-dimensional space corresponding to the surface of any smooth convex three-dimensional object, we can find the shortest path between two points by stretching a string tightly between those points. In a general space, we call the paths that represent the shortest (more technically, the extremal) distance between two points a **geodesic**. A space's geometric characteristics therefore define unique geodesic paths in that space.[1]

---
1. Technically, if two points in a space are separated by a distances large compared to the scale over which the space's curvature becomes significant, one may be able to find more than one geodesic connecting the points. For example, the poles of a sphere can be connected by an infinite number of great circles. But if the points are separated by distances small compared to that scale, the geodesic between a given pair of points is unique. Let's assume this.

1. INTRODUCTION

The **geodesic hypothesis** of general relativity asserts simply that

A free particle follows a geodesic in spacetime.

(where "a free particle" is one free of non-gravitational interactions). According to this hypothesis, a gravitational field shapes spacetime, which in turn specifies the geodesics that particles must follow.

The geodesic hypothesis makes sense only in *spacetime*, not in three-dimensional space. To see this, consider a thrown ball moving in a parabolic trajectory from point A to point B in the space near the earth's surface. But I could also fire a bullet from point A in such a way that it passes through point B: because of its greater speed, such a bullet would follow a much shallower parabola between the points (see figure 1.1). But the definition of a geodesic implies that there should be a *unique* geodesic between points A and B. Therefore the ball and bullet, even though both are "free," cannot both be following a geodesic, contrary to the hypothesis!

However, if the ball and bullet follow geodesics in *spacetime*, the apparent paradox evaporates. Figure 1.2 shows graphs of the ball's and bullet's trajectories in space *and* time. From this graph, we can draw two important conclusions. First, we see that even though the ball and bullet start out from point A at the same time (by hypothesis), they do not end up at point B at the same time, so they are *not* traveling between the same two points in *spacetime*. Two objects that do travel between between A and B *in the same time* would also have to have the same initial velocity and therefore *would* follow exactly the same trajectory in both space and time.

Second, though the paths of the ball and bullet are clearly *different* geodesic paths when plotted in spacetime, if we measure time in meters of light-travel time, then figure 1.2 shows that both have approximately the *same* radius of curvature (roughly 1 light-year = 1 ly). These different geodesics thus share a common curvature that they plausibly get from the spacetime around the earth.

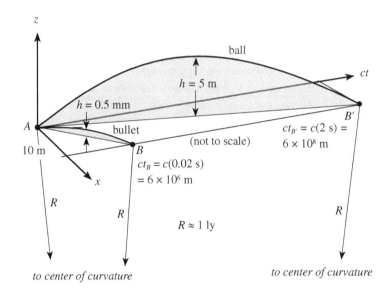

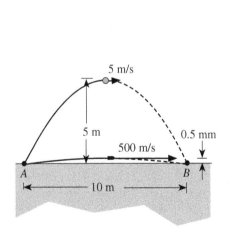

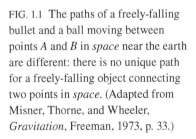

FIG. 1.1 The paths of a freely-falling bullet and a ball moving between points A and B in *space* near the earth are different: there is no unique path for a freely-falling object connecting two points in *space*. (Adapted from Misner, Thorne, and Wheeler, *Gravitation*, Freeman, 1973, p. 33.)

FIG 1.2 Plotted in *spacetime*, the bullet and ball paths have different ending times t_B and $t_{B'}$, so their final points B and B' in spacetime are *not* the same. However, if we express displacements along the time axis of this spacetime graph in terms of ct (where c is the invariant speed of light) so that all axes have the same units of meters, then the two paths *do* turn out to have approximately the same radius of curvature $R \sim 1$ light-year = 1 ly (see problem P1.1). Note that two projectiles moving between the same positions in space *and* time would also have to have the same initial *speed* and thus *would* follow the same (unique) path in spacetime. (Adapted from Misner, Thorne, and Wheeler, *Gravitation*, Freeman, 1973, p. 33.)

Why Gravitational Mass *Is* Inertial Mass. If we accept the geodesic hypothesis, then gravitational and inertial mass are the same thing, as I will now argue. Note that near the earth's surface, the geodesic for an object released from rest is a trajectory where the object accelerates downward at a rate of $g = 9.8$ m/s². According to the geodesic hypothesis, this is the "natural" path for a free object to follow, analogous to the straight-line geodesic an object would "naturally" follow in deep space (far from any gravitating objects). Now in deep space, accelerating an object away from a straight-line geodesic requires one to exert a force on the object. Analogously, if I hold an object at rest near the earth, I must exert an upward force on the object sufficient to accelerate it at a rate of $g = 9.8$ m/s² relative to the downward geodesic it naturally wants to follow. The magnitude of force required, according to Newton's second law, is simply $m_I g$, where m_I is the object's *inertial* mass.

However, it is precisely the magnitude of the *upward* force required to hold an object at rest that scales and balances measure when we "weigh" an object. In Newtonian mechanics, we imagine this upward force to be balanced by (and equal in magnitude to) a "gravitational force" $m_G g$ acting on the object, and thus we imagine the scale to register the object's "weight," which (after division by g) yields the object's gravitational mass m_G. But from the perspective of general relativity, the only real force acting on the object is the upward force (since a net force is required to accelerate an object relative to its geodesic), and that net force has a magnitude of $m_I g$. Therefore, when we *think* we are measuring an object's gravitational mass using a scale, what we are *really* measuring is its resistance to acceleration. So of course $m_G = m_I$: they are really the same thing!

Inertial and Noninertial Reference Frames. The paragraph above implies that in general relativity, we consider an object's "weight" (that is, the gravitational force acting on it) to be fictitious, not real. How can this be? To answer this question, we have to rethink the definition of inertial and noninertial reference frames.

In Newtonian mechanics, we typically define an inertial reference frame (IRF) to be "a frame in which a free object initially at rest remains at rest." However, it seems that we immediately waive this definition when we treat a reference frame at rest on the earth's surface as being even approximately inertial, since a free object initially at rest obviously does *not* remain at rest, but rather accelerates downward at a rate of g! The Newtonian explanation, of course, is that an object near the earth's surface is not "free," but rather subject to a gravitational force exerted on it by the earth, and that is why it accelerates. However, the only evidence for this "force" is the observed acceleration of a dropped object, which is unnatural only if we *assume* that a reference frame at rest on the earth's surface is inertial.

In general relativity, we take the definition of an IRF as given above literally and seriously. A reference frame at rest on the earth's surface is therefore *not* inertial, since a free object does not remain at rest. The only reference frames near the earth that are even approximately inertial are *freely falling frames*. We know, for example, that in a freely falling frame such as an orbiting space shuttle, an object placed at rest in midair remains floating at rest, consistent with the definition of an IRF!

Of course a Newtonian physicist would claim that this is an illusion, because both the shuttle and object happen to fall toward the earth with the same acceleration, and so remain at rest with each other. So is the decision about whether to take a frame at rest on the earth's surface or a freely falling frame as being an IRF merely a matter of perspective? No! One of Einstein's greatest triumphs was to show that this choice has physical consequences that we can examine experimentally.

The Equivalence Principle. Einstein pointed out that if a freely falling reference frame is truly an IRF, then it should be *physically equivalent* to a freely floating frame in deep space (far from any massive bodies), in the sense that any experiment performed in a freely falling frame should yield the same result as in the deep-space frame.

1. INTRODUCTION

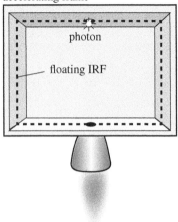

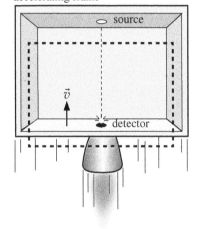

FIG. 1.3 (a) At the instant that an accelerating frame in deep space is at rest with respect to a floating inertial frame (IRF) in deep space, a photon is emitted by a source in the accelerating frame's ceiling. (b) By the time that the photon reaches the detector on the floor, the accelerating frame (and thus the detector) is moving upward relative to the IRF. The detector will therefore measure the photon's frequency to be blue-shifted. Since an accelerating frame in deep space is physically equivalent to a frame at rest on the earth's surface, we would expect to see the photon blue-shifted in a frame on the earth as well.

Similarly, a frame at rest near the earth's surface, since it is accelerating upward at a rate of $g = 9.8$ m/s^2 with respect to the inertial frames in its vicinity, should be *physically equivalent* to a rocket-powered noninertial frame accelerating at a rate of g relative to freely floating frames in deep space. This is one statement of what physicists call Einstein's **Equivalence Principle**.

Consider the following experimental test of this principle. Imagine that light of a certain frequency is emitted by a source attached to the ceiling of a frame at rest on the surface of the earth. This light is detected, and its frequency measured, by a detector on the floor of that frame. If such a frame is really *inertial* (in contradiction to general relativity), then the detected light will be observed to have the same frequency as the emitted light. If such a frame is *noninertial* (as required by the Equivalence Principle), on the other hand, then the light will be observed to be slightly blue-shifted, as I will now argue.

The Equivalence Principle requires that what we observe in a frame at rest near the earth's surface should be the same as what we observe in an accelerating frame in deep space. Imagine that such an accelerating frame happens to be at rest with respect to a floating IRF at the time ($t = 0$) that a certain photon is emitted by the accelerating frame's ceiling source (see figure 1.3). In the floating IRF, that photon's frequency remains constant. But by the time t' the photon reaches the detector on the floor of the accelerating frame, that accelerating frame is moving upward relative to the IRF at a speed of $v = gt'$. Therefore, in the floating IRF, the detector at the time of reception is moving with this speed toward the source (which was at rest in the IRF at the time of emission). Therefore, the detector will observe the light to be slightly Doppler-shifted toward the blue. This shift will be very small (because t' will be small: see problem P1.2 for a calculation of the magnitude), but nonzero.

The effect is indeed so small that it eluded firm experimental verification for more than 50 years after Einstein first predicted it. However, in 1959, R. V. Pound and G. A. Rebka were able to verify this effect in the reference frame corresponding to a 22.5-m high tower in the Jefferson Physical Laboratory at Harvard University (see Pound and Rebka, "Gravitational Redshift in Nuclear Resonance" *Phys. Rev. Lett.* **3**, 439–441, 1959). This experiment used gamma rays from a radioactive sample of Fe-57 as a source, and took advantage of the Mössbauer effect to make a very precise measurement of the frequency shift in the gamma rays as they were absorbed by an Fe-57 sample at the other end of the tower. This experiment verified the predicted blue-shift to an uncertainty of about 10%. Subsequent experiments have verified this in earth-based frames to within one part in 10^4.

We see that experimental evidence firmly supports the conclusion that frames at rest on the earth's surface are really noninertial, and consequently that freely falling frames *are* inertial. This means that the "force of gravity" that *appears* to press us to the floor in a frame at rest near the earth is really a fictitious force, as fictitious as the force that appears to press us to the floor in a frame that is accelerating upward. Indeed, this force vanishes in a truly inertial (that is, freely falling) frame: in such a frame, objects are exposed as being truly "weightless." Any force that appears or disappears depending on our choice of reference frame cannot be real.

The Reality of Gravity. So is gravity *entirely* fictitious? Is there *nothing* about gravity that is real (that is, observable in an inertial frame)? The answer to both of these questions is no. Gravity *is* real, and *does* have an aspect observable in an inertial frame, it is just not the downward force we usually consider "gravity."

To see what this aspect is, consider a large room freely falling toward the earth. Imagine that we place four balls so that they float initially at rest with respect to the room; one near the room's ceiling, one near its floor, one near a wall, and one near the opposite wall (see figure 1.4). What happens to these balls as the room falls?

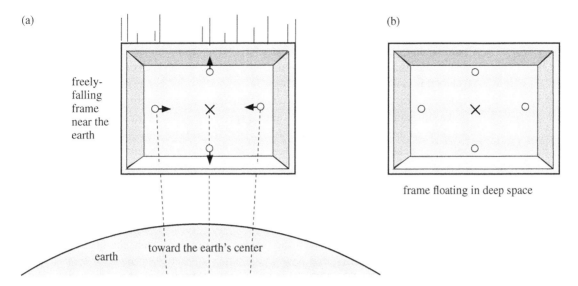

FIG. 1.4 (a) Because the gravitational field of the earth (indeed, any gravitating object) is non-uniform, off-center floating balls in a freely falling frame will experience small accelerations relative to the frame's center of mass. (b) Such accelerations are *not* observed in a frame floating in deep space: floating balls initially at rest remain truly at rest.

To predict what happens, let us retreat for the moment back into the Newtonian picture (which will predict the correct behavior even if it does not provide the correct interpretation). In that picture, the room's center of mass falls toward the earth with a certain acceleration. The ball near the ceiling is just a bit farther from the earth's center than the room's center of mass, so it experiences a slightly smaller acceleration, just as the ball near the floor experiences a slightly larger acceleration. The balls near the walls accelerate toward the earth's *center* and thus along lines angled slightly inward with respect to the direction along which the room accelerates. So as time passes, we will observe the top and bottom balls to accelerate *away* from the room's center, while the side balls will accelerate *toward* the room's center. This is *not* the behavior that we would observe in a frame floating in deep space: in such a frame, the balls would remain strictly at rest.

The relative accelerations of off-center free bodies, then, is something we *do* observe in an inertial (freely falling) frame near a gravitating object, but *not* in an inertial frame in deep space. These relative accelerations therefore represent a frame-independent (and thus real) indication that we must be near a gravitating object.

We call this aspect of gravity the **tidal effect** of a gravitational field, because (as Newton himself first realized) this effect explains tides on the earth's surface. Note that we can consider the earth to be a frame freely falling in the moon's gravitational field. Like the balls in our freely falling room, ocean waters on the sides of the earth closest to and farthest from the moon will accelerate away from the earth's center and thus bulge outward, while ocean waters on the sides will press inward. This explains the 12-hour tidal variation of the depth of the ocean.

Spacetime Is Curved. How do we interpret these tidal effects from the perspective of general relativity? Figure 1.5 shows a spacetime graph of the trajectories of the two side balls in our falling-room experiment. Since these balls are initially at rest with respect to each other, their paths in spacetime are initially parallel (they remain at an initially constant separation as time passes). As time progresses, however, they eventually begin to move toward each other with increasing speed, so the paths curve toward each other as shown.

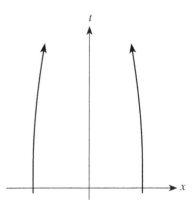

FIG. 1.5 Plotted in spacetime, the geodesics that the balls in figure 1.4 follow (as measured in the freely-falling frame) are initially parallel (the balls have initially constant separation), but gradually bend toward each other (because their separation eventually decreases). This bending of initial parallel lines signals that the underlying spacetime is *curved*.

But remember that these paths are *geodesics* (that is, the straightest possible lines) in spacetime. A fundamental axiom of Euclidean (flat plane) geometry is that initially parallel straight lines remain parallel. Here we see that initially parallel geodesics in spacetime (the straightest possible lines we have in spacetime) do not *remain* parallel. While this violates Euclid's axiom for plane geometry, this behavior is typical of curved space. For example, on the two-dimensional space corresponding to the surface of a globe, lines of longitude are great circles (geodesics). These lines are parallel at the equator but do not remain parallel as one goes toward the poles. We conclude that the relative acceleration of geodesics near a gravitating object indicates that the geometry of spacetime is *curved* (non-Euclidean) there. This curvature of spacetime is the frame-independent signal that a gravitational field is present. Moreover, once we understand exactly *how* spacetime is curved near a gravitating body, we can calculate that spacetime's geodesics and therefore predict how freely-falling bodies will move.

The Einstein Equation. Given the geodesic hypothesis, then, the central task of a theory of gravity is to predict how a gravitating body affects the curvature of spacetime. On November 25, 1915, Einstein completed the theory of general relativity by proposing an equation that linked the presence of matter and energy to the curvature of spacetime, an equation that we now call the **Einstein equation** (or the Einstein Field Equation). This equation reads as follows:

$$G^{\mu\nu} = 8\pi G T^{\mu\nu} \tag{1.4}$$

where $G^{\mu\nu}$ is a 4×4 matrix (tensor) that describes the curvature of spacetime at a given point in space and time, G is the universal gravitational constant, and $T^{\mu\nu}$ is a 4×4 matrix describing the density and flow of matter and energy at the same point in space and time. This equation and the **geodesic equation** used to calculate geodesics in an arbitrarily curved spacetime comprise the core equations of general relativity.

General Relativity in a Nutshell. In the chapters that follow, we will explore in great detail the mathematical meaning of both the Einstein equation and the geodesic equation. For our purposes at the moment, however, it is enough to understand their *physical* meaning. To summarize, if we know how spacetime is curved, we can use the geodesic equation (the mathematical equivalent of stretching a string between two points in spacetime) to calculate how objects will move in that spacetime. If we know the density and flow of matter and energy in spacetime, we can use the Einstein equation to calculate how spacetime is curved. The great physicist John Archibald Wheeler summarized the theory's essence even more briefly this way:

Spacetime tells matter how to move; matter tells spacetime how to curve.

This is general relativity in a nutshell. What could be simpler? Our task in what follows will simply involve unfolding the implications of this simple but profound statement.

HOMEWORK PROBLEMS

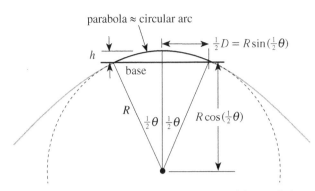

FIG. 1.6 This figure illustrates how we can model a parabola with height h and base D by a circular arc having an appropriate radius R.

P1.1 This problem explores the claims made in figure 1.2.

a. Show that the values of h and t for the bullet and ball trajectories in that figure are consistent with a gravitational acceleration of $g = 10$ m/s^2.

b. The ball's and bullet's paths in the spacetime graph in figure 1.2 are actually *very* gently curved parabolas when you consider how long their bases are in that diagram. As shown in figure 1.6, we can use a circular arc as an excellent approximation for a gently curved parabola. We can calculate that arc's effective radius R, and thus quantify the path's curvature in a direct and reasonably intuitive way. Note from the figure that the peak of a circular arc that spans an angle θ is a distance R away from the circle's center, but the center of the arc's base is only $R\cos(\tfrac{1}{2}\theta)$ from the center. The height of the arc's peak above its base is therefore $h = R[1 - \cos(\tfrac{1}{2}\theta)]$. Since the angle θ will be quite small in this case, we can expand $\cos(\tfrac{1}{2}\theta)$ in a power series and drop higher order terms: $\cos(\tfrac{1}{2}\theta) \approx 1 - \tfrac{1}{2}(\tfrac{1}{2}\theta)^2 = 1 - \tfrac{1}{8}\theta^2$ (assuming that θ is in radians and is much smaller than 1). In the same limit, $\sin(\tfrac{1}{2}\theta) \approx \tfrac{1}{2}\theta$, so the base's length D is thus $2R\sin(\tfrac{1}{2}\theta) \approx R\theta$. Now, in figure 1.2, the length of each curve's base in the spacetime diagram is almost exactly equal to ct, where t is the time required for the projectile to go between spatial points A and B. Combine the approximations above with the given values of h and t for each path to show that effective radii of curvature of the ball's and bullet's paths in figure 1.2 are both approximately $R \approx 10^{16}$ m ≈ 1 ly. Also check that θ is indeed very small for both paths, justifying the approximations we have made.

P1.2 Imagine that a laser on the ceiling of a laboratory on the earth emits a flash of light directed toward a sensor on the floor a distance $d = 25$ m below (the laboratory is in a tower). This lab is equivalent (from the point of view of the gross effects of gravity) to an identical lab accelerating upward in deep space with a uniform acceleration of magnitude g. Imagine that we observe the flash being emitted and detected in an inertial lab surrounding the accelerating lab. For the sake of simplicity, imagine that the two labs are at rest with respect to each other at the instant the flash is emitted. In the time it takes the flash to reach the floor (as measured in the inertial lab), the accelerating lab attains a certain speed v relative to the inertial lab. Thus (according to observers in the inertial lab) the floor detector in the accelerated lab is moving toward the laser with speed v at the time the pulse is detected, so the floor detector measures the laser light's wavelength to be blue-shifted to the value λ given by the relativistic Doppler shift formula $\lambda/\lambda_0 = \sqrt{(1-v/c)/(1+v/c)}$, where λ_0 is the wavelength of the light as emitted by the laser and v is the detector's speed relative to the laser at the time of detection.

a. Argue that the fractional shift in wavelength is

$$\frac{\lambda_0 - \lambda}{\lambda_0} \approx \frac{gd}{c^2} \qquad (1.1)$$

when $gd/c^2 \ll 1$ and $v/c \ll 1$. (*Hint:* You will find the binomial approximation $(1 + x)^n \approx 1 + nx$ helpful. This approximation is accurate to order x^2.)

b. What would be the fractional shift in wavelength in a lab on the earth's surface?

c. What would be the fractional shift in wavelength if the lab were located on the surface of a neutron star having a mass of $M = 3.0 \times 10^{30}$ kg (≈ 1.5 the mass of the sun) and a radius of $R = 12$ km? [*Hint:* First estimate the magnitude of $\vec{g}$ using Newton's law of universal gravitation. You can find the value of the universal gravitational constant G on the inside front cover.]

P1.3 Another consequence of the Equivalence Principle is that light will be bent in a gravitational field. This has never been measured on the surface of the earth, but was verified qualitatively by observing starlight passing near the sun's edge during a total eclipse in 1919. Why can't this experiment be done on the earth's surface? Let's predict how much bending we should see in a laboratory at rest on the earth's surface. What one would observe in such a laboratory should be the same as what one would observe in a laboratory accelerating in deep space with a uniform acceleration of $\vec{a} = -\vec{g}$, where $\vec{g}$ is the local acceleration of gravity on the earth's surface. Imagine that a laser at one end of the laboratory emits a beam of light that originally travels parallel to the laboratory floor.

P1.3 (continued)

This light shines on the opposite wall of the laboratory a horizontal distance $d = 3.0$ m away from the laser.

a. Find the magnitude of this deflection in a laboratory on the surface of the earth.

b. Find the magnitude of this deflection if the laboratory sits on the surface of a neutron star having a mass of $M = 3.0 \times 10^{30}$ kg (≈ 1.5 the mass of the sun) and a radius of $R = 12$ km. [*Hint:* First estimate the magnitude of $\vec{g}$ using Newton's law of universal gravitation. You can find the value of the universal gravitational constant G on the inside front cover.]

P1.4 We can calculate how much starlight passing the sun's edge will be deflected (according to the Equivalence Principle) as follows. Assume that the light's deflection is so small that we can approximate a given photon's trajectory by a straight line along the x axis that grazes the sun's surface, as shown in figure 1.7. Assume that the photon has the same acceleration as any other object as it passes near the sun and that this acceleration has the magnitude $|\vec{a}| = GM/r^2$ predicted by Newtonian physics (where $G = 6.67 \times 10^{-11}$ N·m²/kg², M is the sun's mass, and r is the distance between the photon and the sun's center), but that the photon's speed remains $\approx c$ (the speed of light) during the whole process. As figure 1.7 shows, the sine of the deflection angle δ will be equal to v_y/c, where v_y is the y component of the photon's final velocity $\vec{v}$. We can determine v_y by integrating $a_y dt$ over the entire trajectory, which you can do by expressing a_y as a function of x and R, writing $dt = dx/c$, and integrating from $x = -\infty$ to $+\infty$. Look up the resulting integral and note that for small angles, $\sin \delta \approx \delta$ (when δ is expressed in radians). Show therefore that the predicted deflection is $\delta = 4.2 \times 10^{-6}$ rad = 0.87 seconds of arc (which is small enough to justify the various approximations we have made above).

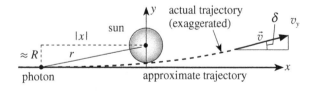

FIG. 1.7 This figure illustrates the trajectory of a photon passing near the sun's edge. Its deflection has been hugely exaggerated. The sun's radius R is about 700,000 km.

[*Note:* Einstein published this prediction in 1907 after first proposing the Equivalence Principle. However, as we will see in chapter 13, the full theory of general relativity actually predicts *twice* this deflection, as Einstein noted in 1915, and the latter result was verified during the 1919 eclipse. It turns out that while the Equivalence Principle works well in reference frames that are small enough compared to the scale over which any enclosed gravitational field varies significantly, in the calculation above we are implicitly using a frame that is large compared to that scale. In such a case, as Einstein himself found, a naive application of the Equivalence Principle yields incorrect results.]

P1.5 Imagine a freely falling reference frame near the surface of the earth. This reference frame has the form of a cube 44 m on a side. Imagine that floating balls are placed at point A in the frame's center, at point B 22 meters above A, and at point C 22 m below A. The frame's center of mass will fall at the same rate as the ball at A which is located at the frame's center. But due to tidal effects, the balls at B and C will fall a bit slower and faster respectively than the frame as a whole. What are the magnitudes of the accelerations of the balls at B and C *relative* to A (that is, what are the magnitudes of $\vec{a}_B - \vec{a}_A$ and $\vec{a}_C - \vec{a}_A$)? Feel free to use Newtonian physics in this calculation: we will show later that general relativity leads to the same results near the surface of the earth to many decimal places. (Hint: You will need to use the binomial approximation $(1 + x)^n \approx 1 + nx$. If you do not, you will find that your calculator does not keep enough digits to yield an accurate result.)

P1.6 Many smart phones (such as the iPhone) have built-in three-axis accelerometers. These accelerometers typically measure the x, y, and z force components per unit mass that must be applied to an internal proof mass to hold it at rest with respect to the rest of the phone. Now, if the phone is freely falling, it represents an inertial reference frame within which the internal proof mass *should* float without requiring *any* forces to hold it at rest. Find an app for your smartphone that is capable of logging acceleration data for all three axes, run the app, and then throw your phone in a nice parabola onto a suitable soft surface. (Note that *you* are responsible for any mishaps!) Use your collected data to argue that the phone is indeed an inertial reference frame during the time interval between leaving your hand and landing on the soft surface.

Two-sided handout for a lecture based on this material ⟶

GENERAL RELATIVITY IN A NUTSHELL

I. FUNDAMENTAL IDEAS
  A. The curious equality of gravitational mass and inertial mass
    1. Illustrating the distinction:
       (a) Compare Coulomb's law with Newton's law of universal gravitation.
       (b) Note how the m's on either side of $GMm_G/r^2 = m_I a$ express *different* things.
    2. Experimental tests show that inertial mass = gravitational mass to 13 decimal places.
    3. It is not credible that this is an accident!
  B. The Geodesic Hypothesis
    1. A plausibility argument: the path followed by a falling object (since it is independent of the object's properties) seems to be a property of the *space*, not the object.
    2. What is a *geodesic*, anyway?
       (a) It is the straightest possible path (or path of "extremal" length) through a space.
       (b) The geometry of the space uniquely specifies such paths.
    3. A statement of the hypothesis: *A free particle follows a geodesic in spacetime.*
    4. Note: This works only if the paths are geodesics in *spacetime* (see figures 1.1 and 1.2).

II. IMPLICATIONS
  A. "Weight" really expresses an object's resistance to acceleration (relative to its geodesic)!
    1. An object's geodesic near the earth accelerates downward at $g = GM/r^2 = 9.8$ m/s^2.
    2. To hold an object at rest, we give it an upward acceleration g relative to its geodesic.
    3. This requires an upward force of $m_I g = m_I GM/r^2$.
    4. So the m's on either side of $GMm_G/r^2 = m_I a$ really express the same thing after all!
  B. Inertial reference frames (IRFs) and freely falling reference frames
    1. Definition of an IRF: a free object at rest (in the frame) remains at rest
    2. Near a gravitating object, only freely falling frames are genuine IRFs.
  C. The Equivalence Principle
    1. A frame on earth's surface is analogous to an accelerating frame in deep space.
    2. Implications:
       (a) bending of light in a gravitational field (see problem P1.3)
       (b) gravitational blue-shift (see figure 1.3)

III. THE REALITY OF GRAVITY
  A. What is real (that is, frame-independent) about gravity?
    1. Is it totally fictitious like "centrifugal force"?
    2. No! *Tidal* effects of gravity cannot be erased by change of reference frame (figure 1.4).
    3. So initially parallel paths of falling objects do not *remain* parallel (figure 1.5).
    4. Implication: *spacetime must be curved* (i.e., have a non-Euclidean geometry).
  B. The Einstein equation
    1. $G^{\mu\nu} = 8\pi G T^{\mu\nu}$, where
       (a) $G^{\mu\nu}$ = 4 × 4 symmetric tensor describing curvature of spacetime at a point
       (b) $T^{\mu\nu}$ = 4 × 4 symmetric tensor describing the density of mass-energy at that point
    2. This is the link between curvature of spacetime and mass that causes it.

IV. GENERAL RELATIVITY IN A NUTSHELL
Spacetime tells matter how to move, matter tells spacetime how to curve (J. A. Wheeler)

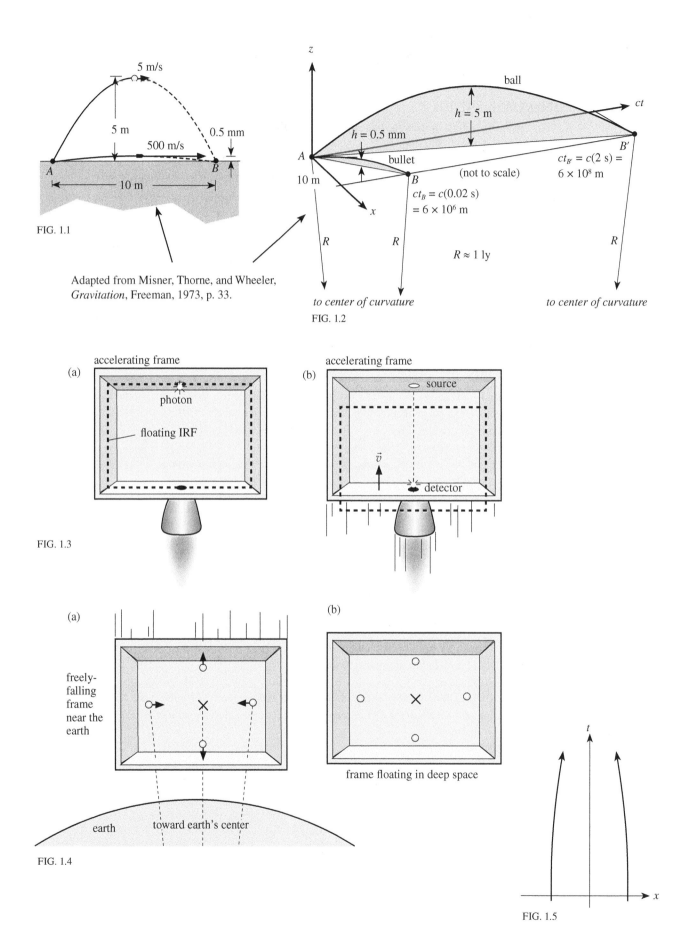

FIG. 1.1

FIG. 1.2

FIG. 1.3

FIG. 1.4

FIG. 1.5

2. REVIEW OF SPECIAL RELATIVITY

INTRODUCTION

FLAT SPACETIME
- Review of Special Relativity
- Four-Vectors
- Index Notation

TENSORS
- Arbitrary Coordinates
- Tensor Equations
- Maxwell's Equations
- Geodesics

SCHWARZSCHILD BLACK HOLES
- The Schwarzschild Metric
- Particle Orbits
- Precession of the Perihelion
- Photon Orbits
- Deflection of Light
- Event Horizon
- Alternative Coordinates
- Black Hole Thermodynamics

THE CALCULUS OF CURVATURE
- The Absolute Gradient
- Geodesic Deviation
- The Riemann Tensor

THE EINSTEIN EQUATION
- The Stress-Energy Tensor
- The Einstein Equation
- Interpreting the Equation
- The Schwarzschild Solution

COSMOLOGY
- The Universe Observed
- A Metric for the Cosmos
- Evolution of the Universe
- Cosmic Implications
- The Early Universe
- CMB Fluctuations & Inflation

GRAVITATIONAL WAVES
- Gauge Freedom
- Detecting Gravitational Waves
- Gravitational Wave Energy
- Generating Gravitational Waves
- Gravitational Wave Astronomy

SPINNING BLACK HOLES
- Gravitomagnetism
- The Kerr Metric
- Kerr Particle Orbits
- Ergoregion and Horizon
- Negative-Energy Orbits

this depends on this

Reference Frames and Events. The study of geometry on a two-dimensional surface starts with the concept of a *point*: other geometric elements (lines, curves, triangles, and so on) are sets of points. The analogous concept for spacetime geometry is an **event**, a physical occurrence (e.g., the collision of two particles) that marks a suitably well-defined point in space and instant of time.

A **reference frame** allows one to quantify an event's location in spacetime by providing a mechanism (real or hypothetical) for assigning **spacetime coordinates** (three spatial coordinates and one time coordinate) to every event. We can visualize a reference frame as being a rigid cubical lattice of measuring sticks with an event-detector and associated clock located at each intersection (see figure 2.1). Any event occurring within the lattice is registered by the nearest detector and is assigned the three spatial coordinates of that detector's lattice position and the time coordinate registered by the detector's associated clock. We can imagine making the spacing between lattice intersections as fine as needed to achieve any desired resolution. An **observer** collects and interprets the data generated by the frame's detectors.

A cubical clock-lattice is usually too simplistic to be a practical technology. However, its simplicity is precisely its value as a visualization: it helps us think clearly about what a reference frame is and what it does. Any valid method of assigning coordinates must yield values equivalent to those obtained from a cubical clock-lattice.

We will often consider reference frames attached to specific objects. We will sometimes refer to a reference frame only obliquely, as in "an observer in the spaceship finds that . . .": since an observer (by definition) interprets measurements obtained in a reference frame, such a phrase presumes a frame attached to the ship.

Inertial Reference Frames. Depending on how it (or the object to which it is attached) moves through spacetime, a given reference frame may be either *inertial* or *noninertial*. In special relativity, we define a **inertial reference frame** (IRF) to be a frame in which a free object moves at a constant velocity (i.e. Newton's first law is obeyed) to some specified accuracy everywhere in the frame. (In general relativity, the best we can do is

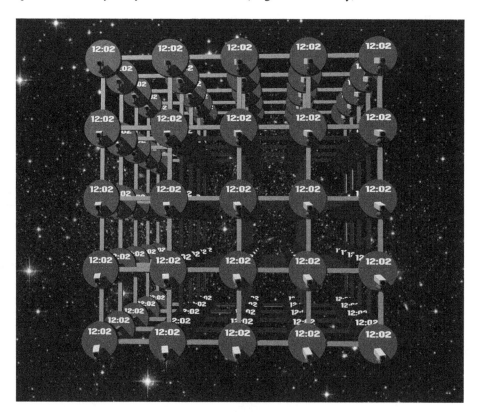

FIG. 2.1 This is the mental image one should have of a reference frame. The video camera attached to each clock records events in the neighborhood. The video records are stamped with the time indicated by the clock. These records are then sent to an observer, who interprets the records.

find frames that are inertial over a sufficiently small region of space and time.) All other reference frames are **noninertial reference frames** (NIRFs). Note that an observer in any frame can determine (by testing Newton's first law) whether the frame is inertial without referring to anything outside the frame.

In general relativity, an object is "free" if it does not participate in non-gravitational interactions with other objects. The definition of an inertial reference frame then implies that the frames near a massive body that most closely approximate ideal IRFs are non-rotating **freely falling reference frames** (FFRFs), sometimes also called **free-float reference frames**. We will see later that the tidal effects of gravity mean that FFRFs are not *exactly* inertial, but we can make this approximation as good as we need by limiting the frame to a sufficiently small region of spacetime.

In the contexts of both special and general relativity, you can easily prove (see box 2.1) that a reference frame will be inertial *if and only if* it moves at a constant velocity and does not rotate relative to an established IRF (as long as both frames are able to assign coordinates to events in a common region of spacetime).

The Principle of Relativity. The **principle of relativity** states that

> The laws of physics are the same in all inertial reference frames (IRFs).

This principle is a *symmetry principle,* akin to the more commonplace principles that the laws of physics are independent of time, position in space, or orientation in space. Since it is a statement about how laws of physics must behave (even laws we might invent in the future), it is more fundamental than the laws themselves.

This principle means that all IRFs are physically equivalent: there is no physical experiment that we could perform entirely within a given IRF that can distinguish it from any other IRF. This does *not* mean that physical quantities (such as a particle's energy or momentum) necessarily have the same *values* in different IRFs, only that all observers find that these values obey the same laws.

On the other hand, Maxwell's equations state that electromagnetic waves (including light waves) move in a given IRF with a certain speed c that appears as a fixed constant in the equations (a constant linked to experimental measurements of the strengths of static electric and magnetic interactions). The value of c is thus an intrinsic part of the laws of electromagnetism and so must be the same in all IRFs.

Relativistic Units. Because the speed of light c is frame-independent, we can connect space and time units in a frame-independent way. The nearly universal convention in general relativity is to express both distance and time in units of meters (rather than in seconds), where we define a **meter of time** to be the time it takes light to travel a meter of distance. Since light travels 299,792,458 m in 1 s (by definition of the SI meter), a meter of time is 1/299,792,458 s = 3.34 ns.

Light moves at a speed of 1 m/m = 1 by definition in this unit system, so all other velocities are expressed as unitless fractions of the speed of light. This implies that mass m, momentum (which must have the same units as mv), and energy (which must have the same units as $\frac{1}{2}mv^2$) all have the same units. For macroscopic objects, I will use the SI kilogram as the common unit for these quantities, but for microscopic objects (molecules and smaller), I will use the electron volt (eV).

Let us call the unit system where time and distance are measured in meters and mass, momentum, and energy are measured in kilograms the **GR (General Relativistic) unit system.** See box 2.2 for conversion factors between SI and GR units.

Clock Synchronization. If the time coordinates assigned by the clocks in a reference frame of the type shown in figure 2.1 are to have any coherent meaning, the clocks must be *synchronized*. Since the speed of light $c = 1$ m/m = 1 must have the same value in all

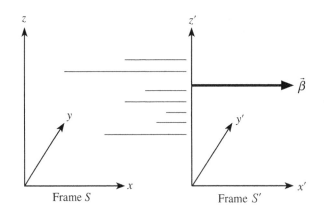

FIG. 2.2 Two reference frames (whose lattices here are represented by simple coordinate axes) whose lattice directions are spatially aligned. (The spatial origins of these lines coincided a short time ago, which we define to be $t = t' = 0$.) We say that such frames are in **standard orientation** with respect to each other.

IRFs (as argued a few paragraphs back), these clocks will be synchronized *if and only if* an observer concludes that the speed of a light flash traveling between any pair of clocks is 1. This means that if a master clock emits a light flash at time t, a clock a distance d from the master will be correctly synchronized with it if it reads time $t + d$ when a signal passes. This provides a straightforward method for synchronizing the clocks in an inertial frame.

(Note that this method does not work in NIRFs, so we will have to use a different approach to assigning events coherent time coordinates in general relativity.)

The Lorentz Transformation. Consider two IRFs S and S' whose spatial coordinate axes are aligned (see figure 2.2). Suppose that S' moves in the $+x$ direction relative to S with speed β and define $t = 0$ in both frames to be the instant when the spatial origins of S and S' coincide. The principle of relativity and the definition of clock synchronization together imply (see box 2.3) that if the coordinates of an event are observed to be t, x, y, and z in frame S, the coordinates of the same event observed in S' are given by the matrix equation

$$\begin{bmatrix} t' \\ x' \\ y' \\ z' \end{bmatrix} = \begin{bmatrix} \gamma & -\gamma\beta & 0 & 0 \\ -\gamma\beta & \gamma & 0 & 0 \\ 0 & 0 & 1 & 0 \\ 0 & 0 & 0 & 1 \end{bmatrix} \begin{bmatrix} t \\ x \\ y \\ z \end{bmatrix} \quad \text{where } \gamma \equiv \frac{1}{\sqrt{1-\beta^2}} \qquad (2.1)$$

This equation is called the **Lorentz transformation equation**, and the square matrix the **Lorentz transformation matrix** or simply the **Lorentz transformation**. (We can generate the **inverse Lorentz transformation** from primed to unprimed coordinates by replacing β with $-\beta$ as you will show in box 2.3.)

In many ways, the Lorentz transformation is analogous to the transformation between two rotated coordinate systems on a two-dimensional plane (see box 2.4).

Because the transformation is linear, the same transformation law applies to coordinate differences Δt, Δx, Δy, Δz between pairs of events:

$$\begin{bmatrix} \Delta t' \\ \Delta x' \\ \Delta y' \\ \Delta z' \end{bmatrix} = \begin{bmatrix} \gamma & -\gamma\beta & 0 & 0 \\ -\gamma\beta & \gamma & 0 & 0 \\ 0 & 0 & 1 & 0 \\ 0 & 0 & 0 & 1 \end{bmatrix} \begin{bmatrix} \Delta t \\ \Delta x \\ \Delta y \\ \Delta z \end{bmatrix} \qquad (2.2)$$

This means that observers in different IRFs will disagree about whether two events are simultaneous, as the first line of equation 2.2 tells us that $\Delta t = 0$ does *not* imply $\Delta t' = 0$ if $\Delta x \neq 0$. Another way to say this is that an observer in one IRF will consider the clocks in a different IRF to be unsynchronized.

The Metric Equation. The **spacetime interval** Δs between two events is defined by the **metric equation**

$$\Delta s^2 = -\Delta t^2 + \Delta x^2 + \Delta y^2 + \Delta z^2 \tag{2.3}$$

The squared spacetime interval Δs^2 corresponds to the negative square of the time interval between two events measured in an IRF where they occur at the same place ($\Delta x = \Delta y = \Delta z = 0$) or the squared spatial distance between those events in an IRF where they occur at the same time ($\Delta t = 0$). The spacetime interval is important because it is a *frame-independent* measure of the events' separation: observers in *all* IRFs agree on the value of Δs between a given pair of events (see box 2.5)! The spacetime interval and the metric equation are to spacetime what *distance* between two points and the Pythagorean theorem are to ordinary plane geometry. This *crucial* equation provides the main link between special and general relativity.

The choice of overall sign in equation 2.3 follows a well-established convention in general relativity that highlights the *spatial* aspect of the interval (and its analogy to distance) over its time aspect.

Categories of Spacetime Intervals. Spacetime intervals are called

$$\begin{aligned} spacelike &\quad \text{if } \Delta s^2 > 0 \;\Rightarrow\; \Delta x^2 + \Delta y^2 + \Delta z^2 > \Delta t^2 \\ lightlike &\quad \text{if } \Delta s^2 = 0 \;\Rightarrow\; \Delta x^2 + \Delta y^2 + \Delta z^2 = \Delta t^2 \\ timelike &\quad \text{if } \Delta s^2 < 0 \;\Rightarrow\; \Delta t^2 > \Delta x^2 + \Delta y^2 + \Delta z^2 \end{aligned}$$

These distinctions arise only because of the minus sign in the metric equation: since there is no minus sign in the corresponding Pythagorean theorem for ordinary space, there is only one kind of distance. Since observers in all IRFs agree on the sign of Δs^2, all will agree on how to classify the interval between a given pair of events.

These classifications are important because events with a spacelike interval *cannot* be causally connected, because the order in time of such events is frame-dependent (see box 2.6). This means that no particle, message, or other causal effect can travel faster than the speed of light in an IRF.

Spacetime Diagrams. A **spacetime diagram** is a convenient way to display the relationship between events in spacetime. A spacetime diagram is a two or possibly three-dimensional graph with a vertical time axis and horizontal spatial axis or axes. (All axes are conventionally given the same scale.) An *event* in spacetime is represented by a *point* on the diagram. We think of a particle's trajectory through spacetime as being a connected sequence of events called the particle's **worldline**, which is represented in a spacetime diagram by a connected sequence of points—i.e., a line (maybe a curve) that shows its spatial position as a function of time t (along the vertical axis). An object's speed at any given instant is the inverse slope of the curve representing its worldline on the diagram, evaluated at that instant. Problems P2.1–P2.4 will give you practice creating and reading such diagrams.

Any light emitted by a given event E will expand (as time passes) into a spherical surface centered on the event. The projection of this sphere on the xy plane is a circle whose radius expands with time. If we draw this on a spacetime diagram, the expanding ring looks like a cone (see figure 2.3). We call this the event's **light cone**. The light cone is important because the worldline of any particle traveling through E must lie within this light cone (because no particle can travel faster than light), and all events that can be caused by E must lie within this cone.

Note that because the metric of spacetime is different than the Pythagorean theorem of ordinary space, the apparent separation of points on a spacetime diagram does not give us any indication of the spacetime interval between them. For example, the spacetime interval between events E and L in figure 2.3 is zero!

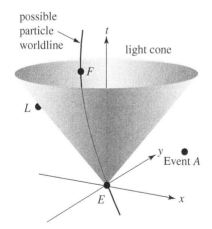

FIG. 2.3 A spacetime diagram that shows the light cone for event E. Note that event A lies outside event E's light cone, so it is causally unrelated to E, but events L and F could be caused by E. Particle worldlines that pass through event E must lie inside its light cone.

Proper Time. We can calculate the **proper time** $\Delta\tau$ measured between two events A and B by a clock traveling along a given worldline connecting those events using a method analogous to how we would calculate the path length along a curve in ordinary space. (1) We first imagine dividing the worldline into infinitesimal segments. (2) We use the metric equation to calculate what an imaginary *inertial* clock would measure while traversing each segment. If the segment is small enough to be reasonably straight, the time $d\tau$ registered by the actual clock should be essentially the same. (3) We then sum $d\tau$ over all segments. The result (see box 2.7) is that

$$\Delta\tau = \int \sqrt{-ds^2} = \int_{t_A}^{t_B} \sqrt{1-v^2}\, dt \tag{2.4}$$

Length Contraction. A moving object's length L in a given IRF is defined to be the distance between two events located at the object's ends that are simultaneous in the frame. Because observers in different IRFs disagree about which pairs of events are simultaneous, they will disagree about the object's length. If an object's length along a given direction is L_R in its own IRF, its length in an IRF where it is observed to move with speed v in that direction is (see box 2.8.)

$$L = L_R\sqrt{1-v^2} \tag{2.5}$$

The Einstein Velocity Transformation. One can use the Lorentz transformation equations to prove (as you will show in box 2.9) that if a particle's velocity in a given IRF is $\vec{v} = [v_x, v_y, v_z]$, its velocity components in another IRF in standard orientation with respect to the first (see figure 2.2) are

$$v'_x = \frac{v_x - \beta}{1 - \beta v_x}, \qquad v'_y = \frac{v_y\sqrt{1-\beta^2}}{1 - \beta v_x}, \qquad v'_z = \frac{v_z\sqrt{1-\beta^2}}{1 - \beta v_x} \tag{2.6}$$

(Again, replacing β with $-\beta$ yields the inverse transformation.) Note that this is *not* a simple linear transformation (unlike the Lorentz transformation).

For Further Reference. A great many books provide excellent introductions to special relativity in more depth and at a more leisurely pace than the whirlwind tour presented in this chapter. In particular, each of the following books offers a much more extensive treatment of special relativity that is especially compatible with the conceptual approach taken in this book:

Taylor and Wheeler, *Spacetime Physics,* 2nd edition, Freeman, 1992.
Moore, *Six Ideas That Shaped Physics: Unit R,* 2nd edition, McGraw-Hill, 2003.

BOX 2.1 Overlapping IRFs Move with Constant Relative Velocities

In this box, our goal is to prove that a rigid frame S' is inertial *if and only if* it moves at a constant velocity relative to an established IRF S (assuming that both frames can assign coordinates to events in a common region of spacetime). We will start with a proof for the "if" clause: that is, we will *assume* that S' moves at a constant velocity and *prove* that it is inertial.

Imagine that there is a free object that in S is observed to be at rest. Since S is inertial by hypothesis, Newton's first law implies that such an object will remain at rest relative to S as time passes. Now consider observing this object in frame S'. First note that *all* observers agree about whether an object is or is not "free"; if the object is uncharged and not magnetic, everyone will concede that it cannot participate in long-range electromagnetic interactions, and if it is not touching anything, everyone will agree that it cannot participate in any other kinds of interactions. Therefore, the observer in S' will agree with the observer in S that the object in question is free. However, the observer in S' also observes the *free object* to move at a constant velocity, because S moves at a constant velocity relative to S' by hypothesis, and because the object is at rest in the S frame, the observer in S' must see it as moving with the same velocity as S. Therefore, this free object is observed to move at a constant velocity in S'. Since the argument does not depend on where the object is located in either frame (as long as it is somewhere in the region of spacetime that can be observed in both frames), it follows that a free object moves with a constant velocity everywhere in S'. Therefore S' is inertial (at least in the commonly observed region of spacetime). Q. E. D.

Exercise 2.1.1. In the space below, prove the "only if" clause; that is, assume that S' is inertial and show that it must move at a constant velocity relative to S. Start by considering a free object at rest in S'.

BOX 2.2 Unit Conversions Between SI and GR Units

The time unit in the GR unit system is the *meter*: this is the only change from SI units. The SI meter of distance (by international agreement) is *defined* to be the distance that light travels in (1/299,792,458) s. Therefore, the GR meter of time is equivalent to

$$1 \text{ m} = \left(\frac{1}{299,792,458}\right) \text{s} = 3.33564095 \times 10^{-9} \text{ s} \approx 3.34 \text{ ns} \quad (2.7)$$

From this basic definition, you can show that (to four significant figures)

$$1 \text{ μs} = 299.8 \text{ m (of time)} \quad (2.8a)$$
$$1 \text{ ms} = 299.8 \text{ km} \quad (2.8b)$$
$$1 \text{ s} = 299,800 \text{ km} \quad (2.8c)$$
$$1 \text{ min} = 17.99 \times 10^6 \text{ km} \quad (2.8d)$$
$$1 \text{ hr} = 1.079 \times 10^9 \text{ km} \quad (2.8e)$$
$$1 \text{ day} = 25.90 \times 10^9 \text{ km} \quad (2.8f)$$
$$1 \text{ y} = 9.461 \times 10^{15} \text{ m} \quad (2.8g)$$
$$\text{age of universe} = 13.7 \text{ Gy} = 1.30 \times 10^{26} \text{ m} \quad (2.8h)$$

We can convert any quantity in SI units into GR units by multiplying by as many factors of the conversion factor $c = 1 = (2.99792458 \times 10^8 \text{ m} / 1 \text{ s})$ as are required to eliminate all units of seconds from the quantity.

$$1 \text{ J} = 1 \text{ kg} \frac{\cancel{m}^2}{\cancel{s}^2}\left(\frac{1 \cancel{s}}{299,792,458 \cancel{m}}\right)^2 = 1.1126501 \times 10^{-17} \text{ kg (energy)} \quad (2.9a)$$

$$1 \text{ kg (energy)} = 8.98755179 \times 10^{16} \text{ J} \quad (2.9b)$$
$$1 \text{ kg (momentum)} = 299,792,458 \text{ kg·m/s} \quad (2.9c)$$
$$1 \text{ eV} = 1.602 \times 10^{-19} \text{ J} = 1.782 \times 10^{-36} \text{ kg (energy)} \quad (2.9d)$$
$$1 \text{ eV (momentum)} = 5.34 \times 10^{-28} \text{ kg·m/s} \quad (2.9e)$$

The eV as a unit for mass, momentum, and energy has a much more convenient size than the kilogram when dealing with subatomic particles, atoms, and molecules.

Here are some constants in GR units that will be useful to us later:

$$g = 1.09 \times 10^{-16} \text{ m}^{-1} = 1 / (9.17 \times 10^{15} \text{ m}) \approx 1 / (1 \text{ ly}) \quad (2.10a)$$
$$G = 7.426 \times 10^{-28} \text{ m/kg} = 1477 \text{ m} / \text{(solar mass)} \quad (2.10b)$$

Exercise 2.2.1. In the space below, verify the last two equations.

BOX 2.3 One Derivation of the Lorentz Transformation

In this box, we will derive the Lorentz transformation (LT) equations from the principle of relativity, including the requirement that the speed of light should be measured to be 1 in any IRF. The literature is full of different ways to derive the LT; I have chosen a method (from Alan Macdonald, private communication) that I think is both conceptually straightforward and exposes the most important conceptual issues. We will assume IRFs S and S' in standard orientation, as shown in figure 2.2.

Distances perpendicular to the line of relative motion. Our first task is to argue that all observers must agree on the values of distances measured perpendicular to the line of the frames' relative motion (the x axis here). Here is an argument by contradiction (from Taylor and Wheeler, *Spacetime Physics*, 2nd edition, Freeman, 1992, p. 65). *Assume*, for the sake of argument, that an observer in one IRF *does* observe the distance between two objects to be different (say, smaller) in the direction perpendicular to the line of their motion than an observer in the objects' own IRF. We will show that this assumption and the principle of relativity together lead to an absurd conclusion, contradicting the original assumption.

Imagine a train car whose wheels are designed to sit directly on the track rails when the car is at rest. Given our assumption, the ground observer will conclude that as the train's speed grows, the distance between its wheels will decrease until the wheels slip off the rails' inside edges, as shown in figure 2.4a. The principle of relativity, however, requires that the effect described neutrally in the previous paragraph must apply to both frames equivalently. An observer on the train must observe the distance between the moving *rails* to decrease (as the speed of the ground increases) until the train's wheels fall off the rails' *outside* edges, as shown in figure 2.4b.

But this is absurd. In the aftermath of the train wreck predicted by both observers, investigators cannot find that the wheels left gouges in the track ties both inside (figure 2.4a) and outside (figure 2.4b) the rails! Our original assumption must therefore be false: observers in two IRFs must in fact *agree* about the values of distances measured perpendicular to the line of relative motion. (Note that if both observers find that the distance between the two wheels is always equal to that between rails, the absurdity is avoided.) For two frames in standard orientation, then we must have $y' = y$ and $z' = z$ if the principle of relativity holds.

Comparing coordinate times. Now we will show that a clock at rest in S' that is *present* at two events registers their time coordinate difference to be $\Delta t' = \Delta t \sqrt{1 - \beta^2}$, where Δt is the coordinate time difference between those same events as registered by a pair of synchronized clocks in frame S.

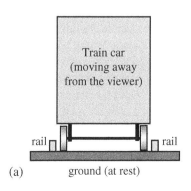
(a)
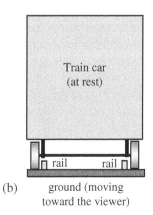
(b)

FIG. 2.4 An illustration of the train-car argument for the frame-independence of distances measured perpendicular to the direction of the relative velocity of two IRFs.

22   2. REVIEW OF SPECIAL RELATIVITY

BOX 2.3 (continued)   One Derivation of the Lorentz Transformation

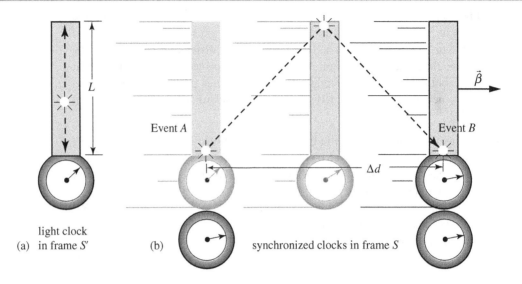

FIG. 2.5 (a) A light clock as viewed in its rest frame S'. (b) The same clock (and the zigzag path of its light flash) as viewed in frame S through which the clock is moving at speed β.

A **light clock** is a type of clock that uses a flash of light bouncing between two mirrors as its timekeeping mechanism (see figure 2.5a). Since the speed of light is 1 in the clock's frame, the clock reads time correctly if its display advances by $2L$ for every complete cycle of the flash, where L is the distance between the mirrors.

Now imagine this clock moving (along with the whole frame S') with speed β through frame S, as illustrated in figure 2.5b. Let events A and B be successive events of the light clock's flash hitting the bottom mirror. During the coordinate time Δt between the events measured in frame S, the clock covers a distance $\Delta d = \beta \Delta t$. Since the speed of light is also 1 in frame S, Δt must be the time required for light to travel the zigzag path shown in figure 2.5b, which (in meters) is simply equal to the distance traveled. By the Pythagorean theorem, this distance is simply

$$\Delta t = 2\sqrt{L^2 + (\tfrac{1}{2}\Delta d)^2} = \sqrt{(2L)^2 + \Delta d^2} = \sqrt{(\Delta t')^2 + \Delta d^2}$$
$$\Rightarrow \quad \Delta t'^2 = \Delta t^2 - \Delta d^2 = \Delta t^2 - (\beta \Delta t)^2 \quad \Rightarrow \quad \Delta t' = \Delta t\sqrt{1-\beta^2} \quad (2.11)$$

The Inverse Lorentz Transformation. Now consider the spacetime diagram shown in figure 2.6. The worldline from event O to event F is the worldline of a clock Q at rest at $x' = 0$ in frame S': event O defines $t = 0$ and $x = 0$ in both frames. Event E is an arbitrary event, and the dashed lines are the worldlines of two light flashes (which have nothing to do with the flash shown in figure 2.5): one is a right-going flash that departs from clock Q at event D and goes to event E, and the other a left-going flash that departs from event E and returns to clock Q at event F.

Clock Q is at $x' = 0$ by definition, so events D and F occur at $x'_D = x'_F = 0$ in frame S'. Since clock Q moves (with S') with speed β in the $+x$ direction in S and starts at $x = 0$ at $t = 0$, we have $x_D = \beta t_D$ and $x_F = \beta t_F$. Finally, since clock Q is present at events O and D, equation 2.11 applies, so we know that $t'_D = t_D\sqrt{1-\beta^2}$, which implies that $t_D = \gamma t'_D$ where $\gamma \equiv [1-\beta^2]^{-1/2}$. Similarly, $t_F = \gamma t'_F$. Therefore,

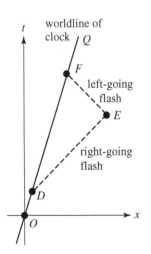

FIG. 2.6 A spacetime diagram displaying an arbitrary event E and light flashes connecting this event to the worldline of clock Q.

BOX 2.3 (continued) One Derivation of the Lorentz Transformation

$$t_D - x_D = t_D - \beta t_D = \gamma(1-\beta)t'_D = \gamma(1-\beta)(t'_D - x'_D) \qquad (2.12a)$$

$$t_F + x_F = t_F + \beta t_F = \gamma(1+\beta)t'_F = \gamma(1+\beta)(t'_F + x'_F) \qquad (2.12b)$$

Now, since the speed of light is 1 in frame S', x' increases along the right-going light-flash worldline at the same rate as t' increases, so the quantity $t' - x'$ will have the same numerical value at all points along that light flash's worldline. The same is true in frame S. Specifically, $t_E - x_E = t_D - x_D$ and $t'_E - x'_E = t'_D - x'_D$, so we have

$$t_E - x_E = \gamma(1-\beta)(t'_E - x'_E) \qquad (2.13a)$$

Similarly, in both frames, the value of x decreases as t increases along the worldline of the left-going flash, so we also have

$$t_E + x_E = \gamma(1+\beta)(t'_E + x'_E) \qquad (2.13b)$$

By adding and subtracting these equations, you can show that

$$t_E = \gamma t'_E + \gamma\beta x'_E \quad \text{and} \quad x_E = \gamma\beta t'_E + \gamma x'_E \qquad (2.14)$$

Exercise 2.3.1. In the space below, verify equations 2.14.

Since the event E was arbitrary, equations 2.14 apply to the coordinates of *any* event. Moreover, as long as events O, D, E, and F have the same y and z coordinates, this argument is independent of the y and/or z coordinates of event E. Therefore, the following equations connect the coordinates of an arbitrary event in frames S and S':

$$\begin{aligned} t &= \gamma(t' + \beta x') \\ x &= \gamma(\beta t' + x') \\ y &= y' \\ z &= z' \end{aligned} \quad \text{or, equivalently,} \quad \begin{bmatrix} t \\ x \\ y \\ z \end{bmatrix} = \begin{bmatrix} \gamma & \gamma\beta & 0 & 0 \\ \gamma\beta & \gamma & 0 & 0 \\ 0 & 0 & 1 & 0 \\ 0 & 0 & 0 & 1 \end{bmatrix} \begin{bmatrix} t' \\ x' \\ y' \\ z' \end{bmatrix} \qquad (2.15)$$

This is the *inverse* Lorentz transformation, which takes coordinates in S' and converts them to coordinates in S.

BOX 2.3 (continued) One Derivation of the Lorentz Transformation

Lorentz Transformation. Equation 2.1 claims that the *direct* Lorentz transformation from S coordinates to S' coordinates is

$$\begin{bmatrix} t' \\ x' \\ y' \\ z' \end{bmatrix} = \begin{bmatrix} \gamma & -\gamma\beta & 0 & 0 \\ -\gamma\beta & \gamma & 0 & 0 \\ 0 & 0 & 1 & 0 \\ 0 & 0 & 0 & 1 \end{bmatrix} \begin{bmatrix} t \\ x \\ y \\ z \end{bmatrix} \quad (2.16)$$

You can show that this is correct using either of two different methods: (1) you can solve the first two rows of equation 2.15 for t' and x', or (2) you can show that the matrix in 2.16 is the matrix inverse of the one in 2.15, and therefore is the transformation that takes you in the reverse direction.

Here, for your convenience, is equation 2.15 again:

$$\begin{aligned} t &= \gamma(t' + \beta x') \\ x &= \gamma(\beta t' + x') \\ y &= y' \\ z &= z' \end{aligned} \quad \text{or, equivalently,} \quad \begin{bmatrix} t \\ x \\ y \\ z \end{bmatrix} = \begin{bmatrix} \gamma & \gamma\beta & 0 & 0 \\ \gamma\beta & \gamma & 0 & 0 \\ 0 & 0 & 1 & 0 \\ 0 & 0 & 0 & 1 \end{bmatrix} \begin{bmatrix} t' \\ x' \\ y' \\ z' \end{bmatrix} \quad (2.15r)$$

Note that you can convert either matrix to the other by substituting $-\beta$ for β. This makes sense, because the only difference between the S and S' frames is that one moves in the $+x$ direction and the other moves in the $-x$ direction.

Exercise 2.3.2. In the space below, work through either one of the two methods enumerated above. (If you choose the first, start by multiplying the bottom equation by β and subtracting the result from the top equation and use the definition of γ.)

BOX 2.4 Lorentz Transformations and Rotations

The purpose of this box is to illustrate some similarities between a two-dimensional rotation transformation and a Lorentz transformation. Consider the S and S' coordinate systems illustrated in figure 2.7. The transformation that converts the coordinates of a given point in the S frame to those of the same point in the S' frame is given by the matrix equation

$$\begin{bmatrix} x' \\ y' \end{bmatrix} = \begin{bmatrix} \cos\theta & \sin\theta \\ -\sin\theta & \cos\theta \end{bmatrix} \begin{bmatrix} x \\ y \end{bmatrix} \tag{2.17}$$

You can check that this is correct pretty easily by looking at points along the x and y axes (for example, $x = r$, $y = 0$) and verify that they are correctly transformed.

The Lorentz transformation can be expressed in a similar way using hyperbolic trigonometric functions. Consider the right triangle shown in figure 2.8. The ordinary trigonometric functions are defined for this triangle as follows:

$$\sin\theta \equiv \frac{y}{\sqrt{x^2 + y^2}}, \quad \cos\theta \equiv \frac{x}{\sqrt{x^2 + y^2}}, \quad \tan\theta \equiv \frac{y}{x} \tag{2.18}$$

The hyperbolic trigonometric functions are defined analogously:

$$\sinh\theta \equiv \frac{y}{\sqrt{x^2 - y^2}}, \quad \cosh\theta \equiv \frac{x}{\sqrt{x^2 - y^2}}, \quad \tanh\theta \equiv \frac{y}{x} \tag{2.19}$$

For ordinary trigonometric functions, the denominator is the length of the triangle's hypotenuse and the angle is the actual angle between the hypotenuse and the horizontal leg, but for the hyperbolic functions, the denominator is not the triangle's hypotenuse and θ is no longer the angle of anything simple, but is rather just an abstract parameter. Still, you can see that the definitions have a similar form.

Now, imagine that we define a "velocity parameter" θ so that $\tanh\theta \equiv \beta$. If we set $y = \beta$ and $x = 1$ in the expressions above, we see that for this θ,

$$\cosh\theta = \frac{1}{\sqrt{1 - \beta^2}} \equiv \gamma, \quad \sinh\theta = \frac{\beta}{\sqrt{1 - \beta^2}} = \gamma\beta \tag{2.20}$$

and the Lorentz transformation can be written

$$\begin{bmatrix} t' \\ x' \\ y' \\ z' \end{bmatrix} = \begin{bmatrix} \cosh\theta & -\sinh\theta & 0 & 0 \\ -\sinh\theta & \cosh\theta & 0 & 0 \\ 0 & 0 & 1 & 0 \\ 0 & 0 & 0 & 1 \end{bmatrix} \begin{bmatrix} t \\ x \\ y \\ z \end{bmatrix} \tag{2.21}$$

The similarity between the part of the Lorentz transformation that involves t and x and the two-dimensional rotation is pretty clear. It is sometimes conceptually useful to think of observers in different IRFs as having different "angles of view" on the region of spacetime they both observe.

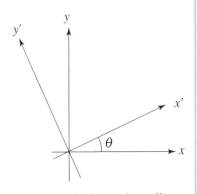

FIG. 2.7 A pair of rotated coordinate systems.

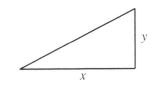

FIG. 2.8 A right triangle.

BOX 2.5 Frame-Independence of the Spacetime Interval

Consider two events whose coordinate separations in an IRF S are given by Δt, Δx, Δy, Δz. Because the Lorentz transformation is linear, the coordinate differences in a frame S' in standard orientation relative to S will be

$$\begin{bmatrix} \Delta t' \\ \Delta x' \\ \Delta y' \\ \Delta z' \end{bmatrix} = \begin{bmatrix} \gamma & -\gamma\beta & 0 & 0 \\ -\gamma\beta & \gamma & 0 & 0 \\ 0 & 0 & 1 & 0 \\ 0 & 0 & 0 & 1 \end{bmatrix} \begin{bmatrix} \Delta t \\ \Delta x \\ \Delta y \\ \Delta z \end{bmatrix} \quad \text{where} \quad \gamma \equiv \frac{1}{\sqrt{1-\beta^2}} \quad (2.22)$$

One can show by direct calculation that

$$\Delta s'^2 \equiv -\Delta t'^2 + \Delta x'^2 + \Delta y'^2 + \Delta z'^2 = -\Delta t^2 + \Delta x^2 + \Delta y^2 + \Delta z^2 \equiv \Delta s^2 \quad (2.23)$$

independent of the values of the coordinate differences or the value of β.

Exercise 2.5.1. Show this in the space below.

BOX 2.6 Frame-Dependence of the Time Order of Events

If event A causes event B, then it follows that event A must occur *before* event B in all IRFs; otherwise, in some IRFs, observers would see an event preceding its cause, which is absurd. For the sake of simplicity in our argument, let us define coordinates so that both events occur on the $+x$ axis (that is, at $y = z = 0$, with $x > 0$) in a certain frame S.

Exercise 2.6.1. In the space below, use an appropriate Lorentz transformation equation for coordinate differences to show that if $\Delta t > 0$ in frame S, but $\Delta s^2 > 0$, (i.e., the interval between the events is spacelike), then it is possible to find a frame S' moving with speed $\beta < 1$ relative to S where $\Delta t' < 0$ (i.e., the time order is different). Also show that this *not* possible if $\Delta s^2 < 0$ (i.e., the interval is timelike).

BOX 2.7 Proper Time Along a Path

Imagine that we divide up a particle's worldline into many infinitesimal steps, as illustrated in figure 2.9. Consider a pair of events A and B on this worldline whose coordinate separations are dt, dx, dy, and dz in the inertial frame S in which the particle is observed. If these separations are truly infinitesimal, the section of worldline between A and B will be essentially straight. An *inertial* clock whose straight worldline goes through both events will therefore follow essentially the same path as the particle. In this clock's frame (call it S'), the events occur at the same place, so $dx' = dy' = dz' = 0$. The squared spacetime interval between the events is thus

$$ds^2 = -dt'^2 + dx'^2 + dy'^2 + dz'^2 = -dt'^2 \qquad (2.24)$$

A clock carried by the particle should read the same time as this inertial clock. You can show, therefore, that

$$d\tau = dt' = \sqrt{-ds^2} = \sqrt{dt^2 - dx^2 - dy^2 - dz^2} = dt\sqrt{1 - v^2} \qquad (2.25)$$

where v is the speed measured in frame S. Integrating this over the worldline yields the result displayed in equation 2.4.

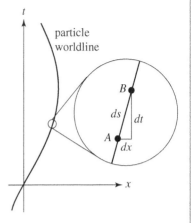

FIG. 2.9 A particle worldline showing a close-up shot of infinitesimal portion.

Exercise 2.7.1. Fill in the missing step in equation 2.25.

BOX 2.8 Length Contraction

Consider an object moving with speed v relative to an inertial frame S. We define a moving object's length L along the direction of its motion in that frame to be the distance between two events that occur simultaneously (in S) at the object's ends. Let the moving object's frame be S'. We can define coordinates so that the object moves along the $+x$ direction in S, and align the axes of S' to be in standard orientation relative to S. This frame will move at a speed $\beta = v$ relative to S.

Let events F and B happen simultaneously in the S frame ($\Delta t = 0$) at the two ends of the object. The distance between these events in frame S is therefore defined to be the object's length in that frame: $L = \Delta x$. The distance $\Delta x'$ between these same events in the object's own frame S' will be its rest length L_R.

Exercise 2.8.1. In the space below, use the Lorentz transformation for coordinate differences to show that $L = L_R\sqrt{1 - \beta^2}$ (equation 2.5).

BOX 2.9 The Einstein Velocity Transformation

Consider two IRFs S and S' in standard orientation. Imagine that an object moves through both frames. We can use the Lorentz transformation to determine the object's velocity components in S' if we know its velocity components in frame S. For example, consider two infinitesimally separated events along the particle's worldline. The Lorentz transformation tells us that

$$v'_y \equiv \frac{dy'}{dt'} = \frac{dy}{\gamma(dt - \beta\, dx)} = \frac{dy/dt}{\gamma[1 - \beta(dx/dt)]} = \frac{v_y\sqrt{1-\beta^2}}{(1-\beta v_x)} \quad (2.26)$$

where in the last step I used the definition of γ.

Exercise 2.9.1. In the space below, use the same approach to calculate v'_x.

HOMEWORK PROBLEMS

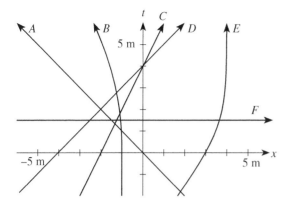

FIG. 2.10 A spacetime diagram showing the worldlines of various particles.

P2.1 The spacetime diagram in figure 2.10 shows the worldlines of various particles moving along the x axis in space. Note that (as is conventional in spacetime diagrams), I have calibrated the t and x axes in this diagram to have the same scales in the same units: this ensures that the worldline of a photon (which has a speed of 1) makes a 45° angle with each axis. For each labeled worldline, describe (1) where the particle is at time $t = 0$, (2) about how fast and in what direction (in the $+x$ direction or the $-x$ direction) the particle is moving at that time, and (3) whether the particle's speed subsequently increases or decreases as time passes. Also identify the one worldline that is physically impossible, and explain why.

P2.2 Imagine that a spaceship is docked at a space station floating in deep space. Assume that the space station defines the origin in its own frame. At $t = 0$ (call this event A) the spaceship starts accelerating in the $+x$ direction away from the space station at a constant rate (as measured in the station frame). The spaceship reaches a cruising speed of 0.50 after 8 Tm of time as measured in the station frame (call this event B). (1 Tm of time = 10^{12} m ≈ 0.93 h.) At this event, the spaceship sends a laser flash signal back to the station, reporting that it has reached its cruising speed. This signal reaches the station at event C. The technician on the space station sends a laser flash response to this message 0.5 Tm later after returning from a lunch break: call this event D. Some time later, the spaceship receives this acknowledgement (event E). Use graph paper to draw a quantitatively accurate two-dimensional spacetime diagram of the situation as an observer in the station's frame would draw it, showing (and labeling) the worldlines of the space station, the spaceship, and the laser flashes (treated as if they were particles). Include labeled points representing the events A through E. Note that it is conventional in spacetime diagrams to calibrate the t and x axes so that they have scales of the same size expressed in the same units (in this case, Tm). Also answer the following questions:

a. *Where* does event B occur in the station frame? Explain your reasoning.

b. Compute the magnitude of the ship's acceleration in GR units (m^{-1}) and in gs (where $1\ g \equiv 9.8\ m/s^2$). Note that a shockproof watch can tolerate an acceleration of about 50 g.

c. When and where does event C occur in the station frame? When and where does event E occur?

P2.3 At $t = 0$, an alien spaceship passes by the earth: let this be event A. At $t = 260$ Gm (according to synchronized clocks on earth and Mars) the spaceship passes by Mars, which is 100 Gm from the earth at the time: let this be event B. (Note that 18 Gm of time = 1.8×10^{10} m ≈ 1 minute). Radar tracking indicates that the spaceship moves at a constant velocity between the earth and Mars. Just after the ship passes the earth, people on the earth launch a probe whose purpose is to catch up with and investigate the spaceship. This probe accelerates away from the earth, moving slowly at first, but moves faster and faster as time passes, eventually catching up with and passing the alien ship just as it passes Mars. In all parts of this problem, you can ignore the effects of gravity and the relative motion of the earth and Mars (which are small) and treat the earth and Mars as if they were both at rest in the inertial reference frame of the solar system. The probe takes some pictures of the alien spacecraft at event B, and immediately sends them encoded as in a burst of laser light back to the earth. The burst arrives at the earth at event C.

a. Use graph paper to draw a quantitatively accurate two-dimensional spacetime diagram of the situation in the solar system frame, showing the worldlines of the earth, Mars, the alien spacecraft, the probe, and the flash of laser light (which you can treat as if it were a particle). Let the position of the earth define $x = 0$ in that frame. Note that it is conventional in spacetime diagrams to calibrate the t and x axes so that they have scales of the same size expressed in the same units (in this case, steps of 20 Gm will probably be appropriate). Also note that the probe's worldline can never have a slope less than 1 (because it cannot move faster than light).

b. Explain how you determined the time coordinate of event C.

c. Use the metric equation to calculate the time that elapses between events A and B in the alien spaceship's frame. (*Hint:* Note that since the spaceship is present at both events, they occur at the same *place*, the spaceship's location, in the spaceship's frame.)

P2.4 Imagine that your boss is on the earth-Pluto shuttle, which travels at a constant velocity of 0.60 straight from the earth to Pluto, a distance of 5.0 Tm (5×10^{12} m) in an inertial frame attached to the sun. Let event A be the shuttle's departure from the earth. About 1.0 Tm into the flight (according to your boss' watch), your boss

P2.4 (continued)
sends a laser message back to you on earth asking you to send a wake-up call appropriately timed so that your boss can catch a 1.0-Tm (56-minute) nap (as measured on your *boss's* watch). Let this be event B. When you receive this message at event C, you immediately reply with the wake-up call and an apology that the call is late, claiming in your defense that the laws of physics prevented a timely response. Your boss receives your message at event D.

a. Use graph paper to draw a quantitatively accurate spacetime diagram of the situation *in the solar system frame,* showing the worldlines of the earth and Pluto, the shuttle, and the two message flashes. Also indicate and label events A, B, C, and D. Note that it is conventional in spacetime diagrams to calibrate the t and x axes so that they have scales of the same size expressed in the same units. (*Hint:* You will need to use the metric equation to determine the time between events A and B in the solar system frame: it is *not* 1.0 Tm).

b. Explain in words why it is impossible to carry out your boss' request.

c. How long has your boss slept (according to your boss's watch) by the time your message is received at the shuttle? Explain carefully. (*Hint:* Again, use the metric equation, and note that the events B and D occur at the same place in the shuttle frame.)

d. Draw a spacetime diagram of the situation as viewed in the shuttle frame, and show that you get the same result (much more easily).

P2.5 In a certain inertial frame S, two events are observed to occur at the same place, but $\Delta t = 60$ m apart in time. In a different inertial frame S' the same two events are observed to occur 45 m apart in space.

a. Is the spacetime interval between these events timelike, lightlike, or spacelike?

b. What is the coordinate time interval between the events in frame S'?

c. Compute the speed of frame S as measured by observers in frame S'. (*Hint:* The events occur at the same place in S. So how far has S moved in the time between the events as seen in frame S'? What is the time between the events in frame S'?)

P2.6 A train traveling with speed $\beta = 3/5$ flashes by a train station. Let event A be the event of the train's front end passing the east end of the station. Let event B be the train's back end passing by the east end of the station. An observer on the train station measures that 100 m of time has passed between these events, and therefore concludes that the train is $\beta \Delta t = 60$ m long.

a. Use the Lorentz transformation equations to determine the time and distance between these events in the train frame.

b. How is the distance between events A and B in the train frame related to the train's length in the train's frame? Explain your answer.

P2.7 A train passes a pair of lights that blink simultaneously in the ground frame. In the train frame, which light blinks first, the one nearer the front end of the train, or the one nearer the back end? Explain your answer.

P2.8 A train 100 m long in its own frame is moving at a speed of 0.80 relative to the ground. Just as the train's rear end passes a signal light on the ground, that light blinks.

a. When do the blink photons reach the train's front according to clocks on the train?

b. When do the blink photons reach the train's front according to clocks on the ground?

P2.9 Two spacecraft, A and B, travel between space stations Alpha and Beta, which are at rest with respect to each other in deep space 6.0 Tm apart (1 Tm = 10^{12} m). Alpha and Beta together comprise an inertial reference frame. Both spacecraft leave station Alpha at noon, as read by clocks on Alpha, and both arrive at Beta 13 Tm (about 12 h) later, as read by clocks on Beta (whose clocks are synchronized with those on Alpha in their common frame). Spaceship A travels straight from Alpha to Beta at a constant velocity, while B travels at a constant speed on a semicircular path of radius 3.0 Tm (note that B has to constantly accelerate to do this).

a. Calculate the trip time measured by clocks on A.

b. Calculate the trip time measured by clocks on B.

P2.10 Consider the situation shown in figure 2.5. Imagine we turn the clock on its side so that the light flash moves parallel or anti-parallel to the clock's motion. Show that equation 2.11 still describes the relationship between $\Delta t'$ in the clock frame and Δt in the ground frame.

P2.11 Imagine that a train is moving at a speed of 4/5. A passenger points a laser out the train window perpendicular to the tracks, and the laser emits a brief flash of light. What angle does the velocity of this light flash make with the tracks in the ground frame?

P2.12 Imagine that in its own reference frame, an object emits light uniformly in all directions. Imagine also that this object moves with a speed β in the +x direction with respect some inertial reference frame S.

a. Show that the half of the light emitted into the forward hemisphere in the object's own frame is in frame S observed to be concentrated in a cone that makes an angle of $\sin^{-1}(\sqrt{1-\beta^2})$ with respect to the x axis.

b. Show that if $\beta = 0.99$, the angle within which half the object's light is concentrated is only 8.1°. (This forward concentration of the radiation emitted by a moving object is called the **headlight effect**.)

3. FOUR-VECTORS

INTRODUCTION

FLAT SPACETIME
Review of Special Relativity
Four-Vectors
Index Notation

TENSORS
Arbitrary Coordinates
Tensor Equations
Maxwell's Equations
Geodesics

SCHWARZSCHILD BLACK HOLES
The Schwarzschild Metric
Particle Orbits
Precession of the Perihelion
Photon Orbits
Deflection of Light
Event Horizon
Alternative Coordinates
Black Hole Thermodynamics

THE CALCULUS OF CURVATURE
The Absolute Gradient
Geodesic Deviation
The Riemann Tensor

THE EINSTEIN EQUATION
The Stress-Energy Tensor
The Einstein Equation
Interpreting the Equation
The Schwarzschild Solution

COSMOLOGY
The Universe Observed
A Metric for the Cosmos
Evolution of the Universe
Cosmic Implications
The Early Universe
CMB Fluctuations & Inflation

GRAVITATIONAL WAVES
Gauge Freedom
Detecting Gravitational Waves
Gravitational Wave Energy
Generating Gravitational Waves
Gravitational Wave Astronomy

SPINNING BLACK HOLES
Gravitomagnetism
The Kerr Metric
Kerr Particle Orbits
Ergoregion and Horizon
Negative-Energy Orbits

this depends on this

3. FOUR-VECTORS

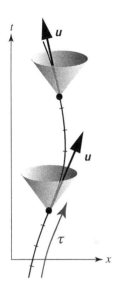

FIG. 3.1 A spacetime diagram displaying an object's worldline. We can label events along the worldline according to the proper time τ measured by the object from some suitable starting event. An object's four-velocity $\boldsymbol{u}$ is tangent to the worldline at each event and inside the event's light cone.

Describing Motion in Terms of Proper Time. In both special and general relativity, it turns out to be convenient to describe an object's motion in a given reference frame (or general coordinate system in GR) by stating how its spacetime coordinates depend on the object's own proper time τ [e.g., by specifying $t(\tau), x(\tau), y(\tau), z(\tau)$] rather than describing how its position coordinates depend on the time coordinate [e.g. by specifying $x(t), y(t), z(t)$]. We will see why this is crucial shortly, but for now, note that this at least treats all the spacetime coordinates symmetrically, and that the object's proper time τ is a frame-independent parameter, so all observers agree about its value at a given event along the object's worldline.

Four-Displacement and Four-Velocity. Consider now an infinitesimal step in proper time $d\tau$ along the object's worldline centered on a certain event E. We define an object's **four-displacement** $d\boldsymbol{s}$ during this infinitesimal proper time interval and its **four-velocity** at event E to be the four-component vectors

$$d\boldsymbol{s} \equiv \begin{bmatrix} dt \\ dx \\ dy \\ dz \end{bmatrix} \quad \text{and} \quad \boldsymbol{u} \equiv \frac{d\boldsymbol{s}}{d\tau} \equiv \begin{bmatrix} dt/d\tau \\ dx/d\tau \\ dy/d\tau \\ dz/d\tau \end{bmatrix} \equiv \begin{bmatrix} u^t \\ u^x \\ u^y \\ u^z \end{bmatrix} \qquad (3.1)$$

Note that u^t should be read "the time component of the four-velocity $\boldsymbol{u}$," not as "u raised to the power t." The reason for writing this component label as a superscript will become clear in the next chapter.

Because the components of $\boldsymbol{u}$ are the same as the components of $d\boldsymbol{s}$ multiplied by a common frame-independent scalar, all observers will agree that $\boldsymbol{u}$ is tangent to the object's worldline at event E (see figure 3.1).

Four-Vectors. Defining the four-velocity this way is valuable because if we know the components of $\boldsymbol{u}$ in one IRF, we can compute its components in a new IRF simply by using the Lorentz transformation (LT)! Let's see why. We know that the components of the displacement $d\boldsymbol{s}$ transform according to the LT:

$$\begin{bmatrix} dt' \\ dx' \\ dy' \\ dz' \end{bmatrix} = \begin{bmatrix} \gamma & -\gamma\beta & 0 & 0 \\ -\gamma\beta & \gamma & 0 & 0 \\ 0 & 0 & 1 & 0 \\ 0 & 0 & 0 & 1 \end{bmatrix} \begin{bmatrix} dt \\ dx \\ dy \\ dz \end{bmatrix} \qquad (3.2)$$

(assuming frames in standard orientation). Because $d\tau$ is frame-independent, the components of an object's four-velocity $\boldsymbol{u}$ transform then as follows:

$$\begin{bmatrix} u'^t \\ u'^x \\ u'^y \\ u'^z \end{bmatrix} = \frac{1}{d\tau}\begin{bmatrix} dt' \\ dx' \\ dy' \\ dz' \end{bmatrix} = \frac{1}{d\tau}\begin{bmatrix} \gamma & -\gamma\beta & 0 & 0 \\ -\gamma\beta & \gamma & 0 & 0 \\ 0 & 0 & 1 & 0 \\ 0 & 0 & 0 & 1 \end{bmatrix}\begin{bmatrix} dt \\ dx \\ dy \\ dz \end{bmatrix} = \begin{bmatrix} \gamma & -\gamma\beta & 0 & 0 \\ -\gamma\beta & \gamma & 0 & 0 \\ 0 & 0 & 1 & 0 \\ 0 & 0 & 0 & 1 \end{bmatrix}\begin{bmatrix} u^t \\ u^x \\ u^y \\ u^z \end{bmatrix} \qquad (3.3)$$

This is a nice, simple, and *linear* transformation law. (Contrast the complicated and *nonlinear* transformation for ordinary velocity components given by equation 2.6.)

We define a **four-vector** to be any four-component quantity whose components transform according to the LT when we change IRFs. Because four-vector components transform exactly like the components of $d\boldsymbol{s}$, if any *one* observer finds an arbitrary four-vector $\boldsymbol{A}$ to be parallel to some four-displacement $d\boldsymbol{s}$, then *all* observers will. This means that a simple four-displacement $d\boldsymbol{s}$ can serve as a universal indicator for the direction of $\boldsymbol{A}$ in spacetime in the same way that the simple spatial displacement represented by a pointing hand in ordinary 3-space can serve as a meaningful direction indicator for other vectors (force, velocity, etc.) in 3-space.

In printing, four-vectors are denoted by symbols set in ***italic bold sans-serif*** type. In handwriting, one can use a squiggly line under the symbol—e.g., $\underset{\sim}{A}$ (this is the standard copy-editor's mark indicating that a character should be bold).

CONCEPT SUMMARY

The Scalar Product and Magnitude. In special relativity, the scalar (dot) product $\mathbf{A} \cdot \mathbf{B}$ of two arbitrary four-vectors $\mathbf{A}$ and $\mathbf{B}$ is defined to be

$$\mathbf{A} \cdot \mathbf{B} \equiv -A^t B^t + A^x B^x + A^y B^y + A^z B^z \tag{3.4}$$

and the squared magnitude of an arbitrary four-vector $\mathbf{A}$ is defined to be

$$A^2 \equiv \mathbf{A} \cdot \mathbf{A} = -(A^t)^2 + (A^x)^2 + (A^y)^2 + (A^z)^2 \tag{3.5}$$

These quantities are interesting and important because their values are *frame-independent* (see box 3.1), just as the squared spacetime interval between the events at the two ends of an infinitesimal four-displacement

$$d\mathbf{s} \cdot d\mathbf{s} = ds^2 = -dt^2 + dx^2 + dy^2 + dz^2 \tag{3.6}$$

is frame-independent. The squared magnitude of an object's four-velocity is

$$\mathbf{u} \cdot \mathbf{u} \equiv -(u^t)^2 + (u^x)^2 + (u^y)^2 + (u^z)^2 = -1 \tag{3.7}$$

no matter how fast the object is moving through space (see box 3.2). This is an important result that we will use often in what follows.

Four-Velocity and Ordinary Velocity. We saw in box 2.7 that an infinitesimal interval of proper time $d\tau$ along an object's worldline is linked to the coordinate time dt for that same interval in a given IRF by

$$d\tau = dt\sqrt{1-v^2} \tag{3.8}$$

where v is the object's ordinary speed in the same IRF. We can use this to express an object's four-velocity components in a given IRF in terms of its speed in that IRF:

$$\begin{bmatrix} u^t \\ u^x \\ u^y \\ u^z \end{bmatrix} = \begin{bmatrix} \dfrac{dt}{dt\sqrt{1-v^2}} \\ \dfrac{dx}{dt\sqrt{1-v^2}} \\ \dfrac{dy}{dt\sqrt{1-v^2}} \\ \dfrac{dz}{dt\sqrt{1-v^2}} \end{bmatrix} = \begin{bmatrix} \dfrac{1}{\sqrt{1-v^2}} \\ \dfrac{v_x}{\sqrt{1-v^2}} \\ \dfrac{v_y}{\sqrt{1-v^2}} \\ \dfrac{v_z}{\sqrt{1-v^2}} \end{bmatrix} \tag{3.9}$$

Therefore, if we know $\vec{v}$, we can calculate $\mathbf{u}$. If we know $\mathbf{u}$, we can calculate $\vec{v}$ by dividing all components of equation 3.9 by u^t:

$$\begin{bmatrix} v_x \\ v_y \\ v_z \end{bmatrix} = \begin{bmatrix} u^x/u^t \\ u^y/u^t \\ u^z/u^t \end{bmatrix} \tag{3.10}$$

According to equation 3.9, the four-velocity of an object at rest is

$$\mathbf{u} = \begin{bmatrix} 1 \\ 0 \\ 0 \\ 0 \end{bmatrix} \text{ when } v = 0 \tag{3.11}$$

When an object's speed v is $\ll 1$, the square root is very nearly equal to 1, so

$$\mathbf{u} = \begin{bmatrix} u^t \\ u^x \\ u^y \\ u^z \end{bmatrix} \approx \begin{bmatrix} 1 \\ v_x \\ v_y \\ v_z \end{bmatrix} \text{ when } v \ll 1 \tag{3.12}$$

(see box 3.3). We will find the four-velocity a very useful quantity in the future, so it is wise to be very familiar with the equations in this section.

Four-Momentum. If an object has nonzero mass m, we define its **four-momentum** p to be

$$\boldsymbol{p} = m\boldsymbol{u} \quad \Rightarrow \quad \begin{bmatrix} p^t \\ p^x \\ p^y \\ p^z \end{bmatrix} = \begin{bmatrix} mu^t \\ mu^x \\ mu^y \\ mu^z \end{bmatrix} = \begin{bmatrix} \dfrac{m}{\sqrt{1-v^2}} \\ \dfrac{mv_x}{\sqrt{1-v^2}} \\ \dfrac{mv_y}{\sqrt{1-v^2}} \\ \dfrac{mv_z}{\sqrt{1-v^2}} \end{bmatrix} \quad (3.13)$$

In modern special relativity and general relativity, we consider an object's mass m to be frame-independent. Thus the four components of $\boldsymbol{p}$ transform according to the Lorentz transformation when we change frames (just multiply equation 3.3 by m to see that this is true), meaning that $\boldsymbol{p}$ is also a four-vector.

Equations 3.5 and 3.7 imply that the squared magnitude of an object's four-momentum $\boldsymbol{p}$ is simply the negative square of its frame-independent mass:

$$\boldsymbol{p} \cdot \boldsymbol{p} = (m\boldsymbol{u}) \cdot (m\boldsymbol{u}) = m^2(\boldsymbol{u} \cdot \boldsymbol{u}) = -m^2 \quad (3.14)$$

Conservation of Four-Momentum. At sufficiently small speeds, equation 3.9 implies that the spatial components of an object's four-momentum reduce to the corresponding components of the object's Newtonian momentum:

$$p^x \approx mv_x, \quad p^y \approx mv_y, \quad p^z \approx mv_z \quad (3.15)$$

so an object's four-momentum represents a reasonable relativistic generalization of the concept of Newtonian momentum.

In Newtonian physics, we assume that an isolated system's total Newtonian momentum is conserved. But what is *really* conserved, its total Newtonian momentum or its total four-momentum? It turns out that conservation of Newtonian momentum is *not* consistent with the principle of relativity: because of the nature of the Einstein velocity transformation (equation 2.6), if a system's Newtonian momentum is observed to be conserved in an IRF, it will *not* be conserved in any other IRF. On the other hand, the linear character of the Lorentz transformation law means that a law of conservation of four-momentum works in all IRFs (as long as *all four* components of $\boldsymbol{p}$ are conserved). Therefore, if anything is conserved, it must be *the system's total four-momentum*. This is explored in depth in box 3.4.

Relativistic Energy and Momentum. The time component of an object's four-momentum $\boldsymbol{p}$ is called its **relativistic energy** $E \equiv p^t$. It is also conventional to define the object's **relativistic momentum** $\vec{p}$ to be the three-dimensional vector whose components are the *spatial* components of $\boldsymbol{p}$, and $\vec{p}^2 = (p^x)^2 + (p^y)^2 + (p^z)^2$ as the squared magnitude of $\vec{p}$ (the notation $\vec{p}^2$ distinguishes it from $p^2 \equiv \boldsymbol{p} \cdot \boldsymbol{p}$). Equations 3.5, 3.10, and 3.14 then imply that

$$m^2 = E^2 - \vec{p}^2 \quad \text{and} \quad \vec{v} = \frac{\vec{p}}{E} \quad (3.16)$$

E is called the object's relativistic energy not only because it is conserved (along with the other components of $\boldsymbol{p}$) but because when $v \ll 1$,

$$E \approx m + \tfrac{1}{2}mv^2 + \ldots \quad (\text{when } v \ll 1) \quad (3.17)$$

(see box 3.3). We see that at such velocities, an object's conserved relativistic energy E reduces to its Newtonian kinetic energy plus a term equal to its mass. In any process that does not involve significant changes in mass (e.g., most everyday processes), this conservation of E thus boils down to conservation of kinetic energy, and thus to (at least a simple form of) the Newtonian law of conservation of energy.

However, equation 3.17 claims that an object's total energy is not simply its kinetic energy, but rather a sum of its kinetic energy and its mass. It is this *sum* that is conserved, not either part separately. Therefore, we see that an object's mass is just another form of energy that might in principle be converted to kinetic energy or vice versa. The expression $E = m$ that describes an object's energy when it is at rest, when written in SI units, becomes $E = mc^2$, the famous cultural icon for the idea that mass is energy (and, more generally, for Einstein's genius in realizing this). Though this idea was shocking in Einstein's time, we now know of many physical processes that convert mass energy to kinetic energy and vice versa.

Equation 3.17 applies only when $v \ll 1$. More generally, we do *not* define an object's kinetic energy K to be $\frac{1}{2}mv^2$ but rather the part of the object's relativistic energy that depends on its speed:

$$E = m + K \quad \Rightarrow \quad K = m\left(\frac{1}{\sqrt{1 - v^2}} - 1\right) \tag{3.18}$$

Four-Momentum of Light. Photons of light clearly carry energy. However, $E = p^t$, so a photon must have a four-momentum vector $\boldsymbol{p}$. But the proper time τ measured along a photon worldline is zero (see equation 3.8), so we cannot define the photon's four-velocity using equation 3.1 or its four-momentum using equation 3.13. Note, however, that equation 3.13 implies that for ordinary objects,

$$\boldsymbol{p} \equiv \begin{bmatrix} E \\ Ev_x \\ Ev_y \\ Ev_z \end{bmatrix} \tag{3.19}$$

We will assume that this applies to photons as well, and take this as the *definition* of the four-momentum of light. This definition ensures that $\boldsymbol{p}$ will be parallel to the photon's worldline, just as it is for an ordinary particle (see figure 3.2, which illustrates the special case $v_x = -1$, $v_y = v_z = 0$). It also implies that the mass of a photon (according to equation 3.16) is

$$m^2 = E^2 - \vec{p}^{\,2} = E^2 - E^2 v^2 = 0 \tag{3.20}$$

You can think of the photon's energy as being purely kinetic energy, with no associated mass energy.

Energy in a Given Observer's Frame. An observer at rest in a given IRF S has a four-velocity $\boldsymbol{u}_{obs}$ with components $u^t = 1$, $u^x = u^y = u^z = 0$ in that frame (see equation 3.11). Let a passing object's four-momentum be $\boldsymbol{p}$. Note that

$$-\boldsymbol{p} \cdot \boldsymbol{u}_{obs} = -(-p^t \cdot 1 + p^x \cdot 0 + p^y \cdot 0 + p^z \cdot 0) = +p^t = E \tag{3.21}$$

Therefore, the negative scalar product of the *object's* four-momentum and the *observer's* four-velocity yields the object's energy as measured in the observer's frame S. Since the value of the scalar product is frame-independent, we can compute $-\boldsymbol{p} \cdot \boldsymbol{u}_{obs}$ in a different (possibly more convenient) frame (where the components of $\boldsymbol{p}$ and $\boldsymbol{u}_{obs}$ may be completely different than those given above) and we will *still* get the object's energy E as measured in the observer's frame. We will find this very handy in future calculations. Box 3.5 presents an example calculation that illustrates the general utility of the frame-independence of the scalar product.

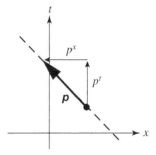

FIG. 3.2 This diagram shows the (dashed) worldline of a photon moving in the $-x$ direction with speed 1. Note that a four-momentum vector parallel to the worldline has $p^x = -p^t$, consistent with equation 3.18 (note that $v_x = -1$ here).

BOX 3.1 The Frame-Independence of the Scalar Product

If **A** and **B** are four-vectors, then when we go from inertial frame S to another inertial frame S' in standard orientation with respect to S, their components transform as follows:

$$\begin{bmatrix} A'^t \\ A'^x \\ A'^y \\ A'^z \end{bmatrix} = \begin{bmatrix} \gamma & -\gamma\beta & 0 & 0 \\ -\gamma\beta & \gamma & 0 & 0 \\ 0 & 0 & 1 & 0 \\ 0 & 0 & 0 & 1 \end{bmatrix} \begin{bmatrix} A^t \\ A^x \\ A^y \\ A^z \end{bmatrix}, \quad \begin{bmatrix} B'^t \\ B'^x \\ B'^y \\ B'^z \end{bmatrix} = \begin{bmatrix} \gamma & -\gamma\beta & 0 & 0 \\ -\gamma\beta & \gamma & 0 & 0 \\ 0 & 0 & 1 & 0 \\ 0 & 0 & 0 & 1 \end{bmatrix} \begin{bmatrix} B^t \\ B^x \\ B^y \\ B^z \end{bmatrix} \quad (3.22)$$

You can show by explicit calculation that

$$-A'^t B'^t + A'^x B'^x + A'^y B'^y + A'^z B'^z = -A^t B^t + A^x B^x + A^y B^y + A^z B^z \quad (3.23)$$

for all $\beta < 1$. This implies that the dot product $\mathbf{A} \cdot \mathbf{B}$ is frame independent.

Exercise 3.1.1. In the space below, verify equation 3.23.

BOX 3.2 The Invariant Magnitude of the Four-Velocity

As discussed in box 2.7, the proper time interval $d\tau$ for an infinitesimal step along an object's worldline is equal to

$$d\tau = \sqrt{-ds^2} = \sqrt{dt^2 - dx^2 - dy^2 - dz^2} \quad (3.24)$$

One can use this, the definition of **u** given by equation 3.1, and equation 3.6 to show rather directly that $\mathbf{u} \cdot \mathbf{u} = -1$.

Exercise 3.2.1. In the space below, verify this.

BOX 3.3 The Low-Velocity Limit of *u*

When |x| < 1, one can expand $(1 + x)^a$ in a power series as follows:

$$(1 + x)^a = 1 + ax + \frac{a(a-1)}{2!}x^2 + \frac{a(a-1)(a-2)}{3!}x^3 + \ldots \quad (3.25)$$

This is called the *binomial expansion* (the right side is essentially a Taylor series expansion of the left side). This means that

$$(1 - v^2)^{-1/2} = 1 + \tfrac{1}{2}v^2 + \tfrac{3}{8}v^4 + \ldots \quad (3.26)$$

Therefore, the four-velocity can be written

$$\begin{bmatrix} u^t \\ u^x \\ u^y \\ u^z \end{bmatrix} = \begin{bmatrix} 1 + \tfrac{1}{2}v^2 + \tfrac{3}{8}v^4 + \ldots \\ v_x(1 + \tfrac{1}{2}v^2 + \tfrac{3}{8}v^4 + \ldots) \\ v_y(1 + \tfrac{1}{2}v^2 + \tfrac{3}{8}v^4 + \ldots) \\ v_z(1 + \tfrac{1}{2}v^2 + \tfrac{3}{8}v^4 + \ldots) \end{bmatrix} \quad (3.27)$$

When $v \ll 1$, we can neglect terms involving higher powers of v, so

$$\begin{bmatrix} u^t \\ u^x \\ u^y \\ u^z \end{bmatrix} = \begin{bmatrix} 1 + O(v^2) \\ v_x[1 + O(v^2)] \\ v_y[1 + O(v^2)] \\ v_z[1 + O(v^2)] \end{bmatrix} \quad (3.28)$$

where $O(v^2)$ means that the correction terms are of the order of magnitude of v^2 or smaller. When $v < 0.1$, this means that the components of *u* will be 1, v_x, v_y, v_z with an error of less than about 1% in each term. This justifies equation 3.12.

Note that in absolute terms, the spatial components of *u* will be at least a factor of v smaller than u^t: the correction term for u^t will be $O(v^2)$, while the correction terms in the spatial components will be $O(v^3)$.

Note also that in the limit that $v \ll 1$,

$$E \equiv p^t \equiv mu^t = m(1 + \tfrac{1}{2}v^2 + \tfrac{3}{8}v^4 + \ldots) = m + \tfrac{1}{2}mv^2 + O(v^4) \quad (3.29)$$

This justifies equation 3.17.

Exercise 3.3.1. In the space below, show that equation 3.26 follows from 3.25.

BOX 3.4 Conservation of Momentum or Four-momentum?

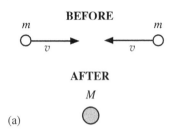

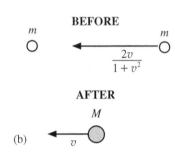

FIG. 3.3 (a) The example collision as viewed in frame S. (b) The same collision viewed in a frame S' moving to the right at the same speed as the left particle.

A single counterexample suffices to show that conservation of Newtonian momentum is inconsistent with the principle of relativity. Consider the following collision as an example. Imagine that in frame S, two identical particles of mass m are moving with the same speed v but opposite velocities $\vec{v}_1$ and $\vec{v}_2$ (see figure 3.3a). Let us define coordinates so that $\vec{v}_1$ points in the $+x$ direction. The particles then collide and form a single particle at rest ($v_3 = 0$) with mass $M = 2m$. The total Newtonian momentum of this system is clearly zero before and after the collision and so it *is* conserved in frame S. Now view this collision in a reference frame S' that moves in the $+x$ direction with speed $\beta = v$. You can use the Einstein velocity transformation to show (as illustrated in figure 3.3b) that in this frame

$$v'_{1x} = 0, \quad v'_{2x} = \frac{-2v}{1+v^2}, \quad v'_{3x} = -v \tag{3.30}$$

while the other components of all three velocities remain zero. In frame S', then, the total Newtonian x-momentum is $-2mv/(1+v^2)$ before the collision and $-2mv$ after the collision, so Newtonian momentum is *not* conserved in this frame. This contradicts the principle of relativity, which requires that any valid law of physics work in all inertial frames. So conservation of Newtonian momentum cannot be correct.

Exercise 3.4.1. In the space below, show that equation 3.30 is correct.

Now we will argue that a law of conservation of four-momentum is consistent with the principle of relativity. The argument depends only on the *linearity* of the Lorentz transformation. One of the basic characteristics of a linear transformation is that the sum of a set of transformed vectors is the same as the transformation of the sum of the original vectors: in matrix notation this looks like

$$\left[\,LT\,\right]\left[V_1\right] + \left[\,LT\,\right]\left[V_2\right] + \ldots = \left[\,LT\,\right]\left(\left[V_1\right] + \left[V_2\right] + \ldots\right) \tag{3.31}$$

where $[LT]$ is the matrix representing the linear transformation and $[V_1], [V_2], \ldots$ are column vectors. Now consider the particular case where $[LT]$ is a Lorentz transformation and the vectors are four-momentum vectors. Assume that four-momentum is conserved in some frame S. This means that if we subtract the total final four-momentum from the total initial four-momentum, we will get zero in frame S:

$$\boldsymbol{p}_{1i} + \boldsymbol{p}_{2i} + \ldots - \boldsymbol{p}_{1f} - \boldsymbol{p}_{2f} - \ldots = 0 \tag{3.32}$$

Now let us use the Lorentz transformation to calculate the total four-momentum in another inertial frame S'. Equation 3.31 means that

BOX 3.4 (continued) Conservation of Momentum or Four-momentum?

$$\left[\mathbf{p}'_{1i}\right] + \left[\mathbf{p}'_{2i}\right] + \ldots - \left[\mathbf{p}'_{1f}\right] - \left[\mathbf{p}'_{2f}\right] - \ldots$$
$$= \left[\, LT\, \right] \left(\left[\mathbf{p}_{1i}\right] + \left[\mathbf{p}_{2i}\right] + \ldots - \left[\mathbf{p}_{1f}\right] - \left[\mathbf{p}_{2f}\right] - \ldots \right) \quad (3.33)$$

But the four-vector quantity in round parentheses is equal to zero, because we are assuming that four-momentum is conserved in the unprimed frame. Another fundamental characteristic of a linear transformation is that it maps the zero vector to the zero vector (a matrix multiplied by a vector whose components are all zero yields zeros as a result). Therefore, the left side of equation 3.33 is also zero, implying that four-momentum is conserved in the primed frame as well.

Let's see how the specifics work out for our particular example collision. Note that in frame S, conservation of four-momentum requires (since $v_1 = v_2 = v$) that

$$\frac{1}{\sqrt{1-v^2}} \begin{bmatrix} m \\ mv \\ 0 \\ 0 \end{bmatrix} + \frac{1}{\sqrt{1-v^2}} \begin{bmatrix} m \\ -mv \\ 0 \\ 0 \end{bmatrix} = \begin{bmatrix} M \\ 0 \\ 0 \\ 0 \end{bmatrix} \quad (3.34)$$

The spatial components of this equation are clearly satisfied, and the time component is also if the final particle's mass is $M = 2m/\sqrt{1-v^2}$ (not $2m$). Therefore, with this value of M, four-momentum is conserved in this frame.

Now, is four-momentum also conserved in frame S'? If so, then we must have

$$\begin{bmatrix} m \\ 0 \\ 0 \\ 0 \end{bmatrix} + \frac{1}{\sqrt{1-(v'_2)^2}} \begin{bmatrix} m \\ -mv'_2 \\ 0 \\ 0 \end{bmatrix} \stackrel{?}{=} \frac{1}{\sqrt{1-v^2}} \begin{bmatrix} M \\ -Mv \\ 0 \\ 0 \end{bmatrix} \quad (3.35)$$

Exercise 3.4.2. First show that given the value of v'_{2x} specified in equation 3.30, $1/\sqrt{1-(v'_2)^2} = (1+v^2)/(1-v^2)$. Then use the frame-independent final particle mass M given above to show that four-momentum is indeed conserved in S'.

Note that the argument in this box does not necessarily imply that four-momentum *is* conserved, only that a law of conservation of four-momentum is consistent with the principle of relativity (in contrast to a law of conservation of Newtonian momentum). Abundant experimental verification (billions of tests per year in particle-physics research) provide the ultimate justification for conservation of four-momentum.

3. FOUR-VECTORS

BOX 3.5 Example: The GZK Cosmic-Ray Energy Cutoff

(Adapted from Hartle, *Gravity*, Addison-Wesley, 2003, p. 94.) Cosmic rays are high-energy elementary particles or nuclei created by violent astrophysical processes. Protons are common cosmic-ray particles, and protons with energies in excess of 3×10^{20} eV (≈ 20 J!) have been observed entering the earth's upper atmosphere.

One of the puzzling aspects about such extremely energetic particles is that they should not be able to move very far even through intergalactic space. In 1966, Griesen, Zatsepin, and Kuz'min noted that even intergalactic space contains a significant density ($\approx 4 \times 10^8$ m^{-3}) of photons from the cosmic microwave background (CMB). These photons were emitted by the Big Bang and have subsequently been red-shifted by the expansion of the universe (as we will see) to a blackbody spectrum with an average photon energy of about 6.4×10^{-4} eV. They also noted that a sufficiently energetic proton colliding with such a photon can initiate the reaction

$$p + \gamma \rightarrow p + \pi^0 \quad \text{or} \quad p + \gamma \rightarrow n + \pi^+ \tag{3.36}$$

among others, where p here stands for the proton, γ for the CMB photon, n for a neutron, π^0 is a neutral pion, and π^+ is a positive pion. Such reactions, they argued, can greatly reduce the energy of the protons before they ever reach earth.

At what proton energies do such collisions become a problem? Conservation of four-momentum for the first of the reactions above requires that

$$\boldsymbol{p}_p + \boldsymbol{p}_\gamma = \boldsymbol{p}_f + \boldsymbol{p}_\pi \tag{3.37}$$

where $\boldsymbol{p}_p, \boldsymbol{p}_\gamma, \boldsymbol{p}_f$ and $\boldsymbol{p}_\pi$ are the four-momenta of the initial proton, the CMB photon, the final proton, and the pion, respectively (the analysis for the reaction that produces a neutron will be almost exactly the same). Consider now observing this process in the "center of mass" (CM) frame where the spatial momenta of the incoming particles are equal and opposite, so that the spatial components of the initial four-momenta add to zero. The reaction will not be possible unless the incoming particles carry in an energy (in that frame) at least equal to that of a proton and pion at rest. When we just *barely* have enough energy, we have (in the CM frame)

$$\boldsymbol{p}_f + \boldsymbol{p}_\pi = \begin{bmatrix} m_p \\ 0 \\ 0 \\ 0 \end{bmatrix} + \begin{bmatrix} m_\pi \\ 0 \\ 0 \\ 0 \end{bmatrix} = \begin{bmatrix} m_p + m_\pi \\ 0 \\ 0 \\ 0 \end{bmatrix} \tag{3.38}$$

The magnitude of this total momentum four-vector for this threshold reaction is

$$(\boldsymbol{p}_f + \boldsymbol{p}_\pi)^2 = (\boldsymbol{p}_f + \boldsymbol{p}_\pi) \cdot (\boldsymbol{p}_f + \boldsymbol{p}_\pi) = -(m_p + m_\pi)^2 \tag{3.39}$$

and since equation 3.37 implies that $(\boldsymbol{p}_p + \boldsymbol{p}_\gamma)^2 = (\boldsymbol{p}_f + \boldsymbol{p}_\pi)^2$, we have

$$(\boldsymbol{p}_p + \boldsymbol{p}_\gamma)^2 = -(m_p + m_\pi)^2 \tag{3.40}$$

This will in fact be the value of $(\boldsymbol{p}_p + \boldsymbol{p}_\gamma)^2$ in *any* reference frame, because the scalar product yields a frame-independent value. So now let us evaluate this in the earth's reference frame, where we know that the CMB photons have an energy $E \approx 6 \times 10^{-4}$ eV. Writing out the squares on both sides yields

$$\cancel{p_p^2} + 2\boldsymbol{p}_p \cdot \boldsymbol{p}_\gamma + \cancel{p_\gamma^2} = -\cancel{m_p^2} - 2m_p m_\pi - m_\pi^2 \tag{3.41}$$

(note that $p_\gamma^2 = [\text{photon mass}]^2 = 0$). Let us set up coordinate axes in the earth frame so that the proton and photon move in the $+x$ and $-x$ directions respectively. Then

$$\boldsymbol{p}_p = \begin{bmatrix} E_p \\ p_{px} \\ 0 \\ 0 \end{bmatrix}, \quad \boldsymbol{p}_\gamma = \begin{bmatrix} E \\ -E \\ 0 \\ 0 \end{bmatrix} \tag{3.42}$$

BOX 3.5 (continued) Example: The GZK Cosmic-Ray Energy Cutoff

Let's *assume* that the proton energy is very large in our frame (so that $E_p \gg m_p$), do the calculation, and then check that our answer is consistent with this assumption. Since $m_p^2 = E_p^2 - p_{px}^2$, this approximation implies that

$$p_{px}^2 = E_p^2 - m_p^2 \approx E_p^2 \quad \text{so} \quad p_{px} \approx E_p \tag{3.43}$$

If you substitute this into equation 3.42, write out the scalar product in 3.41, and solve for E_p, you will find that

$$E_p \approx \frac{m_\pi(2m_p + m_\pi)}{4E} \tag{3.44}$$

Since $m_p = 938$ MeV and $m_\pi = 135$ MeV, you can show that $E_p \approx 10^{20}$ eV for CMB photons with $E \approx 6.4 \times 10^{-4}$ eV. This is the GZK cutoff for proton energies: protons with energies greater than this will be able to lose energy to collisions with CMB photons when traveling between galaxies. (Note that the extremely large value justifies our approximation that $E_p \gg m_p$.)

Exercise 3.5.1. Verify equation 3.44 and the stated value for E_p.

Of course, since the CMB spectrum is that of a blackbody, there are some CMB photons that have significantly higher energies than the average, and one can see from equation 3.44 that higher-energy photons in the CMB will cause problems for even lower-energy cosmic-ray protons. A more refined calculation that takes account of the spread of photon energies suggests that the cutoff should become significant at a proton energy of about 6×10^{19} eV. An article from the HiRes cosmic ray observatory team published in 2008 in *Physical Review Letters* (**100** (10), 101101) claims indeed to observe such a cutoff in cosmic ray energies.

The problem with this result is as follows. One can show that the mean free path for protons with such energies in intergalactic space is on the order of tens of millions of light years. So if this calculation is correct (as the HiRes data suggests), the protons with higher energies that we observe must have come from the handful of galaxies within this distance from the earth. The difficulty is that it is hard to imagine astrophysical processes that could generate such energetic protons at all, much less processes that could be taking place in the rather staid group of our local galaxies without our knowing about them from other forms of radiation. Moreover, the trajectories of observed cosmic ray protons with such energies do not seem to point to any particular nearby galaxy. But the GZK calculation clearly implies that these protons must come from local sources. This problem is known as the **GZK paradox**. A number of experiments are currently in progress to try to resolve this apparent paradox. Solving it may bring to light some new and interesting physics!

HOMEWORK PROBLEMS

P3.1 Imagine that the function
$$x(\tau) = \frac{1}{g}[\cosh(g\tau) - 1] \qquad (3.45)$$
describes the worldline of an object moving along the x axis of some inertial frame, where $x(\tau)$ is the object's x-position as a function of its proper time τ. Note that g has units of m^{-1} in GR units. (This function happens to describe the worldline of a object whose acceleration has the constant value g in its own instantaneous rest frame.) Note also that $x = 0$ at $\tau = 0$, because $\cosh(0) = 1$.

a. Calculate u^x as a function of τ for this object.
b. Use the requirement that $\boldsymbol{u} \cdot \boldsymbol{u} = -1$ to determine u^t as a function of τ for this object.
c. What is this object's speed $v = dx/dt$ in the inertial frame? Is this speed ever greater than 1?
d. Show that $gt = \sinh(g\tau)$, where t is the coordinate time measured in the inertial reference frame.
e. Use the result of the previous part to find expressions for u^x, u^t, and v in terms of gt.

[Hints: Hyperbolic trigonometric functions satisfy the following identities: $\cosh^2 x - \sinh^2 x = 1$, $d(\sinh x)/dx = \cosh x$, $d(\cosh x)/dx = +\sinh x$, $\tanh x \equiv \sinh x/\cosh x$.]

P3.2 Imagine that an object is moving in the $+x$ direction in a certain inertial frame with a speed
$$v = \sqrt{1 - \frac{1}{(gt+1)^2}} \qquad (3.46)$$
where g is a constant with units of m^{-1} and $t \geq 0$ is the time in that frame. (Note that $0 < v < 1$ for all times $t \geq 0$.)

a. Find u^t as a function of t.
b. Show that $g\tau = \ln(gt + 1)$.
c. Express u^t as a function of τ.
d. Express u^x as a function of both t and τ.
e. Find expressions for $x(\tau)$ and $t(\tau)$. (You may look up the required integral.)

P3.3 What are the components (in MeV) of p for a neutral pion ($m_\pi = 135$ MeV) moving with velocity $v = 3/5$ in the xy plane direction 38.7° counterclockwise from the x axis? [Hint: $\cos(38.7°) = 4/5$, $\sin(38.7°) = 3/5$.]

P3.4 A positive pion π^+ (with mass 140 MeV) at rest will decay after an average lifetime of 26 ns to an antimuon μ^+ (mass 106 MeV) and a muon neutrino ν_μ with negligible mass. Use conservation of four-momentum to determine the outgoing neutrino's energy. (Hint: Treat the neutrino as if it were a photon.)

P3.5 An electron and a positron (anti-electron), each with mass m, approach each other along the lab frame's x axis with equal speeds v. They collide and annihilate to form two photons with equal energies E that move in opposite directions along the x axis.

a. Show that four-momentum is conserved in the lab frame as long as $E = m/\sqrt{1 - v^2}$.
b. Use the Einstein velocity transformation to find the positron's speed in the *electron's* frame in terms of v, and use the Lorentz transformation to find the photon energies in that frame in terms of m and v. Then show that four-momentum is conserved in the electron's frame if it is conserved in the lab frame. [Hint: Remember that $1 - v^2 = (1 + v)(1 - v)$.]

P3.6 A spacecraft (with full fuel tanks) of mass M is at rest in the solar system, preparing to disembark on a grand mission to Alpha Centauri. Its engines work by combining matter and antimatter and directing the hard gamma rays that result out the spacecraft's rear. The engines fire for a brief time, bringing the ship's speed to $v = 0.95$ with respect to the solar system.

a. What is the mass m of the ship after the engines have fired, expressed as a fraction of M? (Hint: Treat the gamma-ray exhaust as a single giant photon.)
b. If the ship is supposed to go to Alpha Centauri, decelerate to rest, do some research there, and then return, what will the ship's total initial mass M have to be if the ship's empty mass is m_0 when it reaches earth? (Express your result as a multiple of m_0.)

P3.7 A photon's energy is related to its wavelength λ by the expression $E = h/\lambda$ (in GR units) where $h = 1240$ eV·nm. Imagine that a photon has a wavelength of λ_0 in the frame of the source. An observer is moving with speed v away from the source.

a. Use a Lorentz transformation of the photon's four-momentum to find the photon's energy and wavelength in the frame of the observer. Express your answer as a fraction of λ_0.
b. Show that equation 3.21 yields the same result.

P3.8 Use conservation of four-momentum to show that an electron-positron pair can *never* collide and annihilate to form a single photon, no matter how the original particles are moving. Show that it is possible, though, for the electron-positron pair to decay to a pair of photons. (Hints: A positron is an anti-electron: it has the same mass as an electron. First argue that we can always find a "center of mass" frame where the electron and positron are moving toward each other on the x axis with equal speeds.)

P3.9 Use the method discussed in box 3.5 to determine the minimum energy a moving electron striking another electron at rest must have if it is to be able to create an electron-positron pair during the collision (in addition to the original two electrons).

P3.10 Use the method discussed in box 3.5 to determine the minimum energy a photon must have if it is to be able to create an electron-positron pair after colliding with (and being absorbed by) an electron at rest. (Note that there are three particles, two electrons and one positron, after the interaction.)

4. INDEX NOTATION

INTRODUCTION

FLAT SPACETIME
| Review of Special Relativity |
| Four-Vectors |
| **Index Notation** |

TENSORS
| Arbitrary Coordinates |
| Tensor Equations |
| Maxwell's Equations |
| Geodesics |

SCHWARZSCHILD BLACK HOLES
| The Schwarzschild Metric |
| Particle Orbits |
| Precession of the Perihelion |
| Photon Orbits |
| Deflection of Light |
| Event Horizon |
| Alternative Coordinates |
| Black Hole Thermodynamics |

THE CALCULUS OF CURVATURE
| The Absolute Gradient |
| Geodesic Deviation |
| The Riemann Tensor |

THE EINSTEIN EQUATION
| The Stress-Energy Tensor |
| The Einstein Equation |
| Interpreting the Equation |
| The Schwarzschild Solution |

COSMOLOGY
| The Universe Observed |
| A Metric for the Cosmos |
| Evolution of the Universe |
| Cosmic Implications |
| The Early Universe |
| CMB Fluctuations & Inflation |

GRAVITATIONAL WAVES
| Gauge Freedom |
| Detecting Gravitational Waves |
| Gravitational Wave Energy |
| Generating Gravitational Waves |
| Gravitational Wave Astronomy |

SPINNING BLACK HOLES
| Gravitomagnetism |
| The Kerr Metric |
| Kerr Particle Orbits |
| Ergoregion and Horizon |
| Negative-Energy Orbits |

this depends on this

The Lorentz Transformation. We have seen that the Lorentz transformation of the components of an arbitrary four-vector **A** can be written in matrix form as

$$\begin{bmatrix} A'^t \\ A'^x \\ A'^y \\ A'^z \end{bmatrix} = \begin{bmatrix} \Lambda^t{}_t & \Lambda^t{}_x & \Lambda^t{}_y & \Lambda^t{}_z \\ \Lambda^x{}_t & \Lambda^x{}_x & \Lambda^x{}_y & \Lambda^x{}_z \\ \Lambda^y{}_t & \Lambda^y{}_x & \Lambda^y{}_y & \Lambda^y{}_z \\ \Lambda^z{}_t & \Lambda^z{}_x & \Lambda^z{}_y & \Lambda^z{}_z \end{bmatrix} \begin{bmatrix} A^t \\ A^x \\ A^y \\ A^z \end{bmatrix} \qquad (4.1)$$

and the inverse LT (that takes one from primed to unprimed coordinates) as

$$\begin{bmatrix} A^t \\ A^x \\ A^y \\ A^z \end{bmatrix} = \begin{bmatrix} (\Lambda^{-1})^t{}_t & (\Lambda^{-1})^t{}_x & (\Lambda^{-1})^t{}_y & (\Lambda^{-1})^t{}_z \\ (\Lambda^{-1})^x{}_t & (\Lambda^{-1})^x{}_x & (\Lambda^{-1})^x{}_y & (\Lambda^{-1})^x{}_z \\ (\Lambda^{-1})^y{}_t & (\Lambda^{-1})^y{}_x & (\Lambda^{-1})^y{}_y & (\Lambda^{-1})^y{}_z \\ (\Lambda^{-1})^z{}_t & (\Lambda^{-1})^z{}_x & (\Lambda^{-1})^z{}_y & (\Lambda^{-1})^z{}_z \end{bmatrix} \begin{bmatrix} A'^t \\ A'^x \\ A'^y \\ A'^z \end{bmatrix} \qquad (4.2)$$

(We will see why I have written row labels as superscripts and column labels as subscripts shortly.) In the very special case where the primed frame is in standard orientation and moves in the $+x$ direction with speed β relative to the unprimed frame, the nonzero components of the direct LT matrix are $\Lambda^t{}_t = \Lambda^x{}_x = \gamma \equiv [1 - \beta^2]^{-1/2}$, $\Lambda^t{}_x = \Lambda^x{}_t = -\gamma\beta$, and $\Lambda^y{}_y = \Lambda^z{}_z = 1$, and the nonzero components of the inverse LT matrix are the same except that $(\Lambda^{-1})^t{}_x = (\Lambda^{-1})^x{}_t = +\gamma\beta$. For more general frame orientations and relative velocities, the components of these matrices are different, but the transformations are *always* linear, so they can always be expressed in this form.

Abstract Indices. Even for simple cases, the matrix notation used above can involve a lot of writing, and we will be using equations like this all the time in GR. Moreover, we will eventually need to write equations (for example) where we multiply a 4 × 4 × 4 × 4 matrix by three vectors to get a vector, which is impossible to write in matrix notation. Our goal in this section is to develop a notation for such equations that is even more elegant and flexible than matrix notation.

Instead of writing A^t, A^x, A^y, A^z each time we want to refer to the components of a four-vector **A**, let us simply write A^μ, where the Greek-letter superscript (we could have used any Greek letter) stands for any *one* of the coordinates in the coordinate system we are using. So, when we use flat-space coordinates $t, x, y,$ and z,

$$A^\mu \text{ stands (abstractly) for } A^t, A^x, A^y, \text{ or } A^z \qquad (4.3)$$

Think of A^μ as representing the "μth component of **A**," with μ unspecified. Similarly, we can represent any *one* of the LT matrix components $\Lambda^t{}_t, \Lambda^t{}_x, \Lambda^t{}_y,$ etc. abstractly as $\Lambda^\mu{}_\nu$, where the abstract indices μ and ν can each take whichever one of the coordinate names (in this case, $t, x, y,$ or z) we please.

Using this notation, we can more compactly write equation 4.1 as

$$A'^\mu = \sum_{\nu = t,x,y,z} \Lambda^\mu{}_\nu A^\nu \qquad (4.4)$$

The instructions under the summation symbol tell us to assign the values t, x, y, z to the index ν and sum the four terms that result. The value of μ is left unspecified: if $\mu = t$, then this equation corresponds to the first row of equation 4.1; if $\mu = x$, it corresponds to the second line of 4.1, and so on. Equations 4.4 and 4.1 are equivalent.

Equation 4.4 can be made even *more* compact if we adopt the rule known as the **Einstein summation convention**:

> If the same Greek-letter index appears exactly once as a superscript *and* exactly once as a subscript in any single *term* of an equation, we will *assume* that term is to be summed over all four possible values of that index.

Using this convention, equations 4.1 and 4.2 can be written simply as

$$A'^\mu = \Lambda^\mu{}_\nu A^\nu \quad \text{and} \quad A^\mu = (\Lambda^{-1})^\mu{}_\nu A'^\nu \qquad (4.5)$$

respectively. Since the index ν appears once as a superscript and once as a subscript in the term on the right side of this equation, the kind of summation shown explicitly in equation 4.4 is assumed. (In contrast, we do *not* assume that μ is summed, because it appears only once in each term, and each time only as a superscript.)

We now see part of the reason I have written some component labels as superscripts: the positioning of the indices specifies exactly *how* we are to sum terms. For example, since the *column* label for the LT matrix components is in the subscript position, equation 4.5 unambiguously implies that we are summing *along the row* of the LT matrix (and down the column of the four-vector), consistent with the implied sum in the analogous matrix equation. (The component label's position has a deeper meaning as well, which we will discover in a few chapters.).

The Metric Tensor. We can write the metric equation using this notation if we define a square matrix whose components are

$$\begin{bmatrix} \eta_{tt} & \eta_{tx} & \eta_{ty} & \eta_{tz} \\ \eta_{xt} & \eta_{xx} & \eta_{xy} & \eta_{xz} \\ \eta_{yt} & \eta_{yx} & \eta_{yy} & \eta_{yz} \\ \eta_{zt} & \eta_{zx} & \eta_{zy} & \eta_{zz} \end{bmatrix} \equiv \begin{bmatrix} -1 & 0 & 0 & 0 \\ 0 & 1 & 0 & 0 \\ 0 & 0 & 1 & 0 \\ 0 & 0 & 0 & 1 \end{bmatrix} \quad (4.6)$$

Here the component labels are both written as subscripts, unlike the Lorentz transformation matrix. (However, just as in equations 4.1 and 4.2, the first subscript is a row label, while the second index is a column label.) Using this matrix, which we call the **metric tensor**, we can write the metric equation very compactly as

$$ds^2 = \eta_{\mu\nu} dx^\mu dx^\nu \quad (4.7)$$

(I will define the term "tensor" more precisely in chapter 6). There are *two* implied sums here, since both μ and ν appear once as a superscript and once as a subscript. Writing out the terms in the implied sums yields a sum with $4 \times 4 = 16$ terms ($\eta_{tt} dt\, dt + \eta_{tx} dt\, dx + \eta_{ty} dt\, dy + \ldots$), but since only the four diagonal components of the metric tensor are nonzero, equation 4.7 boils down to

$$\begin{aligned} ds^2 &= \eta_{tt} dt^2 + \eta_{xx} dx^2 + \eta_{yy} dy^2 + \eta_{zz} dz^2 \\ &= -dt^2 + dx^2 + dy^2 + dz^2 \end{aligned} \quad (4.8)$$

We can similarly write the dot product of two four-vectors as

$$\mathbf{A} \cdot \mathbf{B} = \eta_{\mu\nu} A^\mu B^\nu \quad (4.9)$$

and the squared magnitude of a four-vector as

$$A^2 = \mathbf{A} \cdot \mathbf{A} = \eta_{\mu\nu} A^\mu A^\nu \quad (4.10)$$

These equations are not all that more compact than their counterparts written normally, so the benefit of the new notation may not be obvious, but be patient....

The Kronecker Delta. The **Kronecker delta** $\delta^\mu{}_\nu$ is a 16-component object whose components are equal to 1 if $\mu = \nu$ and 0 otherwise. Expressed as a matrix,

$$\begin{bmatrix} \delta^t{}_t & \delta^t{}_x & \delta^t{}_y & \delta^t{}_z \\ \delta^x{}_t & \delta^x{}_x & \delta^x{}_y & \delta^x{}_z \\ \delta^y{}_t & \delta^y{}_x & \delta^y{}_y & \delta^y{}_z \\ \delta^z{}_t & \delta^z{}_x & \delta^z{}_y & \delta^z{}_z \end{bmatrix} = \begin{bmatrix} 1 & 0 & 0 & 0 \\ 0 & 1 & 0 & 0 \\ 0 & 0 & 1 & 0 \\ 0 & 0 & 0 & 1 \end{bmatrix} \quad (4.11)$$

we see that the Kronecker delta expresses the identity matrix in abstract-index form. For example, the matrix equation $[\Lambda^{-1}][\Lambda] = [I]$ (which states that a LT to the primed frame followed by an inverse LT back to the unprimed frame should be equivalent to no transformation) can be written in abstract-index notation as

$$(\Lambda^{-1})^\mu{}_\alpha \Lambda^\alpha{}_\nu = \delta^\mu{}_\nu \quad (4.12)$$

with an implicit sum over all possible values of α.

Summing over either index of the Kronecker delta is equivalent to replacing the value of the summed index with the value of the Kronecker delta's other index. For example, you can show (see box 4.1) that

$$\delta^\mu{}_\nu A^\nu = A^\mu, \quad \delta^\mu{}_\alpha \eta_{\mu\nu} = \eta_{\alpha\nu}, \text{ and so on} \tag{4.13}$$

The Electromagnetic Field Tensor. Another interesting quantity in special relativity is the electromagnetic field tensor, whose components are

$$\begin{bmatrix} F^{tt} & F^{tx} & F^{ty} & F^{tz} \\ F^{xt} & F^{xx} & F^{xy} & F^{xz} \\ F^{yt} & F^{yx} & F^{yy} & F^{yz} \\ F^{zt} & F^{zx} & F^{zy} & F^{zz} \end{bmatrix} = \begin{bmatrix} 0 & E_x & E_y & E_z \\ -E_x & 0 & B_z & -B_y \\ -E_y & -B_z & 0 & B_x \\ -E_z & B_y & -B_x & 0 \end{bmatrix} \tag{4.14}$$

where $[E_x, E_y, E_z]$ are the components of the electric field vector $\vec{E}$ and $[B_x, B_y, B_z]$ are the same for the magnetic field vector $\vec{B}$ (in GR units, both vectors have the same units of $kg \cdot C^{-1} m^{-1}$: see box 4.2). Using this tensor, we can write a relativistically valid version of the Lorentz force law (which describes the total electromagnetic force acting on a particle with charge q moving through an electromagnetic field),

$$\frac{dp^\mu}{d\tau} = qF^{\mu\nu}\eta_{\nu\alpha}u^\alpha \tag{4.15}$$

and Gauss's law and the Ampere-Maxwell relation become the single equation

$$\frac{\partial F^{\mu\nu}}{\partial x^\nu} = 4\pi k J^\mu \tag{4.16}$$

where k is the Coulomb constant (in GR units), and $J^t = \rho$, $J^x = \rho v_x$, $J^y = \rho v_y$, and $J^z = \rho v_z$ are the components of the four-current $\boldsymbol{J}$ of charge flowing at the event (ρ is the density of charge at the event in question and v_x, v_y, and v_z are the usual velocity components of the flowing charge). In equation 4.16, the superscript ν in the denominator of the derivative is considered to be equivalent to a subscript in the numerator, so there *is* an implicit sum over ν. See box 4.3 for a discussion of these equations. However, you can see how the abstract component notation here yields very compact versions of these important equations.

Free and Bound Indices. Consider the following example equation regarding the total four-momentum of a two-particle system:

$$p'^\mu_{\text{tot}} = p'^\mu_1 + p'^\mu_2 = \Lambda^\mu{}_\nu p^\nu_1 + \Lambda^\mu{}_\alpha p^\alpha_2 \tag{4.17}$$

If $\mu = t$, this equation tells us that the primed-frame t component of the system's four-momentum $\boldsymbol{p}_{\text{tot}}$ is the sum of the primed-frame t components of particle 1's four-momentum $\boldsymbol{p}_1$ and particle 2's four-momentum $\boldsymbol{p}_2$, which in turn is the sum of the t rows of the matrix equations for the Lorentz transformation of $\boldsymbol{p}_1$'s and $\boldsymbol{p}_2$'s unprimed-frame components. If $\mu = x$, the equation makes the corresponding statement about the x components and so on. This abstract equation therefore actually represents four equations about the four components of the total momentum. Note that there are implicit sums over the indices ν and α.

What I want you to focus on at the moment is the *structure* of this equation. The index μ appears in *every* term of the equation, and it is not summed in *any* term. We are free to assign any value to μ that we like in order to specify which component of the equation we are talking about. We therefore call μ a **free index**.

In contrast, the sole purpose of the pair of indices labeled ν in the term $\Lambda^\mu{}_\nu p^\nu_1$ is to indicate a sum. We are *not* free to specify a value for these indices: they must together take on all four possible values and the four resulting terms summed if the equation is to make any sense. We call ν a **bound index** (some texts call this a **dummy index**). The index α in the next term is also a bound index.

Do the exercises in box 4.4 to practice identifying free and bound indices.

The Rules. Because abstract-index notation is so abstract, it is easy to write equations that superficially look good but make no sense. Similarly, when manipulating such equations, it is all too easy to convert a perfectly good equation into nonsense. To help you avoid making such errors, let me list some rules to follow.

1. **The number of free indices.** One cannot add or equate quantities with different numbers of components. Therefore every term in an equation must have the same number of free indices (in the same vertical positions), and you must use the same symbols for these free indices across all terms.

    Bad: $A^2 = \eta_{\alpha\beta} A^\mu A^\nu$ Bad: $A^\mu = B^\nu$ Bad: $A_\mu = B^\mu$

    The only exception is that one can (by convention) set a quantity with any number of components equal to zero: for example, $A^\mu = 0$ is allowed, meaning that each component of the four-vector $\mathbf{A}$ is equal to zero.

2. **Renaming Indices.** One can rename any index with a different letter (the choice of Greek letters is arbitrary) as long as (1) you rename *every* occurrence of that letter in the equation, and (2) (when renaming *bound* indices in a given term) you avoid using a letter already used *in that same term*. (It is OK to rename bound indices in one term to have the same letter as bound indices in a different term: indeed, this is often useful.)

    Bad: $A'^\mu = \Lambda^\mu{}_\nu A^\nu \rightarrow A'^\mu = \Lambda^\mu{}_\mu A^\mu$ or $A'^\alpha = \Lambda^\mu{}_\nu A^\nu$

Otherwise, you can treat abstract-index equations as if they were ordinary symbolic equations: you can multiply both sides by the same quantity, pull out common factors, and/or distribute common factors into a sum of terms. The point of the rules above is to help ensure that any manipulations that you do preserve the nature of any implicit sums and help you recognize errors when you make them.

However, if you are ever anxious about this notation, there is a foolproof (if tedious) way to check that a given manipulation does not do something illegal:

3. **The Fundamental Rule.** When in doubt, write it out!

In other words, if you are unsure that what you are doing is legal, write out any implied sums and make sure that what you are doing is legal for the expanded equation (where the rules of ordinary algebra apply).

Do the exercises in box 4.5 to practice using these rules.

Some Useful Identities. We can use the abstract notation to work out some nice proofs that would otherwise require a lot of writing. For example, we can use the frame-independence of the spacetime interval to show that any valid LT matrix must satisfy the identity

$$\eta_{\alpha\beta} = \eta_{\mu\nu} \Lambda^\mu{}_\alpha \Lambda^\nu{}_\beta \tag{4.18}$$

and that the same applies to the inverse transformation

$$\eta_{\alpha\beta} = \eta_{\mu\nu} (\Lambda^{-1})^\mu{}_\alpha (\Lambda^{-1})^\nu{}_\beta \tag{4.19}$$

We can also prove the following useful result for an arbitrary four-vector $\mathbf{A}$:

$$\frac{d}{d\tau}(A^2) = 2\eta_{\mu\nu} A^\mu \frac{dA^\nu}{d\tau} \tag{4.20}$$

We can also use the fact that the components of the electromagnetic field tensor obey the relation $F^{\mu\nu} = -F^{\nu\mu}$ to show that

$$F^{\mu\nu} \eta_{\mu\alpha} \eta_{\nu\beta} u^\alpha u^\beta = 0 \tag{4.21}$$

no matter what a particle's four-velocity $\mathbf{u}$ might be.

Working through these proofs is very instructive practice in using abstract index notation. See box 4.6 for details.

BOX 4.1 Behavior of the Kronecker Delta

Exercise 4.1.1. Write out the implicit sum in equation $\delta^\mu{}_\nu A^\nu = A^\mu$ for all four possible values of μ and show that the equality is satisfied in all four cases. Do the same for $\delta^\mu{}_\nu \eta_{\mu\alpha} = \eta_{\nu\alpha}$ for all four values of ν.

BOX 4.2 EM Field Units in the GR Unit System

Exercise 4.2.1. The SI units for the electric field are N/C. Show that the equivalent GR units are kg/(m·C).

Exercise 4.2.2. The SI unit for the magnetic field is the tesla, where 1 T = N·s/(m·C). Show that the equivalent GR units are kg/(m·C), just as for the electric field.

BOX 4.3 Electromagnetic Equations in Index Notation

You may recall that the **Lorentz force law** for the Newtonian electromagnetic force on a particle with charge q moving at a velocity $\vec{v}$ through a point where the electric and magnetic fields are $\vec{E}$ and $\vec{B}$ is

$$\vec{F}_{em} = q(\vec{E} + \vec{v} \times \vec{B}) \tag{4.22}$$

If no other forces act on this particle, the force is equal to the particle's mass m times its acceleration $\vec{a}$. Now, in the low-velocity limit, the components of this particle's four-velocity are $u^t \approx 1$, $u^x \approx v_x$, $u^y \approx v_y$, and $u^z \approx v_z$, and those of its four-momentum are $p^t = m + K$, $p^x \approx mv_x$, $p^y \approx mv_y$, and $p^z \approx mv_z$.

Exercise 4.3.1. Write out the implied sum in $dp^\mu/d\tau = qF^{\mu\nu}\eta_{\nu\alpha}u^\alpha$ for $\mu = x$ and show that it is equivalent to the x component of equation 4.22 at low velocities.

You may also recall that Gauss's law and the Ampere-Maxwell relation in differential form are typically written in classical E&M as

$$\text{div}(\vec{E}) = \frac{\partial E_x}{\partial x} + \frac{\partial E_y}{\partial y} + \frac{\partial E_z}{\partial z} = \frac{\rho}{\varepsilon_0} \quad \text{and} \quad \text{curl}(\vec{B}) - \varepsilon_0\mu_0\frac{\partial \vec{E}}{\partial t} = \mu_0\vec{J} \tag{4.23}$$

where ρ is the local charge density, $\vec{J} = \rho\vec{v}$ is the local current density, and the constants ε_0 and μ_0 are related to the Coulomb constant k by $4\pi k = 1/\varepsilon_0 = \mu_0$ (in GR units). Note that $\text{curl}(\vec{B})$ is a vector with components $\partial B_z/\partial y - \partial B_y/\partial z$, $\partial B_x/\partial z - \partial B_z/\partial x$, and $\partial B_y/\partial x - \partial B_x/\partial y$, respectively.

Exercise 4.3.2. Show that when $\mu = t$, $\partial F^{\mu\nu}/\partial x^\nu = 4\pi k J^\mu$ becomes Gauss's law, and when $\mu = x$, it becomes the x component of the Ampere-Maxwell relation.

BOX 4.4 Identifying Free and Bound Indices

The following are all valid equations in index notation:

a) $\eta_{\mu\nu} \dfrac{du^\mu}{d\tau} u^\nu = 0$ (4.24a)

b) $\eta_{\mu\nu}(\Lambda^{-1})^\nu{}_\alpha = \eta_{\mu\beta}\Lambda^\beta{}_\alpha$ (4.24b)

c) $\dfrac{dp^\mu}{d\tau} = 0$ for a free particle (4.24c)

d) $\eta_{\alpha\mu}\eta_{\beta\nu}\dfrac{\partial F^{\mu\nu}}{\partial x^\sigma} + \eta_{\sigma\mu}\eta_{\alpha\nu}\dfrac{\partial F^{\mu\nu}}{\partial x^\beta} + \eta_{\beta\mu}\eta_{\sigma\nu}\dfrac{\partial F^{\mu\nu}}{\partial x^\alpha} = 0$ (4.24d)

e) $F^{\mu\nu}\eta_{\mu\alpha}\eta_{\nu\beta}u^\alpha u^\beta = 0$ (4.24e)

f) $\eta_{\mu\nu}\eta^{\mu\nu} = 4$ (4.24f)

Exercise 4.4.1. Circle the free indices in the equations above (or write "none"). How many component equations does each equation represent?

BOX 4.5 Rule Violations

Consider the following equations:

a) $p^\mu = mu^\alpha \delta^\mu{}_\beta$ (4.25a)

b) $p^\mu \eta_{\mu\nu} p^\nu + m^2 = 0$ (4.25b)

c) $\dfrac{dp^\mu}{d\tau} = 0$ for a free particle (4.25c)

d) $\eta_{\alpha\beta} = \eta_{\mu\nu}\Lambda^\mu{}_\alpha \Lambda^\nu{}_\beta$ (4.25d)

e) $\dfrac{dp^\alpha}{d\tau} = qF^{\mu\nu}u^\alpha$ (4.25e)

Exercise 4.5.1. Which of these equations violate Rule 1 about free indices? Write "OK" or "violates" next to each equation, as appropriate.

Consider the following examples of renaming an index:

a) $\Lambda^\alpha{}_\beta A^\beta = A'^\alpha$ renamed to $\Lambda^\alpha{}_\beta A^\beta = A'^\mu$ (4.26a)

b) $\eta_{\mu\nu}A^\mu B^\nu = 0$ renamed to $\eta_{\mu\mu}A^\mu B^\mu = 0$ (4.26b)

c) $\dfrac{dp^\mu}{d\tau} = 0$ renamed to $\dfrac{dp^\alpha}{d\tau} = 0$ (4.26c)

d) $\eta_{\alpha\beta} = \eta_{\mu\nu}\Lambda^\mu{}_\alpha \Lambda^\nu{}_\beta$ renamed to $\eta_{\mu\nu} = \eta_{\alpha\beta}\Lambda^\alpha{}_\mu \Lambda^\beta{}_\nu$ (4.26d)

e) $\eta_{\mu\nu}A^\mu B^\nu + \eta_{\alpha\beta}A^\alpha C^\beta = 0$ renamed to $\eta_{\mu\nu}A^\mu B^\nu + \eta_{\mu\nu}A^\mu C^\nu = 0$ (4.26e)

f) $\delta^\alpha{}_\mu \delta^\beta{}_\nu F^{\mu\nu} = F^{\alpha\beta}$ renamed to $\delta^\mu{}_\mu \delta^\nu{}_\nu F^{\mu\nu} = F^{\mu\nu}$ (4.26f)

Exercise 4.5.2. Which of these equations violate Rule 2 about renaming indices? Write "OK" or "violates" next to each equation, as appropriate.

BOX 4.6 Example Derivations

The purpose of this box is to illustrate some example derivations that one might do with abstract index notation. First consider equation 4.18. Because the spacetime interval is frame invariant, note that

$$ds^2 = \eta_{\mu\nu} dx^\mu dx^\nu = \eta_{\mu\nu} dx'^\mu dx'^\nu \qquad (4.27)$$

Using the definition of the Lorentz transformation, we can write this

$$\eta_{\mu\nu} dx^\mu dx^\nu = \eta_{\mu\nu}(\Lambda^\mu{}_\alpha dx^\alpha)(\Lambda^\nu{}_\beta dx^\beta) = \eta_{\mu\nu}\Lambda^\mu{}_\alpha \Lambda^\nu{}_\beta dx^\alpha dx^\beta \qquad (4.28)$$

(the last step follows because multiplication is commutative for the factors in any given term of the implied sums). Now we can rename the indices on the right side so that $\mu \to \alpha$, $\nu \to \beta$, $\alpha \to \mu$, and $\beta \to \nu$ (we can do this swapping of names as long as we preserve the distinctiveness of each implied sum and don't mix it up with the other implied sums).

$$\eta_{\mu\nu} dx^\mu dx^\nu = \eta_{\alpha\beta}\Lambda^\alpha{}_\mu \Lambda^\beta{}_\nu dx^\mu dx^\nu \qquad (4.29)$$

Now subtract the right side from the left, and pull out the common factor $dx^\mu dx^\nu$:

$$(\eta_{\mu\nu} - \eta_{\alpha\beta}\Lambda^\alpha{}_\mu \Lambda^\beta{}_\nu) dx^\mu dx^\nu = 0 \qquad (4.30)$$

This must be true for any possible four displacement, whose components dx^μ can be completely arbitrary (though infinitesimal). This is possible only if the quantity in parentheses is itself zero. For example, equation 4.30 must be true when only one component (say dx) is nonzero, which establishes that the quantity in parentheses is zero when $\mu = \nu = x$. Equation 4.30 must also be true when *two* components are nonzero, which (in combination with the first result) implies that it must be true when μ and ν are different. The bottom line is that we must have

$$\eta_{\mu\nu} - \eta_{\alpha\beta}\Lambda^\alpha{}_\mu \Lambda^\beta{}_\nu = 0 \quad \Rightarrow \quad \eta_{\mu\nu} = \eta_{\alpha\beta}\Lambda^\alpha{}_\mu \Lambda^\beta{}_\nu \qquad (4.31)$$

Exercise 4.6.1. Use a similar approach (involving renaming indices and grouping like terms) to prove equation 4.20.

We can prove equation 4.21 as follows. A fundamental property of the electromagnetic field tensor is that $F^{\mu\nu} = -F^{\nu\mu}$. Therefore

$$F^{\mu\nu}\eta_{\mu\alpha}\eta_{\nu\beta}u^\alpha u^\beta = -F^{\nu\mu}\eta_{\mu\alpha}\eta_{\nu\beta}u^\alpha u^\beta \qquad (4.32)$$

On the right side, rename μ to ν and vice versa, and α to β and vice versa:

$$F^{\mu\nu}\eta_{\mu\alpha}\eta_{\nu\beta}u^\alpha u^\beta = -F^{\mu\nu}\eta_{\nu\beta}\eta_{\mu\alpha}u^\beta u^\alpha = -F^{\mu\nu}\eta_{\mu\alpha}\eta_{\nu\beta}u^\alpha u^\beta \qquad (4.33)$$

The last step follows because the order of multiplication is unimportant. But we see that the two sides are now completely equivalent except for the minus sign. Since there are no free indices, this quantity is a simple scalar. The only scalar that is equal to negative itself is zero. Therefore, the structure of the EM field tensor means that

$$F^{\mu\nu}\eta_{\mu\alpha}\eta_{\nu\beta}u^\alpha u^\beta = 0 \qquad (4.34)$$

HOMEWORK PROBLEMS

P4.1 Which of the following are validly constructed index equations? (Consider only the equation's structure, not its meaning.) If they are not, what is the problem?

a. $0 = m^2 + (p^\mu)^2$

b. $dF^{\mu\nu}/d\tau = 0$

c. $dp^\mu/d\tau = g$ (where g is some constant)

d. $F_{\alpha\beta} = \eta_{\alpha\mu}\eta_{\beta\nu}F^{\mu\sigma}$

e. $A^{\alpha\beta} = \eta_{\alpha\mu}\eta_{\beta\nu}F^{\mu\nu}$ (where [A] is some matrix)

f. $A^\mu = \delta^\mu{}_\alpha A^\alpha$ (where is some four-vector)

g. $0 = A^\mu + B^\nu$ (where and are four-vectors)

h. $qF^{\mu\nu} = dp^\mu/d\tau$

P4.2 In which of the following cases have I renamed indices in a valid way? (Consider only the equation's structure, not its meaning.)

a. $A^2 = \eta_{\alpha\beta}A^\alpha A^\beta \Rightarrow A^2 = \eta_{\mu\nu}A^\alpha A^\beta$

b. $0 = \eta_{\alpha\beta}A^\beta + \eta_{\alpha\mu}B^\mu \Rightarrow 0 = \eta_{\alpha\beta}(A^\beta + B^\beta)$

c. $\eta_{\mu\nu} = \eta_{\alpha\beta}\Lambda^\alpha{}_\mu \Lambda^\beta{}_\nu \Rightarrow \eta_{\mu\nu} = \eta_{\alpha\alpha}\Lambda^\alpha{}_\mu \Lambda^\alpha{}_\nu$

d. $dp^\mu/d\tau = qF^{\mu\nu}\eta_{\nu\alpha}u^\alpha \Rightarrow dp^\mu/d\tau = qF^{\mu\nu}\eta_{\nu\mu}u^\mu$

e. $(\Lambda^{-1})^\alpha{}_\mu \eta_{\alpha\nu} = \eta_{\mu\beta}\Lambda^\beta{}_\nu \Rightarrow (\Lambda^{-1})^\beta{}_\mu \eta_{\beta\nu} = \eta_{\mu\alpha}\Lambda^\alpha{}_\nu$

P4.3 Show that $(\Lambda^{-1})^\alpha{}_\mu \eta_{\alpha\nu} = \eta_{\mu\beta}\Lambda^\beta{}_\nu$.

P4.4 Prove equation 4.19.

P4.5 What is the value of $\delta^\mu{}_\mu$? Explain your reasoning. (*Hint:* It is *not* equal to 1. Knowing this will be very important to us later.)

P4.6 Use equation 4.21 to prove that the square magnitude of a charged particle's four-momentum (which is its mass) is conserved (as it must be) when it moves in an electromagnetic field. In other words, show that if equations 4.15 and 4.21 are true, then

$$\frac{d}{d\tau}(\mathbf{p}\cdot\mathbf{p}) = 0 \tag{4.35}$$

P4.7 Write out equation 4.15 in the low-velocity limit when $\mu = t$. What is the physical meaning of the resulting equation? (*Hint:* Remember that the time component of the four-momentum is relativistic energy.)

P4.8 Show that $\eta_{\mu\nu}F^{\mu\nu} = 0$ if $F^{\mu\nu}$ is the electromagnetic field tensor.

P4.9 Evaluate $\eta_{\mu\alpha}\eta_{\nu\beta}F^{\mu\nu}F^{\alpha\beta}$ in terms of the components of $\vec{E}$ and $\vec{B}$.

P4.10 As we will see in chapter 6, the components of the electromagnetic field tensor in a primed inertial reference frame are related to its components in an unprimed reference frame according to the following rule:

$$F'^{\mu\nu} = \Lambda^\mu{}_\alpha \Lambda^\nu{}_\beta F^{\alpha\beta} \tag{4.36}$$

where [Λ] is the Lorentz transformation matrix from the unprimed to the primed frame. This is a generalization of the transformation rule for four-vectors given by equation 4.5: here there is one Lorentz transformation matrix factor for *each* upper index in .

Assuming that this is correct, use the result of problem P4.3 to prove that the quantity defined in problem P4.9 has the same numerical value in every inertial reference frame, and therefore represents something that is frame-independent about an electromagnetic field.

P4.11 Use index notation to argue that for any particle having nonzero rest mass, $\mathbf{u}\cdot\mathbf{a} = 0$, where = four-acceleration $\equiv d\mathbf{u}/d\tau$. (*Hint:* Note that according to equation 3.7, $\mathbf{u}\cdot\mathbf{u} = -1$ for all particles that have nonzero rest mass.)

5. ARBITRARY COORDINATES

INTRODUCTION

FLAT SPACETIME
Review of Special Relativity
Four-Vectors
Index Notation

TENSORS
Arbitrary Coordinates
Tensor Equations
Maxwell's Equations
Geodesics

SCHWARZSCHILD BLACK HOLES
The Schwarzschild Metric
Particle Orbits
Precession of the Perihelion
Photon Orbits
Deflection of Light
Event Horizon
Alternative Coordinates
Black Hole Thermodynamics

THE CALCULUS OF CURVATURE
The Absolute Gradient
Geodesic Deviation
The Riemann Tensor

THE EINSTEIN EQUATION
The Stress-Energy Tensor
The Einstein Equation
Interpreting the Equation
The Schwarzschild Solution

COSMOLOGY
The Universe Observed
A Metric for the Cosmos
Evolution of the Universe
Cosmic Implications
The Early Universe
CMB Fluctuations & Inflation

GRAVITATIONAL WAVES
Gauge Freedom
Detecting Gravitational Waves
Gravitational Wave Energy
Generating Gravitational Waves
Gravitational Wave Astronomy

SPINNING BLACK HOLES
Gravitomagnetism
The Kerr Metric
Kerr Particle Orbits
Ergoregion and Horizon
Negative-Energy Orbits

this depends on → this

5. ARBITRARY COORDINATES

Introduction to Arbitrary Coordinates. As we saw at the beginning of the course, general relativity tells us that gravity results from curved spacetime. We have seen in the past few chapters how to describe the flat spacetime of special relativity using cartesian spatial coordinates and a time coordinate defined by synchronized clocks in an inertial frame. But in curved spacetimes, we cannot use cartesian coordinates. Moreover, since our eventual goal is to calculate *how* matter curves the spacetime around it, we often do not know the spacetime's geometry *a priori*, and therefore do not know what kind of coordinate system to use.

Our goal in the next few chapters is to develop mathematical techniques for writing physical equations in a way that is completely independent of the coordinate system we actually end up using. This will generalize the principle of relativity: we will learn how to express the laws of physics in a way that is not only independent of our choice of inertial reference frame, but in fact entirely independent of our choice of coordinates!

A **coordinate system** is ultimately simply some kind of organized scheme for attaching numbers (**coordinates**) to points in space and/or events in spacetime. The clock-lattice scheme that we considered in chapter 2 is one way, but by no means the only way, to attach coordinates to events. The *only* assumptions that we will make here about our coordinate systems are that (1) our space is not so horribly curved that we cannot treat a sufficiently small patch of it as if it were flat, and (2) our coordinates vary smoothly so that neighboring points have nearly the same coordinates.

To make things simple and easy to visualize, we will in this chapter be primarily working with arbitrary coordinates in a flat two-dimensional (2D) space. However, the *methods* we develop for handling arbitrary coordinates will end up working just as well for curved spaces in any number of dimensions.

No matter how we construct our coordinate system, the distance ds between two infinitesimally separated points is a coordinate-independent quantity, because we can measure it directly with a ruler without having to define a coordinate system at all. The fundamental way that we connect arbitrary coordinates to physical reality is by specifying how the distance between two infinitesimally-separated points depends on their coordinate separations. A **cartesian** x, y coordinate system is one in which the distance ds between two infinitesimally-separated points is given by $ds^2 = dx^2 + dy^2$ everywhere in the 2D plane. A **curvilinear** coordinate system is any non-cartesian coordinate system where this simple Pythagorean relationship is not true. How can we connect the coordinate-independent distance between two points with their coordinate separations in such a case?

Definition of a Coordinate Basis. Consider arbitrary coordinates u, w for a 2D space. When using index notation, we will interpret dx^u as being equivalent to du, and dx^w as being equivalent to dw, and we will assume that Greek indices have two possible values u and w. (In the last chapter, in the context of cartesian coordinates in flat spacetime, I stated that indices could represent either t, x, y, or z, but when we use arbitrary coordinates, the indices represent whatever the index names might be.) I will also represent 2D vectors with the same bold-face notation as we used for four-vectors in the previous chapters. This will keep the notation from changing when we generalize to 4D spacetimes.

Now, no matter how our u, w coordinate system is defined, at each point $\mathcal{P}$ in the space, we can define a pair of basis vectors $\mathbf{e}_u, \mathbf{e}_w$ such that

1. $\mathbf{e}_u$ points tangent to the w = constant curve toward increasing u.
2. $\mathbf{e}_w$ points tangent to the u = constant curve toward increasing w.
3. The lengths of $\mathbf{e}_u, \mathbf{e}_w$ are defined so that the displacement vector $d\mathbf{s}$ between the point $\mathcal{P}$ at coordinates u, w and any infinitesimally separated neighboring point $\mathcal{Q}$ at coordinates $u + du, w + dw$ can be written

$$d\mathbf{s} = du\,\mathbf{e}_u + dw\,\mathbf{e}_w = dx^\mu \mathbf{e}_\mu \tag{5.1}$$

Figure 5.1 illustrates how these basis vectors are defined.

CONCEPT SUMMARY

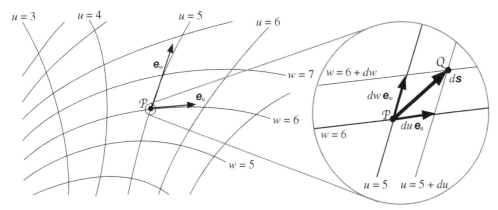

FIG. 5.1 This drawing shows an arbitrary coordinate system, a point $\mathcal{P}$, the basis vectors $\mathbf{e}_u$ and $\mathbf{e}_w$ at that point, and a close-up view of how we describe an infinitesimal displacement $d\mathbf{s}$ as a sum of the basis vectors multiplied by the corresponding changes in the coordinate values. Ensuring that $dw\,\mathbf{e}_w$ and $du\,\mathbf{e}_u$ add up to the actual displacement $d\mathbf{s}$ defines the lengths of $\mathbf{e}_u$ and $\mathbf{e}_w$.

If we define basis vectors this way, then du and dw become the components of $d\mathbf{s}$ in that basis, and we call the set of basis vectors $\mathbf{e}_u, \mathbf{e}_w$ a **coordinate basis**. We do not *have* to define the basis vectors this way, but it proves *very* convenient, as we will see. A coordinate basis is generally *different* than the cartesian coordinate basis vectors $\mathbf{e}_x, \mathbf{e}_y$ (more commonly written $\mathbf{i}, \mathbf{j}$ or $\hat{x}, \hat{y}$) in that (1) $\mathbf{e}_u \cdot \mathbf{e}_w$ may be nonzero, (2) $\mathbf{e}_u$ and $\mathbf{e}_w$ may not have unit length, and (3) $\mathbf{e}_u$ and $\mathbf{e}_w$ may change in magnitude and/or direction as one moves from point to point.

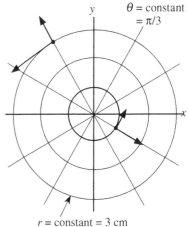

FIG. 5.2 Coordinate basis vectors for a polar coordinate system.

An Example. Consider r and θ coordinates for a flat 2D space. The coordinate basis for this coordinate system consists of the vectors $\mathbf{e}_r$ and $\mathbf{e}_\theta$, whose directions vary from point to point ($\mathbf{e}_r$ pointing radially away from the origin, and $\mathbf{e}_\theta$ perpendicular to it) and whose magnitudes are given by $\text{mag}(\mathbf{e}_r) = 1$ and $\text{mag}(\mathbf{e}_\theta) = r$ (see figure 5.2). This definition ensures that we can write $d\mathbf{s} = dr\,\mathbf{e}_r + d\theta\,\mathbf{e}_\theta$. If we were to use conventional polar-coordinate *unit* vectors $\mathbf{e}_{\hat{r}}$ and $\mathbf{e}_{\hat{\theta}}$, which both have unit magnitude by definition, we would have to write $d\mathbf{s} = dr\,\mathbf{e}_{\hat{r}} + r d\theta\,\mathbf{e}_{\hat{\theta}}$ instead. Therefore, the conventional polar-coordinate basis vectors $\mathbf{e}_{\hat{r}}$ and $\mathbf{e}_{\hat{\theta}}$ do *not* comprise a "coordinate basis." This example is discussed more fully in box 5.1.

General Vectors. Once we have established a coordinate basis, then we can *define* the components A^u, A^w of an arbitrary vector $\mathbf{A}$ at the point $\mathcal{P}$ so that

$$\mathbf{A} \equiv A^\mu \mathbf{e}_\mu = A^u \mathbf{e}_u + A^w \mathbf{e}_w \tag{5.2}$$

The Metric Tensor. The scalar product of $d\mathbf{s}$ with itself is the square of the physical distance between the endpoints of $d\mathbf{s}$:

$$\begin{aligned} ds^2 = d\mathbf{s} \cdot d\mathbf{s} &= (du\,\mathbf{e}_u + dw\,\mathbf{e}_w) \cdot (du\,\mathbf{e}_u + dw\,\mathbf{e}_w) \\ &= du^2\,\mathbf{e}_u \cdot \mathbf{e}_u + du\,dw\,\mathbf{e}_u \cdot \mathbf{e}_w + dw\,du\,\mathbf{e}_w \cdot \mathbf{e}_u + dw^2\,\mathbf{e}_w \cdot \mathbf{e}_w \\ &= dx^\alpha dx^\beta\,\mathbf{e}_\alpha \cdot \mathbf{e}_\beta \equiv g_{\alpha\beta}\,dx^\alpha dx^\beta \end{aligned} \tag{5.3}$$

The set of four components $g_{\alpha\beta} \equiv \mathbf{e}_\alpha \cdot \mathbf{e}_\beta$ comprises the **metric tensor** for our 2D coordinate basis. The equation above therefore specifies the relationship between the coordinate separations and the physical distance between two points, and represents a generalization of the Pythagorean theorem for our arbitrary coordinate system. Note that $g_{\alpha\beta}$ is generally a function of position in space.

This is the generalization of the metric tensor $\eta_{\alpha\beta}$ introduced in the last section. In *flat* spacetime, we can always find a cartesian coordinate basis where the basis vectors are mutually orthogonal ($\mathbf{e}_\alpha \cdot \mathbf{e}_\beta = 0$ when $\alpha \neq \beta$) and have unit magnitude ($\mathbf{e}_\alpha \cdot \mathbf{e}_\beta = \pm 1$ when $\alpha = \beta$, with the negative value indicating a time coordinate) at all events. This is *not* generally possible in a curved spacetime.

Transformation of Coordinates. Now consider a general coordinate transformation in two dimensions between coordinates u, w and new coordinates $p(u,w)$ and $q(u,w)$. The chain rule for partial derivatives implies that infinitesimal changes in the new coordinates are related to changes in the old coordinates as follows:

$$dp = \frac{\partial p}{\partial u} du + \frac{\partial p}{\partial w} dw \quad \text{and} \quad dq = \frac{\partial q}{\partial u} du + \frac{\partial q}{\partial w} dw \tag{5.4}$$

If we consider the p, q coordinate system the primed coordinates and u, w the unprimed coordinates, then we can write this compactly in index notation as follows:

$$dx'^{\mu} = \frac{\partial x'^{\mu}}{\partial x^{\nu}} dx^{\nu} \tag{5.5}$$

(with an implicit sum over the ν index, since we will consider the superscript in the *denominator of a derivative* to be equivalent to a subscript).

But if we use a coordinate basis in both coordinate systems, then dp and dq are the *actual components* of the infinitesimal displacement $d\mathbf{s}$ in the primed system, and du and dw play the same role in the unprimed system. Since by definition the components of an arbitrary vector $\mathbf{A}$ must transform in the same way that the components of the displacement do, the components of $\mathbf{A}$ must transform as follows:

$$A'^{\mu} = \frac{\partial x'^{\mu}}{\partial x^{\nu}} A^{\nu} \tag{5.6}$$

This is the *general* transformation law for the components of a vector when we are using a coordinate basis. The simplicity of this transformation law is precisely why coordinate bases are so useful. From this point on, we will *assume* that we will use coordinate bases unless we explicitly state otherwise.

It follows from the argument above that the reverse transformation of vector components from the primed to the unprimed system is simply

$$A^{\mu} = \frac{\partial x^{\mu}}{\partial x'^{\nu}} A'^{\nu} \tag{5.7}$$

An Important Identity. Basic partial differential calculus implies that

$$\frac{\partial x'^{\mu}}{\partial x^{\alpha}} \frac{\partial x^{\alpha}}{\partial x'^{\nu}} = \delta^{\mu}{}_{\nu} \tag{5.8}$$

If we write this out explicitly for our p, q and u, w coordinate systems, this says that

$$\frac{\partial p}{\partial u}\frac{\partial u}{\partial p} + \frac{\partial p}{\partial w}\frac{\partial w}{\partial p} = \frac{dp}{dp} = 1, \quad \frac{\partial p}{\partial u}\frac{\partial u}{\partial q} + \frac{\partial p}{\partial w}\frac{\partial w}{\partial q} = \frac{dp}{dq} = 0, \tag{5.9a}$$

$$\frac{\partial q}{\partial u}\frac{\partial u}{\partial p} + \frac{\partial q}{\partial w}\frac{\partial w}{\partial p} = \frac{dq}{dp} = 0, \quad \frac{\partial q}{\partial u}\frac{\partial u}{\partial q} + \frac{\partial q}{\partial w}\frac{\partial w}{\partial q} = \frac{dq}{dq} = 1 \tag{5.9b}$$

We will find this identity *very* useful in what follows.

The Transformation of the Metric Tensor. The fact that the magnitude of $d\mathbf{s}$ is a frame-independent quantity by definition directly implies that

$$ds^2 = g'_{\mu\nu} dx'^{\mu} dx'^{\nu} = g_{\alpha\beta} dx^{\alpha} dx^{\beta} \tag{5.10}$$

This directly implies (see box 5.2) that

$$g'_{\mu\nu} = \frac{\partial x^{\alpha}}{\partial x'^{\mu}} \frac{\partial x^{\beta}}{\partial x'^{\nu}} g_{\alpha\beta} \quad \text{and} \quad g_{\alpha\beta} = \frac{\partial x'^{\mu}}{\partial x^{\alpha}} \frac{\partial x'^{\nu}}{\partial x^{\beta}} g'_{\mu\nu} \tag{5.11}$$

(Note again that we assume there to be implicit sums over the α and β indices in the first equation and the μ and ν indices in the second.)

Equations 5.11 provides a handy way to find the components of the metric tensor in a new coordinate system if you know how the new coordinates depend on the old coordinates and you know the metric tensor in latter system. Box 5.3 illustrates such a calculation for a simple 2D coordinate system.

Coordinate Transformations in Flat Spacetime. The cartesian-like coordinates that we use for an inertial reference frame in the flat spacetimes of special relativity are an example of a coordinate basis. The Lorentz transformation equations in fact represent a special case of the general transformation rule given above. To see this, consider the full coordinate transformations between two inertial frames in standard orientation and where the primed frame is moving with speed β in the $+x$ direction relative to the unprimed frame are

$$
\begin{aligned}
t' &= \gamma(t - \beta x) \\
x' &= \gamma(-\beta t + x) \\
y' &= y \\
z' &= z
\end{aligned}
\quad \text{and} \quad
\begin{aligned}
t &= \gamma(t' + \beta x') \\
x &= \gamma(\beta t' + x') \\
y &= y' \\
z &= z'
\end{aligned}
\tag{5.12}
$$

By taking the partial derivatives of these equations, you can easily show that

$$\frac{\partial x'^\mu}{\partial x^\nu} = \Lambda^\mu{}_\nu = \begin{bmatrix} \gamma & -\gamma\beta & 0 & 0 \\ -\gamma\beta & \gamma & 0 & 0 \\ 0 & 0 & 1 & 0 \\ 0 & 0 & 0 & 1 \end{bmatrix} \tag{5.13a}$$

$$\frac{\partial x^\mu}{\partial x'^\nu} = (\Lambda^{-1})^\mu{}_\nu = \begin{bmatrix} \gamma & \gamma\beta & 0 & 0 \\ \gamma\beta & \gamma & 0 & 0 \\ 0 & 0 & 1 & 0 \\ 0 & 0 & 0 & 1 \end{bmatrix} \tag{5.13b}$$

(See box 5.4.) Because the transformation equations 5.12 are linear, the partial derivatives in equations 5.13 are *constant*, which is *not* generally the case for arbitrary coordinate transformations.

Equations 5.13 imply that for cartesian-like coordinates in flat spacetime, the general transformation law for the components of an arbitrary four-vector **A** given in equation 5.6 is the same as the Lorentz transformation law we saw earlier:

$$A'^\mu = \frac{\partial x'^\mu}{\partial x^\nu} A^\nu = \Lambda^\mu{}_\nu A^\nu \tag{5.14}$$

(compare with equation 4.5). We also know that Lorentz transformations obey the identity specified in equation 5.8,

$$(\Lambda^{-1})^\mu{}_\alpha \Lambda^\alpha{}_\nu = \frac{\partial x'^\mu}{\partial x^\alpha} \frac{\partial x^\alpha}{\partial x'^\nu} = \delta^\mu{}_\nu \tag{5.15}$$

(compare with equation 4.12), and that the transformation law for the metric tensor of flat spacetime is

$$\eta'_{\alpha\beta} = \frac{\partial x^\mu}{\partial x'^\alpha} \frac{\partial x^\nu}{\partial x'^\beta} \eta_{\mu\nu} = \eta_{\mu\nu} (\Lambda^{-1})^\mu{}_\alpha (\Lambda^{-1})^\nu{}_\beta = \eta_{\alpha\beta} \tag{5.16}$$

according to equation 4.19. This means that the components of the metric tensor for flat spacetime have *the same numerical value* in all cartesian-like coordinate systems connected by Lorentz transformations. You can check that equation 5.16 is true component by component (see box 5.5).

The Metric for a Spherical Surface. In curved spaces and spacetimes, we are stuck with curvilinear coordinates. As an example, box 5.6 discusses the curvilinear θ, ϕ (latitude-longitude) coordinate system for the 2D surface of a sphere of radius R. We see there that the metric for this space in this coordinate system is

$$g_{\mu\nu} = \begin{bmatrix} g_{\theta\theta} & g_{\theta\phi} \\ g_{\phi\theta} & g_{\phi\phi} \end{bmatrix} = \begin{bmatrix} R^2 & 0 \\ 0 & R^2 \sin^2\theta \end{bmatrix} \tag{5.17}$$

This result will be very valuable to us later as a simple example of a curved space and when we seek to construct metrics for spherically symmetric spacetimes.

58   5. ARBITRARY COORDINATES

BOX 5.1 The Polar Coordinate Basis

Consider ordinary polar coordinates r and θ (see figure 5.3). Note that the distance between two points with the same r coordinate but separated by an infinitesimal step $d\theta$ in θ is $r\,d\theta$ (by the definition of angle). So there are (at least) two ways to define a basis vector for the θ direction (which we define to be tangent to the r = constant curve): (1) we could define a basis vector $\mathbf{e}_{\hat{\theta}}$ with a unit magnitude, in which case the differential displacement vector for the step we are considering would be $d\mathbf{s} = r\,d\theta\,\mathbf{e}_{\hat{\theta}}$, or (2) we can define a basis vector $\mathbf{e}_\theta$ with magnitude r, so that we can write $d\mathbf{s} = d\theta\,\mathbf{e}_\theta$. In each case, the magnitude of the displacement will be $r\,d\theta$, but in the second case, the coordinate change $d\theta$ itself becomes the component of $d\mathbf{s}$, which is convenient. This latter choice is the one that defines the "coordinate basis" vector for the θ direction in polar coordinates.

The length of an infinitesimal step dr in the r direction (tangent to the θ = constant curve) is simply dr, so if we define $\mathbf{e}_r$ to have unit magnitude, we have $d\mathbf{s} = dr\,\mathbf{e}_r$ for such a step. Here, the basis vector with unit length is (in this case) the appropriate choice for a "coordinate basis" vector in the r direction.

Once we have established these basis vectors, we can write the components of an arbitrary infinitesimal displacement in any direction as

$$d\mathbf{s} = dr\,\mathbf{e}_r + d\theta\,\mathbf{e}_\theta \tag{5.18}$$

Note carefully that this equation does not apply to finite displacements, but only displacements small enough so that the basis vectors $\mathbf{e}_r$ and $\mathbf{e}_\theta$ do not change significantly over the distance spanned by the displacement. (See the exercise below.)

Note that by the nature of polar coordinates, basis vectors that point tangent to the θ = constant and r = constant curves are perpendicular to each other at all points, but $\mathbf{e}_r$ (for example) does not point in the same direction at one point as it does at another, as illustrated in figure 5.3. The metric for the polar coordinate basis is

$$g_{\mu\nu} \equiv \mathbf{e}_\mu \cdot \mathbf{e}_\nu = \begin{bmatrix} \mathbf{e}_r \cdot \mathbf{e}_r & \mathbf{e}_r \cdot \mathbf{e}_\theta \\ \mathbf{e}_\theta \cdot \mathbf{e}_r & \mathbf{e}_\theta \cdot \mathbf{e}_\theta \end{bmatrix} = \begin{bmatrix} 1 & 0 \\ 0 & r^2 \end{bmatrix} \tag{5.19}$$

Important note: We can always specify the components of a metric tensor either by listing them in a matrix (as above) or by writing out the metric equation. For example, if we compare the abstract and concrete versions of the metric equation

$$ds^2 = g_{\mu\nu}\,dx^\mu\,dx^\nu = dr^2 + r^2\,d\theta^2 \tag{5.20}$$

we can immediately infer that $g_{rr} = 1$, $g_{\theta\theta} = r^2$, and $g_{r\theta} = g_{\theta r} = 0$ (because terms involving $dr\,d\theta$ and $d\theta\,dr$ do not appear). The latter approach is often very convenient.

Exercise 5.1.1. By drawing on the diagram below, show that the displacement $\Delta\mathbf{s}$ between the two points with coordinates of $r = 1$ cm, $\theta = 0°$ and $r = 2$ cm, $\theta = 90°$ is *not* accurately given by equation 5.18 (because it is not infinitesimal).

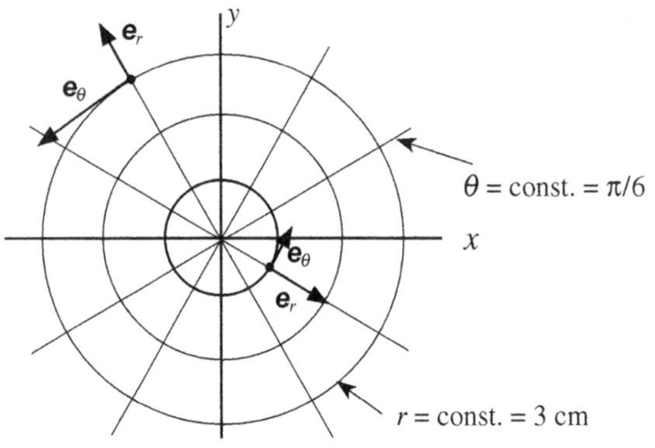

FIG. 5.3. A diagram that displays the r = constant and θ = constant curves for polar coordinates and the polar coordinate basis vectors at selected points.

BOX 5.2 Proof of the Metric Transformation Law

One can prove equation 5.11 as follows. Start with equation 5.10, repeated here for convenience:

$$g'_{\mu\nu} dx'^{\mu} dx'^{\nu} = g_{\alpha\beta} dx^{\alpha} dx^{\beta} \qquad (5.10r)$$

Use the inverse transformation law for the components of an infinitesimal displacement (equation 5.7) to rewrite the above as

$$g'_{\mu\nu} dx'^{\mu} dx'^{\nu} = g_{\alpha\beta} \frac{\partial x^{\alpha}}{\partial x'^{\mu}} \frac{\partial x^{\beta}}{\partial x'^{\nu}} dx'^{\mu} dx'^{\nu} \qquad (5.21)$$

Now you can follow the mode of argument used in box 4.6 to show that

$$g'_{\mu\nu} = g_{\alpha\beta} \frac{\partial x^{\alpha}}{\partial x'^{\mu}} \frac{\partial x^{\beta}}{\partial x'^{\nu}} \qquad (5.22)$$

This is equation 5.11.

Exercise 5.2.1. Fill in the gap between equation 5.21 and 5.22. Note that because of the implicit sums in equation 5.21, this is more complicated than saying "divide both sides by $dx'^{\mu} dx'^{\nu}$"!

BOX 5.3 A 2D Example: Parabolic Coordinates

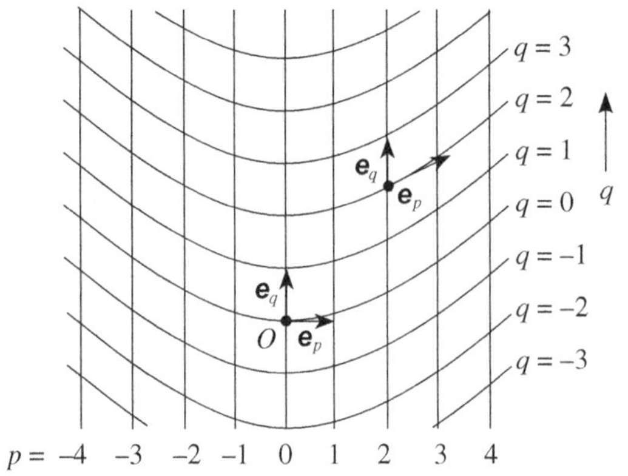

FIG. 5.4. A diagram that displays the $p = $ constant and $q = $ constant curves for parabolic coordinates and the parabolic coordinate basis vectors at selected points.

Consider the parabolic coordinate system p, q shown in figure 5.4. The transformation functions from ordinary cartesian coordinates x, y to these coordinates are

$$p(x,y) = x \quad \text{and} \quad q(x,y) = y - cx^2 \tag{5.23}$$

where c is a constant. The inverse transformation functions are

$$x(p,q) = p \quad \text{and} \quad y(p,q) = cp^2 + q \tag{5.24}$$

Exercise 5.3.1. Show that equations 5.24 are the correct inverse transformations.

Exercise 5.3.2. Calculate all eight partial derivatives $\partial x'^\mu / \partial x^\nu$ and $\partial x^\mu / \partial x'^\nu$.

The metric equation for the cartesian coordinates x, y is $ds^2 = dx^2 + dy^2$, so the metric tensor for these coordinates must be

$$g_{\alpha\beta} = \begin{bmatrix} 1 & 0 \\ 0 & 1 \end{bmatrix} \tag{5.25}$$

You can use equation 5.11 to show that the metric for the p, q system is

$$g'_{\mu\nu} = \begin{bmatrix} 1 + 4c^2p^2 & 2cp \\ 2cp & 1 \end{bmatrix} \tag{5.26}$$

BOX 5.3 (continued) A 2D Example: Parabolic Coordinates

For example, if we choose the coordinate indices $\mu = p, \nu = q$, we see that

$$g_{pq} = \frac{\partial x^\alpha}{\partial p}\frac{\partial x^\beta}{\partial q} g_{\alpha\beta} = \frac{\partial x}{\partial p}\frac{\partial x}{\partial q} g_{xx} + \frac{\partial x}{\partial p}\frac{\partial y}{\partial q} g_{xy} + \frac{\partial y}{\partial p}\frac{\partial x}{\partial q} g_{yx} + \frac{\partial y}{\partial p}\frac{\partial y}{\partial q} g_{yy}$$

$$= 1 \cdot 0 \cdot 1 + 0 + 0 + 2cp \cdot 1 = 2cp \tag{5.27}$$

Exercise 5.3.3. Use the same technique to verify the other components of equation 5.26. Does the fact that this metric has off-diagonal components make sense?

Exercise 5.3.4. Let a vector $\mathbf{A}$ have p, q components $A^p = 1, A^q = 0$.
a) Find this vector's components in the x, y coordinate system.
b) Do these components make sense? (*Hint:* Sketch $\mathbf{e}_p, \mathbf{e}_q$ at a typical point.)
c) Show that $A^2 = \mathbf{A} \cdot \mathbf{A}$ of this vector has the same value in both systems.

BOX 5.4 The LTEs as an Example General Transformation

Notice that for the Lorentz transformation

$$\frac{\partial x'^t}{\partial x^t} = \frac{\partial t'}{\partial t} = \frac{\partial}{\partial t}\gamma(t - \beta x) = \gamma + 0 = \gamma = \Lambda^t{}_t \qquad (5.28)$$

Exercise 5.4.1. Similarly, check that $\partial x'^\mu/\partial x^\nu = \Lambda^\mu{}_\nu$ when $\mu = x$ and $\nu = t$, and when $\mu = \nu = y$.

BOX 5.5 The Metric Transformation Law in Flat Space

Let's check equation 5.16 for $\alpha = \beta = t$.

$$\begin{aligned}\eta'_{tt} &= (\Lambda^{-1})^\mu{}_t(\Lambda^{-1})^\nu{}_t \eta_{\mu\nu} \\ &= (\Lambda^{-1})^t{}_t(\Lambda^{-1})^\nu{}_t \eta_{t\nu} + (\Lambda^{-1})^x{}_t(\Lambda^{-1})^\nu{}_t \eta_{x\nu} \\ &\quad + (\Lambda^{-1})^y{}_t(\Lambda^{-1})^\nu{}_t \eta_{y\nu} + (\Lambda^{-1})^z{}_t(\Lambda^{-1})^\nu{}_t \eta_{z\nu}\end{aligned} \qquad (5.29)$$

Now, $\eta_{t\nu}$ is only nonzero when $\nu = t$, $\eta_{x\nu}$ only when $\nu = x$, and so on. Moreover $(\Lambda^{-1})^y{}_t = (\Lambda^{-1})^z{}_t = 0$. Therefore,

$$\begin{aligned}\eta'_{tt} &= (\Lambda^{-1})^t{}_t(\Lambda^{-1})^t{}_t \eta_{tt} + (\Lambda^{-1})^x{}_t(\Lambda^{-1})^x{}_t \eta_{xx} \\ &= \gamma^2(-1) + (\gamma\beta)^2(+1) = \frac{-1 + \beta^2}{1 - \beta^2} = -1 = \eta_{tt}\end{aligned} \qquad (5.30)$$

The other components are analogous.

Exercise 5.5.1. Check the cases where $\alpha = t$ and $\beta = x$, and where $\alpha = \beta = x$.

BOX 5.6 A Metric for a Sphere

Consider the 2D surface of a sphere of radius R. The most commonly used coordinate system for a spherical surface is a latitude-longitude system using angular coordinates θ and ϕ. As illustrated in figure 5.5, curves of constant longitude ϕ are great circles that intersect at both poles. The coordinate ϕ labels these curves by the angle each makes at the north pole with a longitude curve arbitrarily chosen to have $\phi = 0$ (in the case of the earth's surface, the great circle going through Greenwich, England). The curves of constant latitude are circles a constant distance from the pole (as measured along the sphere's surface). On the earth's surface, we conventionally label these circles by the angle θ that a line drawn from any point on the circle to the earth's center makes with the earth's equatorial plane (so that the equator has $\theta = 0$). However, in physics, we usually define θ to be the angle measured down from the north pole, so that $\theta = 0$ at the north pole, $\pi/2$ at the equator, and π at the south pole. I will use the physics definition throughout this book.

A nice feature of this coordinate system is that the curves of longitude and latitude are always perpendicular to each other. This means that $g_{\theta\phi} = g_{\phi\theta} = \mathbf{e}_\theta \cdot \mathbf{e}_\phi = 0$, i.e., the matrix for this coordinate system's metric is diagonal. We can determine the other metric components as follows. Consider first the infinitesimal displacement corresponding to an infinitesimal change in latitude $d\theta$ along a curve of constant longitude. Since that curve is a great circle, its radius is R, so the arclength along the sphere's surface subtended by the angle $d\theta$ is $R\,d\theta$. Similarly, since the diagram shows that a circle of latitude θ has a radius of $R\sin\theta$, the length of the infinitesimal displacement corresponding to an infinitesimal change $d\phi$ along a circle of constant latitude must have a length $R\sin\theta\,d\phi$. Because these displacements are perpendicular, and because in the infinitesimal limit, the patch of area spanned by these displacements is almost flat, we can use the Pythagorean theorem to determine the squared length of the displacement $d\mathbf{s}$ that is the sum of such displacements:

$$ds^2 = (R\,d\theta)^2 + (R\sin\theta\,d\phi)^2 = R^2\,d\theta^2 + R^2\sin^2\theta\,d\phi^2 \tag{5.31}$$

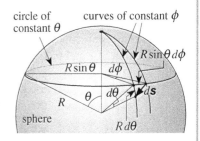

FIG. 5.5: A drawing of the surface of a sphere, showing curves of constant longitude ϕ and latitude θ, and an infinitesimal displacement $d\mathbf{s}$ comprised of infinitesimal steps in both the θ and ϕ directions.

Comparing this to the abstract form of the metric equation $ds^2 = g_{\mu\nu}\,dx^\mu\,dx^\nu$, we see that $g_{\theta\theta} = R^2$, $g_{\phi\phi} = R^2\sin^2\theta$, and $g_{\phi\theta} = g_{\theta\phi} = 0$.

Exercise 5.6.1. What would the metric components be if we were to measure θ up from the equator rather than down from the pole?

HOMEWORK PROBLEMS

P5.1 Consider the coordinate basis discussed in box 5.1 for polar coordinates r, θ.

a. Find the transformation equations that take one from 2D cartesian coordinates x, y to r, θ and vice versa.

b. Evaluate the partial derivatives of x and y with respect to r and θ and vice versa.

c. Find the metric for polar coordinates by transforming the metric from cartesian coordinates. Is your result consistent with equation 5.19?

P5.2 (Do problem P5.1 first.) Consider the coordinate basis discussed in box 5.1 for polar coordinates r, θ.

a. We know that an object in uniform circular motion has a constant radius, so it must have a velocity **v** such that $v^r = 0$. Use the polar-coordinate metric to show that if we assume this velocity has a constant (and coordinate-system-independent!) squared magnitude of $\mathbf{v} \cdot \mathbf{v} = v^2$, then we must have $v^\theta = \pm v/r$ (where the sign depends on which way the object moves around the circle).

b. Find this object's velocity components v^x and v^y in cartesian coordinates. Express your result as a function of r and θ. Use a sketch to show that these components do indeed describe a vector tangent to a circle of radius r, and also show that the squared magnitude of this vector in the cartesian system is indeed v^2.

P5.3 (Do problem 5.1 first.) Consider the coordinate basis discussed in box 5.1 for polar coordinates r, θ.

a. Consider an object moving at a constant speed v in the $+y$ direction of the cartesian coordinate system, so that $v^y = v$, $v^x = 0$. Find the components v^r and v^θ of this object's velocity in the polar coordinate system. Express your results both purely in terms of r and θ and purely in terms of x and y.

b. Imagine that the object starts at $x = b$, $y = 0$ at time $t = 0$. Its subsequent y position at later times t will therefore be simply $y = vt$. Use this to express both the object's r and θ position and its polar coordinate velocity components v^r and v^θ at all times $t > 0$ in terms of v, b, r, and t. Does your result make sense? (In particular, if you sketch the object's path, you should be able to see that its velocity will be mostly in the θ direction at early times, but mostly in the r direction at late times. Is this consistent with your mathematical expressions?)

P5.4 We can define "semilog" coordinates p, q for a flat 2D plane by the relations $p = x$ and $q = e^{by}$, where b is a constant. For the sake of argument, let $b = 0.40$ cm^{-1}.

a. Sketch what the "curves" of constant p and constant q look like in a cartesian x, y coordinate system.

b. An object at $y = 2.0$ cm has an acceleration **a** whose components in the cartesian coordinate system are $a^x = 0.2$ cm/s^2 and $a^y = -0.5$ cm/s^2. What are the components of **a** in the semilog system? (Be careful with units!)

c. What is the metric of the semilog coordinate system? Is this metric diagonal?

d. Show that **a** as defined in part *b* has the same magnitude in both the cartesian and semilog systems.

e. What is the length of the basis vector $\mathbf{e}_q$?

P5.5 We can define "sinusoidal" coordinates u, w on a flat 2D plane by the relations $u = x$ and $w = y - A\sin(bx)$, where A and b are constants. For the sake of concreteness, let $A = 1.0$ cm and $b = \pi/2$ cm^{-1}.

a. Sketch what the "curves" of constant u and constant w look like in a cartesian x, y coordinate system.

b. What is the metric of the sinusoidal coordinate system? Is this metric diagonal?

c. Imagine that an object moves with constant velocity **v** such that $v^x = v$ and $v^y = 0$. Such an object's position will be $x = vt$ (assuming $x = 0$ at $t = 0$) and $y = $ constant. Find the object's velocity components v^u and v^w in the u, w coordinate system. Express your results in terms of v, t, A, and b.

d. Show that the squared magnitude of **v** is still the constant v^2 in this coordinate system, even though the velocity component v^w is *not* constant in time. Explain *why* v^w is not constant, even though the vector **v** in abstract always points in the same direction and always has the same magnitude.

e. Argue therefore that dv^w/dt cannot be equal to the component a^w of the object's acceleration vector **a** in the u, w coordinate system. (*Hint:* Note that $a^x = a^y = 0$ in the cartesian system.) We will learn in a later chapter how to take derivatives *correctly* in an curvilinear coordinate system.

P5.6 Consider polar-coordinate-like "radial-longitude" coordinates r, ϕ for the 2D surface of a sphere of radius R, where r is the distance along the sphere's surface measured from the north pole and ϕ is an angular longitude coordinate measured around the pole. Note that curves of constant r and curves of constant ϕ are always perpendicular to each other everywhere on the sphere. Therefore (as we did in Box 5.1 for polar coordinates), by considering displacements on the sphere's surface that lie purely in the r and ϕ directions, infer the components $g_{\mu\nu}$ of the metric for this coordinate system (assuming we use a coordinate basis). Express these components purely in terms of R and r. (We will later find a similar coordinate system helpful in describing the spatial geometry of the universe.)

P5.7 Consider the two-dimensional surface of a paraboloid defined by the relation $z = br^2$ (where b is some constant and $r^2 = x^2 + y^2$) in a 3D flat (Euclidean) space.

a. Sketch this surface in a 3D xyz plot.

b. Define coordinates r, ϕ for this surface, where the r coordinate of a point on the surface is defined as above and ϕ is an angle measured around the surface's axis of symmetry (the z axis), like a longitudinal coordinate on the earth. Determine the metric components $g_{\mu\nu}$ for these coordinates on the paraboloid's surface, assuming that we use a coordinate basis. (*Hint:* Note that a step toward larger r on the surface means not only moving *away* from the z axis in the 3D space but also moving *upward* to more positive z. What is the distance ds along the surface involved in a step of dr along a curve of constant ϕ?)

6. TENSOR EQUATIONS

INTRODUCTION		
FLAT SPACETIME	**TENSORS**	**SCHWARZSCHILD BLACK HOLES**
Review of Special Relativity	Arbitrary Coordinates	The Schwarzschild Metric
Four-Vectors	**Tensor Equations**	Particle Orbits
Index Notation	Maxwell's Equations	Precession of the Perihelion
	Geodesics	Photon Orbits
		Deflection of Light
		Event Horizon
		Alternative Coordinates
		Black Hole Thermodynamics
THE CALCULUS OF CURVATURE	**THE EINSTEIN EQUATION**	
The Absolute Gradient	The Stress-Energy Tensor	
Geodesic Deviation	The Einstein Equation	
The Riemann Tensor	Interpreting the Equation	
	The Schwarzschild Solution	
COSMOLOGY	**GRAVITATIONAL WAVES**	**SPINNING BLACK HOLES**
The Universe Observed	Gauge Freedom	Gravitomagnetism
A Metric for the Cosmos	Detecting Gravitational Waves	The Kerr Metric
Evolution of the Universe	Gravitational Wave Energy	Kerr Particle Orbits
Cosmic Implications	Generating Gravitational Waves	Ergoregion and Horizon
The Early Universe	Gravitational Wave Astronomy	Negative-Energy Orbits
CMB Fluctuations & Inflation		

this depends on this

6. TENSOR EQUATIONS

Introduction. We have seen that a (four-)vector is *defined* to be a quantity whose components transform in a specific way (i.e., like the components dx^μ of the displacement vector $d\mathbf{s}$) when we change our coordinate system:

$$A'^\mu = \frac{\partial x'^\mu}{\partial x^\nu} A^\nu \tag{6.1}$$

In this section, we will see how this definition fits into a larger scheme of quantities called *tensors*, and we will finally discover the fundamental reason why some component labels are written as superscripts and others as subscripts.

Covectors. A **covector** is *defined* to be a quantity whose components transform as follows when we change coordinate systems:

$$B'_\nu = \frac{\partial x^\mu}{\partial x'^\nu} B_\mu \tag{6.2}$$

Compare equations 6.1 and 6.2 carefully. Note that the partial derivative for the covector transformation has the primed coordinate in the *denominator* of the partial derivative, not the numerator, and that the implied summation is over the coordinates in the numerator. Finally, note that in order to make the summation work, the index on the covector quantity must be a subscript, not a superscript. Putting a component label in the subscript position instead of the superscript position distinguishes covectors from ordinary vectors.

Covectors are also called **1-forms**, **covariant vectors**, and **dual vectors**. Those who call them "covariant vectors" also typically call ordinary vectors "contravariant vectors" to make the contrast clear.

The Gradient as Covector. There are two important types of quantities that transform as covectors. The first is the **gradient** of a scalar function. Consider an arbitrary scalar function of position $\Phi(t,x,y,z) \equiv \Phi(x^\mu)$ whose value at any given event is frame-independent. The gradient of such a function is defined to be

$$\partial_\mu \Phi \equiv \frac{\partial \Phi}{\partial x^\mu} \tag{6.3}$$

This object has one component for each of the possible values of the free index μ (so in 4D spacetime, it has four components). According to the chain rule of partial differential calculus, the values of these components transform as follows:

$$\partial'_\nu \Phi \equiv \frac{\partial \Phi}{\partial x'^\nu} = \frac{\partial x^\mu}{\partial x'^\nu} \frac{\partial \Phi}{\partial x^\mu} = \frac{\partial x^\mu}{\partial x'^\nu} \partial_\mu \Phi \tag{6.4}$$

If you compare this with equation 6.2, you will see that the components of the gradient do transform as the components of a covector. Box 6.1 discusses simple 2D examples of gradient covectors.

Lowering an Index. The second important quantity that transforms like a covector is the quantity we get when we multiply a vector by the metric tensor and sum over the vector's components. For example, if A^μ is a vector, then the quantity

$$A_\mu \equiv g_{\mu\nu} A^\nu \tag{6.5}$$

is a covector. We can prove this as follows. Applying the definition 6.5 in the primed coordinate system and using equations 5.11 and 5.8, we find that

$$A'_\mu = g'_{\mu\nu} A'^\nu = \left[\frac{\partial x^\alpha}{\partial x'^\mu} \frac{\partial x^\beta}{\partial x'^\nu} g_{\alpha\beta}\right]\left[\frac{\partial x'^\nu}{\partial x^\sigma} A^\sigma\right] = \frac{\partial x^\alpha}{\partial x'^\mu}\left[\frac{\partial x^\beta}{\partial x'^\nu} \frac{\partial x'^\nu}{\partial x^\sigma}\right] g_{\alpha\beta} A^\sigma$$

$$= \frac{\partial x^\alpha}{\partial x'^\mu} \delta^\beta_{\;\sigma} g_{\alpha\beta} A^\sigma = \frac{\partial x^\alpha}{\partial x'^\mu}(g_{\alpha\beta} A^\beta) \equiv \frac{\partial x^\alpha}{\partial x'^\mu} A_\alpha \tag{6.6}$$

Again, if we compare this with equation 6.2, we find that the components of the quantity defined by equation 6.5 do indeed transform like those of a covector.

The process of multiplying by the metric tensor and summing is called **lowering the index** of a vector's components. Indeed, you can think of the components A^μ and A_μ as being two different ways to represent the core physical quantity that lies behind **A**: A^μ are **A**'s raw vector components and A_μ as the components of the "covector version" of **A**. See box 6.2 for some examples.

Note that the metric tensor for ordinary cartesian coordinates in an ordinary 2D or 3D flat space is simply the identity matrix: for example, the Pythagorean theorem in a flat 2-space reads $ds^2 = dx^2 + dy^2$, so

$$ds^2 \equiv g_{\mu\nu} dx^\mu dx^\nu = dx^2 + dy^2 \quad \Rightarrow \quad g_{\mu\nu} = \begin{bmatrix} 1 & 0 \\ 0 & 1 \end{bmatrix} \tag{6.7}$$

Therefore, in such a space, there is *no distinction* between the vector and covector components of a vector **A**. This is why you probably have not heard of covectors before: in the types of orthonormal coordinate systems you have considered up to now, the concept represents an unnecessary complication.

The Scalar Product of a Vector and Covector. The scalar product of a vector with components A^μ and a covector with components B_μ is defined to be $A^\mu B_\mu$ (with the implied sum over μ). This scalar product is interesting because its value is independent of coordinate system. The proof is direct:

$$A'^\mu B'_\mu = \left[\frac{\partial x'^\mu}{\partial x^\alpha} A^\alpha\right]\left[\frac{\partial x^\beta}{\partial x'^\mu} B_\beta\right] = \frac{\partial x^\beta}{\partial x'^\mu}\frac{\partial x'^\mu}{\partial x^\alpha} A^\alpha B_\beta = \delta^\beta{}_\alpha A^\alpha B_\beta = A^\alpha B_\alpha \tag{6.8}$$

In fact, equation 6.5 means that this is really the same as $A^\mu g_{\mu\nu} B^\nu$, which we have already defined as the scalar product of two vectors, and which we have already seen is frame independent for Lorentz transformations. We now see that $A^\mu B_\mu = A^\mu g_{\mu\nu} B^\nu$ has a coordinate-independent value not only in special relativity, but in *any arbitrary coordinate system* we use (as long as we use a coordinate basis)!

The Inverse Metric. We have already seen that the components of the metric tensor transform as follows (see equation 5.11):

$$g'_{\mu\nu} = \frac{\partial x^\alpha}{\partial x'^\mu}\frac{\partial x^\beta}{\partial x'^\nu} g_{\alpha\beta} \tag{6.9}$$

Compare this to the transformation equation 6.2 for a covector. Note that $g_{\alpha\beta}$, which has *two* subscript indices, has a transformation law that is essentially a doubled version of equation 6.2: there is a partial derivative factor associated with each lower index that has the same form as the single factor associated with the single lower index in equation 6.2.

We define the **inverse metric tensor** $g^{\mu\nu}$ to be the matrix inverse of the normal metric. In abstract component notation, this condition requires that

$$g^{\mu\alpha} g_{\alpha\nu} = \delta^\mu{}_\nu \tag{6.10}$$

You can prove (see box 6.3) that when we change coordinates, the components of the inverse metric transform according to

$$g'^{\mu\nu} = \frac{\partial x'^\mu}{\partial x^\alpha}\frac{\partial x'^\nu}{\partial x^\beta} g^{\alpha\beta} \tag{6.11}$$

Compare this to the transformation equation 6.1 for a vector. Note that $g^{\alpha\beta}$, which has *two* superscript indices, has a transformation law that again has a partial derivative factor associated with each upper index having the same form as the single factor associated with the single upper index in equation 6.1. Are we beginning to sense a pattern here?

Tensors. With this in mind, we *define* an **nth-rank tensor T** to be a quantity with D^n components $T^{\mu\nu\cdots}{}_{\alpha\cdots}$ (where D is the number of dimensions) that transform when one changes coordinate systems according to the transformation rule

6. TENSOR EQUATIONS

TABLE 6.1 Examples of tensors and their transformation rules. (The number of components is for 4D spacetime.)

Rank n	# Components	Other Name	Symbol	Transformation Law	Example Quantities
0	$4^0 = 1$	(invariant scalar)	Φ	$\Phi' = \Phi$	Rest energy m
1	$4^1 = 4$	vector	A^μ	$A'^\mu = \dfrac{\partial x'^\mu}{\partial x^\nu} A^\nu$	four-velocity u^μ four-momentum p^μ gradient $\partial_\mu \Phi$ of a scalar
		covector	A_μ	$A'_\mu = \dfrac{\partial x^\nu}{\partial x'^\mu} A_\nu$	
2	$4^2 = 16$	tensor	$T^{\mu\nu}$	$T'^{\mu\nu} = \dfrac{\partial x'^\mu}{\partial x^\alpha} \dfrac{\partial x'^\nu}{\partial x^\beta} T^{\alpha\beta}$	Inverse metric $g^{\mu\nu}$ EM field tensor $F^{\mu\nu}$ Stress-energy tensor $T^{\mu\nu}$ Kronecker delta $\delta^\mu{}_\nu$ Metric tensor $g_{\mu\nu}$ Ricci tensor $R_{\mu\nu}$ Einstein tensor $G_{\mu\nu}$
			$T^\mu{}_\nu$	$T'^\mu{}_\nu = \dfrac{\partial x'^\mu}{\partial x^\alpha} \dfrac{\partial x^\beta}{\partial x'^\nu} T^\alpha{}_\beta$	
			$T_{\mu\nu}$	$T'_{\mu\nu} = \dfrac{\partial x^\alpha}{\partial x'^\mu} \dfrac{\partial x^\beta}{\partial x'^\nu} T_{\alpha\beta}$	
3	$4^3 = 64$	tensor	$M^{\mu\nu}{}_\alpha$	$M'^{\mu\nu}{}_\alpha = \dfrac{\partial x'^\mu}{\partial x^\beta} \dfrac{\partial x'^\nu}{\partial x^\gamma} \dfrac{\partial x^\sigma}{\partial x'^\alpha} M^{\beta\gamma}{}_\sigma$	(no obvious examples)
4	$4^4 = 256$	tensor	$R^\alpha{}_{\beta\mu\nu}$	$R'^\alpha{}_{\beta\mu\nu} = \dfrac{\partial x'^\alpha}{\partial x^\gamma} \dfrac{\partial x^\delta}{\partial x'^\beta} \dfrac{\partial x^\sigma}{\partial x'^\mu} \dfrac{\partial x^\lambda}{\partial x'^\nu} R^\gamma{}_{\delta\sigma\lambda}$	Riemann tensor $R^\alpha{}_{\beta\mu\nu}$

$$T'^{\mu\nu\cdots}{}_{\alpha\cdots} = \frac{\partial x'^\mu}{\partial x^\beta} \frac{\partial x'^\nu}{\partial x^\gamma} \cdots \frac{\partial x^\sigma}{\partial x'^\alpha} \cdots T^{\beta\gamma\cdots}{}_{\sigma\cdots} \tag{6.12}$$

that is, each superscript index on the tensor has a corresponding partial-derivative factor like that for a vector index, and each subscript index gets a corresponding partial-derivative factor like that for covector index.

According to this definition, a coordinate-independent scalar is a zeroth-rank tensor ($4^0 = 1$ component, no indices, no transformation factors):

$$\Phi' = \Phi \tag{6.13}$$

Both vectors and covectors are first-rank tensors. The metric tensor $g_{\mu\nu}$ is a second-rank tensor with two lower indices, the Kronecker delta $\delta^\mu{}_\nu$ is a second-rank tensor with one upper and one lower index (see box 6.4), and the electromagnetic field tensor $F^{\mu\nu}$ is a second-rank tensor with two upper indices. Table 6.1 provides a summary of tensor transformation rules up to 4th-rank tensors.

Tensor Equations. Why are tensor quantities useful? The most important reason is that if we can express a law of physics in the form of (nth-rank tensor **A**) = (nth-rank tensor **B**), then we know it will have exactly the same form in *every* coordinate system, because the tensor components on both sides of this equation transform in exactly the same way when we change coordinate systems. We say that such a law is **manifestly covariant.**

A law expressed in this form is therefore *automatically* consistent with the principle of relativity, because if it is true in any IRF, it will be true in all IRFs, and indeed in *any* meaningful coordinate system (even those attached to accelerating frames)! So any law cast in this form will satisfy a generalized principle of relativity: if it is true in an IRF, it will be true in *any other arbitrary coordinate system*.

Einstein assumed that this tensor language was tied in such an especially intimate way with the fundamental structure of the universe that true laws of physics would have a simple form when expressed in this language. This proved to be a valuable guide for constructing the equations of general relativity.

TABLE 6.2 Basic tensor operations that yield new tensors.

Operation	Example	Result	Comments
tensor sum	$p_1^\mu + p_2^\mu = p_{tot}^\mu$	tensor of same rank	Addition is defined only between tensors having the same rank and index positions, and whose corresponding components have the same units.
tensor product	$A^\mu B_\nu = M^\mu{}_\nu$	tensor of rank $n_1 + n_2$	The components of the resulting tensor are the products of all possible pairings of the input tensor components. The values of n_1 and n_2 are the ranks of the input tensors
contraction over an upper and lower index	$R^\alpha{}_{\mu\alpha\nu} = R_{\mu\nu}$	tensor of rank $n - 2$	Starting tensor must have a rank ≥ 2, and the sum must be over an upper and lower index
raising an index	$g^{\mu\alpha} R_{\alpha\nu} = R^\mu{}_\nu$	tensor of same rank	
lowering an index	$g_{\mu\alpha} T^{\alpha\nu} = T_\mu{}^\nu$	tensor of same rank	

Tensor Operations. Table 6.2 describes a set of basic operations that convert tensors to other tensors: in box 6.5, you will prove that each operation does in fact yield a tensor of the rank specified. We can construct more complicated operations from a sequence of the basic operations noted above. For example, the scalar product **A · B** of two vectors A^μ and B^ν is a tensor product of A^μ, B^ν, and $g_{\alpha\beta}$, followed by a pair of contractions over the two indices of $g_{\alpha\beta}$ to yield $A^\mu g_{\mu\nu} B^\nu$. (Alternatively, we could lower the index of B^ν, take tensor product with A^μ to form $A^\mu B_\nu$, and then contract over the upper and lower indices, yielding $A^\mu B_\mu$.)

If we start with a physically valid tensor equation, we can apply any of the operations described above and get a valid tensor equation. These operations therefore provide building blocks for tensor equations that might describe physical laws.

The deep meaning of the superscript and subscript indices should now finally be clear: they indicate how the components of the tensor to which they are attached transform when we change coordinate systems. We sum only over an upper and lower index because only such a sum (and not, for example, the sum over two upper indices or two lower indices) yields a new tensor quantity. If we follow the rules carefully, we are able to manipulate tensor equations in a way that ensures that what we have at the end is a valid tensor equation.

The Gradient of a Tensor Is Not a Tensor. Note that while the gradient of a scalar invariant yields a covector, the gradient of any *other* tensor does *not* yield another tensor. For example, imagine that A^μ is a four-vector. The components of the gradient of A^μ transform as follows when we change coordinate systems:

$$\partial'_\nu A'^\mu \equiv \frac{\partial A'^\mu}{\partial x'^\nu} = \frac{\partial}{\partial x'^\nu}\left(\frac{\partial x'^\mu}{\partial x^\alpha} A^\alpha\right) = \frac{\partial x^\beta}{\partial x'^\nu} \frac{\partial}{\partial x^\beta}\left(\frac{\partial x'^\mu}{\partial x^\alpha} A^\alpha\right)$$

$$= \frac{\partial x^\beta}{\partial x'^\nu} \frac{\partial^2 x'^\mu}{\partial x^\beta \partial x^\alpha} A^\alpha + \frac{\partial x^\beta}{\partial x'^\nu} \frac{\partial x'^\mu}{\partial x^\alpha} (\partial_\beta A^\alpha) \qquad (6.14)$$

by the product rule. The second term appearing in the bottom line, if it were alone, would be the transformation rule for a tensor. However, if the coordinate transformation factors $\partial x'^\mu / \partial x^\nu$ are not constants, then the second derivative appearing in the first term will not be zero, meaning that the gradient of a vector does *not* transform as a tensor, and therefore is not a tensor. The same issue arises with derivatives with respect to a particle's proper time τ (see problem P6.7). This is an issue we must address in the future, because most physical equations involve such derivatives.

(However, the components of the Lorentz transformation *are* constants, so if we limit ourselves to transformations between IRFs in cartesian coordinates, the first term in equation 6.14 is zero and the gradient of a vector *does* transform like a tensor. This also applies to the gradients of higher-rank tensors.)

BOX 6.1 EXAMPLE Gradient Covectors

Consider polar coordinates on a flat plane. The transformation equations between the polar coordinates r, θ (the primed coordinate system) and cartesian coordinates x, y (the unprimed coordinate system) are

$$x = r\cos\theta \quad \text{and} \quad r = \sqrt{x^2 + y^2} \qquad (6.15a)$$

$$y = r\sin\theta \quad \text{and} \quad \theta = \tan^{-1}\left(\frac{y}{x}\right) \qquad (6.15b)$$

Consider also the scalar function $\Phi = bxy = br^2\cos\theta\sin\theta$.

Exercise 6.1.1. Calculate the four transformation partials $\partial x^\mu/\partial x'^\nu$ for the transformations given above. (If you have done problem P5.1, just copy the result.)

Exercise 6.1.2. Use the basic definition of the gradient to calculate the components of the gradient of Φ in both coordinate systems. (*Hint:* You should find that $\partial_r\Phi = 2br\cos\theta\sin\theta$ and $\partial_\theta\Phi = br^2[\cos^2\theta - \sin^2\theta]$).

Exercise 6.1.3. Now show that if you take the components of the gradient of Φ in the unprimed system and transform them using the covector transformation rule, you get the components of the gradient that you calculated in 6.1.2.

BOX 6.2 Lowering Indices

Again consider polar coordinates on a flat plane. The transformation equations between the polar coordinates r, θ (the primed coordinate system) and cartesian coordinates x, y (the unprimed coordinate system) are given by equation 6.15. Consider also the vector **v** whose components are $v^x = 1$ and $v^y = 0$.

Exercise 6.2.1. Lower the indices of this vector in the cartesian coordinate system to show that $v_x = 1$ and $v_y = 0$.

Exercise 6.2.2. Use the partials you calculated in exercise 6.1.1 to transform these components to the polar coordinate system. You should find that $v_r = \cos\theta$ and $v_\theta = -r\sin\theta$.

Exercise 6.2.3. One can show that in the polar coordinate system, $v^r = \cos\theta$ and $v^\theta = -\sin\theta/r$ (see problem P6.1). Show that $v'^\mu v'_\mu = 1$. Does this make sense?

BOX 6.3 The Inverse Metric

If equation 6.10 is to be the definition of the inverse metric, then it must be true in all coordinate systems:

$$g^{\mu\alpha} g_{\alpha\nu} = \delta^{\mu}{}_{\nu} \quad \text{and also} \quad g'^{\mu\beta} g'_{\beta\nu} = \delta^{\mu}{}_{\nu} \tag{6.16}$$

We also know that 6.9 is true. Assume that we have found the inverse metric $g^{\mu\alpha}$ in the unprimed frame. What I will show is that the quantity $g'^{\mu\nu}$ given by the transformation law in equation 6.11 is indeed the matrix inverse of $g'_{\mu\nu}$. Since there is only one possible matrix inverse to $g'_{\mu\nu}$, this will mean that the transformation law for $g'^{\mu\nu}$ is indeed given by equation 6.11.

We can start with the product $g'^{\mu\beta} g'_{\beta\nu}$ and use equations 6.11 and 6.9 to express this in terms of the metric and inverse metric in the unprimed frame:

$$g'^{\mu\beta} g'_{\beta\nu} = \left[\frac{\partial x'^{\mu}}{\partial x^{\alpha}} \frac{\partial x'^{\beta}}{\partial x^{\sigma}} g^{\alpha\sigma} \right] \left[\frac{\partial x^{\gamma}}{\partial x'^{\beta}} \frac{\partial x^{\delta}}{\partial x'^{\nu}} g_{\gamma\delta} \right] = \frac{\partial x'^{\mu}}{\partial x^{\alpha}} \frac{\partial x'^{\beta}}{\partial x^{\sigma}} \frac{\partial x^{\gamma}}{\partial x'^{\beta}} \frac{\partial x^{\delta}}{\partial x'^{\nu}} g^{\alpha\sigma} g_{\gamma\delta} \tag{6.17}$$

Now, the fundamental identity given by equation 5.8 implies that we can replace the middle pair of partials with a Kronecker delta:

$$g'^{\mu\beta} g'_{\beta\nu} = \frac{\partial x'^{\mu}}{\partial x^{\alpha}} \delta^{\gamma}{}_{\sigma} \frac{\partial x^{\delta}}{\partial x'^{\nu}} g^{\alpha\sigma} g_{\gamma\delta} = \frac{\partial x'^{\mu}}{\partial x^{\alpha}} \frac{\partial x^{\delta}}{\partial x'^{\nu}} g^{\alpha\sigma} g_{\sigma\delta} \tag{6.18}$$

If you apply the first of equations 6.16 and do the sum over the resulting Kronecker delta, you should find that

$$g'^{\mu\beta} g'_{\beta\nu} = \frac{\partial x'^{\mu}}{\partial x^{\alpha}} \frac{\partial x^{\alpha}}{\partial x'^{\nu}} \tag{6.19}$$

But according to the fundamental identity, this is equivalent to $g'^{\mu\beta} g'_{\beta\nu} = \delta^{\mu}{}_{\nu}$. We see that $g'^{\mu\beta}$ is indeed the unique matrix inverse to $g'_{\beta\nu}$ in the primed coordinate system. Therefore by the logic outlined in the first paragraph, equation 6.11 must be the correct transformation law for the inverse metric.

Exercise 6.3.1. Fill in the gap between equations 6.18 and 6.19.

BOX 6.4 The Kronecker Delta Is a Tensor

The Kronecker delta is defined to have the *same* components ($\delta^\mu{}_\nu = 1$ if $\mu = \nu$ and zero otherwise) in every coordinate system. In spite of this, the Kronecker delta does (surprisingly) satisfy the tensor transformation rule for a tensor with one upper and one lower index:

$$\frac{\partial x'^\mu}{\partial x^\alpha} \frac{\partial x^\beta}{\partial x'^\nu} \delta^\alpha{}_\beta = \delta'^\mu{}_\nu \equiv \delta^\mu{}_\nu \qquad (6.20)$$

Exercise 6.4.1. Prove this. (*Hint:* Use the fundamental identity.)

BOX 6.5 Tensor Operations

Let's look at each of the five operations in table 6.2 in turn. Here is the proof that the sum of two tensors with the same number of indices in the same positions yields a tensor. For the sake of concreteness, consider the following sum of third-rank tensors: $C^{\mu\nu}{}_\alpha = A^{\mu\nu}{}_\alpha + B^{\mu\nu}{}_\alpha$. The transformed components of **C** will be

$$C'^{\mu\nu}{}_\alpha \equiv A'^{\mu\nu}{}_\alpha + B'^{\mu\nu}{}_\alpha = \frac{\partial x'^\mu}{\partial x^\beta}\frac{\partial x'^\nu}{\partial x^\gamma}\frac{\partial x^\sigma}{\partial x'^\alpha} A^{\beta\gamma}{}_\sigma + \frac{\partial x'^\mu}{\partial x^\beta}\frac{\partial x'^\nu}{\partial x^\gamma}\frac{\partial x^\sigma}{\partial x'^\alpha} B^{\beta\gamma}{}_\sigma$$

$$= \frac{\partial x'^\mu}{\partial x^\beta}\frac{\partial x'^\nu}{\partial x^\gamma}\frac{\partial x^\sigma}{\partial x'^\alpha} (A^{\beta\gamma}{}_\sigma + B^{\beta\gamma}{}_\sigma) = \frac{\partial x'^\mu}{\partial x^\beta}\frac{\partial x'^\nu}{\partial x^\gamma}\frac{\partial x^\sigma}{\partial x'^\alpha} C^{\beta\gamma}{}_\sigma \qquad (6.21)$$

So this tensor does indeed satisfy the appropriate tensor transformation law.

Now let's look at the tensor product. For the sake of concreteness, consider the the following product $C_{\mu\nu}{}^\alpha = A_{\mu\nu} B^\alpha$. You can easily show that the product **C** satisfies the correct tensor transformation rule for a third-rank tensor with two lower and one upper index.

Exercise 6.5.1. Do this (it will take only about two steps).

Now consider the contraction $A_\mu = C^\alpha{}_{\alpha\mu}$. The transformation goes like this:

$$A'_\mu \equiv C'^\alpha{}_{\alpha\mu} = \frac{\partial x'^\alpha}{\partial x^\beta}\frac{\partial x^\gamma}{\partial x'^\alpha}\frac{\partial x^\sigma}{\partial x'^\mu} C^\beta{}_{\gamma\sigma} = \delta^\gamma{}_\beta \frac{\partial x^\sigma}{\partial x'^\mu} C^\beta{}_{\gamma\sigma} \qquad (6.22)$$

by the fundamental identity. If we sum over the Kronecker delta, we get

$$A'_\mu = \frac{\partial x^\sigma}{\partial x'^\mu} C^\beta{}_{\beta\sigma} = \frac{\partial x^\sigma}{\partial x'^\mu} A_\sigma \qquad (6.23)$$

BOX 6.5 (continued) Tensor Operations

We have already seen the proof that lowering the index on a vector yields a covector. The proof for lowering the index of a more general tensor quantity is analogous. In a very similar way, one can use equation 6.11 to show that if we raise the first index of, say, $C_{\mu\nu}{}^{\alpha}$ the resulting quantity $C^{\mu}{}_{\nu}{}^{\alpha}$ transforms like a tensor with two upper indices and one lower index, in other words, that

$$C'^{\mu}{}_{\nu}{}^{\alpha} \equiv g'^{\mu\sigma} C'_{\sigma\nu}{}^{\alpha} = \frac{\partial x'^{\mu}}{\partial x^{\beta}} \frac{\partial x^{\gamma}}{\partial x'^{\nu}} \frac{\partial x'^{\alpha}}{\partial x^{\delta}} C^{\beta}{}_{\gamma}{}^{\delta} \qquad (6.24)$$

Exercise 6.5.2. Do the proof.

Exercise 6.5.3. Finally, show that $C^{\mu}{}_{\mu}{}^{\alpha}$ transforms like a four-vector C^{α}.

HOMEWORK PROBLEMS

P6.1 Consider polar coordinates in a two-dimensional flat space (see boxes 6.1 and 6.2).

a. Find the transformation partials $\partial x'^\mu / \partial x^\nu$ (where the primed coordinates are polar coordinates and the unprimed coordinates are cartesian coordinates).

b. Use these partials to transform the cartesian components of the vector **v** described in box 6.2 to polar components, and so verify the claim made in exercise 6.2.3.

c. We have seen that the polar coordinate metric is

$$g'_{\mu\nu} = \begin{bmatrix} 1 & 0 \\ 0 & r^2 \end{bmatrix} \quad (6.25)$$

Show that the inverse metric is:

$$g'^{\mu\nu} = \begin{bmatrix} 1 & 0 \\ 0 & r^{-2} \end{bmatrix} \quad (6.26)$$

d. Use this to raise the polar-coordinate components of the covector v_μ. You should get the same components as you found in part b.

P6.2 Consider the situation described in boxes 6.1 and 6.2. Show that $v^\mu \partial_\mu \Phi$ has the same numerical value in both the polar and cartesian coordinate systems.

P6.3 Given an arbitrary second-rank tensor $F^{\mu\nu}$, prove that the contraction $F \equiv F^\mu{}_\mu \equiv g_{\mu\nu} F^{\mu\nu}$ (which we call the trace of $F^{\mu\nu}$) transforms like a scalar invariant.

P6.4 Consider four-vectors **A** and **B** whose components in a certain abstract coordinate system are $A^0 = 1$ m, $A^1 = 2$ m, $A^2 = -1$ m, $A^3 = 0$ m and $B^0 = 3$ s^{-1}, $B^1 = -1$ s^{-1}, $B^2 = 0$ s^{-1}, and $B^3 = -2$ s^{-1}, where 0, 1, 2, 3 are abstract index names.

a. Can we add these four-vectors? If so, what are the components of the sum? If not, why not?

b. Find the components of the second-rank tensor product $M^{\mu\nu} = A^\mu B^\nu$.

c. Is the tensor product commutative? (That is, does $A^\mu B^\nu$ yield the same matrix as $B^\mu A^\nu$?)

d. If the metric for this coordinate system is the flat-space metric $\eta_{\mu\nu}$, what is $M^\mu{}_\mu$? How is this related to **A** · **B**?

P6.5 What is the numerical value of the scalar $g_{\mu\nu} g^{\mu\nu}$ in an arbitrary coordinate system?

P6.6 We say that a tensor **M** is **symmetric** if $M^{\mu\nu} = M^{\nu\mu}$ or $M_{\mu\nu} = M_{\nu\mu}$ (the indices exchanged *must* be in the same position). The metric $g_{\mu\nu}$ is an example of such a tensor. We say that a tensor **F** is **antisymmetric** if $F^{\mu\nu} = -F^{\nu\mu}$ or $F_{\mu\nu} = -F_{\nu\mu}$. The electromagnetic field tensor is an example of an antisymmetric tensor.

a. Show that if a tensor is either symmetric or antisymmetric in one coordinate system, it has the same property in *all* coordinate systems.

b. Prove that if $F^{\mu\nu}$ is antisymmetric then $F_{\mu\nu}$ is also antisymmetric and vice versa. Similarly, show that if $M^{\mu\nu}$ is symmetric, then $M_{\mu\nu}$ is symmetric.

c. Prove that $M_{\mu\nu} F^{\mu\nu} = 0$ if **M** is symmetric and **F** is antisymmetric.

d. We call the scalar $F^\mu{}_\mu$ the **trace** of a second-rank tensor **F**. Prove that $F^\mu{}_\mu = 0$ if **F** is antisymmetric.

e. How many independent components does a second-rank symmetric tensor have in a 4D spacetime? How many independent components does an antisymmetric second-rank tensor have?

P6.7 Consider an arbitrary four-vector **A** that depends on a particle's position in spacetime (examples of such vectors include the particle's four-velocity **u**, its four-momentum **p**, and vectors we might construct from these). The particle's position x^μ in spacetime is generally a function of the particle's proper time τ: $x^\mu = x^\mu(\tau)$, so we can consider such a vector to depend on τ.

a. Show that $dA^\mu/d\tau$ is not a four-vector unless the coordinate transformation partials are independent of position in spacetime (which will rarely be the case in general coordinate systems).

b. Argue, however, that if $\Phi(x^\mu)$ is a scalar that depends on position, then $d\Phi/d\tau$ evaluated along a particle's worldline *is* a valid scalar.

Note: Since the coordinate transformation partials *are* independent of position in spacetime for the particular case of cartesian coordinates in flat spacetime, $dA^\mu/d\tau$ *is* a four-vector in that particular context.

P6.8 Assume that we are using cartesian coordinates in the flat spacetime of special relativity. Which one of the following equations might possibly express the correct tensor generalization of Newton's second law for a particle with mass m and charge q moving in an electromagnetic field $F^{\mu\nu}$ with four-velocity u^μ? Also explain why each of the other proposed equations is *not* credible. (Hints: Focus more on form than on content. Note that t is the coordinate time measured in our inertial frame, where τ is the proper time measured by the particle itself along its worldline.)

A. $m\dfrac{du^\mu}{dt} = qF^{\mu\nu}$ 　　B. $m\dfrac{du^\mu}{d\tau} = qF^{\mu\nu}$

C. $m\dfrac{du^\mu}{dt} = qF^{\mu\nu} u_\nu$ 　　D. $m\dfrac{du^\mu}{d\tau} = qF^{\mu\nu} u_\nu$

E. $m\dfrac{du^\mu}{dt} = qF^{\mu\nu} u^\nu$ 　　F. $m\dfrac{du^\mu}{d\tau} = qF^{\mu\nu} u^\nu$

P6.9 The Poisson equation for the Newtonian gravitational potential Φ at a given point in space is

$$\nabla^2 \Phi \equiv \frac{\partial^2 \Phi}{\partial x^2} + \frac{\partial^2 \Phi}{\partial y^2} + \frac{\partial^2 \Phi}{\partial z^2} = 4\pi G \rho \quad (6.27)$$

where ρ is the density of mass at that point and G is the universal gravitational constant. This equation, which is the gravitational analogue of Gauss's law for the electrostatic potential, accurately describes how mass affects the Newtonian gravitational potential as long as the mass distribution does not change with time. The Newtonian potential Φ in turn affects a particle's motion as follows:

P6.9 (continued)

$$\frac{d\vec{v}}{dt} = -\vec{\nabla}\Phi, \quad \text{where} \quad \vec{\nabla}\Phi \equiv \left[\frac{\partial\Phi}{\partial x}, \frac{\partial\Phi}{\partial y}, \frac{\partial\Phi}{\partial z}\right] \quad (6.28)$$

and $\vec{v}$ is the particle's velocity. This is also works very well when the particle's speed is much smaller than that of light. Our goal in this problem is to construct (and critique) a possible relativistic theory of gravity by converting these equations into tensor equations assuming that Φ and ρ are true relativistic scalars.

a. In the context of cartesian coordinates in the flat spacetime of special relativity (where the gradient of a tensor is a tensor), find a tensor equation that reduces to equation 6.27 when Φ is time-independent. You will find that the tensor equation makes physical claims beyond what equation 6.27 implies because it describes how Φ and ρ are connected in time-dependent situations.

b. Find the simplest tensor equation that reduces to equation 6.28 when the particle's speed $\ll 1$ and Φ is time-independent. Again, you should find that this equation implies new physics about how a time-varying potential Φ affects a particle's motion.

c. However, this theory has problems. Show that the requirement that $\mathbf{u} \cdot \mathbf{u} = -1$ implies that

$$u^\mu \frac{du_\mu}{d\tau} = 0 \quad (6.29)$$

where τ is the particle's proper time. Then show that if you multiply both sides of your answer to part b by u^μ, you find that we must have $u^\mu \partial_\mu \Phi = 0$, which is true only for an arbitrary particle if $\partial_\mu \Phi = 0$. Finally, show that this contradicts equation 6.27.

d. In 1912, Gunnar Nordström proposed that the tensor replacement for equation 6.28 should instead be

$$\frac{du_\mu}{d\tau} = -\partial_\mu \Phi - u_\mu u^\alpha \partial_\alpha \Phi \quad (6.30)$$

Show that this fixes the problem, and that it still reduces to equation 6.28 in the limit that the particle's speed is small and Φ does not depend on time.

e. Argue that this theory predicts that Φ satisfies the wave equation in empty space and so predicts the existence of gravitational waves. (*Hint:* Look up "wave equation" if you aren't familiar with the term.)

f. However, this whole lovely theory breaks down because the mass density ρ is *not* a genuine relativistic scalar. To see this, consider an imaginary box that has volume V in its rest frame and encloses N particles of mass m, so that $\rho = Nm/V$. Now look at this box in an inertial frame where it is moving with some speed β. Explain carefully why the mass density for the particles in the box is *not* the same in the new inertial frame. (Note that m really *is* a genuine scalar.)

P6.10 In the context of cartesian coordinates in the flat spacetime of special relativity (where the gradient of a tensor is a tensor), the tensor equation

$$\partial_\alpha F_{\mu\nu} + \partial_\nu F_{\alpha\mu} + \partial_\mu F_{\nu\alpha} = 0 \quad (6.31)$$

is true (where $F_{\mu\nu}$ is the fully lowered version of the electromagnetic field tensor). This equation actually represents 64 component equations, but most are identically zero.

a. In the context of cartesian coordinates in special relativity, what are the components of $F_{\mu\nu}$?

b. Show that because $F_{\mu\nu} = -F_{\nu\mu}$, equation 6.31 is identically zero when any two indices are the same.

c. Faraday's law in GR units reads

$$\text{curl}(\vec{E}) + \frac{\partial \vec{B}}{\partial t} = 0 \quad (6.32)$$

Show that equation 6.31 with $\mu = x$, $\nu = y$, and $\alpha = t$ gives you a component of this law. (Which one?)

d. Show that equation 6.31 with $\mu = x$, $\nu = y$, and $\alpha = z$ gives you Gauss's law for the magnetic field.

P6.11 While we have been focusing so far on tensors in either two spatial dimensions or four spacetime dimensions, tensors exist in ordinary three-dimensional space as well. An example is the moment of inertia tensor I, which specifies how a rotating object's angular momentum $\vec{L}$ is related to its angular velocity $\vec{\omega}$. The tensor equation is $L^i = I^{ij}\omega_j = I^{ij}g_{jk}\omega^k$, where g_{jk} is the metric, which in 3D cartesian coordinates is simply the identity matrix. (The convention in general relativity is to use Latin letters i, j, k, l, m, and n instead of Greek letters for indices that range over three purely spatial coordinates.) $\vec{L}$ is parallel to $\vec{\omega}$ only when the latter happens to be parallel to an axis of symmetry for an object.

a. Show that the transformation partials that take us from one cartesian coordinate system x^i to another x'^i whose axes have been rotated counterclockwise by an angle θ around the z axis are

$$R^i{}_j \equiv \frac{\partial x'^i}{\partial x^j} = \begin{bmatrix} \cos\theta & \sin\theta & 0 \\ -\sin\theta & \cos\theta & 0 \\ 0 & 0 & 1 \end{bmatrix} \quad (6.33)$$

(*Hint:* Consider a unit vector in each axis direction.)

b. We can often *calculate* a symmetric object's moment of inertia most easily in a coordinate system aligned with its symmetry axes (its "principal axes"). Imagine that in such a coordinate system S, we find that

$$I^{ij} = \begin{bmatrix} I_1 & 0 & 0 \\ 0 & I_2 & 0 \\ 0 & 0 & I_3 \end{bmatrix} \quad (6.34)$$

However, suppose that we actually want to *use* a coordinate system S' rotated an angle θ about the z axis relative to the principle axis system. Use the tensor transformation rule to find the moment of inertia's components in the coordinate system S'.

c. Show that we can write this transformation in matrix form as $[R][I][R^T]$, where $[R^T]$ is the transpose of the matrix in equation 6.33. Why the transpose?

7. MAXWELL'S EQUATIONS

INTRODUCTION

FLAT SPACETIME
Review of Special Relativity
Four-Vectors
Index Notation

TENSORS
Arbitrary Coordinates
Tensor Equations
Maxwell's Equations
Geodesics

SCHWARZSCHILD BLACK HOLES
The Schwarzschild Metric
Particle Orbits
Precession of the Perihelion
Photon Orbits
Deflection of Light
Event Horizon
Alternative Coordinates
Black Hole Thermodynamics

THE CALCULUS OF CURVATURE
The Absolute Gradient
Geodesic Deviation
The Riemann Tensor

THE EINSTEIN EQUATION
The Stress-Energy Tensor
The Einstein Equation
Interpreting the Equation
The Schwarzschild Solution

COSMOLOGY
The Universe Observed
A Metric for the Cosmos
Evolution of the Universe
Cosmic Implications
The Early Universe
CMB Fluctuations & Inflation

GRAVITATIONAL WAVES
Gauge Freedom
Detecting Gravitational Waves
Gravitational Wave Energy
Generating Gravitational Waves
Gravitational Wave Astronomy

SPINNING BLACK HOLES
Gravitomagnetism
The Kerr Metric
Kerr Particle Orbits
Ergoregion and Horizon
Negative-Energy Orbits

this depends on this

7. MAXWELL'S EQUATIONS

Introduction. In the last chapter, we saw that if we can write a law of physics in the form of a tensor equation, it will have exactly the same form in all reference frames and coordinate systems. Such a *manifestly covariant* equation automatically satisfies the principle of relativity.

The tensor formalism actually provides a powerful tool for finding the appropriate relativistic generalizations of pre-relativistic laws of physics. In this chapter, I will illustrate this approach by showing how the Lorentz force law and *all* of Maxwell's equations of electromagnetism naturally emerge when we seek to express the definition of the electric field and Gauss's law in tensor form. By going through this process, you will not only see why Maxwell's equations are needed to make electrostatics consistent with relativity and practice using and reading tensor equations, but also you will learn techniques and concepts that we will find useful when we seek later to do a similar thing with the Newtonian laws of gravitation.

In the last chapter, we saw that the gradient of a tensor is not generally a tensor (see equation 6.14). However, in this chapter we will confine our attention exclusively to inertial reference frames (IRFs) described in Cartesian coordinates. In this case, as discussed in the parenthetical paragraph below equation 6.14, the gradient of a tensor *is* a tensor. This will be important in what follows.

Basic Electrostatics. The fundamental law that connects an electric charge distribution to the electric field that it creates is **Gauss's law** (which in the case of static fields follows directly from Coulomb's law and the superposition principle). In differential form, Gauss's law says that at every point in space, we have

$$\vec{\nabla} \cdot \vec{E} \equiv \frac{\partial E_x}{\partial x} + \frac{\partial E_y}{\partial y} + \frac{\partial E_z}{\partial z} = 4\pi k \rho \tag{7.1}$$

where $\vec{E}(x,y,z)$ is the electric field and $\rho(x,y,z)$ is the charge density (both considered as functions of position) and k is the Coulomb constant. (See box 7.1 for how this equation is connected to the maybe more familiar *integral* form of Gauss's law.) The definition of the electric field $\vec{E} \equiv \vec{F}_e/q$ and Newton's second law imply that

$$\frac{d\vec{p}}{dt} = \vec{F}_e = q\vec{E} \tag{7.2}$$

where $\vec{F}_e$ is the electrostatic force on the particle and $\vec{p}$ is its momentum. We will start by assuming only that these equations are empirically true in the static and non-relativistic limit. Our task in what follows is to find tensor versions of equations 7.1 and 7.2 that reduce to these simple equations in the non-relativistic limit.

The Transformation Properties of the Charge Density. The first step in generalizing such equations is to determine the transformation properties of quantities involved. For example, is the charge density ρ a relativistic scalar, a component of a four-vector, a component of a second-rank tensor, or what?

One of the deepest principles of electrostatics is that *charge* is a relativistic scalar invariant. To see that this must be so, consider the observed fact that all atoms are exactly electrically neutral. The protons in an atom's nucleus are essentially at rest, while the innermost electrons orbit the nucleus with increasingly relativistic speeds as we go higher on the periodic table. Yet the total negative charge of the electrons exactly balances the total electric charge of the protons no matter how fast the electrons are moving. This implies that charge is a frame-independent quantity.

The Four-Current J. Consider now a set of particles with total charge q at rest inside a tiny box of volume V' that is also at rest in a certain inertial reference frame S'. Now imagine looking at the same box in a frame S in which frame S' (and thus the box) moves with speed β in the $+x$ direction. The box in S contains the same amount of charge q as in S' (because charge is invariant) but is observed to have a smaller volume $V = V'\sqrt{1-\beta^2}$, because the box's length in the x direction is observed to be Lorentz-contracted by a factor of $\sqrt{1-\beta^2}$ (see figure 7.1). Therefore the charge *density* observed in frame S will be

$$\rho \equiv \frac{q}{V} = \frac{q}{V'\sqrt{1-\beta^2}} = \frac{\rho'}{\sqrt{1-\beta^2}} = \gamma\rho' \qquad (7.3a)$$

Moreover, in this frame, the charge moves with speed v in the $+x$ direction, so even though there was zero current density in frame S', in S there is a current density of

$$\vec{J} \equiv \rho\vec{v} = \rho\vec{\beta} \qquad (7.3b)$$

If we define the **four-current** J in any inertial reference frame to have components $J^t \equiv \rho$, $J^x \equiv \rho v_x$, $J^y \equiv \rho v_y$, and $J^z \equiv \rho v_z$, then we see that equations 7.3 are the same as inverse Lorentz transformation equations (see equation 5.12) that transform the components of J from the primed frame S' (where its components are $J'^t = \rho'$ and $J'^x = J'^y = J'^z = 0$) to the unprimed frame:

$$\rho \equiv J^t = \gamma(J'^t + \beta J'^x) = \gamma(\rho' + 0) = \gamma\rho' \qquad (7.4a)$$

$$J^x = \gamma(\beta\rho' + J'^x) = \gamma(\beta\rho' + 0) = \gamma\beta\rho' \qquad (7.4b)$$

$$J^y = J'^y = 0 \quad \text{and} \quad J^z = J'^z = 0 \qquad (7.4c)$$

As long as we are confining ourselves to using cartesian-like coordinates in inertial reference frames, the Lorentz transformation equations *are* the transformation equations for the components of a four-vector (see equations 5.7 and 5.13b). Therefore, *we conclude that the charge density is the time component of a four-vector J.*

The Generalization of Gauss's Law. Since the right side of the generalization of equation 7.1 is a four-vector, the left side must be also. Now, the sum on the left side superficially looks as if it might be something like $\partial_\nu E^\nu$, where E^ν is a hypothetical electric field four-vector with components $[?, E_x, E_y, E_z]$. But this can't be right, because that would make the left side a scalar, not a four-vector. On the other hand, if we operate on a second-rank tensor $F^{\mu\nu}$ with ∂_ν and sum over the ν index, the result *is* a four-vector. So the simplest tensor generalization of Gauss's law is

$$\partial_\nu F^{\mu\nu} = 4\pi k J^\mu \qquad (7.5)$$

with the electric field being components of a second-rank field tensor

$$F^{\mu\nu} = \begin{bmatrix} ? & E_x & E_y & E_z \\ ? & ? & ? & ? \\ ? & ? & ? & ? \\ ? & ? & ? & ? \end{bmatrix} \qquad (7.6)$$

(We know that the electric field must be components along the $\mu = t$ row of the tensor because equation 7.1 corresponds to the $\mu = t$ component of equation 7.5.)

Generalizing Equation 7.2. Equation 7.5 provides us with a manifestly covariant equation that connects a field tensor $F^{\mu\nu}$ to the four-current J^μ that creates the field. But what is the physical meaning of all of the extra components of $F^{\mu\nu}$? Equation 7.2 describes the physical effects of the electric field components, so generalizing it might tell us about what the other components mean.

The obvious generalization of $d\vec{p}/dt$ on the left side is $dp^\mu/d\tau$ (where p^μ is the test particle's four-momentum and τ is the proper time measured along its worldline). This is a four-vector whose spatial components reduce to $d\vec{p}/dt$ in the non-relativistic limit. If the left side of equation 7.2 becomes a four-vector, the right side must be as well. The charge q is a frame-independent scalar, so what q multiplies must be a four-vector. But the field tensor $F^{\mu\nu}$ that is the generalization of the electric field is a second-rank tensor, not a four-vector. How can we possibly convert this field tensor into a four-vector?

The only way to convert a second-rank tensor to a four-vector is to multiply it by a four-vector and contract over one shared index, as we did in equation 7.5. But the right side in this case can't be something like $q\partial_\nu F^{\mu\nu}$ because that would introduce partial

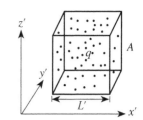

FIG. 7.1a In the S' frame, the charge in this box has an average density of $\rho' = q/V' = q/L'A$.

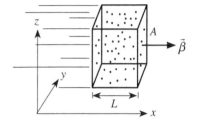

FIG. 7.1b In the S frame, where the box and its S' frame move in the $+x$ direction with speed β, the charge in the same box is still q, and the box's area perpendicular to the x axis is still A, but its length has been Lorentz contracted to $L = \sqrt{1-\beta^2}\, L' = L'/\gamma$. The density of charge in the box in this frame is thus $\rho = q/LA = (q/L'A)\gamma = \gamma\rho'$.

derivatives that don't appear in equation 7.2. The only other four-vector possibly connected with this situation is the test particle's four-velocity u^μ. So the generalization of equation 7.2 should plausibly be something like

$$\frac{dp^\mu}{d\tau} = qF^{\mu\nu}u_\nu \qquad (7.7)$$

where $u_\nu \equiv \eta_{\nu\alpha}u^\alpha$ is the covector version of the four-velocity.

If this is so, then the fact that a test particle's invariant squared mass $m^2 = -p^\mu p_\mu$ must remain fixed as it responds to the field constrains the form of $F^{\mu\nu}$. As you will show in box 7.2, m being fixed requires that for all possible fields

$$0 = \frac{d}{d\tau}(\mathbf{p}\cdot\mathbf{p}) = 2p_\mu\frac{dp^\mu}{d\tau} = 2qm\,u_\mu u_\nu F^{\mu\nu} \qquad (7.8)$$

where in the last step, I used equation 7.7 and $\mathbf{p} \equiv m\mathbf{u}$. This will be *automatically* true if $F^{\mu\nu}$ is antisymmetric (i.e., $F^{\nu\mu} = -F^{\mu\nu}$), because then

$$u_\mu u_\nu F^{\mu\nu} = -u_\mu u_\nu F^{\nu\mu} = -u_\nu u_\mu F^{\nu\mu} = -u_\mu u_\nu F^{\mu\nu} \qquad (7.9)$$

where in the third step I used the fact that $u_\mu u_\nu = u_\nu u_\mu$ (since multiplication is commutative) and in the fourth step I renamed the bound indexes $\nu \to \mu$ and vice versa. But zero is the only number that can be equal to negative itself, so $u_\mu u_\nu F^{\mu\nu} = 0$ no matter what $F^{\mu\nu}$ might otherwise be. This also implies that the diagonal elements of $F^{\mu\nu}$ must be zero, since (for example) $F^{tt} = -F^{tt}$ (switching indices), so the diagonal elements must be their own negatives. So the field tensor must have the form

$$F^{\mu\nu} = \begin{bmatrix} 0 & E_x & E_y & E_z \\ -E_x & 0 & ? & ? \\ -E_y & -? & 0 & ? \\ -E_z & -? & -? & 0 \end{bmatrix} \qquad (7.10)$$

Now, for a particle at rest, $u^t = 1$ and $u^x = u^y = u^z = 0$, so we have $u_t \equiv g_{t\alpha}u^\alpha = \eta_{t\alpha}u^\alpha = \eta_{tt}u^t = -1$ and $u_x = u_y = u_z = 0$. Therefore, when the particle is at rest, the $\mu = x$ component of equation 7.7 reduces to

$$\frac{dp^x}{d\tau} = qF^{x\nu}u_\nu = qF^{xt}u_t = -qF^{xt} = +qE_x \qquad (7.11)$$

and the $\mu = y$ and $\mu = z$ components are similar. The spatial components of equation 7.7 thus nicely reduce to $d\vec{p}/dt = q\vec{E}$ (equation 7.2) for a test particle at rest.

However, we see that the relativistic generalization of the definition of the electric field *requires* the existence of three other field components (corresponding to the three question marks above the diagonal in equation 7.10) that do not happen to exert a force on a charged test particle at rest. Let's name these components (arbitrarily, but for the sake of argument), B_z, $-B_y$, and B_x, respectively:

$$F^{\mu\nu} = \begin{bmatrix} 0 & E_x & E_y & E_z \\ -E_x & 0 & B_z & -B_y \\ -E_y & -B_z & 0 & B_x \\ -E_z & B_y & -B_x & 0 \end{bmatrix} \qquad (7.12)$$

If our charged test particle is moving, then in the non-relativistic limit $u^t \approx 1$ and $u^x \approx v_x, u^y \approx v_y$, and $u^z \approx v_z$. Lowering the index in cartesian coordinates changes the time component's sign but leaves the spatial components intact (see box 7.3), so the x component of equation 7.7 for a particle moving non-relativistically is

$$\frac{dp^x}{d\tau} \approx q(-E_x)(-1) + 0 + qv_y B_z - qv_z B_y = q\left[E_x + (\vec{v}\times\vec{B})_x\right] \qquad (7.13)$$

We see from this equation that the mystery field tensor components are none other than the components of the empirical magnetic field, which is now revealed to be the relativistically necessary sibling of the electric field. If the magnetic field had been unknown before Einstein, he could have predicted its existence this way!

Consequences. Now that we know the full form of the field tensor, we can predict how the electric and magnetic fields transform when we change reference frames (see problem P7.1). We can also see immediately that since Gauss's law (equation 7.1) is the complete $\mu = t$ component of the tensor equation $\partial_\nu F^{\mu\nu} = 4\pi k J^\mu$, it will be accurate in its original form even in dynamic and highly relativistic circumstances. The spatial components of this equation (as we saw in box 4.3) are equivalent to the three components of the Ampere-Maxwell relation,

$$\vec{\nabla} \times \vec{B} - \frac{\partial \vec{E}}{\partial t} = 4\pi k \vec{J} \quad (7.14)$$

(in GR units) and we can now be confident that this relation is also valid in dynamic and highly relativistic situations. Finally, note that if we take the gradient of both sides of $\partial_\nu F^{\mu\nu} = 4\pi k J^\mu$ and sum over the index μ, we find that

$$4\pi k \partial_\mu J^\mu = \partial_\mu \partial_\nu F^{\mu\nu} = 0 \quad \Rightarrow \quad \partial_\mu J^\mu = 0 \quad (7.15)$$

because the field tensor is antisymmetric (see box 7.5). The equation $\partial_\mu J^\mu = 0$ expresses the law of *conservation of charge* (see box 7.4).

The point is that knowing only Gauss's law for static electric fields, the definition of the electric field, and relativistic tensor mathematics, one can predict (1) the existence of the magnetic field, (2) the form of the Lorentz force law for electromagnetism, (3) how the fields transform when we change frames, (4) the fully relativistic form of half of Maxwell's equations, and (5) the fact that charge is conserved.

The Other Two Maxwell Equations. Now consider the relativistic generalization of the equation that relates a *static* electric field to the electric potential

$$\vec{E} = -\vec{\nabla}\phi \quad (7.16)$$

We know that the left side of this equation generalizes to the field tensor $F^{\mu\nu}$, so the right side must generalize to the gradient of some four-vector **A**, something like $F^{\mu\nu} = \partial^\mu A^\nu$, where $\partial^\mu = \eta^{\mu\alpha}\partial_\alpha$ is the gradient with an index raised (as shown in box 7.3, raising the index changes only the sign of the time component). But $F^{\mu\nu} = \partial^\mu A^\nu$ cannot be correct, because the field tensor must be antisymmetric for all fields. On the other hand, the definition

$$F^{\mu\nu} \equiv \partial^\mu A^\nu - \partial^\nu A^\mu \quad (7.17)$$

automatically *ensures* that $F^{\mu\nu}$ is antisymmetric without putting any restrictions on the four-potential **A**. The $\mu = x$ and $\nu = t$ component of this equation reads

$$F^{xt} = -E_x = \partial^x A^t - \partial^t A^x = \frac{\partial \phi}{\partial x} + \frac{\partial A^x}{\partial t} \quad (7.18)$$

if we identify $A^t = \phi$. The $\mu = y$ and $\mu = z$ components are analogous. These reduce to the components of $\vec{E} = -\vec{\nabla}\phi$ in the *static* limit (where the time derivatives of all fields are zero), the only case where the latter is valid. Equation 7.17 provides a fully relativistic generalization of $\vec{E} = -\vec{\nabla}\phi$ even when the fields are *not* static.

We see that the electric potential $\phi = A^t$, but what are the other components of the four-potential **A**? You can show from equations 7.17 and 7.12 (see box 7.6) that one can generate the *magnetic* field from the four-potential's spatial components $\vec{A}$ using the following vector-like equation

$$\vec{B} = \vec{\nabla} \times \vec{A} \quad (7.19)$$

providing a non-obvious (and fully relativistic) generalization of $\vec{E} = -\vec{\nabla}\phi$.

You can also show (see box 7.7) that requiring that the field tensor be constructed from potentials as described by equation 7.17 implies that

$$\partial^\alpha F^{\mu\nu} + \partial^\nu F^{\alpha\mu} + \partial^\mu F^{\nu\alpha} = 0 \quad (7.20)$$

identically. These provide the other two Maxwell equations (see problem P7.2). So we see that the other two Maxwell equations follow from the relativistic generalization of $\vec{E} = -\vec{\nabla}\phi$ and the requirement that the field tensor be antisymmetric.

BOX 7.1 Gauss's Law in Integral and Differential Form

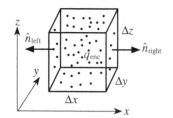

FIG. 7.2 A tiny box enclosing charge q_{enc}. The box is immersed in an electric field $\vec{E}$ which is nearly uniform over the tiny box. Normal vectors are shown for the box's left and right faces.

Gauss's law in integral form tells us that

$$\oint \vec{E} \cdot d\vec{A} = \frac{q_{\text{enc}}}{\varepsilon_0} = 4\pi k q_{\text{enc}} \tag{7.21}$$

where the flux integral is over some closed surface and q_{enc} is the charge enclosed by that surface. Consider now a tiny rectangular box oriented with faces parallel to the three cartesian coordinate axes whose lengths along those axes are Δx, Δy, and Δz, respectively. Assume that box's center has coordinates x, y, and z (see figure 7.2).

If the box is sufficiently small, the electric field will be essentially constant in magnitude and direction over any face of the box, so to a good approximation, the integral of the flux over that face will be equal to the dot product of the electric field evaluated at the face's center and the area vector for the entire face:

$$\int_{\text{face}} \vec{E} \cdot d\vec{A} \approx \vec{E}_{\text{center}} \cdot \int_{\text{face}} d\vec{A} = \vec{E}_{\text{center}} \cdot A\hat{n} \tag{7.22}$$

where A is the face's area and $\hat{n}$ an outward unit vector perpendicular to its surface. This approximation becomes exact in the limit that Δx, Δy, and Δz go to zero.

Consider now the two faces of the box that are perpendicular to the x direction. The unit vector $\hat{n}$ for the left face points in the $-x$ direction, so the flux through this face is approximately $\vec{E}(x - \tfrac{1}{2}\Delta x, y, z) \cdot \hat{n}\, \Delta y \Delta z = -E_x(x - \tfrac{1}{2}\Delta x, y, z)\, \Delta y \Delta z$ (note that the coordinates $x - \tfrac{1}{2}\Delta x, y, z$ specify the point at the center of this face). Similarly, the flux through the right face is approximately $+E_x(x + \tfrac{1}{2}\Delta x, y, z)\, \Delta y \Delta z$. Therefore, the net flux through these two faces is

$$\left[E_x(x + \tfrac{1}{2}\Delta x, y, z) - E_x(x - \tfrac{1}{2}\Delta x, y, z) \right] \Delta y \Delta z$$

$$= \left[\frac{E_x(x + \tfrac{1}{2}\Delta x, y, z) - E_x(x - \tfrac{1}{2}\Delta x, y, z)}{\Delta x} \right] \Delta x \Delta y \Delta z \approx \frac{\partial E_x}{\partial x} \Delta x \Delta y \Delta z \tag{7.23}$$

since the expression in brackets in the second line is the definition of the partial derivative $\partial E_x / \partial x$ in the limit that $\Delta x \to 0$. Again, the latter expression for the net flux through these faces becomes exact in the limit that Δx, Δy, and Δz go to zero.

Exercise 7.1.1. Check that the corresponding expressions for the net flux through the pairs of faces perpendicular to the y axis and z axis reduce to $(\partial E_y / \partial y)\, \Delta x \Delta y \Delta z$ and $(\partial E_z / \partial z)\, \Delta x \Delta y \Delta z$, respectively, in the same limit.

So if we substitute the results of equation 7.23 and exercise 7.1.1 into equation 7.21, divide both sides by the box's volume $\Delta x \Delta y \Delta z$, and then take the limit as this volume goes to zero, we get

$$\frac{\partial E_x}{\partial x} + \frac{\partial E_y}{\partial y} + \frac{\partial E_z}{\partial z} = \lim_{\Delta x \Delta y \Delta z \to 0} \left[4\pi k \frac{q_{\text{enc}}}{\Delta x \Delta y \Delta z} \right] = 4\pi k \rho \tag{7.24}$$

This is Gauss's law in differential form.

BOX 7.2 The Derivative of m^2

Exercise 7.2.1. Use the product rule and the definitions of raised and lowered indices (in an IRF in cartesian coordinates) to prove that $0 = d(p_\mu p^\mu)/d\tau = 2p_\mu(dp^\mu/d\tau)$. *Hint:* Start with $-m^2 = p_\mu p^\mu = p^\mu \eta_{\mu\nu} p^\nu$ and use the product rule.

BOX 7.3 Raising and Lowering Indices in Cartesian Coordinates

It is *very important* (not only here but in the future) to know how raising and lowering indices affects components when we are using cartesian coordinates. Consider an arbitrary four-vector $A^\mu = [A^t, A^x, A^y, A^z]$. If we lower its index in the standard way ($A_\mu = g_{\mu\nu} A^\nu$), we find that in cartesian coordinates this becomes

$$A_t = \eta_{t\nu} A^\nu = \eta_{tt} A^t + \eta_{tx} A^x + \eta_{ty} A^y + \eta_{tz} A^z = (-1)A^t + 0 + 0 + 0 = -A^t \quad (7.25)$$

$$A_x = \eta_{x\nu} A^\nu = \eta_{xt} A^t + \eta_{xx} A^x + \eta_{xy} A^y + \eta_{xz} A^z = 0 + (+1)A^x + 0 + 0 = +A^x \quad (7.26)$$

and similarly for the y and z components. Therefore, we see that lowering an index *in cartesian coordinates* flips the sign of the time component's value but does not change the spatial components' values.

In cartesian coordinates in flat spacetime, it turns out that the components $\eta^{\mu\nu}$ of the inverse metric are the same as those of the metric $\eta_{\mu\nu}$:

$$\eta^{\mu\nu} = \begin{bmatrix} -1 & 0 & 0 & 0 \\ 0 & 1 & 0 & 0 \\ 0 & 0 & 1 & 0 \\ 0 & 0 & 0 & 1 \end{bmatrix}, \text{ as } \eta^{\mu\alpha}\eta_{\alpha\nu} = \begin{bmatrix} -1 & 0 & 0 & 0 \\ 0 & 1 & 0 & 0 \\ 0 & 0 & 1 & 0 \\ 0 & 0 & 0 & 1 \end{bmatrix}\begin{bmatrix} -1 & 0 & 0 & 0 \\ 0 & 1 & 0 & 0 \\ 0 & 0 & 1 & 0 \\ 0 & 0 & 0 & 1 \end{bmatrix} = \begin{bmatrix} 1 & 0 & 0 & 0 \\ 0 & 1 & 0 & 0 \\ 0 & 0 & 1 & 0 \\ 0 & 0 & 0 & 1 \end{bmatrix} \quad (7.27)$$

as is required for matrix inverses. You can use this to show that if we raise the index of an arbitrary covector B_μ in the standard way ($B^\mu = g^{\mu\nu} B_\nu$), we find the same result: raising an index flips the sign of the time component's value but does not change the spatial components' values.

Exercise 7.3.1. Verify the last statement.

BOX 7.4 The Tensor Equation for Conservation of Charge

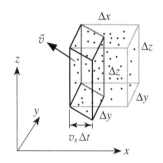

FIG. 7.3 All of the charge in the dark-edged parallelepiped will move through the left face of the gray box in time Δt if the charge in the vicinity of that face has velocity $\vec{v}$. (In this figure, $v_y = 0$.)

Consider a tiny box like that shown in figure 7.2 in box 7.1. Imagine that this box is immersed in a sea of charged particles that could be individually moving randomly but in the infinitesimal region about a given point x, y, z have an *average* velocity $\vec{v}(x, y, z)$ that varies smoothly with position, like the water in a river. Imagine that our box is so small that this average velocity evaluated over any given face of the box is approximately uniform.

Focus now on the two faces of the box that are perpendicular to the x axis. As figure 7.3 illustrates, if $v_x < 0$ in the vicinity of the left-most of these faces, then all of the charges in a parallelepiped with volume $\Delta y \Delta z (-v_x) \Delta t$ will move *out* of the box during the (very short) time interval Δt. The total amount of charge that moves *out* of the box through the left face during this time interval is therefore roughly

$$\Delta q_{\text{left}} \approx -\rho(x - \tfrac{1}{2}\Delta x, y, z) v_x(x - \tfrac{1}{2}\Delta x, y, z) \Delta t \, \Delta y \, \Delta z$$
$$\equiv -J^x(x - \tfrac{1}{2}\Delta x, y, z) \Delta t \, \Delta y \, \Delta z \quad (7.28)$$

where $\rho(x - \tfrac{1}{2}\Delta x, y, z)$, $v_x(x - \tfrac{1}{2}\Delta x, y, z)$ and $J^x(x - \tfrac{1}{2}\Delta x, y, z)$ are the average charge density, average charge x-velocity, and the x component of the current density $\vec{J} \equiv \rho \vec{v}$ evaluated at the center of that left face, respectively. Note that if $v_x > 0$, then Δq_{left} will be negative (indicating that charge has moved *into* the box). Similarly, the amount of charge flowing out of the right face of this box during this time interval will be approximately $\Delta q_{\text{right}} = +J^x(x + \tfrac{1}{2}\Delta x, y, z) \Delta t \, \Delta y \, \Delta z$ (positive because charge will flow out if $v_x(x + \tfrac{1}{2}\Delta x, y, z) > 0$ in this case). Therefore the net amount of charge flowing out of these two faces during Δt will be

$$\Delta q_{\text{left}} + \Delta q_{\text{right}} \approx \left[J^x(x + \tfrac{1}{2}\Delta x, y, z) - J^x(x - \tfrac{1}{2}\Delta x, y, z) \right] \Delta y \, \Delta z \, \Delta t$$

$$\Rightarrow \quad \frac{\Delta q_{\text{left}} + \Delta q_{\text{right}}}{\Delta x \Delta y \Delta z \Delta t} \approx \left[\frac{J^x(x + \tfrac{1}{2}\Delta x, y, z) - J^x(x - \tfrac{1}{2}\Delta x, y, z)}{\Delta x} \right] \quad (7.29)$$

Note that this approximation becomes exact in the limit that Δx, Δy, Δz, and Δt go to zero, and in that limit, we have, by definition of the partial derivative,

$$\lim_{\Delta x \to 0} \left[\frac{J^x(x + \tfrac{1}{2}\Delta x, y, z) - J^x(x - \tfrac{1}{2}\Delta x, y, z)}{\Delta x} \right] = \frac{\partial J^x}{\partial x} \quad (7.30)$$

Exercise 7.4.1. Find expressions analogous to equation 7.29 for the net charge escaping through the front and back faces and through the top and bottom faces.

BOX 7.4 (continued) The Tensor Equation for Conservation of Charge

Now, if we add equation 7.30 to the expressions you get for exercise 7.4.1, you should find we have a term on the left side equal to

$$\frac{\Delta q_{\text{left}} + \Delta q_{\text{right}} + \Delta q_{\text{front}} + \Delta q_{\text{back}} + \Delta q_{\text{top}} + \Delta q_{\text{bottom}}}{\Delta x \, \Delta y \, \Delta z \, \Delta t} \tag{7.31}$$

But by conservation of charge, the total charge that leaves the box through all six faces must come at the expense of charge *inside* the box, so the numerator of this expression becomes simply $-\Delta q_{\text{inside}}$, where q_{inside} is the total charge enclosed by the box. Now, note that when we take the limit of this expression as Δx, Δy, Δz, and Δt all go to zero we find that

$$\lim_{\Delta x, \Delta y, \Delta z, \Delta t \to 0} \left[-\frac{\Delta q_{\text{inside}}}{\Delta x \, \Delta y \, \Delta z \, \Delta t} \right] = -\frac{\partial \rho}{\partial t} \tag{7.32}$$

If you combine this expression with equations 7.29 and 7.30 and their equivalents for the other pairs of faces, you can show that in this limit, we indeed have

$$\partial_\mu J^\mu = 0 \tag{7.33}$$

Exercise 7.4.2. Verify equation 7.33.

BOX 7.5 The Antisymmetry of *F* Implies Charge Conservation

Exercise 7.5.1. Use the antisymmetry of $F^{\mu\nu}$ and the fact that the order of partial derivatives is irrelevant to show that $\partial_\mu \partial_\nu F^{\mu\nu} = 0$. (*Hint:* Study the argument surrounding equation 7.9.)

BOX 7.6 The Magnetic Potential

According to equation 7.12, the components of the electromagnetic field tensor $F^{\mu\nu}$ that yield the components of the classical magnetic field vector are $F^{xy} = B_z$, $F^{zx} = B_y$, and $F^{yz} = B_x$. The generalized EM potential equation $F^{\mu\nu} \equiv \partial^\mu A^\nu - \partial^\nu A^\mu$ (equation 7.17) then implies that

$$B_x = F^{yz} = \partial^y A^z - \partial^z A^y = \partial_y A^z - \partial_z A^y = \frac{\partial A^z}{\partial y} - \frac{\partial A^y}{\partial z} \qquad (7.34a)$$

$$B_y = F^{zx} = \partial^z A^x - \partial^x A^z = \partial_z A^x - \partial_x A^z = \frac{\partial A^x}{\partial z} - \frac{\partial A^z}{\partial x} \qquad (7.34b)$$

$$B_z = F^{xy} = \partial^x A^y - \partial^y A^x = \partial_x A^y - \partial_y A^x = \frac{\partial A^y}{\partial x} - \frac{\partial A^x}{\partial y} \qquad (7.34c)$$

where in the next-to-last step of each line, I used the fact (see box 7.3) that in cartesian coordinates in flat space, raising or lowering a spatial index does nothing to a vector component. Now, the notation "$\vec{\nabla} \times \vec{A}$" is meant to imply that you should calculate the "cross product" of $\vec{\nabla}$ and $\vec{A}$, treating the gradient operator $\vec{\nabla}$ as if it were a "vector" $\vec{\nabla} \equiv [\partial/\partial x, \partial/\partial y, \partial/\partial z]$ and considering the components of $\vec{A}$ to be the spatial components of the four-vector **A**.

Exercise 7.6.1. Show (using your favorite method for evaluating the cross product) that the x component of $\vec{B} = \vec{\nabla} \times \vec{A}$ is equivalent to equation 7.34a above. (The other components are analogous.)

By the way, we have seen that the expression $\vec{E} = -\vec{\nabla}\phi$ applies only in the case of static fields. You can show that $F^{\mu\nu} \equiv \partial^\mu A^\nu - \partial^\nu A^\mu$ with $\mu = t$ implies that a fully correct version of this equation can be written in "vector" form as

$$\vec{E} = -\frac{\partial \vec{A}}{\partial t} - \vec{\nabla}\phi \qquad (7.35)$$

Exercise 7.6.2. Check that $F^{\mu\nu} \equiv \partial^\mu A^\nu - \partial^\nu A^\mu$ with $\mu = t$ and $\nu = x$ yields the x component of equation 7.35. (The other components are analogous.) Also, what equation does $\mu = \nu = t$ yield?

BOX 7.7 Proof of the Source-Free Maxwell Equations (Equation 7.20).

Exercise 7.7.1. By inserting $F^{\mu\nu} \equiv \partial^\mu A^\nu - \partial^\nu A^\mu$ (equation 7.17) into the left side of $\partial^\alpha F^{\mu\nu} + \partial^\nu F^{\alpha\mu} + \partial^\mu F^{\nu\alpha} = 0$ (equation 7.20), show that the latter is indeed identically zero. (Remember that the order of partial derivatives does not matter.)

HOMEWORK PROBLEMS

P7.1 By applying the general equation for the transformation properties of a tensor (see equation 6.12) to the electromagnetic field tensor $F^{\mu\nu}$ and remembering that the Lorentz transformation is a special case of such a transformation (see equations 5.13), show that if we know a electric field's components in one IRF (frame S), we can calculate its components in a second IRF (frame S') in standard orientation with respect to the first as follows:

$$E'_x = E_x \quad (7.36a)$$
$$E'_y = \gamma(E_y - \beta B_z) \quad (7.36b)$$
$$E'_z = \gamma(E_z + \beta B_y) \quad (7.36c)$$

Also use the same procedure to find the analogous transformation equations for the components of the magnetic field.

P7.2 a. Argue that equation 7.20 has 64 components.
b. Show that in fact the left side of the equation is identically zero unless the indices represented by α, μ, and ν are all different (e.g., $\alpha = t$, $\mu = x$ and $\nu = z$). (*Hint:* The antisymmetric character of $F^{\mu\nu}$ is crucial here.)
c. Pick any *three* sets of distinct choices for α, μ, and ν such that the three index values are all different, and show that equation 7.20 reduces to either Gauss's law for the magnetic field ($\vec{\nabla} \cdot \vec{B} = 0$) or to a component of Faraday's law (in GR units):

$$\frac{\partial \vec{B}}{\partial t} + \vec{\nabla} \times \vec{E} = 0 \quad (7.37)$$

P7.3 a. Re-express equation 7.5 in terms of the electromagnetic four-potential A^{μ}.
b. Show that a new four-potential $A^{\mu}_{\text{new}} = A^{\mu} + \partial^{\mu}\Lambda$, where Λ is any arbitrary function of the spacetime coordinates, generates the same field tensor $F^{\mu\nu}$ as does A^{μ}. Therefore adding such a term does not change anything observable about the electromagnetic field. For obscure historical reasons, physicists call such a change a "gauge transformation."
c. It turns out that this degree of freedom allows us always to choose a four-potential A^{μ} such that $\partial_{\mu}A^{\mu} = 0$ (physicists call this the "Lorenz gauge.") Show that with such a choice, your answer to the first part of this problem becomes much simpler.

Note: Your answer to part c, the Lorenz gauge condition $\partial_{\mu}A^{\mu} = 0$, and the definition of $F^{\mu\nu}$ in equation 7.17 together are equivalent to all of Maxwell's equations!

P7.4 The result we found in box 7.3 applies to tensor indices in general, not just four-vectors. Show that (in cartesian coordinates in a flat spacetime) $T^{\cdots}{}_{\mu}{}^{\cdots} = -T^{\cdots \mu \cdots}$ if $\mu = t$ and $T^{\cdots}{}_{\mu}{}^{\cdots} = +T^{\cdots \mu \cdots}$ otherwise, where the dots stand for an arbitrary number of indices.

P7.5 Argue that the Lorentz force equation (in GR units)

$$\frac{d\vec{p}}{dt} \equiv \vec{F}_{em} = q(\vec{E} + \vec{v} \times \vec{B}) \quad (7.38)$$

is correct even in the extreme relativistic limit, as long as we define $\vec{p}$ to be the particle's *relativistic* momentum (i.e., the spatial components of $\mathbf{p}$) in the reference frame where $t, \vec{v}, \vec{E}$, and $\vec{B}$ are being measured).

P7.6 Carefully write out a solution for exercise 7.7.1.

P7.7 According to the tensor transformation rules discussed in chapter 6, the quantity $F^{\mu\nu}F_{\mu\nu}$ must be a frame-independent scalar. This tells us (intriguingly) that there is some aspect of the electromagnetic field that is frame-independent. Express this frame-independent quantity in terms of the electric and magnetic field components. (*Hint:* This quantity is *not* zero. You may find the result of problem P7.4 helpful.)

P7.8 In any introductory physics textbook, you can find the equation for the density of energy stored in an electromagnetic field. In GR units, this equation reads

$$\rho_E = \frac{1}{8\pi k}(E^2 + B^2) \quad (7.39)$$

In this problem, we will seek a tensor generalization of this equation.
a. We argued in this chapter that electric charge was a scalar, but charge *density* was the time component of a four-vector $\mathbf{J}$. Argue that since energy is the time component of a four-vector (the total four-momentum $\mathbf{p}$ enclosed by our differential volume), energy *density* (in general) must be the T^{tt} component of a second-rank tensor $\mathbf{T}$, a tensor that physicists call the "stress-energy" tensor. (*Hints:* Adapt the argument that in the text leads to the conclusion below equations 7.4, but replace charge q with total energy E. Ultimately show that the energy density ρ_E of a bunch of particles transforms like the time-time component of a second-rank tensor. If this applies to the energy density of particles, it ought also to the energy density of the field as well.)
b. This means that the right-hand side of the tensor version of equation 7.39 must also be a second-rank tensor. This tensor must be constructed from only $F^{\mu\nu}$ and $\eta_{\mu\nu}$ (since no other tensors have anything to do with the electromagnetic field or the spacetime in which we find that field). Find a tensor equation whose t-t component is equation 7.39. (*Hint:* You might find the answer to problem P7.7 useful.)
c. If T^{tt} is the density of energy p^t in the field, then T^{tx} is plausibly the density of x-momentum p^x in the field. Express T^{tx} in terms of components of $\vec{E}$ and $\vec{B}$. Does this expression ring any bells? (*Hint:* Look up "Poynting vector.")

(Problem P6.9 is also a good problem for this chapter.)

8. GEODESICS

INTRODUCTION

FLAT SPACETIME
- Review of Special Relativity
- Four-Vectors
- Index Notation

TENSORS
- Arbitrary Coordinates
- Tensor Equations
- Maxwell's Equations
- **Geodesics**

SCHWARZSCHILD BLACK HOLES
- The Schwarzschild Metric
- Particle Orbits
- Precession of the Perihelion
- Photon Orbits
- Deflection of Light
- Event Horizon
- Alternative Coordinates
- Black Hole Thermodynamics

THE CALCULUS OF CURVATURE
- The Absolute Gradient
- Geodesic Deviation
- The Riemann Tensor

THE EINSTEIN EQUATION
- The Stress-Energy Tensor
- The Einstein Equation
- Interpreting the Equation
- The Schwarzschild Solution

COSMOLOGY
- The Universe Observed
- A Metric for the Cosmos
- Evolution of the Universe
- Cosmic Implications
- The Early Universe
- CMB Fluctuations & Inflation

GRAVITATIONAL WAVES
- Gauge Freedom
- Detecting Gravitational Waves
- Gravitational Wave Energy
- Generating Gravitational Waves
- Gravitational Wave Astronomy

SPINNING BLACK HOLES
- Gravitomagnetism
- The Kerr Metric
- Kerr Particle Orbits
- Ergoregion and Horizon
- Negative-Energy Orbits

this depends on → this

8. GEODESICS

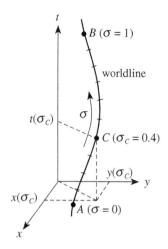

FIG. 8.1 We can describe a worldline between two events A and B by labeling events along the worldline with an arbitrary parameter σ and then specifying the spacetime coordinates of these events as a function of σ. The diagram illustrates the coordinates of the event C corresponding to a parameter value of $\sigma_C = 0.4$.

Introduction. A **geodesic** through spacetime can be defined in one of two ways: (1) as *the worldline of longest proper time* between a given pair of events, or (2) as *the straightest possible worldline*. In this section, we will focus on the first definition, and derive from that definition an expression for calculating geodesics in any arbitrary coordinate system.

A geodesic in an ordinary space would be the path of *shortest distance* between two points. So why is a geodesic in spacetime the worldline of *longest* proper time? The ultimate reason is the minus sign in the metric equation, which often screws up our expectations. However, here is a clue that this is correct. In the flat spacetime of special relativity, a free particle follows an *inertial* worldline (that is, it moves at a constant speed in a straight line). Moreover, in special relativity, an *inertial* clock passing through two events measures the *longest possible* proper time between those events (see box 8.1 for a discussion). If the geodesic hypothesis is true (i.e., if free particles follow geodesics), then in flat spacetime at least, the geodesic between two events is thus indeed the worldline of longest proper time between those events. We are simply extending this equivalence to possibly curved spacetimes.

Timelike Geodesics. Consider two events A and B that have a timelike separation, and consider possible timelike worldlines between them. We can describe such a worldline by specifying the spacetime coordinates of events along it as a function of a parameter σ that goes from 0 at event A to 1 at event B, i.e., by specifying $x^\mu(\sigma)$ (see figure 8.1). The proper time measured along this worldline is then

$$\tau_{AB} = \int \sqrt{-ds^2} = \int_0^1 \sqrt{-g_{\mu\nu}(x^\alpha(\sigma))\frac{dx^\mu}{d\sigma}\frac{dx^\nu}{d\sigma}}\, d\sigma \tag{8.1}$$

where the notation $g_{\mu\nu}(x^\alpha(\sigma))$ reminds us that the metric is a function of the coordinates of events along the worldline, which in turn are functions of the parameter σ.

Finding the worldline of longest proper time is analogous to the goal of finding the trajectory $q_i(t)$ of extreme action using the variational principle in mechanics. In that case, we define a system's action S to be the time-integral of a Lagrangian L:

$$S \equiv \int_{t_A}^{t_B} L(q_i, \dot{q}_i)\, dt \tag{8.2}$$

where $L(q_i, \dot{q}_i)$ indicates that the Lagrangian is a function of the generalized position coordinates q_i and the generalized velocities $\dot{q}_i$ (where i ranges over however many such coordinates you have). In classical mechanics we found that the trajectory $q_i(t)$ physically followed by the system was the one that made the action S extreme, which in turn was the trajectory that satisfied the Euler-Lagrange equations

$$0 = \frac{d}{dt}\left(\frac{\partial L}{\partial \dot{q}_i}\right) - \frac{\partial L}{\partial q_i} \tag{8.3}$$

The situation here is exactly analogous: we have a Lagrangian function that depends on coordinates x^μ and "velocities" $\dot{x}^\alpha \equiv dx^\alpha/d\sigma$,

$$L(x^\alpha, \dot{x}^\alpha) \equiv \sqrt{-g_{\mu\nu}(x^\alpha)\dot{x}^\mu \dot{x}^\nu} \tag{8.4}$$

and we hope to find the worldline $x^\mu(\sigma)$ that makes the integral of this Lagrangian extreme. By analogy, the worldline we want will be the one that satisfies the Euler-Lagrange equations for this Lagrangian,

$$0 = \frac{d}{d\sigma}\left[\frac{\partial L}{\partial \dot{x}^\alpha}\right] - \frac{\partial L}{\partial x^\alpha} \tag{8.5}$$

For an N-dimensional space, this corresponds to N separate differential equations (one for each coordinate and therefore for each value of the abstract index α). In the particular case of 4D spacetime, we will have four differential equations.

If you do not recall the variational principle of mechanics or find the analogy difficult, box 8.2 gives a full derivation of equation 8.5 from first principles.

We can simplify this equation and make its meaning clearer in our specific context as follows. Note that according to the chain rule

$$\frac{\partial L}{\partial \dot{x}^\alpha} = \frac{\partial}{\partial \dot{x}^\alpha}\sqrt{-g_{\mu\nu}\dot{x}^\mu \dot{x}^\nu} = \frac{1}{2\sqrt{-g_{\mu\nu}\dot{x}^\mu \dot{x}^\nu}}(-g_{\mu\nu}\delta^\mu{}_\alpha \dot{x}^\nu - g_{\mu\nu}\dot{x}^\mu \delta^\nu{}_\alpha) \quad (8.6)$$

because $\partial \dot{x}^\mu / \partial \dot{x}^\alpha$ is 1 if $\mu = \alpha$ and zero otherwise, and because the metric does not depend on $\dot{x}^\alpha$ but rather only on x^α. If we note that the square root in the denominator is just L, and do the sums over the Kronecker deltas, we get

$$\frac{\partial L}{\partial \dot{x}^\alpha} = \frac{1}{2L}(-g_{\alpha\nu}\dot{x}^\nu - g_{\mu\alpha}\dot{x}^\mu) = \frac{1}{2L}(-g_{\alpha\mu}\dot{x}^\mu - g_{\mu\alpha}\dot{x}^\mu)$$
$$= \frac{1}{2L}(-g_{\alpha\mu}\dot{x}^\mu - g_{\alpha\mu}\dot{x}^\mu) = \frac{1}{L}(-g_{\alpha\mu}\dot{x}^\mu) = -\frac{1}{L}g_{\alpha\mu}\frac{dx^\mu}{d\sigma} \quad (8.7)$$

But notice that

$$\tau(\sigma) = \int_0^{\tau(\sigma)} d\tau = \int_0^\sigma \sqrt{-g_{\mu\nu}\frac{dx^\mu}{d\sigma}\frac{dx^\nu}{d\sigma}}\,d\sigma \Rightarrow \frac{d\tau}{d\sigma} = \sqrt{-g_{\mu\nu}\frac{dx^\mu}{d\sigma}\frac{dx^\nu}{d\sigma}} = L \quad (8.8)$$

If we plug this into equation 8.7, we find that

$$\frac{\partial L}{\partial \dot{x}^\alpha} = -g_{\alpha\mu}\frac{dx^\mu}{d\sigma}\frac{d\sigma}{d\tau} = -g_{\alpha\mu}\frac{dx^\mu}{d\tau} \quad (8.9)$$

Similarly,

$$\frac{\partial L}{\partial x^\alpha} = -\frac{1}{2L}\frac{\partial g_{\mu\nu}}{\partial x^\alpha}\frac{dx^\mu}{d\sigma}\frac{dx^\nu}{d\sigma} = -\frac{1}{2}\frac{\partial g_{\mu\nu}}{\partial x^\alpha}\frac{dx^\mu}{d\sigma}\frac{d\sigma}{d\tau}\frac{dx^\nu}{d\sigma} = -\frac{1}{2}\frac{\partial g_{\mu\nu}}{\partial x^\alpha}\frac{dx^\mu}{d\tau}\frac{dx^\nu}{d\sigma} \quad (8.10)$$

Therefore, equation 8.5 becomes

$$0 = \frac{d}{d\sigma}\left[-g_{\alpha\mu}\frac{dx^\mu}{d\tau}\right] + \frac{1}{2}\frac{\partial g_{\mu\nu}}{\partial x^\alpha}\frac{dx^\mu}{d\tau}\frac{dx^\nu}{d\sigma} \quad (8.11)$$

Let's rename $\mu \to \beta$ in the first term to make it clear that the implied sums in the first and second terms are independent. If we now multiply through by $-d\sigma/d\tau$ and use our compact expression for the partial derivative of the metric, we get

$$0 = \frac{d}{d\tau}\left(g_{\alpha\beta}\frac{dx^\beta}{d\tau}\right) - \tfrac{1}{2}\partial_\alpha g_{\mu\nu}\frac{dx^\mu}{d\tau}\frac{dx^\nu}{d\tau} \quad (8.12)$$

This is the **geodesic equation** in the form we will most commonly use. It is simpler to use than equation 8.5 because (1) this equation *explicitly* refers to the spacetime's metric tensor $g_{\alpha\beta}$ instead of burying it in the Lagrangian L, and (2) all references to the arbitrary worldline parameter σ have vanished: solving this set of differential equations yields the coordinates $x^\alpha(\tau)$ of events along the geodesic as a function of the proper time τ that a clock traveling along the geodesic would measure. This makes the solutions easier to interpret physically.

Note that the factor of $g_{\alpha\beta}$ in the parentheses in equation 8.12 depends *implicitly* on τ because $g_{\alpha\beta}$ generally is a function of the coordinates, which in turn are functions $x^\mu(\tau)$ for events along the geodesic worldline. Expanding this term yields

$$\frac{d}{d\tau}\left(g_{\alpha\beta}\frac{dx^\beta}{d\tau}\right) = \left(\frac{\partial g_{\alpha\beta}}{\partial x^\gamma}\frac{dx^\gamma}{d\tau}\right)\frac{dx^\beta}{d\tau} + g_{\alpha\beta}\frac{d^2x^\beta}{d\tau^2} \quad (8.13)$$

by the product and chain rules. If we rename $\gamma \to \mu$ and $\beta \to \nu$ in the first term on the right, substitute the result into equation 8.12, multiply by the inverse metric $g^{\gamma\alpha}$, and sum over α (see box 8.3), we can express the geodesic equation in the form

$$0 = \frac{d^2x^\gamma}{d\tau^2} + g^{\gamma\alpha}(\partial_\mu g_{\alpha\nu} - \tfrac{1}{2}\partial_\alpha g_{\mu\nu})\frac{dx^\mu}{d\tau}\frac{dx^\nu}{d\tau} \quad (8.14)$$

This form of the geodesic equation is a bit more complicated than equation 8.12 in that we have to calculate the inverse metric as well as knowing the metric, but it does more *explicitly* expose the implications of the metric varying with position.

In particular, note that if the metric is everywhere *independent* of position (as it is when we use cartesian coordinates in flat spacetime, where $g_{\mu\nu} = \eta_{\mu\nu} =$ constant), then a free particle's coordinate acceleration $d^2x^\gamma/d\tau^2$ is *zero*, i.e., it moves in a straight line at a constant velocity relative to our reference frame's coordinate grid (as we in fact observe). However, if the metric *does* vary with position, then a free particle following a geodesic will appear to accelerate relative to the coordinate grid ($d^2x^\gamma/d\tau^2 \neq 0$), whether the metric's position-dependence is due to the intrinsic curvature of spacetime or simply due to our use of curvilinear coordinates.

Whether we use equation 8.12 or equation 8.14, I hope it is clear that if we know a spacetime's metric $g_{\mu\nu}$ as a function of position, we can (in principle) solve the set of four differential equations implied by either equation for $x^\gamma(\tau)$ and thus find the geodesic worldline in whatever coordinate system we choose to use.

This equation also works for geodesics in 2D spaces if we substitute pathlength s along the path for proper time τ. See box 8.4 and box 8.5 for examples.

A Hint for Solving the Geodesic Equation. When solving these equations, it is often advantageous to note that the definition of proper time implies that

$$\mathbf{u} \cdot \mathbf{u} = g_{\mu\nu} \frac{dx^\mu}{d\tau} \frac{dx^\nu}{d\tau} = -\left(\frac{d\tau}{d\tau}\right)^2 = -1 \qquad (8.15)$$

(see equation 3.7) for any timelike worldline, including geodesics. Since we have implicitly used this equation in deriving the geodesic equation (see equation 8.8), it really does not provide any *independent* information about the geodesic except about the scale of the proper time parameter τ (the geodesic equation by itself cannot fix the scale of the parameter τ for reasons discussed in box 8.6). Indeed, one will often find that taking the τ-derivative of equation 8.15 simply yields one of the geodesic equations. However since equation 8.15 is a *first*-order differential equation in $dx^\mu/d\tau$, using it generally enables one to avoid integrating one of the second-order geodesic equation components while simultaneously fixing the scale of τ. This can be a *huge* payoff, so using it is highly recommended in most cases.

In a 2D ordinary space (with no time coordinate), the equivalent equation is

$$g_{\mu\nu} \frac{dx^\mu}{ds} \frac{dx^\nu}{ds} = +\left(\frac{ds}{ds}\right)^2 = +1 \qquad (8.16)$$

where s is the pathlength along the curve.

Lightlike Geodesics. The whole line of argument in the previous section breaks down when we consider photon worldlines, because the proper time measured along a photon worldline is always $\tau_{AB} = 0$, so τ is not a good parameter for a lightlike geodesic and the whole idea of finding the worldline of longest proper time becomes absurd (as τ for a light geodesic is already as small as possible).

So how can we find the geodesic equation for a photon? We can in principle use a parameter other than τ. In practice, it is easier to use an approach that extends the geodesic equation we have already derived for ordinary particles.

As an ordinary particle approaches the speed of light in special relativity, it becomes increasingly photon-like. Specifically, its energy $E = m/\sqrt{1-v^2}$ and its momentum $p = mv/\sqrt{1-v^2}$ become huge compared to its mass m. Eventually, the particle's mass becomes completely negligible, so its geodesic should be almost the same as a photon's (whose mass is strictly zero). Equivalently, a photon's geodesic *should* correspond to the limit of the family of geodesics for ordinary particles as we take the limit $m \to 0$ while holding the particle energy E fixed (as taking $m \to 0$ with $E = m/\sqrt{1-v^2}$ fixed means we are requiring $v \to 1$).

So, one way to determine a photon's geodesic is to re-express the implications of the geodesic equation for an ordinary particle in terms of quantities that don't behave badly in the limit that $m \to 0$. This usually means dividing one of the geodesic equation components by another. See box 8.7 for a simple illustration.

BOX 8.1 The Worldline of Longest Proper Time in Flat Spacetime

We can calculate in any IRF the proper time a clock measures when traveling along a given worldline between two events A and B using equation 2.4:

$$\Delta \tau_{BA} = \int_{t_A}^{t_B} \sqrt{1 - v^2}\, dt \tag{8.17}$$

where t is the coordinate time and v is the coordinate speed *in that frame*. While we use frame-dependent quantities t and $v(t)$ to calculate this proper time, the resulting proper time $\Delta \tau_{BA}$ is frame-*independent*, because all observers must agree about what a clock following that worldline will read as it travels between events A and B. Even so, it should also be clear that the proper time will generally depend on the particular worldline the clock follows between A and B.

Consider now a clock attached to a free particle, and consider two events A and B along that particle's worldline. In special relativity, a free particle moves in a straight line at a constant velocity by definition of an IRF. Since it does not matter which IRF we choose to evaluate proper time along any worldline, let us choose, for the sake of convenience, to do our analysis in the IRF in which our free particle is at rest, and to choose the origin of our frame to coincide with the particle's position. In a spacetime diagram based in such a frame, this particle's worldline will be a straight line up the t axis, as shown in figure 8.2. The proper time $\Delta \tau_{AB,F}$ that the free particle's clock will measure between A and B is, according to equation 8.17,

$$\Delta \tau_{BA,F} = \int_{t_A}^{t_B} \sqrt{1 - 0^2}\, dt = t_B - t_A = \Delta t_{BA} \tag{8.18}$$

Now consider the proper time $\Delta \tau_{BA,NF}$ measured by a clock attached to a particle compelled to travel along any other worldline going between events A and B, such as the curved worldline illustrated in figure 8.2. (Since the worldline of a *free* particle traveling between A and B must be straight up the t axis in our frame, a particle following a curved worldline will *not* be free: forces must be exerted on it to keep it following this noninertial worldline. Hence the "NF" subscript.) For at least some part of the time interval between the events, this particle will have a nonzero speed in our chosen frame. Therefore, the proper time its clock measures along this worldline will be

$$\Delta \tau_{BA,NF} = \int_{t_A}^{t_B} \sqrt{1 - [v(t)]^2}\, dt \tag{8.19}$$

where $v(t) \neq 0$ for at least some of the time interval. But since $\sqrt{1-v^2} \leq 1$ for all possible values of v, the integrand in equation 8.19 will always be less than or equal to the integrand in equation 8.18 (and strictly less than that integrand at least some of the time). This implies that

$$\Delta \tau_{BA,NF} = \int_{t_A}^{t_B} \sqrt{1 - v^2}\, dt < \int_{t_A}^{t_B} \sqrt{1}\, dt = \Delta \tau_{BA,F} \tag{8.20}$$

Therefore, we conclude that in this IRF, the proper time measured along the free particle's worldline is larger than that measured along any other worldline.

But since the values of these proper times are frame independent, observers in all IRFs must agree with this conclusion. Moreover, since the argument can be applied to any pair of events A and B, any free-particle worldline between those events, and any other worldline connecting those events, the result is completely general: *the worldline of a free particle moving between two events A and B is the worldline of longest proper time between those events.* Q. E. D.

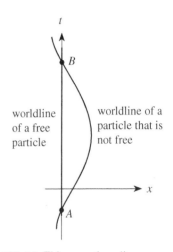

FIG. 8.2 This spacetime diagram shows the worldline of a free particle moving between two events A and B (shown in an IRF where the particle is at rest), and the worldline of a particle following a noninertial worldline between the same events. Since the x-velocity of the latter at any time is equal to the inverse slope of its worldline on this diagram at that time, the noninertial particle's speed must at some time be nonzero if its worldline is to be distinct from the free particle's worldline.

BOX 8.2 Derivation of the Euler-Lagrange Equation

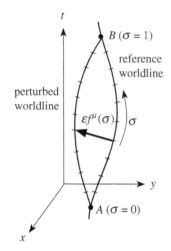

FIG. 8.3 For any given value of the parameter σ, the spacetime coordinates $x_{pt}^\mu(\sigma)$ of the event on the perturbed worldline corresponding to σ is generated by adding four "disturbance functions" $\varepsilon f^\mu(\sigma)$ to the coordinates $x^\mu(\sigma)$ of the corresponding event on the reference worldline. This displaces the perturbed worldline away from the reference worldline at every point.

Consider two events A and B and a particular worldline between them, which we will call the "reference worldline." Let us describe this worldline by specifying the spacetime coordinates of events along it as a function of a parameter σ that goes from 0 at event A to 1 at event B, i.e., by specifying $x^\mu(\sigma)$. Let us also consider small perturbations of this worldline, which we can describe by

$$x_{pt}^\mu(\sigma) = x^\mu(\sigma) + \varepsilon f^\mu(\sigma) \tag{8.21}$$

where $x_{pt}^\mu(\sigma)$ are the coordinates of events along the perturbed worldline as a function of our parameter σ, $f^\mu(\sigma)$ are four arbitrary "disturbance" functions of σ, and ε is a small number. Figure 8.3 illustrates such a reference worldline and one possible perturbation of it. We will assume that the disturbance functions $f^\mu(\sigma)$ are 0 at both $\sigma = 0$ and $\sigma = 1$ so that all the perturbed worldlines go through events A and B just as the reference worldline does.

Our goal is to find the reference worldline that has the longest possible proper time, that is, the largest possible value of

$$\tau_{AB} \equiv \int_0^1 L(\sigma)\, d\sigma, \quad \text{where} \quad L = \sqrt{-g_{\mu\nu} \dot{x}^\mu \dot{x}^\nu} \tag{8.22}$$

The signal that we have succeeded is when an arbitrary infinitesimal perturbation of the reference worldline yields a proper time that is the *same* as that of the reference worldline, just as we know that we have reached the point x_{ex} where a function $f(x)$ is extreme when the value of $f(x)$ remains the same for points in the neighborhood of $f(x_{ex})$, that is, its slope is zero at that point.

In the limit that ε is infinitesimal, we can evaluate L_{pt} on the perturbed worldline by expanding the value of L at the corresponding point on the reference worldline in a Taylor series to first order in ε:

$$L_{pt}(\sigma) = L(\sigma) + \frac{\partial L}{\partial x^\alpha} \varepsilon f^\alpha(\sigma) + \frac{\partial L}{\partial \dot{x}^\alpha} \varepsilon \frac{df^\alpha}{d\sigma} \tag{8.23}$$

by the chain rule of partial differential calculus, since L depends on only σ through its dependence on $x^\mu(\sigma)$ and $\dot{x}^\mu(\sigma)$. Note the implicit sums over α. So

$$\frac{dL_{pt}}{d\varepsilon} = \lim_{\varepsilon \to 0} \frac{L_{pt} - L}{\varepsilon} = \frac{\partial L}{\partial x^\alpha} f^\alpha(\sigma) + \frac{\partial L}{\partial \dot{x}^\alpha} \frac{df^\alpha}{d\sigma} \tag{8.24}$$

Therefore

$$\frac{d\tau_{pt}}{d\varepsilon} = \frac{d}{d\varepsilon} \int_0^1 L_{pt}\, d\sigma = \int_0^1 \frac{dL_{pt}}{d\varepsilon}\, d\sigma = \int_0^1 \left[\frac{\partial L}{\partial x^\alpha} f^\alpha(\sigma) + \frac{\partial L}{\partial \dot{x}^\alpha} \frac{df^\alpha}{d\sigma} \right] d\sigma \tag{8.25}$$

We can simplify the second term in the integrand by integrating by parts

$$\int_0^1 \frac{\partial L}{\partial \dot{x}^\alpha} \frac{df^\alpha}{d\sigma}\, d\sigma = -\int_0^1 f^\alpha \frac{d}{d\sigma}\left(\frac{\partial L}{\partial \dot{x}^\alpha} \right) d\sigma \tag{8.26}$$

Exercise 8.2.1. Fill in the missing steps in equation 8.26. In particular, where does the minus sign come from?

BOX 8.2 (continued) Derivation of the Euler-Lagrange Equation

If we plug equation 8.26 into equation 8.25, we find that

$$\frac{d\tau_{pt}}{d\varepsilon} = \int_0^1 \left[\frac{\partial L}{\partial x^\alpha} - \frac{d}{d\sigma}\left(\frac{\partial L}{\partial \dot{x}^\alpha}\right) \right] f^\alpha(\sigma) d\sigma \qquad (8.27)$$

The condition that our *reference* worldline is a geodesic is the requirement that $d\tau_{pt}/d\varepsilon = 0$: that is that the value of τ be a maximum at $\varepsilon = 0$ for the family of worldlines generated by varying the perturbation scale parameter ε. Moreover, this must be true no matter what the perturbation functions $f^\alpha(\sigma)$ are. These two conditions together then imply that the reference worldline will have the longest proper time compared to all possible neighboring worldlines. The first condition states

$$0 = \int_0^1 \left[\frac{\partial L}{\partial x^\alpha} - \frac{d}{d\sigma}\left(\frac{\partial L}{\partial \dot{x}^\alpha}\right) \right] f^\alpha(\sigma) d\sigma \qquad (8.28)$$

Applying the second condition simultaneously requires that

$$\frac{\partial L}{\partial x^\alpha} - \frac{d}{d\sigma}\left(\frac{\partial L}{\partial \dot{x}^\alpha}\right) = 0 \qquad (8.29)$$

This is the same as equation 8.5.

Exercise 8.2.2. Why does this follow? *Hint:* $f^\alpha(\sigma)$ are *arbitrary* functions.

BOX 8.3 Deriving the Second Form of the Geodesic Equation

Exercise 8.3.1. Show that substituting equation 8.13 into equation 8.12 (and doing what is suggested in the paragraph below equation 8.13) yields equation 8.14.

BOX 8.4 Geodesics for Flat Space in Parabolic Coordinates

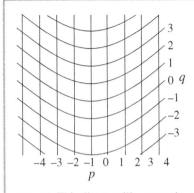

FIG. 8.4 This diagram illustrates the curves of constant p and q coordinates in a 2D flat space.

Consider the parabolic coordinate system p, q for 2D flat space discussed in box 5.3 (see figure 8.4). The coordinate transformations between these coordinates and cartesian x, y coordinates are given by equations 5.23 and 5.24:

$$p(x,y) = x, \quad q(x,y) = y - cx^2 \tag{8.30}$$

$$x(p,q) = p, \quad y(p,q) = cp^2 + q \tag{8.31}$$

According to equation 5.26, the metric for this coordinate system is

$$g_{\mu\nu} = \begin{bmatrix} 1+4c^2p^2 & 2cp \\ 2cp & 1 \end{bmatrix} \tag{8.32}$$

Our goal in this box is to use the geodesic equation to calculate the geodesics in this coordinate system and show that they correspond to straight lines.

In 2D space, we replace the proper time τ in the geodesic equation 8.12 with the pathlength s. If we then write out the implicit sums over β, μ, and ν for the $\alpha = p$ component of that equation, we get

$$0 = \frac{d}{ds}\left\{g_{pp}\frac{dp}{ds} + g_{pq}\frac{dq}{ds}\right\} - \frac{1}{2}\left[\frac{\partial g_{pp}}{\partial p}\left(\frac{dp}{ds}\right)^2 + 2\frac{\partial g_{pq}}{\partial p}\left(\frac{dp}{ds}\right)\left(\frac{dq}{ds}\right) + \frac{\partial g_{qq}}{\partial p}\left(\frac{dq}{ds}\right)^2\right] \tag{8.33}$$

since $g_{pq} = g_{qp}$. If you substitute in the metric components from equation 8.32, evaluate the partial and s derivatives, and cancel some terms, you should find that

$$0 = [1 + 4c^2p^2]\frac{d^2p}{ds^2} + 4c^2p\left(\frac{dp}{ds}\right)^2 + 2cp\frac{d^2q}{ds^2} \tag{8.34}$$

Exercise 8.4.1. Verify that equation 8.34 is correct.

Doing the same for the $\alpha = q$ component of the geodesic equation yields

$$0 = 2cp\frac{d^2p}{ds^2} + 2c\left(\frac{dp}{ds}\right)^2 + \frac{d^2q}{ds^2} \tag{8.35}$$

BOX 8.4 (continued) Geodesics for Flat Space in Parabolic Coordinates

Exercise 8.4.2. Verify equation 8.35.

If we multiply equation 8.35 by $2cp$ and subtract it from equation 8.34, we get

$$0 = \frac{d^2p}{ds^2} \quad \Rightarrow \quad \frac{dp}{ds} = b \quad \Rightarrow \quad p = bs \tag{8.36}$$

where b is an undetermined constant and I have defined $s = 0$ to be where $p = 0$. If we then plug these results back into equation 8.35, we find that

$$0 = 2cb^2 + \frac{d^2q}{ds^2} \quad \Rightarrow \quad \frac{d^2q}{ds^2} = -2cb^2 = \text{constant} \tag{8.37}$$

This is also easy to integrate:

$$q = -cb^2 s^2 + hs + q_0 = -cp^2 + \frac{h}{b}p + q_0 \tag{8.38}$$

where h and q_0 are constants and in the last step I have used $p = bs$ to express q for the geodesic curve in terms of the other coordinate p. If we add cp^2 to both sides and use equation 8.31 to convert to x, y coordinates, we find that

$$q + cp^2 = y = \frac{h}{b}x + q_0 \tag{8.39}$$

In x, y coordinates we recognize this as being a straight line of slope h/b. Therefore, the geodesic equation produces a straight line in this 2D flat space as expected.

In this case, all components of the geodesic equation are so easy to integrate that using the definition of pathlength (equation 8.16) is unnecessary. However, note that our solution yields three arbitrary constants (b, h, q_0), but only *two* constants are required to determine everything we need to know about a straight line, including the pathlength between two points. You can show that equation 8.16 in this situation yields $b^2 + h^2 = 1$, so b and h are not really independent. But the geodesic equation does not supply this constraint on b (which amounts to specifying the scale of the parameter s): it comes only from equation 8.16.   *(continued)*

BOX 8.4 (continued) Geodesics for Flat Space in Parabolic Coordinates

Exercise 8.4.3. Verify that equation 8.16 in p, q coordinates yields $b^2 + h^2 = 1$ (note that equations 8.36 and 8.38 imply that $dp/ds = b$ and $dq/ds = -2cb^2 s + h$).

BOX 8.5 Geodesics for the Surface of a Sphere

A very important example of a two-dimensional curved space is the curved surface of a sphere. Consider latitude-longitude coordinates θ, ϕ for this surface, where the latitude coordinate θ is measured (in the conventional physics way) down from the north pole instead of up from the equator. As discussed in box 5.6, the metric for the spherical surface in these coordinates is given by

$$g_{\mu\nu} = \begin{bmatrix} R^2 & 0 \\ 0 & R^2 \sin^2\theta \end{bmatrix} \tag{8.40}$$

where R is the sphere's radius. Using this metric, you can show that the $\alpha = \theta$ and $\alpha = \phi$ components of the geodesic equation in 2D space yield, respectively,

$$0 = \frac{d^2\theta}{ds^2} - \sin\theta\cos\theta\left(\frac{d\phi}{ds}\right)^2 \tag{8.41}$$

$$0 = \frac{d}{ds}\left(\sin^2\theta\frac{d\phi}{ds}\right) \tag{8.42}$$

Exercise 8.5.1. Check that equation 8.41 and 8.42 are correct.

BOX 8.5 (continued) Geodesics for the Surface of a Sphere

We can integrate equation 8.42 quite easily to get

$$\sin^2\theta \frac{d\phi}{ds} = \text{const.} \equiv \frac{c}{R} \quad \Rightarrow \quad \frac{d\phi}{ds} = \frac{c}{R\sin^2\theta} \tag{8.43}$$

where c is a *unitless* but otherwise arbitrary constant (since R has units of s, including R as part of the constant of integration makes it clearer how the units work out). Substituting this into equation 8.41 yields

$$\frac{d^2\theta}{ds^2} = \cos\theta \sin\theta \left(\frac{c}{R\sin^2\theta}\right)^2 = \frac{c^2}{R^2}\frac{\cos\theta}{\sin^3\theta} \tag{8.44}$$

This differential equation would be very difficult to solve directly.

In this case, using the definition of the pathlength (equation 8.16) really helps us with the integration. In this metric, the definition reads

$$1 = R^2\left(\frac{d\theta}{ds}\right)^2 + R^2\sin^2\theta\left(\frac{d\phi}{ds}\right)^2 \tag{8.45}$$

Substituting $d\phi/ds$ from equation 8.43 and solving for $d\theta/ds$ yields

$$\frac{d\theta}{ds} = \pm\frac{1}{R\sin\theta}\sqrt{\sin^2\theta - c^2} \tag{8.46}$$

Exercise 8.5.2. Verify that equation 8.46 is correct.

This actually *can* be integrated analytically. Problem P8.2 discusses how to solve the problem from here. My goal in this box is only to display the basic outline for how one might arrive at a solution in this case (which will be very similar to cases in curved spacetime that we will consider later).

In this case, though, we can take advantage of the symmetry of a sphere to show that the geodesics must be great circles. Consider first the equator, the curve where $\theta = \pi/2$. Since $d^2\theta/ds^2$ and $\cos\theta$ are zero at all points on this curve, equation 8.41 is satisfied no matter what $d\phi/ds$ is. So the equator is a geodesic. Now consider a geodesic going from an arbitrary point A on the sphere to some other arbitrary point B. Without changing anything physical, we can rotate the coordinates around until both A and B lie on the equator. Since the equator is the *unique* geodesic connecting any two points on the equator, the geodesic between A and B must then lie entirely along the new equator. Since any closed curve on the sphere's surface that we can match to the equator in a rotated coordinate system is a great circle by definition, it follows that the geodesic between A and B in their original positions must have been part of a great circle.

BOX 8.6 The Geodesic Equation Does Not Determine the Scale of τ

Exercise 8.6.1. Show that if $x^\mu(\tau)$ satisfies the geodesic equation 8.12, then so will $x^\mu(\bar{\tau})$, where $\bar{\tau} \equiv b\tau$ and b is an arbitrary constant. Show, however, that the same is *not* true of the definition of τ given by equation 8.15. This implies that while the geodesic equation does not uniquely fix the scale of the parameter τ, equation 8.15 does.

BOX 8.7 Light Geodesics in Flat Spacetime

In the case of cartesian coordinates in flat spacetime, the metric $g_{\mu\nu} = \eta_{\mu\nu}$ is constant, so the geodesic equation in the form given in equation 8.14 implies that

$$0 = \frac{d^2 x^\mu}{d\tau^2} \quad \Rightarrow \quad \frac{dx^\mu}{d\tau} = c^\mu \qquad (8.47)$$

where the c^μ are constants of integration (one for each value of μ). The equation for $\mu = t$ implies that $dt/d\tau = c^t$. We can use this to eliminate τ from the others:

$$\frac{dx^i}{dt} = \frac{dx^i}{d\tau}\frac{d\tau}{dt} = \frac{c^i}{c^t} \equiv v^i \quad (\text{where } i = x, y, z) \qquad (8.48)$$

where the v^i are three new constants (one for each value of i). This tells us that a particle following a geodesic in flat spacetime moves at a constant (cartesian) coordinate velocity.

Now, as we take the limit where the particle mass $m \to 0$ for constant energy $E = m\, dt/d\tau$, $d\tau \to 0$ and $c^\mu \to \infty$, so the geodesic equation solutions given in equation 8.47 become undefined. But the solutions described by equation 8.48 remain perfectly well defined in this limit. So we conclude that a *photon* following a geodesic in flat spacetime will *also* move at constant cartesian coordinate velocity. The metric equation tells us what the speed is.

Exercise 8.7.1. For light, we have $0 = ds^2 = -dt^2 + dx^2 + dy^2 + dz^2$. Show that this in conjunction with equation 8.48 implies that the speed of light is 1.

HOMEWORK PROBLEMS

P8.1 In this problem, we will explore a simplified argument (developed by Richard Feynman) that illustrates why the worldline of a projectile near the earth's surface could indeed be a worldline of longest proper time. (For a different version of this problem, see problem P9.8.)

a. As we saw qualitatively in the first chapter and quantitatively in problem P1.2, the equivalence principle implies that an observer at rest on the earth's surface at $z = 0$ will observe light from a source at a higher position z in the earth's field to be blue-shifted as follows:

$$\frac{\lambda_e - \lambda}{\lambda_e} \approx gz \quad \Rightarrow \quad \frac{\lambda}{\lambda_e} \approx 1 - gz \quad (8.49)$$

where λ is the wavelength the observer measures, λ_e is the wavelength observed where it was emitted, and g is the acceleration of gravity (in GR units). This equation is a good approximation as long as $gz \ll 1$.

Since successive crests of the light wave travel the same distance from the source to the floor through exactly the same spacetime in the earth frame, the earth-frame observer can only conclude that if the crests are *arriving* more rapidly than expected, the source must be *emitting* them more rapidly than expected. This can only mean that a clock at altitude z must run faster than one at $z = 0$. Argue that if an observer at $z = 0$ measures an infinitesimal coordinate time dt between two events, a clock at rest at z will measure a time $dT \approx (1 + gz)dt$ between those events. (*Hint:* Recall the binomial approximation: $(1 + x)^n \approx 1 + nx$ as long as $x \ll 1$).

b. Now imagine that we launch a clock from $z = 0$ as a projectile. At altitude z, assume that the clock has speed v. The proper time that the projectile clock reads between two infinitesimally separated events along its worldline at height z will be shorter than the time measured by a clock at rest at z by the time dilation factor $\sqrt{1 - v^2}$. Argue therefore that if $v \ll 1$ and $gz \ll 1$, then the proper time $d\tau$ measured by the projectile clock between such events will be related to the time dt measured between those events by an observer at $z = 0$ as follows:

$$d\tau \approx dt(1 - \tfrac{1}{2}v^2 + gz) \quad (8.50)$$

(*Hint:* Again, recall the binomial approximation.)

c. If we want to choose a trajectory for the projectile clock that has maximum proper time between two given events A and B in spacetime, we see that we must contend with two competing effects: on the one hand, we would like to minimize the speed, but on the other, we would like to maximize the altitude. What trajectory provides the optimal trade-off between these effects? To find out, define a Lagrangian

$L = 1 - \tfrac{1}{2}(\dot{x}^2 + \dot{y}^2 + \dot{z}^2) + gz$ (where $\dot{x} \equiv dx/dt = v_x$ and so on) and use normal Lagrangian methods to find the equations for $x(t), y(t), z(t)$ that maximize

$$\tau_B - \tau_A = \int_{t_A}^{t_B} L\, dt \quad (8.51)$$

You should find that the projectile clock's optimal trajectory is the standard parabola of projectile motion.

P8.2 Our goal in this problem is to finish solving the geodesic equation for θ, ϕ coordinates on the surface of a sphere and show that the solutions are great circles.

a. We begin where we left off in box 8.5. Show that we can rearrange equation 8.46 to read

$$ds = \pm \frac{R \sin\theta\, d\theta}{\sqrt{a^2 - \cos^2\theta}} \quad (8.52)$$

where $a \equiv \sqrt{1 - c^2}$ is an undetermined constant. Note that for this equation to be real, we must have $|\cos\theta| \leq a$ and $0 \leq a \leq 1$.

b. Define a new variable $u \equiv \cos\theta$, rewrite equation 8.52 in terms of u, and integrate both sides (you can find the integral of the right side in any decent table of integrals). Choose the constant of integration so that the pathlength s is 0 when the curve crosses the equatorial plane (where $u = 0$). You should find that

$$u \equiv \cos\theta = \pm a \sin\psi \quad \text{where} \quad \psi \equiv \frac{s}{R} \quad (8.53)$$

c. Now we can integrate equation 8.43. Note that

$$\sin^2\theta = 1 - \cos^2\theta = 1 - a^2\sin^2\psi \quad (8.54)$$

If you substitute this into equation 8.43 and integrate with respect to ψ (again, you should be able to find the integral in an integral table), you should get

$$\tan(\phi - \phi_0) = c \tan\psi \quad (8.55)$$

where ϕ_0 is a constant of integration specifying the value of ϕ when $\psi = 0$, i.e., when $s = 0$ (which is where the geodesic crosses the equator according to part b) and c is the same constant as in part a ($c = \sqrt{1 - a^2}$). We now have completely specified the geodesic curve as a function of $\psi \equiv s/R$.

d. But is this curve a great circle? A great circle is the intersection of a sphere's surface and a plane containing the sphere's center. For simplicity's sake, let us choose our plane to be the one whose equation in cartesian coordinates is $z = my$: this plane will have a slope of m and will cross the equatorial plane along the x axis. The equation for the surface of a sphere in cartesian coordinates is

$$x^2 + y^2 + z^2 = R^2 \quad (8.56)$$

You can find the equation for the intersection of these surfaces by substituting $z = my$ into equation 8.56 to eliminate z and then using the definitions of x and y in spherical coordinates

$$x = R\sin\theta\cos\phi, \quad y = R\sin\theta\sin\phi \quad (8.57)$$

to get an equation in terms of θ and ϕ. You should find the result to be

$$\sqrt{\frac{1}{\sin^2\theta} - 1} = \cot\theta = \pm m\sin\phi \quad (8.58)$$

e. Let us now see if this great circle coincides with the solution to the geodesic equation implied by equations 8.53 and 8.55. First, note that since the plane defined by $z = my$ crosses the equatorial plane along the x axis, and since equation 8.57 implies that $\phi = 0$ on that axis, the great circle we found in part d should correspond to a solution where $\phi_0 = 0$. Second, note that $\sin\phi = \tan\phi/\sqrt{1 + \tan^2\phi}$ (imagine a right triangle whose opposite and adjacent sides have length $\tan\phi$ and 1, respectively). Use this and equation 8.55 to express $\sin\phi$ in terms of $\tan\psi$. Then multiply top and bottom of the resulting expression by $\cos\psi$, use a trig identity to get an expression purely in terms of $\sin\psi$, and use equations 8.53 and 8.54 to simplify. You should find that

$$\cot\theta = \pm\frac{a}{c}\sin\phi = \pm\frac{a}{\sqrt{1-a^2}}\sin\phi \quad (8.59)$$

This indeed coincides with equation of the great circle found in part d if the plane's slope m coincides with the value of $a/\sqrt{1-a^2}$. *So our geodesic equation yields solutions that are great circles*, as expected. (**Comment:** The value of a is actually the sine of the angle that the plane of the great circle makes with the sphere's equatorial plane.)

P8.3 Our goal in this problem is to show that solutions to the geodesic equation in 2D polar coordinates on a flat plane are indeed straight lines.

a. We can specify an arbitrary line on a flat 2D surface by specifying the distance b between the origin O and the point B on the line that is closest, and the angle α that the line OB makes with the x axis (see figure 8.5). Let us also define the pathlength s to be zero at B and increase as we go along the line in the direction counterclockwise around the origin. Using figure 8.5, argue that such a line can be described in polar coordinates as a function of s by

$$r^2 = s^2 + b^2 \quad (8.60)$$

and

$$\theta = \alpha + \tan^{-1}\left(\frac{s}{b}\right) \quad (8.61)$$

b. Show that the geodesic equations for the polar coordinate system imply

$$\frac{d\theta}{ds} = \frac{c}{r^2} \quad (8.62)$$

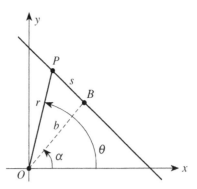

FIG. 8.5 A straight line described in terms of the distance b of closest approach to the origin O and the angle α that the perpendicular OB makes with the x axis.

$$\frac{d^2r}{ds^2} = \frac{c^2}{r^3} \quad (8.63)$$

where c is some constant of integration.

c. In this case we cannot easily integrate equation 8.63 directly, but we can use the definition of pathlength (equation 8.16) to determine dr/ds. Show in this case that equations 8.16 and 8.62 together imply that

$$\frac{dr}{ds} = \pm\sqrt{1 - \frac{c^2}{r^2}} \quad (8.64)$$

d. This equation is essentially an integral of equation 8.63. Prove this by showing that taking the s-derivative of equation 8.64 yields equation 8.63.

e. Note that equation 8.64 makes sense only if $r \geq c$. Show that if you isolate the r-dependent terms on one side and the ds on the other side and integrate, you get $r^2 = s^2 + c^2$, which is the same as our original equation 8.60 if we identify the constant of integration c as being the distance b in figure 8.5.

f. Finally, show that if you substitute $r^2 = s^2 + c^2$ into the geodesic equation 8.62, set $c = b$, and integrate, you get $\theta = \alpha + \tan^{-1}(s/b)$ (where α is another constant of integration), which is our original equation 8.61. We have therefore shown that solutions to the geodesic equation for polar coordinates are simply straight lines of arbitrary orientation expressed in polar coordinates (as we might expect).

P8.4 a. Show that the geodesic equation implies that

$$0 = g_{\alpha\beta}\frac{du^\beta}{d\tau} + (\partial_\sigma g_{\alpha\beta})u^\sigma u^\beta - \tfrac{1}{2}(\partial_\alpha g_{\mu\nu})u^\mu u^\nu \quad (8.65)$$

where $u^\alpha \equiv dx^\alpha/d\tau$ is the four-velocity of the object following the geodesic. (*Hint:* Remember that the metric depends implicitly on τ. Why?)

b. Use this result to prove that the geodesic equation implies that $\boldsymbol{u}\cdot\boldsymbol{u}$ is constant for any object following a geodesic (as must be, since $\boldsymbol{u}\cdot\boldsymbol{u} = -1$ by definition of $\boldsymbol{u}$). [*Hint:* Write out $d(\boldsymbol{u}\cdot\boldsymbol{u})/d\tau$, multiply equation 8.65 by u^α and sum over α, and compare.]

8. GEODESICS

P8.5 Consider a 2D spacetime whose metric is

$$ds^2 = -dt^2 + [f(q)]^2 dq^2 \qquad (8.66)$$

where $f(q)$ is any function of the spatial coordinate q.

a. Show that the t component of the geodesic equation implies that $dt/d\tau$ is constant for a geodesic.
b. The q component of the geodesic equation is hard to integrate, but show that requiring $\mathbf{u} \cdot \mathbf{u} = -1$ implies that $f\,dq/d\tau$ is a constant for a geodesic.
c. Argue then that the trajectory $q(t)$ of a free particle in this spacetime is such that $dq/dt = \text{constant}/f$.
d. Imagine that we transform to a new coordinate system with coordinates t and x, where $x = F(q)$ and $F(q)$ is the antiderivative of $f(q)$. Show that the metric in the new coordinate system is the metric for flat spacetime, so the spacetime described by equation 8.66 is simply flat spacetime in disguise. We know geodesics in flat spacetime obey $dx/d\tau = (f\,dq)/d\tau = \text{constant}$, so the result of part b is not surprising.

P8.6 Consider a 2D spacetime whose metric is

$$ds^2 = -e^{-x/a} dt^2 + dx^2 \qquad (8.67)$$

where a is a constant with units of meters. (Unlike the spacetime in problem P8.5, the underlying geometry of this spacetime is *not* flat.)

a. Show that the t component of the geodesic equation implies that $dt/d\tau = c\,e^{x/a}$, where c is some constant.
b. The x component of the geodesic equation is hard to integrate, but show that requiring $\mathbf{u} \cdot \mathbf{u} = -1$ yields

$$\frac{dx}{d\tau} = \pm\sqrt{c^2 e^{x/a} - 1} \qquad (8.68)$$

c. We can calculate the trajectory $x(t)$ for a geodesic in this spacetime by calculating

$$\frac{dx}{dt} = \frac{dx/d\tau}{dt/d\tau} \qquad (8.69)$$

and integrating. Your integral will yield $t(x)$: you have to invert this function to find $x(t)$. Express your result in terms of a, c, and the initial conditions $x = x_0$ and $dx/d\tau = u_0$ at time $t = 0$. You should find that

$$x(t) = a \ln\left[\left(\frac{t}{2a} + \frac{u_0}{c}\right)^2 + \frac{1}{c^2}\right] \qquad (8.70)$$

Also verify that $x(0) = x_0$.
d. Find the trajectory $x(t)$ for a freely-falling particle starting from rest at the origin (i.e., $x_0 = 0$ and $u_0 = 0$). Plot a graph of x/a versus t/a for this geodesic worldline, making the t axis vertical if possible. (*Hint:* Note that although I left the constant c in equation 8.70 for the sake of simplicity, it is completely determined by the initial conditions.)

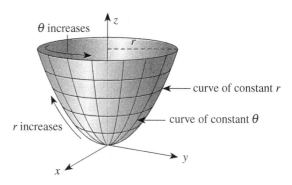

FIG. 8.6 Polar-like coordinates r and θ on the surface of a paraboloid such that $z = ar^2 = a(x^2 + y^2)$. The origin is at the paraboloid's base.

e. We often also describe a worldline by specifying $x^\mu(\tau)$. Find expressions for $x(\tau)$ and $t(\tau)$ for the case of a particle starting from rest at the origin (make sure that you set $\tau = 0$ when $t = 0$). Argue that the proper time τ measured along this worldline asymptotically approaches a maximum as $t \to \infty$ and calculate that maximum. (The constant a should be the only undetermined constant in your final expressions.)

P8.7 Consider polar-like coordinates r and θ not on a flat surface but rather on the 2D surface of a paraboloid whose equation in cartesian coordinates (in a 3D flat space) is $z = ar^2$, where $r^2 = x^2 + y^2$ and a is some constant (see figure 8.6). The origin ($r = 0$) is the paraboloid's bottom. The coordinate r labels points on the surface that are the same distance from the origin, but the distance *along* the surface between the origin and those points is not r. The coordinate θ measures the angle around the paraboloid's axis.

a. Argue that the metric for the distance ds between infinitesimally separated (but otherwise arbitrary) points on the surface is

$$ds^2 = (1 + 4a^2 r^2) dr^2 + r^2 d\theta^2 \qquad (8.71)$$

(*Hint:* The projection of ds along a $\theta = $ constant curve will be $\sqrt{dr^2 + dz^2}$, and $z = ar^2$.)
b. Show that the geodesic equation for this metric implies the following:

$$r^2 \frac{d\theta}{ds} = c \qquad (8.72a)$$

$$0 = 4a^2 r \left(\frac{dr}{ds}\right)^2 + (1 + 4a^2 r^2)\frac{d^2r}{ds^2} - \frac{c^2}{r^3} \qquad (8.72b)$$

where c is a constant of integration.
c. A purely radial curve on this surface is a curve such that $d\theta/ds = 0$ everywhere along it. For such a curve, equation 8.71 implies that $dr/ds = (4a^2 r^2 + 1)^{-1/2}$. Argue that such a curve is a geodesic (that is, it satisfies equations 8.72) with $c = 0$.

9. THE SCHWARZSCHILD METRIC

INTRODUCTION

FLAT SPACETIME
Review of Special Relativity
Four-Vectors
Index Notation

TENSORS
Arbitrary Coordinates
Tensor Equations
Maxwell's Equations
Geodesics

SCHWARZSCHILD BLACK HOLES
The Schwarzschild Metric
Particle Orbits
Precession of the Perihelion
Photon Orbits
Deflection of Light
Event Horizon
Alternative Coordinates
Black Hole Thermodynamics

THE CALCULUS OF CURVATURE
The Absolute Gradient
Geodesic Deviation
The Riemann Tensor

THE EINSTEIN EQUATION
The Stress-Energy Tensor
The Einstein Equation
Interpreting the Equation
The Schwarzschild Solution

COSMOLOGY
The Universe Observed
A Metric for the Cosmos
Evolution of the Universe
Cosmic Implications
The Early Universe
CMB Fluctuations & Inflation

GRAVITATIONAL WAVES
Gauge Freedom
Detecting Gravitational Waves
Gravitational Wave Energy
Generating Gravitational Waves
Gravitational Wave Astronomy

SPINNING BLACK HOLES
Gravitomagnetism
The Kerr Metric
Kerr Particle Orbits
Ergoregion and Horizon
Negative-Energy Orbits

this depends on this

9. THE SCHWARZSCHILD METRIC

Introduction. This chapter begins a long set of chapters that explore the physical meaning of the *Schwarzschild metric*. This metric tensor is a spherically symmetric, time-independent solution to the Einstein equation in a vacuum, and is thus suitable for describing the spacetime in the empty space *surrounding* a spherical, static object. This metric therefore plays much the same role in general relativity that the formula for the gravitational field of a point mass does in Newtonian gravity and the formula for the electric field of a point charge does in electrostatics.

Spherical Coordinates for Flat Spacetime. According to box 5.6, the metric equation for latitude-longitude coordinates θ, ϕ on a 2D spherical surface is given by

$$ds^2 = R^2 d\theta^2 + R^2 \sin^2\theta \, d\phi^2 \tag{9.1}$$

where R is the radius of the spherical surface (in three dimensions). We can label any event in an ordinary flat spacetime with spherical coordinates r, θ, and ϕ, where r (now a variable) specifies the radius of the sphere around the origin on which the event lies, and θ and ϕ specify the event's latitude and longitude coordinates on that surface. Therefore, the metric for spherical coordinates in flat spacetime should be

$$ds^2 = -dt^2 + dr^2 + r^2 d\theta^2 + r^2 \sin^2\theta \, d\phi^2 \tag{9.2}$$

The Schwarzschild Metric. In 1916, Karl Schwarzschild (who at the time was fighting in the trenches of World War I and who died shortly thereafter) discovered that the following metric satisfied the Einstein field equations in empty space:

$$ds^2 = -\left(1 - \frac{r_s}{r}\right) dt^2 + \left(1 - \frac{r_s}{r}\right)^{-1} dr^2 + r^2 d\theta^2 + r^2 \sin^2\theta \, d\phi^2 \tag{9.3}$$

where r_s is a constant with units of length called the **Schwarzschild radius**. (We will verify that this metric is indeed a solution to the Einstein equation later in the course.) This metric is *spherically symmetric* (note that surfaces of constant r have the same metric as that for latitude-longitude coordinates on the surface of a sphere and so have the geometry of a sphere, and the other components of the metric depend only on r), *time-independent*, and becomes the flat space metric in the limit as $r \to \infty$. This metric is therefore a suitable candidate for that describing the spacetime in the empty space surrounding a spherical, static object.

The Meaning of the Radial Coordinate. The Schwarzschild metric labels events in spacetime using coordinates t, r, θ, and ϕ that look very much like the coordinates we would use to describe flat spacetime in spherical coordinates. But what do these coordinates really mean physically?

It is very important to remember that in general relativity, coordinates *have no intrinsic meaning*, no matter how evocatively they might be named: they are simply numbers for labeling events in spacetime. The *only* thing that gives coordinates meaning is the metric equation, which links coordinate differences to physically measurable distances and/or time intervals. This means that we can discover what coordinates really mean only by examining the metric equation.

Let's begin by considering the r coordinate. Consider a circle of constant r (meaning that $dr = 0$ for all steps around the circle) in the equatorial plane ($\theta = \pi/2$, so $\sin\theta = 1$, and $d\theta = 0$) at an instant of time (meaning that $dt = 0$ for all events on the circle). The physical distance (as measured, say, by a tape measure) around this circle can be found by integrating the spacetime interval ds between infinitesimally separated points around the curve, where

$$ds^2 = 0 + 0 + 0 + r^2 d\phi^2 \quad \Rightarrow \quad ds = r \, d\phi \quad \text{for such a circle} \tag{9.4}$$

If we integrate this all the way around the circle, we get

$$\text{circumference} \equiv C = \oint ds = \int_0^{2\pi} r \, d\phi = 2\pi r \quad \Rightarrow \quad r = \frac{C}{2\pi} \tag{9.5}$$

We see that the Schwarzschild metric equation defines the r coordinate of a spherical surface in that spacetime to be $(1/2\pi)$ times the circumference of that sphere's equator. (Since the orientation of the equatorial plane is arbitrary, this really applies to any circle around the object.)

But the r coordinate is *not* equivalent to radial distance here. To see this, integrate ds along a radial line (i.e., a line made up of steps where $dt = d\theta = d\phi = 0$):

$$ds^2 = \frac{dr^2}{(1 - r_s/r)} \quad \Rightarrow \quad ds = \frac{dr}{\sqrt{1 - r_s/r}} \quad \text{for such a line} \quad (9.6)$$

You can show that the radial distance between any two events A and B along such a line is *greater* than $|r_B - r_A|$ (see box 9.1).

So though the value of r sequentially labels spheres of various sizes around the origin, it does not actually specify their *distance* from the origin. We call the Schwarzschild r coordinate a **circumferential radial coordinate**.

The discrepancy between the r coordinate and the radial distance (which means that the ratio of the circumference of a circle of constant r to its radius is $\neq 2\pi$) is an indication that the Schwarzschild metric describes a curved (non-Euclidean) spacetime. To see a different example of the same phenomenon, consider the curved surface of a sphere, and consider a circle around the sphere's north pole, as shown in figure 9.1. Such a circle will correspond to a curve of fixed latitude angle θ if we use latitude-longitude coordinates for the sphere. Define this circle's "circumferential radius" r to be $C/2\pi$, where C is the circle's circumference. According to the metric for a sphere in latitude-longitude coordinates (equation 9.1), the length of a differential step around this circle is $ds^2 = R^2 d\theta^2 + R^2 \sin^2\theta\, d\phi^2 \Rightarrow ds = R\sin\theta\, d\phi$, where R is the sphere's radius in three dimensions. The circumference of the circle is thus

$$C = \int ds = R\sin\theta \int d\phi = 2\pi R \sin\theta \quad \Rightarrow \quad r \equiv \frac{C}{2\pi} = R\sin\theta \quad (9.7)$$

But from figure 9.1, it should be clear that the "radius" of this circle (i.e., the rim's distance from the center as measured on the sphere's surface) is $R\theta$, which is larger (one can also prove this directly from the metric: see problem P9.3). Though the curvature of Schwarzschild spacetime is mathematically different from the curvature of a spherical surface, they both display this discrepancy between the circumferential radius and the actual radial distance.

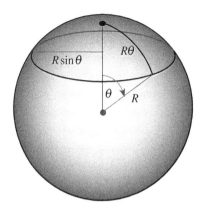

FIG. 9.1 The circumferential radius of a circle on a sphere is not the same as the radial distance from the circle's center as measured on the sphere's surface because the surface of the sphere is curved. In an analogous way, the Schwarzschild radial coordinate does not coincide with radial distance measured within the spacetime from $r = 0$.

The Meaning of r_s. Consider now the "Schwarzschild radius" constant r_s that appears in the Schwarzschild metric. The geodesic equation implies that the initial radial acceleration of an object released from rest at a given r coordinate is

$$\frac{d^2r}{d\tau^2} = -\frac{\frac{1}{2} r_s}{r^2} \quad (9.8)$$

where τ is the object's own proper time (see box 9.2). Now, if the object is at rest, its proper time will be the same as the time measured by an observer who remains at rest and watches the object fall, and while we have seen that the r coordinate does not exactly correspond to a radial distance, the distinction between r and radial distance becomes negligible as r becomes very large compared to r_s.

The Newtonian Law of Universal Gravitation tells us that an observer watching an object fall from rest toward a gravitating object of mass M will observe its radial acceleration to be

$$\frac{d^2r}{d\tau^2} = -\frac{GM}{r^2} \quad (9.9)$$

where r here is strictly a *radial* coordinate, $d\tau$ is the time measured by the observer, and G is the universal gravitational constant. If we compare these equations, we see that the predicted Schwarzschild acceleration will agree with the Newtonian result at large radii ($r \gg r_s$) if and only if $r_s = 2GM$, where M is the mass of whatever is the source of the gravitational field described by the Schwarzschild coordinates.

Note that in GR units, G has units of m/kg (see equation 2.10b), so GM has units of meters. GM for the sun is about 1.5 km (1.477 km to be more precise), and GM for the earth is about 4.45 mm (see box 9.3). This means that $r \gg r_s = 2GM$ for any point in the empty space outside either the earth or sun (since the radius of the earth is 6.38×10^9 mm and the radius of the sun is about 700,000 km). The Newtonian approximation in equation 9.9 should be excellent in such circumstances. On the other hand, a typical **neutron star** has $GM \approx 2.2$ km and a surface r-coordinate $R \approx 10$ to 12 km, so $R \approx 5GM$ (not *too* much larger than $2GM$), so we would expect to see significant deviations from the predictions of Newtonian gravity in the empty space surrounding a neutron star. We also have strong empirical evidence to suggest that objects exist whose surface (if it exists at all) is inside $r_s = 2GM$; such objects are called **black holes** for reasons that will become clear in future chapters. We will see in fact that (non-rotating) black holes can have no fixed surface at all, so the Schwarzschild solution applies to all radial coordinates $r > 0$.

If we substitute $r_s = 2GM$ into equation 9.3, we get the Schwarzschild metric equation in the form that we will use from now on:

$$ds^2 = -\left(1 - \frac{2GM}{r}\right)dt^2 + \left(1 - \frac{2GM}{r}\right)^{-1} dr^2 + r^2 d\theta^2 + r^2 \sin^2\theta \, d\phi^2 \quad (9.10)$$

This makes it very clear how this metric equation is linked to the mass of the object that creates this curved spacetime.

The Meaning of the Time Coordinate. Now consider the Schwarzschild time coordinate t. Consider a clock at rest at a given r coordinate ($dr = d\theta = d\phi = 0$). According to the Schwarzschild metric equation, the proper time $\Delta\tau$ that this clock measures between two events at its location is related to the coordinate difference Δt between those events as follows:

$$\Delta\tau = \int d\tau = \int \sqrt{-ds^2} = \int \sqrt{\left(1 - \frac{2GM}{r}\right)dt^2 + 0 + 0 + 0}$$

$$= \sqrt{1 - \frac{2GM}{r}} \int dt = \sqrt{1 - \frac{2GM}{r}} \, \Delta t \quad (9.11)$$

We see that the clock's proper time between the events agrees with the t coordinate difference between those events if and only if the clock is at $r = \infty$. The t coordinate difference between two events therefore corresponds to the time between them *as measured by a clock at $r = \infty$*.

But how can we really *measure* the t coordinate difference between two events that occur at a given finite r far from the clock a $r = \infty$? Imagine constructing a spherical lattice around the origin. At each lattice intersection, imagine placing a clock that measures proper time at that location, and also a "t-meter" that registers the t coordinate. (The latter is not a "clock," because it does not measure the correct local time.) The t-meter receives light flashes sent out at 1.0-s time intervals from a master clock at a very large r ($\approx \infty$). Each time that the t-meter receives such a flash, it advances its reading by 1.0 s. We can even synchronize the t-meter with the distant master clock by having it send a flash back to the master clock each time it receives a flash. Because the metric equation is time-independent (nothing depends on t) and time-symmetric (replacing dt by $-dt$ does not change anything) it should take the flash as much time to go back to the master clock as it did to come down, so the "t-meter" will be synchronized with the master clock if upon reception of its flash it reads halfway between the times of master clock's emission and reception of the flashes. The point is that we can in fact set up a t-meter at every lattice intersection that displays the t coordinate ("time at infinity") at that location.

Equation 9.11 tells us that the clock at rest at a given lattice intersection measures less time between two events than the t-meter at that intersection does, and that the discrepancy in the measurements gets larger as r decreases: the lower the clock, the slower it runs compared to the *t*-meter. This discrepancy is another indication that the spacetime described by the Schwarzschild metric is a curved spacetime.

The Gravitational Redshift. I argued in the first chapter that an observer in a room at rest in a gravitational field would see light that moves toward the floor to be blue-shifted and light that moves toward the ceiling to be red-shifted. The Schwarzschild metric implies the same thing.

To see this, consider two successive crests of a light wave emitted by a source at r_E. A clock at r_E measures the proper time $\Delta\tau_E$ between the crest emission events. In GR units, this time is measured in meters and is equal to the wavelength λ_E that would be measured at r_E, since the speed of light is 1 in a local frame attached to the clock. Let the *t*-coordinate difference between these events (as measured by the *t*-meter at r_E) be Δt.

Now let the light move out to a radius $r_R > r_E$. Figure 9.2 shows a spacetime diagram of the worldlines of two successive crests in r, t coordinates. The worldlines of these crests will not necessarily be straight lines at 45° any more (because of the curvature of spacetime), but because the Schwarzschild metric does not depend on t, the *shapes* of the worldlines must be the same. Therefore, if the *t*-coordinate separation of the crests was Δt at r_E, the *t*-coordinate separation at the reception radius r_R must have the *same* value Δt (as measured by the *t*-meter there).

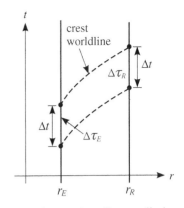

FIG. 9.2 A spacetime diagram displaying the worldlines of two successive crests of a light wave being emitted at radial coordinate r_E and received at radial coordinate r_R.

This does not mean that the proper time $\Delta\tau_R$ between these events measured by a clock at rest at r_R is the same as $\Delta\tau_E$. Indeed, equation 9.11 implies that

$$\lambda_E = \Delta\tau_E = \sqrt{1 - \frac{2GM}{r_E}}\,\Delta t \quad \text{and} \quad \lambda_R = \Delta\tau_R = \sqrt{1 - \frac{2GM}{r_R}}\,\Delta t \qquad (9.12a)$$

$$\Rightarrow \quad \frac{\lambda_R}{\lambda_E} = \sqrt{\frac{1 - 2GM/r_R}{1 - 2GM/r_E}} \qquad (9.12b)$$

This is greater than 1 if $r_R > r_E$, so the light is red-shifted as it climbs upward. This is the gravitational redshift formula. (Note also that the redshift becomes *infinite* as r_E approaches $2GM$: this indicates that there is something peculiar about that radial coordinate, which we will explore in a later chapter.)

When the radial coordinates are very large compared to $2GM$, then $2GM/r \ll 1$ and we can use the binomial approximation to simplify the expression:

$$\frac{\lambda_R}{\lambda_E} = \sqrt{\frac{1 - 2GM/r_R}{1 - 2GM/r_E}} \approx \left(1 - \frac{GM}{r_R}\right)\left(1 + \frac{GM}{r_E}\right)$$

$$\Rightarrow \quad \frac{\lambda_R}{\lambda_E} \approx 1 + \frac{GM}{r_E} - \frac{GM}{r_R} \quad \left(\text{if } \frac{2GM}{r} \ll 1\right) \qquad (9.13)$$

where in the last step I dropped a term of order of magnitude $(2GM/r)^2$. If, in addition, the difference $h \equiv r_R - r_E \ll r_E$, you can show (box 9.4) that

$$\frac{\lambda_R}{\lambda_E} \approx 1 + gh \quad \left(\text{if } \frac{2GM}{r} \ll 1 \text{ and } h \ll r_E\right) \qquad (9.14)$$

where $g \equiv GM/r_E^2$ is the gravitational acceleration measured at the emission radius (see equations 9.8 and 9.9). This is the redshift expected from a simple argument involving the equivalence of a frame at rest in a gravitational field and an accelerating frame in deep space (see box 9.4).

Conclusion. We see that the Schwarzschild metric equation displays all of the effects that we would expect to see for a gravitational field: (1) objects released from rest accelerate to smaller r with the expected Newtonian acceleration, and (2) we observe the gravitational redshift expected from the Equivalence Principle. It seems, therefore, that this metric equation could indeed accurately describe the spacetime in the vacuum surrounding a spherical and static gravitating object of mass M.

BOX 9.1 Radial Distance

Equation 9.6 (with $r_s = 2GM$) implies that the actual distance (as measured by a ruler) between infinitesimally separated events along a purely radial worldline is

$$ds = \frac{dr}{\sqrt{1 - 2GM/r}} \qquad (9.15)$$

The total radial distance between two points with r coordinates r_A and r_B is thus

$$\Delta s = \int ds = \int_{r_A}^{r_B} \frac{dr}{\sqrt{1 - 2GM/r}} \qquad (9.16)$$

In the limit that $2GM \ll r$, we can use the binomial approximation to do this integral approximately. If you do this, you should find that

$$\Delta s \approx (r_B - r_A) + GM \ln\left(\frac{r_B}{r_A}\right) \quad \text{if} \quad 2GM \ll r \qquad (9.17)$$

Exercise 9.1.1. Verify equation 9.17.

Exercise 9.1.2. Argue that the condition applies for all points above the surface of the earth. Then calculate the extra distance beyond 100 km that one would measure between $r_A = 6380$ km ($\approx$ the earth's surface) and $r_B = 6480$ km. (*Hint:* For the earth, $GM = 4.44$ mm.)

One can also do the integral exactly, by changing variables $u \equiv 2GM/r$, which means that $du = -2GM\,dr/r^2$, or $dr = -2GM\,du/u^2$. The integral then becomes something that one can look up in a good integral table. If you do this, you should find (see problem P9.4) that (as long as $r > 2GM$)

$$\Delta s = \left[r\sqrt{1 - \frac{2GM}{r}} + 2GM \tanh^{-1}\sqrt{1 - \frac{2GM}{r}} \right]_{r_A}^{r_B} \qquad (9.18)$$

(I am not going to make you check this here.)

BOX 9.2 Falling from Rest in Schwarzschild Spacetime

The geodesic equation for any spacetime is (see equation 8.14):

$$\frac{d^2 x^\gamma}{d\tau^2} = -g^{\gamma\alpha}(\partial_\nu g_{\alpha\mu} - \tfrac{1}{2}\partial_\alpha g_{\mu\nu}) u^\mu u^\nu \tag{9.19}$$

In order to evaluate this, we need to know the four-velocity u^μ of an object at rest. The spatial components of u^μ are zero, but the time component $u^t \equiv dt/d\tau \ne 1$ here because the coordinate time t in this situation is not the same as the time τ measured by the object's clock, even if the object is at rest. We can most easily evaluate what u^t is in these coordinates using the tensor equation $\boldsymbol{u} \cdot \boldsymbol{u} = u^\mu g_{\mu\nu} u^\nu = -1$, which applies in all coordinate systems. In this case, since the spatial components of u^μ are all zero, you can show that this implies that

$$u^t = \left(1 - \frac{2GM}{r}\right)^{-1/2} \tag{9.20}$$

(This is a valuable technique that we will use often to evaluate the t component of the four-velocity of an object at rest in a given spacetime.)

Exercise 9.2.1. Verify equation 9.20.

Now return to equation 9.19. First, note that when the metric is diagonal (as it is in this case), the definition of the metric inverse $g_{\alpha\gamma} g^{\gamma\beta} = \delta_\alpha^{\;\beta}$ simply implies that the metric inverse is also diagonal and each diagonal element of $g^{\alpha\beta}$ is simply equal to $1/g_{\alpha\beta}$. Note also in this case that since only the t component of the four-velocity of our object is nonzero, the only nonzero term in the implicit sums over μ and ν will be the one where $\mu = \nu = t$. Using this information, you can easily show that the geodesic equation for $\gamma = r$ in this case implies that

$$\frac{d^2 r}{d\tau^2} = -\frac{GM}{r^2} \tag{9.21}$$

Exercise 9.2.2. Verify equation 9.21.

BOX 9.3 *GM* for the Earth and the Sun

Exercise 9.3.1. Use the information in box 2.2 and the masses of the earth and sun (which you can look up almost anywhere) to show that $GM = 4.44$ mm for the earth and 1.477 km for the sun.

BOX 9.4 The Gravitational Redshift for Weak Fields

According to equation 9.13, if we define $h = r_R - r_E$, then as long as both r_E and r_R are much larger than $2GM$, the ratio of the received wavelength to the emitted wavelength will be

$$\frac{\lambda_R}{\lambda_E} \approx 1 + \frac{GM}{r_E} - \frac{GM}{r_E + h} = 1 + \frac{GM}{r_E}\left(1 - \frac{1}{1 + h/r_E}\right) \quad (9.22)$$

Let's assume now that $h \ll r_E$. From here, the binomial approximation leads fairly directly to the result that

$$\frac{\lambda_R}{\lambda_E} \approx 1 + \frac{GM}{r_E^2}h = 1 + gh \quad (9.23)$$

where $g = GM/r_E^2$ is the gravitational acceleration at $r = r_E$ that we would compute from Newtonian theory.

Exercise 9.4.1. Verify equation 9.23.

BOX 9.4 (continued) The Gravitational Redshift for Weak Fields

Now how does this compare to what we would find from the idea that a frame at rest in a gravitational field with strength g is equivalent to a frame that is accelerating at a rate g upward in deep space? Consider a room of height h in deep space that is accelerating at a rate of g upward relative to an inertial frame in the same vicinity. Imagine that at $t = 0$ the accelerating room is instantaneously at rest with respect to the inertial frame, and at that very instant a source on the floor emits a flash of light that travels vertically upward until it reaches the ceiling of the accelerating room. In the inertial frame, this flash has the same energy E_E and wavelength $\lambda_E = h_P/E_E$ (where h_P is Planck's constant) when it is received as it does when it is emitted. However, by the time $t \approx h$ in the inertial frame that the flash reaches the room's ceiling, the accelerating room is moving upward with speed $v = gt = gh$ relative to the inertial frame. Since this will be pretty small, the four-velocity of the receiver in the accelerating frame's roof will have components $u_R^t \approx 1, u_R^x = 0, u_R^y = 0$, and $u_R^z \approx +v \approx gh$ in the inertial frame. The light has four-momentum $p^t = E_E$, $p^x = 0$, $p^y = 0$, and $p^z = E_E$ in that frame. As we saw in chapter 3 (see equation 3.21), the energy of the light as observed by the receiver is $E_R = -\boldsymbol{p} \cdot \boldsymbol{u}_R$. If you use this, the fact that the room's speed $gh \ll 1$, and the binomial approximation, you will find that

$$\frac{\lambda_R}{\lambda_E} = \frac{E_E}{E_R} \approx 1 + gh \tag{9.24}$$

This is consistent with equation 9.23.

Exercise 9.4.2. Verify equation 9.24.

9. THE SCHWARZSCHILD METRIC

HOMEWORK PROBLEMS

P9.1 Imagine that you are observing a neutron star whose mass is $M = 3.0 \times 10^{30}$ kg and whose surface is at a Schwarzschild radial coordinate $R_S = 12$ km from a robot spacecraft hovering at a Schwarzschild radial coordinate of $R_R = 17.0$ km. A flash of light emitted by a surface event is observed by spacecraft.

a. Compute the fractional redshift $(\lambda_R - \lambda_E)/\lambda_E$ of this light using the approximate method discussed in box 9.4. The result won't be very accurate, because h here is not small relative to R_S. Still, it probably will yield a reasonably good answer if you evaluate g halfway between the surface and the spaceship.

b. Calculate $(\lambda_R - \lambda_E)/\lambda_E$ exactly using the Schwarzschild metric and compare with the result of part a.

P9.2 Recent Hubble measurements (see Barstow et al., *Mon. Not. R. Astron. Soc.* **362**, 1134–1142, 2005) of Sirius' white dwarf companion show that the white dwarf plausibly has a mass of $1.02\, M_\odot$ and a radius of 5640 km (with uncertainties of roughly 2%) and that spectral features of its light have a fractional redshift of about 2.68×10^{-4} (with an uncertainty of about 6%). Are these results consistent with the redshift prediction of general relativity?

P9.3 Using the methods illustrated in this chapter, show *from the spherical metric* (equation 9.1) that the radial distance along the surface of a sphere of radius R from its north pole to a circle at latitude θ is $R\theta$.

P9.4 Verify the formula (equation 9.18) for the actual radial distance between two radial coordinates r_A and r_B in Schwarzschild spacetime. You should be able to find the integral you need in a good table of integrals.

P9.5 An advanced civilization constructs two concentric massless shells around a neutron star of mass M. The inner shell has a circumference of $6\pi GM$ and the outer shell has a circumference of $20\pi GM$. What is the (exact) physical distance between the shells?

P9.6 An observer stationed at a fixed Schwarzschild radial coordinate R near a spherical star of mass M observes a photon moving radially away from the star and measures its energy to be E. What are the components of this photon's four-momentum $\mathbf{p}$ in the Schwarzschild coordinate basis? (*Hints:* $\mathbf{p} \cdot \mathbf{p} = 0$ for a photon. Also note that equation 9.20 applies to the observer.)

P9.7 Consider the following metric equation: (9.25)

$$ds^2 = -dt^2 + dr^2 + R^2 \sinh^2(r/R)(d\theta^2 + \sin^2\theta\, d\phi^2)$$

where R is a constant with units of length.

a. What kinds of clocks and in what positions register the coordinate time t?

b. Does this metric describe a spherically symmetric spacetime? Carefully justify your response.

c. Is the r coordinate a radial coordinate or a circumferential coordinate? Is the circumference of a circle bigger than, equal to, or less than $2\pi r$? Explain.

d. If we were to write this metric tensor as a matrix, it would have no off-diagonal terms. What does this imply about the coordinate system?

P9.8 (For a different version of this problem, see P8.1.) If we put the result from equation 9.14 together with the fact that $\lambda_E = \Delta\tau_E$ and $\lambda_R = \Delta\tau_R$, we see that the proper time $\Delta\tau$ between successive passing light flashes as measured by a clock at a vertical position $z = h$ is related to the proper time $\Delta\tau_0$ measured between the same flashes at a vertical position $z_0 = 0$ by

$$\frac{\Delta\tau}{\Delta\tau_0} \approx 1 + gh = 1 + gz \qquad (9.26)$$

when $gh \ll 1$. This essentially tells us that clocks higher in a gravitational field run somewhat faster than lower clocks. In box 9.4, we saw that this result is what we would expect (both qualitatively and quantitatively) from the Equivalence Principle. In this problem we will see that this result (ignoring the Schwarzschild metric) combined with the proper time formula from special relativity (equation 2.25) implies that (in the weak-field, slow-object limit) an object following a worldline of maximal proper time follows the trajectory we expect from Newtonian mechanics.

To begin, consider two clocks: a movable clock M and a clock F fixed at a vertical position $z_0 = 0$. Imagine that we want to move clock M vertically from clock F's position to some higher position and back during a set time interval T as measured by the fixed clock F. The question is, what trajectory maximizes the proper time τ that clock M measures for this trip?

a. According to equation 9.26, clock M runs faster as z increases, so it seems that we would want to move it very quickly to a high position and then leave it there as long as we can. Explain qualitatively why equation 2.25 describes a competing effect that reduces the benefit of doing that.

b. Show indeed that during an infinitesimal time interval when clock M has is at vertical position z and has vertical speed $|dz/dt| = v$, equations 9.26 and 2.25 imply that the proper time $d\tau$ it measures is related to the time dT that clock F measures by

$$d\tau \approx dT(1 + gz - \tfrac{1}{2}v^2) \qquad (9.27)$$

assuming both $gz \ll 1$ and $v \ll 1$.

c. So the *total excess* time measured by clock M is

$$\tau - T = \int_0^T \left[gz - \frac{1}{2}\left(\frac{dz}{dt}\right)^2\right] dT \qquad (9.28)$$

Treating the quantity in brackets as a Lagrangian, show that clock M's proper time is maximized when $d^2z/dt^2 = -g$, that is, when it follows the trajectory predicted by Newtonian mechanics.

10. PARTICLE ORBITS

INTRODUCTION

FLAT SPACETIME
Review of Special Relativity
Four-Vectors
Index Notation

TENSORS
Arbitrary Coordinates
Tensor Equations
Maxwell's Equations
Geodesics

SCHWARZSCHILD BLACK HOLES
The Schwarzschild Metric
Particle Orbits
Precession of the Perihelion
Photon Orbits
Deflection of Light
Event Horizon
Alternative Coordinates
Black Hole Thermodynamics

THE CALCULUS OF CURVATURE
The Absolute Gradient
Geodesic Deviation
The Riemann Tensor

THE EINSTEIN EQUATION
The Stress-Energy Tensor
The Einstein Equation
Interpreting the Equation
The Schwarzschild Solution

COSMOLOGY
The Universe Observed
A Metric for the Cosmos
Evolution of the Universe
Cosmic Implications
The Early Universe
CMB Fluctuations & Inflation

GRAVITATIONAL WAVES
Gauge Freedom
Detecting Gravitational Waves
Gravitational Wave Energy
Generating Gravitational Waves
Gravitational Wave Astronomy

SPINNING BLACK HOLES
Gravitomagnetism
The Kerr Metric
Kerr Particle Orbits
Ergoregion and Horizon
Negative-Energy Orbits

this depends on this

10. PARTICLE ORBITS

Introduction. In chapter 8, we discussed the general geodesic equation for arbitrary coordinates in arbitrary spacetimes. The geodesic equation, in the form that will be most useful in this chapter, reads

$$0 = \frac{d}{d\tau}\left(g_{\mu\nu}\frac{dx^\nu}{d\tau}\right) - \tfrac{1}{2}\partial_\mu g_{\alpha\beta}\frac{dx^\alpha}{d\tau}\frac{dx^\beta}{d\tau} \qquad (10.1)$$

In this chapter, we will use the geodesic equation to explore the trajectories that particles with nonzero rest mass follow in Schwarzschild spacetime and develop a variety of tools for visualizing and modeling those trajectories.

The Schwarzschild Metric Tensor. The Schwarzschild metric equation is

$$ds^2 = -\left(1 - \frac{2GM}{r}\right)dt^2 + \left(1 - \frac{2GM}{r}\right)^{-1}dr^2 + r^2 d\theta^2 + r^2\sin^2\theta\, d\phi^2 \qquad (10.2)$$

Comparing this with the general metric equation $ds^2 = g_{\mu\nu}dx^\mu dx^\nu$, we can read off the nonzero components of the Schwarzschild metric tensor:

$$g_{tt} = -\left(1 - \frac{2GM}{r}\right), \quad g_{rr} = \left(1 - \frac{2GM}{r}\right)^{-1}, \quad g_{\theta\theta} = r^2, \quad g_{\phi\phi} = r^2\sin^2\theta \qquad (10.3)$$

Conserved Quantities. The $\mu = t$ component of equation 10.1 tells us that

$$0 = \frac{d}{d\tau}\left(g_{t\nu}\frac{dx^\nu}{d\tau}\right) - \frac{1}{2}\partial_t g_{\alpha\beta}\frac{dx^\alpha}{d\tau}\frac{dx^\beta}{d\tau} \qquad (10.4)$$

The Schwarzschild metric is both diagonal and time-independent, so this becomes

$$0 = \frac{d}{d\tau}\left(g_{tt}\frac{dt}{d\tau}\right) + 0 \quad \Rightarrow \quad \text{constant} = -g_{tt}\frac{dt}{d\tau} = \left(1 - \frac{2GM}{r}\right)\frac{dt}{d\tau} \equiv e \qquad (10.5)$$

The quantity e is therefore conserved along all geodesic trajectories in Schwarzschild spacetime. We can interpret this quantity to be the *relativistic energy per unit mass that the object would have at infinity*, because if we substitute $r = \infty$ into equation 10.5, then e reduces to $dt/d\tau$. Since t is the time measured by a clock at infinity, this is the same as the object's four-velocity component u^t as measured by the observer at infinity, which in turn is p^t/m = relativistic energy per mass.

The $\mu = \phi$ component of equation 10.1 tells us that

$$0 = \frac{d}{d\tau}\left(g_{\phi\nu}\frac{dx^\nu}{d\tau}\right) - \frac{1}{2}\partial_\phi g_{\alpha\beta}\frac{dx^\alpha}{d\tau}\frac{dx^\beta}{d\tau} \qquad (10.6)$$

The Schwarzschild metric is both diagonal and independent of ϕ, so this becomes

$$0 = \frac{d}{d\tau}\left(g_{\phi\phi}\frac{d\phi}{d\tau}\right) \quad \Rightarrow \quad \text{constant} = g_{\phi\phi}\frac{d\phi}{d\tau} = r^2\sin^2\theta\frac{d\phi}{d\tau} \equiv \ell \qquad (10.7)$$

The quantity ℓ is therefore also conserved along all geodesic trajectories in Schwarzschild spacetime. For a trajectory on the equatorial plane, $\sin^2\theta = 1$, so this quantity reduces to being $\ell = r^2(d\phi/d\tau)$, which we can interpret as being relativistic angular momentum per unit mass [in Newtonian mechanics, $L/m = r^2\omega = r^2(d\phi/dt)$.]

As discussed in box 10.1, symmetry requires that each geodesic in Schwarzschild spacetime lies on a plane through the origin. We can, therefore, without loss of generality, choose our coordinates so that any given orbit of interest lies in the equatorial ($\theta = \pi/2$) plane. We will assume this in what follows.

The Radial Equation of Motion. Since all of the Schwarzschild metric components depend on r, the r component of the geodesic equation is a bit complicated. This is another case where using $-1 = \mathbf{u}\cdot\mathbf{u} = g_{\mu\nu}u^\mu u^\nu$ really pays off. If you substitute in the metric components from equation 10.3 and use the results in equations 10.5 and 10.7, the result (see box 10.2) is that

$$\tfrac{1}{2}(e^2 - 1) = \frac{1}{2}\left(\frac{dr}{d\tau}\right)^2 - \frac{GM}{r} + \frac{\ell^2}{2r^2} - \frac{GM\ell^2}{r^3} \equiv \tilde{E} \qquad (10.8)$$

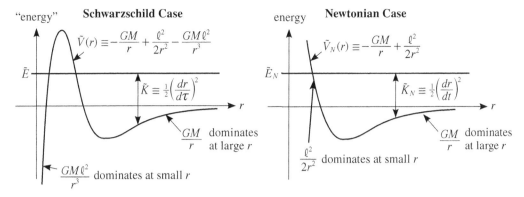

FIG. 10.1 These graphs show the effective potential-energy-per-unit-mass functions governing radial motion in the Schwarzschild case (left) and Newtonian case (right).

We can interpret this as being structurally equivalent to a conservation-of-energy equation, with $\tilde{K} \equiv \frac{1}{2}(dr/d\tau)^2$ serving as an effective "radial kinetic energy per unit mass," $\tilde{E}$ as an effective conserved "energy per unit mass," and

$$\tilde{V}(r) \equiv -\frac{GM}{r} + \frac{\ell^2}{2r^2} - \frac{GM\ell^2}{r^3} \tag{10.9}$$

as an effective "potential energy per unit mass" that depends only on r. With these definitions, equation 10.8 becomes simply $\tilde{E} = \tilde{K} + \tilde{V}(r)$.

Indeed, one can show (see box 10.3) that one can use conservation of Newtonian energy to derive an analogous equation for Newtonian orbits:

$$\frac{1}{2}\left(\frac{dr}{dt}\right)^2 + \tilde{V}_N \equiv \tilde{E}_N \quad \text{where} \quad \tilde{V}_N(r) \equiv -\frac{GM}{r} + \frac{\ell^2}{2r^2} \tag{10.10}$$

and $\tilde{E}_N$ is the actual Newtonian total energy per unit mass. The only formal difference between the two equations is that the Schwarzschild equation has an extra inverse-r^3 term in the $\tilde{V}(r)$ function. (But also note that in the Schwarzschild case, r is the *circumferential* radius, and e, not $\tilde{E}$, is the true relativistic energy per unit mass.)

Graphical Interpretation of the Possible Orbits. Figure 10.1 shows graphs of the effective potential-energy-per-unit-mass curves for the Schwarzschild and Newtonian cases. One can read these graphs to determine characteristics of the possible trajectories much the way that one interprets potential energy graphs for one-dimensional motion in ordinary mechanics. For both cases:

- $\tilde{K} = 0$ where $\tilde{E} = \tilde{V}(r)$: such points are "turning points" where outgoing radial motion becomes ingoing motion and vice versa.

- $\tilde{E} < 0$ orbits correspond to bound orbits (ellipses in the Newtonian case) that have maximal and minimal radial coordinates [at radii where $\tilde{E} = \tilde{V}(r)$].

- $\tilde{E} > 0$ orbits correspond to unbound orbits (hyperbolas in the Newtonian case).

- Radii where $d\tilde{V}/dr = 0$ are possible circular-orbit radii (see box 10.4).

However, there are special features of the curve for Schwarzschild spacetime:

- The effective potential energy goes to negative infinity as $r \to 0$ instead of going to positive infinity.

- For sufficiently high $\tilde{E}$, spiral orbits exist that go into $r = 0$.

- As long as $\ell > \sqrt{12}\,GM$ (see box 10.4 and box 10.6), we have an unstable circular orbit (at the radial coordinate where $\tilde{V}(r)$ is maximum) *and* a stable circular orbit (at the radial coordinate where $\tilde{V}(r)$ is minimum).

Circular Orbits. Indeed, by solving for the radii where $d\tilde{V}/dr = 0$, one can show (see box 10.4) that for a given value of ℓ, the radial coordinates r_c of possible circular orbits in Schwarzschild spacetime are

$$r_c = \frac{6GM}{1 \pm \sqrt{1 - 12(GM/\ell)^2}} \qquad (10.11)$$

with the inner orbit being unstable and the other orbit being stable. The radius of the single (stable) Newtonian circular orbit is $r = \ell^2/GM$. One can also prove (see box 10.5) that Kepler's third law

$$\Omega^2 = \frac{GM}{r_c^3} \quad \Rightarrow \quad T^2 = \left|\frac{2\pi}{\Omega}\right|^2 = \frac{4\pi^2}{GM}r_c^3 \qquad (10.12)$$

still applies in Schwarzschild spacetime as long as we take T to be the period of the orbit as measured by an observer at infinity ($\Omega = d\phi/dt$ is the angular speed of the orbiting object as determined by an observer at infinity). However, it does not quite *mean* the same thing, as the Schwarzschild radial coordinate is not equal to the Newtonian radial coordinate.

Equations for Radial Acceleration. By taking the τ-derivative of both sides of equation 10.8, you can find the following equation for the radial acceleration of an object in Schwarzschild spacetime:

$$0 = \tfrac{1}{2}2\frac{dr}{d\tau}\left(\frac{d^2r}{d\tau^2}\right) + \frac{GM}{r^2}\frac{dr}{d\tau} - \frac{2\ell^2}{2r^3}\frac{dr}{d\tau} + \frac{3GM\ell^2}{r^4}\frac{dr}{d\tau}$$

$$\Rightarrow \quad \frac{d^2r}{d\tau^2} = -\frac{GM}{r^2} + \frac{\ell^2}{r^3} - \frac{3GM\ell^2}{r^4} \quad \text{(Schwarzschild)} \qquad (10.13)$$

The corresponding Newtonian equation (found by taking t-derivative of both sides of equation 10.10) is

$$\frac{d^2r}{dt^2} = -\frac{GM}{r^2} + \frac{\ell^2}{r^3} \qquad \text{(Newtonian)} \qquad (10.14)$$

Note that again the Newtonian equation lacks the final term that appears in the Schwarzschild case. Note that for an object moving radially, $\ell \equiv r^2 d\phi/d\tau = 0$, so the radial acceleration is precisely $d^2r/d\tau^2 = -GM/r^2$ in such a case.

These equations turn out to be more useful for constructing computer models of trajectories than equations 10.8 or 10.10.

Astrophysical Implications. For Schwarzschild spacetime, one can show (see box 10.6) that contrary to the predictions of Newtonian gravity (where stable circular orbits exist for all radii), there are *no* stable circular orbits with $r \leq 6GM$. Therefore, $r = 6GM$ represents the "innermost stable circular orbit" (ISCO) for anything orbiting a highly compact (non-rotating) object.

This has astrophysical relevance, because compact objects do exist whose outer radii are smaller than $6GM$. Neutron stars, which are created by catastrophic stellar core collapse during supernova, have masses typically on the order of 1.4 solar masses and radii of roughly 10 km $\approx 5GM$. The spacetime in the vacuum outside this radius will be Schwarzschild if the star is not rotating (we will study rotating objects later). A non-rotating black hole (which we will also discuss more later) is entirely Schwarzschild spacetime all the way down to $r = 0$. Astrophysicists have strong evidence that both neutron stars and black holes exist in the universe.

Part of the evidence comes from observations of point-like X-ray sources in our local group of galaxies. The graphs shown in figure 10.1 mean that particles with any significant angular momentum per unit mass ℓ falling toward a compact object will "bounce off" of the potential barrier in $\tilde{V}(r)$ before reaching the object. Eventually, such particles will organize themselves into a flat "accretion disk" around the object. Friction between particles in the disk radiates energy away, rapidly circularizing their orbits (note that stable circular orbits have the lowest energy for a given angular momentum, since they correspond to the bottom point of a valley in the "potential energy" graph). Since there is no easy way for particles to "radiate" angular momentum out of the disk, it might seem the disk's total angular momentum should be conserved, thus keeping the circularized orbits stable. However, it is now widely believed that magnetic interactions between charged particles create turbulence that allows a few particles to carry angular momentum outward, slowly decreasing the angular momentum of the rest. This allows most particles in the disk to slowly spiral inward. Once particles pass through the ISCO, there is no longer any barrier in $\tilde{V}(r)$ (see box 10.6), so particles rapidly fall into or accrete onto the object from there.

How much energy can be released by particles in the accretion disk? Consider a particle starting essentially at rest at $r = \infty$. According to equation 10.5, its relativistic energy per unit rest mass at infinity is $e = 1$ (meaning that it has its mass energy and nothing else). At the ISCO radius $r = 6GM$, equation 10.11 implies that the particle's angular momentum per unit mass ℓ in its last circular orbit is such that

$$12\left(\frac{GM}{\ell}\right)^2 = 1 \quad \Rightarrow \quad \ell = \sqrt{12}\,GM \tag{10.15}$$

If you plug this, $r = 6GM$, and $dr/d\tau = 0$ into the energy equation 10.8, you can show (see box 10.7) that the particle's final energy-per-unit-mass just as it drifts past the ISCO has decreased by about

$$\Delta e = \sqrt{\frac{8}{9}} - 1 \approx -0.057 \tag{10.16}$$

This means that in order to make it from infinity to the ISCO, the particle must radiate away energy equivalent to 5.7% of its rest mass. Even if the particle then falls into a black hole (where its remaining energy is entirely absorbed), the energy released just by the disk is *enormous*. For comparison, the fusion reaction used by most stars converts only about 0.7% of their hydrogen fuel's mass-energy into radiated energy (and nuclear fission is more than an order of magnitude *less* efficient).

One can get an order-of-magnitude estimate of how the inner disk's temperature depends on its luminosity as follows. The inner disk is where particle velocities are highest and the disk will be hottest. Let's crudely estimate that essentially all of the energy comes from the portion of the disk between its inner radius R and $2R$, that the disk's temperature T is constant over this region, that it radiates energy like an ideal black body, and that the rest of the disk emits nothing. The Stefan-Boltzmann law says that the luminosity L of a blackbody of area A at temperature T is

$$L \equiv \frac{\text{energy radiated}}{\text{time}} = \sigma A T^4 \tag{10.17}$$

where σ is the Stefan-Boltzmann constant = 5.67×10^{-8} W/(m²K⁴). Given the assumptions above, you can show (see problem P10.1) that the temperature of the disk will be of order of magnitude

$$T \sim \left(\frac{L}{L_\odot}\right)^{1/4} \left(\frac{M_\odot}{M}\right)^{1/2} (2 \times 10^6 \text{ K}) \tag{10.18}$$

where $L_\odot$ and $M_\odot$ are the sun's luminosity and mass, respectively. Note that although the temperature goes up with increased luminosity and decreases with increasing mass, T varies as small powers of these quantities, so for any stellar-sized source radiating a stellar-like energy, T will not be much different than 10^6 K. Since such temperatures are about 500 times higher than the sun's surface temperature, the typical wavelengths of light emitted will be about 500 times shorter, or on the order of magnitude of 1 nm, in the X-ray region (photon energy ≈ 1.2 keV).

In fact, we observe a number of point X-ray sources in our galaxy and in neighboring galaxies having luminosities up to $10^6 L_\odot$. Emission from accretion disks around highly compact objects is the most reasonable explanation for such sources (as other energy sources would require an implausible rate of fuel consumption).

BOX 10.1 Schwarzschild Orbits Must Be Planar

A simple symmetry argument provides the proof that all Schwarzschild orbits must be planar. Consider an object whose initial velocity lies in the equatorial plane. Its subsequent trajectory *must also* lie in the equatorial plane, because in this spherically symmetric spacetime, one side of the equatorial plane is identical to the other, so there is no reason for a free object (whose motion is completely determined by the spacetime) to leave the plane and thus choose one side over the other. Also in a spherically symmetric spacetime, the equatorial plane is no different than any other plane going through the origin. No matter what an object's initial velocity might be, that velocity and the origin define a plane through the origin, so the object's trajectory will be confined to that plane. Therefore *all geodesic trajectories in Schwarzschild spacetime must be planar*.

You can show that this statement is consistent with the geodesic equation. The $\mu = \theta$ component of equation 10.1 tells us that

$$0 = \frac{d}{d\tau}\left(g_{\theta\nu}\frac{dx^\nu}{d\tau}\right) - \frac{1}{2}\partial_\theta g_{\alpha\beta}\frac{dx^\alpha}{d\tau}\frac{dx^\beta}{d\tau} \tag{10.19}$$

You can show that this component of the geodesic equation becomes

$$0 = r^2\frac{d^2\theta}{d\tau^2} + 2r\frac{dr}{d\tau}\frac{d\theta}{d\tau} - r^2\sin\theta\cos\theta\left(\frac{d\phi}{d\tau}\right)^2 \tag{10.20}$$

and that $\theta = \pi/2$ = constant is a solution to this equation. This means that orbits in the equatorial plane are allowed by the geodesic equation. Moreover, the fact that the value of θ does not accelerate when $d\theta/d\tau = 0$ and $\theta = \pi/2$ also means that if the object's trajectory initially lies in the equatorial plane, it cannot not curve away from that plane. Again, since the equatorial plane is no different than any other plane through the origin, this proves (in a different way) that geodesic trajectories in Schwarzschild spacetime must be planar.

Exercise 10.1.1. Verify equation 10.20 and show that $\theta = \pi/2$ = constant is a solution to that equation.

BOX 10.2 The Schwarzschild "Conservation of Energy" Equation

The Schwarzschild metric tensor is diagonal, so the implied sums in the equation $-1 = g_{\mu\nu} u^\mu u^\nu$ yield only four nonzero terms:

$$-1 = g_{tt}\left(\frac{dt}{d\tau}\right)^2 + g_{rr}\left(\frac{dr}{d\tau}\right)^2 + g_{\theta\theta}\left(\frac{d\theta}{d\tau}\right)^2 + g_{\phi\phi}\left(\frac{d\phi}{d\tau}\right)^2 \quad (10.21)$$

If you substitute in the value of the metric components and use equations 10.5 and 10.7, you should be able to show that this can be written

$$1 = \left(1 - \frac{2GM}{r}\right)^{-1} e^2 - \left(1 - \frac{2GM}{r}\right)^{-1}\left(\frac{dr}{d\tau}\right)^2 - \frac{\ell^2}{r^2} \quad (10.22)$$

for orbits in the equatorial plane ($\theta = \pi/2$ = constant).

Exercise 10.2.1. Verify equation 10.22.

From equation 10.22, it is only a few steps to equation 10.8.

Exercise 10.2.2. Work out the steps between equation 10.22 and equation 10.8.

BOX 10.3 Deriving Conservation of Newtonian Energy for Orbits

Consider an object following a Newtonian orbit in the r, ϕ plane. We can write conservation of Newtonian energy for such an orbit in the following form:

$$E = \tfrac{1}{2}m(v_r^2 + v_\phi^2) - \frac{GMm}{r} = \tfrac{1}{2}m\left[\left(\frac{dr}{dt}\right)^2 + r^2\left(\frac{d\phi}{dt}\right)^2\right] - \frac{GMm}{r} \quad (10.23)$$

where E is the object's total Newtonian energy, m is its mass, M is the mass of the primary at the origin, v_r is the radial component of the object's velocity, and v_ϕ is the component perpendicular to the radial component in the direction in which ϕ increases. But notice that the object's angular momentum around the origin is

$$L = mrv_\phi = mr^2\frac{d\phi}{dt}, \quad \text{so} \quad \ell \equiv \frac{L}{m} = r^2\frac{d\phi}{dt} \quad (10.24)$$

Exercise 10.3.1. Show that these equations together imply equation 10.10, as long as we define $\tilde{E}_N \equiv E/m$.

BOX 10.4 The Radii of Circular Orbits

Note that if we take the τ-derivative of both sides of equation 10.8, we get

$$0 = \frac{d^2r}{d\tau^2}\frac{dr}{d\tau} + \frac{d\tilde{V}}{dr}\frac{dr}{d\tau} \quad \Rightarrow \quad \frac{d^2r}{d\tau^2} = -\frac{d\tilde{V}}{dr} \quad (10.25)$$

This means that points where $d\tilde{V}/dr = 0$ will be points where an object will experience no radial acceleration. If the object's effective energy $\tilde{E} = \tilde{V}(r)$ at such a point, then the object will have no radial velocity and no radial acceleration, so it will remain at constant r. Such a particle must have nonzero angular momentum (figure 10.1 makes it clear that if $\ell = 0$, then we will never have $d\tilde{V}/dr = 0$), so it will therefore follow a circular orbit around the origin.

So the radii r_c of possible circular orbits correspond to values of the radial coordinate r where $d\tilde{V}/dr = 0$. Setting the r-derivative of equation 10.9 to zero yields

$$0 = +\frac{GM}{r_c^2} - \frac{\ell^2}{r_c^3} + \frac{3GM\ell^2}{r_c^4} \quad (10.26)$$

Exercise 10.4.1. Verify equation 10.26.

BOX 10.4 (continued) The Radii of Circular Orbits

The fastest way to equation 10.11 (which is the most useful form for the equation for r_c) is to define $u_c \equiv 1/r_c$. If you substitute this into equation 10.26, divide both sides by u_c^2, and solve the resulting quadratic equation, you will get equation 10.11, which (for the sake of convenience) I reproduce here:

$$r_c = \frac{6GM}{1 \pm \sqrt{1 - 12(GM/\ell)^2}} \qquad (10.11r)$$

Exercise 10.4.2. Verify this.

Note that equation 10.11 implies that as long as the square root is real, there will be a circular orbit outside $r = 6GM$ (corresponding to the negative sign in the denominator) and one inside that radius (corresponding to the positive sign). From this equation, you can also find the smallest value for $|\ell|$ for which circular orbit solutions exist at all.

Exercise 10.4.3. What is this value of $|\ell|$?

BOX 10.5 Kepler's Third Law

The component of the geodesic equation 10.1 with $\mu = r$ implies that for the diagonal Schwarzschild metric

$$0 = \frac{d}{d\tau}\left(g_{rr}\frac{dr}{d\tau}\right) - \frac{1}{2}\frac{\partial g_{\alpha\beta}}{\partial r}\frac{dx^\alpha}{d\tau}\frac{dx^\beta}{d\tau} \qquad (10.27)$$

For a circular orbit ($dr/d\tau = 0$) in the equatorial plane ($d\theta/d\tau = 0$), you can show that this reduces to

$$0 = \frac{\partial g_{tt}}{\partial r}\left(\frac{dt}{d\tau}\right)^2 + \frac{\partial g_{\phi\phi}}{\partial r}\left(\frac{d\phi}{d\tau}\right)^2 \qquad (10.28)$$

Exercise 10.5.1. Verify this.

Define $\Omega \equiv d\phi/dt$. This is the angular speed that an observer at infinity (whose time is equal to the coordinate time t) will consider the orbiting particle to have. The orbital period T this observer measures is simply $T = |2\pi/\Omega|$. If we multiply equation 10.28 through by $(d\tau/dt)^2$ and apply the chain rule, we get

$$0 = \frac{\partial g_{tt}}{\partial r} + \frac{\partial g_{\phi\phi}}{\partial r}\left(\frac{d\phi}{d\tau}\frac{d\tau}{dt}\right)^2 = \frac{\partial g_{tt}}{\partial r} + \frac{\partial g_{\phi\phi}}{\partial r}\left(\frac{d\phi}{dt}\right)^2 = \frac{\partial g_{tt}}{\partial r} + \frac{\partial g_{\phi\phi}}{\partial r}\Omega^2 \qquad (10.29)$$

From this, you can prove that $\Omega^2 = GM/r^3$ and from that the other part of equation 10.12.

Exercise 10.5.2. Do this.

BOX 10.6 The Innermost Stable Circular Orbit (ISCO)

Stable circular-orbit radii correspond to local *minima* of the effective potential energy function $\tilde{V}(r)$ displayed in figure 10.1 (see if you can remember why). To determine whether an extremum is a minimum, we need to see whether $d^2\tilde{V}/dr^2$ is positive (i.e., the curve is concave up) at the extremum. Equation 10.26 implies that

$$\frac{d\tilde{V}}{dr} = \frac{GM}{r^2} - \frac{\ell^2}{r^3} + \frac{3GM\ell^2}{r^4} \tag{10.30}$$

$$\Rightarrow \quad \frac{d^2\tilde{V}}{dr^2} = -\frac{2GM}{r^3} + \frac{3\ell^2}{r^4} - \frac{12GM\ell^2}{r^5} \tag{10.31}$$

At a local *minimum*, we must have $d\tilde{V}/dr = 0$ and $d^2\tilde{V}/dr^2 > 0$.

Exercise 10.6.1. Set the first expression to 0, multiply it by $2/r$, and add it to the second to prove that minima can exist only for $r > 6GM$.

For the record, figure 10.2 shows a graph of $\tilde{V}(r)$ when $\ell/GM = \sqrt{12}$ (the value of ℓ that leads to a circular orbit of radius $r = 6GM$ according to equation 10.11). You can see that $r = 6GM$ in this case corresponds to an inflection point, not a minimum, and that there is no barrier preventing a particle with slightly more energy than the circular-orbit energy from falling in.

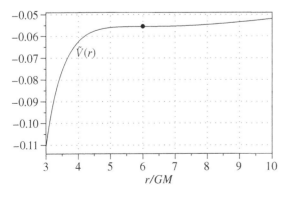

FIG. 10.2 A plot of $\tilde{V}(r)$ when $\ell = \sqrt{12}\,GM$ (the value of ℓ that an object has in the ISCO).

BOX 10.7 The Energy Radiated by an Inspiraling Particle

Exercise 10.7.1. Substitute $\ell/GM = \sqrt{12}$, $r = 6GM$, and $dr/d\tau = 0$ into equation 10.8 and solve for e to find a particle's energy per unit rest mass as measured at infinity when it is in the ISCO. Then subtract from the value of e for the particle when it *was* at infinity to find its *change* in e during the inspiral process. Note that it is e that is the physically relevant energy here, not $\tilde{E}$ (which was invented to make it easier to compare the Schwarzschild and Newtonian cases.)

HOMEWORK PROBLEMS

P10.1 Verify that equation 10.18 is correct. (*Hint:* Calculate the approximate area A of the disk between $R = 6GM$ and $2R$, substitute this into equation 10.17, multiply top and bottom by $L_\odot$ and $M_\odot$, and solve for T. Note that $GM_\odot = 1477$ m and $L_\odot = 3.9 \times 10^{26}$ W.)

P10.2 An object falls radially inward toward a black hole with mass M, starting at rest at infinity. How much time will a clock on the object register between the events of the object passing through the Schwarzschild radial coordinates $r = 10GM$ and $r = 2GM$? (*Hint:* Argue that an object released from rest at infinity will have $\tilde{E} = 0$, i.e., $e = 1$.)

P10.3 Two objects fall radially in from infinity, one having $e = 1$ and the other having $e = 2$. An observer at rest at $r = 6GM$ watches these objects pass. How much faster is the second object moving than the first object according to this observer? (*Hints:* Remember that the observer will measure each object's energy to be $E = -\mathbf{p} \cdot \mathbf{u}_{\text{obs}}$. Calculate this in Schwarzschild coordinates: you may find equation 9.20 helpful. Then one can infer the speed the observer will measure using $E = m/\sqrt{1-v^2}$.) (Adapted from problem 9.7 in Hartle, *Gravity*, Addison Wesley, 2003.)

P10.4 a. Find a general expression for dr/dt for a geodesic in the equatorial plane as a function of r, GM, e, and ℓ. What does this equation say happens as r approaches $2GM$? (As we will see in chapter 14, this proves to be an artifact of the Schwarzschild coordinate system.)
b. Find expressions for both $dr/d\tau$ and dr/dt if we drop an object from rest in the equatorial plane at radial coordinate r_0. (*Hint:* You should be able to determine specific values of e and ℓ in this case.)

P10.5 Imagine we launch an object radially from $r = r_0$ with sufficient speed so that it comes to rest at $r = r_1 > r_0$ before falling back to $r = r_0$. Find an expression (in terms of GM, r_0, and r_1) for the proper time measured by the object during this trajectory. (*Hints:* Determine e in terms of r_1, and change variables to $u \equiv r/r_1$. Note that $u \le 1$ for the entire trajectory. Feel free to look up a fairly nasty integral.)

P10.6 Imagine that an object in a stable circular orbit around a neutron star ($GM = 2.2$ km) has an angular momentum per unit mass of $\ell = 6GM = 13.2$ km.
a. Calculate the radius of the orbit.
b. Calculate the period of orbit as measured by a clock traveling with the object. Express your answer in milliseconds. (*Hint:* You can very easily calculate $d\phi/d\tau$, which is constant for the orbit.)
c. Calculate the period of the orbit as measured by an observer at infinity. Express your answer in milliseconds.

P10.7 a. Use equation 10.11 to show that for a circular orbit around a gravitating object of mass M,

$$\ell^2 = \frac{GMr_c^2}{(r_c - 3GM)} \tag{10.32}$$

for both signs of the original equation, where r_c is the circular orbit's radial coordinate. Note that this equation implies that for objects with nonzero mass, no circular orbits of any kind exist for $r \le 3GM$.
b. Use this to show that the effective energy-per-unit-mass $\tilde{E}$ for a circular orbit as a function of r_c is

$$\tilde{E} = -\frac{GM}{2r_c}\left(1 - \frac{3GM}{r_c}\right)^{-1}\left[1 - \frac{4GM}{r_c}\right] \tag{10.33}$$

and compare to the Newtonian result $E/m = -GM/2r_c$.
c. Find e as a function of r_c alone and check that $e = \sqrt{8/9}$ when $r_c = 6GM$.

P10.8 Consider an object starting essentially at rest at infinity, but with an infinitesimal tangential velocity sufficient to give it an angular-momentum-per-unit-mass ℓ. Argue that if ℓ has the appropriate value, this particle can spiral in to an unstable circular orbit at $r_c = 4GM$, and find that appropriate value of ℓ in terms of GM. (*Hint:* Use the results of problem P10.7.)

P10.9 A spaceship is in a stable circular orbit at a Schwarzschild radial coordinate of $r = 10GM$ around a supermassive black hole whose mass is 10^6 solar masses.
a. What is this orbit's circumference in kilometers?
b. What is the effective energy per unit mass $\tilde{E}$ and angular momentum per unit mass ℓ for this object? (*Hint:* Use the results of problem P10.7.)
c. What is the period of the spaceship's orbit according to its own clock?

P10.10 Find the relation between $\omega \equiv d\phi/d\tau$ and r for a circular orbit. How does this compare to the relationship $\Omega^2 = GM/r^3$ found in box 10.5?

P10.11 Using the method displayed in box 10.4, calculate the expression that for Newtonian mechanics is analogous to equation 10.11. Also show that the Newtonian result is the large-ℓ, large-r limit of equation 10.11.

P10.12 Consider three observers, one in a spaceship in a circular orbit of radius r, one stationary at radius r, and one effectively at infinity. Calculate the period of the orbit measured by each observer as a function of r, and from that period, infer the speed at which each would consider the spacecraft to be moving if we define that speed to be the circumference of the orbit $2\pi r$ divided by the observer's time. Rank these speeds from smallest to greatest, and explain why this ranking makes sense physically. Are any of the speeds (so calculated) possibly greater than 1? If so, also explain how that is possible. (*Hint:* Equation 10.32 may be helpful.)

P10.13 In chapter 12 we will see that photons can orbit a Schwarzschild black hole at a constant radial coordinate of $r = 3GM$. Consider a photon in such an orbit.

a. The definition of the Schwarzschild r coordinate implies that if the photon moves through an angular displacement of $d\phi$ in a certain coordinate time dt, the physical distance the photon moves is $r\,d\phi$. Therefore, an observer at infinity (whose clock measures time dt) will conclude that the photon's speed is $r\,d\phi/dt$. Use the fact that $ds = 0$ along a photon worldline to show that its speed (so defined) is $V = 0.577$.

b. An observer at $r = 3GM$ observes this same photon orbit exactly once in a time T. Use the Schwarzschild metric to compute the time T this stationary observer's clock measures between the two events of the photon passing once and then passing a second time as a fraction of the coordinate time Δt between these events. Use this to calculate the photon's speed v as measured by that stationary observer.

c. Explain qualitatively and physically why v measured by the observer at $r = 3GM$ is not the same the value of V measured by the observer at $r \approx \infty$.

P10.14 A comet with mass m comes in from essentially rest at infinity but with sufficient angular momentum so that it approaches a black hole, loops partway around it, then recedes back to to infinity. Our goal in this problem is to determine the comet's speed as measured by a stationary observer at the comet's point of closest approach.

a. Argue that as r goes to infinity, $d\phi/d\tau$ must go to zero for any finite ℓ. Then use the metric equation and the definition of e to argue that $e \approx 1$ for an comet having $dr/d\tau \approx 0$ at large r, even if it has finite ℓ.

b. Show that at any radial coordinate r

$$\frac{d\phi}{d\tau} = \left(1 - \frac{2GM}{r}\right)\frac{\ell}{r^2}\frac{dt}{d\tau} \quad (10.34)$$

c. Write out the relation $-1 = \mathbf{u} \cdot \mathbf{u}$ at the point of closest approach and use equation 10.34 to show that this comet's four-momentum $\mathbf{p} \equiv m\mathbf{u}$ at its point of closest approach has the following time component in the Schwarzschild coordinate system:

$$p^t = \frac{m}{\sqrt{\left(1 - \frac{2GM}{R}\right) - \left(1 - \frac{2GM}{R}\right)^2 \frac{\ell^2}{R^2}}} \quad (10.35)$$

where R is the (unknown) radial coordinate of the comet's closest approach.

d. We have seen in previous contexts that an object's energy as measured by an observer moving with four-velocity $\mathbf{u}_{obs}$ will be $E_{obs} = -\mathbf{p} \cdot \mathbf{u}_{obs}$. Since dot products have the same value in every coordinate system, we can use Schwarzschild coordinates to calculate this dot product, but the result will still be the energy that the observer would measure. Find $\mathbf{u}_{obs}$ in Schwarzschild coordinates for a stationary observer at R, and evaluate E_{obs} for the comet in terms of ℓ, GM, and R.

e. In the observer's orthonormal frame $E = m/\sqrt{1-v^2}$. Use this to evaluate the comet's speed v according to a stationary observer at R in terms of ℓ, GM, and R.

f. Use equation 10.8 to find the radial coordinate of closest approach R in terms of ℓ. Explain why there are two solutions, and argue which one you want. Also, show that in the large-ℓ limit, your desired solution approaches the result we would get if gravity were Newtonian, which is $\ell^2/2GM$. Is the point of closest approach closer or farther than the Newtonian result? Does this make sense? (*Hint:* Study figure 10.1 to help you answer the question about why there are two solutions and answer the last question.)

P10.15 As you may know from discussions of the so-called twin paradox, one can effectively travel to the future by getting into a spaceship and traveling to and back from some distant point at nearly the speed of light. However, if you have a local black hole, you can do this *much* less expensively as follows. Put yourself in an orbit with the correct $\tilde{E}$ and ℓ at your starting point at approximately infinite r so that (subsequently without using any fuel) you spiral into an unstable circular orbit near to the black hole, hang out there for a while, and then spiral back out to your starting place.

a. If you start essentially at rest at a very large radius, but give yourself just the right tiny bump in the tangential direction to give yourself the right ℓ, show that you can spiral in to an unstable circular orbit at $r = 4GM$ and hang out there for a while, before spiraling back out again. Calculate the correct value of ℓ. (*Hint:* See problem P10.7.) Also, for the portion of your trajectory where your orbit is approximately circular at $r = 4GM$, by what factor does your clock run slower than one at approximately infinite r?

b. You can improve this performance by giving yourself enough radially inward velocity at very large r to end up in an unstable circular orbit at a closer radius. Imagine that for the portion of your trajectory where your orbit is approximately circular, you want your clock to run 10 times more slowly than a clock at very large r. Calculate the value of ℓ that you would need and what your speed v at very large r needs to be to put yourself into the required orbit. You should find that the required v will be relativistic, but that traveling such a speed in flat spacetime would give you *much* less of a slowdown.

11. PRECESSION OF THE PERIHELION

- **INTRODUCTION**

- **FLAT SPACETIME**
  - Review of Special Relativity
  - Four-Vectors
  - Index Notation

- **TENSORS**
  - Arbitrary Coordinates
  - Tensor Equations
  - Maxwell's Equations
  - Geodesics

- **SCHWARZSCHILD BLACK HOLES**
  - The Schwarzschild Metric
  - Particle Orbits
  - **Precession of the Perihelion**
  - Photon Orbits
  - Deflection of Light
  - Event Horizon
  - Alternative Coordinates
  - Black Hole Thermodynamics

- **THE CALCULUS OF CURVATURE**
  - The Absolute Gradient
  - Geodesic Deviation
  - The Riemann Tensor

- **THE EINSTEIN EQUATION**
  - The Stress-Energy Tensor
  - The Einstein Equation
  - Interpreting the Equation
  - The Schwarzschild Solution

- **COSMOLOGY**
  - The Universe Observed
  - A Metric for the Cosmos
  - Evolution of the Universe
  - Cosmic Implications
  - The Early Universe
  - CMB Fluctuations & Inflation

- **GRAVITATIONAL WAVES**
  - Gauge Freedom
  - Detecting Gravitational Waves
  - Gravitational Wave Energy
  - Generating Gravitational Waves
  - Gravitational Wave Astronomy

- **SPINNING BLACK HOLES**
  - Gravitomagnetism
  - The Kerr Metric
  - Kerr Particle Orbits
  - Ergoregion and Horizon
  - Negative-Energy Orbits

this depends on this

11. PRECESSION OF THE PERIHELION

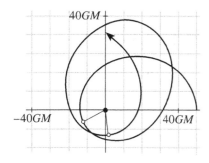

FIG. 11.1 This drawing is a tracing of a computer plot showing an object orbiting a black hole in a non-circular orbit. The "perihelion" is the point in each orbit that is closest to the black hole (white dots). The perihelion shift is about 60° per cycle in this extreme case. (The word *perihelion* is from the Greek *peri* "near" + *helios* "sun," so maybe we should say "periholion" in this case).

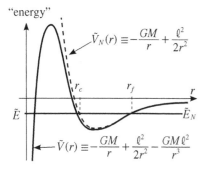

FIG. 11.2 A graph of the effective potential energy curves governing the radial motion of an object orbiting a star of mass M according to Newtonian physics (dashed curve) and general relativity (solid curve). This figure overlays the two curves displayed separately in figure 10.1.

Introduction. When Einstein published a draft of his theory of general relativity on November 25 of 1915, one of his greatest announcements was an explanation of the anomalous precession of the perihelion of Mercury, a puzzle that had bothered astronomers since the 1850s. Careful astronomical observations had demonstrated well before that time that Mercury's orbital ellipse precessed with time. Until the 1850s, though, people assumed that this was caused by the gravitational effects of the other planets. The bulk of Mercury's observed precession is indeed explained this way, but once all effects had been accounted for, a small amount remained. The modern result is 43.11 ± 0.45 arc-seconds/century (Weinberg, *Gravitation and Cosmology*, Wiley, 1972, p. 198), but this value was known to an uncertainty of better than about ±15% since 1882 (see Pais, *Subtle Is the Lord*, Oxford, 1982, p. 254). Einstein's successful resolution of this puzzle gave his theory great credibility. The purpose of this chapter is to demonstrate how general relativity explains this classic result.

Why Do Orbits Precess? General relativity predicts that non-circular orbits in Schwarzschild spacetime are not closed. Figure 11.1 shows a computer solution for an extreme orbit close to a black hole.

Why does this happen? We can understand this qualitatively as follows. One can show that Newtonian orbits with total Newtonian energy $E < 0$ are in fact closed ellipses: the orbiting object returns exactly to its initial position after one complete orbit. We will see a mathematical proof of this shortly, but for the moment let's accept this as given. Figure 11.2 shows the effective potential energy curves that govern the radial part of the motion of an object orbiting a star with mass M in according to Newtonian physics (the dashed curve) and general relativity (the solid curve). Note that the $-GM\ell^2/r^3$ term causes the GR curve to deviate downward from the Newtonian curve as r becomes smaller.

The radial coordinate of an object in an elliptical Newtonian orbit with energy $\tilde{E}_N$ will oscillate back and forth between a minimum radius r_c and a maximum radius r_f. In Newtonian mechanics, the time required for a complete oscillation happens to be exactly the time required for the object to complete one orbit around the star, so the orbit is closed. But you can see that an object having effective energy $\tilde{E}$ subject to the GR potential will have essentially the same far radius r_f, but will get closer to the star than in the Newtonian case and spend extra time at small radii. This is where the object's angular speed is the greatest, so during a complete oscillation, an object subject to the GR potential covers a bit *more* angle than in the Newtonian case, so that when the object returns to the same point in its radial oscillation, it has covered a bit more than 2π radians in angle. This causes the orbit to precess.

Calculating the Shape of an Orbit. Predicting the angle of precession analytically is generally impossible, but the calculation becomes tractable if we consider the limit where $r \gg GM$ and a nearly circular orbit.

Equations 10.7 and 10.8 in chapter 10 express the consequences of the geodesic equation for equatorial orbits in Schwarzschild spacetime:

$$r^2 \frac{d\phi}{d\tau} = \ell \tag{11.1}$$

$$\tilde{E} = \frac{1}{2}\left(\frac{dr}{d\tau}\right)^2 - \frac{GM}{r} + \frac{\ell^2}{2r^2} - \frac{GM\ell^2}{r^3} \tag{11.2}$$

where ℓ and $\tilde{E}$ are constants. Remember that the analogous Newtonian equations are the same except that $\tau \to t$ and the last term in equation 11.2 does not appear.

We can recast these equations into an equation that describes the shape $r(\phi)$ of the orbit as follows. First note that we can write

$$\frac{dr}{d\tau} = \frac{dr}{d\phi}\frac{d\phi}{d\tau} = \frac{dr}{d\phi}\frac{\ell}{r^2} \tag{11.3}$$

where in the last step I have used equation 11.1. If we substitute this into equation 11.2, we get a differential equation for the shape of the orbit:

$$\tilde{E} = \frac{1}{2}\left(\frac{dr}{d\phi}\right)^2 \frac{\ell^2}{r^4} - \frac{GM}{r} + \frac{\ell^2}{2r^2} - \frac{GM\ell^2}{r^3} \quad (11.4)$$

We could integrate this directly to find $r(\phi)$, but the equations become much easier to understand if we define

$$u \equiv \frac{1}{r} \quad \Rightarrow \quad \frac{du}{d\phi} = -\frac{1}{r^2}\frac{dr}{d\phi} \quad \Rightarrow \quad \frac{dr}{d\phi} = -r^2\frac{du}{d\phi} = -\frac{1}{u^2}\frac{du}{d\phi} \quad (11.5)$$

If we substitute these results into equation 11.4, we can eliminate all references to r. If you do this and take the derivative of the resulting equation with respect to ϕ, you will find (see box 11.1) that equation 11.4 becomes

$$\frac{d^2u}{d\phi^2} + u = \frac{GM}{\ell^2} + 3GMu^2 \quad (11.6)$$

The corresponding Newtonian equation is the same (see box 11.2) except that (again), the final term is missing

$$\frac{d^2u}{d\phi^2} + u = \frac{GM}{\ell^2} \quad \text{(Newtonian)} \quad (11.7)$$

Notice that equation 11.7 is the equation of a harmonic oscillator with a constant driving force. Equation 11.6 is the harmonic oscillator equation with a somewhat more complicated driving force.

Perturbation Calculation. We cannot solve equation 11.6 analytically, but we can solve it using an approach known as a perturbation calculation. This is a powerful technique for solving difficult equations that is often used in physics and is therefore a useful addition to your repertoire as a working physicist.

Consider first a circular orbit with a constant radius r_c. The value of $u_c = 1/r_c$ must satisfy equation 11.6, so we have

$$\frac{d^2u_c}{d\phi^2} + u_c = \frac{GM}{\ell^2} + 3GMu_c^2 \quad (11.8)$$

Now consider a perturbation of this orbit that is slightly non-circular. Define

$$u(\phi) = u_c + u_c w(\phi) \quad (11.9)$$

where the unitless "wobble" function $w(\phi) \ll 1$. If you plug equation 11.9 into equation 11.6 and use equation 11.8 to eliminate some terms involving u_c, you should find (see box 11.3) that the perturbation ("wobble") function must satisfy

$$\frac{d^2w}{d\phi^2} + w = 6GMu_c w + 3GMu_c w^2 \quad (11.10)$$

Now, so far, the calculation has been exact. However, since $w(\phi) \ll 1$ by hypothesis, the second term on the right will be much smaller than the first term. So

$$\frac{d^2w}{d\phi^2} + (1 - 6GMu_c)w \approx 0 \quad \text{or} \quad \frac{d^2w}{d\phi^2} + \left(1 - \frac{6GM}{r_c}\right)w \approx 0 \quad (11.11)$$

This is formally the same as the harmonic oscillator equation $d^2x/dt^2 + \omega^2 x = 0$ with the constant $(1 - 6GM/r_c)$ playing the role of ω^2. We know that the solution to the harmonic oscillator equation is $x(t) = A\cos(\omega t + \theta_0)$ (where A and θ_0 are constants). So by analogy, the solution to equation 11.11 must be

$$w(\phi) = A\cos(\omega\phi + \phi_0) \quad \text{where} \quad \omega \equiv \left(1 - \frac{6GM}{r_c}\right)^{1/2} \quad (11.12)$$

and where A and ϕ_0 are constants. (You can easily verify that 11.12 is the solution by substituting it back into equation 11.11.)

The minimum value of r occurs when $w(\phi)$ is a maximum, i.e., when the argument of the cosine function in equation 11.12 is $2\pi n$, where n is an integer. Successive points

11. PRECESSION OF THE PERIHELION

FIG. 11.3 This surface has the same metric as the spatial equatorial plane of Schwarzschild spacetime. The circles on the surface correspond to r in steps of $2GM$. (The "throat" becomes vertical at $r = 2GM$, and it is not possible to represent the geometry *inside* $r = 2GM$ by a curved surface in a three-dimensional Euclidean space.)

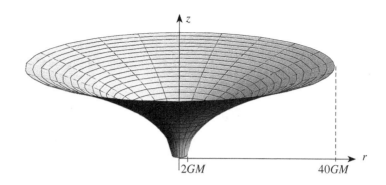

of closest approach therefore occur when the argument of the function $A\cos(\omega\phi + \phi_0)$ changes by 2π, i.e., when

$$\omega\Delta\phi = \left(1 - \frac{6GM}{r_c}\right)^{1/2} \Delta\phi = 2\pi$$

$$\Rightarrow \quad \Delta\phi = \left(1 - \frac{6GM}{r_c}\right)^{-1/2} 2\pi \approx 2\pi\left(1 + \frac{3GM}{r_c}\right) = 2\pi + \frac{6\pi GM}{r_c} \quad (11.13)$$

where in the next-to-last step I used the binomial approximation, assuming that r_c is large enough so that $GM/r_c \ll 1$. This implies that the angular coordinate must go once around from the previous angle of closest approach and then $6\pi GM/r_c$ radians more before we are again at minimum r. When applied to Mercury's orbit, this perihelion advance is roughly 43 arc-seconds per century (see box 11.4).

The Effect of Spatial Curvature. Physically, this precession arises both because $g_{tt} = -(1 - 2GM/r)$ (clock rates depend on r) and because $g_{rr} = (1 - 2GM/r)^{-1}$ (the spatial part of the Schwarzschild geometry is curved). The goal of this section is to show that the latter effect contributes 1/3 of the total precession.

Since our orbit remains in the equatorial plane, we will focus on the set of events on the equatorial plane $\theta = \pi/2$ at some arbitrary instant of time t. The geometry for this set of events is found by setting $dt = 0$, $\theta = \pi/2$, and $d\theta = 0$ in the Schwarzschild metric: the metric for this particular spatial plane is thus

$$ds^2 = \frac{dr^2}{(1 - 2GM/r)} + r^2 d\phi^2 \quad (11.14)$$

As discussed in chapter 9, we consider the geometry described by this metric to be "curved" because the radial distance $s = \int(1 - 2GM/r)^{-1} dr$ between two concentric circles is not equal to the difference Δr in their circumferential radii, indicating that the two-dimensional space described by this metric is not Euclidean.

However, this is all a bit abstract. We can actually *see* the curvature of a spherical surface's geometry because we can draw a diagram that displays a surface having the geometry described by the spherical metric $ds^2 = R^2 d\theta^2 + R^2 \sin^2\theta d\phi^2$ as a two-dimensional curved surface in three-dimensional Euclidean space.

It turns out that we can do the same thing for the Schwarzschild equatorial plane. Consider a three-dimensional Euclidean space labeled by polar coordinates r and ϕ on the horizontal plane and a vertical z coordinate. You can show (see box 11.5) that the surface described by the function

$$z(r) = \sqrt{8GM(r - 2GM)} \quad (11.15)$$

has the same metric in r, ϕ coordinates as given in equation 11.14 and thus the same geometry as the two-dimensional t = constant, $\theta = \pi/2$ slice through Schwarzschild spacetime. Figure 11.3 displays this surface, which is called **Flamm's paraboloid**. This diagram, though it has become the iconic representation of a black hole, captures only one limited aspect of the four-dimensional Schwarzschild geometry. However, we can

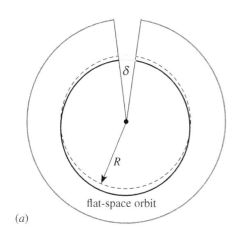
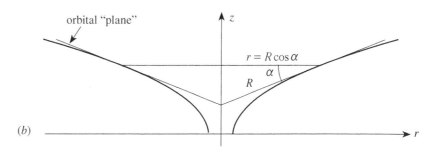

FIG. 11.4 (*a*) This diagram shows the orbit as it would look in flat space. (*b*) This diagram shows Flamm's paraboloid in a side view. To most closely match the surface, we have to convert the flat diagram of part *a* into a cone whose surface is tangent to the paraboloid at the orbit's average radial coordinate *r*.

see immediately that radial distances between circles on this surface will be larger than Δr and that the discrepancies get worse as $r \to 2GM$.

A diagram that displays the geometry implied by a given two-dimensional metric as a two-dimensional curved surface in a three-dimensional Euclidean space is called an **embedding diagram**. Though it is not always possible to construct an embedding diagram for a given metric, such diagrams can yield helpful insights.

For example, we can use figure 11.3 to display why the spatial curvature alone yields a perihelion shift of $2\pi GM/r_c$ (1/3 of that given in equation 11.13) for nearly circular orbits at large radii.[1] Figure 11.4*a* shows a circle of radius R on a flat plane (dashed curve), and an elliptical orbit generated by adding a small wobble to this circle (solid curve). Let this represent the orbit a particle would follow if space were flat. However, when space is curved, we should (as a first approximation) consider the "plane" of the orbit to be more like a cone that is tangent to Flamm's paraboloid at a radial coordinate of $r = R\cos\alpha$, where $\tan\alpha$ is equal to the paraboloid's slope dz/dr at that radius (see figure 11.4*b*). Such a conical surface will closely approximate the paraboloid's actual surface for the narrow range of radii between the orbital extremes. To convert the flat drawing of figure 11.4*a* to a cone with the appropriate angle, we need to cut a wedge of angle δ out of the drawing, so that

$$(2\pi - \delta)R = 2\pi r = 2\pi R\cos\alpha \quad \text{where} \quad \tan\alpha = \left[\frac{dz}{dr}\right]_{\text{at } r} \quad (11.16)$$

Now, in the limit that $r \gg 2GM$, space should nearly be flat, so α should be very small. If you use the small-angle approximation, evaluate dz/dr at the specified radius, expand the cosine in a power series for small α, and solve equation 11.16 for δ, you should find (see box 11.6) that the wedge angle is

$$\delta \approx \frac{2\pi GM}{r} = \frac{1}{3}\left(\frac{6\pi GM}{r}\right) \quad (11.17)$$

Cutting out this wedge advances the perihelion by the same amount. (In problem P11.5, you can prove that the remaining 2/3 of the shift comes from the g_{tt} metric component alone.)

Interpreting the Embedding Diagram. Popular treatments of general relativity sometimes use a "rubber sheet" analogy to describe what massive objects do to spacetime. In this analogy, a massive object placed on a rubber sheet deforms it into a shape that look like Flamm's paraboloid. Small objects on the sheet are then "attracted" by the depression and can even be put into "orbit" around it.

If we push this analogy even the least bit beyond the most qualitative level, its flaws become clear. First, *spatial* curvature is not the most important factor even in

---

[1]. This argument comes from Rindler, *Relativity*, 2/e, Oxford Press, 2006, pp. 234–5)

determining the perihelion shift. The curvature in *spacetime* due to g_{tt}'s dependence on r is much more important, and indeed it is the *only* effect that matters for particles at rest (*it* is responsible for the apparent "attraction" of gravity). There is no good way to visualize this curvature, and the rubber sheet only misdirects.

Second, it should also be clear from the above that the orbit a particle follows is *not* a geodesic on Flamm's paraboloid, but rather a geodesic in the entire curved spacetime. Flamm's paraboloid, for all its stature as an icon, is really useful *only* for (1) vividly displaying the difference between the Schwarzschild circumferential r coordinate and radial distance, and (2) estimating the effects of the purely spatial part of the spacetime curvature (as we have just done).

Computer Models of Orbits. The calculation we have done in this chapter applies only to nearly circular orbits with $r \gg 2GM$. To handle orbits that violate these conditions, one must resort to computer solutions of the geodesic equations. Equation 10.13, which says that $d^2r/d\tau^2 = -GM/r^2 + \ell^2/r^3 - 3GM\ell^2/r^4$, provides the most helpful starting place. Imagine dividing proper time into discrete steps of size $\Delta\tau$, and let r_n be the value of r at the nth time step. Using a suitable difference approximation for the double derivative (see box 11.7) yields

$$r_{n+1} = 2r_n - r_{n-1} + \Delta\tau^2 \left(-\frac{GM}{r_n^2} + \frac{\ell^2}{r_n^3} - \frac{3GM\ell^2}{r_n^4} \right) \tag{11.18}$$

This nicely allows us to calculate the next value of r given its values at the two previous time steps. Using equation 11.1 and a suitably centered approximation for the single derivative (again, see box 11.7), we can then calculate ϕ into the future:

$$\phi_{n+1} = \phi_n + \Delta\tau \frac{\ell}{[\frac{1}{2}(r_{n+1} + r_n)]^2} \tag{11.19}$$

So if we choose a suitably small value for $\Delta\tau$ and two initial values for r and one for ϕ and then iterate equations 11.18 and 11.19, we can calculate the orbit's coordinates r and ϕ as far into the future as desired.

BOX 11.1 Verifying the Orbital Equation for $u(\phi)$

Exercise 11.1.1. Substitute the results from equation 11.5 into equation 11.4 and take the derivative with respect to ϕ to get equation 11.6. The equations are repeated below for your convenience.

$$\tilde{E} = \frac{1}{2}\left(\frac{dr}{d\phi}\right)^2 \frac{\ell^2}{r^4} - \frac{GM}{r} + \frac{\ell^2}{2r^2} - \frac{GM\ell^2}{r^3} \qquad (11.4r)$$

$$u \equiv \frac{1}{r} \quad \Rightarrow \quad \frac{du}{d\phi} = -\frac{1}{r^2}\frac{dr}{d\phi} \quad \Rightarrow \quad \frac{dr}{d\phi} = -r^2\frac{du}{d\phi} = -\frac{1}{u^2}\frac{du}{d\phi} \qquad (11.5r)$$

$$\frac{d^2u}{d\phi^2} + u = \frac{GM}{\ell^2} + 3GMu^2 \qquad (11.6r)$$

BOX 11.2 Verifying the Newtonian Orbital Equation

The Newtonian equations corresponding to equations 11.1 and 11.2 are

$$\ell = r^2\frac{d\phi}{dt} \quad \text{and} \quad \tilde{E}_N = \frac{1}{2}\left(\frac{dr}{dt}\right)^2 - \frac{GM}{r} + \frac{\ell^2}{2r^2} \qquad (11.20)$$

The same line of reasoning that leads from equations 11.1 and 11.2 to equation 11.6 yields equation 11.7 (i.e., $d^2u/d\phi^2 + u = GM/\ell^2$) from equations 11.20.

Exercise 11.2.1. Verify that this is correct.

BOX 11.3 Verifying the Equation for the Orbital "Wobble"

Exercise 11.3.1. Fill in the missing steps between equations 11.6, 11.8, and 11.9 and the final equation for the "wobble" given by equation 11.10. The equations are repeated below for your convenience.

$$\frac{d^2u}{d\phi^2} + u = \frac{GM}{\ell^2} + 3GMu^2 \tag{11.6r}$$

$$u_c = \frac{GM}{\ell^2} + 3GMu_c^2 \tag{11.8r}$$

$$u(\phi) = u_c + u_c w(\phi) \tag{11.9r}$$

$$\frac{d^2w}{d\phi^2} + w = 6GMu_c w + 3GMu_c w^2 \tag{11.10r}$$

BOX 11.4 Application to Mercury

Exercise 11.4.1. The mean orbital radius of Mercury is about 57.9×10^6 km, and its period is roughly 0.241 y. Remember that GM for the sun is 1.477 km. Show that the predicted perihelion shift of $6\pi GM/r_c$ radians per orbit corresponds to roughly 43 arc-seconds/century (where 1 arc-second = 1 as = $1°/3600$). (Note that the "mean orbital radius" is not exactly the same as r_c, so this calculation will be a bit of an approximation.) When Einstein calculated this result and realized that it explained the hitherto unexplained precession of Mercury's orbit, it gave him "palpitations of the heart," and "for a few days, I was beside myself with joyous excitement." (Pais, *Subtle Is the Lord*, Oxford, 1982, p. 253).

BOX 11.5 Constructing the Schwarzschild Embedding Diagram

Our goal in this box is to construct in a three-dimensional Euclidean space a two-dimensional surface having the same metric as a $t = \text{const}$, $\theta = \pi/2$ slice through Schwarzschild spacetime, which is given by equation 11.14 (repeated here):

$$ds^2 = \frac{dr^2}{1 - 2GM/r} + r^2 d\phi^2 \qquad (11.14r)$$

Consider a three-dimensional Euclidean space described by polar coordinates r and ϕ on the horizontal plane and a vertical coordinate z. The metric for such a space in this coordinate system is

$$ds^2 = dr^2 + r^2 d\phi^2 + dz^2 \qquad (11.21)$$

Note that a horizontal circle in this space has circumference $s = \int r\, d\phi = 2\pi r$, so r is a circumferential coordinate for such circles, as it is in the Schwarzschild metric.

In this three-dimensional space, we can describe a two-dimensional surface by specifying its height z as a function of the horizontal coordinates r and ϕ. Since the Schwarzschild geometry is symmetric under rotations in ϕ, we would expect the surface that represents it to display the same symmetry, so in this particular case, we expect the height of our surface to be a function of r alone: $z = z(r)$. The metric on this surface can be found by substituting $z(r)$ into equation 11.21 to eliminate dz and thus find a metric that depends only on r and ϕ. Doing this yields

$$ds^2 = dr^2 + r^2 d\phi^2 + \left(\frac{dz}{dr}\right)^2 dr^2 = \left[1 + \left(\frac{dz}{dr}\right)^2\right] dr^2 + r^2 d\phi^2 \qquad (11.22)$$

Our goal is to find the function $z(r)$ that makes the metric of this surface match the metric in equation 11.14, that is, $z(r)$ such that

$$1 + \left(\frac{dz}{dr}\right)^2 = \frac{1}{1 - 2GM/r} \qquad (11.23)$$

This is a first-order differential equation that you can solve.

Exercise 11.5.1. Solve equation 11.23 for dz/dr and integrate the resulting equation. You should find that $z(r) = \pm\sqrt{8GM(r - 2GM)}$ plus an irrelevant constant of integration that merely moves the surface up or down vertically. We choose (arbitrarily) the positive solution to represent our surface.

One can work through the same process to find embedding diagrams for other two-dimensional metrics. Problem P11.6 shows that the embedding diagram for the spherical surface metric does indeed yield the equation for a spherical surface.

BOX 11.6 Calculating the Wedge Angle δ

Exercise 11.6.1. If α is small, then $\tan\alpha \approx \alpha$ and $\cos\alpha \approx 1 - \frac{1}{2}\alpha^2$. Use these approximations and equation 11.16 to show that in the limit that $r \gg 2GM$, the wedge angle δ that we must remove from the circular orbit is $\approx 2\pi GM/r$.

BOX 11.7 A Computer Model for Schwarzschild Orbits

We want to find a computer algorithm for evaluating r and ϕ at discrete proper times separated by a fixed (small) $\Delta\tau$ so that their values represent good approximations (at the same discrete times) to the solutions to

$$\frac{d^2r}{d\tau^2} = -\frac{GM}{r^2} + \frac{\ell^2}{r^3} - \frac{3GM\ell^2}{r^4} \quad \text{and} \quad \frac{d\phi}{d\tau} = \frac{\ell}{r^2} \quad (11.24)$$

respectively. The main step in creating such an algorithm is to replace the derivatives in the differential equations with finite differences that represent good approximations to the required derivatives. To find a suitable difference for the double-derivative in the first equation, consider these Taylor series expansions:

$$r(\tau_n + \Delta\tau) = r(\tau_n) + \left[\frac{dr}{d\tau}\right]_{\text{at }\tau_n}\Delta\tau + \frac{1}{2}\left[\frac{d^2r}{d\tau^2}\right]_{\text{at }\tau_n}\Delta\tau^2 + \frac{1}{6}\left[\frac{d^3r}{d\tau^3}\right]_{\text{at }\tau_n}\Delta\tau^3 + O(\Delta\tau^4) \quad (11.25)$$

$$r(\tau_n - \Delta\tau) = r(\tau_n) - \left[\frac{dr}{d\tau}\right]_{\text{at }\tau_n}\Delta\tau + \frac{1}{2}\left[\frac{d^2r}{d\tau^2}\right]_{\text{at }\tau_n}\Delta\tau^2 - \frac{1}{6}\left[\frac{d^3r}{d\tau^3}\right]_{\text{at }\tau_n}\Delta\tau^3 + O(\Delta\tau^4) \quad (11.26)$$

where τ_n is the value of the proper time at the nth time step and $O(\Delta\tau^4)$ means that the remaining terms are of order of $\Delta\tau^4$ or smaller. Define $r_{n+1} \equiv r(\tau_n + \Delta\tau)$, $r_n \equiv r(\tau_n)$, and $r_{n-1} \equiv r(\tau_n - \Delta\tau)$. Note if we add the two power series above, the odd powers cancel out, and if we then subtract $2r_n$ to kill the first even term, we get

$$r_{n+1} - 2r_n + r_{n-1} = \left[\frac{d^2r}{d\tau^2}\right]_{\text{at }\tau_n}\Delta\tau^2 + 2O(\Delta\tau^4) \quad (11.27)$$

Note that if we drop the terms summarized by $2O(\Delta\tau^4)$, we will only be making an error that is roughly $\Delta\tau^2$ smaller than the second-derivative term: we therefore say that the difference expression on the left is a "second-order-accurate" approximation to $[d^2r/d\tau^2]\Delta\tau^2$ evaluated at τ_n. As long as we choose $\Delta\tau$ to be sufficiently small, this should be good enough. It also turns out that using *at least* second-order-accurate approximations greatly increases the probability that the numerical solution will remain stable as the calculation proceeds.

So if we drop the $O(\Delta\tau^4)$ term in equation 11.27 and substitute the right side of the first of equations 11.24 for second derivative, we get

$$r_{n+1} - 2r_n + r_{n-1} = \Delta\tau^2\left[-\frac{GM}{r^2} + \frac{\ell^2}{r^3} - \frac{3GM\ell^2}{r^4}\right]_{\text{at }\tau_n} = \Delta\tau^2\left[-\frac{GM}{r_n^2} + \frac{\ell^2}{r_n^3} - \frac{3GM\ell^2}{r_n^4}\right] \quad (11.28)$$

Solving this for r_{n+1} yields equation 11.18.

BOX 11.7 (continued) A Computer Model for Schwarzschild Orbits

Now consider the following Taylor series for r and ϕ expanded about the instant $\tau_{n+1/2} \equiv \tau_n + \frac{1}{2}\Delta\tau$ halfway between τ_n and τ_{n+1}:

$$r_{n+1} = r_{n+1/2} + \left[\frac{dr}{d\tau}\right]_{\text{at }\tau_{n+1/2}}(\tfrac{1}{2}\Delta\tau) + \frac{1}{2}\left[\frac{d^2r}{d\tau^2}\right]_{\text{at }\tau_{n+1/2}}(\tfrac{1}{2}\Delta\tau)^2 + O(\Delta\tau^3) \quad (11.29a)$$

$$r_n = r_{n+1/2} - \left[\frac{dr}{d\tau}\right]_{\text{at }\tau_{n+1/2}}(\tfrac{1}{2}\Delta\tau) + \frac{1}{2}\left[\frac{d^2r}{d\tau^2}\right]_{\text{at }\tau_{n+1/2}}(\tfrac{1}{2}\Delta\tau)^2 - O(\Delta\tau^3) \quad (11.29b)$$

$$\phi_{n+1} = \phi_{n+1/2} + \left[\frac{d\phi}{d\tau}\right]_{\text{at }\tau_{n+1/2}}(\tfrac{1}{2}\Delta\tau) + \frac{1}{2}\left[\frac{d^2\phi}{d\tau^2}\right]_{\text{at }\tau_{n+1/2}}(\tfrac{1}{2}\Delta\tau)^2 + O(\Delta\tau^3) \quad (11.29c)$$

$$\phi_n = \phi_{n+1/2} - \left[\frac{d\phi}{d\tau}\right]_{\text{at }\tau_{n+1/2}}(\tfrac{1}{2}\Delta\tau) + \frac{1}{2}\left[\frac{d^2\phi}{d\tau^2}\right]_{\text{at }\tau_{n+1/2}}(\tfrac{1}{2}\Delta\tau)^2 - O(\Delta\tau^3) \quad (11.29d)$$

where $r_{n+1} \equiv r(\tau_n + \Delta\tau) = r(\tau_{n+1/2} + \tfrac{1}{2}\Delta\tau)$, $r_{n+1/2} \equiv r(\tau_{n+1/2})$, and $r_n \equiv r(\tau_n) \equiv r(\tau_{n+1/2} - \tfrac{1}{2}\Delta\tau)$, and similarly for ϕ.

Exercise 11.7.1. Use these Taylor series to show that $\frac{1}{2}(r_{n+1} + r_n)$ is a second-order-accurate approximation for $r_{n+1/2}$ and that $\phi_{n+1} - \phi_n$ is a second-order-accurate approximation for $[d\phi/d\tau]\Delta\tau$ evaluated at $\tau_{n+1/2}$.

Therefore, a second-order-accurate approximation for $d\phi/d\tau = \ell/r^2$ is

$$\phi_{n+1} - \phi_n = \frac{\ell\,\Delta\tau}{r_{n+1/2}^2} = \frac{\ell\,\Delta\tau}{\tfrac{1}{4}(r_{n+1} + r_n)^2} \quad (11.30)$$

Adding ϕ_n to both sides yields equation 11.19.

The common element in these approximations is that the differences (or the sum in the case of $r_{n+1/2}$) are symmetrical about the evaluation point. We call these **centered difference** approximations.

All that remains now to complete the algorithm is to choose appropriate values for $\Delta\tau$ and ℓ and initial values of r and ϕ. We fundamentally need to choose (as in the case of Newtonian mechanics) an initial position and velocity for both coordinates to define the orbit. It is easiest to take our initial point to be at an extreme point in the orbit (i.e., to be a point where $dr/d\tau = 0$) and define ϕ to be zero there. Mindful that we want to choose centered approximations, let's define this point in the orbit to occur at time $\tau_{1/2}$ a half step after τ_0. We can choose our initial radius $r_{1/2}$ arbitrarily, but requiring that $dr/d\tau = 0$ at this time means that $r_1 - r_0 \approx 0$,

BOX 11.7 (continued) A Computer Model for Schwarzschild Orbits

implying that $r_1 = r_0 = r_{1/2}$. Choosing an initial angular speed $[d\phi/d\tau]_{1/2}$ at radius $r_{1/2}$ allows us to calculate the constant $\ell = r_{1/2}^2 [d\phi/d\tau]_{1/2}$, and since we are taking $\phi_{1/2} = 0$, we also have $\phi_1 = +\frac{1}{2}[d\phi/d\tau]_{1/2}\Delta\tau$. Given a good choice for $\Delta\tau$, we have what we need to calculate r and ϕ at future time steps.

The easiest way to choose $\Delta\tau$ is to choose it to be a small fraction (something like 1/500) of the time required for a circular orbit of radius $r_{1/2}$. This will ensure that for all but the most extremely non-circular orbits, $\Delta\tau$ will be small enough so that neither r nor ϕ changes much during a time step, which in turn means that ignoring high-order terms in the Taylor series we have used will be a sound approximation.

One can calculate the proper time period for a circular orbit with a given radius r_c as follows. For a circular orbit, $d^2 r_c/d\tau^2 = 0$. If you substitute this into the first relation in equation 11.24 and solve for ℓ, you can show that

$$\frac{\ell_c}{GM} = \frac{r_c/GM}{\sqrt{(r_c/GM) - 3}} \tag{11.31}$$

Exercise 11.7.2. Verify this.

We can therefore calculate a circular orbit period for all $r > 3GM$ (even though orbits for $r < 6GM$ cannot be stable). Since the angular speed $d\phi/d\tau$ for a circular orbit is constant, the proper time required for a complete orbit at our initial radius $r_{1/2}$ is, then,

$$\tau_c = \left[\frac{d\tau}{d\phi}\right]_c 2\pi = 2\pi \frac{r_{1/2}^2}{\ell_c} \tag{11.32}$$

We can therefore set $\Delta\tau = \tau_c / 500$ or so, and be pretty comfortable that this will be small enough.

If we go this route, we have to calculate the angular speed $[d\phi/d\tau]_c = \ell_c/r_{1/2}^2$ for a circular orbit anyway, it is also pretty convenient to specify the object's initial angular speed $[d\phi/d\tau]_{1/2}$ as a fraction of that speed. This saves us from randomly guessing what an appropriate angular speed might be.

Finally, a computer can deal only with unitless numbers, so it is convenient to express r, τ, and ℓ in units of the distance unit GM. This amounts to nothing more than setting $GM = 1$ in all of the equations in this box that explicitly refer to GM.

HOMEWORK PROBLEMS

P11.1 Calculate the approximate perihelion shifts we would expect for the orbits of Venus, the earth, and Mars. These planets have periods of 0.615 y, 1 y, and 1.881 y, respectively, and mean orbital radii of 108.2 Gm, 149.6 Gm, and 227.9 Gm, respectively, where 1 Gm = 10^9 m = 10^6 km. (It is more difficult to measure these planets' perihelion shifts than for Mercury because they orbit farther from the sun, because the earth's and Venus's orbits are more nearly circular than Mercury's, and because the confounding perturbations due to the large outer planets are larger.)

P11.2 Find the precession rate for the perigee for a nearly circular low-earth orbit. (The "perigee" for an orbit around the earth is the same as the "perihelion" for an orbit around the sun.) The radius of a low earth orbit will be about 6500 km and its period is about 90 minutes. Express your answer in arc-seconds per century.

P11.3 Find the precession rate for a nearly circular orbit's periastron if the orbit has a mean Schwarzschild radial coordinate of $r = 400$ km and is going around a neutron star whose mass is such that $GM = 2.0$ km. Calculate the result per orbit and per time as measured by someone at infinity. (*Hints:* The periastron is to an orbit around a star what the perihelion is to an orbit around the sun. To calculate the rate of the precession as measured by the observer at infinity, you will have to find the period of this orbit as measured by someone at infinity. There are results from the last chapter that will help you.)

P11.4 In 1993, Russell Hulse and Joseph Taylor Jr. won the Nobel Prize for their 1974 discovery of the binary system PSR B1913+16, a pair of very tightly orbiting neutron stars. This was a prize-worthy discovery because one of the stars is a pulsar (a neutron star that radiates radio pulses tied to its rotation rate). By monitoring the Doppler shifts of the radio signals from the pulsar in this pair, astrophysicists have been able to determine characteristics of the stars' orbits to extraordinary accuracy, allowing a number of tests of general relativity that were not possible before. For example, one can use this data to measure empirically the periastron shift of the stars' orbits ("astron" = star).

The result we derived in this chapter does not apply to this system, because it violates most of our assumptions and approximations. The neutron stars in this case have comparable masses, so we don't have the case of a particle following a geodesic in the spacetime created by much more massive (and therefore essentially stationary) object. As we will see later, the gravitational field of two objects is *not* the same as the simple sum of the fields of each one separately, so we can't just "superpose" two Schwarzschild spacetimes somehow. The neutron stars are rapidly rotating, so even if each were alone, their surrounding spacetimes would not be correctly modeled by the Schwarzschild metric. Their orbits are also highly elliptical, so our "small wobble" approximation breaks down.

One can, however, make an approximate prediction for the periastron shift using what is called a **post-Newtonian** approach, where one adds the lowest-order corrections predicted by general relativity to an otherwise Newtonian calculation. Such a calculation is outlined in Damour and Deruelle, "General relativistic celestial mechanics of binary systems. I. The post-Newtonian motion," *Annales de l'Institut Henri Poincaré*, vol. 43, no. 1, 1985, pp. 107–132. The result of this calculation is that (to lowest order) the periastron shift (per orbit) for a binary pair is

$$\phi = 6\pi GM_{\text{tot}}\bar{u} \quad \text{where} \quad \bar{u} \equiv \frac{1}{2}\left(\frac{1}{r_c} + \frac{1}{r_f}\right) \quad (11.33)$$

and where M_{tot} is the total mass of the binary system, r_c is the stars' smallest separation, and r_f is their largest.

The relevant data for PSR B1913+16 (which you can easily find online) are as follows:

mass of larger star	= $1.441 M_\odot$
mass of smaller star	= $1.387 M_\odot$
least separation	= 746,000 km
greatest separation	= 3,153,600 km
orbital period	= 7.7524 h

where $M_\odot$ is a solar mass (= 1.477 km in GR units). Using this information, calculate the periastron shift for the PSR B1913+16 system and compare with the measured shift of 4.22° per year.

P11.5 Consider the following spacetime metric (11.34)

$$ds^2 = -\left(1 - \frac{2GM}{r}\right)dt^2 + dr^2 + r^2 d\theta^2 + r^2 \sin^2\theta\, d\phi^2$$

This is the same as the Schwarzschild metric except that the spatial part of the metric is *flat*. Our goal is to find the precession of the perihelion for this metric.

a. Show that the t component of the geodesic equation yields $(1 - 2GM/r)(dt/d\tau) = e$ and the ϕ component yields $d\phi/d\tau = \ell/r^2$ in the equatorial plane (as we found for the Schwarzschild metric).

b. Show that $-1 = \mathbf{u}\cdot\mathbf{u}$ for motion in the equatorial plane of this metric yields

$$1 = \left(1 - \frac{2GM}{r}\right)^{-1} e^2 - \left(\frac{dr}{d\tau}\right)^2 - \frac{\ell^2}{r^2} \quad (11.35)$$

c. Follow the procedure outlined in this chapter to convert this to an equation in terms of $dr/d\phi$ and then $du/d\phi$ (where $u = 1/r$). Take the ϕ derivative of the result to get

$$\frac{d^2u}{d\phi^2} + u = \frac{GMe^2}{\ell^2(1 - 2GMu)^2} \quad (11.36)$$

d. In this case, it is easiest to take the large-radius approximation $2GMu = 2GM/r \ll 1$ now rather than later. Use the binomial approximation to show that

$$\frac{d^2u}{d\phi^2} + \left[1 - 4\left(\frac{GMe}{\ell}\right)^2\right]u \approx \frac{GMe^2}{\ell^2} \quad (11.37)$$

continued...

e. Now consider a circular orbit $u = u_c$ = constant. For this orbit, equation 11.37 becomes

$$\left[1 - 4\left(\frac{GMe}{\ell}\right)^2\right]u_c \approx \frac{GMe^2}{\ell^2} \qquad (11.38)$$

Solve this for $(GMe/\ell)^2$.

f. Now define $u(\phi) = u_c + u_c w(\phi)$ and find the differential equation for w. You can then use the result of the last part to find the perihelion shift for this metric in the large-r limit. You should find it to be 2/3 of what we found for the Schwarzschild metric.

P11.6 In this problem, we will find the embedding diagram for a surface with the metric

$$ds^2 = R^2 d\theta^2 + R^2 \sin^2\theta \, d\phi^2 \qquad (11.39)$$

where R is constant. Don't presume at this point that we know anything about what these variables mean: this metric simply describes a two-dimensional geometry in some unknown coordinates.

a. Define a new variable $r \equiv R \sin\theta$. Use this to find $d\theta$ in terms of r and dr and show that we can rewrite the metric above as follows:

$$ds^2 = \frac{dr^2}{1 - r^2/R^2} + r^2 d\phi^2 \qquad (11.40)$$

Now, if we interpret ϕ as an angular coordinate, then we can interpret r as a circumferential radial coordinate. This space is curved, since the radial distance from the origin to a circle of radius r is not r.

b. Now follow the process outlined in box 11.5 to embed a surface with this metric in a three-dimensional Euclidean space described by cylindrical coordinates r, ϕ, and z. You should find that the surface is described by the equation $z(r) = \pm\sqrt{R^2 - r^2}$.

c. Argue that this is the surface of a sphere. Therefore, the proper embedding diagram for the metric described by equations 11.39 and 11.40 is a spherical surface in the Euclidean three-space.

P11.7 Consider the following metric for a 2D space:

$$ds^2 = \cosh^2(r/R) \, dr^2 + r^2 d\phi^2 \qquad (11.41)$$

where R is a constant with units of length. Follow the process outlined in box 11.5 to embed a surface with this metric in a three-dimensional Euclidean space described by cylindrical coordinates r, ϕ, and z. (Remember that $\cosh^2 x - \sinh^2 x = 1$ and that the derivative of $\cosh x$ is $\sinh x$ and vice versa.) In your solution, specify the function $z(r)$ analytically and submit a printout of a computer-drawn graph of $z(r)$.

P11.8 Consider the following metric for a 2D space:

$$ds^2 = \frac{dr^2}{\cos^2(r/R)} + r^2 d\phi^2 \qquad (11.42)$$

where R is a constant with units of length. Follow the process outlined in box 11.5 to embed a surface with this metric in a three-dimensional Euclidean space described by cylindrical coordinates r, ϕ, and z. (You should be easily able to do the required integral in this case.) In your solution, specify the function $z(r)$ analytically, and submit a printout of a computer-drawn graph of $z(r)$.

P11.9 In this problem, we will find the embedding diagram for a surface described by the metric

$$ds^2 = d\rho^2 + (\rho^2 + b^2) \, d\phi^2 \qquad (11.43)$$

where b is a constant. Don't presume at this point that we know anything about what these variables mean: this metric simply describes a two-dimensional geometry in some unknown coordinates.

a. Define a new variable $r \equiv \sqrt{\rho^2 + b^2}$. Use this to find ρ and $d\rho$ in terms of r and dr, and show that we can rewrite the metric above as

$$ds^2 = \frac{r^2 dr^2}{r^2 - b^2} + r^2 d\phi^2 \qquad (11.44)$$

Now, if we interpret ϕ as an angular coordinate, then we can interpret r as a circumferential radial coordinate. This space is curved, since the radial distance from the origin to a circle of radius r is not r.

b. Now follow the process outlined in box 11.5 to embed a surface with this metric in a three-dimensional Euclidean space described by cylindrical coordinates r, ϕ, and z. You should find that the surface is described by the equation $z(r) = b \cosh^{-1}(r/b)$ (you may have to look up an integral).

c. Graph or sketch this curve in the rz plane.

P11.10 In this problem we will use the geodesic equation to show that circular (constant-r) trajectories are *not* geodesics on Flamm's paraboloid.

a. Use the metric equation (equation 11.14) for the surface of Flamm's paraboloid to prove that for a circular orbit, we must have $d\phi/ds = 1/r$.

b. Show that circular orbit solutions for which $dr/ds = 0$ at all times (and thus $d^2r/ds^2 = 0$), and where $d\phi/ds = 1/r$, do *not* satisfy the r component of the geodesic equation (equation 8.12) for this metric.

P11.11 Implement the computer model described in box 11.7 using either a spreadsheet, a computer mathematics tool such as Mathematica, or a computer programming language such as C. Set your initial radius to $50GM$ and your initial angular speed to about 0.75 times the circular orbit speed, graph $r(\phi)$, and see if you can reproduce the results shown in figure 11.1. When your implementation seems to be working well, experiment with some other initial conditions. How does changing the initial radius affect your results? Changing the initial angular speed? Submit some kind of printout or file that shows how you implemented the computer model, and printouts of graphs of $r(\phi)$ for at least three cases you examined.

12. PHOTON ORBITS

INTRODUCTION

FLAT SPACETIME
Review of Special Relativity
Four-Vectors
Index Notation

TENSORS
Arbitrary Coordinates
Tensor Equations
Maxwell's Equations
Geodesics

SCHWARZSCHILD BLACK HOLES
The Schwarzschild Metric
Particle Orbits
Precession of the Perihelion
Photon Orbits
Deflection of Light
Event Horizon
Alternative Coordinates
Black Hole Thermodynamics

THE CALCULUS OF CURVATURE
The Absolute Gradient
Geodesic Deviation
The Riemann Tensor

THE EINSTEIN EQUATION
The Stress-Energy Tensor
The Einstein Equation
Interpreting the Equation
The Schwarzschild Solution

COSMOLOGY
The Universe Observed
A Metric for the Cosmos
Evolution of the Universe
Cosmic Implications
The Early Universe
CMB Fluctuations & Inflation

GRAVITATIONAL WAVES
Gauge Freedom
Detecting Gravitational Waves
Gravitational Wave Energy
Generating Gravitational Waves
Gravitational Wave Astronomy

SPINNING BLACK HOLES
Gravitomagnetism
The Kerr Metric
Kerr Particle Orbits
Ergoregion and Horizon
Negative-Energy Orbits

this → depends on → this

12. PHOTON ORBITS

The Problem with Photons. Since the proper time τ along a photon worldline is zero, we cannot *directly* apply the geodesic equation 10.1 or any of the equations of motion derived from it in chapter 10 to photons. However (as discussed in chapter 8), ordinary particles become more and more photon-like as their mass m becomes negligible compared to their energy. To find equations of motion for photons in Schwarzschild spacetime, we will therefore use the equations in chapter 10 in combinations that remain well-defined in the limit that $m \to 0$.

The Impact Parameter b. With this in mind, let us define

$$b = \frac{\ell}{e} = \frac{r^2(d\phi/d\tau)}{(1 - 2GM/r)(dt/d\tau)} = r^2\left(1 - \frac{2GM}{r}\right)^{-1}\frac{d\phi}{dt} \quad (12.1)$$

This is a conserved quantity for any geodesic, and it remains perfectly well-defined as we take $m \to 0$. For a photon, the quantity b corresponds to perpendicular distance between the photon's trajectory at very large r and the radial line initially parallel to that trajectory, as shown in figure 12.1 (see box 12.1). In classical physics, we call this distance the trajectory's **impact parameter**, and we will continue that usage here.

Note that in flat space, this expression would simply be $b = r^2 d\phi/dt$.

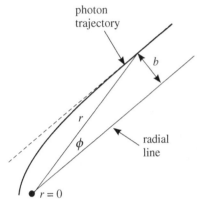

FIG. 12.1 This figure illustrates the meaning of the impact parameter b.

The Equation of Radial Motion for a Photon. According to the Schwarzschild metric equation, the coordinate differences dt, dr, and $d\phi$ between two infinitesimally separated events along an equatorial *photon* worldline will be related by

$$0 = ds^2 = -\left(1 - \frac{2GM}{r}\right)dt^2 + \left(1 - \frac{2GM}{r}\right)^{-1}dr^2 + r^2 d\phi^2 \quad (12.2)$$

If we divide both sides by $(1 - 2GM/r)dt^2$, use equation 12.1, and rearrange things a bit (see box 12.2), we get

$$1 = \left(1 - \frac{2GM}{r}\right)^{-2}\left(\frac{dr}{dt}\right)^2 + \frac{b^2}{r^2}\left(1 - \frac{2GM}{r}\right) \quad (12.3)$$

Equations 12.1 and 12.3 provide a complete set of equations of motion for the photon in that, given b, we could (in principle) solve equation 12.3 for $r(t)$ and then substitute this into equation 12.1 and to find $\phi(t)$. This parameterizes the photon's motion not in terms of proper time τ (as we would with an ordinary particle) but in terms of the Schwarzschild time coordinate t.

We can put this equation into a more evocative form if we divide both sides of equation 12.3 by b^2 to get

$$\frac{1}{b^2} = \left[\frac{1}{b}\left(1 - \frac{2GM}{r}\right)^{-1}\frac{dr}{dt}\right]^2 + \frac{1}{r^2}\left(1 - \frac{2GM}{r}\right) \quad (12.4)$$

This equation has the form of a conservation of energy equation, where $1/b^2$ plays the role of the conserved energy, the first quantity on the right side is a complicated quantity we might consider a "radial kinetic energy," and the last term plays the role of an effective potential energy. With the help of this equation, we can read characteristics of a photon's motion off a potential energy graph like the one shown in figure 12.2 (see box 12.3 for a discussion of its features). In particular, note that for $b > \sqrt{27}\,GM$ (that is, for $1/b^2 < 1/27[GM]^2$), a photon coming in from infinity will rebound to infinity, but for impact parameters less than this critical value, the photon will spiral in to $r = 0$. Photons with exactly this critical value will enter an (unstable) circular orbit at $r = 3GM$.

The equivalent equation of motion for flat spacetime (see box 12.4) is

$$\frac{1}{b^2} = \left[\frac{1}{b}\frac{dr}{dt}\right]^2 + \frac{1}{r^2} \quad \text{(flat spacetime)} \quad (12.5)$$

Note that in this case, the effective potential energy function $1/r^2$ has no peak, but rather increases to infinity as r becomes small. This means that a photon coming in from infinity will always go back out to infinity, as one would expect in flat space.

CONCEPT SUMMARY

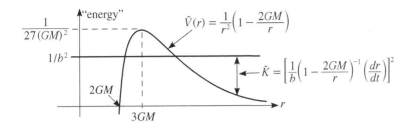

FIG. 12.2 This shows a graph of the effective potential energy function for the radial equation of motion for a photon. Larger values on the "energy" axis correspond to smaller impact parameters.

Orthonormal Coordinate Bases. As stated, equations 12.1 and 12.3 (or 12.4) allow us to plot the trajectory of a photon in Schwarzschild coordinates r and ϕ as a function of the Schwarzschild time coordinate t. But how would an observer at rest in Schwarzschild spacetime characterize the photon's direction of motion?

Answering a question like this requires thinking carefully about how observers in a gravitational field will measure quantities. This is a good opportunity to digress for a moment from studying the motion of photons in particular to address the more general question of how an ordinary observer would *interpret* four-vector quantities whose components we know in a coordinate system like Schwarzschild coordinates. This general approach will be valuable to us in a number of contexts in the future.

Just as we can use cartesian coordinates on any sufficiently small patch of a curved surface (such as the surface of the earth), we can set up a *local* cartesian-like coordinate system in any sufficiently small region of spacetime. Indeed, any observer who is trying to measure physical quantities in his or her local region will do precisely this if they set up a clock lattice (like the one described in chapter 2) or something equivalent. Observers in a gravitational field will have trouble *synchronizing* lattice clocks because clocks higher in the field run faster than those lower, but such problems will be negligible in a sufficiently small frame.

Such an observer would use the mutually orthogonal lattice directions to define orthonormal basis four-vectors $\mathbf{o}_x, \mathbf{o}_y$, and $\mathbf{o}_z$ such that

$$\mathbf{o}_x \cdot \mathbf{o}_x = \mathbf{o}_y \cdot \mathbf{o}_y = \mathbf{o}_z \cdot \mathbf{o}_z = 1 \text{ and } \mathbf{o}_x \cdot \mathbf{o}_y = \mathbf{o}_x \cdot \mathbf{o}_z = \mathbf{o}_y \cdot \mathbf{o}_z = 0 \quad (12.6)$$

(I am using the notation $\mathbf{o}_\alpha$ instead of the more general notation $\mathbf{e}_\alpha$ for these basis four-vectors to remind us that these are a certain **o**bserver's **o**rthonormal basis vectors.) The observer will say that these four-vectors have components $[0, 1, 0, 0]$, $[0, 0, 1, 0]$, and $[0, 0, 0, 1]$ respectively, in his or her local cartesian-like coordinate system. The observer's own four-velocity $\mathbf{u}_{\text{obs}} \equiv \mathbf{o}_t$ defines the basis vector in the time direction, since the observer will consider his or her *own* four-velocity to have components $[1, 0, 0, 0]$. This observer's metric tensor will have components

$$g_{\mu\nu,\text{obs}} \equiv \begin{bmatrix} \mathbf{o}_t \cdot \mathbf{o}_t & \mathbf{o}_t \cdot \mathbf{o}_x & \mathbf{o}_t \cdot \mathbf{o}_y & \mathbf{o}_t \cdot \mathbf{o}_z \\ \mathbf{o}_x \cdot \mathbf{o}_t & \mathbf{o}_x \cdot \mathbf{o}_x & \mathbf{o}_x \cdot \mathbf{o}_y & \mathbf{o}_x \cdot \mathbf{o}_z \\ \mathbf{o}_y \cdot \mathbf{o}_t & \mathbf{o}_y \cdot \mathbf{o}_x & \mathbf{o}_y \cdot \mathbf{o}_y & \mathbf{o}_y \cdot \mathbf{o}_z \\ \mathbf{o}_z \cdot \mathbf{o}_t & \mathbf{o}_z \cdot \mathbf{o}_x & \mathbf{o}_z \cdot \mathbf{o}_y & \mathbf{o}_z \cdot \mathbf{o}_z \end{bmatrix} = \begin{bmatrix} -1 & 0 & 0 & 0 \\ 0 & 1 & 0 & 0 \\ 0 & 0 & 1 & 0 \\ 0 & 0 & 0 & 1 \end{bmatrix} = \eta_{\mu\nu} \quad (12.7)$$

Call such a frame a **LOF** ("locally orthogonal frame" or "local observer's frame").

Since in *any* coordinate basis, a vector $\mathbf{A} \equiv \mathbf{e}_\mu A^\mu$ and $g_{\mu\nu} \equiv \mathbf{e}_\mu \cdot \mathbf{e}_\nu$, we have

$$\mathbf{e}_\mu \cdot \mathbf{A} = \mathbf{e}_\mu \cdot \mathbf{e}_\nu A^\nu = g_{\mu\nu} A^\nu = A_\mu \quad (12.8)$$

In our particular observer's LOF, $g_{\mu\nu} = \eta_{\mu\nu}$, so

$$\mathbf{o}_t \cdot \mathbf{A} = \eta_{t\mu} A^\mu_{\text{obs}} = \eta_{tt} A^t_{\text{obs}} = -A^t_{\text{obs}}, \quad \mathbf{o}_x \cdot \mathbf{A} = \eta_{x\mu} A^\mu_{\text{obs}} = \eta_{xx} A^x_{\text{obs}} = +A^x_{\text{obs}} \quad (12.9)$$

(you will explicitly check this in box 12.5). Similarly, we see that $\mathbf{o}_y \cdot \mathbf{A} = A^y_{\text{obs}}$ and $\mathbf{o}_z \cdot \mathbf{A} = A^z_{\text{obs}}$. We have already seen that $-\mathbf{u}_{\text{obs}} \cdot \mathbf{p} = -\mathbf{o}_t \cdot \mathbf{p}$ yields the energy $E \equiv p^t_{\text{obs}}$ the observer would measure for an object with four-momentum $\mathbf{p}$: equation 12.9 simply extends this technique to cover *all* components of *any* four-vector. Once we know these components, we can interpret them just as we would in cartesian coordinates in special relativity (for example, $v_{x,\text{obs}} = p^x_{\text{obs}}/p^t_{\text{obs}}$, and so on).

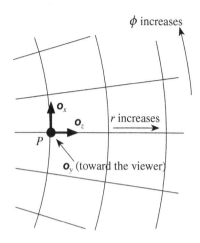

FIG. 12.3 The three spatial basis vectors for a LOF (locally orthonormal frame) that an observer at rest at point P might use. We are looking down on the Schwarzschild equatorial plane. Note that the direction in which θ increases is away from us, but to have a right-handed coordinate system, we need $\mathbf{o}_y$ to point *toward* us, in the $-\theta$ direction.

Now, suppose that we know the components A^μ of a four-vector **A** in some *global* coordinate system (such as the Schwarzschild coordinate system) that covers all of spacetime. If we also know the components $(\mathbf{o}_\alpha)^\mu$ of the observer's basis four-vectors $\mathbf{o}_\alpha$ in that global coordinate system, then we can evaluate the inner products $A^t_{\text{obs}} = -\mathbf{o}_t \cdot \mathbf{A}$, $A^x_{\text{obs}} = \mathbf{o}_x \cdot \mathbf{A}$, etc., in the *global* coordinate system. Since an inner product's numerical value is the same in all coordinate systems, the result will be the same as the observer would obtain. This is generally a *much* easier way to find **A**'s components in the observer's coordinate system than trying to use the general tensor transformation rule $A'^\mu = (\partial x'^\mu/\partial x^\nu)A^\nu$, which requires knowing exactly how the observer's coordinates x'^μ depend on the global coordinates x^μ.

The general process for finding the global system components $(\mathbf{o}_\alpha)^\mu$ of an observer's basis four-vectors $\mathbf{o}_\alpha$ is as follows: (1) Find the components u^μ_{obs} of the observer's four velocity $\mathbf{u}_{\text{obs}} \equiv \mathbf{o}_t$. (2) Decide how to orient the observer's *spatial* basis vectors relative to the global coordinate system. Good alignment usually allows us to set most of the global components of each spatial vector to zero. (3) Use the orthonormality condition $\mathbf{o}_\mu \cdot \mathbf{o}_\nu = \eta_{\mu\nu}$ to determine the remaining components.

For example, consider an observer at rest in Schwarzschild coordinates. As the observer is at *rest*, the three spatial components of $(\mathbf{o}_t)^\mu \equiv u^\mu_{\text{obs}} \equiv dx^\mu/d\tau$ are zero. If we align the $\mathbf{o}_x, \mathbf{o}_y,$ and $\mathbf{o}_z$ vectors with the ϕ, $-\theta$, and r Schwarzschild directions, respectively (see figure 12.3), then each spatial basis vector also has only one nonzero Schwarzschild component (the ϕ, θ, and r component, respectively). You can then show (see box 12.6) that requiring that $\mathbf{o}_\mu \cdot \mathbf{o}_\nu = \eta_{\mu\nu}$ easily determines the one nonzero Schwarzschild component in each of the observer's basis vectors, yielding

$(\mathbf{o}_t)^\mu$	$(\mathbf{o}_x)^\mu$	$(\mathbf{o}_y)^\mu$	$(\mathbf{o}_z)^\mu$
$\begin{bmatrix} \dfrac{1}{\sqrt{1-2GM/r}} \\ 0 \\ 0 \\ 0 \end{bmatrix}$	$\begin{bmatrix} 0 \\ 0 \\ 0 \\ \dfrac{1}{r\sin\theta} \end{bmatrix}$	$\begin{bmatrix} 0 \\ 0 \\ -\dfrac{1}{r} \\ 0 \end{bmatrix}$	$\begin{bmatrix} 0 \\ \sqrt{1-2GM/r} \\ 0 \\ 0 \end{bmatrix}$

(12.10)

The Four-Momentum of Light. In the case of photons, the four-vector of interest is the photon's four-momentum **p**, so let's find that four-momentum. Now, the Schwarzschild components of **p** for a particle with *nonzero* rest mass m are

$$p^\mu = m\frac{dx^\mu}{d\tau} = m\frac{dt}{d\tau}\frac{dx^\mu}{dt} = \frac{me}{1-2GM/r}\frac{dx^\mu}{dt} = \frac{E}{1-2GM/r}\frac{dx^\mu}{dt} \quad (12.11)$$

where I have used equation 10.5 to eliminate the $dt/d\tau$ term and $E = me$ is the particles's relativistic energy at infinity. The last expression in equation 12.11 is well-defined in the limit $m \to 0$, so we will take it to be the *definition* of the Schwarzschild four-momentum components for a photon having energy E at infinity. If the photon moves in the equatorial plane, then equations 12.1 and 12.3 imply that

$$p^t = \frac{E}{(1-2GM/r)}, \quad p^\phi = \frac{E}{(1-2GM/r)}\frac{d\phi}{dt} = E\frac{b}{r^2} \quad (12.12a)$$

$$p^\theta = 0, \quad \text{and} \quad p^r = \frac{E}{(1-2GM/r)}\frac{dr}{dt} = \pm E\sqrt{1 - \frac{b^2}{r^2}\left(1 - \frac{2GM}{r}\right)} \quad (12.12b)$$

(The last is positive for an outgoing photon, negative for an ingoing photon.)

The Critical Escape Angle. Imagine now that an observer at rest at Schwarzschild coordinate r emits a photon in the Schwarzschild equatorial plane at an angle ψ (as measured by the observer) relative to the radially outward direction. At what critical angle ψ_c will this light be just barely captured by the gravitating object?

The x (sideward) component of the light's ordinary velocity in the observer's cartesian system is (as usual) the ratio of the x and t components of its four-momentum: $v_{x,\text{obs}} = p^x_{\text{obs}}/p^t_{\text{obs}}$. Since the photon's speed is 1 in any local observer's frame, the sine of the angle of emission, as measured by the observer, must therefore be

$$\sin \psi = \frac{v_{x,\text{obs}}}{1} = \frac{p^x_{\text{obs}}}{p^t_{\text{obs}}} = \frac{\mathbf{o}_x \cdot \mathbf{p}}{-\mathbf{o}_t \cdot \mathbf{p}} \quad (12.13)$$

The critical angle will correspond to when $b^2 = 27(GM)^2$: according to figure 12.2, any impact parameter smaller than this will spiral in to $r = 0$. You can show by using equations 12.10, 12.12, and 12.13 (see box 12.7) that the critical angle is

$$\sin \psi_c = \frac{GM\sqrt{27}}{r} \sqrt{1 - \frac{2GM}{r}} \quad (12.14)$$

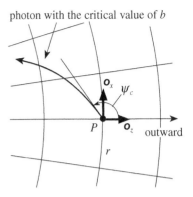

FIG. 12.4 The definition of the critical angle ψ_c at a point P in Schwarzschild spacetime. We are looking down on the Schwarzschild equatorial plane. Photons emitted from P at an angle $\psi > \psi_c$ (that is, in a more inward direction) according to an observer at P will spiral into the black hole; those emitted with $\psi < \psi_c$ will escape. Those emitted exactly at ψ_c will spiral into an unstable circular orbit at $r = 3GM$.

What does this result mean? As r becomes very large, the value of $\sin \psi_c$ approaches zero. But does this mean that the critical angle approaches zero or 180° (both of which have sines of zero)? Far from the gravitating object, we would expect to be able to emit light in almost any direction without having it be captured, so the angle (which is measured from the outward direction) should be 180° in this limit. We see that as r decreases, the sine increases, so the critical angle decreases *from* 180° (see figure 12.4). When r reaches $3GM$, the sine of the critical angle is 1, so the critical angle is 90°, meaning that photons emitted even slightly inward will be captured, those emitted slightly outward will escape, and those emitted exactly at 90° will have exactly the critical value of b and will therefore (according to figure 12.2) move in a circular orbit around the object at that radius, and therefore circle around and illuminate the observer from behind!

As r decreases to $2GM$, the sine of the critical angle decreases to zero, so the critical angle also decreases to zero: even light emitted directly outward is captured. Below that radius, the square root becomes undefined, so some assumption we have made is breaking down. (We will see *what* assumption in chapter 14.)

The Observer Looks Around. The Schwarzschild metric is both t-independent (no metric component depends on t) and t-symmetric (replacing dt by $-dt$ does not change anything). This means that the spatial trajectory of a photon should be the same whether it moves away from the observer or toward the observer. Thus equation 12.14 also applies to photons *arriving* at the observer's location. If the gravitating object that creates the Schwarzschild spacetime is a black hole, then photons from distant stars that reach the observer must have angles relative to the outward direction that are *smaller* than the critical angle ψ_c, because any photons reaching the observer with an angle greater than ψ_c would have to have come *from* $r = 0$, which is impossible for a black hole. Therefore, observers closer and closer to a black hole will see the hole's blackness occupy a greater and greater fraction of the sky. At $r = 3GM$, the hole fills exactly half the sky, and at smaller radii, the hole occupies more than half the sky. Observers very close to $r = 2GM$ will see but a tiny cone of light arriving from the external universe.

Such an observer will also see light from the outside universe to be *blue-shifted*. This is because to the observer, a photon's energy is $E_{\text{obs}} = -\mathbf{o}_t \cdot \mathbf{p}$. Using equations 12.10 and 12.12 to evaluate the dot product in Schwarzschild coordinates yields

$$E_{\text{obs}} = -\mathbf{o}_t \cdot \mathbf{p} \equiv -(\mathbf{o}_t)^\mu g_{\mu\nu} p^\nu = -(\mathbf{o}_t)^t g_{tt} p^t + (15 \text{ zeros})$$

$$= -\left(1 - \frac{2GM}{r}\right)^{-1/2}\left[-\left(1 - \frac{2GM}{r}\right)\right] E\left(1 - \frac{2GM}{r}\right)^{-1} = \frac{E}{\sqrt{1 - 2GM/r}} \quad (12.15)$$

Thus, the energy E_{obs} measured by any stationary observer at any finite radial coordinate r in Schwarzschild spacetime is greater than that photon's energy E at infinity (and infinitely so at $r = 2GM$). (Similarly, observers at infinity will find light emitted by observers at finite r to be red-shifted.) This is fundamentally because clocks at finite r run slower than clocks at infinity (see problem P12.5).

BOX 12.1 The Meaning of the Impact Parameter b

Consider a photon moving in the equatorial plane of Schwarzschild spacetime. Imagine that in the limit of very large r, the photon is moving in a trajectory parallel to a certain radial line from the origin, but a distance d to one side of that line. In the limit that r becomes very large, equations 12.1 and 12.3 become

$$b \approx r^2 \frac{d\phi}{dt} \quad \text{and} \quad \frac{dr}{dt} \approx -1 \tag{12.16}$$

But since spacetime is almost flat at large radii, $\phi \approx \sin^{-1}(d/r) \approx d/r$ (see figure 12.1). This means that

$$\frac{d\phi}{dt} = \frac{d\phi}{dr}\frac{dr}{dt} \approx -\frac{d(d/r)}{dr} = \frac{d}{r^2} \quad \Rightarrow \quad d \approx r^2 \frac{d\phi}{dt} \tag{12.17}$$

If we compare this with the first of equations 12.16, we see that $b = d$. So b is indeed the distance between the photon's trajectory and the parallel radial line at very large distances.

BOX 12.2 Derivation of the Equation of Motion for a Photon

Exercise 12.2.1. Carry out the missing steps leading from equations 12.1 and 12.2 to equation 12.3 (the equation of motion for a photon).

BOX 12.3 Features of the Effective Potential Energy Function for Light

The effective potential energy function for light described in figure 12.2 is

$$\tilde{V}(r) = \frac{1}{r^2}\left(1 - \frac{2GM}{r}\right) \tag{12.18}$$

Note that this function goes to zero at $r = 2GM$ and as $r \to \infty$. You can show that it has a single extremum at $r = 3GM$ and that $\tilde{V}(3GM) = 1/[27(GM)^2]$.

Exercise 12.3.1. Verify both claims in the previous sentence.

BOX 12.4 Photon Motion in Flat Spacetime

In flat spacetime, the equations that correspond to equations 12.1 and 12.2 are

$$b = \frac{\ell}{e} = \frac{r^2 d\phi/d\tau}{dt/d\tau} = r^2 \frac{d\phi}{dt} \tag{12.19}$$

$$0 = -dt^2 + dr^2 + r^2 d\phi^2 \tag{12.20}$$

Exercise 12.4.1. Show that equation 12.5 (the equation of motion for a photon in flat spacetime) follows from these equations.

BOX 12.5 Evaluating 4-Vector Components in an Observer's Frame

Consider the first equation $A^t_{\text{obs}} = -\mathbf{o}_t \cdot \mathbf{A}$. In the observer's coordinate system, the four-vector $\mathbf{A}$ (by definition) has components

$$A^\mu_{\text{obs}} = \begin{bmatrix} A^t_{\text{obs}} \\ A^x_{\text{obs}} \\ A^y_{\text{obs}} \\ A^z_{\text{obs}} \end{bmatrix} \quad (12.21)$$

The components of the observer's four-velocity $\mathbf{u}_{\text{obs}} \equiv \mathbf{o}_t$ in the observer's own reference frame (whose metric is the same as that for flat space) is

$$(\mathbf{o}_t)^\mu_{\text{obs}} = \begin{bmatrix} 1 \\ 0 \\ 0 \\ 0 \end{bmatrix} \quad (12.22)$$

Therefore, the inner product between these vectors in the observer's frame is

$$\mathbf{o}_t \cdot \mathbf{A} \equiv \eta_{\mu\nu}(\mathbf{o}_t)^\mu_{\text{obs}} A^\nu_{\text{obs}} = \eta_{t\nu}(1) A^\nu_{\text{obs}} = \eta_{tt} A^t_{\text{obs}} = -A^t_{\text{obs}} \quad (12.23)$$

where the middle step follows because only $(\mathbf{o}_t)^t = 1$ is nonzero in the sum over μ and the next step follows because the metric is diagonal. Since the inner product of two four-vectors is a scalar, its value will be the same in all coordinate systems.

Exercise 12.5.1. Show in a similar way that $A^x_{\text{obs}} = \mathbf{o}_x \cdot \mathbf{A}$.

BOX 12.6 An Orthonormal Basis in Schwarzschild Coordinates

Consider the metric component $\mathbf{o}_t \cdot \mathbf{o}_t = \eta_{tt} = -1$. If the observer is at rest in Schwarzschild coordinates, then the spatial Schwarzschild components of $\mathbf{o}_t = \mathbf{u}_{\text{obs}}$ must all be zero. Therefore,

$$-1 = \mathbf{o}_t \cdot \mathbf{o}_t = g_{\mu\nu}(\mathbf{o}_t)^\mu(\mathbf{o}_t)^\nu = g_{tt}(\mathbf{o}_t)^t(\mathbf{o}_t)^t$$

$$\Rightarrow \quad (\mathbf{o}_t)^t = \frac{1}{\sqrt{-g_{tt}}} = \frac{1}{\sqrt{1 - 2GM/r}} \quad (12.24)$$

where $g_{\mu\nu}$ is the Schwarzschild metric here. Since the other components are zero, this leads to the first of equations 12.10.

Because this vector has no spatial components, the condition

$$0 = \eta_{tx} = \mathbf{o}_t \cdot \mathbf{o}_x = g_{\mu\nu}(\mathbf{o}_t)^\mu(\mathbf{o}_x)^\nu = \frac{g_{t\nu}}{\sqrt{-g_{tt}}}(\mathbf{o}_x)^\nu = \frac{g_{tt}}{\sqrt{-g_{tt}}}(\mathbf{o}_x)^t \quad (12.25)$$

implies that $\mathbf{o}_x$ has zero Schwarzschild time component. The same argument applies to the other spatial basis vectors. (Note that if the observer were moving in Schwarzschild spacetime, then $\mathbf{o}_t$ would have nonzero spatial components, and the spatial basis vectors *would* possibly have a nonzero time component.)

Exercise 12.6.1. Argue in the space provided on the next page that the rest of equations 12.10 are correct if $\mathbf{o}_x, \mathbf{o}_y,$ and $\mathbf{o}_z$ point in the $\phi, -\theta,$ and r directions, respectively, in Schwarzschild spacetime.

BOX 12.6 (continued) An Orthonormal Basis in Schwarzschild Coordinates

BOX 12.7 Derivation of the Critical Angle for Photon Emission

Equation 12.13 tells us that the angle ψ an emitted photon's path makes with the outward direction is such that

$$\sin \psi = \frac{v_{x,\text{obs}}}{1} = \frac{p^x_{\text{obs}}}{p^t_{\text{obs}}} = \frac{\mathbf{o}_x \cdot \mathbf{p}}{-\mathbf{o}_t \cdot \mathbf{p}} \qquad (12.13r)$$

You can use equations 12.10 and 12.12 (too lengthy to repeat here) to show that

$$\sin \psi_c = \frac{GM\sqrt{27}}{r}\sqrt{1 - \frac{2GM}{r}} \qquad (12.14r)$$

Exercise 12.7.1. Do this.

HOMEWORK PROBLEMS

P12.1 Imagine that the beam from a laser at infinity would miss a black hole's center by $6GM$ if spacetime were flat. Will the beam be absorbed by the black hole?

P12.2 Imagine that we shine an initially cylindrical beam of photons (with initially parallel velocities) from infinity directly toward a gravitating object. What is the largest radius R that the beam can have at infinity and still be entirely absorbed by the object? **Note:** The absorption cross-section of the object for light will then be πR^2. (Adapted from Hartle, *Gravity*, Addison Wesley, 2003, problem 9.16.)

P12.3 By calculating $v_x = p_x/p_t = \mathbf{o}_x \cdot \mathbf{p}/(-\mathbf{o}_t \cdot \mathbf{p})$ and $v_z = p_z/p_t = \mathbf{o}_z \cdot \mathbf{p}/(-\mathbf{o}_t \cdot \mathbf{p})$, show that light moving in the equatorial plane in Schwarzschild spacetime has speed 1 as measured by all observers using orthonormal frames at rest at radii $r > 2GM$.

P12.4 a. Show that an observer at rest at $r = 6GM$ in the Schwarzschild spacetime around the black hole would see the black hole occupy a region 45° around the directly downward direction.

b. We might represent the part of the sky covered by the black hole using a diagram like the one shown below for $r = 6GM$. You should imagine the observer standing at the circle's center and looking left directly toward the black hole. The black region shows the range of angles above and below the radial direction toward the hole for which the observer will see only blackness. Draw similar diagrams for observers at rest at $r = 2.5GM, 3GM, 4GM,$ and $5GM$.

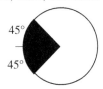

P12.5 Argue that the results from equation 9.12 (which were derived directly from the r-dependence of clock rates in Schwarzschild spacetime) are consistent with equation 12.15. (*Hint:* Remember that for a photon, $E = hc/\lambda = h/\lambda$ in GR units, where h is Planck's constant.)

P12.6 In this problem, we will show that equations 12.5 and 12.19 together imply that photon trajectories in flat spacetime are straight.

a. Show that these equations imply

$$\frac{dr}{dt} = \pm\sqrt{1 - \frac{b^2}{r^2}}, \quad \frac{d\phi}{dt} = \frac{b}{r^2} \quad (12.26)$$

b. Show that if we divide the second of these equations by the first and integrate with respect to r, we find that $\phi = \pm\cos^{-1}(b/r) + \alpha$, where α is some constant of integration.

c. Argue that this is the equation of a straight line in polar coordinates, with b corresponding to the smallest distance between the line and the origin.

P12.7 (Important!) Consider an observer who is falling radially into a black hole from rest at infinity.

a. Use results from chapter 10 to argue that the components of this observer's four-velocity must be $dt/d\tau = (1 - 2GM/r)^{-1}$, $dr/d\tau = -\sqrt{2GM/r}$, $d\theta/d\tau = 0$, and $d\phi/d\tau = 0$. These, therefore, are also the Schwarzschild components of the observer's basis vector $\mathbf{o}_t$.

b. Let us align the observer's spatial basis vectors so that the $\mathbf{o}_z$ vector has no θ or ϕ components, the $\mathbf{o}_y$ vector has no r or ϕ components, and the $\mathbf{o}_x$ vector has no r or θ components (i.e., the spatial projections of these basis vectors point purely in the $r, \theta,$ and ϕ directions, respectively). The observer's metric is $\mathbf{o}_\mu \cdot \mathbf{o}_\nu = \eta_{\mu\nu}$. Prove that satisfying this relation for all μ and ν necessarily implies that

$$(\mathbf{o}_z)^\mu = \begin{bmatrix} \dfrac{-\sqrt{2GM/r}}{1 - 2GM/r} \\ 1 \\ 0 \\ 0 \end{bmatrix} \quad (12.27)$$

and that the $\mathbf{o}_y$ and $\mathbf{o}_x$ vectors are as given in equation 12.10. (Do not just show that these results work: prove that these basis vector *must* have these components given our assumptions.)

c. Find the critical angle for this observer as a function of r, and draw diagrams like the one shown in problem P12.4 for $r = 4GM, 3GM, 2GM,$ and GM.

d. Imagine that such a falling observer receives radial signals from infinity. Will these signals be red-shifted or blue-shifted? What is the fractional change in wavelength of these signals?

P12.8 Consider an observer who is falling radially from rest at infinity, as in problem P12.7. Imagine that just as the observer falls through $r = 6GM$, an object in a circular orbit at that radius happens to pass through the observer's coordinate system. Use the results of problem P12.7 to determine that object's ordinary velocity components and its speed in the observer's frame.

P12.9 Use the method outlined in problem P12.7 and results from chapter 10 to find the Schwarzschild components of a set of orthonormal basis vectors for an observer in a circular orbit of radius r, assuming that we choose the observer's $\mathbf{o}_x, \mathbf{o}_y,$ and $\mathbf{o}_z$ vectors so that they have *spatial* components only in the $\phi, -\theta,$ and r directions (one must have a nonzero t component, though). Express your results entirely in terms of r and GM (eliminate e and ℓ). Show that your calculation falls apart for $r \leq 3GM$, and explain why physically.

13. DEFLECTION OF LIGHT

INTRODUCTION

FLAT SPACETIME
- Review of Special Relativity
- Four-Vectors
- Index Notation

TENSORS
- Arbitrary Coordinates
- Tensor Equations
- Maxwell's Equations
- Geodesics

SCHWARZSCHILD BLACK HOLES
- The Schwarzschild Metric
- Particle Orbits
- Precession of the Perihelion
- Photon Orbits
- **Deflection of Light**
- Event Horizon
- Alternative Coordinates
- Black Hole Thermodynamics

THE CALCULUS OF CURVATURE
- The Absolute Gradient
- Geodesic Deviation
- The Riemann Tensor

THE EINSTEIN EQUATION
- The Stress-Energy Tensor
- The Einstein Equation
- Interpreting the Equation
- The Schwarzschild Solution

COSMOLOGY
- The Universe Observed
- A Metric for the Cosmos
- Evolution of the Universe
- Cosmic Implications
- The Early Universe
- CMB Fluctuations & Inflation

GRAVITATIONAL WAVES
- Gauge Freedom
- Detecting Gravitational Waves
- Gravitational Wave Energy
- Generating Gravitational Waves
- Gravitational Wave Astronomy

SPINNING BLACK HOLES
- Gravitomagnetism
- The Kerr Metric
- Kerr Particle Orbits
- Ergoregion and Horizon
- Negative-Energy Orbits

this depends on this

154   13. DEFLECTION OF LIGHT

Orbit Equation for Photons. In this section, we will use the photon equations of motion we found in chapter 12 to find equations for the shape $r(\phi)$ of photon worldlines in Schwarzschild spacetime. According to equations 12.1 and 12.3, the basic equations of motion for photons in terms of coordinate time t are

$$\frac{d\phi}{dt} = \frac{b}{r^2}\left(1 - \frac{2GM}{r}\right), \quad 1 = \left(1 - \frac{2GM}{r}\right)^{-2}\left(\frac{dr}{dt}\right)^2 + \frac{b^2}{r^2}\left(1 - \frac{2GM}{r}\right) \quad (13.1)$$

As we found in chapter 11, it is actually easier to find an equation for $u(\phi)$, where $u \equiv 1/r$. You can show using the first of the equations above that

$$\frac{dr}{dt} = \frac{dr}{du}\frac{du}{d\phi}\frac{d\phi}{dt} = -\frac{du}{d\phi}b(1 - 2GMu) \quad (13.2)$$

(see box 13.1). Substituting this into the second of equations 13.1 yields

$$\left(\frac{du}{d\phi}\right)^2 + u^2 = \frac{1}{b^2} + 2GMu^3 \quad (13.3)$$

(see box 13.2). Taking the derivative of both sides with respect to ϕ yields

$$2\frac{du}{d\phi}\frac{d^2u}{d\phi^2} + 2u\frac{du}{d\phi} = 0 + 6GMu^2\frac{du}{d\phi} \quad \Rightarrow \quad \frac{d^2u}{d\phi^2} + u = 3GMu^2 \quad (13.4)$$

A Perturbation Solution. As in chapter 11, we can find a solution to this otherwise nasty nonlinear differential equation using a perturbation approach. Consider first what a photon's orbital path would be if there were no gravitating object at the origin (i.e., if $M = 0$ in the equations above). Then equation 13.4 becomes a simple harmonic oscillator equation $d^2u/d\phi^2 + u = 0$ whose solution is

$$u_{zg}(\phi) = u_c \sin\phi = \frac{\sin\phi}{r_c} \quad \Rightarrow \quad r_{zg}(\phi) = \frac{r_c}{\sin\phi} \quad (13.5)$$

where u_c is a constant, $r_c = 1/u_c$ and I have oriented the coordinate system so that $\phi = \pi/2$ at the point where this zero-gravity path is closest to the origin. Note that r_c is the radius of the zero-gravity trajectory's closest approach to the star. Figure 13.1 shows a picture of the actual photon orbit and the zero-gravity trajectory.

As long as $r_c \gg 2GM$, the photon's path will not be *much* affected by the gravitational field, so we can try to approximate the actual solution for the photon's path as a perturbation of the zero-gravity straight-line path. Define

$$u(\phi) = u_c \sin\phi + u_c w(\phi) \quad (13.6)$$

where $w(\phi)$ is a unitless "wobble function" away from the straight-line zero-gravity limit. We will assume that $w(\phi) \ll 1$. If you substitute this into equation 13.4 and throw away all but the largest term on the right side after squaring the u, you can show that the equation for w becomes

$$\frac{d^2w}{d\phi^2} + w = \varepsilon\sin^2\phi = \frac{\varepsilon}{2}(1 - \cos 2\phi) \quad \text{where} \quad \varepsilon \equiv \frac{3GM}{r_c} \quad (13.7)$$

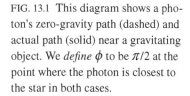

FIG. 13.1 This diagram shows a photon's zero-gravity path (dashed) and actual path (solid) near a gravitating object. We *define* ϕ to be $\pi/2$ at the point where the photon is closest to the star in both cases.

(see box 13.3). This looks like the differential equation for a driven harmonic oscillator. The general approach in that case is to attempt a solution with the same frequency as the driving term. Therefore, let's attempt a solution of the form $A + B\cos 2\phi$. By direct substitution into equation 13.7, you can show (see box 13.4) that this solution works as long as $A = 3B = \frac{1}{2}\varepsilon$, and that the solution for $u(\phi)$ is therefore

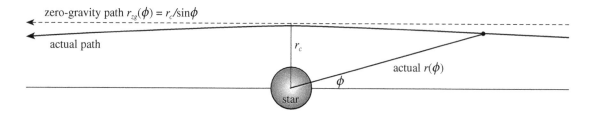

FIG. 13.2 This diagram is essentially the same as figure 13.1 except viewed from much farther away. At very large distances from the star (i.e., as $u \to 0$), the angle that the photon's path makes with the horizontal axis approaches ϕ_0.

$$u(\phi) \approx \frac{1}{r_c}\left[\sin\phi + \frac{3GM}{2r_c}(1 + \tfrac{1}{3}\cos 2\phi)\right] \tag{13.8}$$

Note that our assumption that $w(\phi) \ll 1$ will be satisfied as long as $3GM/r_c \ll 1$.

What does this solution mean? Note that for our zero-gravity solution, $u(\phi)$ became zero (and thus r became infinite) at $\phi = 0$. This is no longer true for equation 13.8; rather, $u(\phi) = 0$ (r becomes infinite) at some angle ϕ_0 such that

$$0 = \sin\phi_0 + \frac{3GM}{2r_c}(1 + \tfrac{1}{3}\cos 2\phi_0) \tag{13.9}$$

In the small-angle approximation, $\sin\phi_0 \approx \phi_0$ and $\cos 2\phi_0 \approx 1$, so this becomes

$$0 \approx \phi_0 + \frac{3GM}{2r_c}\frac{4}{3} \quad \Rightarrow \quad \phi_0 \approx -\frac{2GM}{r_c} \tag{13.10}$$

(Note that if $3GM/r_c \ll 1$, then this will indeed be a small angle, justifying the approximation.) Note that equation 13.9 implies that $\sin\phi_0$ will be a small negative number, so the photon's path will go to infinite radius in the 3rd and 4th quadrants, just below $\phi = 0$ and $\phi = 180°$, as shown in figure 13.2. The total deflection angle is therefore

$$\delta \approx 2|\phi_0| \approx \frac{4GM}{r_c} \tag{13.11}$$

The quantity r_c turns out to be equal (to the level of our approximation) to the impact parameter b that a distant observer would compute from the photon's actual initial or final direction of motion, and is also very nearly equal to the actual distance of the photon's closest approach to the origin (see problem P13.1).

A Historical Note. You can show (see box 13.5) that for starlight just grazing the sun's surface, the deflection angle should be 1.74 as (arc-seconds). This had not ever been observed or measured by 1915, when Einstein published his general theory of relativity. However, the British Astronomer Royal Sir Frank Dyson sent physicists Arthur Eddington and Andrew Crommelin to Principe off the coast of western Africa and Sobral in northern Brazil, respectively, to observe total solar eclipse of May 29, 1919. This eclipse was unusually long and was to take place against an especially rich backdrop of stars, and Dyson hoped that during the eclipse it would be possible to photograph and measure the deflection of stars' images near the sun. Both expeditions ran into difficulties (technical problems in Sobral, bad weather in Principe), but Eddington managed to produce results of 1.98 ($\pm$ 0.16) as and 1.61 ($\pm$ 0.4) as from the data taken from Sobral and Principe, respectively. These were within two standard deviations of Einstein's prediction and comfortably excluded either zero deflection or a pseudo-Newtonian result of 0.87 as (derived assuming that photons fall like other particles in Newtonian mechanics: see problem P1.4). Though a few physicists were critical of Eddington's statistics, most accepted these results as validating Einstein's prediction.[1]

A report was published in the *London Times* on November 7, 1919, and a (lengthy, multi-part) headline in the *New York Times* on November 10, 1919, gushed "LIGHTS ALL ASKEW IN THE HEAVENS — Men of Science More or Less Agog Over Results of Eclipse Observations. — EINSTEIN THEORY TRIUMPHS — Stars Not Where They Seemed or Were Calculated to be, but Nobody Need Worry. — A BOOK FOR 12 WISE MEN — No More in All the World Could Comprehend it, Said Einstein When His Daring Publishers Accepted It." The story of German and British scientists collaborating on a significant scientific advance played very well in the aftermath of World War I. Before these articles, Einstein was not well known to physicists outside of Germany, but afterward, Einstein became the contemporary equivalent of a rock star: his iconic status as the archetype of "genius" starts here.

---

1. Most of the information in this paragraph comes from P. Coles, "Einstein, Eddington and the 1919 Eclipse" from *Historical Development of Modern Cosmology*, ASP Conference Proceedings Vol. 252, Martínez, Trimble, Pons-Bordería. eds. This article is available online at http://adsabs.harvard.edu/full/2001ASPC..252...21C.

13. DEFLECTION OF LIGHT

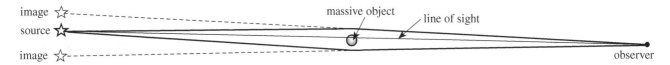

FIG. 13.3 This figure illustrates how a gravitating object might create two deflected images of a single distant source.

Gravitational Lensing. Figure 13.3 shows that if both the source and observer are sufficiently far from a central gravitating object, the bending of light by the object's field can create *two* deflected images of the source, one from bent light traveling above the gravitating body and one from bent light traveling below. Indeed, if the source were directly behind a spherical object, the observer would see the source imaged as a ring around the lensing object. Multiple images and ring segments are in fact observed (the first were observed in 1979), and their analysis has now become an important tool for astronomers. In this section we will develop mathematics that will help us understand and analyze such images.

In what follows, we will assume that the gravitating object is sufficiently spherical so that the deflection result in equation 13.11 applies, that the gravitating object is small enough compared to the distance to either source or observer that the bending takes place over a negligibly small distance and light otherwise travels in straight lines, and that the angles involved are very small. Figure 13.4a shows a diagram showing the basic geometry of the lensing process for *one* of the two images under these assumptions. Note that if the angles are small, sines and tangents of any of the angles are essentially equal to the angle itself, and cosines of those angles are essentially 1. Therefore, for example, the total horizontal distance on the diagram $\equiv D_S =$ (distance to the source) $\cos \beta \approx$ distance to the source, and the distance between O' and the source S is $D_S \tan \beta \approx D_S \beta$, where β is the angle between the line of sight to the source S (absent gravity) and that to the gravitating object ("lens") L.

Given these approximations, the distances along the figure's left edge satisfy

$$D_S \theta = D_S \beta + D_{LS} \delta \tag{13.12}$$

where θ is angle of the line of sight to the deflected image S' relative to that to the lens, δ is the angle through which light from the source is deflected by the lens, and D_{LS} is the horizontal projection of the distance between the lens and source (which is essentially the same as that distance). If the lensing object has mass M, then the deflection angle is $\delta \approx 4GM/r_c \approx 4GM/b$ by equation 13.11. Since the impact parameter is $b \approx D_L \theta$ in this case (where D_L is the distance between the observer and lens), you can show (see box 13.6) that equation 13.12 can be written

$$0 = \theta^2 - \beta\theta - \theta_E^2 \quad \text{where} \quad \theta_E \equiv \sqrt{4GM \frac{D_{LS}}{D_L D_S}} \tag{13.13}$$

Note that if $\beta = 0$ (corresponding to the source being directly behind the lens), then the image of the source is a so-called **Einstein ring** (by symmetry, light is bent equally around all sides of the lens) at the angle θ_E. However, if the source is off-center ($\beta \neq 0$), then equation 13.13 is a simple quadratic with the two solutions

$$\theta_\pm = \tfrac{1}{2}\left(\beta \pm \sqrt{\beta^2 + 4\theta_E^2}\right) \tag{13.14}$$

corresponding to the two possible images, the positive one outside the Einstein ring angle θ_E and the negative one inside (see box 13.6 again). Since the amount of bending is the same for all light paths with the same impact parameter but at different azimuthal angles ϕ *around* the lens, the source's images will have the same azimuthal angular width $\Delta\phi$ around the lens as the source, as shown in figure 13.4b (which shows the situation as the observer sees it). The radial thickness $\Delta\theta_\pm$ of each image can be found by taking the β-derivative of equation 13.14:

$$\Delta\theta_\pm \approx \frac{d\theta_\pm}{d\beta}\Delta\beta = \frac{1}{2}\left(1 \pm \frac{\beta}{\sqrt{\beta^2 + 4\theta_E^2}}\right)\Delta\beta \tag{13.15}$$

Since the quantity in parentheses is greater than zero but less than 2, both images will be somewhat compressed in this direction. Therefore, the two images of the source will

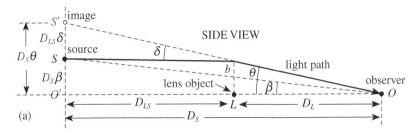

FIG. 13.4 (*a*) This diagram shows a side view of a lensing situation for one of the two images. (All angles have been grossly exaggerated.) Note that we are using the "thin lens" approximation where the light is bent suddenly in a small region near the source. (*b*) This diagram shows both images as viewed by the observer. Note that each image is stretched horizontally and compressed vertically.

be distorted into ring segments that are thinner radially and (generally) wider than the source itself, as illustrated in figure 13.4*b*. Figure 13.5 (on the next page) shows the distorted images of at least one background galaxy created by a cluster of galaxies in the foreground. Many ring-segment-like images are displayed.

Image Brightness. This distortion affects the brightness of the images as well. Consider each of the images to be a "target" in a plane through the lens L and perpendicular to the line OO'. All light from the source that enters this target will be directed toward the observer. Since the source is approximately the same distance from each part of both targets, the amount of light that each target collects will be proportional to the cross-sectional area that each presents to the source. Keeping in mind the small-angle approximation, the top target on this plane will have a radial thickness of $D_L \Delta\theta_+$ and a length (along its arc) of $(D_L \theta_+)\Delta\phi$, so it presents a total area of $D_L^2 \theta_+ \Delta\theta_+ \Delta\phi$ to the source. The argument for the other image and the source itself (in the absence of the lens) is analogous, so the ratio of the brightness $I_\pm$ of either image (i.e., the total energy received per unit time) to the energy per unit time I_s that *would* be received from the source if the gravitational lens were not there is

$$\frac{I_\pm}{I_s} = \frac{D_L^2 |\theta_\pm| \Delta\theta_\pm \Delta\phi}{D_L^2 \beta \Delta\beta \Delta\phi} = \frac{|\theta_\pm| \Delta\theta_\pm}{\beta \Delta\beta} \tag{13.16}$$

You can show (see box 13.7) that this and equations 13.14 and 13.15 imply that

$$\frac{I_\pm}{I_s} = \frac{1}{4}\left(\frac{\beta}{\sqrt{\beta^2 + 4\theta_E^2}} + \frac{\sqrt{\beta^2 + 4\theta_E^2}}{\beta} \pm 2\right) \tag{13.17}$$

and that the first two terms add up to something more than 2, so this ratio is always positive (as it must be). The image inside the Einstein ring angle θ_E (i.e., the negative solution) is dimmer than the source, but the image outside is brighter.

Astronomers have used this brightening effect to detect stellar-mass objects that are too dim to image directly (such as white dwarfs, neutron stars, black holes) in the halo surrounding our own galaxy. Such objects are called **massive compact halo objects** (MACHOs). If a MACHO passes in front of a visible star, it creates two images of the star, as discussed above. These images are too close together to be resolved as separate images, but their combined brightness is given by

$$\frac{I_{tot}}{I_s} = \frac{I_+ + I_-}{I_s} = \frac{1}{2}\left(\frac{\beta}{\sqrt{\beta^2 + 4\theta_E^2}} + \frac{\sqrt{\beta^2 + 4\theta_E^2}}{\beta}\right) \tag{13.18}$$

which is always greater than 1. Therefore, as the MACHO passes in front of the star, it will appear to brighten in a characteristic way (problem P13.6 discusses what such an event will look like). Researchers in the 1990s monitored hundreds of thousands of stars over several hundred days. A number of MACHOs were discovered this way, providing an estimate of the number of MACHOs in our galaxy's halo. One such event is shown in figure 13.6.

FIG. 13.5 This Hubble photograph shows multiple distorted images of background galaxies due to gravitational lensing by the galactic cluster CL0024+1654 in the foreground (the large globe-like galaxies near the center of the photograph). The distinctive twisted and beaded shape of some of the probable images leads researchers to think that the distorted images at 4, 8, 9, and 10 o'clock are actually all images of one *single* background galaxy (more than two images are formed because the cluster is not spherically symmetric). The cluster is about 5 billion light years from us, and the imaged galaxy might be twice as distant. This is direct visual evidence of the effect of gravitational fields on light. **Credits:** W. N. Colley and E. Turner (Princeton University), J. A. Tyson (AT&T Bell Labs, Lucent Technologies), Hubble Space Telescope, and NASA. To see the image in (artificially enhanced) color that helps one distinguish the images from the foreground cluster, visit hubblesite.org/newscenter/archive/releases/1996/10.

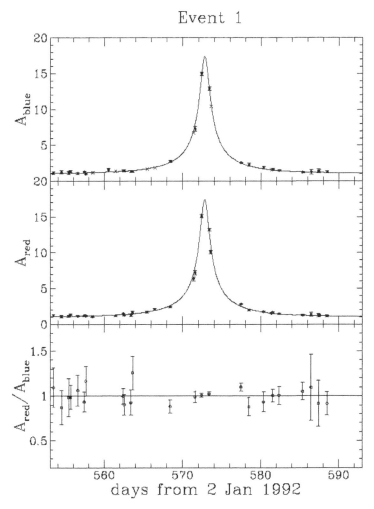

FIG. 13.6 This diagram shows the light curve for one lensing event observed by the MACHO Project. The solid line is a best fit to the data points based on equation 13.18 and a parameterized model for how the lensing object passed in front of the imaged star. The point of the bottom graph is to demonstrate that the pulse in brightness was the same at the red and blue ends of the spectrum, as would be expected from a purely gravitational effect. Most other candidate explanations for such a burst in brightness (such as a stellar explosion of some type) would be brighter in one end of the spectrum. Credit: Alcock, et al. (the MACHO Collaboration), courtesy Douglas Welch. You can find the image online at http://wwwmacho.anu.edu.au/Results/Bulge/fig4a.gif.

BOX 13.1 Checking Equation 13.2

Exercise 13.1.1. Verify that $d\phi/dt = b(1 - 2GM/r)/r^2$ (from equation 13.1) and $u \equiv 1/r$ together imply that $dr/dt = -(du/d\phi)b(1 - 2GMu)$ (equation 13.2).

13. DEFLECTION OF LIGHT

BOX 13.2 The Differential Equation for the Shape of a Photon Orbit

Exercise 13.2.1. Show that $dr/dt = -(du/d\phi)b(1 - 2GMu)$ (equation 13.2) and

$$1 = \left(1 - \frac{2GM}{r}\right)^{-2}\left(\frac{dr}{dt}\right)^2 + \frac{b^2}{r^2}\left(1 - \frac{2GM}{r}\right) \quad \text{(from equation 13.1)}$$

together imply that $(du/d\phi)^2 + u^2 = (1/b^2) + 2GMu^3$ (equation 13.3).

BOX 13.3 The Differential Equation for the Photon "Wobble"

Exercise 13.3.1. Show that substituting $u(\phi) = [\sin\phi + w(\phi)]/b$ (equation 13.6, noting that $u_c = 1/b$) into $d^2u/d\phi^2 + u = 3GMu^2$ (equation 13.4) yields

$$\frac{d^2w}{d\phi^2} + w = \frac{\varepsilon}{2}(1 - \cos 2\phi) \quad \text{where} \quad \varepsilon \equiv \frac{3GM}{b} \tag{13.7r}$$

after keeping only the leading term on the right and using a trigonometric identity to express that leading term in terms of $\cos 2\phi$. This is the differential equation for the photon's wobble.

BOX 13.4 The Solution for $u(\phi)$ in the Large-r Limit

Exercise 13.4.1. Show that $w(\phi) = A + B\cos 2\phi$ solves equation 13.7 as long as the constants A and B have certain values (which you should determine).

BOX 13.5 The Maximum Angle of Light Deflection by the Sun

A photon from a distant star that reaches earth after just grazing the surface of the sun will have $r_c \approx$ distance of closest approach $\approx$ the sun's radius (696,000 km). Note also that for the sun, $GM \approx 1.477$ km.

Exercise 13.5.1. Show that such a photon will be deflected through an angle of about $1.74''$ (where $1'' = 1$ arc-second $= 1°/3600$).

BOX 13.6 The Lens Equation

Figure 13.4 and the small-angle approximation imply that $D_S\theta = D_S\beta + D_{LS}\delta$, (equation 13.12), where $\delta \approx 4GM/r_c \approx 4GM/b$ (equation 13.11) and $b \approx D_L\theta$ (from the diagram). The summary claims that together these imply that

$$0 = \theta^2 - \beta\theta - \theta_E^2 \quad \text{where} \quad \theta_E \equiv \sqrt{4GM\frac{D_{LS}}{D_L D_S}} \quad (13.13r)$$

for the angle θ of images formed by a gravitational lens.

Exercise 13.6.1. Verify that this is true.

Exercise 13.6.2. The solutions to this equation are $\theta_\pm = \tfrac{1}{2}\left(\beta \pm \sqrt{\beta^2 + 4\theta_E^2}\right)$. The θ_+ solution is clearly outside the Einstein ring angle θ_E. Show that $|\theta_-| \leq \theta_E$, meaning that the θ_- solution is inside the ring. (*Hint:* Argue that $\theta_- \leq 0$, and then use this information in equation 13.13, above.)

BOX 13.7 The Ratio of Image Brightness to the Source Brightness

Equations 13.16 and 13.15 imply that

$$\frac{I_\pm}{I_s} = \frac{|\theta_\pm|\Delta\theta_\pm}{\beta\,\Delta\beta} \quad \text{where} \quad \Delta\theta_\pm = \frac{1}{2}\left(1 \pm \frac{\beta}{\sqrt{\beta^2 + 4\theta_E^2}}\right)\Delta\beta \qquad (13.19)$$

We also have $\theta_\pm = \frac{1}{2}\left(\beta \pm \sqrt{\beta^2 + 4\theta_E^2}\right)$ from equation 13.14. The claim in equation 13.17 is that the above imply that

$$\frac{I_\pm}{I_s} = \frac{1}{4}\left(\frac{\beta}{\sqrt{\beta^2 + 4\theta_E^2}} + \frac{\sqrt{\beta^2 + 4\theta_E^2}}{\beta} \pm 2\right) \qquad (13.17r)$$

Exercise 13.7.1. Verify that this is correct.

Exercise 13.7.2. Verify that this ratio is always positive. (*Hint:* Argue that for any positive x, $x + (1/x) - 2 \geq 0$.)

13. DEFLECTION OF LIGHT

HOMEWORK PROBLEMS

P13.1 Equation 13.8 states that

$$u(\phi) \approx \frac{1}{r_c}\left[\sin\phi + \frac{3GM}{2r_c} + \frac{GM}{2r_c}\cos 2\phi\right] \quad (13.20)$$

Because we dropped terms of order $(GM/r_c)^2$ from equation 13.8, the solution will also neglect terms of that order. Now, according to equation 13.3, this $u(\phi)$ should satisfy

$$\left(\frac{du}{d\phi}\right)^2 + u^2 - 2GMu^3 = \frac{1}{b^2} \quad (13.21)$$

a. We can use this to evaluate b in terms of r_c for our solution for $u(\phi)$. Substitute equation 13.20 into the above, multiply through by r_c^2, and throw away all terms of order $(GM/r_c)^2$ (don't even bother to calculate them) to show that

$$\left(\frac{r_c}{b}\right)^2 \approx 1 + \frac{GM}{r_c}(-2\cos\phi\sin 2\phi + 3\sin\phi$$
$$+ \sin\phi\cos 2\phi - 2\sin^3\phi) \quad (13.22)$$

Since the left side of this expression is a constant, it must have the same value at all angles, so we can evaluate this at any angle we choose. Obviously, at $\phi = 0$, the entire term in parentheses vanishes, so we can conclude that $b = r_c$ (to this level of approximation, anyway).

b. Figure 13.1 implies that the photon's closest approach to the star happens at $\phi = 90°$. Use equation 13.20 and the binomial approximation to show that

$$r_{min} \approx r_c\left(1 - \frac{GM}{r_c}\right) \quad (13.23)$$

c. (Optional) Use trigonometric identities to show that the quantity in parentheses on the right side of equation 13.22 is in fact zero for *all* ϕ. (This is a useful check on the validity of equation 13.22, because the left side must be a constant independent of ϕ.)

P13.2 Imagine that a massive galaxy lies between a distant quasar and the earth in such a position that $\beta = \frac{1}{2}\theta_E$. Assume that we can treat the galaxy as a point mass.
a. What are the angles of the two quasar images relative to the direction of the galaxy (in terms of θ_E)?
b. How much brighter or dimmer is each image than the quasar would appear if the galaxy were not present?

P13.3 The "double quasar" Q0957+561 was the first set of gravitational lens images discovered (Walsh et al., *Nature* **279**, 381–384, 31 May 1979). In this system, a lensing galaxy about 3.7×10^9 ly from the earth creates two images of a quasar about 8.7×10^9 ly from the earth. The images are (roughly) on opposite sides of the lensing galaxy, about 5.0 as and 1.0 as from the galaxy's center, respectively. Let us (naïvely) assume that we can treat the lensing galaxy as a point mass.
a. Determine the values of β and θ_E for this situation.
b. Find the mass of the lensing galaxy in solar masses.

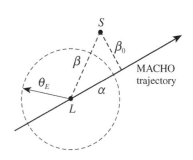

FIG. 13.7 The path of a MACHO L as it passes near a light source S, as an observer might project it all against the sky.

P13.4 Consider the solutions $\theta_\pm = \frac{1}{2}\left(\beta \pm \sqrt{\beta^2 + 4\theta_E^2}\right)$ (see equation 13.14) for the angles of the images created by a gravitational lens. It is clear from this that as β becomes small, the image angles become $\theta_\pm = \pm\theta_E$, meaning that they approach the Einstein ring.
a. What happens in the limit where β becomes large compared to θ_E? Express both image angles in terms of a power series in θ_E/β but use the binomial approximation to keep only the leading nonzero power of θ_E/β.
b. Express the brightness ratio $I_\pm/I_s$ for these images in this limit, keeping only the leading nonzero power of θ_E/β in your calculation. Does your result make sense?
c. Show in the case of starlight deflected by the sun that the outer image is deflected through an angle of about $4GM_\odot/R_\odot$, where $M_\odot$ and $R_\odot$ are the mass and radius of the sun. (Hints: Argue that $D_{LS} \approx D_S \gg D_L$ in this case, and that $\beta \approx R_\odot/D_L$.)
d. Explain why we can't see the second image in the case of starlight deflected by the sun.

P13.5 As a MACHO moves in front of a star, the star will brighten significantly only when the angular separation β between the MACHO and the star is of the order of magnitude of θ_E, the Einstein ring angle. We can therefore estimate the duration of the "blip" in brightness by estimating how long it takes the MACHO to move an angular distance of order of magnitude θ_E.
a. For a solar-mass MACHO at a distance $D_L \sim 30{,}000$ ly (roughly our distance from the galactic center) moving at a typical galactic orbital speed of $v \sim 200$ km/s, what is this time? (Make a plausible estimate for D_S.) Compare with the blip's width in figure 13.6.
b. Using the estimates for D_L and v given in part *a*, estimate the mass of the MACHO involved in the detection event recorded in figure 13.6.

P13.6 Figure 13.7 (above) shows a MACHO L (as we would view it projected on the sky) traveling along a straight line that takes it within an angle β_0 of a distant source S. At any given time, the angle between L and S is β, and the angular distance between L and its point of closest approach is α. Because the angles involved here are all very small, we can model the patch of sky involved as a flat surface and treat the angles as distances

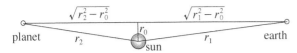

FIG. 13.8 A Newtonian diagram of an experiment to measure the Shapiro delay.

that obey the Pythagorean theorem: $\beta^2 = \alpha^2 + \beta_0^2$. Assume that the MACHO moves at a steady angular speed of $d\alpha/dt = \theta_E/t_E$ across the sky, where $t_E \equiv$ the time required for the MACHO to move an angular distance of θ_E, and define $t = 0$ to be the time of the MACHO's closest approach to S. Also define $u_0 \equiv \beta_0/\theta_E$ (this is the angle of closest approach in units of θ_E).

a. Show that the magnification of the source's brightness as a function of time is

$$\frac{I_{\text{tot}}}{I_s} = \frac{1}{2}\left[\frac{1}{q(t)} + q(t)\right]$$

$$\text{where } q(t) \equiv \sqrt{1 + \frac{4}{(t/t_E)^2 + u_0^2}} \quad (13.24)$$

b. Use a computer graphing tool to show what this function looks like as a function of t/t_E for various values of u_0. Submit a printout of your results.

c. Use your graphing tool to estimate the values of u_0 and t_E for the event shown in figure 13.6.

P13.7 A gravitating object not only *bends* a light beam passing near the object, it also *delays* it, an effect physicists call the **Shapiro delay,** after the physicist who first measured this effect in the solar system by bouncing radar signals off of Mercury and Venus as they passed behind the sun (Shapiro et al., *Phys. Rev. Lett.* **20**, 1968, 1265–1269). Recent measurements using the Cassini spacecraft (Bertotti et al., *Nature* **425**, 2003, 374–376) have verified the delay predicted by general relativity to about 3 parts in 10^5.

A typical experiment for measuring the Shapiro delay in the solar system might go as follows. A transmitter on a planet at Schwarzschild radial coordinate r_2 sends a signal to the earth at Schwarzschild radial coordinate r_1 at such a time that the signal passes within a Schwarzschild radial coordinate r_0 of the center of the sun. If Newtonian physics were true, the signal would follow a straight path as shown in figure 13.8, and the light travel time along this path would be the same as the length of this line, which by the Pythagorean theorem is $\sqrt{r_1^2 - r_0^2} + \sqrt{r_2^2 - r_0^2}$. However, in general relativity, the path is not *quite* straight, space is not *quite* flat (meaning the Pythagorean theorem does not quite apply), and the coordinate speed of light is not *quite* exactly 1 during the whole trip, so this Newtonian analysis is not *quite* right.

We can do a correct analysis (according to general relativity) by solving equation 12.3 for dr/dt, and then integrating the result to find the time that it takes light to get from r_0 to a general coordinate r.

a. First, use equation 12.3 to show that if we define r_0 to be the radial coordinate of the signal's closest approach to the sun, then

$$b = r_0\left(1 - \frac{2GM}{r_0}\right)^{-1/2} \quad (13.25)$$

b. Use this to eliminate b in equation 12.3 and show that

$$dt = \frac{r\,dr}{\left(1 - \frac{2GM}{r}\right)\sqrt{r^2 - r_0^2 \frac{1 - 2GM/r}{1 - 2GM/r_0}}} \quad (13.26)$$

c. If we could integrate this from $r = r_0$ to r, we could calculate the time $t(r, r_0)$ that it takes a light signal to go from radial coordinates r_0 to r. Since this also has to be the time the signal takes to go from r to r_0, the total coordinate time t required for the signal to go from the planet to the earth in figure 13.8 would then simply be $t(r_2, r_0) + t(r_1, r_0)$. Sadly, we cannot analytically integrate equation 13.26 in its present form. However, in this situation, $2GM/r \ll 1$, so we can use this to greatly simplify the integral. However, we don't want to simplify it *too* much. Show that if we completely ignore the $2GM/r$ and $2GM/r_0$ terms, we simply get $t(r, r_0) = \sqrt{r^2 - r_0^2}$, which is the Newtonian result.

d. So let's instead keep terms to first order in $2GM/r$ and $2GM/r_0$. Use the binomial approximation inside the square root, define $u \equiv r/r_0$, and rearrange things to get

$$dt \approx \frac{r_0 u\,du}{\sqrt{u^2 - 1}\left(1 - \frac{2GM}{r_0 u}\right)\sqrt{1 - \frac{2GM}{r_0 u(u+1)}}} \quad (13.27)$$

to first order in $2GM/r$ and $2GM/r_0$.

e. Note that $u \geq 1$ always, so both terms involving $2GM$ in the above are less than or equal to $2GM/r_0$ and so are much smaller than 1. Therefore, you can use the binomial approximation to convert the binomials involving these terms to binomials in the numerator. Multiplying out those binomials and keeping only terms to first order in GM/r converts the single expression in equation 13.27 into the sum of three simpler expressions, which we can integrate separately. Find those three expressions, evaluate the integrals from $u = 1$ to arbitrary u (looking up integrals if necessary), and substitute $u \equiv r/r_0$ back into the results to show that

$$t(r, r_0) \approx \sqrt{r^2 - r_0^2} + 2GM\ln\left(\frac{r + \sqrt{r^2 - r_0^2}}{r_0}\right)$$

$$+ GM\sqrt{\frac{r - r_0}{r + r_0}} \quad (13.28)$$

Note that the first term is the Newtonian result and the two correction terms are positive.

f. Assume that the planet is Venus ($r_2 = 0.723$ AU). Calculate the *extra* time (above the Newtonian result) required for a signal to get from Venus to earth after passing close to the sun's surface ($r_0 = 696{,}000$ km). Your result should be on the order of a hundred microseconds.

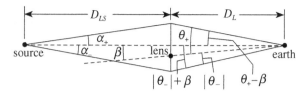

FIG. 13.9 Schematic diagram for determining the difference in the two pathlengths for light traveling to the earth from a lensed source. I have hugely exaggerated the angles for clarity. I have also defined D_L, D_{LS}, and the points where each beam bends slightly differently here than in figure 13.4, but as the angles are so small, the differences are negligible.

P13.8 Consider a situation where light from a distant quasar is bent by a nearer galaxy into two images. Variations in the quasar's light output may appear in the images at different times, because the time required for light to travel along each path to the earth may not be the same. Determining the time required for light to follow either bent path involves calculating the "Shapiro delay" for that path (see problem P13.7), which is partly due to the increased geometrical length of the bent path compared to a straight path, and partly due to effects related to to the curvature of spacetime near the lensing galaxy. For the sake of simplicity, let's neglect the latter. Figure 13.9 shows a simplified diagram of the two paths that the quasar light follows to get to the earth. We will assume that all angles are very small.

a. Argue that

$$\alpha_+ \approx (\theta_+ - \beta)\frac{D_L}{D_{LS}} \quad \text{and} \quad \alpha_- \approx (|\theta_-| + \beta)\frac{D_L}{D_{LS}} \quad (13.29)$$

b. Show that

$$\theta_+ - \beta = \theta_0 - \tfrac{1}{2}\beta \quad \text{and} \quad |\theta_-| + \beta = \theta_0 + \tfrac{1}{2}\beta$$

where $\theta_0 \equiv \tfrac{1}{2}\sqrt{\beta^2 + 4\theta_E^2}$ (13.30)

(Note that I have written $|\theta_-|$ instead of θ_- because the latter is technically defined to be negative in equation 13.14, and we need the *magnitude* of that angle here.)

c. Show that if we keep terms only through the order of the angles squared, the *difference* between the pathlengths corresponding to the two images is

$$\Delta D = D_L\left(1 + \frac{D_L}{D_{LS}}\right)\beta\theta_0 \quad (13.31)$$

(*Hint:* Expand cosines in a power series, and use the binomial approximation.)

d. The "double quasar" object named Q0957+561 is actually two images of a single quasar 8.7 Gy from earth created by a lensing galaxy 3.7 Gy from the earth. For this system, $\beta \approx 1.8\theta_E$ and $\theta_E = 1.08 \times 10^{-5}$ radians. Calculate ΔD for this situation and compare with the measured delay of 417 ± 3 days. (*Hints:* Don't expect very good agreement here, as our model is far too crude to yield a trustworthy prediction. Also note that 1 Gy $\equiv 10^9$ y = 10^9 ly in the GR unit system.)

P13.9 Thinkers associated with Project Icarus (www.icarusinterstellar.org/) have speculated about using the gravitational field of the sun as a giant lens to help astronomers image planets in nearby solar systems. In this problem, we will explore that possibility. The idea is to place a satellite in a position where it would see light from a distant object imaged as an Einstein ring around the sun. The immense size of the sun as a lens would mean a huge magnification of the object's apparent brightness as well as extraordinary resolution (though the image would have to be processed by a computer to remove the distortion created by the gravitational lens).

a. Show that the closest that the satellite could be to the sun for the sun's gravitational field to focus light from a distant object on the satellite is about 550 AU. (For comparison, note that Pluto's average distance from the sun is 39.5 AU.) Also carefully explain why we could (and maybe would *want* to) locate the satellite significantly *farther* from the sun, but not closer. Note that the radius of the sun is 696,000 km.

b. Calculate the Einstein angle θ_E for this system, assuming that the satellite is at 550 AU and we are imaging something essentially infinitely far away. Compare with the angular radius of the sun as seen from the satellite.

c. Imagine that we want to image a point source on a planet 1 AU from Alpha Centauri, which is 4.3 ly away. Calculate the distance to Alpha Centauri in AU, and compute the angular separation δ between the star and planet as viewed from the satellite. Compare with θ_E.

d. Equation 13.18 predicts that the intensity magnification goes to infinity as $\beta \to 0$ (though only for an ideal point source, as our derivation assumed that the source's angular width $\Delta\beta$ is much smaller than β). Assuming that we are able to position our spacecraft to within at least 1 km of the ideal position, determine the minimum intensity magnification for an ideal point source at the planet's position.

e. Even if we are focused on the planet, we will also get some interfering light from the star. Treating the star as a point source, argue that there will be only one image visible, and compute the intensity magnification for that image. Considering that the star is likely to be billions of times brighter than the planet, will we have to come up with some scheme to shield our detector from the star's image to see the planet's?

f. Argue that a 1-km shift of our spacecraft's position corresponds to a shift of about 500 km of the ideally imaged ($\beta = 0$) spot on the planet. Argue, therefore, that most of the planet's light will *not* be magnified by the same factor as calculated in part d (that is, that the planet is not really an ideal point source).

14. EVENT HORIZON

INTRODUCTION

FLAT SPACETIME
- Review of Special Relativity
- Four-Vectors
- Index Notation

TENSORS
- Arbitrary Coordinates
- Tensor Equations
- Maxwell's Equations
- Geodesics

SCHWARZSCHILD BLACK HOLES
- The Schwarzschild Metric
- Particle Orbits
- Precession of the Perihelion
- Photon Orbits
- Deflection of Light
- **Event Horizon**
- Alternative Coordinates
- Black Hole Thermodynamics

THE CALCULUS OF CURVATURE
- The Absolute Gradient
- Geodesic Deviation
- The Riemann Tensor

THE EINSTEIN EQUATION
- The Stress-Energy Tensor
- The Einstein Equation
- Interpreting the Equation
- The Schwarzschild Solution

COSMOLOGY
- The Universe Observed
- A Metric for the Cosmos
- Evolution of the Universe
- Cosmic Implications
- The Early Universe
- CMB Fluctuations & Inflation

GRAVITATIONAL WAVES
- Gauge Freedom
- Detecting Gravitational Waves
- Gravitational Wave Energy
- Generating Gravitational Waves
- Gravitational Wave Astronomy

SPINNING BLACK HOLES
- Gravitomagnetism
- The Kerr Metric
- Kerr Particle Orbits
- Ergoregion and Horizon
- Negative-Energy Orbits

this ➡ depends on ➡ this

14. EVENT HORIZON

Focus on Black Holes. Let me emphasize again that the Schwarzschild metric applies only in the vacuum *outside* a spherical and static gravitating object, not *inside* such an object. Most gravitating objects (such as normal stars and galaxies) have surface r-coordinates that are large compared to their Schwarzschild radii $r_s = 2GM$, so we will not observe for such objects the strongly non-Newtonian behaviors that we have discussed that happen at small r-coordinates. In this and the next few chapters, however, we will focus specifically on the physics of *black holes*, i.e., objects which do not have a surface outside of $2GM$. Black holes display most vividly the differences between general relativity and Newtonian gravitational theory. In this chapter, we will focus specifically on the strange physics of the surface at $r = 2GM$ that we call the Schwarzschild spacetime's *event horizon*.

A Catalog of Pathologies at $r = 2GM$. The Schwarzschild metric is

$$ds^2 = -\left(1 - \frac{2GM}{r}\right)dt^2 + \left(1 - \frac{2GM}{r}\right)^{-1}dr^2 + r^2 d\theta^2 + r^2 \sin^2\theta \, d\phi^2 \quad (14.1)$$

We also saw in chapter 9 (see equation 9.12b) that the light emitted at r-coordinate r_E and received at r_R is red-shifted or blue-shifted according to

$$\frac{\lambda_R}{\lambda_E} = \sqrt{\frac{1 - 2GM/r_R}{1 - 2GM/r_E}} \quad (14.2)$$

Table 14.1 summarizes the equations of motion for both ordinary and massless particles in the equatorial plane. It is clear that strange things happen at the Schwarzschild radius $r_s = 2GM$ in almost all of the equations. Specifically, we observe the following pathologies:

1. A clock at rest at $r = 2GM$ registers *no* time (i.e., when $r = 2GM$ and $dr = 0$, $d\theta = 0$, and $d\phi = 0$, then $d\tau^2 = (1 - 2GM/r) dt^2 = 0$).
2. (Related to this,) light emitted from rest at $r = 2GM$ is infinitely red-shifted when observed at any (larger) radius. Moreover, the redshift formula simply does not work for $r < 2GM$.
3. All particles (even photons) falling inward appear to a distant observer to "freeze" at $r = 2GM$ (see the expressions for dr/dt and $d\phi/dt$ in table 14.1).
4. Worst of all, g_{rr} goes to infinity at $r = 2GM$, implying that the derivative ds/dr of the radial distance s with respect to radial coordinate r diverges.

Possible Roots of These Pathologies. These pathologies (particularly the infinities) signal that something bad is going on at $r = 2GM$. There are two possible explanations for what is going wrong:

TABLE 14.1 This table summarizes the equations of motion for particles in the Schwarzschild equatorial plane.

	Particles with mass $m > 0$	Particles with mass $m = 0$ (e.g., photons)
Conserved quantities	$e = \left(1 - \frac{2GM}{r}\right)\frac{dt}{d\tau}, \quad \ell = r^2 \frac{d\phi}{d\tau}$	$b \equiv \frac{\ell}{e} = \frac{r^2}{(1 - 2GM/r)}\frac{d\phi}{dt}$
Equation for $d\tau^2$	$d\tau^2 = \left(1 - \frac{2GM}{r}\right)dt^2 - \left(1 - \frac{2GM}{r}\right)^{-1}dr^2 - r^2 d\phi^2$	$0 = \left(1 - \frac{2GM}{r}\right)dt^2 - \left(1 - \frac{2GM}{r}\right)^{-1}dr^2 - r^2 d\phi^2$
Equation for $dr/d\tau$	$\Rightarrow \frac{dr}{d\tau} = \pm\sqrt{e^2 - \left(1 - \frac{2GM}{r}\right)\left(1 + \frac{\ell^2}{r^2}\right)}$	$\Downarrow$
Equation for dr/dt	$\Rightarrow \frac{dr}{dt} = \pm\left(1 - \frac{2GM}{r}\right)\sqrt{1 - \frac{1}{e^2}\left(1 - \frac{2GM}{r}\right)\left(1 + \frac{\ell^2}{r^2}\right)}$	$\Rightarrow \frac{dr}{dt} = \pm\left(1 - \frac{2GM}{r}\right)\sqrt{1 - \left(1 - \frac{2GM}{r}\right)\frac{b^2}{r^2}}$
Equation for $d\phi/dt$	$\Rightarrow \frac{d\phi}{dt} = \frac{\ell}{er^2}\left(1 - \frac{2GM}{r}\right)$	$\Rightarrow \frac{d\phi}{dt} = \frac{b}{r^2}\left(1 - \frac{2GM}{r}\right)$

1. The spacetime has a *geometric* pathology at $r = 2GM$.
2. The Schwarzschild *coordinate system* is broken at $r = 2GM$.

A *geometric* pathology occurs when the physical characteristics of the spacetime are such that we cannot describe it at all using the mathematics we have developed. One of the most fundamental assumptions we made in chapter 5 was that our spacetime was not so horribly curved that we could not model a sufficiently small patch around any point as being flat. The apex of a cone is an example of a geometric pathology: since we cannot model even an infinitesimal region around the apex as being flat, our mathematics breaks down and no coordinate system will adequately describe the surface of the cone at that point.

On the other hand, a *coordinate* pathology occurs when the underlying geometry of the spacetime is perfectly reasonable but we happen to be using a coordinate system that describes that geometry poorly at one or more events or locations. For example, the latitude-longitude coordinate system on the surface of a sphere exhibits coordinate pathologies at the poles, because the $g_{\phi\phi}$ component of the metric $ds^2 = R^2 d\theta^2 + R^2 \sin^2\theta \, d\phi^2$ goes to zero there. This is not because the pole is geometrically different than any other location on the spherical surface, but rather because in the coordinate system, all the lines of longitude come together at the poles, meaning that the pole has no well-defined ϕ coordinate.

This kind of pathology seems to be analogous to the way that the Schwarzschild metric component g_{tt} goes to zero at $r = 2GM$; by analogy, we might conclude that we cannot assign well-defined t coordinates to events at $r = 2GM$ due to some problem with the way the coordinate system is defined. But the fact that $g_{rr} \to \infty$ at that radius *seems* qualitatively different.

Evidence for a Coordinate Pathology. But is it really so? First of all, note that even though $g_{rr} \to \infty$ at $r = 2GM$, it does not necessarily follow that the physical distance between a given initial r-coordinate R and $r = 2GM$ is infinite (though an infinite physical distance to $r = 2GM$ from any point outside *would* be strong evidence of a geometric pathology). In fact, you can show (see box 14.1) that for a purely radial displacement, the physical distance between $r = R$ and $r = 2GM$ is well defined and equal to

$$\Delta s = R\sqrt{1 - \frac{2GM}{R}} + 2GM \tanh^{-1}\sqrt{1 - \frac{2GM}{R}} \qquad (14.3)$$

For example, the total physical distance between $r = 3GM$ and $r = 2GM$ turns out to be $3.05GM$, which is obviously finite.

You can also show (see box 14.2) that the proper time measured by a particle falling from rest at $r = R$ to $r = 0$ (passing through $r = 2GM$ on the way) is

$$\Delta \tau = \frac{\pi R^{3/2}}{\sqrt{8GM}} \qquad (14.4)$$

which is also obviously finite. For example, the proper time required to fall from $R = 1000GM$ (1500 km) to $r = 0$ for a solar-mass black hole is about 0.18 s.

The point is that we do *not* observe the kinds of problems that we would expect if there were a real geometric divergence at $r = 2GM$. This suggests that the problem is a *coordinate* pathology.

The Problem with the Time Coordinate. Let's see if we can figure out where the coordinate problem is. Consider again the discussion in chapter 9 of how we might set up a reference frame around a black hole to assign Schwarzschild coordinates to events. Imagine that we first construct a reference spherical surface with an extremely large radius centered on the black hole and mark it with a grid consisting of curves of constant latitude and longitude. At each grid intersection, we attach a perpendicular girder that we allow to hang radially toward the star, and then we construct (at various levels down toward the star) girders perpendicular to the radial girders that (when connected) form new spherical surfaces at various (smaller) r coordinates, thus creating a latticework of nested spheres and radial lines. Each intersection in this lattice is formed by a radial girder (corresponding to a specific value of θ and ϕ) meeting horizontal girders that are part of a spherical surface (whose r-coordinate we can determine by measuring the surface's equatorial circumference and dividing by 2π). At each lattice intersection, we can then put (1) a little plaque specifying its r, θ, and ϕ coordinates, (2) a clock (for measuring local time), and (3) a "t-meter" that is like a clock, except that we adjust its ticking rate by turning a knob until it matches the received rate of pulses sent at 1-second intervals by a master clock at infinity. The Schwarzschild coordinates of any event in the lattice can be registered by recording the r, θ, and ϕ coordinates of the nearest lattice intersection and the time indicated by the t-meter at that intersection.

However, this whole scheme breaks down at $r = 2GM$. First of all, as we climb down a radial girder toward $r = 2GM$, signals from the master clock at infinity become more and more blue-shifted, so we have to crank up t-meter rates higher and higher to match the master clock's rate. If we could get down to $r = 2GM$, this rate would become infinite compared to local physical clocks, so we could no longer crank up the t-meter high enough to match the master clock's signal rate. Thus we cannot assign meaningful t coordinates to events on this surface. This is a consequence of g_{tt} going to zero at $r = 2GM$.

Worse yet, we cannot even construct the lattice to $r = 2GM$ or inward. Note that the Schwarzschild metric implies that a hypothetical girder particle at rest ($dr = 0$, $d\theta = 0$, and $d\phi = 0$) at $r = 2GM$ measures zero proper time along its worldline:

$$d\tau^2 = -ds^2 = \left(1 - \frac{2GM}{r}\right)dt^2 + 0 + 0 + 0 = 0 \quad \text{at } r = 2GM \qquad (14.5)$$

This means that the girder particle's worldline is lightlike, and only zero-mass particles (such as photons) can follow lightlike worldlines. Therefore, while we can (in principle) construct fixed girders at any $r > 2GM$, no material girder particle (no matter how strong the girder is) can remain at rest at $r = 2GM$. So as we lower radial girders down toward the black hole, the girder's bottom end will inevitably rip apart and fall into the black hole before it reaches $r = 2GM$.

Time and Space Coordinates Inside $r = 2GM$. Indeed, *no* particle (even a photon) inside $r = 2GM$ can remain at constant r. As we will see in this section, this is because the *physical meaning* of the r and t coordinates changes inside $r = 2GM$.

Consider for the moment an arbitrary *diagonal* metric for spacetime in arbitrary coordinates, and consider a worldline along which one coordinate (call it the q coordinate) varies but the other three are constant. The sign of the corresponding diagonal metric component g_{qq} tells us whether q is a time coordinate or a spatial coordinate. If $g_{qq} < 0$, then $ds^2 < 0$ (and $d\tau^2 > 0$) along the worldline, so the worldline is *timelike* and q is a *time* coordinate. On the other hand, if $g_{qq} > 0$, then $ds^2 > 0$, so the worldline is *spacelike* and q is a *spatial* coordinate. Any *spacetime* will have three spatial coordinates and one time coordinate. There is also a physical distinction between space and time that is deeper even than relativity: a material object can move around freely in space, but only toward the future in time.

Now, for $r > 2GM$ in Schwarzschild coordinates, $g_{tt} < 0$ and g_{rr}, $g_{\theta\theta}$, and $g_{\phi\phi}$ are all positive, so the t coordinate is a time coordinate and the others are spatial coordinates. But for $r < 2GM$, the t and r coordinates reverse roles: $g_{tt} > 0$ and $g_{rr} < 0$, meaning that the r coordinate becomes a *time* coordinate, while t becomes a spatial coordinate. (Remember that the coordinate *names* mean nothing: the *only* physical meaning the coordinates have comes from the metric.) So a particle following a timelike or lightlike worldline inside $r < 2GM$ *cannot* remain at a fixed coordinate r any more than a particle in flat spacetime could remain at a fixed coordinate time t.

Even so, the Schwarzschild metric also implies that the coordinate r retains its original meaning as a *circumferential* coordinate even for $r < 2GM$. A surface of constant r and t still has the metric (and thus the geometry) of a spherical surface, and the value of r is still equal to a surface's equatorial circumference divided by 2π. We can still *imagine* (though we can no longer construct) a nested set of spherical surfaces of constant r and t inside $r = 2GM$, and these surfaces have the same geometrical meaning as they do outside that radius. It is simply that a particle can no longer remain at rest with respect to these surfaces, but rather must move *through* them to go forward in time.

Inside $r = 2GM$, the Future Is $r = 0$. Indeed, a particle must inevitably move inward toward $r = 0$. We can easily show that this is true for geodesics. As we saw in chapter 10, the geodesic equation for a massive particle implies that

$$\frac{d^2r}{d\tau^2} = -\frac{GM}{r^2} + \frac{\ell^2}{r^3}\left(1 - \frac{3GM}{r}\right) \tag{14.6}$$

(see equation 10.13). Note that nothing unusual happens to this equation at $r = 2GM$ and our derivation of it nowhere assumed that r was a spacelike coordinate, so it should apply for all r. For $r < 3GM$, the final quantity in parentheses is negative, so any freely-falling particle's radial acceleration will be negative, independent of the value of ℓ. So no matter how a freely-falling particle is moving initially, it will end up moving inward eventually. So if inside $r = 2GM$, a freely-falling particle is constrained to move in only one direction through nested constant-r surfaces, the only plausible direction is *inward*. Moreover, the future cannot lie in a different direction for freely-falling particles than for other particles, so *the future of any particle within the Schwarzschild radius is inward*, and the ultimate terminus of any particle's worldline will be $r = 0$.

Box 14.3 shows that the longest possible proper time that *any* object can survive once it passes $r = 2GM$ is πGM, no matter how it might struggle. So not only will any object's future terminate at $r = 0$, it will inevitably end a relatively short time after passing the event horizon.

The Event Horizon. We see that the surface where $r = 2GM$ functions as a one-way membrane. A spaceship at a given r outside this surface might (with sufficient expenditure of fuel) return to larger r, but any spaceship or other object that crosses this boundary will move inward with the same inevitability as moving forward in time. Timelike or lightlike worldlines can cross this surface going inward, but no such worldline can emerge from smaller r going outward. We call such a one-way surface an **event horizon**.

The Problem with Schwarzschild Coordinates. The core problem with the Schwarzschild coordinate system is that it does not deal well with the existence of this event horizon. The key advantage of Schwarzschild coordinates is that they become ordinary space and time coordinates at large r. But this convenience is also their downfall. When we define the coordinate t to be time as registered by a clock at infinity, we have chosen a definition that we cannot apply even in principle to events inside the event horizon: since it is *impossible* for an observer at infinity (or even just outside the event horizon) to receive information from such events, the observer cannot meaningfully assign a coordinate time t to those events. It is also very problematic to have coordinates that change from spatial to time coordinates (or vice versa), as the very method of measuring coordinate displacements must change (we measure spacelike displacements with a ruler, but time displacements with a clock). This means that we can never come up with a single physical definition for a coordinate that changes meaning this way.

The pathologies we listed at the beginning of this chapter *all* ultimately have their origin in these problems with the Schwarzschild-coordinate definitions. The ultimate solution to this problem is to find coordinates that are well defined both inside and outside the event horizon and which retain their spacelike or timelike identity everywhere. In the next chapter, we will discuss two different coordinate systems that do just that.

14. EVENT HORIZON

BOX 14.1 Finite Distance to $r = 2GM$

According to the Schwarzschild metric, the differential physical distance ds corresponding to a purely radial differential displacement dr is

$$ds = \frac{dr}{\sqrt{1 - 2GM/r}} \quad \Rightarrow \quad \Delta s = \int \frac{dr}{\sqrt{1 - 2GM/r}} \tag{14.7}$$

This is not an integral that one can look up directly in most integral tables (and WolframAlpha gave me something really ugly). But let's define

$$u \equiv \sqrt{1 - \frac{2GM}{r}} \quad \Rightarrow \quad r = \frac{2GM}{1 - u^2} \tag{14.8}$$

Exercise 14.1.1. Verify that the last equality is correct.

With this substitution, the total distance from radial coordinate $r = 2GM$ to $r = R$ is

$$\Delta s = 4GM \int_0^{u(R)} \frac{du}{(1 - u^2)^2} \tag{14.9}$$

Exercise 14.1.2. Check that this is correct.

BOX 14.1 (continued) Finite Distance to $r = 2GM$

This integral is easy to find in a good printed table of integrals: the indefinite integral (from Dwight, *Tables of Integrals and Other Mathematical Data*, 4th edition, Macmillan, item 140.2) is

$$\int \frac{du}{(1-u^2)^2} = \frac{u}{2(1-u^2)} + \frac{1}{4}\ln\left|\frac{1+u}{1-u}\right| \tag{14.10}$$

Using this and the identity (Dwight, item 702),

$$\tanh^{-1} u = \frac{1}{2}\ln\left|\frac{1+u}{1-u}\right| \tag{14.11}$$

you can easily show that $\Delta s = R\sqrt{1 - 2GM/R} + 2GM\tanh^{-1}\sqrt{1 - 2GM/R}$, which is equation 14.3 (as claimed).

Exercise 14.1.3. Check that this is correct.

Exercise 14.1.4. Check that the physical distance from $r = 3GM$ to $r = 2GM$ is indeed $3.05GM$. (If your calculator cannot handle inverse hyperbolic functions, use equation 14.11 to eliminate $\tanh^{-1} u$.)

BOX 14.2 Proper Time for Free Fall from $r = R$ to $r = 0$

Consider a freely-falling object that is initially at rest on the equatorial plane at $r = R$. An object initially at rest has no angular momentum (implying that $\ell = 0$) and is not moving radially (meaning $dr/d\tau = 0$), so according to the equation for $dr/d\tau$ in table 14.1, the energy-per-unit-mass e for such a particle must be

$$0 = \left(\frac{dr}{d\tau}\right)^2 = e^2 - \left(1 - \frac{2GM}{R}\right)\left(1 + \frac{0^2}{R^2}\right) \quad\Rightarrow\quad e^2 = 1 - \frac{2GM}{R} \quad (14.12)$$

The same equation then implies that at *other* radii during the object's fall,

$$\frac{dr}{d\tau} = -\sqrt{\left(1 - \frac{2GM}{R}\right) - \left(1 - \frac{2GM}{r}\right)} = -\sqrt{\frac{2GM}{r} - \frac{2GM}{R}} \quad (14.13)$$

Therefore, the elapsed proper time required to go from $r = R$ to zero is

$$\Delta\tau = \int_R^0 \frac{-dr}{\sqrt{2GM/r - 2GM/R}} = \sqrt{\frac{R}{2GM}} \int_0^R \frac{r^{1/2} dr}{\sqrt{R - r}} \quad (14.14)$$

where in the last step I reversed the integral's endpoints (thus reversing its sign) and multiplied top and bottom by $\sqrt{rR/2GM}$. One might be able to find this integral in a table, but you may have better luck defining $u^2 \equiv r \Rightarrow dr = 2u\,du$ and $u_0^2 \equiv R$. With this substitution, the integral becomes

$$\Delta\tau = 2\sqrt{\frac{R}{2GM}} \int_0^{u_0} \frac{u^2 du}{\sqrt{u_0^2 - u^2}} \quad (14.15)$$

Exercise 14.2.1. Look either integral up in a table, evaluate the result at the endpoints, and thus show that $\Delta\tau = (\pi/2)\sqrt{R^3/2GM}$ (equation 14.4).

BOX 14.3 The Future Is Finite Inside the Event Horizon

Inside the event horizon, we can rewrite the Schwarzchild metric as follows:

$$d\tau^2 = -ds^2 = \left(\frac{2GM}{r} - 1\right)^{-1} dr^2 - \left(\frac{2GM}{r} - 1\right) dt^2 - r^2(d\theta^2 + \sin^2\theta \, d\phi^2) \quad (14.16)$$

Note that the quantities inside the parentheses are now all positive for $r < 2GM$.

Now consider an arbitrary infinitesimal segment of an arbitrary worldline inside the event horizon for an object with mass $m > 0$. Since r is the time coordinate inside the event horizon, dr^2 must be nonzero for any timelike worldline. If dt, $d\theta$, or $d\phi$ are nonzero for a given segment, the resulting proper time along that segment will be shorter than if $dt = d\theta = d\phi = 0$. So the worldline with longest proper time between any two given r-coordinate values will be a *purely radial* worldline. (This is analogous to the fact that in flat spacetime, the longest proper time between two given t coordinates in a given IRF will be measured by an object that moves purely in the t direction, i.e., an object at rest with respect to the other coordinates.)

Moreover, this radial worldline *must* be a geodesic. Here's the argument. Consider now a pair of events A and B along a purely radial worldline line, and consider other worldlines that go through these events but which wander away from the radial worldline between A and B. Again, equation 14.16 implies that the proper time measured along any infinitesimal segment of such a wandering worldline will be smaller than (or at best equal to) the corresponding purely radial segment with the same dr. Therefore, the purely radial worldline is the worldline of *longest proper time* between the events A and B, which is the *definition* of a geodesic. Q. E. D.

This means that if you fall inside a black hole's event horizon, you will survive longest if you use your rockets to kill any speed that you might have in the t, θ, or ϕ directions, and then give up struggling and fall freely toward $r = 0$. Though it seems counter-intuitive, trying to escape by firing your rockets in any direction will only hasten your inevitable demise!

You will also live longest by maximizing the distance that you freely drop, i.e., by starting with $dt = d\theta = d\phi = 0$ as soon as possible after $r = 2GM$. So, according to equation 14.16 (and recognizing that $dr/d\tau$ is negative) the *absolute maximum* proper time that any object can survive inside the event horizon is given by

$$\Delta\tau = \int_{2GM}^{0} \frac{-dr}{\sqrt{2GM/r - 1}} = \int_{0}^{2GM} \frac{r^{1/2} dr}{\sqrt{2GM - r}} \quad (14.17)$$

Exercise 14.3.1. Evaluate this integral using whichever method you used in exercise 14.2.1 and show that the answer is πGM.

HOMEWORK PROBLEMS

P14.1 You are in a spaceship originally at rest at $r = 10GM$ near a supermassive black hole at the center of a galaxy. Assume that the black hole has a mass equal to a million solar masses. If your engines fail and you begin to fall radially inward, about how long do you have to live (as measured by your own watch)?

P14.2 Imagine that you are at rest at $r = 16GM$ near a supermassive black hole whose mass is a million solar masses. You launch a clock radially outward at such a speed so it comes to rest at $r = 32GM$. The clock subsequently falls radially back into the black hole.
  a. How much time does the launched clock register between its launch and the time that it comes to rest?
  b. How much time does it register between coming to rest and crossing the event horizon?
  c. How much time does it register between crossing the event horizon and its destruction at the origin?

P14.3 As discussed in the chapter, a falling object appears to a distant observer to "freeze" at the event horizon. This, however, is purely an artifact of the coordinates. Consider an inward-falling object moving in the equatorial plane with arbitrary e and ℓ. Show that an observer at rest at a given R will measure its squared speed to be

$$v_{\text{obs}}^2 = 1 - \frac{1}{e^2}\left(1 - \frac{2GM}{R}\right) \qquad (14.18)$$

Note that this approaches 1 (the speed of light) as R approaches $2GM$, no matter what the particle's energy-per-unit-mass-at-infinity e might be. So compared to any *local* observer, the particle does *not* "freeze." [*Hints:* Use the equations in table 14.1 to evaluate the particle's four-velocity $\boldsymbol{u} = [dt/d\tau, dr/d\tau, 0, d\phi/d\tau]$ at R for arbitrary e and ℓ. Then use the techniques discussed in chapter 12 to evaluate the components of $\boldsymbol{u}$ in the orthonormal coordinate system at rest at R, and finally use $v_{\text{obs},x} = u^x/u^t$ and $v_{\text{obs},z} = u^z/u^t$ in that frame.]

P14.4 Figure 14.1 shows a spacetime diagram for Schwarzschild coordinates t and r. At $r = 4GM$, I have drawn a light cone for a particle with nonzero rest mass moving radially. I have determined this cone as follows. The proper time measured by such a particle must be real, so according to the Schwarzschild metric equation, a particle moving radially ($d\theta = d\phi = 0$) must satisfy

$$0 < d\tau^2 = -ds^2 = \left(1 - \frac{2GM}{r}\right)dt^2 - \frac{dr^2}{(1 - 2GM/r)}$$

$$\Rightarrow 0 < \tfrac{1}{2}dt^2 - 2\,dr^2 \Rightarrow dt^2 > 4\,dr^2 \qquad (14.19)$$

when $r = 4GM$. Outside the event horizon, we know that the forward direction in proper time corresponds to

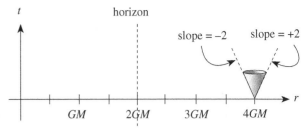

FIG. 14.1 This drawing shows the light cone for particles with nonzero rest mass moving radially at $r = 4GM$. Worldlines for such particles must lie within this cone.

increasing coordinate time t, so dt must be positive and the slope dt/dr of an outgoing object's worldline on the diagram must be more vertical than 2, and that for an ingoing particle must be more vertical than –2.

Draw analogous (and quantitatively accurate) light cones at $r = 3GM, \tfrac{5}{2}GM, \tfrac{3}{2}GM, GM$, and $\tfrac{1}{2}GM$. Also describe in words what happens to the light cones as we approach $r = 2GM$ from either side. (Note that for $r < 2GM$, forward in proper time corresponds to $dr < 0$.)

P14.5 Imagine that an observer at rest at a given $R > 2GM$ launches an object with nonzero mass radially outward with an initial velocity v_0 *as measured by the observer*. Show that the escape speed as measured by the observer is the same as in Newtonian mechanics for a Newtonian radius equal to R, and show that this speed approaches the speed of light as $R \to 2GM$. [*Hints:* First find the value of e that a particle must have to just barely make it to rest at infinity. Then use the equations in table 14.1 evaluate the particle's four-velocity $\boldsymbol{u} = [dt/d\tau, dr/d\tau, 0, 0]$ at R for that e. Then use the techniques discussed in chapter 12 to evaluate the components of $\boldsymbol{u}$ in an orthonormal coordinate system at rest at R, and finally argue that $v_0 = u^z/u^t$ in that frame.]

P14.6 Imagine an observer falling from rest at infinity into a black hole. Imagine that this observer measures the wavelength of light passing on its way radially inward. In problem P12.7, you should have found that the observer will see this light to be red-shifted by the factor

$$\frac{\lambda_r}{\lambda_e} = 1 + \sqrt{\frac{2GM}{r}} \qquad (14.20)$$

Even though in that problem, we derived the expression assuming a falling observer outside the event horizon, a falling observer inside the event horizon should still be able to receive an incoming photon, and this equation exhibits no bizarre behavior at $r = 2GM$ (unlike the corresponding formula for the blue-shift observed by an observer at rest). Go over the argument in that problem again, this time using the metric in the form given by equation 14.16, and verify that this expression is correct.

P14.7 (Based on a discussion in Rindler, *Relativity: Special, General, and Cosmological*, 2nd edition, Oxford, 2006, pp. 267–272.) In this problem, we will see that even in a completely flat spacetime, bad coordinates can lead to weird problems similar to those seen in Schwarzschild coordinates. Consider an ordinary flat spacetime labeled in some inertial reference frame by cartesian coordinates t, x, y, and z. Consider new coordinates R and T for this flat spacetime that are defined implicitly by the equations

$$t(T,R) = \pm b\sqrt{2R-1}\,\sinh T \quad (14.21a)$$
$$x(T,R) = \pm b\sqrt{2R-1}\,\cosh T \quad (14.21b)$$

when $R > \tfrac{1}{2}$, and

$$t(T,R) = \pm b\sqrt{1-2R}\,\cosh T \quad (14.21c)$$
$$x(T,R) = \pm b\sqrt{1-2R}\,\sinh T \quad (14.21d)$$

when $R < \tfrac{1}{2}$, where b is a positive constant with units of meters, R and T are unitless, and we choose the signs in front of the expressions for t and x to be either both positive or both negative together. Assume that y and z are the same in both coordinate systems.

a. Argue that in a spacetime diagram based on t and x coordinates, curves of constant R are hyperbolas (i.e., they satisfy the relation $t^2 - x^2 = $ constant) and that they face rightward or leftward when $R > \tfrac{1}{2}$ (depending on our choice of sign) and upward or downward when $R < \tfrac{1}{2}$ (depending on our choice of sign). (*Hint:* Note that $\cosh^2 q - \sinh^2 q = 1$ for all q.)

b. Argue that curves of constant T are straight lines with constant slope whose magnitude is less than 1 when $R > \tfrac{1}{2}$ and greater than 1 when $R < \tfrac{1}{2}$. [*Hint:* $\tanh q \equiv \sinh q / \cosh q \equiv \tfrac{1}{2}(e^q - e^{-q}) / \tfrac{1}{2}(e^q + e^{-q})$.]

c. Figure 14.2 shows the constant-coordinate curves for the T, R coordinate system. Show that the metric for this coordinate system is *everywhere* given by

$$ds^2 = -(2R-1)b^2\,dT^2 + \frac{b^2\,dR^2}{2R-1} + dy^2 + dz^2 \quad (14.22)$$

Be sure to consider all regions of the graph.

d. Argue that when $R > \tfrac{1}{2}$, R is a spatial coordinate and T is a time coordinate, but when $R < \tfrac{1}{2}$, the situation is reversed, and that the "future" for all particles in the upper of these regions is toward smaller R.

e. Use figure 14.2 to argue that the positive-x half of the line where $R = \tfrac{1}{2}$ and $T = \infty$ is an event horizon; i.e., particles can cross from larger R to smaller R but not the reverse. Also, why do we consider this to be an "event horizon" in T, R coordinates but not in t, x coordinates?

f. Use the metric equation to argue that photons are at rest at $R = \tfrac{1}{2}$, but that no particle with nonzero rest mass can be at rest at that position. Argue that this is consistent with the spacetime diagram in figure 14.2.

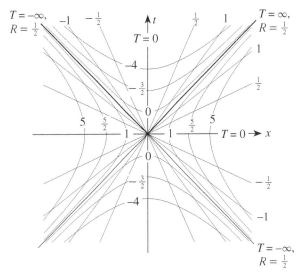

FIG. 14.2 This is a spacetime diagram of the T, R coordinate system for ordinary flat spacetime discussed in problem P14.7. I have labeled each of the hyperbolas with their corresponding value of R and each line of constant T with its corresponding value of T, using the symbols R and T only where there might be ambiguity. Note that the extra-dark $R = \tfrac{1}{2}$ light-flash worldlines flanking the upper quarter of the graph are event horizons in the T, R coordinate system: particles may move *into* the upper region where $R < \tfrac{1}{2}$, but they cannot escape.

g. Solve the T component of the geodesic equation to show that $dT/d\tau = eb^{-2}(2R-1)^{-1}$, where e is a constant of integration. Then use the metric equation to find an expression for $dR/d\tau$ in terms of e, b, and R for a particle with no y or z velocity. Combine this with the result for $dT/d\tau$ to find dR/dT. Now consider a particle released from rest from $R = 1$ at $T = 0$. Determine e for such a particle and integrate both the expression for $dR/d\tau$ and that for dR/dT to find the coordinate time T and the proper time τ required for such a particle reach $R = \tfrac{1}{2}$. (Feel free to look up integrals.) Argue that this trip requires infinite time T, but finite time τ, and argue that these results are consistent with the diagram.

Note: The R and T coordinates described in this problem are not complete fantasy: they are related to a coordinate system that a uniformly accelerated observer might plausibly use to describe a flat spacetime. Such an observer would really observe an event horizon at the fixed position corresponding to $R = \tfrac{1}{2}$, that light is at rest there, and that objects require an infinite coordinate time T to reach that horizon. However, since we are talking about ordinary flat spacetime here, these odd behaviors are all exposed as being purely coordinate artifacts. See Rindler for a more detailed discussion of how this coordinate system is related to coordinates an accelerated observer might use.

P14.8 Is a black hole really black? Consider a shell of incandescent gas falling into a black hole. According to the equation for dr/dt in table 14.1, such matter will appear to an observer at infinity to "freeze" at the event horizon. So shouldn't an observer see the event horizon always covered with hot glowing gas?

The answer is no. To make the problem simpler, consider instead a laser that falls radially into the black hole from rest at infinity. The laser is pointed radially outward and emits light with a fixed wavelength λ_e in its own frame. By the same method used in problem P12.7 to arrive at equation 14.20, one can show that the redshift for light emitted by a radially falling object and detected at infinity (instead of the other way around) is

$$\frac{\lambda_e}{\lambda_r} = 1 - \sqrt{\frac{2GM}{r}} = \frac{E_r}{E_e} \equiv u \quad (14.23)$$

where E_e and E_r are the energies of the emitted and received photons, respectively ($E = h/\lambda$ in GR units). The quantity u is thus the fraction of a photon's energy (as observed in the laser's frame) that an observer receives at infinity. Let's accept that this formula is correct (you do not need to verify it unless you do the optional part c).

a. Granting this, use the formula for dr/dt in table 14.1 to show that the expression for the rate at which u decreases with time is

$$\frac{du}{dt} = -\frac{GM}{r^2}\left(1 - \frac{2GM}{r}\right) \quad (14.24)$$

(*Hint:* The laser has to start at rest at infinity.)

b. Note that the laser's r coordinate appears to freeze at $r = 2GM$ and remains close to that value for a long time as viewed from infinity. Note also

$$1 - \frac{2GM}{r} = \left(1 + \sqrt{\frac{2GM}{r}}\right)\left(1 - \sqrt{\frac{2GM}{r}}\right) \quad (14.25)$$

Use this and the definition of u to argue that once the laser's r coordinate is sufficiently close to $2GM$, $du/dt \to -u/2GM$, implying that the received photon energy decreases exponentially with coordinate time $\propto e^{-t/2GM}$. For a solar-mass black hole, this e-folding time $2GM$ is on the order of 10 μs, so the laser's light is very rapidly red-shifted to the point where it is not detectable. So a black hole really does become black.

c. (Optional) Use the methods discussed in problem P12.7 to verify that equation 14.23 is correct for light emitted radially outward by an object falling from infinity.

15. ALTERNATIVE COORDINATES

INTRODUCTION

FLAT SPACETIME
- Review of Special Relativity
- Four-Vectors
- Index Notation

TENSORS
- Arbitrary Coordinates
- Tensor Equations
- Maxwell's Equations
- Geodesics

SCHWARZSCHILD BLACK HOLES
- The Schwarzschild Metric
- Particle Orbits
- Precession of the Perihelion
- Photon Orbits
- Deflection of Light
- Event Horizon
- **Alternative Coordinates**
- Black Hole Thermodynamics

THE CALCULUS OF CURVATURE
- The Absolute Gradient
- Geodesic Deviation
- The Riemann Tensor

THE EINSTEIN EQUATION
- The Stress-Energy Tensor
- The Einstein Equation
- Interpreting the Equation
- The Schwarzschild Solution

COSMOLOGY
- The Universe Observed
- A Metric for the Cosmos
- Evolution of the Universe
- Cosmic Implications
- The Early Universe
- CMB Fluctuations & Inflation

GRAVITATIONAL WAVES
- Gauge Freedom
- Detecting Gravitational Waves
- Gravitational Wave Energy
- Generating Gravitational Waves
- Gravitational Wave Astronomy

SPINNING BLACK HOLES
- Gravitomagnetism
- The Kerr Metric
- Kerr Particle Orbits
- Ergoregion and Horizon
- Negative-Energy Orbits

this depends on this

15. ALTERNATIVE COORDINATES

Introduction. The best way to show that the fact that $g_{rr} \to \infty$ in the Schwarzschild coordinate system at $r = 2GM$ is a problem with the Schwarzschild coordinate system (and not with the underlying geometry) is to find other coordinate systems that describe the same spacetime but do not exhibit the same pathology. This chapter will discuss several alternate coordinate systems we can use to describe Schwarzschild spacetime. *Global Rain Coordinates* have a clear physical meaning both inside and outside the event horizon, but involve a non-diagonal metric that can be tricky to interpret. *Kruskal-Szekeres* coordinates, in contrast, have no obvious physical meaning, but have a purely diagonal and easily interpreted metric and make it easy to draw meaningful and useful spacetime diagrams for particle trajectories both inside and outside a black hole. These two coordinate systems provide complementary pictures of the physical nature of a black hole.

Global Rain Coordinates. The core problem with the Schwarzschild coordinate system is that the stationary coordinate lattice we use to define the r and t coordinates cannot exist inside the event horizon. The **global rain coordinate** system gets around this problem by using an array of *steadily infalling* clocks to assign coordinates to events.

We begin as before by setting up a sphere of extremely large radius centered on a black hole, marking latitude and longitude coordinates on its surface. We synchronize clocks on this sphere's surface: these clocks will then all (approximately) read Schwarzschild time t. Then (tapping into our infinite research budget), we start dropping robot observers *from rest* at a steady rate from each location on the sphere, making sure that each robot's clock is synchronized with the clock on the sphere from which it is dropped. (Each robot clock will thus *initially* read the Schwarzschild time t at which it was dropped.) These robot observers will drop radially along worldlines of fixed (and thus known) θ and ϕ. Moreover, each observer can continually determine its current circumferential r coordinate by extending a ruler to the neighboring observer with the same θ but at $\phi + \Delta\phi$, where $\Delta\phi$ is small. Each observer will then see, overlapping its position, a ruler attached to the observer with coordinate $\phi - \Delta\phi$ (see figure 15.1). Since the observers' angular separation $\Delta\phi$ is constant, the circumferential radial coordinate r at any instant is $r \approx \Delta s / \sin\theta \Delta\phi$, where Δs is the distance between the observers displayed on the ruler. Therefore each observer can keep a running record of its current r coordinate. Finally, we define an event's coordinate time $\mathring{t}$ (note the raindrop) to be the time registered on the clock of whichever robot observer is passing that event's location as it occurs, the event's r coordinate to be that particular robot's r coordinate at that instant, and the event's θ and ϕ coordinates to be that robot's (known) θ and ϕ values. The observing robot can report this event and its coordinates to other observers.

These coordinates, because they are operationally defined by the process above, have a clear and immediate physical interpretation. The time coordinate $\mathring{t}$ is also clearly timelike both outside and inside the horizon, since (by definition) it is measured by a clock. Therefore, this coordinate system gets around the problem of coordinates changing meaning at the event horizon.

The Metric for Global Rain Coordinates. The new time coordinate $\mathring{t}$ must be a function of the original Schwarzschild coordinates. The transformation cannot depend on θ or ϕ, because spherical symmetry means that all angular directions are equivalent, so $\mathring{t}$ must be a function $\mathring{t}(t, r)$ of the coordinates t and r alone.

Now, the metric involves *differentials* of coordinate quantities. We can express the differential $d\mathring{t}$ in terms of dt and dr as

$$d\mathring{t} = \left(\frac{\partial \mathring{t}}{\partial t}\right) dt + \left(\frac{\partial \mathring{t}}{\partial r}\right) dr \tag{15.1}$$

We can actually determine the partial derivatives directly from the description of the coordinate system given in the previous section. Consider the first partial derivative. Physically, this derivative poses the following question: For two events separated by an infinitesimal Schwarzschild coordinate time dt but *at fixed radial coordinate r*, what is

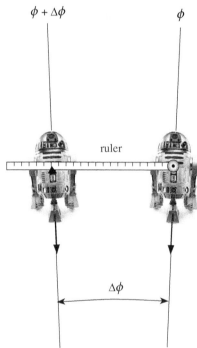

FIG. 15.1 This figure shows two falling robot observers. The one on the left can read its radial coordinate r by measuring the distance Δs on the ruler held by the one on the right and dividing by $\sin\theta \, d\phi$. (This figure is adapted from Taylor and Wheeler, *Exploring Black Holes,* Addison-Wesley, 2000, p. B-8.)

the rain coordinate time difference $d\mathring{t}$ between those events? To answer this question, note that the rain time between these events is measured by the internal clocks of two falling robots, one passing radius r at the time of the first event, and the other falling past the same radius at the time of the second event dt later. Now, each robot requires the same Schwarzschild time to fall from infinity to a given r, so if the second robot passed r a time dt after the first, it must have been *dropped* a time dt after the first. Since each robot clock was set to Schwarzschild time when it was dropped, the second robot's clock will initially be set a time dt later than the first robot's clock. The additional proper times that each robot registers *during* the drop to the same r must be the same. Therefore, the difference $d\mathring{t}$ in the times they register between the *arrival* events at r must be the *same* as the Schwarzschild time difference dt between the *drop* events, so $d\mathring{t} = dt$ for these events, implying that

$$\frac{\partial \mathring{t}}{\partial t} = 1 \qquad (15.2)$$

Now consider the partial derivative $\partial \mathring{t}/\partial r$. Physically, this derivative poses the following question: For two events separated by an infinitesimal radial coordinate difference dr but occurring *at the same Schwarzschild time*, what will be the rain time $d\mathring{t}$ between the events? To answer this question, note that the rain time between these events is measured by the internal clocks of two falling robots that are dr apart when both events occur at the same Schwarzschild time instant. The lower one must have been dropped a bit earlier (which affects the time it reads, as discussed above) but also fell a bit farther (which *also* affects the time it reads). Calculating these competing effects is a bit more complicated than the previous case, but is not that difficult. The result (see box 15.1) is

$$\frac{\partial \mathring{t}}{\partial r} = \sqrt{\frac{2GM}{r}}\left(1 - \frac{2GM}{r}\right)^{-1} \qquad (15.3)$$

Substituting this and equation 15.2 into equation 15.1 yields

$$d\mathring{t} = dt + \frac{\sqrt{2GM/r}}{1 - 2GM/r}\,dr \qquad (15.4)$$

If you solve this for dt and substitute the result into the Schwarzschild metric, you will see (see box 15.2) that you get

$$ds^2 = -\left(1 - \frac{2GM}{r}\right)d\mathring{t}^2 + 2\sqrt{\frac{2GM}{r}}\,d\mathring{t}\,dr + dr^2 + r^2(d\theta^2 + \sin^2\theta\,d\phi^2) \qquad (15.5)$$

This is the metric for global rain coordinates. Since there is nothing in the arguments above that does not apply equally well inside the event horizon as outside it, this metric applies everywhere in Schwarzschild spacetime. Note also that the global rain metric components are finite everywhere except $r = 0$.

"But wait!" you cry. "Doesn't $g_{\mathring{t}\mathring{t}}$ become positive for $r < 2GM$, implying that the $\mathring{t}$ coordinate is spacelike inside the event horizon?" No. Remember that we have defined $\mathring{t}$ to be measured by clocks in objects that are following timelike worldlines, so $\mathring{t}$ must *always* be a time coordinate. The argument we developed in the last chapter about how to interpret signs of metric components applies only to *diagonal* metrics. The metric for global rain coordinates is *not* diagonal: it has off-diagonal terms $g_{r\mathring{t}} = g_{\mathring{t}r} = \sqrt{2GM/r}$. This means that signs of the metric components are not quite so easy to interpret.

In fact, note that for $r < 2GM$ it appears that *all* of the metric components are positive, so at first glance it might seem like the metric has *no* time coordinate! However, let's take seriously the idea that $\mathring{t}$ is a genuine time component that always moves toward the future for any timelike worldline. The only way that we can get a timelike displacement ($ds^2 < 0$) out of the metric above for $d\mathring{t} > 0$ is when dr is *negative*. We conclude, therefore, that *objects following timelike worldlines inside the event horizon must move inward* (as we argued in the last chapter).

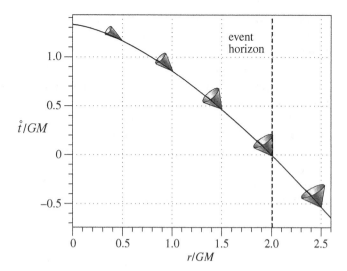

FIG. 15.2 This spacetime diagram, drawn in rain coordinates, shows the worldline of a freely-falling robot observer and the light cones implied by equation 15.6 for such an observer. (In this diagram, I have set the origin of time to be such that the robot's clock measures $\mathring{t} = 0$ when crossing the event horizon.)

We can in fact be more quantitative. You can show (see box 15.3) that at any given $r < 2GM$, dr for purely radial worldlines must be sufficiently negative so that

$$-\left(\sqrt{\frac{2GM}{r}} + 1\right) \leq \frac{dr}{d\mathring{t}} \leq -\left(\sqrt{\frac{2GM}{r}} - 1\right) \qquad (15.6)$$

The limits (which are the rain-coordinate velocities of radial photon worldlines) are symmetric around the robot observers' own inward coordinate velocity

$$\frac{dr}{d\mathring{t}} \equiv \frac{dr}{d\tau} = -\sqrt{\frac{2GM}{r}} \quad \text{(for robot observer)} \qquad (15.7)$$

Therefore, after a robot observer falls past the event horizon, a photon emitted by its taillight does not move actually outward but rather inward less rapidly than the observer (though it appears to the freely-falling robot observer to move outward). Figure 15.2 shows light cones for this observer both inside and outside the horizon.

Kruskal-Szekeres Coordinates. While we can work with global rain coordinates, its non-diagonal metric can be awkward and misleading. Kruskal-Szekeres (henceforth KS) coordinates have a nice, diagonal metric, and are probably the most useful coordinate system for understanding and displaying the behavior of photons. Start by defining KS coordinates u and v such that

$$\left(\frac{r}{2GM} - 1\right)e^{r/2GM} = u^2 - v^2 \quad \text{and} \quad t = 2GM \ln\left|\frac{u+v}{u-v}\right| \qquad (15.8)$$

where the r coordinate is now considered to be a function $r(u,v)$ that is implicitly defined by the first of equations 15.8. If you take differentials of both sides of these expressions and substitute the results for dr and dt into the Schwarzschild metric, you will obtain the KS metric for the empty spacetime around a spherically symmetric star or black hole (see box 15.4).

$$ds^2 = -\frac{32(GM)^3}{r}e^{-r/2GM}(dv^2 - du^2) + r^2(d\theta^2 + \sin^2\theta \, d\phi^2) \qquad (15.9)$$

Here are some of the features of this coordinate system:
- No metric component is infinite at $r = 2GM$.
- There is a pathology at $r = 0$ (which turns out to be geometric).
- The v coordinate is always timelike and the u coordinate is always spacelike (since $g_{vv} < 0$ and $g_{uu} > 0$ always for this diagonal metric).
- Radial light worldlines follow paths such that $du = \pm dv$.
- Massive particles must follow paths such that $dv > |du|$.

To see the last two points, note that $ds^2 = 0$ along any photon worldline, and that *radial* photon worldlines also have $d\theta = d\phi = 0$. Substituting this into equation 15.9 yields

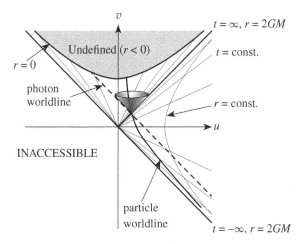
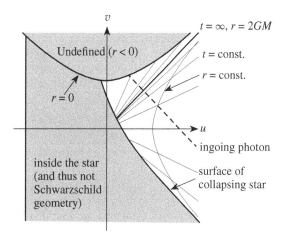

FIG. 15.3 A spacetime diagram of Schwarzschild spacetime using Kruskal-Szekeres coordinates.

FIG. 15.4 A more realistic diagram that takes account of how a black hole forms from a collapsing star.

$du^2 - dv^2 = 0$ for radial photon worldlines, implying that $du = \pm dv$. Massive particles, however, must have $ds^2 \equiv -d\tau^2 < 0$, which is possible only if $dv^2 > du^2$. Non-radial worldlines make ds^2 *less* negative, so the requirement that $dv^2 > du^2$ applies to the worldlines of *all* massive particles.

Now imagine that we draw a spacetime diagram using v as the vertical axis and u as the horizontal axis (see figure 15.3). On such a diagram, radially moving photons follow worldlines with slopes ± 1, and all light cones face toward the top of the diagram. According to the first of equations 15.8, an object having a fixed $r > 2GM$ will follow a hyperbolic worldline $u^2 - v^2 =$ positive constant. Note also that the second of equations 15.8 tells us that a worldline such that $u = \pm av$ (with a constant) will be a worldline of constant t, because

$$\frac{t}{2GM} = \ln\left|\frac{u+v}{u-v}\right| = \ln\left|\frac{\pm av + v}{\pm av - v}\right| = \ln\left|\frac{\pm a + 1}{\pm a - 1}\right| = \text{constant} \qquad (15.10)$$

Such worldlines will be simple lines with slope $\pm 1/a$. The event horizon $r = 2GM$ corresponds to worldlines such that $u^2 - v^2 = 0$, i.e., lines with slope ± 1 through the origin. These lines also correspond with $t = \pm\infty$ ($a = \pm 1$ in equation 15.10).

The beauty of drawing a spacetime diagram in KS coordinates is that habits developed when drawing spacetime diagrams in flat space still apply. Photon worldlines always have slope ± 1, and the worldlines of massive particles always have slopes whose absolute values are greater than 1. So a typical worldline for a particle traveling into the black hole might look as shown in figure 15.3.

We can also read directly from the diagram facts that were hard to wring from the Schwarzschild and global rain metrics. For example, it is obvious from the diagram that no particle that moves inward past $r = 2GM$ can ever hope to come out again. It is also clear that the $r = 0$ hyperbola spans the complete future of any such particle: the particle's arrival at $r = 0$ is inevitable, and no valid particle worldline can escape this fate.

Note also that viewed from outside, an infalling particle does not reach $r = 2GM$ until $t = \infty$. This is corroborated by the equation for dr/dt in table 14.1, which tells us that an observer at infinity observes such a particle to "freeze" at the event horizon.

It is important to remember that Schwarzschild coordinates (and thus KS coordinates) describes the spacetime only in the vacuum *outside* a spherical object. If a star collapses to form a black hole, the geometry inside the star is entirely different. So the *complete* KS graph, which describes a spacetime that existed at $t = -\infty$, is a bit fictional. A more realistic KS graph would look as shown in figure 15.4. The lower gray region refers to the region of spacetime not described by KS coordinates.

BOX 15.1 Calculating $\partial \mathring{t}/\partial r$

As discussed in the text, to calculate $\partial \mathring{t}/\partial r$, we need to consider a pair of events separated by a differential circumferential coordinate difference dr, but that occur *at the same Schwarzschild coordinate time*. Our goal is to calculate the difference $d\mathring{t}$ between the clocks carried by robot observers dropping past those events. Let's measure the steps dr and $d\mathring{t}$ relative to the lower event, so that dr is positive.

First, we should note that an object *dropped from rest at infinity* will have energy per unit mass at infinity $e = 1$, since the energy of an object at rest is simply equal to its mass. Moreover, such an object will have zero angular momentum per unit mass ℓ. Substituting these constants into the equations for $dr/d\tau$ and dr/dt in table 14.1, we find that for our radially dropping robot observers

$$\frac{dr}{d\tau} = -\sqrt{\frac{2GM}{r}} \quad \text{and} \quad \frac{dr}{dt} = -\left(1 - \frac{2GM}{r}\right)\sqrt{\frac{2GM}{r}} \tag{15.11}$$

(You don't have to show this, but make sure that you see how these follow from the equations in the table: these relations are crucial in what follows.)

Now, the upper robot must have been dropped *slightly later* in Schwarzschild time than the lower one (if they had been dropped at the same time, they would have arrived at the same radius after the same Schwarzschild times had passed). Indeed it must have been later by precisely the Schwarzschild time required for a robot to cover the extra distance $-dr$ at our radius. Call this difference in drop times dt_d. According to the second of equations 15.11, then, we have

$$dt_d = \frac{-dr}{(dr/dt)} = \frac{dr}{(1 - 2GM/r)\sqrt{2GM/r}} \tag{15.12}$$

Since the robot's clocks are set to Schwarzschild coordinate time t when they are dropped, this will represent their initial difference when they were dropped.

After being dropped from $r_0 \approx \infty$, each robot measures a proper time that, according to the first of equations 15.11, will be

$$\tau = \int_{r_0}^{r_f} \frac{d\tau}{dr} dr = -\int_{r_0}^{r_f} \sqrt{\frac{r}{2GM}} dr = \frac{2(r_0^{3/2} - r_f^{3/2})}{3\sqrt{2GM}} \tag{15.13}$$

where r_f is each robot's final position. Since the two robots fall to slightly different final positions, these elapsed proper times will be somewhat different. Since the differential in r_f is dr, the differential in the elapsed proper times will be

$$d\tau = 0 - \frac{2}{3\sqrt{2GM}} \frac{3}{2} r^{1/2} dr = -\sqrt{\frac{r}{2GM}} dr \tag{15.14}$$

($d\tau$ is negative here, as the upper robot registers less time for its shorter fall).

The total difference $d\mathring{t}$ in the times the robot clocks display when they arrive at the two events will be the sum of these two times, meaning that $\partial \mathring{t}/\partial r = (dt_d + d\tau)/dr$. Using equations 15.12 and 15.14, you can show that $\partial \mathring{t}/\partial r = \sqrt{2GM/r}/(1 - 2GM/r)$, as claimed in equation 15.3.

Exercise 15.1.1. Check that this is true.

BOX 15.2 The Global Rain Metric

Equation 15.4 tells us that $d\mathring{t} = dt + \sqrt{2GM/r}\,(1 - 2GM/r)^{-1}\,dr$. If you solve this for dt and substitute the result into the Schwarzschild metric

$$ds^2 = -\left(1 - \frac{2GM}{r}\right)dt^2 + \frac{dr^2}{1 - 2GM/r} + r^2(d\theta^2 + \sin^2\theta\,d\phi^2) \quad (15.15)$$

you will get the metric in global rain coordinates $\mathring{t}, r, \theta,$ and ϕ.

Exercise 15.2.1. Calculate the metric and compare with equation 15.5.

BOX 15.3 The Limits on $dr/d\mathring{t}$ Inside the Event Horizon

For purely radial worldlines, $d\theta = d\phi = 0$. Under these circumstances, if we divide the global rain metric in equation 15.5 by $d\mathring{t}^2$, we find that the condition for a timelike worldline inside the event horizon becomes

$$0 \geq \left(\frac{ds}{d\mathring{t}}\right)^2 = \left(\frac{2GM}{r} - 1\right) + 2\sqrt{\frac{2GM}{r}}\,\frac{dr}{d\mathring{t}} + \left(\frac{dr}{d\mathring{t}}\right)^2 \quad (15.16)$$

This is a parabolic function of $dr/d\mathring{t}$, which (because the coefficient of $[dr/d\mathring{t}]^2$ is positive) is concave upward. It will therefore be negative between its two roots, which you can find by setting the function equal to zero and solving for $dr/d\mathring{t}$ using the quadratic formula. The roots will apply to radially moving photons: the region between the roots will represent possible values of $dr/d\mathring{t}$ for particles following timelike worldlines. If $d\theta$ and/or $d\phi$ are nonzero, the additional positive terms in the metric will push the parabola in the positive direction, making the roots closer together. Therefore, the roots of equation 15.16 represent the absolute maximum range for particle worldlines inside the event horizon.

Exercise 15.3.1. Find equation 15.16's roots and compare with equation 15.6.

BOX 15.4 Transforming to Kruskal-Szekeres Coordinates

We can transform the metric equation from Schwarzschild coordinates to Kruskal-Szekeres (KS) coordinates most quickly by using equations 15.8 to express the differentials dr and dt in terms of du and dv, and then substituting those expressions for dr and dt into the Schwarzschild metric equation. This is equivalent to doing a formal transformation using the transformation law for the components of the metric (but is usually faster). To begin, consider the first of equations 15.8:

$$u^2 - v^2 = \left(\frac{r}{2GM} - 1\right)e^{r/2GM} \tag{15.17}$$

If you take the differential of both sides of this equation, you should find that

$$dr = \frac{8(GM)^2}{r}e^{-r/2GM}(u\,du - v\,dv) \tag{15.18}$$

Exercise 15.4.1. Verify equation 15.18.

Now consider the second of equations 15.8:

$$t = 2GM \ln\left|\frac{u+v}{u-v}\right| = 2GM\ln|u+v| - 2GM\ln|u-v| \tag{15.19}$$

Because of the absolute values, when we evaluate the differential of this equation, we actually have to consider four separate cases:

- $u + v > 0, u - v > 0 \Rightarrow t = 2GM\ln(u+v) - 2GM\ln(u-v)$
- $u + v > 0, u - v < 0 \Rightarrow t = 2GM\ln(u+v) - 2GM\ln(v-u)$
- $u + v < 0, u - v > 0 \Rightarrow t = 2GM\ln(-u-v) - 2GM\ln(u-v)$
- $u + v < 0, u - v < 0 \Rightarrow t = 2GM\ln(-u-v) - 2GM\ln(v-u)$

Consider, for the sake of argument, the second of these cases. We have

$$dt = \frac{2GM}{u+v}(du+dv) - \frac{2GM}{v-u}(dv-du) = 2GM\left(\frac{du+dv}{u+v} - \frac{du-dv}{u-v}\right) \tag{15.20}$$

where in the last step, I multiplied both the numerator and the denominator of the second term by -1.

Exercise 15.4.2. Show that any other case (pick one) also yields equation 15.20.

BOX 15.4 (continued) Transforming to Kruskal-Szekeres Coordinates

The point is that it turns out that all four cases lead to the same differential expression given by equation 15.20. If we gather the terms in this equation that involve du and separate them from those that involve dv, we find that

$$dt = \left(\frac{2GM}{u+v} - \frac{2GM}{u-v}\right)du + \left(\frac{2GM}{u+v} + \frac{2GM}{u-v}\right)dv$$

$$= 2GM\left(\frac{u-v-u-v}{u^2-v^2}\right)du + 2GM\left(\frac{u-v+u+v}{u^2-v^2}\right)dv$$

$$= \frac{4GM}{u^2-v^2}(u\,dv - v\,du) = \frac{4GMe^{-r/2GM}}{(r/2GM-1)}(u\,dv - v\,du) \qquad (15.21)$$

where in the last step, I used equation 15.17 to get rid of the $u^2 - v^2$.

If you now square equations 15.18 and 15.21 and use equation 15.17 to eliminate the factor of $u^2 - v^2$ that arises, you should find that (as equation 15.9 claims)

$$-\left(1 - \frac{2GM}{r}\right)dt^2 + \left(1 - \frac{2GM}{r}\right)^{-1}dr^2 = -\frac{32(GM)^3}{r}e^{-r/2GM}(dv^2 - du^2) \qquad (15.22)$$

Exercise 15.4.3. Verify this result.

HOMEWORK PROBLEMS

P15.1 Another way to arrive at the expression for $d\mathring{t}$ in terms of dt and dr is to use the methods of chapter 12. Note that $d\mathring{t}$ for any pair of events separated by a differential four-displacement $d\mathbf{s}$ is the projection $-\mathbf{o}_{\hat{t}} \cdot d\mathbf{s}$ of this four-displacement on the timelike unit vector $\mathbf{o}_{\hat{t}}$ of the falling robot's orthonormal coordinate system. As discussed in chapter 12, we can calculate this dot product in the Schwarzschild coordinate system.

a. (Do this if you have not done problem P12.7.) Using the methods of chapter 12 and equation 15.11, show that the robot observer's time basis vector expressed in Schwarzschild coordinates has components $(\mathbf{o}_{\hat{t}})^\mu = [(1-2GM/r)^{-1}, -\sqrt{2GM/r}, 0, 0]$.

b. Use this to calculate $-\mathbf{o}_{\hat{t}} \cdot d\mathbf{s}$ in Schwarzschild coordinates, and compare with equation 15.4.

P15.2 Discussion problem. Present at least one argument for and one argument against the following assertion: A freely falling observer has a speed greater than 1 inside the event horizon.

P15.3 Find an expression for $\mathring{t}(r)$ for radial photon worldlines. Plot a graph of your resulting expression for light starting inward from $r = 2GM$ at $\mathring{t} = 0$, and compare to the trajectory of a falling robot observer that passes through the same starting event. (*Hints:* You will have to look up an integral: use the substitution $u = \sqrt{2GM/r}$ to put the integral into a more tractable form. You should find that the photon reaches $r = 0$ at $\mathring{t} \approx 0.77 GM$.)

P15.4 Imagine an observer falling into a black hole. An observer at infinity imagines that the observer remains stuck at $r = 2GM$. Does this mean that the observer can see the future of the outside universe essentially to infinity before being crushed at $r = 0$? Alternatively, is the observer completely cut off from the outside universe after falling through the event horizon? Use a KS diagram to definitively answer both questions.

P15.5 Imagine that your spaceship accidently falls through the event horizon of a black hole while shooting photon torpedoes radially outward at a pursuing alien spaceship. You know that any photon torpedoes you fired after falling through the event horizon will eventually end up moving toward smaller r just as you will. Is there any danger that your ship will be hit by your own radially moving photon torpedoes before you are crushed at $r = 0$? Answer using a KS spacetime diagram. (Assume that photon torpedoes follow photon worldlines.)

P15.6. (Adapted from Hartle, *Gravity*, Addison Wesley, 2003, problem 12.13.) Is it dark inside a black hole? Assume that the glowing surface of the collapsing star that creates the black hole continues to radiate as it collapses to $r = 0$. Will other observers that fall through the event horizon later be able to see light from this surface? Answer using a KS spacetime diagram.

P15.7 Imagine that you are hovering in your spaceship at a fixed radial coordinate R above a black hole, and (while fixing an antenna on the outside hull) you drop a valuable wrench, which falls toward the black hole. Someone at infinity would claim that the wrench never actually crosses the event horizon. Does that mean that you can use your spaceship to go rescue the wrench at any later time, or do you have a limited time (according to your watch) to start going after the wrench if you are going to catch it and get out alive? Answer using a KS spacetime diagram.

P15.8. Imagine a spaceship hovering above a black hole, tail downward. The engines fail and the ship begins to fall vertically downward into the black hole. Is there any time when observers at the ship's front end cannot see or receive signals from the tail end? What about when the tail end has passed the event horizon but the front end has not? Will the observers be able to see the tail end of the ship hit $r = 0$? Answer using a KS spacetime diagram.

P15.9 Your spaceship is hovering above a Schwarzschild black hole at a fixed radius R such that
$$\left(\frac{R}{2GM} - 1\right)e^{+R/2GM} = 1$$
At $t = 0$, a shuttle leaves this position, intending to travel closer to the event horizon and back. However, a problem with the shuttle's engines makes them so weak that the best the shuttle can do is to follow a vertical worldline on a KS spacetime diagram (which is not a geodesic, but still does not provide much hope for the shuttle).

a. On a KS spacetime diagram, sketch the paths of both spacecraft as well as the event horizon.

b. What is the latest Schwarzschild time t (as a multiple of GM) after the shuttle departs that the other ship could depart from its position at R to go rescue the shuttle?

16. BLACK HOLE THERMODYNAMICS

INTRODUCTION

FLAT SPACETIME
Review of Special Relativity
Four-Vectors
Index Notation

TENSORS
Arbitrary Coordinates
Tensor Equations
Maxwell's Equations
Geodesics

SCHWARZSCHILD BLACK HOLES
The Schwarzschild Metric
Particle Orbits
Precession of the Perihelion
Photon Orbits
Deflection of Light
Event Horizon
Alternative Coordinates
Black Hole Thermodynamics

THE CALCULUS OF CURVATURE
The Absolute Gradient
Geodesic Deviation
The Riemann Tensor

THE EINSTEIN EQUATION
The Stress-Energy Tensor
The Einstein Equation
Interpreting the Equation
The Schwarzschild Solution

COSMOLOGY
The Universe Observed
A Metric for the Cosmos
Evolution of the Universe
Cosmic Implications
The Early Universe
CMB Fluctuations & Inflation

GRAVITATIONAL WAVES
Gauge Freedom
Detecting Gravitational Waves
Gravitational Wave Energy
Generating Gravitational Waves
Gravitational Wave Astronomy

SPINNING BLACK HOLES
Gravitomagnetism
The Kerr Metric
Kerr Particle Orbits
Ergoregion and Horizon
Negative-Energy Orbits

this depends on this

16. BLACK HOLE THERMODYNAMICS

Introduction. Up to this point, we have been thinking of black holes entirely from the perspective of classical physics. As a classical object, a black hole is very simple. We have seen that the future of any matter or energy that falls inside the event horizon is $r = 0$. This applies even to the matter that originally formed the black hole: once a spherical object has collapsed beyond its own event horizon, it must *all* end up at $r = 0$. So a black hole consists of an infinitely-dense point of mass at $r = 0$ surrounded by a curved but spherically symmetric Schwarzschild spacetime, whose most significant feature is the spherical event horizon at $r = 2GM$.

The infinite concentration of mass-energy at $r = 0$ creates a singularity in the geometry of spacetime (a point of infinite curvature) that cannot be removed by a coordinate transformation (note that the Schwarzschild, global rain, and Kruskal-Szekeres metrics all have infinite components at $r = 0$). This singularity signals a breakdown in the fundamental assumptions of general relativity,

Presumably, a future quantum field theory of gravitation will "blur" this singularity over a small but finite volume. But even so, one might think that any quantum-mechanical features of a black hole would be observable only deep inside the event horizon, and thus would be completely irrelevant for any external observers.

In the 1970s, however, Stephen Hawking and his students discovered that black holes have quantum-mechanical aspects that *can* be observed by observers at infinity. They first discovered and proved (quite generally) that the area of a black hole's event horizon must always increase. Now, the second law of thermodynamics states that an isolated system's *entropy* must always increase. The similarity in these laws led them to speculate that a black hole might have an entropy proportional to its event horizon's area. But according to thermodynamic theory, an object that has an entropy S must also have a temperature $T = (\partial S/\partial U)^{-1}$ where U is the object's internal energy (for a black hole $U = M$). Hawking then was able to show in 1974 that a black hole will radiate photons as if it were a black body with a certain temperature T, in spite of our classical argument that nothing can emerge from a black hole!

Our goal in this chapter is to explore these thermodynamic aspects of black holes, starting with the blackbody radiation. In addition to being physically interesting in their own right, these quantum-mechanical features of black holes provide important clues about what an eventual quantum theory of gravity must look like.

Particles with Negative Energy at Infinity. We will start by noting a purely classical aspect of the Schwarzschild geometry. Particles in Schwarzschild spacetime can have a negative energy per unit mass at infinity e inside the event horizon. To see this, remember that for any particle with nonzero mass

$$e \equiv \left(1 - \frac{2GM}{r}\right)\frac{dt}{d\tau} \tag{16.1}$$

(see the first equation in table 14.1). *Outside* the event horizon, $(1 - 2GM/r) > 0$ (since $r > 2GM$) and $dt/d\tau > 0$ (since t is a time coordinate outside the horizon and so must increase along a timelike worldline), so e must be positive. But *inside* the event horizon, $(1 - 2GM/r) < 0$ and $dt/d\tau$ can have *either* sign (as the Schwarzschild t coordinate is a spatial coordinate inside the horizon). So, though any particle that falls through the event horizon must have positive e, once it is inside, it is possible (in principle) for it to interact with another particle and give up enough energy to end up with *negative e* while still satisfying equation 16.1. Of course, neither particle will be able to escape after this interaction, so the point might seem moot.

Why a Black Hole Radiates. However, this provides the key to understanding how a black hole can radiate energy. Relativistic quantum field theory predicts that even in a vacuum, "quantum fluctuations" constantly occur that produce particle-antiparticle pairs that exist for very short periods of time before recombining. Since energy has to be conserved, one of these particles must have a positive energy E and the other a negative energy $-E$. Quantum field theory predicts that in flat spacetime, the particle with negative energy can exist for only a short time that (by sheer dimensional analysis) must be of order $\Delta t \sim \hbar/E$ before it must recombine with its antiparticle to form a vacuum with zero energy again. (This is analogous to the way that a particle's quantum wavefunction can penetrate a very short *distance* into a region where its potential energy is greater than its total energy and thus its momentum $p = \sqrt{2mK} = \sqrt{2m(E-V)}$ is imaginary.) To put it another way, because a particle with negative energy cannot travel a significant distance through a vacuum, it will inevitably recombine with its partner after a very short time.

Now imagine that such a fluctuation occurs very close to the event horizon of a Schwarzschild black hole, and imagine further that during its short lifetime, the negative-energy particle happens to travel through the event horizon. Since it is possible for even a classical particle to have negative energy inside the event horizon, the negative-energy particle is no longer classically forbidden, and thus is no longer forced to eventually recombine with its partner. In fact, the negative-energy particle (since it still must follow a timelike worldline) *must* end up at $r = 0$ (thereupon reducing the black hole's total mass-energy somewhat), while the positive-energy particle might escape to infinity. This process allows the black hole to radiate particles that effectively drain away its mass-energy!

An Estimate of the Typical Energy of Radiated Particles. A fully correct calculation of the energy spectrum of photons radiated by a black hole requires detailed knowledge of quantum field theory beyond our scope. But we can very crudely estimate the typical photon energy as follows.

Let us observe the quantum fluctuation in a frame that is freely falling. By the equivalence principle, the fluctuation should behave in this freely falling frame as it would in flat spacetime. Specifically, imagine that our frame's origin is instantaneously at rest at the position $r = 2GM + \varepsilon$ (where $\varepsilon \ll 2GM$) where the fluctuation occurs. You can show (see box 16.1) that the frame's origin will reach the event horizon in a proper time of roughly $\Delta\tau \approx 2\sqrt{2GM\varepsilon}$ if $\varepsilon \ll 2GM$. The radiation process can work only if the negative-energy partner survives long enough to make it to the event horizon, that is, for a time in the freely falling frame that is comparable to the time (just calculated) the frame's origin needs to reach the event horizon. So the positive-energy partner's characteristic energy E in the freely falling frame will be such that

$$\frac{\hbar}{E} \sim \Delta\tau \quad \Rightarrow \quad E \sim \frac{\hbar}{\Delta\tau} = \frac{\hbar}{2\sqrt{2GM\varepsilon}} \quad (16.2)$$

(If E is much larger than this, the negative-energy particle will probably not survive long enough to make it to the event horizon; if E is much smaller, the particles will likely survive long enough for *both* to fall into the event horizon).

This, therefore, is the characteristic energy of the outgoing particle in the freely falling frame. What is this particle's energy if and when it reaches infinity? We can calculate this using the techniques of chapter 12. In our falling frame's local orthonormal coordinate system, the particle's energy is given by $E = -\mathbf{o}_t \cdot \mathbf{p}$, where $\mathbf{o}_t \equiv \mathbf{u}_{\text{obs}}$ is the freely falling frame's four-velocity and $\mathbf{p}$ is the particle's four-momentum. Since the dot product is a frame-independent scalar, we can evaluate it using Schwarzschild coordinates. The result (see box 16.2) is

$$E = \frac{me}{\sqrt{1-2GM/r}} = \frac{E_\infty}{\sqrt{1-2GM/r}} \quad (16.3)$$

where $E_\infty \equiv me$ is the particle's energy measured at infinity. Solving for this quantity and using $r = 2GM + \varepsilon$ (the frame's location at the fluctuation event), we get

$$E_\infty = \sqrt{1-\frac{2GM}{r}}\, E \sim \sqrt{1-\frac{2GM}{2GM+\varepsilon}}\, \frac{\hbar}{2\sqrt{2GM\varepsilon}} \quad (16.4)$$

This formula makes sense: note that the quantity in the big square root is very small, so the particle's energy at infinity (after it has climbed out through the gravitational field) will be much smaller than it was in the freely falling frame near the horizon.

Now, note that if $\varepsilon \ll 2GM$, then

$$\sqrt{1 - \frac{2GM}{2GM + \varepsilon}} = \sqrt{1 - \frac{1}{1 + \varepsilon/2GM}} \approx \sqrt{1 - \left(1 - \frac{\varepsilon}{2GM}\right)} = \sqrt{\frac{\varepsilon}{2GM}} \quad (16.5)$$

where I have used the binomial approximation in the next-to-last step. Substituting this into equation 16.4 yields

$$E_\infty \sim \frac{\hbar}{4GM} \quad (16.6)$$

Note that the ε has dropped out! So the energy at infinity E_∞ of emitted particles has a *unique* characteristic value that does not depend on how far from the event horizon the fluctuation occurred (as long as $\varepsilon \ll 2GM$).

Even so, quantum fluctuations are much more *probable* if the falling-frame energy E involved in the quantum fluctuation is small, because the violation of the usual rules of classical physics is not so gross. A photon is its own antiparticle, and photons have zero mass, so fluctuations can produce photon pairs with arbitrarily low energies. However, a fluctuation that produces at least one member of a massive particle-antiparticle pair with sufficient energy to exist classically must involve $E \geq m$, which is much less likely. Moreover, a particle with nonzero mass must have $E_\infty \geq m$ (not just $E \geq m$) to make it to infinity instead of eventually falling back into the black hole. This is virtually impossible considering that equation 16.6 implies that the *typical* energy-at-infinity for particles produced near the event horizon of a solar-mass black hole is $E_\infty \sim 6 \times 10^{-47}$ kg (see problem P16.1), more than 16 orders of magnitude smaller than an electron's mass!

Therefore, we would expect a black hole of any astronomical mass to radiate photons almost exclusively (even neutrinos have *some* rest mass). Though our derivation of equation 16.3 assumed that the particles formed by the fluctuation had nonzero mass m, this mass drops out, so our results do apply to photons (we can take the limit $m \to 0$ without changing the result). Therefore, equation 16.6 should be a good rough estimate of the typical energies of photons emitted.

Results of Hawking's Calculation. Since the quantum fluctuations create particle-antiparticle pairs with a *range* of energies, we would expect the energies of particles reaching infinity to have some kind of distribution that peaks near the characteristic energy estimated above. Hawking's more careful calculation shows, in fact, that the energy spectrum of photons radiated from a black hole is exactly the same as that of a blackbody whose temperature T is such that

$$k_B T = \frac{\hbar}{8\pi GM} \quad \Rightarrow \quad T = \frac{\hbar}{8\pi k_B GM} \quad \text{(in GR units)} \quad (16.7)$$

where k_B is Boltzmann's constant. Note that the characteristic photon energy given by $k_B T$ differs from our estimate above by only a factor of 2π, not bad for an order-of-magnitude estimate. In GR units, you can show (see box 16.3) that the values of Boltzmann's constant, Planck's constant, and the temperature are

$$k_B = 1.536 \times 10^{-40} \text{ kg/K}, \quad \hbar \approx 3.518 \times 10^{-43} \text{ kg} \cdot \text{m}$$

$$\Rightarrow T = \frac{\hbar}{8\pi k_B GM_\odot}\left(\frac{M_\odot}{M}\right) = \frac{6.17 \times 10^{-8} \text{ K}}{M/M_\odot} \quad (16.8)$$

This means that a solar-mass black hole has a temperature of about 60 nK, and larger black holes have even lower temperatures. This is significantly lower than the background temperature of the universe, which is 2.73 K at the present.

The Lifetime of a Black Hole. The rate at which any blackbody radiates energy is $dE/dt = A\sigma T^4$, where A is the object's surface area and σ is the Stefan-Boltzmann constant ($\sigma = 2.105 \times 10^{-33}$ kg·m^{-3}K^{-4}). As a black hole radiates energy, its mass decreases, which increases its temperature, which makes it radiate energy even more rapidly. You can show (see box 16.4) that an isolated black hole (surrounded by vacuum at absolute zero) would survive for

$$\tau_{\text{life}} = (2.095 \times 10^{67} \text{ y})\left(\frac{M}{M_\odot}\right)^3 \qquad (16.9)$$

A solar-mass black hole will thus survive for about 2×10^{67} y after the background temperature of the universe falls below 60 nK. On the other hand, if we were to be able to create a 1-kg black hole in the laboratory, it would last only 0.083 fs before evaporating (the evaporation would actually be pretty catastrophic, as it would release about as much energy as a large thermonuclear bomb during this time).

Black Hole Entropy. Historically, the idea of black hole entropy was developed first, and was followed by the discovery that a black hole should radiate. However, it makes more sense in hindsight to develop the ideas in the other order. In this section, we will see that the fact that a black hole radiates photons as if it were a blackbody implies that it must have an entropy of a certain value.

In statistical mechanics, one can show from very fundamental principles that an object's temperature T must be related to its entropy S as follows:

$$\frac{1}{T} \equiv \frac{\partial S}{\partial U} \qquad (16.10)$$

where U is the object's internal energy. In the case of a black hole, its internal energy is equal to its mass M, and $1/T = 8\pi k_B GM/\hbar$. Integrating the expression for $1/T$ with respect to M, and assuming that a black hole having no mass also has no entropy, we find that a black hole's entropy must be

$$S = \frac{4\pi k_B GM^2}{\hbar} = \frac{k_B 4\pi(2GM)^2}{4G\hbar} = \frac{k_B}{4G\hbar} A \qquad (16.11)$$

where $A \equiv 4\pi(2GM)^2$ is the area of the black hole's horizon.

The second law of thermodynamics then implies that when black holes interact, the total area of their event horizons must increase. This is clearly true for two black holes with equal masses M colliding to form a single black hole: the total event-horizon area before the collision is $2 \cdot 4\pi(2GM)^2 = 32\pi(GM)^2$, while the total after combination is $4\pi(2G[2M])^2 = 64\pi(GM)^2$. You can easily generalize this for arbitrary combinations of Schwarzschild black-hole masses (see problem P16.7).

In statistical mechanics, an object's entropy is $S = k_B \ln \Omega$, where Ω is the number of microstates (quantum states) consistent with object's macroscopic parameters. In the absence of a clear quantum theory of gravity, we have no idea what kinds of quantum states we are talking about in the case of a black hole. However, this result puts an important constraint on quantum theories of gravity: any valid theory must yield equation 16.11 for the entropy. People have been trying for more than a decade to generate this result from string theory, spin network theory, and other possible quantum theories of gravity, with results that are tantalizing but not definitive (see Susskind, *The Black Hole War*, Little/Brown, 2008 for a popular discussion).

Thermal Equilibrium. The temperature of a normal object typically *increases* when you add energy. As you can see from equation 16.7, the *reverse* is true for black holes (a behavior that black holes share with other gravitationally bound systems such as stars, clusters, etc.). This means that black holes behave oddly when brought into contact with normal thermodynamic objects.

For example, consider a black hole bathed by the cosmic background radiation, which we can treat as a reservoir (i.e., a thermodynamic object so large that it can absorb or provide large amounts of energy without changing its temperature). Given enough time, a *normal* object in contact with a reservoir will come into a stable equilibrium with it at the same temperature. But this is not true for a black hole. A black hole initially colder than the background will absorb more energy from the background than it emits. But this makes it even colder, driving it *away* from equilibrium. Similarly, if it is initially hotter than the background, it emits more energy than it receives, and thus gets still hotter. A black hole can therefore never be in a stable equilibrium with a reservoir (see problem P16.8 for more discussion).

BOX 16.1 Free-Fall Time to the Event Horizon from $r = 2GM + \varepsilon$

One of the fundamental equations of motion for particles with nonzero mass in Schwarzschild spacetime (see table 14.1) is

$$\frac{dr}{d\tau} = \pm\sqrt{e^2 - \left(1 - \frac{2GM}{r}\right)\left(1 + \frac{\ell^2}{r^2}\right)} \qquad (16.12)$$

From this equation, one can show that the proper time required for an object to fall from $r = 2GM + \varepsilon$ to $r = 2GM$ is

$$\Delta\tau = -\int_{2GM+\varepsilon}^{2GM} \frac{dr}{\sqrt{\frac{2GM}{r} - \frac{2GM}{2GM+\varepsilon}}} = +\int_0^\varepsilon \frac{d\rho}{\sqrt{\frac{2GM}{2GM+\rho} - \frac{2GM}{2GM+\varepsilon}}} \qquad (16.13)$$

where $\rho \equiv r - 2GM$.

Exercise 16.1.1. Verify both steps of this equation.

In the limit that $\varepsilon \ll 2GM$, you can apply the binomial approximation to both terms in the square root in equation 16.13, making the integral easy to calculate.

Exercise 16.1.2. Do this and show that $\Delta\tau \approx 2\sqrt{2GM\varepsilon}$ in this limit.

BOX 16.2 Calculating E_∞

In the freely falling frame, the positive-energy particle has energy $E = -\mathbf{o}_t \cdot \mathbf{p}$, where $\mathbf{p}$ is the particle's four-momentum and $\mathbf{o}_t \equiv \mathbf{u}_{obs}$ is the frame's four-velocity. Our job is to calculate this dot product in the Schwarzschild coordinate system and use this to connect the particle's energy E in the falling frame to E_∞, the energy a Schwarzschild observer would measure it to have when it arrives at infinity.

Since the freely falling frame is at rest in the Schwarzschild coordinate system at the time the particle-antiparticle pair is formed, the spatial components of the frame's four-velocity will be zero at that time. The definition of the four-velocity also implies that $-1 = \mathbf{u} \cdot \mathbf{u} = g_{\mu\nu} u^\mu u^\nu$. You can use this information to show that, expressed in Schwarzschild coordinates,

$$u^t = \frac{1}{\sqrt{1 - 2GM/r}} \tag{16.14}$$

Exercise 16.2.1. Verify that this is correct.

You can use this and the definition of the particle's 4-momentum $p^\mu \equiv m(dx^\mu/d\tau)$ to evaluate the dot product in the Schwarzschild coordinate system, yielding

$$E = \sqrt{1 - \frac{2GM}{r}}\, m \frac{dt}{d\tau} \tag{16.15}$$

Exercise 16.2.2. Check that this is right.

BOX 16.2 (continued) Calculating E_∞

But according to equation 16.1,

$$e \equiv \left(1 - \frac{2GM}{r}\right)\frac{dt}{d\tau} \quad (16.16)$$

where e is the particle's conserved energy per unit mass at infinity. Therefore, the particle's energy at infinity is $E_\infty \equiv me$. You can use this, in combination with equations 16.16 and 16.15, to prove equation 16.3 (repeated here for convenience):

$$E = \frac{E_\infty}{\sqrt{1 - 2GM/r}} \quad (16.17)$$

Exercise 16.2.3. Check that this is correct.

BOX 16.3 Evaluating k_B, $\hbar$, and T for a Solar-Mass Black Hole

Exercise 16.3.1. The SI values of k_B and $\hbar$ are (respectively) 1.3807×10^{-23} J/K and 1.0546×10^{-34} J·s. Use these values and the fact that $GM_\odot = 1477$ m (where $M_\odot$ is the mass of the sun) to calculate $\hbar/8\pi k_B GM_\odot$ in K. Compare your results with those shown in equation 16.8. (Be careful with units!)

BOX 16.4 Lifetime of a Black Hole

The rate at which a black hole's mass-energy M is radiated away by its blackbody radiation is given by the Stefan-Boltzmann formula

$$-\frac{dM}{dt} = \frac{dE_{\text{rad}}}{dt} = A\sigma T^4 \tag{16.18}$$

where A is the area of the black hole's event horizon, T is its temperature (which is $\hbar/8\pi k_B GM$), and σ is the Stefan-Boltzmann constant = 2.105×10^{-33} kg·m^{-3}K^{-4}. You can easily integrate this expression to show that a black hole that initially has mass M will evaporate to nothing after a time

$$\tau_{\text{life}} = \frac{256\pi^3 k_B^4}{3G\sigma\hbar^4}(GM)^3 \tag{16.19}$$

Exercise 16.4.1. Verify that this is correct.

Exercise 16.4.2. Evaluate these constants to verify equation 16.9. (Note that 1 y = 9.461×10^{15} m, and that $G = 7.426 \times 10^{-28}$ m/kg in GR units.)

HOMEWORK PROBLEMS

P16.1 Use equation 16.6 to show that the typical relativistic energy at infinity of particles emitted by the event horizon of a solar-mass black hole is about 6×10^{-47} kg. Also express this in electron volts (eV).

P16.2 Imagine that black holes with a spectrum of masses were formed during the Big Bang 13.7 Gy ago.
a. How massive would a black hole have been at the Big Bang if it is just evaporating today?
b. How much energy would such a black hole release in the last second of its life? Compare to the energy released by an atomic bomb (which is $\sim 4 \times 10^{14}$ J).

P16.3 Certain speculative string theory models suggested that the Large Hadron Collider might create microscopic black holes with mass-energies of ~ 1 TeV. Some in the popular media worried that these black holes might be dangerous because they could eventually gobble up the earth. But what would the lifetime of such a black hole be (if general relativity still applies to black holes of such size)?

P16.4 (Adapted from Ohanian and Ruffini, *Gravitation and Spacetime*, 2/e, Norton, 1994, pp. 481–482). A heat engine takes a thermal energy Q_H from a hot object at absolute temperature T_H, extracts a bit of work W, and dumps the remaining energy into a cold object at temperature T_C. The second law of thermodynamics implies that the efficiency of such an engine *must* be less than

$$\epsilon \equiv \frac{W}{Q_H} = 1 - \frac{T_C}{T_H} \quad (16.20)$$

Now, imagine the following heat engine. We collect in a mirrored box some thermal radiation from a glowing object at temperature T_H. We can think of this box as containing an extra mass equal (in GR units) to the energy enclosed. The energy per unit mass of this radiation is therefore $e = 1$ at this time. We then lower this box with a winch to bring it to rest just outside a black hole's event horizon, where the box's energy per unit mass approaches $e = 0$ (see equation 10.8). The energy difference must show up as work done on the winch. We then open the box, dump the radiation into the black hole, and then haul the box back up. Of course, we have to put energy back into the empty box to do this, but we have essentially converted all of the thermal energy *inside* the box to work, in violation of the second law of thermodynamics.

However, let's look at this argument a bit more closely. Wien's law says that the wavelength of thermal radiation with temperature T_H is $\lambda \sim \hbar/k_B T_H$ in GR units. The box must therefore have at least this size to contain the radiation. We can therefore lower the box's center only to within a physical distance Δs of $\lambda/2$ above the event horizon.

a. Use equation 14.3 (and the fact that $\tanh^{-1} x \approx x$ for $x \ll 1$) to argue that the smallest r coordinate that the box's center can have in its final position will be such that

$$\left(\frac{\lambda}{2}\right)^2 \approx (4GM)^2 \left(1 - \frac{2GM}{r}\right) \quad (16.21)$$

b. Use equation 10.8 to argue that the average value of e for the thermal radiation in the box at this r is not zero but in fact about $\lambda/8GM$.

c. Argue that the engine's efficiency is $1 - e$, and that the second law of thermodynamics is satisfied if the black hole has a temperature of $T_C \sim \hbar/8GMk$.

P16.5 Consider an observer at a fixed radius R. How much hotter does this observer consider the black hole to be than an observer at infinity? What happens as R approaches $2GM$? How close (in R) would you have to be to a solar-mass black hole's event horizon to observe its temperature to be ~ 300 K (room temperature)? (*Hints:* The temperature of any radiating black body is proportional to the most likely photon energy observed. Use the methods of chapter 12 to find the most probable photon energy E an observer at R will measure compared if the same at infinity is E_∞.)

P16.6 Find an expression for the mass of a black hole as a function of time t, its initial mass M_0, and its initial lifetime t_0 (no other physical constants). Submit a computer-generated plot of M/M_0 versus t/t_0.

P16.7 Prove that the total entropy for two black holes of arbitrary masses M_1 and M_2 is less than that of the black hole they form were they to coalesce.

P16.8 The entropy of a thermal "reservoir" (an object large enough to exchange significant amounts of energy without changing its temperature) is $S_R = U/T_R + C$, where U is the reservoir's internal energy, T_R is its fixed temperature, and C is some constant. Imagine a black hole exchanging thermal radiation with a reservoir. Let the combined system have some total conserved energy U_{tot}. Note that if the black hole has mass M, the energy of the reservoir is then $U_{\text{tot}} - M$. A *stable* equilibrium is when energy is distributed in such a way that the total entropy of the combined system is a local maximum. Argue that the total entropy of the black hole/reservoir combination does not *have* a local maximum except at the extremes ($M = 0$ or $M = U_{\text{tot}}$).

P16.9 The entropy S of most everyday thermodynamic systems is on the order of magnitude of Nk_B, where N is the number of particles in the system and k_B is Boltzmann's constant. Even in stars, where the matter is very hot, the ratio $S/Nk_B \sim 10$. What is the ratio S/Nk_B for a solar-mass black hole (where N is the number of particles that have fallen into the black hole)? How does this compare to the ratio for normal matter? (*Hint:* For the sake of simplicity, assume that the black hole was formed out of pure ionized hydrogen, i.e., protons and separated electrons, one electron per proton).

17. THE ABSOLUTE GRADIENT

INTRODUCTION		

FLAT SPACETIME	TENSORS	SCHWARZSCHILD BLACK HOLES
Review of Special Relativity	Arbitrary Coordinates	The Schwarzschild Metric
Four-Vectors	Tensor Equations	Particle Orbits
Index Notation	Maxwell's Equations	Precession of the Perihelion
	Geodesics	Photon Orbits
		Deflection of Light
		Event Horizon
		Alternative Coordinates
		Black Hole Thermodynamics

THE CALCULUS OF CURVATURE	THE EINSTEIN EQUATION	
The Absolute Gradient	The Stress-Energy Tensor	
Geodesic Deviation	The Einstein Equation	
The Riemann Tensor	Interpreting the Equation	
	The Schwarzschild Solution	

COSMOLOGY	GRAVITATIONAL WAVES	SPINNING BLACK HOLES
The Universe Observed	Gauge Freedom	Gravitomagnetism
A Metric for the Cosmos	Detecting Gravitational Waves	The Kerr Metric
Evolution of the Universe	Gravitational Wave Energy	Kerr Particle Orbits
Cosmic Implications	Generating Gravitational Waves	Ergoregion and Horizon
The Early Universe	Gravitational Wave Astronomy	Negative-Energy Orbits
CMB Fluctuations & Inflation		

this depends on → this

17. THE ABSOLUTE GRADIENT

Introduction. In the past few chapters, we have thoroughly explored the Schwarzschild geometry for the empty space surrounding a static, spherical object. But what makes us think that this should be the *correct* exterior geometry for such an object? With this chapter we begin a journey that will culminate in the Einstein equation that connects a gravitational field to its source. With this equation, we will be able to show that the Schwarzschild geometry is indeed correct.

In this chapter, we will start by addressing a core mathematical problem left over from chapter 6: How can we compute the *tensor* gradient of a tensor field?

A Review of the Problem. Consider a scalar field $\Phi(x^\alpha)$ (the notation simply means that Φ is potentially a function of all four coordinates). Back in chapter 6, we found that the *gradient* $\partial_\mu \Phi \equiv \partial \Phi / \partial x^\mu$ of such a scalar function is a covector in every coordinate system. This is because if we transform to a new coordinate system, the chain rule of partial differential calculus implies that

$$(\partial'_\mu \Phi) \equiv \frac{\partial \Phi}{\partial x'^\mu} = \frac{\partial x^\nu}{\partial x'^\mu} \frac{\partial \Phi}{\partial x^\nu} = \frac{\partial x^\nu}{\partial x'^\mu}(\partial_\nu \Phi) \tag{17.1}$$

and this coincides with the transformation law of a covector. However, we also saw that this only applies to scalars: in general curvilinear coordinates, the gradient of a more complicated tensor is not a tensor. To repeat the example in equation 6.14:

$$(\partial'_\nu A'^\mu) \equiv \frac{\partial A'^\mu}{\partial x'^\nu} = \frac{\partial}{\partial x'^\nu}\left(\frac{\partial x'^\mu}{\partial x^\alpha} A^\alpha\right) = \frac{\partial x^\beta}{\partial x'^\nu} \frac{\partial}{\partial x^\beta}\left(\frac{\partial x'^\mu}{\partial x^\alpha} A^\alpha\right)$$

$$= \frac{\partial x^\beta}{\partial x'^\nu} \frac{\partial^2 x'^\mu}{\partial x^\beta \partial x^\alpha} A^\alpha + \frac{\partial x^\beta}{\partial x'^\nu} \frac{\partial x'^\mu}{\partial x^\alpha}\left(\frac{\partial A^\alpha}{\partial x^\beta}\right) \tag{17.2}$$

The second term, if it were to appear alone, would be the transformation law for a second-rank tensor with one upper and one lower index. But the presence of the first term, which is nonzero any time that the transformation functions are not linear, means that this quantity is not a tensor.

On the other hand, we saw in the chapter on Maxwell's equations that space and time derivatives of vectors and tensors are useful for expressing physical laws. How can we write equations involving space-time derivatives in a way that preserves the manifestly coordinate-independent character of tensor equations?

Our goal in this chapter is to develop a way of calculating the true gradient, or **absolute gradient**, of a tensor. Unlike the ordinary partial derivative, the absolute gradient of any tensor yields another tensor quantity, making the absolute gradient suitable for constructing tensor equations that are valid in every coordinate system.

Exposing the Problem. To see more clearly why the simple gradient of a vector field is not a tensor, consider the simple case of a constant vector field $\mathbf{A}(x^\sigma)$ in a flat two-dimensional space. The true gradient of such a vector field should be zero (since it doesn't change as we go from place to place). In a cartesian coordinate system, the components A^μ of such a vector field are constant, so $\partial_\alpha A^\mu = 0$ as expected. But in a curvilinear coordinate system, the components A^μ of even a truly constant vector field may *not* be constant, because the basis vectors used to evaluate the vector components change as one moves from point to point. We need to find a way to correct $\partial_\alpha A^\mu$ to remove the part due to changes in the basis vectors if we hope to find the true change in the vector function $\mathbf{A}(x^\sigma)$.

Christoffel Symbols. Let's define a set of coefficients $\Gamma^\nu_{\mu\alpha}$ (which are called **Christoffel symbols**) at a given point or event P such that

$$\frac{\partial \mathbf{e}_\alpha}{\partial x^\mu} \equiv \Gamma^\nu_{\mu\alpha} \mathbf{e}_\nu \tag{17.3}$$

The partial derivative here is the differential change in the basis vector $\mathbf{e}_\alpha$ as we move from P to a point a differential displacement dx^μ along a curve where the other coordinates are constant, divided by that differential displacement dx^μ. The change $\partial \mathbf{e}_\alpha$ in

$\mathbf{e}_\alpha$ is a vector, so we can write it as a sum over the basis vectors $\mathbf{e}_\nu$ at that point (the sum is over ν in equation 17.3). The four coefficients $\Gamma^\nu_{\mu\alpha}$ appearing in that sum also depend on the component direction of the displacement (specified by μ) and which particular basis vector $\mathbf{e}_\alpha$ is being examined, so each coefficient has three indices. In a four-dimensional spacetime, there are therefore a total of $4 \times 4 \times 4 = 64$ Christoffel symbols for a given coordinate system.

The True Change in a Vector A. Now consider a vector field $\mathbf{A}(x^\sigma)$. The product rule implies that the *true* amount that $\mathbf{A}$ changes when we move an arbitrary infinitesimal displacement d (whose components are dx^α) is

$$d\mathbf{A} = d(A^\mu \mathbf{e}_\mu) = \left(\frac{\partial A^\mu}{\partial x^\sigma} dx^\sigma\right) \mathbf{e}_\mu + A^\mu \frac{\partial \mathbf{e}_\mu}{\partial x^\alpha} dx^\alpha \tag{17.4}$$

You can use equation 17.3 and rename indices to show that this can be written

$$d\mathbf{A} = \left[\frac{\partial A^\mu}{\partial x^\alpha} + \Gamma^\mu_{\alpha\nu} A^\nu\right] \mathbf{e}_\mu dx^\alpha \equiv (\nabla_\alpha A^\mu) \mathbf{e}_\mu dx^\alpha \tag{17.5}$$

(see box 17.1) The quantity

$$\nabla_\alpha A^\mu \equiv \frac{\partial A^\mu}{\partial x^\alpha} + \Gamma^\mu_{\alpha\nu} A^\nu \tag{17.6}$$

is called the **absolute gradient** of the vector field $\mathbf{A}(x^\sigma)$: the term involving Christoffel symbols corrects the partial derivative for the variations in the basis vectors. The absolute gradient $\nabla_\alpha A^\mu$ must be a (2nd-rank) tensor, because equation 17.5 implies that $(\nabla_\alpha A^\mu) dx^\alpha$ yields the components of a four-vector. However, neither the partial derivative alone nor the Christoffel symbol alone are tensors: only the combination given by equation 17.6 is a tensor.

The Absolute Gradient of a General Tensor. You can show that the absolute gradient of a covector B_μ must be

$$\nabla_\alpha B_\mu = \frac{\partial B_\mu}{\partial x^\alpha} - \Gamma^\nu_{\alpha\mu} B_\nu \tag{17.7}$$

(see box 17.2). Using a similar approach to the one shown in that box, one can show from this that the absolute gradient of a general tensor (e.g., $T^{\mu\nu}{}_\sigma$) is given by

$$\nabla_\alpha T^{\mu\nu}{}_\sigma = \frac{\partial T^{\mu\nu}{}_\sigma}{\partial x^\alpha} + \Gamma^\mu_{\alpha\beta} T^{\beta\nu}{}_\sigma + \Gamma^\nu_{\alpha\delta} T^{\mu\delta}{}_\sigma - \Gamma^\gamma_{\alpha\sigma} T^{\mu\nu}{}_\gamma \tag{17.8}$$

The general rule is as follows: the absolute gradient of a tensor is the sum of the ordinary gradient of the tensor plus a positive Christoffel symbol (times the tensor) for each upper index (summing over that index and the second lower Christoffel symbol index), plus a negative Christoffel symbol (times the tensor) for each lower index (summing over that index and the upper Christoffel symbol index).

Symmetry of the Christoffel Symbol. Though it is not at all obvious from the definition of the Christoffel symbols (equation 17.3), the Christoffel symbols are symmetric in their lower two indices:

$$\Gamma^\alpha_{\mu\nu} = \Gamma^\alpha_{\nu\mu} \tag{17.9}$$

(See box 17.3 for the proof.) This means that while there are 64 Christoffel symbols for every coordinate system, only 40 are independent, since there are only 10 unique ways that we can arrange the four coordinate values among the two lower indices if permutations don't count (e.g., for Schwarzschild coordinates, the 10 possibilities are tt, tr, $t\theta$, $t\phi$, rr, $r\theta$, $r\phi$, $\theta\theta$, $\theta\phi$, and $\phi\phi$).

Calculating the Christoffel Symbols. By taking the dot product of both sides of equation 17.3 with another basis vector and using equation 17.9, you can prove that the Christoffel symbols can be calculated from the metric as follows:

$$\Gamma^{\alpha}_{\mu\nu} = \tfrac{1}{2}g^{\alpha\sigma}\left[\partial_\mu g_{\nu\sigma} + \partial_\nu g_{\sigma\mu} - \partial_\sigma g_{\mu\nu}\right] \tag{17.10}$$

(see box 17.4). This is easier to remember than it looks. A factor of the inverse metric generates the Christoffel symbol's superscript index. The negative term has the symbol's lower indices as the indices of the metric, and the other two terms in the bracket are simply cyclic permutations of this last term ($\mu \to \nu \to \sigma \to \mu \ldots$).

A Geodesic Is a "Locally Straight" Curve. Previously, we defined a geodesic as the curve of extremal distance between two points. We are now in a position to define a curve that is locally straight. How can we tell if a line is locally straight? An object's four-velocity $\boldsymbol{u} = d\boldsymbol{s}/d\tau$ is always tangent to its path. The path will therefore be locally straight if this four-vector does not change as we move an infinitesimal step along the path:

$$0 = \frac{d\boldsymbol{u}}{d\tau} = \frac{d}{d\tau}(u^\mu \boldsymbol{e}_\mu) = \frac{du^\mu}{d\tau}\boldsymbol{e}_\mu + u^\mu \frac{d\boldsymbol{e}_\mu}{d\tau} \tag{17.11}$$

But the basis vector $\boldsymbol{e}_\mu$ changes with proper time only because it depends on position and the object's position depends with time. Therefore, using the chain rule in equation 17.11 and the definition $u^\mu \equiv dx^\mu/d\tau$, we see that equation 17.11 becomes

$$0 = \frac{d^2 x^\mu}{d\tau^2}\boldsymbol{e}_\mu + \frac{dx^\mu}{d\tau}\frac{dx^\nu}{d\tau}\frac{\partial \boldsymbol{e}_\mu}{\partial x^\nu} = \frac{d^2 x^\mu}{d\tau^2}\boldsymbol{e}_\mu + \frac{dx^\mu}{d\tau}\frac{dx^\nu}{d\tau}\Gamma^\alpha_{\mu\nu}\boldsymbol{e}_\alpha$$

$$= \left[\frac{d^2 x^\mu}{d\tau^2} + \Gamma^\mu_{\alpha\beta}\frac{dx^\alpha}{d\tau}\frac{dx^\beta}{d\tau}\right]\boldsymbol{e}_\mu \;\;\Rightarrow\;\; 0 = \frac{d^2 x^\mu}{d\tau^2} + \Gamma^\mu_{\alpha\beta}\frac{dx^\alpha}{d\tau}\frac{dx^\beta}{d\tau} \tag{17.12}$$

where in the beginning of the second line, I renamed some bound indices so I could pull out a common factor of $\boldsymbol{e}_\mu$. You can use equation 17.10 to show that the last equation is completely equivalent to the geodesic equation we have been using up to now (see box 17.5). Indeed, comparing the results of the version of the geodesic equation given by equation 8.12 with equation 17.12 provides the fastest way to actually calculate the Christoffel symbols. This nice trick is illustrated in box 17.6.

In the next chapter (box 18.3), we will see how to calculate $d/d\tau$ for *any* four-vector that varies at a particle's location as the particle moves along a worldline.

Locally Inertial Reference Frames. Perhaps you have been wondering what exactly I mean by the "true change" in a four-vector. I have been talking about quantities such as d and d as if it was clear what we mean by these quantities in a curved spacetime. In a certain sense, such quantities should be clear at least qualitatively: a four-vector is an arrow whose magnitude and length we can define in spacetime without reference to any coordinate system, so we can in principle talk about what changes in these vectors mean without reference to any coordinate system: d for example, is the arrow that is the vector difference $\boldsymbol{u}(\tau + d\tau) - \boldsymbol{u}(\tau)$. This is really no different than subtracting vectors in a Euclidean 3-space, which we do comfortably all the time without using coordinate systems. However, we always know that in a Euclidean 3-space, if we wanted to calculate the change in a vector, we could revert to a coordinate system where we can calculate component differences instead of drawing arrow pictures. But in an arbitrary curved spacetime, the whole problem is that we *cannot* calculate the difference between two vectors located at different points in spacetime simply by subtracting their components, because the basis vectors are different at the different points, introducing spurious changes in the components that do not reflect the true difference. Equation 17.5 claims that the true change d in a vector field as we travel through a differential displacement d is given by $d\boldsymbol{A} = dx^\alpha(\nabla_\alpha A^\mu)\boldsymbol{e}_\mu$, but what does this physically mean?

We can answer this question most transparently using the important concept of a **locally inertial reference frame** (LIF). We have seen in chapter 12 how we can at a given event define a locally orthogonal frame (LOF) whose metric is the same as the flat-space metric $\eta_{\mu\nu}$ over a small region of spacetime. We have also discussed the intuitive idea that any curved surface is locally flat over a small enough patch of area, so any curved spacetime should be locally flat over a small enough region of spacetime. In this section, we will explore this concept more mathematically.

It turns out that in any non-singular spacetime, we can *always* find a coordinate system centered on an event P whose metric $g_{\mu\nu}$ at P reduces to the flat spacetime metric $\eta_{\mu\nu}$ (a LOF) *and* whose first derivatives all vanish at P: $\partial_\alpha g_{\mu\nu}(\text{at } P) = 0$. We do *not* have enough coordinate freedom to make all of the *second* derivatives of the metric vanish, though: these second derivatives contain information about the actual curvature of spacetime. These issues are explored in more depth in box 17.7.

Now, note that in a coordinate system where the first derivatives of the metric vanish, the Christoffel symbols also vanish (see equation 17.10). This means that in such a coordinate system the geodesic equation becomes simply $d^2x^\mu/d\tau^2 = 0$, the equation of a straight line. We call any LOF where the Christoffel symbols also vanish a "locally inertial frame" because particles following geodesics will look as if they obey Newton's first law in such a frame. A non-rotating and freely falling orthogonal reference frame is a LIF.

Now we are in a position to answer the question with which I started the section. Because the Christoffel symbols are zero at event P in a LIF, the absolute gradient in a LIF reduces to being the same thing as the ordinary gradient (see equations 17.6, 17.7, and 17.8). Therefore, a vector field is "truly" constant at a point if its ordinary gradient vanishes at that event in a LIF, and "true" differences between vectors at neighboring events can be evaluated as simple component differences in a LIF located at one of the events. A LIF therefore provides a backup coordinate system in which we can easily evaluate what "true" changes in a vector mean.

But note that we can do this only *locally*. Because we cannot make the second derivatives of the metric vanish, the metric of even the best LIF will begin to deviate from $\eta_{\mu\nu}$ as we move far enough away from the event P. Therefore, while everything looks simple in a LIF, it is a valid description of spacetime only in a small region around the event P.

The Absolute Gradient of the Metric. The LIF concept is very handy for writing proofs for equations involving gradients. For example, consider the absolute gradient of the metric $\nabla_\alpha g_{\mu\nu}$ in an arbitrary coordinate system in an arbitrary spacetime. In a LIF centered on any given event, the absolute gradient reduces to the ordinary gradient, and we see that its value is

$$\nabla_\alpha g_{\mu\nu} = \partial_\alpha g_{\mu\nu} = 0 \tag{17.13}$$

by definition of the LIF. But the equation $\nabla_\alpha g_{\mu\nu} = 0$ is a tensor equation, so if it holds in any coordinate system, it must hold in all. Therefore, we conclude directly that the absolute gradient of the metric must be zero in *every* coordinate system. This is a handy and powerful result, because it means that we can treat the metric as if it were a constant in any equation involving the absolute gradient.

One can prove this using equations 17.8 and 17.10 (see problem P17.9), but the method described above is *much* faster.

Physics Equations in Curved Spacetime. We can also take advantage of the equivalence principle applied to LIFs to convert physics equations valid in flat spacetime to versions that work in arbitrary coordinates. For example, in an inertial reference frame, Maxwell's equations at any given event in spacetime are

$$\partial_\nu F^{\mu\nu} = 4\pi k J^\mu, \qquad \partial^\mu F^{\alpha\beta} + \partial^\beta F^{\mu\alpha} + \partial^\alpha F^{\beta\mu} = 0 \tag{17.14}$$

By the equivalence principle, this should also be the form of these equations in a LIF centered on that event in a curved spacetime. But in that LIF, these equations are equivalent to the tensor equations

$$\nabla_\nu F^{\mu\nu} = 4\pi k J^\mu, \qquad \nabla^\mu F^{\alpha\beta} + \nabla^\beta F^{\mu\alpha} + \nabla^\alpha F^{\beta\mu} = 0 \tag{17.15}$$

where $\nabla^\mu \equiv g^{\mu\nu}\nabla_\nu$, etc. Therefore, these latter equations should apply in any arbitrary coordinate system in any arbitrary spacetime.

BOX 17.1 Absolute Gradient of a Vector

You can use the definition $\partial \mathbf{e}_\alpha / \partial x^\mu \equiv \Gamma^\nu_{\mu\alpha} \mathbf{e}_\nu$ (equation 17.3) to show that equation 17.4 implies equation 17.5, i.e., that

$$d\mathbf{A} = d(A^\mu \mathbf{e}_\mu) = \left(\frac{\partial A^\mu}{\partial x^\sigma} dx^\sigma\right)\mathbf{e}_\mu + A^\mu \frac{\partial \mathbf{e}_\mu}{\partial x^\alpha} dx^\alpha = \left[\frac{\partial A^\mu}{\partial x^\sigma} + \Gamma^\mu_{\sigma\nu} A^\nu\right] \mathbf{e}_\mu dx^\sigma \quad (17.16)$$

Exercise 17.1.1. Verify this (thus establishing the absolute gradient of $\mathbf{A}$).

BOX 17.2 Absolute Gradient of a Covector

We can prove that equation 17.7 correctly gives the absolute gradient of a covector as follows. Consider the scalar product $A^\mu B_\mu$ of an arbitrary vector field $\mathbf{A}$ and an arbitrary covector field $\mathbf{B}$. By the product rule,

$$\nabla_\alpha(A^\mu B_\mu) = (\nabla_\alpha A^\mu) B_\mu + A^\mu (\nabla_\alpha B_\mu) = \left[\frac{\partial A^\mu}{\partial x^\alpha} + \Gamma^\mu_{\alpha\nu} A^\nu\right] B_\mu + A^\mu (\nabla_\alpha B_\mu) \quad (17.17)$$

But because a scalar's value does not depend on any basis vectors, the absolute gradient of a scalar is the same as its ordinary gradient. Therefore

$$\nabla_\alpha(A^\mu B_\mu) = \partial_\alpha(A^\mu B_\mu) = \frac{\partial A^\mu}{\partial x^\alpha} B_\mu + A^\mu \frac{\partial B_\mu}{\partial x^\alpha} \quad (17.18)$$

If you set equation 17.18 equal to the last version of equation 17.17 and rename some bound indices, you can show that

$$0 = \left[-\frac{\partial B_\mu}{\partial x^\alpha} + \Gamma^\sigma_{\alpha\mu} B_\sigma + (\nabla_\alpha B_\mu)\right] A^\mu \quad (17.19)$$

Since this must be true for totally arbitrary $\mathbf{A}$, the quantity in brackets must be equal to zero. Rearranging this gives equation 17.7.

Exercise 17.2.1. Show that equation 17.19 follows from equations 17.18 and 17.17.

BOX 17.3 Symmetry of the Christoffel Symbols

Consider the double absolute gradient of a scalar function $\Phi(x^\alpha)$. According to equation 17.7, this is equal to

$$\nabla_\mu \nabla_\nu \Phi = \nabla_\mu (\partial_\nu \Phi) = \partial_\mu \partial_\nu \Phi - \Gamma^\alpha_{\mu\nu}(\partial_\alpha \Phi) \qquad (17.20)$$

Now, note that in a LIF, this reduces to

$$\nabla_\mu \nabla_\nu \Phi = \partial_\mu \partial_\nu \Phi \quad \text{(in a LIF)} \qquad (17.21)$$

Note that this expression is symmetric under interchange of the μ and ν indices: Now, the expression $\nabla_\mu \nabla_\nu \Phi = \nabla_\nu \nabla_\mu \Phi$ is a tensor equation, so if it is true in any coordinate system, it must be true in all. Therefore, equation 17.20 must also be symmetric under interchange of the μ and ν indices, which is true only if the Christoffel symbol is symmetric under the interchange of its lower indices. Q. E. D.

BOX 17.4 The Christoffel Symbols in Terms of the Metric

Start by taking the scalar product of the defining equation with $\mathbf{e}_\rho$:

$$\Gamma^\nu_{\mu\alpha} \mathbf{e}_\nu \cdot \mathbf{e}_\rho = \frac{\partial \mathbf{e}_\alpha}{\partial x^\mu} \cdot \mathbf{e}_\rho = \frac{\partial(\mathbf{e}_\alpha \cdot \mathbf{e}_\rho)}{\partial x^\mu} - \mathbf{e}_\alpha \cdot \frac{\partial \mathbf{e}_\rho}{\partial x^\mu} = \frac{\partial g_{\alpha\rho}}{\partial x^\mu} - \Gamma^\beta_{\mu\rho} \mathbf{e}_\alpha \cdot \mathbf{e}_\beta$$

$$\Rightarrow \quad \Gamma^\nu_{\mu\alpha} g_{\nu\rho} + \Gamma^\beta_{\mu\rho} g_{\alpha\beta} = \partial_\mu g_{\alpha\rho} \quad \Rightarrow \quad \Gamma^\nu_{\mu\alpha} g_{\nu\rho} + \Gamma^\nu_{\mu\rho} g_{\nu\alpha} = \partial_\mu g_{\alpha\rho} \qquad (17.22)$$

where in the last step, I renamed $\beta \to \nu$ and used the fact that the metric is symmetric to reverse the indices on the middle term. By cyclically renaming the indices μ, α, and ρ, we can generate two more similar equations:

$$\Gamma^\nu_{\alpha\rho} g_{\nu\mu} + \Gamma^\nu_{\alpha\mu} g_{\nu\rho} = \partial_\alpha g_{\rho\mu} \qquad (17.23)$$

$$\Gamma^\nu_{\rho\mu} g_{\nu\alpha} + \Gamma^\nu_{\rho\alpha} g_{\nu\mu} = \partial_\rho g_{\mu\alpha} \qquad (17.24)$$

If you add 17.22 and 17.23 and subtract 17.24, take advantage of the symmetry in the lower indices of the Christoffel symbols, multiply both sides by $\frac{1}{2} g^{\sigma\rho}$, and use $g^{\sigma\rho} g_{\rho\nu} = \delta^\sigma_\nu$, you can show that

$$\Gamma^\sigma_{\mu\alpha} = \tfrac{1}{2} g^{\sigma\rho} [\partial_\mu g_{\alpha\rho} + \partial_\alpha g_{\rho\mu} - \partial_\rho g_{\mu\alpha}] \qquad (17.25)$$

This is equivalent to equation 17.10.

Exercise 17.4.1. Verify equation 17.25.

BOX 17.5 Checking the Geodesic Equation

The geodesic equation given by equation 8.12 is

$$0 = \frac{d}{d\tau}\left(g_{\mu\nu}\frac{dx^\nu}{d\tau}\right) - \tfrac{1}{2}\partial_\mu g_{\alpha\beta}\frac{dx^\alpha}{d\tau}\frac{dx^\beta}{d\tau} \qquad (17.26)$$

To see that this is the same as equation 17.12, first note that $g_{\mu\nu}$ depends on τ because $g_{\mu\nu}$ depends on position, and the object's position depends on τ as it follows the geodesic. So, using the chain rule to expand the derivative in parentheses,

$$0 = \frac{\partial g_{\mu\nu}}{\partial x^\sigma}\frac{dx^\sigma}{d\tau}\frac{dx^\nu}{d\tau} + g_{\mu\nu}\frac{d^2x^\nu}{d\tau^2} - \tfrac{1}{2}\partial_\mu g_{\alpha\beta}\frac{dx^\alpha}{d\tau}\frac{dx^\beta}{d\tau}$$

$$= \tfrac{1}{2}\partial_\sigma g_{\mu\nu}\frac{dx^\sigma}{d\tau}\frac{dx^\nu}{d\tau} + \tfrac{1}{2}\partial_\sigma g_{\mu\nu}\frac{dx^\sigma}{d\tau}\frac{dx^\nu}{d\tau} + g_{\mu\nu}\frac{d^2x^\nu}{d\tau^2} - \tfrac{1}{2}\partial_\mu g_{\alpha\beta}\frac{dx^\alpha}{d\tau}\frac{dx^\beta}{d\tau} \qquad (17.27)$$

In the last step, I simply divided the front term into two equal pieces and wrote the derivative in the more compact notation. Now, you can rename $\sigma \to \alpha, \nu \to \beta$ in the first term and $\sigma \to \beta, \nu \to \alpha$ in the second term, multiply through by $g^{\sigma\mu}$, use the fact that $g^{\sigma\mu}g_{\mu\nu} = \delta^\sigma{}_\nu$, and use equation 17.10 to get equation 17.12.

Exercise 17.5.1. Derive equation 17.12 from equation 17.27.

BOX 17.6 A Trick for Calculating Christoffel Symbols

In the previous box, we saw that equations 17.26 and 17.12 were equivalent. Comparing these equations actually provides a fast way to calculate Christoffel symbols for any metric. For example, consider the Schwarzschild metric. Since the Schwarzschild metric does not depend on time, the time component of the geodesic equation in this case reads

$$0 = \frac{d}{d\tau}\left(g_{t\nu}\frac{dx^\nu}{d\tau}\right) = \frac{\partial g_{t\nu}}{\partial x^\sigma}\frac{dx^\sigma}{d\tau}\frac{dx^\nu}{d\tau} + g_{t\nu}\frac{d^2x^\nu}{d\tau^2} \qquad (17.28)$$

But the metric is diagonal, so in the sum over ν, only the $\nu = t$ term is nonzero. Moreover, $g_{tt} = -(1 - 2GM/r)$, so only its r-derivative is nonzero. Therefore, equation 17.28 simplifies to

$$0 = -\frac{\partial}{\partial r}\left(1 - \frac{2GM}{r}\right)\frac{dr}{d\tau}\frac{dt}{d\tau} - \left(1 - \frac{2GM}{r}\right)\frac{d^2t}{d\tau^2}$$

$$\Rightarrow \quad 0 = \frac{2GM}{r^2}\left(1 - \frac{2GM}{r}\right)^{-1}\frac{dr}{d\tau}\frac{dt}{d\tau} + \frac{d^2t}{d\tau^2} \qquad (17.29)$$

Now let's compare this to equation 17.12:

$$0 = \frac{d^2t}{d\tau^2} + \Gamma^t{}_{\mu\nu}\frac{dx^\mu}{d\tau}\frac{dx^\nu}{d\tau} \qquad (17.12r)$$

We can see immediately that of the 16 Christoffel symbols $\Gamma^t{}_{\mu\nu}$ that have t as a superscript, the only nonzero symbols are

BOX 17.6 (continued) A Trick for Calculating Christoffel Symbols

$$\Gamma^t_{rt} = \Gamma^t_{tr} = \frac{GM}{r^2}\left(1 - \frac{2GM}{r}\right)^{-1} \tag{17.30}$$

because there are no $(dx^\mu/d\tau)(dx^\nu/d\tau)$ terms in equation 17.29 other than the single term involving $(dr/d\tau)(dt/d\tau)$. Moreover, because the Christoffel symbols are symmetric, and because $(dr/d\tau)(dt/d\tau) = (dt/d\tau)(dr/d\tau)$, the single term in equation 17.29 corresponds to the sum of two equal terms in equation 17.12. Therefore each of the Christoffel symbols in equation 17.30 corresponds to half the value of the single term involving $(dr/d\tau)(dt/d\tau)$ in equation 17.29.

Exercise 17.6.1. Use the same method to find all of the Schwarzschild Christoffel symbols with θ as a superscript.

BOX 17.7 The Local Flatness Theorem

In this box, I will first argue that it should be possible to find a LIF as defined in the text. Consider a transformation $x'^\mu(x^\nu)$ that converts unprimed coordinates to primed coordinates. Under this coordinate transformation, the components of the metric transform as follows:

$$g'_{\mu\nu} = \frac{\partial x^\alpha}{\partial x'^\mu}\frac{\partial x^\beta}{\partial x'^\nu}g_{\alpha\beta} \tag{17.31}$$

Our goal is to find a coordinate transformation that makes $g'_{\mu\nu} = \eta_{\mu\nu}$ and $\partial'_\alpha g'_{\mu\nu} = 0$ at some event P.

Let's expand each of the 16 transformation partials $\partial x^\alpha/\partial x'^\mu$ in a Taylor series in x'^ν around some event P with primed coordinates x'^ν_P:

$$\frac{\partial x^\alpha}{\partial x'^\mu} = a^\alpha{}_\mu + b^\alpha{}_{\mu\nu}\Delta x'^\nu + c^\alpha{}_{\mu\nu\beta}\Delta x'^\nu \Delta x'^\beta + \ldots \tag{17.32a}$$

where
$$a^\alpha{}_\mu \equiv \left(\frac{\partial x^\alpha}{\partial x'^\mu}\right)_{\text{at } P} \tag{17.32b}$$

$$b^\alpha{}_{\mu\nu} \equiv \left[\frac{\partial}{\partial x'^\nu}\left(\frac{\partial x^\alpha}{\partial x'^\mu}\right)\right]_{\text{at } P} = \left[\frac{\partial^2 x^\alpha}{\partial x'^\nu \partial x'^\mu}\right]_{\text{at } P} \tag{17.32c}$$

$$c^\alpha{}_{\mu\nu\beta} \equiv \frac{1}{2!}\left[\frac{\partial^2}{\partial x'^\beta \partial x'^\nu}\left(\frac{\partial x^\alpha}{\partial x'^\mu}\right)\right]_{\text{at } P} = \frac{1}{2}\left[\frac{\partial^3 x^\alpha}{\partial x'^\beta \partial x'^\nu \partial x'^\mu}\right]_{\text{at } P} \tag{17.32d}$$

and
$$\Delta x'^\nu \equiv x'^\nu - x'^\nu_P \tag{17.32e}$$

The sets of quantities $a^\alpha{}_\mu, b^\alpha{}_{\mu\nu}$, and $c^\alpha{}_{\mu\nu\beta}$ are constant coefficients that we are free to choose as we seek to find a transformation that takes us to the LIF: different choices for these coefficients simply generate different transformations.

BOX 17.7 (continued) The Local Flatness Theorem

In order to determine how much freedom we have, it is valuable to count how many free choices we have at each step of the power series expansion. In the case of the coefficients $a^\alpha{}_\mu = (\partial x^\alpha / \partial x'^\mu)_P$, we can choose all 16 of the coefficients freely. However, because the order of partial differentiation does not matter, in the case of $b^\alpha{}_{\mu\nu} = (\partial^2 x^\alpha / \partial x'^\mu \partial x'^\nu)_P$, there are 4 choices of α, but only 10 distinct choices for μ and ν (out of the 16 possible choices of these index values, there are 12 choices that give the two indices different values, and 6 of these will be copies of the other 6 with the indices reversed, so there are 16 − 6 = 10 unique values). Therefore the coefficients $b^\alpha{}_{\mu\nu}$ have only 40 values that we can freely choose. Similarly, $c^\alpha{}_{\mu\nu\beta} = (\partial^3 x^\alpha / \partial x'^\mu \partial x'^\nu \partial x'^\beta)_P$ has 80 independent values: 4 choices for α times 20 distinct combinations of the other three indices that are not permutations of others.

Exercise 17.7.1. Counting the 20 distinct ways to distribute values among the three available indices is the same as counting the number of ways that one can distribute the numbers 0, 1, 2, 3 among three slots so that no combination is a permutation of any other combination. For example, 000 is unique, 010 is unique, 001 is a permutation of the second, and so on. List all 20 valid combinations in this format. (*Hint:* Start with the combinations where all three numbers are the same, then go to combinations where two numbers are the same, then to combinations where the three numbers are unique.)

Expanding the unprimed metric in a similar power series yields

$$g_{\mu\nu} = [g_{\mu\nu}]_{\text{at } P} + \Delta x'^\gamma [\partial'_\gamma g_{\mu\nu}]_{\text{at } P} + \tfrac{1}{2} \Delta x'^\gamma \Delta x'^\sigma [\partial'_\gamma \partial'_\sigma g_{\mu\nu}]_{\text{at } P} + \ldots \quad (17.33)$$

Now let's substitute equations 17.32 and 17.33 into equation 17.31 and gather terms involving similar powers of $\Delta x'^\nu$. To make things easier to see, let's suppress all the indices and simply write equations 17.32 and 17.33 this way:

$$\left[\frac{\partial \mathbf{x}}{\partial \mathbf{x}'}\right] = \mathbf{a} + \mathbf{b}(\Delta \mathbf{x}') + \mathbf{c}(\Delta \mathbf{x}')^2 + \ldots \quad (17.34a)$$

$$\mathbf{g} = \mathbf{g}_P + \partial \mathbf{g}_P \Delta \mathbf{x}' + \tfrac{1}{2} \partial^2 \mathbf{g}_P (\Delta \mathbf{x}')^2 + \ldots \quad (17.34b)$$

with the boldface letters reminding us that there are indices that we are suppressing. In the same compact notation, equation 17.31 says that

$$\mathbf{g}' = \left[\frac{\partial \mathbf{x}}{\partial \mathbf{x}'}\right]^2 \mathbf{g} = [\mathbf{a} + \mathbf{b}\,\Delta \mathbf{x}' + \mathbf{c}(\Delta \mathbf{x}')^2 + \ldots]^2 [\mathbf{g}_P + \partial \mathbf{g}_P \Delta \mathbf{x}' + \tfrac{1}{2}\partial^2 \mathbf{g}_P (\Delta \mathbf{x}')^2 + \ldots]$$

$$= \mathbf{a}^2 \mathbf{g}_P + (\mathbf{abg}_P + \mathbf{bag}_P + \mathbf{a}^2 \partial \mathbf{g}_P)\,\Delta \mathbf{x}' + (\mathbf{acg}_P + \mathbf{ab}\partial \mathbf{g}_P + \ldots)(\Delta \mathbf{x}')^2 + \ldots$$

$$(17.35)$$

Exercise 17.7.2. Convince yourself that this is correct, and write the four missing terms (in compact notation) inside the parentheses of the last term.

BOX 17.7 (continued) The Local Flatness Theorem

Now, examine equation 17.35. The first term sets the value of $g'_{\mu\nu}$ at event P. Let's think about whether we have enough freedom to set

$$[g'_{\mu\nu}]_P = a^\alpha{}_\mu a^\beta{}_\nu [g_{\alpha\beta}]_P = \eta_{\mu\nu} \quad \text{(reverting to index notation)} \quad (17.36)$$

Because the metric is symmetric, equation 17.36 only represents 10 independent conditions on the 16 components of $a^\alpha{}_\mu$. So we can find values $a^\alpha{}_\mu$ that satisfy these equations and have 6 degrees of freedom left over! This is because even once we have determined the metric, we still have some freedom to choose coordinates: we can rotate our final coordinate system and boost it to a constant velocity relative to the original coordinate system and still have the flat-space metric. The 6 leftover degrees of freedom allow us to specify a rotation about an arbitrary axis in 3-space and a 3-component boost for our new coordinate system. If we go only this far, we have a LOF, but not yet a LIF.

If we compare equation 17.35 with a normal power series expansion of the primed metric, we see that the factor in parentheses in the second term of equation 17.35 corresponds to the first derivative of the primed metric evaluated at P:

$$a^\alpha{}_\mu b^\beta{}_{\nu\gamma}[g_{\alpha\beta}]_P + b^\alpha{}_{\mu\gamma} a^\beta{}_\nu [g_{\alpha\beta}]_P + a^\alpha{}_\mu a^\beta{}_\nu [\partial'_\gamma g_{\alpha\beta}]_P = [\partial'_\gamma g'_{\mu\nu}]_P \quad (17.37)$$

Since the metric is symmetric, this represents $4 \times 10 = 40$ independent equations. Do we have enough freedom to set $\partial'_\gamma g'_{\mu\nu} = 0$ at event P? Yes, but just barely. The unprimed metric and its derivatives are given, and we have now used up our freedom to choose the values of $a^\alpha{}_\mu$ in the first step. So the 40 independent equations should allow us to find a unique solution for the 40 independent values of $b^\alpha{}_{\mu\nu}$ that will set the metric first derivatives to zero at point P, yielding a LIF.

The factor in parentheses in the third term of equation 17.35 corresponds to the second derivative of the primed metric evaluated at P:

$$a^\alpha{}_\mu c^\beta{}_{\nu\gamma\sigma}[g_{\alpha\beta}]_P + a^\alpha{}_\mu b^\beta{}_{\nu\gamma}[\partial'_\sigma g_{\alpha\beta}]_P + \text{etc.} = \partial'_\gamma \partial'_\sigma g'_{\mu\nu} \quad (17.38)$$

Because both the metric and the partial derivatives are symmetric under the interchange of indices, this single line represents 100 independent equations (isn't index notation wonderful?). Again, the values of the unprimed metric and its derivatives are given, and we have now determined the values of both $a^\alpha{}_\mu$ and $b^\alpha{}_{\mu\nu}$, so our freedom at this point is limited to the 80 independent values of $c^\alpha{}_{\mu\nu\sigma}$. We can therefore use our freedom to set 80 of the metric second-derivatives to zero, but we will have 20 metric second derivatives left over that we *cannot* set to zero. These 20 nonzero derivatives tell us something about the curvature of spacetime. We will in fact shortly see that a quantity called the Riemann tensor, which depends on the second derivatives of the metric, and which describes the curvature of spacetime, has 20 independent components.

So, I have shown that one should in principle be able to find a coordinate system centered at an event P where the metric (at P) is the flat-space metric *and* the first derivatives of the metric (at P) are zero, but we cannot make all of the *second* derivatives of the metric vanish at P unless we are in a truly flat space. This counting argument is not exactly a mathematical proof valid in all circumstances, but it is strong evidence of such an assertion.

Problem P17.10 shows that you can always set the first derivatives of the metric equal to zero at a point P without actually changing the value of the metric at that point as long as the absolute gradient of the metric is zero. We proved that the absolute gradient of the metric is zero using a LIF, but one can also prove it from equation 17.10, which in turn requires only that the Christoffel symbols are symmetric. We can see, therefore, that the existence of a LIF, the absolute gradient of the metric being zero, and the symmetry of the Christoffel symbols are all linked: taking any one of these statements to be true enables us to strictly prove the others.

17. THE ABSOLUTE GRADIENT

HOMEWORK PROBLEMS

P17.1 Use the method described in box 17.6 to evaluate all eight Christoffel symbols for polar coordinates, and check that at least one (nonzero) symbol agrees with the result calculated directly from equation 17.10.

P17.2 Use the method described in box 17.6 to evaluate all the Schwarzschild Christoffel symbols not calculated in that box, and check that at least one (nonzero) symbol agrees with the result calculated directly from equation 17.10.

P17.3 Consider a cylindrical r, θ, z coordinate basis in a flat spacetime. The spatial part of the metric equation for such a coordinate system is the same as for polar coordinates with an extra term for the z coordinate:

$$ds^2 = -dt^2 + dr^2 + r^2 d\theta^2 + dz^2 \qquad (17.39)$$

a. Use the method described in box 17.6 to evaluate the Christoffel symbols for this coordinate basis.
b. Express the four components of the Maxwell equation $\nabla_\nu F^{\mu\nu} = 4\pi k J^\mu$ in the cylindrical coordinate basis. (*Hint:* Define the components of $F^{\mu\nu}$ in analogy to equation 7.12, i.e., $F^{tr} = E_r$, $F^{rz} = -B_\theta$, etc. This will *not* yield the results that you would find by looking up cylindrical-coordinate expressions in an electrodynamics book, because we are here using a cylindrical *coordinate* basis, not the cylindrical *orthornomal* basis typically used in electrodynamics textbooks.)

P17.4 Consider arbitrary tensor quantities $B^\mu{}_\nu$ and A^α in an arbitrary coordinate system. Prove that the absolute gradient of the product of these quantities obeys the product rule; i.e., $\nabla_\beta(B^\mu{}_\nu A^\alpha) = (\nabla_\beta B^\mu{}_\nu) A^\alpha + B^\mu{}_\nu (\nabla_\beta A^\alpha)$. [The product rule was assumed in box 17.2 but never directly proven. There is a quick way to do this involving using a locally inertial frame (LIF), but a brute-force calculation from the definition is also OK.]

P17.5 Consider the semi-log p, q coordinate system described in problem P5.4, where $p = x$ and $q = e^{by}$.
a. Determine which is the only nonzero Christoffel symbol for this metric and calculate its value.
b. Consider the vector field $\mathbf{A}(x^\mu)$ such that $A^x = 0$, $A^y = Cx$, where C is some constant. What is the absolute gradient $\nabla_\nu A^\mu$ of this vector field in cartesian coordinates? What is its absolute gradient in p, q coordinates? Carefully explain why the components of the absolute gradient are *not* the same in both coordinate systems.
c. Treat the absolute gradient $\nabla_\nu A^\mu$ found in part *b* as a second-rank tensor with one upper and one lower index. Transform the tensor from p, q to x, y coordinates, and show that the result is the same as the absolute gradient you directly computed in x, y coordinates.

P17.6 (Requires having done problem P5.5.) Consider the situation described in problem P5.5, where an object was moving in a 2D flat space with velocity components that were clearly constant in cartesian coordinates but *not* constant in a sinusoidal coordinate system. If we define the object's acceleration to be $a^\mu \equiv dv^\mu/dt$, we found that it had different magnitudes in the different coordinate systems, a problem we could not resolve in chapter 5. Using the methods of this chapter, show how to express $\mathbf{a} \equiv d\mathbf{u}/dt$ *correctly* in both coordinate systems, and show that for the object in question, the components of $\mathbf{a}$ are zero in *both* coordinate systems. (Note that in this Newtonian situation, t is a *parameter*, analogous to τ in spacetime, not a coordinate.)

P17.7 An object's four-acceleration is $\mathbf{a} \equiv d\mathbf{u}/d\tau$. As discussed in this chapter, $\mathbf{a} = 0$ for an object following a geodesic. Consider an object at rest at a Schwarzschild coordinate r. Such an object is *not* following a geodesic: a spaceship would have to constantly fire its rocket engines to hover at such a position. Find $\mathbf{a}$'s Schwarzschild components a^μ for such an object, and compute its coordinate-independent magnitude $a \equiv \sqrt{\mathbf{a} \cdot \mathbf{a}}$. [*Hints:* Note that $a^\mu \neq du^\mu/d\tau$. Start by determining the object's four-velocity $\mathbf{u}$ (as we have done before in chapter 12): remember that u^t is neither 0 nor 1. Then adapt equation 17.11 or 17.12 to the case where $du/d\tau \neq 0$. Finally, if you have not done problem P17.2, use some method to evaluate the handful of Schwarzschild Christoffel symbols you need.]

Note: Since a is coordinate-independent, it will be equal to the magnitude of the *ordinary* acceleration $\vec{a} = d\vec{v}/dt$ we would observe in a freely falling frame instantaneously at rest with respect to the object, since $dt = d\tau$ and $\mathbf{u} = [1, v_x, v_y, v_z]$ in such a frame. Knowing a would thus help you determine the thrust you'd need to hover at r. You should find that a goes to infinity as r approaches $2GM$.

P17.8 Imagine a vector field $\mathbf{v}(r)$ near a Schwarzschild black hole that in the Schwarzschild coordinate basis has components $v^\mu = [(1 - 2GM/r), 0, 0, 0]$. Calculate the components of this field's absolute gradient $\nabla_\alpha v^\mu$ in the Schwarzschild coordinate basis.

P17.9 According to the rules articulated below equation 17.8, the absolute gradient of the metric should be

$$\nabla_\alpha g_{\mu\nu} = \partial_\alpha g_{\mu\nu} - \Gamma^\sigma_{\alpha\mu} g_{\sigma\nu} - \Gamma^\sigma_{\alpha\nu} g_{\mu\sigma} \qquad (17.40)$$

Use equation 17.10 to show that this is zero.

P17.10 Consider a coordinate transformation such that

$$\frac{\partial x^\alpha}{\partial x'^\mu} = \delta^\alpha{}_\mu - \Gamma^\alpha_{\mu\nu,P}(x'^\nu - x'^\nu_P) \qquad (17.41)$$

where $\Gamma^\alpha_{\mu\nu,P}$ are constants equal to the Christoffel symbols of the unprimed metric evaluated at point P. Show that since the absolute gradient of the unprimed metric is zero, this transformation, when plugged into equation 17.31, yields a primed metric at point P that has the same components as the unprimed metric at P, but whose first derivatives are all zero.

18. GEODESIC DEVIATION

INTRODUCTION

FLAT SPACETIME
Review of Special Relativity
Four-Vectors
Index Notation

TENSORS
Arbitrary Coordinates
Tensor Equations
Maxwell's Equations
Geodesics

SCHWARZSCHILD BLACK HOLES
The Schwarzschild Metric
Particle Orbits
Precession of the Perihelion
Photon Orbits
Deflection of Light
Event Horizon
Alternative Coordinates
Black Hole Thermodynamics

THE CALCULUS OF CURVATURE
The Absolute Gradient
Geodesic Deviation
The Riemann Tensor

THE EINSTEIN EQUATION
The Stress-Energy Tensor
The Einstein Equation
Interpreting the Equation
The Schwarzschild Solution

COSMOLOGY
The Universe Observed
A Metric for the Cosmos
Evolution of the Universe
Cosmic Implications
The Early Universe
CMB Fluctuations & Inflation

GRAVITATIONAL WAVES
Gauge Freedom
Detecting Gravitational Waves
Gravitational Wave Energy
Generating Gravitational Waves
Gravitational Wave Astronomy

SPINNING BLACK HOLES
Gravitomagnetism
The Kerr Metric
Kerr Particle Orbits
Ergoregion and Horizon
Negative-Energy Orbits

this depends on this

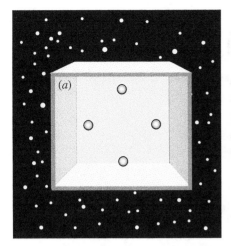
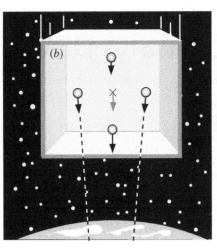

FIG. 18.1 These drawings show a set of four balls floating (initially) at rest in two different reference frames. In (a), the frame floats in deep space. In (b) and (c), the frame is freely falling near the earth. From a Newtonian perspective, as shown in (b), the balls' accelerations (black arrows) do not quite have the same magnitude and direction as the frame center's acceleration (gray arrow), so from the perspective of an observer *in* that frame, the balls will display a tiny acceleration relative to the frame, as shown in (c). This does not happen in deep space.

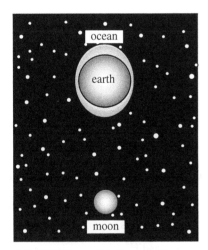

FIG. 18.2 The earth is freely falling in the moon's gravitational field. The oceans therefore experience the same kind of relative acceleration in the earth's frame that the balls do in figure 18.1c. This causes the oceans to become deeper on the sides of the earth facing and opposite to the moon, and shallower on the sides. The sun's field contributes a smaller effect not shown. (This drawing is obviously not drawn to scale.)

Introduction. One of Euclid's basic axioms about *flat* space is that parallel lines (geodesics) never meet. In spacetime, we saw qualitatively in the first chapter that the tidal effects of gravity cause objects following initially parallel geodesics to accelerate relative to each other as time passes, meaning that their geodesics do not remain parallel. Since this contradicts Euclid's axiom, this is a fundamental indicator that spacetime surrounding a gravitating object is *curved*.

Our ultimate goal in this chapter is to construct a tensor called the **Riemann tensor** that quantifies the relative acceleration of initially parallel geodesics. Since such geodesics accelerate relative to each other only in a curved space(time), the Riemann tensor will be zero everywhere in flat space(time) but nonzero in a curved space(time), providing a foolproof way to distinguish curved and flat space(time)s.

To set the stage, we will first review the phenomenon of geodesic deviation and examine it in the context of Newtonian gravitational theory.

The Physics of Tidal Effects. Consider a frame floating in deep space, as shown in figure 18.1a, and imagine that we place balls at rest at various locations in that frame. As such a frame is inertial, these balls will continue to float at rest. Their worldlines drawn on a spacetime diagram would be purely vertical and parallel.

According to Einstein's equivalence principle, a *freely falling frame* is the closest thing we have to an inertial reference frame near the earth's surface. Imagine that we perform the same experiment in such a frame. From a Newtonian perspective (figure 18.1b), the gravitational force exerted by the earth causes the balls and frame to accelerate downward together, and so the balls will initially appear to remain at rest. But balls above the frame's center experience a bit *less* acceleration than the frame's center (because they are a bit farther from the earth's center). Similarly, balls below the center experience a bit *more* acceleration. Balls on either side of the center experience a gravitational acceleration that has a slight inward component, because the direction of the acceleration must always point toward the earth's center.

From the point of view of an observer sitting *in* this freely falling frame (figure 18.1c), balls above and below the center will therefore slowly accelerate *away* from the center, while balls on the sides will slowly accelerate *toward* the center. If we plot these balls' geodesics on a spacetime diagram, their worldlines will be initially vertical and parallel (because the balls are initially at rest), but after enough time has passed, we will see their worldlines curve toward or away from each other.

This relative acceleration of free objects in a freely falling frame demonstrates a gravitational field's **tidal effects**. It is precisely this aspect of the moon's gravitational field that raises tides in the earth's oceans (see figure 18.2). Offering a coherent explanation of the ocean tides was one of Newton's great triumphs.

A Newtonian Analysis of Tidal Effects. In Newtonian mechanics, we can describe a gravitational field either in terms of a **gravitational field vector** $\vec{g} \equiv \vec{F}_g/m$ (where

$\vec{F}_g$ is the gravitational force on a test object of mass m) or in terms of a **gravitational potential** $\Phi \equiv V_g/m$ (where V_g is that test object's gravitational potential energy). The gravitational *field* in empty space a distance r away from the the center of a spherical object of mass M is $\vec{g} = (-GM/r^2)\hat{r}$ (where $\hat{r}$ is a unit vector in the radially outward direction), while the potential is $\Phi = -GM/r$. Just as the electric field and electric potential are linked by the expression $\vec{E} = -\vec{\nabla}\phi$, so the the gravitational field and potential are linked by $\vec{g} = -\vec{\nabla}\Phi$.

The analogy with general relativity is closer if we base our analysis on the gravitational potential. According to Newton's second law, the acceleration of a freely falling particle is $\vec{a} = \vec{F}_g/m \equiv \vec{g} = -\vec{\nabla}\Phi$, or in index notation,

$$\frac{d^2x^i}{dt^2} = -\eta^{ij}\frac{\partial \Phi}{\partial x^j} = -\eta^{ij}[\partial_j \Phi]_{\vec{x}} \tag{18.1}$$

where the Latin indexes i and j are summed only over *spatial* coordinates and η^{ij} represents the spatial components of the flat-space metric. (Following a widely adopted convention, I will use Latin indices instead of Greek indices to indicate a sum *only* over spatial coordinates.) The notation $[\partial_j \Phi]_{\vec{x}}$ reminds us that at every instant of time, we must evaluate the derivative *at the particle's position* $\vec{x}$.

Consider now a reference particle falling according to equation 18.1 and a second particle falling a short displacement $\vec{n}(t)$ away, meaning that its position coordinates at a given instant of time are $x^i(t) + n^i(t)$. We call $\vec{n}(t)$ the **separation vector** between these particles at time t. The relative acceleration of these freely falling particles (the signal of the tidal effect) will then be simply $d^2\vec{n}/dt^2$.

How might we calculate this relative acceleration? Well, by Newton's second law, the acceleration of the second particle is

$$\frac{d^2(x^i + n^i)}{dt^2} = -\eta^{ij}[\partial_j \Phi]_{\vec{x}+\vec{n}} \tag{18.2}$$

where the notation $[\partial_j \Phi]_{\vec{x}+\vec{n}}$ reminds us to evaluate the derivative at the second particle's position, not the reference particle's position. We can estimate this derivative by expanding it in a Taylor series around the reference particle's position:

$$[\partial_j \Phi]_{\vec{x}+\vec{n}} = [\partial_j \Phi]_{\vec{x}} + n^k\left(\frac{\partial}{\partial x^k}[\partial_j \Phi]\right)_{\vec{x}} + \ldots \tag{18.3}$$

If the separation vector is small compared to the scale over which the potential varies significantly, we can ignore higher-order terms in this expansion. If we substitute this into equation 18.2, and subtract equation 18.1 from both sides, we get

$$\cancel{\frac{d^2x^i}{dt^2}} + \frac{d^2n^i}{dt^2} - \cancel{\frac{d^2x^i}{dt^2}} \approx \cancel{-\eta^{ij}[\partial_j \Phi]_{\vec{x}}} - \eta^{ij}n^k[\partial_k\partial_j \Phi]_{\vec{x}} \cancel{+\eta^{ij}[\partial_j \Phi]_{\vec{x}}}$$

$$\Rightarrow \frac{d^2n^i}{dt^2} \approx -\eta^{ij}[\partial_k\partial_j \Phi]n^k \equiv -\eta^{ij}\frac{\partial^2 \Phi}{\partial x^k \partial x^j}n^k \tag{18.4}$$

where we evaluate all derivatives at the *first* particle's position.

Equation 18.4 is the **Newtonian tidal deviation equation**. Given a freely falling reference particle (which we could take to coincide with the origin of our freely falling frame), we can use it to calculate the relative acceleration of any falling particle separated from the reference particle by a small vector $\vec{n}(t)$. For example, if we choose the z coordinate to be radially outward near a spherical object, you can use this equation (see box 18.1) to show that

$$\frac{d^2n^x}{dt^2} = -\frac{GM}{r^3}n^x, \quad \frac{d^2n^y}{dt^2} = -\frac{GM}{r^3}n^y, \quad \frac{d^2n^z}{dt^2} = +\frac{2GM}{r^3}n^z \tag{18.5}$$

Particles vertically separated from the reference particle (or frame origin) will therefore accelerate away from it, and particles horizontally displaced from it will accelerate toward it, consistent with what is shown in figure 18.1c.

Introduction to the Einsteinian Analysis of Geodesic Deviation. From an Einsteinian perspective, the relative accelerations of particles in a freely falling frame are *not* caused by variations in the gravitational force (there is no such thing). Rather, free particles follow geodesics in spacetime, and if these particles accelerate relative to each other, it is because these geodesics are curving relative to each other, indicating that the underlying spacetime is curved. This curvature is the frame-independent indicator of the presence of gravity (see figure 18.1).

In the analysis that follows, we will derive an equation of geodesic deviation in almost exactly the same way that we did in the Newtonian analysis, but using relativistic concepts of curved spacetime and expressing everything in terms of tensors so that our analysis is valid in arbitrary coordinates.

The Situation. Consider two free particles that follow neighboring geodesics described by the two functions $x^\alpha(\tau)$ (reference particle) and $\bar{x}^\alpha(\tau) \equiv x^\alpha(\tau) + n^\alpha(\tau)$ (second particle) where τ is the proper time measured *by the reference particle* and $\boldsymbol{n}$ is the infinitesimal **separation four-vector** that stretches between the two objects at a given τ (see figure 18.3). If these paths are infinitesimally close together and are also parallel at $\tau = 0$, then τ will almost exactly equal the proper time measured by the second particle as well. Note that the separation four-vector $n^\alpha \equiv \bar{x}^\alpha - x^\alpha$ is *not* a true four-vector unless it is infinitesimal.

To say that these geodesics *remain* parallel is to say that the two objects have zero relative acceleration, or (mathematically) $d^2\boldsymbol{n}/d\tau^2 = 0$, where $d^2\boldsymbol{n}/d\tau^2$ is the *absolute* double time-derivative of $\boldsymbol{n}$ (i.e., as computed in a LIF). Our goal is to calculate a tensor expression for the components of $d^2\boldsymbol{n}/d\tau^2$.

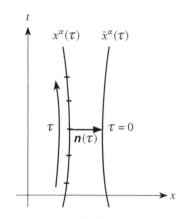

FIG. 18.3 The worldlines of two objects following neighboring geodesics. The vector $\boldsymbol{n}$ connects the two objects at all times τ. The geodesics are parallel at $\tau = 0$.

Implications of the Geodesic Equation. Since each particle follows a geodesic, each will obey the geodesic equation:

$$0 = \frac{d^2 x^\alpha}{d\tau^2} + \Gamma^\alpha_{\mu\nu} \frac{dx^\mu}{d\tau} \frac{dx^\nu}{d\tau}, \quad 0 = \frac{d^2 \bar{x}^\alpha}{d\tau^2} + \bar{\Gamma}^\alpha_{\mu\nu} \frac{d\bar{x}^\mu}{d\tau} \frac{d\bar{x}^\nu}{d\tau} \quad (18.6)$$

This expresses a particle's coordinate acceleration in terms of the Christoffel symbols, which are combinations of first derivatives of the metric, just as Newton's second law expresses it in terms of derivatives of the gravitational potential. Note that in each equation, the Christoffel symbol is evaluated at each object's position: therefore in the left equation $\Gamma^\alpha_{\mu\nu} = \Gamma^\alpha_{\mu\nu}(x^\sigma(\tau))$ and in the right equation, $\bar{\Gamma}^\alpha_{\mu\nu} = \Gamma^\alpha_{\mu\nu}(\bar{x}^\sigma(\tau))$. Since the separation between these positions is infinitesimal, we can evaluate the Christoffel symbol at $\bar{x}^\alpha(\tau)$ using a Taylor-series expansion around its value at $x^\alpha(\tau)$ (analogously to what we did before with $\partial_j \Phi$):

$$\bar{\Gamma}^\alpha_{\mu\nu}(\text{at } \bar{x}^\alpha(\tau)) \approx \Gamma^\alpha_{\mu\nu}(\text{at } x^\alpha(\tau)) + n^\sigma [\partial_\sigma \Gamma^\alpha_{\mu\nu}](\text{at } x^\alpha(\tau)) \quad (18.7)$$

If we plug this equation and $\bar{x}^\alpha(\tau) \equiv x^\alpha(\tau) + n^\alpha(\tau)$ into equation 18.6, we can eliminate all references to the barred function $\bar{x}^\alpha(\tau)$:

$$0 = \frac{d^2(x^\alpha + n^\alpha)}{d\tau^2} + [\Gamma^\alpha_{\mu\nu} + n^\sigma(\partial_\sigma \Gamma^\alpha_{\mu\nu})] \frac{d(x^\mu + n^\mu)}{d\tau} \frac{d(x^\nu + n^\nu)}{d\tau}$$

$$= \frac{d^2 x^\alpha}{d\tau^2} + \frac{d^2 n^\alpha}{d\tau^2} + [\Gamma^\alpha_{\mu\nu} + n^\sigma(\partial_\sigma \Gamma^\alpha_{\mu\nu})]\left(\frac{dx^\mu}{d\tau} + \frac{dn^\mu}{d\tau}\right)\left(\frac{dx^\nu}{d\tau} + \frac{dn^\nu}{d\tau}\right) \quad (18.8)$$

where now both the Christoffel symbol and its derivative are evaluated at $x^\alpha(\tau)$. If you multiply out the quantities in parentheses, drop all terms that are second order or higher in n^α, and then use 18.6 to eliminate some terms, you should find that

$$0 = \frac{d^2 n^\alpha}{d\tau^2} + 2\Gamma^\alpha_{\mu\nu} u^\mu \frac{dn^\nu}{d\tau} + n^\sigma(\partial_\sigma \Gamma^\alpha_{\mu\nu}) u^\mu u^\nu \quad (18.9)$$

where $u^\mu \equiv dx^\mu/d\tau$ is the reference particle's four-velocity (see box 18.2).

The Equation of Geodesic Deviation. The simple derivative $d^2 n^\alpha/d\tau^2$ is not, however, the same as the absolute double tau-derivative of $\boldsymbol{n}$, because the change in n^α as we move forward in proper time may be due to both true changes in $\boldsymbol{n}$ and/or changes in the basis vectors used to determine n^α from $\boldsymbol{n}$. Since $\boldsymbol{n}$ is a four-vector, its absolute single tau-derivative is (by analogy to equation 17.12)

$$\left(\frac{d\mathbf{n}}{d\tau}\right)^\alpha = \frac{dn^\alpha}{d\tau} + \Gamma^\alpha_{\mu\nu}u^\mu n^\nu \tag{18.10}$$

(see box 18.3). This is also a four-vector, so (by analogy) the absolute second tau-derivative of $\mathbf{n}$ must be

$$\left(\frac{d^2\mathbf{n}}{d\tau^2}\right)^\alpha = \left(\frac{d}{d\tau}\left[\frac{d\mathbf{n}}{d\tau}\right]\right)^\alpha = \frac{d}{d\tau}\left(\frac{dn^\alpha}{d\tau} + \Gamma^\alpha_{\mu\nu}u^\mu n^\nu\right) + \Gamma^\alpha_{\sigma\nu}u^\sigma\left(\frac{dn^\nu}{d\tau} + \Gamma^\nu_{\beta\gamma}u^\beta n^\gamma\right)$$

$$= \frac{d^2 n^\alpha}{d\tau^2} + \frac{d\Gamma^\alpha_{\mu\nu}}{d\tau}u^\mu n^\nu + \Gamma^\alpha_{\mu\nu}\frac{du^\mu}{d\tau}n^\nu + \Gamma^\alpha_{\mu\nu}u^\mu\frac{dn^\nu}{d\tau}$$

$$+ \Gamma^\alpha_{\sigma\nu}u^\sigma\frac{dn^\nu}{d\tau} + \Gamma^\alpha_{\sigma\nu}\Gamma^\nu_{\beta\gamma}u^\sigma u^\beta n^\gamma \tag{18.11}$$

Note that if we rename $\sigma \to \mu$ in the fifth term, it becomes the same as the fourth term. Note also that the Christoffel symbol depends on τ only because it depends on the first object's position $x^\alpha(\tau)$, so

$$\frac{d\Gamma^\alpha_{\mu\nu}}{d\tau} = \frac{dx^\sigma}{d\tau}\frac{\partial \Gamma^\alpha_{\mu\nu}}{\partial x^\sigma} = u^\sigma \partial_\sigma \Gamma^\alpha_{\mu\nu} \tag{18.12}$$

Therefore, substituting equation 18.12 into equation 18.11 yields (18.13)

$$\left(\frac{d^2\mathbf{n}}{d\tau^2}\right)^\alpha = \frac{d^2 n^\alpha}{d\tau^2} + (\partial_\sigma \Gamma^\alpha_{\mu\nu})u^\sigma u^\mu n^\nu + \Gamma^\alpha_{\mu\nu}\frac{du^\mu}{d\tau}n^\nu + 2\Gamma^\alpha_{\mu\nu}u^\mu\frac{dn^\nu}{d\tau} + \Gamma^\alpha_{\sigma\nu}\Gamma^\nu_{\beta\gamma}u^\sigma u^\beta n^\gamma$$

Now, if you use equation 18.9 to eliminate $d^2 n^\alpha/d\tau^2$, use the geodesic equation to eliminate $du^\mu/d\tau$, and rename indices to pull out a common factor of $u^\sigma u^\mu n^\nu$ from each of the remaining terms, you should find (see box 18.4) that

$$\left(\frac{d^2\mathbf{n}}{d\tau^2}\right)^\alpha = \left(\partial_\sigma \Gamma^\alpha_{\mu\nu} - \partial_\nu \Gamma^\alpha_{\mu\sigma} + \Gamma^\alpha_{\sigma\gamma}\Gamma^\gamma_{\mu\nu} - \Gamma^\alpha_{\nu\gamma}\Gamma^\gamma_{\mu\sigma}\right)u^\sigma u^\mu n^\nu \tag{18.14}$$

Everything else in this equation is a tensor, so the quantity in parentheses must be a tensor. We in fact define the **Riemann Tensor** to be the negative of that quantity:

$$R^\alpha{}_{\mu\nu\sigma} \equiv \partial_\nu \Gamma^\alpha_{\mu\sigma} - \partial_\sigma \Gamma^\alpha_{\mu\nu} + \Gamma^\alpha_{\nu\gamma}\Gamma^\gamma_{\mu\sigma} - \Gamma^\alpha_{\sigma\gamma}\Gamma^\gamma_{\mu\nu} \tag{18.15}$$

Equation 18.14 therefore becomes simply

$$\left(\frac{d^2\mathbf{n}}{d\tau^2}\right)^\alpha = -R^\alpha{}_{\mu\nu\sigma}u^\sigma u^\mu n^\nu \tag{18.16}$$

This is the **equation of geodesic deviation**. If the Riemann tensor for a given spacetime is zero everywhere, initially parallel geodesics *remain* parallel and spacetime is flat; if not, the spacetime is curved. This tensor thus provides a way to *distinguish* flat and curved spacetimes, something one *cannot* do simply by looking at the metric. Box 18.5 displays a simple example for a two-dimensional space.

Note that the Riemann tensor (which involves second derivatives of the metric) plays the same role in this equation that $\eta^{ij}\partial_k\partial_j\Phi$ does in equation 18.4.

A Mnemonic for the Riemann Tensor. Let's rewrite the definition of the Riemann tensor in terms of coordinate labels that progress in a more orderly way:

$$R^\alpha{}_{\beta\mu\nu} \equiv \partial_\mu \Gamma^\alpha_{\beta\nu} - \partial_\nu \Gamma^\alpha_{\beta\mu} + \Gamma^\alpha_{\mu\gamma}\Gamma^\gamma_{\beta\nu} - \Gamma^\alpha_{\nu\sigma}\Gamma^\sigma_{\beta\mu} \tag{18.17}$$

Note that the positive terms are a derivative of a Christoffel symbol and a product of Christoffel symbols. In both positive terms, the free lower indices go in the order μ, β, ν (i.e., the 3rd, 2nd, 4th Riemann indices), and in the product terms, the final index of the first symbol sums with the first index of the second. The negative terms have exactly the same form, but the μ and ν (3rd and 4th) indices reversed.

So a mnemonic might be "3 to 4 is positive, but 4 to 3 is not. Twins bond inside." Translation: The lower free indices go 324 ("3 to 4", get it?) in the positive terms but 423 in the negative terms, and in the product terms, the bound index on each Christoffel symbol is the one closest to the other, i.e., the innermost indices. If you get these things, the other indices are easy to figure out.

BOX 18.1 Newtonian Tidal Deviation Near a Spherical Object

Our goal in this box is to evaluate the Newtonian tidal deviation when the potential is $\Phi = -GM/r$ and the $+z$ direction in our freely falling reference frame coincides with the radial direction. This will also give you the chance to see the power of using index notation even in a Newtonian calculation!

First, note that in index notation $r \equiv \sqrt{x^2 + y^2 + z^2} = \sqrt{\eta_{mn} x^m x^n}$, where η_{mn} represents the spatial components of the flat-space metric ($\eta_{mn} = 1$ if $m = n$ and zero otherwise). Note also that $\partial x^i / \partial x^j = \delta^i_j$. This means that by the chain rule

$$\partial_j r \equiv \frac{\partial}{\partial x^j} \sqrt{\eta_{mn} x^m x^n} = \frac{\eta_{mn} x^m \delta^n_j + \eta_{mn} x^n \delta^m_j}{2\sqrt{\eta_{mn} x^m x^n}} = \frac{2\eta_{jn} x^n}{2\sqrt{\eta_{mn} x^m x^n}} = \frac{\eta_{jn} x^n}{r} \quad (18.18)$$

where in the second-to-last step, I performed the sums over the Kronecker delta, yielding $\eta_{mj} x^m + \eta_{jn} x^n$ in the numerator, renamed the summed index in the first term n instead of m, and used the fact that $\eta_{nj} = \eta_{jn}$ to add the two terms. This was the hard part: you can use this and the chain and product rules to show that

$$\partial_j \Phi = \frac{GM}{r^3} \eta_{jn} x^n, \quad \partial_k \partial_j \Phi = -\frac{3GM}{r^5} \eta_{km} \eta_{jn} x^m x^n + \frac{GM}{r^3} \eta_{kj} \quad (18.19)$$

Exercise 18.1.1. Verify these results.

The Newtonian tidal deviation equation (equation 18.4) in this case becomes

$$\frac{d^2 n^i}{dt^2} = -\eta^{ij}(\partial_k \partial_j \Phi) n^k = \frac{3GM}{r^5} \eta^{ij} \eta_{km} \eta_{jn} x^m x^n n^k - \frac{GM}{r^3} \eta^{ij} \eta_{kj} n^k \quad (18.20)$$

Remember that we evaluate this at the position of the reference particle. If our reference particle is at the center of our freely falling reference frame and the $+z$ direction corresponds to the radial direction, then the position where we are evaluating this corresponds to $z = r$, $x = y = 0$. Using the above expression (and noting that $\eta^{ij} \eta_{jk} = \delta^i_k$ by definition of the inverse metric), you should be able to show that $d^2 n^x / dt^2 = (-GM/r^3) n^x$ and $d^2 n^z / dt^2 = (+2GM/r^3) n^z$ as in equation 18.5.

Exercise 18.1.2. Check this.

BOX 18.2 Proving Equation 18.9

If we multiply out the terms in equation 18.8, we get

$$0 = \frac{d^2x^\alpha}{d\tau^2} + \frac{d^2n^\alpha}{d\tau^2} + \Gamma^\alpha_{\mu\nu}\left(\frac{dx^\mu}{d\tau}\frac{dx^\nu}{d\tau} + \frac{dx^\mu}{d\tau}\frac{dn^\nu}{d\tau} + \frac{dn^\mu}{d\tau}\frac{dx^\nu}{d\tau} + \cancel{\frac{dn^\mu}{d\tau}\frac{dn^\nu}{d\tau}}\right)$$
$$+ n^\sigma(\partial_\sigma \Gamma^\alpha_{\mu\nu})\left(\frac{dx^\mu}{d\tau}\frac{dx^\nu}{d\tau} + \cancel{\frac{dx^\mu}{d\tau}\frac{dn^\nu}{d\tau}} + \cancel{\frac{dn^\mu}{d\tau}\frac{dx^\nu}{d\tau}} + \cancel{\frac{dn^\mu}{d\tau}\frac{dn^\nu}{d\tau}}\right) \quad (18.21)$$

In this equation, the crosses indicate terms that are second order or higher in n^α that we will drop. Now you can subtract the geodesic equation (the first equation in 18.6) and use $\Gamma^\alpha_{\mu\nu} = \Gamma^\alpha_{\nu\mu}$ to arrive at equation 18.9, repeated here:

$$0 = \frac{d^2n^\alpha}{d\tau^2} + 2\Gamma^\alpha_{\mu\nu}u^\mu\frac{dn^\nu}{d\tau} + n^\sigma(\partial_\sigma \Gamma^\alpha_{\mu\nu})u^\mu u^\nu \quad (18.9r)$$

Exercise 18.2.1. Verify this.

BOX 18.3 The Absolute Derivative of *n*

Note that the true derivative of the separation four-vector **n** is

$$\frac{d\mathbf{n}}{d\tau} = \frac{d}{d\tau}(n^\alpha \mathbf{e}_\alpha) = \frac{dn^\alpha}{d\tau}\mathbf{e}_\alpha + n^\alpha \frac{d\mathbf{e}_\alpha}{d\tau} = \frac{dn^\alpha}{d\tau}\mathbf{e}_\alpha + n^\alpha \frac{dx^\mu}{d\tau}\frac{\partial \mathbf{e}_\alpha}{\partial x^\mu} \quad (18.22)$$

since the basis vectors depend on τ only because the reference object's position does and the basis vectors depend on position. If you use the definition of the Christoffel symbols ($\partial \mathbf{e}_\beta/\partial x^\mu \equiv \Gamma^\alpha_{\beta\mu}\mathbf{e}_\alpha$) and rename some indices, you can show that

$$\frac{d\mathbf{n}}{d\tau} = \left(\frac{dn^\alpha}{d\tau} + \Gamma^\alpha_{\sigma\mu}n^\sigma u^\mu\right)\mathbf{e}_\alpha \Rightarrow \left(\frac{d\mathbf{n}}{d\tau}\right)^\alpha = \frac{dn^\alpha}{d\tau} + \Gamma^\alpha_{\sigma\mu}n^\sigma u^\mu \quad (18.23)$$

Exercise 18.3.1. Verify this.

BOX 18.4 Proving Equation 18.14

Equation 18.13 (repeated here) says that

$$\left(\frac{d^2\mathbf{n}}{d\tau^2}\right)^\alpha = \frac{d^2 n^\alpha}{d\tau^2} + (\partial_\sigma \Gamma^\alpha_{\mu\nu}) u^\sigma u^\mu n^\nu + \Gamma^\alpha_{\mu\nu} \frac{du^\mu}{d\tau} n^\nu + 2\Gamma^\alpha_{\mu\nu} u^\mu \frac{dn^\nu}{d\tau} + \Gamma^\alpha_{\sigma\nu} \Gamma^\nu_{\beta\gamma} u^\sigma u^\beta n^\gamma \quad (18.13r)$$

Equation 18.9 (solved for $d^2 n^\alpha/d\tau^2$) and equation 18.6 (solved for $du^\mu/d\tau \equiv d^2 x^\mu/d\tau^2$) tell us respectively that

$$\frac{d^2 n^\alpha}{d\tau^2} = -2\Gamma^\alpha_{\mu\nu} u^\mu \frac{dn^\nu}{d\tau} - n^\sigma (\partial_\sigma \Gamma^\alpha_{\mu\nu}) u^\mu u^\nu \quad \text{and} \quad \frac{du^\mu}{d\tau} = -\Gamma^\mu_{\alpha\beta} u^\alpha u^\beta \quad (18.24)$$

(Check that you see this.) If you plug equations 18.24 into equation 18.13, cancel some terms, and rename some indices to pull out the common factor, you should be able to arrive at equation 18.14 (repeated here):

$$\left(\frac{d^2\mathbf{n}}{d\tau^2}\right)^\alpha = (\partial_\sigma \Gamma^\alpha_{\mu\nu} - \partial_\nu \Gamma^\alpha_{\mu\sigma} + \Gamma^\alpha_{\sigma\gamma} \Gamma^\gamma_{\mu\nu} - \Gamma^\alpha_{\nu\gamma} \Gamma^\gamma_{\mu\sigma}) u^\sigma u^\mu n^\nu \quad (18.14r)$$

Exercise 18.4.1. Do this.

BOX 18.5 An Example of Calculating the Riemann Tensor

Consider the following metric for a 2-dimensional space in u, w coordinates:

$$ds^2 = a^2 du^2 + \sin^2(u) dw^2 \quad (18.25)$$

where a is a constant. Is this the metric of a flat space or a curved space? We can tell by evaluating the components of the Riemann tensor. But in order to calculate components of the Riemann tensor, we need to compute the Christoffel symbols for this metric. We can do this most easily by using the trick discussed in box 17.6, i.e., by calculating the geodesic equation in the form given by equation 17.26:

$$0 = \frac{d}{d\tau}\left(g_{\mu\nu} \frac{dx^\nu}{d\tau}\right) - \tfrac{1}{2} \partial_\mu g_{\alpha\beta} \frac{dx^\alpha}{d\tau} \frac{dx^\beta}{d\tau} \quad (18.26)$$

and comparing the results with the geodesic equation given by equation 17.12,

$$0 = \frac{d^2 x^\mu}{d\tau^2} + \Gamma^\mu_{\alpha\beta} \frac{dx^\alpha}{d\tau} \frac{dx^\beta}{d\tau} \quad (18.27)$$

to read off the Christoffel symbols.

For our metric, the $\mu = u$ component of equation 18.26 becomes

$$0 = \frac{d}{d\tau}\left(g_{u\nu} \frac{dx^\nu}{d\tau}\right) - \tfrac{1}{2} \partial_u g_{\alpha\beta} \frac{dx^\alpha}{d\tau} \frac{dx^\beta}{d\tau} = \frac{d}{d\tau}\left(g_{uu} \frac{du}{d\tau}\right) - \tfrac{1}{2} \frac{\partial g_{ww}}{\partial u}\left(\frac{dw}{d\tau}\right)^2$$

$$= a^2 \frac{d^2 u}{d\tau^2} - \sin u \cos u \left(\frac{dw}{d\tau}\right)^2 \quad (18.28)$$

BOX 18.5 (continued) An Example of Calculating the Riemann Tensor

Comparing this with equation 18.27, we see that the Christoffel symbols for $\mu = u$ are

$$\Gamma^u_{ww} = -\frac{1}{a^2}\sin u \cos u, \quad \text{other } \Gamma^u_{\alpha\beta} = 0 \tag{18.29}$$

By looking at the $\mu = w$ component, you can show similarly that

$$\Gamma^w_{uw} = \Gamma^w_{wu} = \cot u, \quad \text{other } \Gamma^w_{\alpha\beta} = 0 \tag{18.30}$$

Exercise 18.5.1. Check that this is true.

For a two-dimensional space, the Riemann tensor has $2^4 = 16$ components. However, if *any* of the components are nonzero anywhere, the space is curved. For the sake of argument, let's choose to evaluate $R^u{}_{wuw}$. The definition of the Riemann tensor given in equation 18.17 implies that

$$R^u{}_{wuw} \equiv \partial_u \Gamma^u_{ww} - \partial_w \Gamma^u_{wu} + \Gamma^u_{u\gamma}\Gamma^\gamma_{ww} - \Gamma^u_{w\sigma}\Gamma^\sigma_{wu} \tag{18.31}$$

Note that no Christoffel symbol depends on w, so the second term must be zero. In the third term, note that the first Christoffel symbol is zero no matter what γ is. By evaluating the other terms, you can show that $R^u{}_{wuw} = a^{-2}\sin^2 u$.

Exercise 18.5.2. Show that this is correct.

Since this is (generally) nonzero, this space must be curved. (This should actually not be surprising: if you transform coordinates to θ and ϕ defined so that $u = \theta$ and $w = a\phi$, then the metric becomes $ds^2 = a^2\,d\theta^2 + a^2\sin^2\theta\,d\phi^2$, which is the metric of the surface of a sphere of radius a in latitude-longitude coordinates.)

HOMEWORK PROBLEMS

P18.1 Calculate the magnitude of the relative acceleration of two balls released from rest a vertical distance of 1 m apart in a freely falling frame near the earth's surface, and roughly estimate how long it would take for the separation of the balls to change by a barely measurable 1 nm. (*Hint:* Use Newtonian mechanics and assume that the distance between the reference ball and the earth's center is approximately constant and roughly equal to the earth's radius during that time period. After you get a result, assess the validity of this approximation.)

P18.2 The electric field $\vec{E}$ outside an infinite flat plate with uniformly distributed charge is constant in both magnitude and direction (it points perpendicular to the plane of the plate). By analogy, the Newtonian gravitational field $\vec{g}$ of an infinite flat plate with uniformly distributed mass is also a constant in both magnitude and direction. What conclusion can you draw from the Newtonian tidal deviation formula about the relative acceleration of balls in a freely falling reference frame near such an object? If this result were to hold up in general relativity, would this field be real according to general relativity? Explain your reasoning.

P18.3 A possible candidate for the metric in the empty space outside a static infinite planar slab of mass might be

$$ds^2 = -dt^2 + f(x)\,dx^2 + dy^2 + dz^2 \qquad (18.32)$$

(This metric shares the translational symmetry of the the slab with respect to shifts in the t, y, and z coordinates, though it is not the most general such candidate.) Could this metric describe a real gravitational field in spacetime?
a. Argue that the only nonzero Christoffel symbol for this metric is $\Gamma^x{}_{xx} = (1/2f)(df/dx)$.
b. Argue that the Riemann tensor for this metric is zero, implying that this spacetime is actually *flat* (i.e., it does not describe a gravitational field). (*Hint:* What single component of the Riemann tensor might possibly be nonzero? Is it nonzero?)

P18.4 The metric for a flat space in polar coordinates is

$$ds^2 = dr^2 + r^2\,d\theta^2 \qquad (18.33)$$

Calculate the $R^r{}_{\theta r\theta}$ component of the Riemann tensor for this space. (We will see in the next chapter that this is the *single* independent component of the Riemann tensor in two spatial dimensions: the other 15 components are either identically zero or are the same as ± 1 times this component. Therefore, if this component is zero, the entire Riemann tensor is zero.)

P18.5 Consider the following metric for a 2D space:

$$ds^2 = \frac{dp^2}{1-kp^2} + p^2\,dq^2 \qquad (18.34)$$

Does this metric describe a flat or curved space? (*Hint:* We will see in the next chapter that for a two-dimensional space, components of the Riemann tensor are either identically zero or equal to $\pm R^p{}_{qpq}$. Therefore, if we know the value of $R^p{}_{qpq}$, we know the entire Riemann tensor. Use this in drawing your conclusion.)

P18.6 In this problem, we will examine one component of the Riemann tensor for Schwarzschild spacetime.
a. Use the method of box 17.6 to show that $\Gamma^t{}_{tr} = \Gamma^t{}_{rt}$ are the only nonzero Schwarzschild Christoffel symbols with an upper index of t, and evaluate these symbols.
b. Show that for Schwarzschild spacetime,

$$R^t{}_{rtr} = \frac{2GM}{r^3}\left(1 - \frac{2GM}{r}\right)^{-1} \qquad (18.35)$$

(*Hint:* You will need $\Gamma^r{}_{rr}$ to do this calculation. It turns out that $\Gamma^r{}_{rr} = -\Gamma^t{}_{rt}$: you don't have to show this. See problem P19.8 for a discussion about why the fact that $R^t{}_{rtr}$ blows up at $r = 2GM$ does *not* imply that spacetime is infinitely curved there.)
c. Argue that Schwarzschild spacetime is curved everywhere, but becomes more flat as r increases.

P18.7 Argue that at the center of a freely falling frame (i.e., a locally inertial frame that remains one as time passes), the equation of geodesic deviation for the geodesic of an object at rest in that frame becomes simply

$$\frac{d^2 n^\alpha}{dt^2} = -R^\alpha{}_{t\beta t}n^\beta \quad \text{(in a LIF)} \qquad (18.36)$$

[*Hint:* In a LIF that remains a LIF as time passes (rather than being a LIF only at a single event), argue that the *time* derivatives of the Christoffel symbols are zero.]

Comment: If we compare this to equation 18.4, we see that we should have $R^i{}_{tkt} \approx \eta^{ij}(\partial_j \partial_k \Phi)$ for a gravitational field that is weak enough to be considered Newtonian (assuming that we use cartesian-like coordinates in our LIF).

P18.8 Another interesting relationship involving the Riemann tensor is the following:

$$(\nabla_\mu \nabla_\nu - \nabla_\nu \nabla_\mu)a^\alpha = R^\alpha{}_{\beta\mu\nu}a^\beta \qquad (18.37)$$

where $\mathbf{a}$ is an arbitrary four-vector. Note that in a truly flat space, we can in principle set up globally cartesian coordinates where these absolute double-derivatives reduce to ordinary partial derivatives, and since the order of partial derivatives does not matter, $R^\alpha{}_{\beta\mu\nu}$ would be zero. But in curved space, this equation states that double absolute gradients of a vector do *not* commute (fundamentally because double-derivatives of the metric become involved).
a. Recognizing that $\nabla_\nu a^\alpha$ is a tensor with one upper and one lower index, show that

$$\nabla_\mu(\nabla_\nu a^\alpha) = \partial_\mu \partial_\nu a^\alpha + (\partial_\mu \Gamma^\alpha{}_{\beta\nu})a^\beta + \Gamma^\alpha{}_{\beta\nu}\partial_\mu a^\beta$$
$$- \Gamma^\sigma{}_{\nu\mu}\partial_\sigma a^\alpha - \Gamma^\sigma{}_{\nu\mu}\Gamma^\alpha{}_{\gamma\sigma}a^\gamma$$
$$+ \Gamma^\alpha{}_{\delta\mu}\partial_\nu a^\delta + \Gamma^\alpha{}_{\delta\mu}\Gamma^\delta{}_{\lambda\nu}a^\lambda \qquad (18.38)$$

b. From this, prove equation 18.37.

19. THE RIEMANN TENSOR

INTRODUCTION

FLAT SPACETIME
- Review of Special Relativity
- Four-Vectors
- Index Notation

TENSORS
- Arbitrary Coordinates
- Tensor Equations
- Maxwell's Equations
- Geodesics

SCHWARZSCHILD BLACK HOLES
- The Schwarzschild Metric
- Particle Orbits
- Precession of the Perihelion
- Photon Orbits
- Deflection of Light
- Event Horizon
- Alternative Coordinates
- Black Hole Thermodynamics

THE CALCULUS OF CURVATURE
- The Absolute Gradient
- Geodesic Deviation
- **The Riemann Tensor**

THE EINSTEIN EQUATION
- The Stress-Energy Tensor
- The Einstein Equation
- Interpreting the Equation
- The Schwarzschild Solution

COSMOLOGY
- The Universe Observed
- A Metric for the Cosmos
- Evolution of the Universe
- Cosmic Implications
- The Early Universe
- CMB Fluctuations & Inflation

GRAVITATIONAL WAVES
- Gauge Freedom
- Detecting Gravitational Waves
- Gravitational Wave Energy
- Generating Gravitational Waves
- Gravitational Wave Astronomy

SPINNING BLACK HOLES
- Gravitomagnetism
- The Kerr Metric
- Kerr Particle Orbits
- Ergoregion and Horizon
- Negative-Energy Orbits

this → depends on → this

Introduction. In the last chapter, we introduced the Riemann tensor, which by definition specifies the relative deviation of initially parallel nearby geodesics:

$$\left(\frac{d^2\boldsymbol{n}}{d\tau^2}\right)^\alpha = -R^\alpha{}_{\beta\mu\nu}u^\beta n^\mu u^\nu \tag{19.1}$$

where $\boldsymbol{u}(\tau)$ is the four-velocity of the reference geodesic as a function of proper time τ along that geodesic, and $\boldsymbol{n}(\tau)$ is the separation four-vector at τ between the reference geodesic and a nearby geodesic. Since the relative acceleration of geodesics is zero in flat space or spacetime and not zero in curved spaces or spacetimes (by definition), the Riemann tensor is zero in flat space and spacetimes, and nonzero in curved spaces or spacetimes. The Riemann tensor thus represents a tensor quantity that is able to distinguish flat from curved spaces or spacetimes.

As we saw in the last chapter (see equation 18.17), the Riemann tensor can be expressed in terms of Christoffel symbols as follows:

$$R^\alpha{}_{\beta\mu\nu} \equiv \partial_\mu \Gamma^\alpha_{\beta\nu} - \partial_\nu \Gamma^\alpha_{\beta\mu} + \Gamma^\alpha_{\mu\gamma}\Gamma^\gamma_{\beta\nu} - \Gamma^\alpha_{\nu\sigma}\Gamma^\sigma_{\beta\mu} \tag{19.2}$$

In this chapter we will explore characteristics of this very important tensor, ideas which we will find useful when we develop the Einstein equation, the equation that links the curvature of spacetime to the presence of matter and energy.

(You should be aware, by the way, that there is no firmly established sign convention for the Riemann tensor: some authors take the sign in equation 19.1 to be positive, and thus reverse all the signs in equation 19.2. This is one of the things that you should check when reading other books and/or papers on GR.)

The Riemann Tensor in a LIF. As we saw in chapter 17, the first derivatives of the metric are zero at the origin of a locally inertial frame (LIF), meaning that the Christoffel symbols are zero there. However, *derivatives* of the Christoffel symbols are not necessarily zero at the origin of a LIF, because such derivatives involve *second* derivatives of the metric. You can in fact show from equation 19.2 (see box 19.1) that at the origin of a LIF

$$R_{\alpha\beta\mu\nu} \equiv g_{\alpha\gamma}R^\gamma{}_{\beta\mu\nu} = \tfrac{1}{2}(\partial_\beta\partial_\mu g_{\alpha\nu} + \partial_\alpha\partial_\nu g_{\beta\mu} - \partial_\beta\partial_\nu g_{\alpha\mu} - \partial_\alpha\partial_\mu g_{\beta\nu}) \tag{19.3}$$

A mnemonic for the order of these indices could be "Inner togetherness is positive." Translation: The two positive terms have the inner Riemann indices (indices 2 and 3) together in either the derivatives or the metric. In the negative terms, these indices are separated. **Please note** that equation 19.3 applies *only at the origin of a LIF* while equation 19.2 is true at all events in all coordinate systems.

Symmetries of the Riemann Tensor. Even so, the LIF form of the Riemann tensor is quite handy for proving symmetries of the Riemann tensor. For example, it is quite clear from the form of equation 19.3 that at the origin of a LIF (since "inner togetherness is positive"),

$$R_{\alpha\beta\nu\mu} = \tfrac{1}{2}(\partial_\beta\partial_\nu g_{\alpha\mu} + \partial_\alpha\partial_\mu g_{\beta\nu} - \partial_\beta\partial_\mu g_{\alpha\nu} - \partial_\alpha\partial_\nu g_{\beta\mu})$$

$$= -\tfrac{1}{2}(\partial_\beta\partial_\mu g_{\alpha\nu} + \partial_\alpha\partial_\nu g_{\beta\mu} - \partial_\beta\partial_\nu g_{\alpha\mu} - \partial_\alpha\partial_\mu g_{\beta\nu}) = -R_{\alpha\beta\mu\nu} \tag{19.4a}$$

Since $R_{\alpha\beta\nu\mu} = -R_{\alpha\beta\mu\nu}$ is a tensor equation, if it is true at the origin event in a LIF, it is true in *every* coordinate system at that event. Moreover, we can construct a LIF at any event in spacetime where the metric is non-singular. Therefore the tensor equation $R_{\alpha\beta\nu\mu} = -R_{\alpha\beta\mu\nu}$ must be true at all events in every coordinate system.

In a similar way, you can also show quite easily from equation 19.3 (see box 19.2) that at all events in every coordinate system

$$R_{\beta\alpha\mu\nu} = -R_{\alpha\beta\mu\nu} \tag{19.4b}$$

$$R_{\mu\nu\alpha\beta} = +R_{\alpha\beta\mu\nu} \tag{19.4c}$$

$$R_{\alpha\beta\mu\nu} + R_{\alpha\nu\beta\mu} + R_{\alpha\mu\nu\beta} = 0 \tag{19.4d}$$

(Note the cyclic rotation of the final three indices in equation 19.4d.) These symmetries mean that even though the Riemann tensor has $4 \times 4 \times 4 \times 4 = 256$ components in four-dimensional spacetime, only 20 of them are independent (see box 19.3). You might recall from chapter 17 that an arbitrary coordinate transformation can set all but 20 of the metric second derivatives to zero. The 20 independent Riemann tensor components represent independent combinations of these 20 independent metric second derivatives.

In a two-dimensional space or spacetime, the Riemann tensor has $2^4 = 16$ components, but only *one* component is independent.

The Bianchi Identity. Another very important feature of the Riemann tensor is the so-called Bianchi identity, which states that

$$\nabla_\sigma R_{\alpha\beta\mu\nu} + \nabla_\nu R_{\alpha\beta\sigma\mu} + \nabla_\mu R_{\alpha\beta\nu\sigma} = 0 \tag{19.5}$$

(Note the rotation of the last two indices with the gradient index.) This is also easily proved using a LIF (see box 19.4). We will see that this identity plays a crucial role in the development of the Einstein equation.

Contractions of the Riemann Tensor. In constructing the Einstein equation, we will also find the following contractions of the Riemann tensor useful. The **Ricci tensor** (pronounced "Reechee") is defined to be the contraction of the Riemann tensor over its first and third indices

$$R_{\beta\nu} \equiv R^\alpha{}_{\beta\alpha\nu} \equiv g^{\alpha\mu} R_{\alpha\beta\mu\nu} \tag{19.6}$$

(Again, there is a sign convention problem here: some authors define the Ricci tensor to be the sum over the first and *fourth* indices, which yields the negative of what we have above.) This tensor is symmetric (see box 19.5),

$$R_{\beta\nu} = R_{\nu\beta} \tag{19.7}$$

and so has (at most) 10 independent components in a four-dimensional spacetime.

The **curvature scalar** is the contraction of the Ricci tensor:

$$R \equiv g^{\beta\nu} R_{\beta\nu} \equiv g^{\beta\nu} g^{\alpha\mu} R_{\alpha\beta\mu\nu} \tag{19.8}$$

The curvature scalar is interesting because it is a relativistic *scalar*, meaning that it is independent of one's choice of coordinates. If this quantity is nonzero, it gives a coordinate-independent measure of a space or spacetime's curvature.

The Ricci tensor and the curvature scalar are obviously both zero in flat space or spacetime, but may also be zero in some curved spaces or spacetimes (though a nonzero value clearly signals that the space or spacetime is curved). Only by evaluating the full Riemann tensor can one *conclusively* distinguish flat and curved spaces or spacetimes.

An Example: The Surface of a Sphere. Box 19.6 shows that the metric, the Christoffel symbols, the Riemann tensor, the Ricci tensor, and the curvature scalar for the surface of a sphere of radius r in latitude-longitude coordinates are

$$ds^2 = r^2 d\theta^2 + r^2 \sin^2\theta \, d\phi^2 \tag{19.9}$$

$$\Gamma^\theta_{\phi\phi} = -\sin\theta\cos\theta, \quad \Gamma^\phi_{\theta\phi} = \Gamma^\phi_{\phi\theta} = \cot\theta; \quad \text{other } \Gamma^\mu_{\alpha\beta} = 0 \tag{19.10}$$

$$R_{\theta\phi\theta\phi} = -R_{\theta\phi\phi\theta} = -R_{\phi\theta\theta\phi} = +R_{\phi\theta\phi\theta} = r^2 \sin^2\theta; \quad \text{other } R_{\alpha\beta\mu\nu} = 0 \tag{19.11}$$

$$R_{\theta\theta} = 1, \quad R_{\phi\phi} = \sin^2\theta, \quad R_{\theta\phi} = R_{\phi\theta} = 0 \tag{19.12}$$

$$R = \frac{2}{r^2} \tag{19.13}$$

(Note that r is a constant here, not a variable.) This important example not only illustrates the process of calculating these components but also provides the results that will be useful to us in the future.

BOX 19.1 The Riemann Tensor in a Locally Inertial Frame

Since the Christoffel symbols are all zero at the origin of a LIF, the Riemann tensor at the origin of a LIF becomes

$$R_{\alpha\beta\mu\nu} \equiv g_{\alpha\sigma} R^{\sigma}{}_{\beta\mu\nu} = g_{\alpha\sigma}\left[\partial_\mu \Gamma^{\sigma}_{\beta\nu} - \partial_\nu \Gamma^{\sigma}_{\beta\mu}\right] \text{ in a LIF} \qquad (19.14)$$

Equation 17.10 tells us that the Christoffel symbols are given by

$$\Gamma^{\sigma}_{\beta\nu} = \tfrac{1}{2} g^{\sigma\gamma}\left[\partial_\beta g_{\nu\gamma} + \partial_\nu g_{\gamma\beta} - \partial_\gamma g_{\beta\nu}\right] \qquad (19.15)$$

If you plug this into the above and use the facts that $g_{\alpha\sigma} g^{\sigma\gamma} = \delta^{\gamma}{}_{\alpha}$ and the first derivatives of the metric are all zero at the origin of the LIF, you can show that

$$R_{\alpha\beta\mu\nu} = \tfrac{1}{2}(\partial_\beta \partial_\mu g_{\alpha\nu} + \partial_\alpha \partial_\nu g_{\beta\mu} - \partial_\beta \partial_\nu g_{\alpha\mu} - \partial_\alpha \partial_\mu g_{\beta\nu}) \qquad (19.3r)$$

Exercise 19.1.1. Verify this.

BOX 19.2 Symmetries of the Riemann Tensor

Exercise 19.2.1. Show that $R_{\beta\alpha\mu\nu} = -R_{\alpha\beta\mu\nu}$ at the origin of a LIF. (Remember that if this tensor equation is true in a LIF it is true in any coordinate system.)

Exercise 19.2.2. Show that $R_{\mu\nu\alpha\beta} = +R_{\alpha\beta\mu\nu}$ at the origin of a LIF.

Exercise 19.2.3. Show that $R_{\alpha\beta\mu\nu} + R_{\alpha\nu\beta\mu} + R_{\alpha\mu\nu\beta} = 0$ at the origin of a LIF.

BOX 19.3 Counting the Riemann Tensor's Independent Components

Note that because $R_{\alpha\beta\nu\mu} = -R_{\alpha\beta\mu\nu}$ and $R_{\beta\alpha\mu\nu} = -R_{\alpha\beta\mu\nu}$, a component of the Riemann tensor can be nonzero only if $\alpha \neq \beta$ and $\mu \neq \nu$. Let's abstractly refer to the four spacetime coordinates as 0, 1, 2, 3 (instead of t, x, y, z or t, r, θ, ϕ, or the like). According to the symmetry relations above, the only unique and nonzero index value pairs for $\alpha\beta$ or $\mu\nu$ are 01, 02, 03, 12, 13, and 23 (all other nonzero pairs are reversals of one of these). So let's arrange these nonzero components in a chart:

$$
\begin{array}{c|cccccc}
\mu\nu \rightarrow & 01 & 02 & 03 & 12 & 13 & 23 \\
\alpha\beta \downarrow 01 & \underline{R_{0101}} & R_{0102} & R_{0103} & R_{0112} & R_{0113} & R_{0123} \\
02 & R_{0201} & \underline{R_{0202}} & R_{0203} & R_{0212} & R_{0213} & R_{0223} \\
03 & R_{0301} & R_{0302} & \underline{R_{0303}} & R_{0312} & R_{0313} & R_{0323} \\
12 & R_{1201} & R_{1202} & R_{1203} & \underline{R_{1212}} & R_{1213} & R_{1223} \\
13 & R_{1301} & R_{1302} & R_{1303} & R_{1312} & \underline{R_{1313}} & R_{1323} \\
23 & R_{2301} & R_{2302} & R_{2303} & R_{2312} & R_{2313} & \underline{R_{2323}}
\end{array} \quad (19.16)
$$

There are 36 of these components. But the symmetry relationship $R_{\mu\nu\alpha\beta} = +R_{\alpha\beta\mu\nu}$ means that not all of these are independent.

Exercise 19.3.1. Argue that $R_{\mu\nu\alpha\beta} = +R_{\alpha\beta\mu\nu}$ means that only 21 of the components in the chart above are independent of each other. (*Hint:* Compare components across the underlined diagonal elements in the chart above.)

The final symmetry equation, $R_{\alpha\beta\mu\nu} + R_{\alpha\nu\beta\mu} + R_{\alpha\mu\nu\beta} = 0$, represents $4 \times 4 \times 4 \times 4 = 256$ individual equations, but if *any two* indices have the same value, you can easily show that this equation is *identically* zero on the basis of what we already know.

Exercise 19.3.2. Verify the latter claim.

Therefore, the only component of this equation that tells us anything new is the component $R_{0123} + R_{0312} + R_{0231} = 0$. This allows us to compute, say, $R_{0213} = -R_{0231}$ on the chart given R_{0123} and R_{0312} (which also both appear on the chart). Therefore, we are left with only 20 independent, nonzero components.

BOX 19.3 (continued) Counting the Riemann Tensor's Independent Components

Exercise 19.3.3. Argue that in a two-dimensional space or spacetime, only one component of the Riemann tensor is independent, and show explicitly how all other nonzero components are related to this component.

BOX 19.4 The Bianchi Identity

At the origin of a LIF, the Christoffel symbols are all zero, so by equations 17.5 through 17.8, the absolute gradient of any tensor quantity is the same as its ordinary gradient. The ordinary gradient of the Riemann tensor in any coordinate basis is

$$\partial_\gamma R_{\alpha\beta\mu\nu} = \partial_\gamma [g_{\alpha\sigma}(\partial_\mu \Gamma^\sigma_{\beta\nu} - \partial_\nu \Gamma^\sigma_{\beta\mu} + \Gamma^\sigma_{\mu\delta}\Gamma^\delta_{\beta\nu} - \Gamma^\sigma_{\nu\lambda}\Gamma^\lambda_{\beta\mu})] \quad (19.17)$$

At the origin of a LIF, the first derivatives of the metric and the Christoffel symbols are zero, though derivatives of the Christoffel symbols are not. This means that the term $\partial_\gamma(\Gamma^\sigma_{\mu\delta}\Gamma^\delta_{\beta\nu}) = (\partial_\gamma \Gamma^\sigma_{\mu\delta})\Gamma^\delta_{\beta\nu} + \Gamma^\sigma_{\mu\delta}(\partial_\gamma \Gamma^\delta_{\beta\nu})$ and its similar partner are zero, because each derivative in the expansion is multiplied by a Christoffel symbol. The remaining terms in the brackets are the same as what you found in Box 19.1 to be

$$R_{\alpha\beta\mu\nu} = \tfrac{1}{2}(\partial_\beta\partial_\mu g_{\alpha\nu} + \partial_\alpha\partial_\nu g_{\beta\mu} - \partial_\beta\partial_\nu g_{\alpha\mu} - \partial_\alpha\partial_\mu g_{\beta\nu}) \quad (19.3r)$$

at the origin of a LIF (see equation 19.14). Use this and the fact that the order of partial differentiation is irrelevant to prove that

$$\nabla_\sigma R_{\alpha\beta\mu\nu} + \nabla_\nu R_{\alpha\beta\sigma\mu} + \nabla_\mu R_{\alpha\beta\nu\sigma} = 0 \quad (19.5r)$$

Exercise 19.4.1. Verify this.

BOX 19.5 The Ricci Tensor Is Symmetric

Exercise 19.5.1. Use the symmetries of the Riemann tensor and the fact that the metric tensor is symmetric to prove that the Ricci tensor is symmetric.

BOX 19.6 The Riemann and Ricci Tensors and R for a Sphere

The metric for the surface of a sphere in latitude-longitude coordinates (see equation 19.9) is $ds^2 = r^2 d\theta^2 + r^2 \sin^2\theta \, d\phi^2$, where r is the (fixed) radius of the sphere here. We can evaluate the Christoffel symbols for this metric using our usual trick involving comparing the two forms of the geodesic equation:

$$0 = \frac{d}{d\tau}\left(g_{\mu\nu}\frac{dx^\nu}{d\tau}\right) - \tfrac{1}{2}\partial_\mu g_{\alpha\beta}\frac{dx^\alpha}{d\tau}\frac{dx^\beta}{d\tau}, \quad \text{and} \quad 0 = \frac{d^2 x^\mu}{d\tau^2} + \Gamma^\mu_{\alpha\beta}\frac{dx^\alpha}{d\tau}\frac{dx^\beta}{d\tau} \quad (19.18)$$

If we set $\mu = \phi$ in the first equation above, the second term becomes zero (because no metric component depends on ϕ). Remembering that the metric is diagonal,

$$0 = \frac{d}{d\tau}\left(g_{\phi\nu}\frac{dx^\nu}{d\tau}\right) + 0 = \frac{d}{d\tau}\left(0 + g_{\phi\phi}\frac{d\phi}{d\tau}\right) = \frac{d}{d\tau}\left(r^2\sin^2\theta\frac{d\phi}{d\tau}\right)$$

$$= r^2 2\sin\theta\cos\theta\frac{d\theta}{d\tau}\frac{d\phi}{d\tau} + r^2\sin^2\theta\frac{d^2\phi}{d\tau^2}$$

$$\Rightarrow 0 = \frac{d^2\phi}{d\tau^2} + 2\frac{\sin\theta\cos\theta}{\sin^2\theta}\frac{d\theta}{d\tau}\frac{d\phi}{d\tau} = \frac{d^2\phi}{d\tau^2} + \cot\theta\frac{d\theta}{d\tau}\frac{d\phi}{d\tau} + \cot\theta\frac{d\phi}{d\tau}\frac{d\theta}{d\tau} \quad (19.19)$$

Comparing this with the second of equations 19.18, we see that

$$\Gamma^\phi_{\phi\theta} = \Gamma^\phi_{\theta\phi} = \cot\theta; \quad \text{other } \Gamma^\phi_{\alpha\beta} = 0 \quad (19.20)$$

Exercise 19.6.1. Use the same method but with $\mu = \theta$ to show that the only other nonzero Christoffel symbol for this metric is $\Gamma^\theta_{\phi\phi} = -\sin\theta\cos\theta$.

BOX 19.6 (continued) The Riemann and Ricci Tensors and R for a Sphere

As we found in box 19.3, nonzero components of the Riemann tensor in 2D must have two 0 subscripts and two 1 subscripts, meaning that the only nonzero components are $R_{\theta\phi\theta\phi} = -R_{\theta\phi\phi\theta} = -R_{\phi\theta\theta\phi} = +R_{\phi\theta\phi\theta}$ (the signs can be found using the symmetries in equations 19.4a and 19.4b). So we only need to calculate

$$R_{\theta\phi\theta\phi} = g_{\theta\alpha}R^{\alpha}{}_{\phi\theta\phi} = g_{\theta\theta}R^{\theta}{}_{\phi\theta\phi} + 0$$
$$= r^2 \left(\partial_\theta \Gamma^\theta_{\phi\phi} - \cancel{\partial_\phi \Gamma^\theta_{\phi\theta}} + \cancel{\Gamma^\theta_{\theta\gamma}\Gamma^\gamma_{\phi\phi}} - \Gamma^\theta_{\phi\sigma}\Gamma^\sigma_{\phi\theta} \right) \qquad (19.21)$$

The second term in the above is zero because no Christoffel symbol depends on ϕ, and the third is zero because $\Gamma^\theta_{\theta\theta} = \Gamma^\theta_{\theta\phi} = 0$.

Exercise 19.6.2. Calculate the other two terms to verify that $R_{\theta\phi\theta\phi} = r^2\sin^2\theta$.

Since the metric is diagonal, the inverse metric is also diagonal. Therefore,

$$R_{\theta\theta} = g^{\alpha\mu}R_{\alpha\theta\mu\theta} = g^{\theta\theta}R_{\theta\theta\theta\theta} + g^{\phi\phi}R_{\phi\theta\phi\theta} = 0 + g^{\phi\phi}R_{\phi\theta\phi\theta} = +g^{\phi\phi}R_{\theta\phi\theta\phi} \quad (19.22a)$$
$$R_{\phi\phi} = g^{\alpha\mu}R_{\alpha\phi\mu\phi} = g^{\theta\theta}R_{\theta\phi\theta\phi} + g^{\phi\phi}R_{\phi\phi\phi\phi} = g^{\theta\theta}R_{\theta\phi\theta\phi} \quad (19.22b)$$
$$R_{\phi\theta} = R_{\theta\phi} = g^{\alpha\mu}R_{\alpha\theta\mu\phi} = g^{\theta\theta}R_{\theta\theta\theta\phi} + g^{\phi\phi}R_{\phi\theta\phi\phi} = 0 + 0 = 0 \quad (19.22c)$$
$$R = g^{\alpha\beta}R_{\alpha\beta} = g^{\theta\theta}R_{\theta\theta} + g^{\phi\phi}R_{\phi\phi} \quad (19.23)$$

Note also that the diagonal components of the inverse metric are simply the reciprocals of the corresponding metric components.

Exercise 19.6.3. Use the information above to evaluate the components of the Ricci tensor $R_{\mu\nu}$ and the curvature scalar R.

HOMEWORK PROBLEMS

P19.1 Use the Riemann tensor's identities to show that

$$R^{\alpha}{}_{\alpha\mu\nu} = 0 \qquad (19.24)$$

P19.2 Use the method described in box 19.3 to count the number of independent components of the Riemann tensor in a three-dimensional space or spacetime.

P19.3 Generalize the method described in box 19.3 to show that the number of independent components of the Riemann tensor in n dimensions is $\frac{1}{12}n^2(n^2 - 1)$. (*Hint:* Argue that the number of off-diagonal elements in a $m \times m$ matrix is $m^2 - m$. The really tricky part is to figure out how many constraints equation 19.4d implies. Note that all the indices have to be different, but once chosen, it does not matter how those four indices are arranged in the first term.)

P19.4 Directly calculate $R_{\theta\phi\phi\theta}$, $R_{\phi\theta\theta\phi}$, and $R_{\phi\theta\phi\theta}$ for longitude-latitude coordinates on the surface of a sphere to verify that $R_{\theta\phi\theta\phi} = -R_{\theta\phi\phi\theta} = -R_{\phi\theta\theta\phi} = R_{\phi\theta\phi\theta}$. Also show by direct calculation that at least one other component of the Riemann tensor is zero.

P19.5 Consider "semilog" coordinates p, q whose metric is

$$ds^2 = dp^2 + dq^2/(bq)^2 \qquad (19.25)$$

Find the Christoffel symbols, the only nonzero Riemann tensor component, the Ricci tensor, and the curvature scalar for this metric. Does this metric describe a flat or curved (two-dimensional) space?

P19.6 Consider the following metric for a two-dimensional space: $ds^2 = dp^2 + e^{2p/p_0} dq^2$, where p_0 is a constant with units of length.

a. Use whatever method you like to show that the only nonzero Christoffel symbols for this metric are Γ^p_{qq} and $\Gamma^q_{pq} = \Gamma^q_{qp}$, and evaluate these symbols.

b. Is this space flat or curved? Prove your assertion. (Remember that in a 2D space or spacetime, there is only one independent Riemann tensor component.)

P19.7 Consider the metric for a flat three-dimensional space expressed in spherical coordinates:

$$ds^2 = dr^2 + r^2 d\theta^2 + r^2 \sin^2\theta \, d\phi^2 \qquad (19.26)$$

The last two terms are the same as the metric analyzed in box 19.6, but r is now a variable, not a constant.

a. Show that in addition to the Christoffel symbols calculated in box 19.6, this metric has the following nonzero Christoffel symbols:

$$\Gamma^r_{\theta\theta} = -r, \quad \Gamma^r_{\phi\phi} = -r\sin^2\theta,$$

$$\Gamma^\theta_{\theta r} = \Gamma^\theta_{r\theta} = \Gamma^\phi_{\phi r} = \Gamma^\phi_{r\phi} = \frac{1}{r} \qquad (19.27)$$

b. On the surface of a two-dimensional sphere, we found that $R_{\theta\phi\theta\phi} = r^2 \sin^2\theta$. Show that by virtue of the additional Christoffel symbols above, the expression for $R_{\theta\phi\theta\phi}$ gains a single term over what we had before, and that this term cancels the $r^2\sin^2\theta$ so that $R_{\theta\phi\theta\phi} = 0$ (as is appropriate for a flat space).

P19.8 (Do this only if you have done problem P12.7.) Problem P18.6 claims that in Schwarzschild coordinates

$$R^t{}_{rtr} = \frac{2GM}{r^3}\left(1 - \frac{2GM}{r}\right)^{-1} \qquad (19.28)$$

(see equation 18.35). The fact that $R^t{}_{rtr}$ blows up at $r = 2GM$ does *not* imply that there is a geometric singularity there: a coordinate problem can also cause this behavior. We can see this by evaluating this tensor component in the orthonormal frame of a freely falling observer of the type described in problem P12.7.

a. In equation 12.8 we saw that we could find the covector components of a four-vector $\boldsymbol{A}$ in the frame of an observer using locally orthonormal coordinates by calculating (in *any* coordinate system) the dot products of $\boldsymbol{A}$ with the observer's basis vectors: $A_{\mu,\text{obs}} = \boldsymbol{o}_\mu \cdot \boldsymbol{A} \equiv (\boldsymbol{o}_\mu)^\alpha g_{\alpha\beta} A^\beta$. Generalize this approach to argue that the components of the Riemann tensor evaluated in an observer's orthonormal coordinate system are

$$(R^\alpha{}_{\beta\mu\nu})_{\text{obs}} = \eta^{\alpha\rho} R^\lambda{}_{\gamma\delta\sigma} g_{\lambda\zeta} (\boldsymbol{o}_\rho)^\zeta (\boldsymbol{o}_\beta)^\gamma (\boldsymbol{o}_\mu)^\delta (\boldsymbol{o}_\nu)^\sigma \qquad (19.29)$$

b. For a freely falling observer of the type considered in problem P12.7, the component of the Riemann tensor corresponding to $R^t{}_{rtr}$ is $(R^t{}_{ztz})_{\text{obs}}$. Show that

$$(R^t{}_{ztz})_{\text{obs}} = \frac{2GM}{r^3} \qquad (19.30)$$

Note that this is perfectly finite at $r = 2GM$. It is infinite only at $r = 0$, where there is a *real* geometric singularity. [*Hint:* As a result of symmetries of the Riemann tensor, you only need to know $R^t{}_{rtr}$ to evaluate $(R^t{}_{ztz})_{\text{obs}}$.]

P19.9 A person who falls into a black hole will eventually be torn apart by tidal forces. Imagine a person falling feet-first into a black hole from rest at essentially infinity. Suppose that someone can tolerate a tension force of F_{max} between his or her extremities and center of mass before experiencing serious pain. Determine how long before reaching $r = 0$ that a person might experience pain. Considering that pain impulses travel along nerves at about 1 m/s, is the person likely to suffer? (*Hints:* You should find that the answers to these questions are independent of the black hole's mass. You may use the result of problem P19.8. Also note that when a person's center of mass follows a geodesic, that person's extremities do not. Therefore, forces will have to be applied to constrain a person's extremities to accelerate relative to the geodesics they want to follow.)

20. THE STRESS-ENERGY TENSOR

INTRODUCTION

FLAT SPACETIME
- Review of Special Relativity
- Four-Vectors
- Index Notation

TENSORS
- Arbitrary Coordinates
- Tensor Equations
- Maxwell's Equations
- Geodesics

SCHWARZSCHILD BLACK HOLES
- The Schwarzschild Metric
- Particle Orbits
- Precession of the Perihelion
- Photon Orbits
- Deflection of Light
- Event Horizon
- Alternative Coordinates
- Black Hole Thermodynamics

THE CALCULUS OF CURVATURE
- The Absolute Gradient
- Geodesic Deviation
- The Riemann Tensor

THE EINSTEIN EQUATION
- **The Stress-Energy Tensor**
- The Einstein Equation
- Interpreting the Equation
- The Schwarzschild Solution

COSMOLOGY
- The Universe Observed
- A Metric for the Cosmos
- Evolution of the Universe
- Cosmic Implications
- The Early Universe
- CMB Fluctuations & Inflation

GRAVITATIONAL WAVES
- Gauge Freedom
- Detecting Gravitational Waves
- Gravitational Wave Energy
- Generating Gravitational Waves
- Gravitational Wave Astronomy

SPINNING BLACK HOLES
- Gravitomagnetism
- The Kerr Metric
- Kerr Particle Orbits
- Ergoregion and Horizon
- Negative-Energy Orbits

this depends on this

20. THE STRESS-ENERGY TENSOR

Introduction. In electricity and magnetism, Gauss's law describes the link between the electric field $\vec{E}$ and charge density ρ that creates it:

$$\vec{\nabla} \cdot \vec{E} = 4\pi k \rho \qquad (20.1)$$

When applied to the external field of a motionless spherical charge distribution with total charge Q, this yields Coulomb's law for the electric field $\vec{E} = (kQ/r^2)\hat{r}$, where $\hat{r}$ is a unit vector pointing away from the sphere's center.

The analogous Newtonian equation that links the gravitational field vector $\vec{g} = -\vec{\nabla}\Phi$ (where Φ is the gravitational potential) with *mass* density ρ is

$$-\vec{\nabla} \cdot \vec{g} = -\vec{\nabla} \cdot (-\vec{\nabla}\Phi) = \nabla^2 \Phi = 4\pi G \rho \qquad (20.2)$$

When applied to the external field of a motionless spherical mass distribution with total mass M, this yields $\vec{g} = (-GM/r^2)\hat{r}$ (the fact that gravity is attractive for positive masses rather than repulsive leads to the sign differences). Our goal in the next few chapters is to develop the appropriate tensor generalization of equation 20.2.

While we rarely use equation 20.2 in practice to calculate Newtonian gravitational fields, it is a more appropriate starting place than $\vec{g} = (-GM/r^2)\hat{r}$ because the latter implicitly involves instantaneous action at a distance if taken literally, a concept fundamentally incompatible with relativity. Equation 20.2, in contrast, is a **local field equation:** it links derivatives of the field at a point with the presence of the field's source *at that same point*, ignoring distant sources. A local field equation not only avoids the action-at-a-distance problem, but also allows us to develop the equation first in a locally inertial reference frame (where the physics is easier to understand) before generalizing it to arbitrary coordinates.

Moreover, we have already blazed the trail in chapter 7 by finding the appropriate tensor generalization of Gauss's law. We will follow much the same trajectory here. We have already seen in chapter 18 that the Riemann tensor plays much the same role in general relativity that $\partial_j \partial_k \Phi$ does in Newtonian physics, so we expect that the *left* side of equation 20.2 will generalize to something involving the Riemann tensor: we will work on this in the next chapter. Our goal in *this* chapter is to find the appropriate relativistic generalization of the mass density ρ.

The Source of Gravity Is Relativistic *Energy*. The first question to answer is, does ρ in the Newtonian equation represent the density of an object's *mass* or its relativistic *energy*? In the Newtonian limit, these are essentially the same!

We know of various processes that can convert energy in the form of mass to other forms of energy, so we can in principle create a gravitating object whose internal *mass* content varies with time (its relativistic *energy* content, in contrast, would be fixed because energy is conserved). A simple argument (see box 20.1) shows that if the source of gravity were *mass*, we could use such an object to create energy out of nothing. This problem does not arise if the source of gravity is energy.

The Stress-Energy of "Dust." The next question is this: Is ρ a relativistic scalar, or is it a component of some kind of tensor quantity? In chapter 7, we argued from the invariance of charge that *charge* density was in fact the time component of a current-density four-vector $\mathbf{J}$. In this chapter we will analogously argue from the transformation properties of energy that energy density ρ is the tt component of a second-rank tensor $\mathbf{T}$ called the **stress-energy tensor.**

We can do this most easily by first considering a specific form of distributed matter called **dust**. *Dust* here is not what you find under your bed; it is rather a technical term referring to a fluid whose constituent particles in the neighborhood of a given event in spacetime are essentially at rest with respect to each other (in contrast with gas particles, which are constantly in random motion). In this way, the fluid is somewhat like airborne dust seen in a shaft of sunlight: the dust particles in the neighborhood of a given point are all carried along in the direction of the local direction of air flow without much individual random motion.

The definition of "dust" means that at every event we can find a locally inertial frame (LIF) in which the particles in a given small volume are all essentially at rest. Consider a set of N identical particles each with mass m at rest inside a tiny box of volume V_0 at rest in a certain LIF named S', as shown in figure 20.1a. The number density of particles in the box is $n_0 = N/V_0$, and since each particle is at rest in this box, the energy of each is simply m. The total energy density is thus $\rho_0 = n_0 m$.

Now imagine observing that box in another LIF named S in which the box and the particles inside all move together with velocity $\vec{v}$. The box as viewed in S still has N particles in it, but because its length in direction of motion is Lorentz-contracted (see figure 20.1b), it has a smaller volume $V = V_0 \sqrt{1 - v^2}$. The particle number density in the box is thus $n \equiv N/V = N/(V_0 \sqrt{1 - v^2}) = n_0/\sqrt{1 - v^2}$ in S.

We can express this in tensor form: note that the fluid's four-velocity $\boldsymbol{u}$ in S is

$$u^\alpha = \begin{bmatrix} 1/\sqrt{1-v^2} \\ v_x/\sqrt{1-v^2} \\ v_y/\sqrt{1-v^2} \\ v_z/\sqrt{1-v^2} \end{bmatrix} = \begin{bmatrix} u^t \\ v_x u^t \\ v_y u^t \\ v_z u^t \end{bmatrix} \quad (20.3)$$

(Note that the metric in S, because it is a LIF, is locally the flat-space metric.) Therefore, we can write the particle number density in this frame in the form

$$n \equiv \frac{N}{V} = \frac{N}{V_0 \sqrt{1-v^2}} = n_0 u^t \quad (20.4)$$

Since the energy of each particle in this frame is $p^t \equiv mu^t$, the total energy density as measured in this frame is

$$\rho \equiv n p^t = (n_0 u^t)(m u^t) = (n_0 m) u^t u^t = \rho_0 u^t u^t \quad (20.5)$$

The energy density ρ_0 of the dust in its own rest frame is a *frame-independent scalar*, because all observers know which LIF is at rest with respect to the dust and what an observer in that LIF measures for the dust's energy density. So the quantity in equation 20.5 is a scalar multiplied by the time components of two four-vectors. We can interpret this as the t-t component of a second-rank tensor

$$T^{\mu\nu} = \rho_0 u^\mu u^\nu \quad \text{(for dust)} \quad (20.6)$$

This is the *stress-energy tensor for dust*. Note that it is *symmetric*: $T^{\mu\nu} = T^{\nu\mu}$.

The Physical Meaning of Stress-Energy Components. By construction, T^{tt} in any frame is the energy density ρ. What do the other components mean in a LIF? We can write the component T^{tx} as

$$T^{tx} = \rho_0 u^t u^x = (n_0 m) u^t u^x = (n_0 u^t) m u^x = n p^x = x\text{-momentum density} \quad (20.7)$$

since p^x is the relativistic x-momentum of each particle. Similarly, T^{ty} and T^{tz} correspond to y- and z-momentum density, respectively. Since the stress-energy tensor is symmetric, the same interpretations can be applied to T^{xt}, T^{yt} and T^{zt} respectively.

However, we can look at component T^{tx} in a different way:

$$T^{tx} = \rho_0 u^t u^x = (n_0 m) u^t u^x = (n_0 u^t) m (u^t v_x) = n p^t v_x = \frac{(n A v_x dt) p^t}{A \, dt} \quad (20.8)$$

The quantity $A v_x dt$ is the volume of dust that will move through an area A perpendicular to the x direction during the time interval dt, so $n A v_x dt$ represents the total number of dust particles that move through that area. Therefore, we can interpret $T^{tx} = T^{xt}$ as specifying the *total energy per unit area per unit time* (i.e., **energy flux**) flowing through a surface perpendicular to the x direction. Let's call this the **x-flux of energy**. Similarly, $T^{ty} = T^{yt}$ and $T^{tz} = T^{zt}$ can be interpreted as the y and z-fluxes of energy.

In a similar way, you can show (see box 20.2) that the stress-energy tensor components $T^{ij} = T^{ji}$ with spatial indices i and j can be interpreted as the i-flux of j-momentum or vice versa.

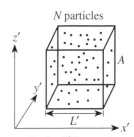

FIG. 20.1a In the S' frame, the number density of particles in the displayed box is $n_0 \equiv N/V_0$, where $V_0 = AL'$ is the box's volume.

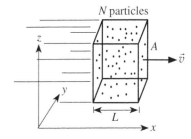

FIG. 20.1b In the S frame, where the particles (and box) move with velocity $\vec{v}$, the box's length is contracted to $L = L'\sqrt{1-v^2}$, so the number density of particles in the box is therefore now $n \equiv N/AL = N/(AL'\sqrt{1-v^2}) = (N/V_0)/\sqrt{1-v^2} = n_0/\sqrt{1-v^2}$.

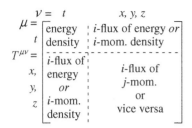

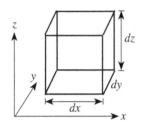

FIG. 20.2 This diagram summarizes the physical meaning of the components of the stress-energy tensor.

FIG. 20.3 An infinitesimal box in a locally inertial reference frame. Dust is moving through the box with a certain four-velocity $\boldsymbol{u}$.

Figure 20.2 summarizes our interpretations for the components of the stress-energy tensor. (Though we have derived them for dust, they apply generally.)

What do these fluxes mean physically? Imagine that we take a plate with area A perpendicular to the x direction and put it in the path of the dust particles, and imagine that the plate absorbs those particles (and thus their energy and momentum). The quantity $T^{tx}A$ yields the *energy per unit (coordinate) time* deposited on the plate, whereas the quantity $T^{yx}A$ yields the *y-momentum per unit (coordinate) time* deposited on the plate, which is the definition of the *y component of the force* the dust exerts on the plate. **Stress** in physics is defined to be force applied per unit area (this is more general than "pressure," because the force in a stress need not act perpendicular to the area). This means that the components of $\boldsymbol{T}$ with two spatial indices specifies *stresses* exerted by the fluid, while T^{tt} specifies *energy* density. This is why this tensor is called the "stress-energy" tensor.

Conservation Law for Dust. Just as $\partial_\mu J^\mu = 0$ expresses the law of conservation of charge in special relativity, the tensor expression

$$\nabla_\nu T^{\mu\nu} = 0 \qquad (20.9)$$

expresses the law of conservation of energy and momentum for a fluid in arbitrary coordinates. We can see that this is true for dust as follows. Consider dust consisting of identical particles of mass m flowing with four-velocity $\boldsymbol{u}$ in an arbitrary spacetime. At a given event, we can always find a locally inertial frame (LIF) in which we can use cartesian-like t, x, y, and z coordinates whose Christoffel symbols are locally zero. Consider an infinitesimal box with dimensions dx, dy, and dz centered on the event's location, as shown in figure 20.3. Assume that the box is small enough so that the dust density and velocity is approximately constant over each face of the box.

Since the particles move a distance $v_x dt$ in an infinitesimal time interval dt, the total energy flowing *into* the box through its left face during dt is

$$\rho v_x \, dt \, dy \, dz = (\rho_0 u^t u^t) v_x \, dt \, dy \, dz = \rho_0 u^t u^x \, dt \, dy \, dz = (T^{tx})_{\text{left}} dt \, dy \, dz \qquad (20.10)$$

where $(T^{tx})_{\text{left}}$ reminds us to evaluate T^{tx} at the box's left face. Similarly, the total energy flowing *out* of the box's right face is $(T^{tx})_{\text{right}} dt \, dy \, dz$. Therefore the net energy that accumulates in the box during dt due to the x component of fluid flow is

$$\left[(T^{tx})_{\text{left}} - (T^{tx})_{\text{right}}\right] dt \, dy \, dz = \left(-\frac{\partial T^{tx}}{\partial x} dx\right) dt \, dy \, dz = -\frac{\partial T^{tx}}{\partial x} dt \, dx \, dy \, dz \qquad (20.11)$$

The expressions for the net energy that has accumulated in the box during dt due to other velocity components of the fluid flow are analogous, so the *total* energy accumulating in the box during time dt is

$$dE_{\text{tot}} = \left[-\frac{\partial T^{tx}}{\partial x} - \frac{\partial T^{ty}}{\partial y} - \frac{\partial T^{tz}}{\partial z}\right] dt \, dx \, dy \, dz \qquad (20.12)$$

But since energy is conserved, any net energy that flows into the box through its sides must reflect itself as an increase in the energy density *inside* the box, so

$$dE_{\text{tot}} = d\rho \, dx \, dy \, dz = \left(\frac{\partial T^{tt}}{\partial t} dt\right) dx \, dy \, dz \qquad (20.13)$$

So if we subtract equation 20.12 from this equation, we find that

$$0 = \left[\frac{\partial T^{tt}}{\partial t} + \frac{\partial T^{tx}}{\partial x} + \frac{\partial T^{ty}}{\partial y} + \frac{\partial T^{tz}}{\partial z}\right] dt \, dx \, dy \, dz \;\Rightarrow\; \partial_\nu T^{t\nu} = 0 \qquad (20.14)$$

This equation therefore expresses the law of conservation of the fluid's energy. Similarly $\partial_\nu T^{i\nu} = 0$ expresses conservation of the fluid's i-momentum ($i = x, y,$ or z).

Because it is always possible to find a LIF at every event in spacetime, and because $\nabla_\nu T^{\mu\nu}$ reduces to $\partial_\nu T^{\mu\nu}$ in such a LIF, and because $\partial_\nu T^{\mu\nu} = 0$ in that LIF as a consequence of conservation of four-momentum, the corresponding *tensor* equation $\nabla_\nu T^{\mu\nu} = 0$ must be true in all coordinate systems as a consequence of conservation of the fluid's four-momentum.

The Stress-Energy of a Perfect Fluid. In most realistic fluids, individual particles exhibit significant random thermal motions even in a frame where the fluid as a whole in a given neighborhood is at rest. We call a fluid a "perfect fluid" when such random motions are significant but the particles remain essentially non-interacting (e.g., viscosity and interparticle potential energies are zero). An ideal gas is an example of a perfect fluid. What is the stress-energy of such a fluid?

Let ρ_0 and p_0 be the energy density and pressure respectively of a perfect fluid at an event measured in a LIF at rest with respect to the fluid at that event. The components of the stress-energy tensor for the gas in the LIF at that event turn out to be

$$T^{\mu\nu} = \begin{bmatrix} \rho_0 & 0 & 0 & 0 \\ 0 & p_0 & 0 & 0 \\ 0 & 0 & p_0 & 0 \\ 0 & 0 & 0 & p_0 \end{bmatrix} \quad \text{(for a perfect fluid at rest in a LIF)} \quad (20.15)$$

The argument for equation 20.15 is discussed in box 20.3: the basic approach involves treating the perfect fluid as comprised of non-interacting subsets of particles that all have essentially the same velocity and can thus be considered dust, and then summing the stress-energy tensors over all those subsets.

Note that in the GR unit system, where time is measured in meters, energy density (kg·m²/m²)/m³ and pressure as force per unit area (kg·m/m²)/m² both end up having units of kg/m³, so the units are consistent along the diagonal.

The appropriate tensor generalization of equation 20.15 that expresses the stress-energy tensor for an arbitrary perfect fluid in arbitrary coordinates is

$$T^{\mu\nu} = (\rho_0 + p_0)u^\mu u^\nu + p_0 g^{\mu\nu} \quad \text{(for a perfect fluid)} \quad (20.16)$$

This is a tensor expression constructed of relevant tensor quantities that (1) reduces to equation 20.15 in a LIF that is instantaneously at rest with respect to the fluid (see box 20.4), and (2) reduces to the dust stress-energy when $p_0 = 0$. Because $\nabla_\nu T^{\mu\nu} = 0$ holds for each of the fluid's dust-like particle subsets, it also must hold for fluid as a whole. You can also easily see that this tensor (like the dust tensor) is symmetric: $T^{\mu\nu} = T^{\nu\mu}$. Because figure 20.2 also correctly describes stress-energy of the dust-like subsets in a LIF, it must also apply to a perfect fluid evaluated in a LIF.

The Conservation Law Specifies Equations of Motion. You can show (see box 20.5) that the conservation law $\nabla_\nu T^{\mu\nu} = 0$ applied to the stress-energy tensor given in equation 20.16 yields the following equations in an arbitrary LIF:

$$\partial_\mu(\rho_0 u^\mu) + p_0 \partial_\mu u^\mu = 0 \quad \text{(equation of continuity)} \quad (20.17)$$

$$(\rho_0 + p_0)(\partial_\mu u^\nu) u^\mu = -(\eta^{\mu\nu} + u^\mu u^\nu)\partial_\mu p_0 \quad \text{(equation of motion)} \quad (20.18)$$

If our fluid is slowly moving in this LIF, then v^2 becomes negligible compared to 1, our fluid's four-velocity becomes approximately $u^\mu \approx [1, v_x, v_y, v_z]$, and we can ignore the distinction between the energy density and pressure ρ_0 and p_0 in the fluid's rest frame and its density and pressure ρ and p measured in the LIF. If the random thermal motions of particles within the fluid are also not relativistic, then the pressure $p \ll \rho$. In this limit, equation 20.17 becomes

$$\partial_\mu(\rho u^\mu) = 0 \quad \Rightarrow \quad 0 = \frac{\partial \rho}{\partial t} + \vec{\nabla} \cdot (\rho \vec{v}) \quad (20.19)$$

and the spatial components of equation 20.18 become

$$\rho(\partial_\mu u^i)u^\mu + \eta^{i\mu}\partial_\mu p = 0 \quad \Rightarrow \quad \rho\left(\frac{\partial}{\partial t} + \vec{u} \cdot \vec{\nabla}\right)\vec{u} = -\vec{\nabla}p \quad (20.20)$$

Those of you who have studied some fluid dynamics will recognize these equations as being the Newtonian law of continuity (expressing mass conservation) and Euler's equation of motion for a perfect fluid (expressing Newton's second law for fluid elements). The point is that $\nabla_\mu T^{\mu\nu} = 0$ specifies all that one needs to know about how fluid elements move in response to internal pressure gradients.

BOX 20.1 Why the Source of Gravity Must Be Energy, Not Mass

Consider a perfectly mirrored box that contains matter and antimatter held in separate magnetic traps. Imagine that an external particle is released essentially from rest a very large distance from the box. The gravitational field of the matter and antimatter will cause the particle to accelerate toward the box, picking up kinetic energy as it approaches. But just as the particle passes the box, we release the magnetic traps, and allow the matter and antimatter particles to annihilate each other, producing photons. If *mass* is the source of a gravitational field, then (since photons have no mass), the box now contains no mass and will thus have no gravitational field. The external particle that is passing the box can therefore carry all the kinetic energy it gained on the inbound trip out to infinity. In other words, this scheme would allow us to fire particles to infinity with energy we create out of nothing, violating the principle of conservation of energy.

Exercise 20.1.1. Explain why this problem does *not* arise if the box's total *energy* is the source of its gravitational field.

BOX 20.2 Interpretation of T^{ij} in a Locally Inertial Frame

Exercise 20.2.1. Use a similar argument to that presented in the text for T^{tx} to argue that $T^{ij} \equiv \rho_0 u^i u^j$ (where i and j are both spatial indices) can be interpreted as the i-flux of j-momentum or vice versa.

BOX 20.3 The Stress-Energy Tensor for a Perfect Fluid in Its Rest LIF

Consider a perfect fluid consisting of randomly moving particles with mass m. We will work in the cartesian-like coordinates of the LIF in which the bulk fluid at a certain event in spacetime is at rest (i.e., where the spatial components of the fluid's total four-momentum are zero). Imagine that this fluid fills a cubical box with sides of length L centered on the event's spatial location.

We can divide the particles in the fluid into subsets of particles having the same four-velocity u^μ. As long as the particles in a perfect fluid do not interact, the particles in a given subset can be treated as "dust," because the particles in the set are at rest with respect to each other. The stress-energy *for a given subset* is

$$\text{(subset)} \quad T^{\mu\nu} = \rho_0 u^\mu u^\nu = n_0 m u^\mu u^\nu \qquad (20.21)$$

where n_0 is the number density of particles in the *subset's rest frame*. According to equation 20.4, the number density of particles in the *fluid's* rest frame (where the particles have four-velocity u^μ) is $n = n_0 u^t$. Thus the *subset's* stress-energy is

$$\text{(subset)} \quad T^{\mu\nu} = \frac{n}{u^t} m u^\mu u^\nu = n \frac{m u^\mu m u^\nu}{m u^t} = n \frac{p^\mu p^\nu}{p^t} = \frac{N}{L^3} \frac{p^\mu p^\nu}{p^t} \qquad (20.22)$$

where N is the number of particles in the subset.

The total stress-energy tensor for the perfect fluid will be the sum of subset stress-energy tensors over all possible subsets. Let $N(|\vec{p}|)\,dp^x dp^y dp^z$ be the number of particles that have a spatial momentum within a certain infinitesimal volume $dp^x dp^y dp^z$ in momentum space centered on a certain value $\vec{p}$. The function $N(|\vec{p}|)$ can depend only on the *magnitude* of $\vec{p}$ because if the motion of molecules is purely random, all *directions* of motion should be equally likely. We can express the perfect fluid's *total* stress-energy tensor as an integral over all possible spatial momenta:

$$\text{(total)} \quad T^{\mu\nu} = \frac{1}{L^3} \iiint dp^x dp^y dp^z \, N(|\vec{p}|) \frac{p^\mu p^\nu}{p^t} \qquad (20.23)$$

Note that the particle energy $p^t = \sqrt{m^2 + |\vec{p}|^2}$, though it is not independent of the spatial momentum $\vec{p}$, also depends only on its magnitude.

This immediately means that all off-diagonal elements in the total perfect-fluid stress-energy tensor must be zero. Consider, for example,

$$T^{xy} = \frac{1}{L^3} \int dp^z \int p^y dp^y \int \frac{N(|\vec{p}|)}{p^t} p^x dp^x \qquad (20.24)$$

When we do the integral over p^x (holding p^y and p^z constant), the quantities $N(|\vec{p}|)$ and p^t depend on $|\vec{p}| = [(p^x)^2 + (p^y)^2 + (p^z)^2]^{1/2} = [(p^x)^2 + \text{const.}]^{1/2}$, implying that $N(|\vec{p}|)/p^t = f[(p^x)^2]$. Therefore, the integral over p^x becomes

$$\int_{-\infty}^{\infty} f([p^x]^2) p^x dp^x = \int_0^{\infty} \{f([p^x]^2) p^x + f([p^x]^2)(-p^x)\} dp^x = 0 \qquad (20.25)$$

Therefore $T^{xy} = 0$. Another way to say this is that for a given p^y and p^z, one is just as likely to have a particle with $+p^x$ as $-p^x$, so the sum over all p^x's will cancel. The same argument applies to any other off-diagonal element of $T^{\mu\nu}$.

Exercise 20.3.1. Explain why this argument does *not* imply that diagonal components such as T^{xx} are zero.

BOX 20.3 (continued) The Stress-Energy Tensor for a Perfect Fluid in Its Rest LIF

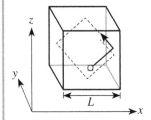

FIG. 20.4 A single fluid particle bouncing around inside a cubical box with side L.

To interpret these diagonal elements, consider a single particle bouncing elastically around the cubical box (see figure 20.4). Since (by hypothesis) it does not interact with other particles, it will hit the right wall once every $\Delta t = 2L/v_x$ (2L because it has to go to the left wall *and back* between collisions). If the collision is elastic, the particle's x-momentum will be reversed, implying that it delivers an x-momentum of $2p^x$ to the right wall every collision. The average force that the bouncing particle delivers to the wall is the x-momentum the wall collects per unit time, which will be $2p^x/\Delta t$. The average *pressure* exerted on the wall is this force divided by the wall's area L^2. The average pressure exerted by this single particle on the wall is

$$\text{pressure for one particle} = \frac{1}{L^2}\frac{2p^x}{\Delta t} = \frac{1}{L^2}\frac{2p^x}{2L/v_x} = \frac{1}{L^3}p^x v^x = \frac{1}{L^3}\frac{p^x p^x}{p^t} \quad (20.26)$$

(The expressions are analogous for walls perpendicular to the y and z directions.) Therefore, we can find the total pressure exerted on this wall by all molecules by integrating this over all possible spatial momenta. Since we have defined $N(|\vec{p}|)\,dp^x dp^y dp^z$ to be the number of particles whose spatial momentum is within a box of volume $dp^x dp^y dp^z$ centered on a spatial momentum $\vec{p}$, then $N(|\vec{p}|)\,dp^x dp^y dp^z\,(p^x p^x/L^3 p^t)$ is the pressure exerted by that subset of particles, and

$$\text{total pressure} = \frac{1}{L^3}\iiint dp^x dp^y dp^z\, N(|\vec{p}|)\frac{p^x p^x}{p^t} = T^{xx} \quad (20.27)$$

where the last step follows from equation 20.23.

Exercise 20.3.2. Argue that $T^{xx} = T^{yy} = T^{zz}$. (Your argument may be either conceptual or mathematical.)

Finally, note that equation 20.23 implies that

$$T^{tt} = \frac{1}{L^3}\iiint dp^x dp^y dp^z\, N(|\vec{p}|)\frac{p^t p^t}{p^t} = \frac{1}{L^3}\iiint dp^x dp^y dp^z\, N(|\vec{p}|)\,p^t \quad (20.28)$$

Exercise 20.3.3. Argue that the right integral is the total energy density ρ.

BOX 20.4 Equation 20.16 Reduces to Equation 20.15

In a LIF where the fluid at a given event is (at least instantaneously) at rest, the metric $g_{\mu\nu} = \eta_{\mu\nu}$ and the fluid's four-velocity at that event is $u^\mu = [1,0,0,0]$.

Exercise 20.4.1. Argue that in such a frame, $T^{\mu\nu} = (\rho_0 + p_0)u^\mu u^\nu + p_0 g^{\mu\nu}$ given in equation 20.16 reduces to $\mathrm{diag}(\rho_0, p_0, p_0, p_0)$ (i.e., equation 20.15).

BOX 20.5 Fluid Dynamics from Conservation of Four-Momentum

In a LIF centered on a certain event, $g_{\mu\nu} = \eta_{\mu\nu}$, the first derivatives of the metric are zero, and absolute gradient becomes the ordinary gradient. Therefore the equation $\nabla_\mu T^{\mu\nu} = 0$ in this frame for $T^{\mu\nu} = (\rho_0 + p_0)u^\mu u^\nu + p_0 g^{\mu\nu}$ becomes

$$0 = \partial_\mu T^{\mu\nu} = u^\mu u^\nu \partial_\mu (\rho_0 + p_0) + (\rho_0 + p_0)[u^\nu \partial_\mu u^\mu + u^\mu \partial_\mu u^\nu] + \eta^{\mu\nu} \partial_\mu p_0 \quad (20.29)$$

This can be simplified as follows. Note that since $-1 = \boldsymbol{u} \cdot \boldsymbol{u} = u^\nu u_\nu$, we have

$$0 = \partial_\mu(u^\nu u_\nu) = \partial_\mu(u^\nu \eta_{\nu\alpha} u^\alpha) = u^\nu \eta_{\nu\alpha} \partial_\mu u^\alpha + u^\alpha \eta_{\nu\alpha} \partial_\mu u^\nu$$

$$= u_\nu \partial_\mu u^\nu + u_\nu \partial_\mu u^\nu = 2 u_\nu \partial_\mu u^\nu \;\Rightarrow\; u_\nu \partial_\mu u^\nu = 0 \quad (20.30)$$

Multiply both sides of equation 20.29 by u_ν, sum over ν, and use $-1 = u^\nu u_\nu$ and equation 20.30 to simplify the result: you should get equation 20.17, repeated here:

$$\partial_\mu(\rho_0 u^\mu) + p_0 \partial_\mu u^\mu = 0 \quad (20.17r)$$

Exercise 20.5.1. Verify that this is true.

BOX 20.5 (continued) Fluid Dynamics from Conservation of Four-Momentum

Now take the equation $\partial_\mu(\rho_0 u^\mu) + p_0 \partial_\mu u^\mu = 0$ you just derived, multiply both sides by u^ν (with no sum), and subtract from equation 20.29 (repeated here):

$$0 = u^\mu u^\nu \partial_\mu(\rho_0 + p_0) + (\rho_0 + p_0)[u^\nu \partial_\mu u^\mu + u^\mu \partial_\mu u^\nu] + \eta^{\mu\nu}\partial_\mu p_0 \quad (20.29r)$$

After canceling terms, you should get equation 20.18 (repeated here):

$$(\rho_0 + p_0)(\partial_\mu u^\nu)u^\mu = -(\eta^{\mu\nu} + u^\mu u^\nu)\partial_\mu p_0 \quad (20.18r)$$

Exercise 20.5.2. Verify this.

HOMEWORK PROBLEMS

P20.1 Find the components (in GR units) of the stress-energy tensor for the ionized hydrogen plasma at the sun's center, where the density is 160 g/cm^3 and the temperature is 1.5×10^7 K. You should find that even under these conditions, $p_0 \ll \rho_0$ and $\rho_0 \approx$ mass density, which is why we generally think of mass density alone as being the source of gravity. (*Hints:* You may find the ideal gas law $pV = Nk_BT$ helpful. Also, 1 g of hydrogen contains Avogadro's number of atoms. In the plasma, hydrogen will be disassociated into protons and electrons. Since even under these conditions, the particles will not be relativistic, you can approximate the relativistic energy of an individual plasma particle as being $p^t \approx m + \text{KE} = m + \frac{3}{2}k_BT$. The value of k_B in GR units is 1.536×10^{-40} kg/K.)

P20.2 Consider a perfect fluid of the type considered in box 20.3. Determine the approximate average speed that randomly moving particles in this fluid must have for the fluid's pressure to be 1% of its energy density in a LIF.

P20.3 Equation 20.16 describes the stress-energy tensor of a perfect fluid moving with four-velocity **u**. Consider a slowly rotating star at the origin of a LIF. Assume that the star has a specified (but arbitrary) radially varying energy density $\rho = \rho(r)$ as measured in the LIF, and is rotating rigidly around the z axis with angular speed ω. Also assume that the star's pressure is negligible compared to its density (see problem P20.1), its mass is small enough so that it does not significantly curve spacetime, and ω is small enough so that the fluid's speed $v = r\omega$ at any point inside the star is $\ll 1$. Express each component of the star's stress-energy tensor $T^{\mu\nu}$ in its interior as a function of the LIF coordinates t, x, y, and z, ignoring terms of order v^2 or smaller.

P20.4 Argue that the stress-energy for an ideal gas of photons in a perfectly-mirrored box in a LIF at rest with respect to the box is

$$T^{\mu\nu} = \begin{bmatrix} \rho & 0 & 0 & 0 \\ 0 & \frac{1}{3}\rho & 0 & 0 \\ 0 & 0 & \frac{1}{3}\rho & 0 \\ 0 & 0 & 0 & \frac{1}{3}\rho \end{bmatrix} \quad (20.31)$$

(*Hints:* Rework the argument in box 20.3 for this case. Note that for photons, we have $p^\mu = [E, Ev_x, Ev_y, Ev_z]$ and $v_x^2 + v_y^2 + v_z^2 = 1$. Note also that exercise 20.3.2 implies that $T^{xx} = T^{yy} = T^{zz}$ for any ideal gas.)

P20.5 (Duplicates much of problem P7.8.) In any introductory physics textbook, you can find the equation for the density of energy stored in an electromagnetic field. In GR units, this equation reads as follows:

$$\rho_E = \frac{1}{8\pi k}(E^2 + B^2) \quad (20.32)$$

Since an electromagnetic field stores energy, it must have a stress-energy tensor. In this problem, we will seek a covariant expression for that tensor.

a. The electromagnetic stress-energy tensor must be constructed from only the antisymmetric EM field tensor $F^{\mu\nu}$ and the symmetric metric $g_{\mu\nu}$ (since no other tensors have anything to do with the electromagnetic field or the spacetime in which we find that field). Find a second-rank tensor constructed of these tensor quantities that is symmetric and whose t-t component in a LIF is equation 20.32. (*Hints:* Since that equation involves products of the field quantities, the tensor you construct will plausibly involve products of the EM field tensor. You will need a sum of two such product terms to get things to come out right. Begin by showing that the scalar $F^{\mu\nu}F_{\mu\nu} = 2(B^2 - E^2)$ in a LIF.)

b. T^{tx} is the density of x-momentum P^x in the field, or equivalently, the x-flux of energy in the field. Express T^{tx} in terms of components of $\vec{E}$ and $\vec{B}$ in a LIF. Does this expression ring any bells? (*Hint:* If it doesn't, look up "Poynting vector.")

P20.6 We define the four-force **F** acting on an object to be the rate $d\mathbf{p}/d\tau$ that four-momentum is delivered to that object per unit proper time τ for that object.

a. Argue that the appropriate covariant expression for the rate $d\mathbf{p}/d\tau$ at which momentum flows *through* a small flat patch with unit normal vector **n** and area A (as measured in its own rest frame) must be

$$\left(\frac{d\mathbf{p}}{d\tau}\right)^\mu = AT^{\mu\nu}g_{\nu\alpha}n^\alpha = AT^{\mu\nu}n_\nu \quad (20.33)$$

(*Hint:* Work this out in a LIF and generalize.) **Note:** This will be equal to a four-force **F** exerted *on* the patch only if the patch *captures* the momentum that would otherwise flow through it. In the case where particles bounce elastically off the patch, the patch captures only the momentum component parallel to **n**.

b. Use the expression above and the equation $T^{\mu\nu} = (\rho_0 + p_0)u^\mu u^\nu + p_0 g^{\mu\nu}$ for a perfect fluid to prove that the magnitude of the four-force exerted on any patch of area A on the wall of a container holding a perfect fluid at rest must be $p_0 A$, no matter what direction **n** points in and no matter what coordinate system we use. (*Hint:* Argue first that if **n** represents a wall in space, and **u** represents the fluid's four-velocity, then we must have $\mathbf{n} \cdot \mathbf{u} = 0$ if the fluid is at rest in the wall's frame.)

P20.7 The tensor $T^{\mu\nu} = -\Lambda g^{\mu\nu}$, where Λ is a positive scalar constant, could *in principle* be a valid stress-energy tensor for some hypothetical fluid in that it is a symmetric tensor, it has a positive energy density in a LIF, and it is consistent with energy conservation ($\nabla_\nu T^{\mu\nu} = -\Lambda \nabla_\nu g^{\mu\nu} = 0$, because the absolute gradient of the metric tensor is zero). In a LIF, how would the pressure and energy density of this hypothetical fluid compare? Adapt the argument of box 20.3 to argue that such a hypothetical fluid could not be constructed of randomly moving, non-interacting particles. (This does turn out to be a physically relevant stress-energy, though, as we will see in the next chapter.)

P20.8 In this problem we will explore how to calculate the stress-energy in an observer's orthonormal frame if we know the stress-energy in any arbitrary coordinate system.

a. In equation 12.8 we saw that we could find the covector components of a four-vector $\mathbf{A}$ in the frame of an observer using locally orthonormal coordinates by calculating (in *any* coordinate system) the dot products of $\mathbf{A}$ with the observer's basis vectors: $A_{\mu,\text{obs}} = \mathbf{o}_\mu \cdot \mathbf{A} \equiv (\mathbf{o}_\mu)^\alpha g_{\alpha\beta} A^\beta$. Generalize this approach to argue that the components of the stress-energy tensor evaluated in an observer's LOF (locally orthonormal frame) are

$$T^{\mu\nu}_{\text{obs}} = \eta^{\mu\alpha} \eta^{\nu\beta} T^{\gamma\delta} g_{\gamma\sigma} (\mathbf{o}_\alpha)^\sigma g_{\delta\lambda} (\mathbf{o}_\beta)^\lambda \quad (20.34)$$

(*Hint:* Consider a second-rank tensor that is the tensor product of two four-vectors.)

b. Let's apply this to the case of an observer moving with four-velocity $\mathbf{u}_{\text{obs}}$ with respect to the local rest frame of a perfect fluid, where $T^{\gamma\delta}$ (assuming we are using cartesian coordinates) is given by equation 20.15. Show that the energy density in the observer's frame is the same as what you would get from equation 20.16 if we assume the fluid in the observer's frame moves at the same speed but in the opposite spatial direction as the observer does in the fluid frame (i.e., $u^x_{\text{fluid}} = -u^x_{\text{obs}}$, $u^y_{\text{fluid}} = -u^y_{\text{obs}}$, and $u^z_{\text{fluid}} = -u^z_{\text{obs}}$).

P20.9 Imagine that the stress-energy of a fluid in a certain LIF is given by

$$T^{\mu\nu} = \begin{bmatrix} \rho & 0 & 0 & 0 \\ 0 & \alpha\rho & 0 & 0 \\ 0 & 0 & \alpha\rho & 0 \\ 0 & 0 & 0 & \alpha\rho \end{bmatrix} \quad (20.35)$$

where α is a unitless constant. Now, according to the result of problem P20.8a, if an observer in a different LIF is moving with four-velocity $\mathbf{u}_{\text{obs}}$ relative to the first, the energy density that observer will measure is

$$\rho_{\text{obs}} = T^{tt}_{\text{obs}} = \eta^{t\sigma} \eta^{t\gamma} T^{\mu\nu} \eta_{\mu\alpha} \eta_{\nu\beta} (\mathbf{o}_\sigma)^\alpha (\mathbf{o}_\gamma)^\beta$$

$$= \eta^{tt} \eta^{tt} T^{\mu\nu} \eta_{\mu\alpha} \eta_{\nu\beta} (\mathbf{o}_t)^\alpha (\mathbf{o}_t)^\beta$$

$$= T^{\mu\nu} \eta_{\mu\alpha} \eta_{\nu\beta} u^\alpha_{\text{obs}} u^\beta_{\text{obs}} \quad (20.36)$$

Note that the observer's four-velocity has the components $u^\alpha_{\text{obs}} = [1, v_x, v_y, v_z]/\sqrt{1-v^2}$ in the original LIF. What constraint on α is imposed by the requirement that observers in *every possible* observer frame find $\rho_{\text{obs}} > 0$?

P20.10 The dominant energy condition (DEC) requires that for *any* causal four-vector $\mathbf{a}$ then $\mathbf{b}$ where

$$b^\beta \equiv -T^{\beta\nu} g_{\nu\alpha} a^\alpha \quad (20.37)$$

is also causal. (A vector $\mathbf{a}$ is "causal" if it is either timelike or lightlike, meaning that $\mathbf{a} \cdot \mathbf{a} \leq 0$, and if it is future-oriented, meaning that $a^t > 0$ in any local orthonormal frame.) The DEC is a proposed constraint on the stress-energy tensors of fluids we might consider "physically reasonable."

a. Argue that the DEC implies that an observer in any LOF will find the fluid's four-momentum *density* $\pi^\nu_{\text{obs}} \equiv T^{t\nu}_{\text{obs}}$ to be causal, which implies that the fluid will not be seen in such a frame to be moving faster than light. (*Hint:* See problem P20.8a. Although π^μ is not a true four-vector, determine whether it is causal as if it were. Note that $\pi^\mu = dp^\mu/dV$ will be causal if and only if the total four-momentum dp^μ in the differential volume dV is.)

b. Argue that for a perfect fluid, the DEC implies that $\rho_0 \geq |p_0|$. (*Hint:* Work in the LOF of an observer at rest with respect to the fluid, and remember that the DEC must hold for *any* $\mathbf{a}$.)

c. Show that a stress-energy of the form $T^{\mu\nu} = -\Lambda g^{\mu\nu}$, where Λ is a positive scalar constant, satisfies the DEC. (In the next chapter we will see that the so-called vacuum energy has a stress-energy of this form.)

d. The **weak energy condition** (WEC) requires that an observer in any LOF observe the fluid's energy density to be positive. Argue that the DEC implies the WEC.

21. THE EINSTEIN EQUATION

INTRODUCTION

FLAT SPACETIME
- Review of Special Relativity
- Four-Vectors
- Index Notation

TENSORS
- Arbitrary Coordinates
- Tensor Equations
- Maxwell's Equations
- Geodesics

SCHWARZSCHILD BLACK HOLES
- The Schwarzschild Metric
- Particle Orbits
- Precession of the Perihelion
- Photon Orbits
- Deflection of Light
- Event Horizon
- Alternative Coordinates
- Black Hole Thermodynamics

THE CALCULUS OF CURVATURE
- The Absolute Gradient
- Geodesic Deviation
- The Riemann Tensor

THE EINSTEIN EQUATION
- The Stress-Energy Tensor
- **The Einstein Equation**
- Interpreting the Equation
- The Schwarzschild Solution

COSMOLOGY
- The Universe Observed
- A Metric for the Cosmos
- Evolution of the Universe
- Cosmic Implications
- The Early Universe
- CMB Fluctuations & Inflation

GRAVITATIONAL WAVES
- Gauge Freedom
- Detecting Gravitational Waves
- Gravitational Wave Energy
- Generating Gravitational Waves
- Gravitational Wave Astronomy

SPINNING BLACK HOLES
- Gravitomagnetism
- The Kerr Metric
- Kerr Particle Orbits
- Ergoregion and Horizon
- Negative-Energy Orbits

this → depends on → this

21. THE EINSTEIN EQUATION

Overview. Our goal in this chapter is to develop the Einstein equation for the gravitational field. This equation is the relativistic generalization of the Newtonian gravitational field equation that is analogous to Gauss's law for the EM field:

$$\vec{\nabla} \cdot \vec{g} = -4\pi G \rho \quad \Rightarrow \quad \nabla^2 \Phi = 4\pi G \rho \tag{21.1}$$

where Φ is the gravitational potential ($\vec{g} = -\vec{\nabla}\Phi$) and ρ is mass density. We saw in the last chapter that the tensor generalization of mass density is the stress-energy tensor $T^{\mu\nu}$. A tensor generalization of $\nabla^2 \Phi = 4\pi G \rho$ thus might have the form

$$G^{\mu\nu} = \kappa T^{\mu\nu} \tag{21.2}$$

where κ is a scalar constant and $G^{\mu\nu}$ is a second-rank tensor that describes something about the curvature of spacetime. We will call $G^{\mu\nu}$ the **Einstein tensor**.

Constraints on $G^{\mu\nu}$. Since $G^{\mu\nu}$ is to describe something about the curvature of spacetime, it is plausible that it should be constructed out of the Riemann tensor, the only tensor quantity that we know that describes curvature. Since $T^{\mu\nu}$ is symmetric, $G^{\mu\nu}$ must also be symmetric. We know that $\nabla_\mu T^{\mu\nu} = 0$ always (because of local conservation of four-momentum) so $\nabla_\mu G^{\mu\nu} = 0$ always (if we take the divergence of both sides of $G^{\mu\nu} = \kappa T^{\mu\nu}$, we should get $0 = 0$). Finally, $G^{\mu\nu} = \kappa T^{\mu\nu}$ should reduce to $\nabla^2 \Phi = 4\pi G \rho$ in the Newtonian limit.

A First Guess. Perhaps the most obvious candidate for a $G^{\mu\nu}$ constructed from the Riemann tensor is the **Ricci tensor**, which we have defined to be

$$R_{\beta\nu} = R^\alpha{}_{\beta\alpha\nu} \quad \text{(note the sum over } \alpha\text{)} \tag{21.3}$$

This is the form of the Ricci tensor with both indexes in the subscript position. The version with both indices raised is

$$R^{\mu\nu} = g^{\mu\beta} g^{\nu\sigma} R_{\beta\sigma} = g^{\mu\beta} g^{\nu\sigma} R^\alpha{}_{\beta\alpha\sigma} = g^{\mu\beta} g^{\nu\sigma} g^{\alpha\gamma} R_{\alpha\beta\gamma\sigma} \tag{21.4}$$

Because of the symmetry $R_{\alpha\beta\gamma\sigma} = R_{\gamma\sigma\alpha\beta}$, the Ricci tensor is symmetric:

$$R^{\nu\mu} = g^{\nu\beta} g^{\mu\sigma} g^{\alpha\gamma} R_{\alpha\beta\gamma\sigma} = g^{\nu\beta} g^{\mu\sigma} g^{\alpha\gamma} R_{\gamma\sigma\alpha\beta} = g^{\nu\sigma} g^{\mu\beta} g^{\gamma\alpha} R_{\alpha\beta\gamma\sigma}$$
$$= g^{\mu\beta} g^{\nu\sigma} g^{\alpha\gamma} R_{\alpha\beta\gamma\sigma} \equiv R^{\mu\nu} \tag{21.5}$$

where in the last step of the first line, I renamed indices $\beta \to \sigma$, $\sigma \to \beta$, $\alpha \to \gamma$, and $\gamma \to \alpha$, and in the first step of the second line, I changed the order of multiplication of the first two metric factors and used the symmetry of the metric to reverse the indices on the third metric factor.

However, this obvious guess has a serious problem: $\nabla_\nu R^{\mu\nu} \neq 0$ in general. This is easiest to see at the origin of a LIF, as discussed in box 21.1.

A Second Guess. What else could $G^{\mu\nu}$ contain besides $R^{\mu\nu}$? The only reasonable things that could go into the construction of $G^{\mu\nu}$ are other contractions of the Riemann tensor and the metric itself: only these quantities reflect the geometric nature of spacetime. Note that $g^{\mu\nu}$ and $g^{\mu\nu} R$, where R is the **curvature scalar**

$$R \equiv g^{\beta\nu} R_{\beta\nu} = g^{\beta\nu} R^\alpha{}_{\beta\alpha\nu} = g^{\beta\nu} g^{\alpha\mu} R_{\alpha\beta\mu\nu} \tag{21.6}$$

are both symmetric second-rank tensors. In fact, $R^{\mu\nu}$, $g^{\mu\nu}$, and $g^{\mu\nu} R$ are the *only* second-rank symmetric tensors that can be constructed out of the metric and the Riemann tensor that are linear in (i.e., do not involve products of) the Riemann tensor. Let's see if we can construct a tensor of the form $R^{\mu\nu} + b g^{\mu\nu} R + \Lambda g^{\mu\nu}$ (where b and Λ are scalar constants) such that

$$\nabla_\nu (R^{\mu\nu} + b g^{\mu\nu} R + \Lambda g^{\mu\nu}) = 0 \tag{21.7}$$

identically. If we can find such a tensor, we will have found the simplest choice for the left side of equation 21.2 (which Occam's Razor suggests will be the *correct* choice). If we cannot, we will have to investigate adding terms that are nonlinear in the Riemann tensor (which we would like to avoid if at all possible).

Since the absolute gradient of the metric is zero and Λ is a constant,

$$\nabla_\nu(\Lambda g^{\mu\nu}) = 0 \tag{21.8}$$

automatically. So the problem reduces to finding the value of b such that

$$\nabla_\nu(R^{\mu\nu} + b g^{\mu\nu} R) = 0 \tag{21.9}$$

You can in fact fairly easily show that the Bianchi identity implies that this quantity will be identically zero if and only if $b = -\frac{1}{2}$ (see box 21.2). Therefore,

$$R^{\mu\nu} - \tfrac{1}{2} g^{\mu\nu} R + \Lambda g^{\mu\nu} = \kappa T^{\mu\nu} \tag{21.10}$$

satisfies all of the constraints we have imposed: both sides are symmetric, second-rank tensors with zero absolute divergence, and the left side is formed only of combinations of the metric and the Riemann tensor linear in the latter.

The Einstein Tensor. For reasons that will become clearer shortly, we define the Einstein tensor $G^{\mu\nu}$ as follows:

$$G^{\mu\nu} \equiv R^{\mu\nu} - \tfrac{1}{2} g^{\mu\nu} R \tag{21.11}$$

The Einstein equation therefore has the most general form

$$G^{\mu\nu} + \Lambda g^{\mu\nu} = \kappa T^{\mu\nu} \tag{21.12}$$

An Alternative Form of the Einstein Equation. If you multiply both sides of equations 21.10 by $g_{\mu\nu}$ you can show that

$$-R + 4\Lambda = \kappa T \tag{21.13}$$

where $T \equiv g_{\mu\nu} T^{\mu\nu} = T^\mu{}_\mu$ (see box 21.3). If we multiply both sides of this by $\tfrac{1}{2} g^{\mu\nu}$, subtract it from equation 21.10, and move the Λ term to the other side, we get

$$R^{\mu\nu} = \kappa(T^{\mu\nu} - \tfrac{1}{2} g^{\mu\nu} T) + \Lambda g^{\mu\nu} \tag{21.14}$$

This version of the Einstein equation is actually more useful in practice than the conceptually more straightforward version given by equation 21.12.

The Newtonian Limit. The next order of business is to determine the value of the constant κ. We can do this by requiring that the Einstein equation reduce to equation 21.1 in the Newtonian limit.

Recall from chapter 18 that the Newtonian equation of tidal deviation is

$$\frac{d^2 n^i}{dt^2} = -\eta^{ij}[\partial_k \partial_j \Phi] n^k \tag{21.15}$$

(this was equation 18.4). The corresponding relativistic equation of geodesic deviation (equation 18.16) states that

$$\left(\frac{d^2 \mathbf{n}}{d\tau^2}\right)^\alpha = -R^\alpha{}_{\mu\nu\sigma} u^\sigma u^\mu n^\nu \tag{21.16}$$

This should reduce to equation 21.15 in the Newtonian limit.

Taking the Newtonian limit involves making three assumptions:

1. The gravitational field is weak (spacetime is almost flat).
2. Objects move with speeds $\ll 1$ in this field.
3. mass-energy density $\rho = T^{tt} \gg$ other $T^{\mu\nu}$ components in the field source.

The first assumption means that we can find a coordinate system that is essentially cartesian (with coordinates we will label as t, x, y, z) but whose metric is insignificantly different from the flat space metric $\eta_{\mu\nu}$ and the difference between absolute and ordinary derivatives becomes negligible. The second assumption means that the four-velocities of freely falling objects following nearby worldlines will have components $u^t \approx 1$, with u^x, u^y, u^z negligible in comparison, and also that the proper time τ along a worldline will be insignificantly different than t.

Therefore, in the Newtonian limit, the spatial components of the relativistic equation of geodesic deviation become

$$\frac{d^2 n^i}{d\tau^2} = -R^i{}_{\mu\nu\sigma} u^\mu n^\nu u^\sigma \approx -R^i{}_{t\nu t} n^\nu = -R^i{}_{tkt} n^k \qquad (21.17)$$

where i and k range over purely spatial indices and the last step follows because $R^i{}_{ttt} = 0$ (one of the symmetries of the Riemann tensor says that $R^\alpha{}_{\beta\mu\nu} = -R^\alpha{}_{\beta\nu\mu}$). Comparing this to the Newtonian equation $d^2 n^i/dt^2 = -\eta^{ij}(\partial_j \partial_k \Phi) n^k$, we see that in the Newtonian limit $\eta^{ij}(\partial_j \partial_k \Phi) \approx R^i{}_{tkt}$. But this means that

$$\nabla^2 \Phi \equiv \eta^{jk}(\partial_j \partial_k \Phi) \approx R^k{}_{tkt} = R^\alpha{}_{t\alpha t} \equiv R_{tt} \qquad (21.18)$$

where the next-to-last step follows because $R^t{}_{ttt} = 0$. Note that when the metric is basically the flat-space metric, $R^{tt} \equiv g^{t\mu} g^{t\nu} R_{\mu\nu} \approx \eta^{t\mu} \eta^{t\nu} R_{\mu\nu} = (-1)(-1) R_{tt} = R_{tt}$, so $R^{tt} \approx \nabla^2 \Phi$ also. The t-t component of the alternative form of the Einstein equation in the Newtonian approximation thus reads

$$\nabla^2 \Phi \approx R^{tt} = \kappa(T^{tt} - \tfrac{1}{2}\eta^{tt} T) + \Lambda g^{tt} \qquad (21.19)$$

But if $\rho = T^{tt} \gg$ other $T^{\mu\nu}$, then $T \equiv g_{\mu\nu} T^{\mu\nu} \approx \eta_{\mu\nu} T^{\mu\nu} \approx \eta_{tt} \rho = -\rho$. Therefore,

$$\nabla^2 \Phi \approx R^{tt} \approx \kappa[\rho - \tfrac{1}{2}(-1)(-\rho)] - \Lambda = \tfrac{1}{2}\kappa\rho - \Lambda \qquad (21.20)$$

This reduces to the Newtonian field equation $\nabla^2 \Phi = 4\pi G \rho$ if and only if our unknown constant $\kappa = 8\pi G$ and our unknown constant $\Lambda \ll 4\pi G \rho$.

The Cosmological Constant. We see from equation 21.20 that the term in the Einstein equation involving Λ acts in the Newtonian limit (if $\Lambda > 0$) as if it were a *negative* energy density that exists even in a vacuum (where $\rho = 0$). This would create a *repulsive* gravitational field. Since we do not observe this effect at the scale of the solar system, Λ must be very small (indeed, accurate observations of the orbits of planets imply that $\Lambda/8\pi G \ll 10^{-7}$ kg/m^3). But is it strictly zero?

When Einstein first proposed the theory of general relativity, the accepted model of the universe was a static model, where the stars and other cosmic objects remained essentially at rest (except for random motions). Einstein included this term in his final field equation because he found that without it, his field equation predicted that the universe would either expand or contract, but could not remain static. He found that a small repulsive term in the field equation could cancel the gravitational attraction of stars and other cosmic objects toward each other and therefore could allow the universe to be static. He called Λ the "cosmological constant."

Not long after Einstein published his field equation in 1915, he determined that the static universe solution that he had found was not stable. Therefore, though a static universe could exist in principle, in *practice*, any small perturbation away from perfect equilibrium would cause either a runaway contraction or a runaway expansion. Moreover, in 1929, Hubble published evidence that the universe was indeed expanding, something that Einstein *could have predicted* had he realized that the static universe solution was untenable. Einstein therefore famously retracted the idea of a cosmological constant, calling it "the worst mistake I ever made." His subsequent work assumed that $\Lambda = 0$.

However, providence may have played a role here. In 1998, astrophysicists discovered that the universe appears to be accelerating in its expansion, which would require a nonzero cosmological constant (as we will see in a few chapters). Refined measurements of the cosmological background radiation have confirmed this result, strongly implying that $\Lambda/8\pi G \approx 0.7 \times 10^{-26}$ kg/m^3. This is small enough that its effects cannot be observed within the solar system, but large enough to have a significant effect on the evolution of the universe. Einstein's original "mistake" has proved instead to be visionary: instead of being wrong, he was merely 80 years ahead of his time!

Physicists currently consider this constant not to be a part of the left side of the Einstein equation (the side that describes the curvature of spacetime) but rather an additional term on the right, as the stress-energy associated with the vacuum:

$$T^{\mu\nu}_{\text{vac}} = -\frac{\Lambda}{8\pi G} g^{\mu\nu} \tag{21.21}$$

Still, it is important to note that this vacuum energy is so small that its effects are completely negligible unless we are dealing with scales larger than the largest galactic superclusters. Therefore, whenever we are dealing with applications of the Einstein equation to situations other than the universe itself, we will assume that $\Lambda = 0 \Rightarrow T^{\mu\nu}_{\text{vac}} = 0$.

The Final Forms of the Einstein Equation. So, the Einstein equation, in the form currently used by physicists, looks like this:

$$G^{\mu\nu} = 8\pi G(T^{\mu\nu} + T^{\mu\nu}_{\text{vac}}) = 8\pi G T^{\mu\nu}_{\text{all}} \tag{21.22}$$

where $T^{\mu\nu}$ is the stress-energy tensor associated with matter and $T^{\mu\nu}_{\text{vac}}$ is the vacuum energy defined by equation 21.21. The form that we will find most *practical* to use, however, is the form given by equation 21.14, modified by including the correct value for κ and using the definition of $T^{\mu\nu}_{\text{vac}}$:

$$R^{\mu\nu} = 8\pi G(T^{\mu\nu}_{\text{all}} - \tfrac{1}{2}g^{\mu\nu} T_{\text{all}}) \tag{21.23}$$

(Note that $T^{\mu\nu}_{\text{vac}} - \tfrac{1}{2}g^{\mu\nu} T_{\text{vac}} = (-\Lambda g^{\mu\nu} + \tfrac{1}{2}g^{\mu\nu} 4\Lambda)/8\pi G = +\Lambda g^{\mu\nu}/8\pi G$.)

The Einstein equation implies that gravity cannot even *exist* in spacetimes having fewer than four dimensions (see problems P21.5 and P21.6). In four dimensions, the Einstein equation has 10 potentially independent components, because the tensors appearing on both sides are symmetric. Though $R^{\mu\nu}$ and $G^{\mu\nu}$ involve only linear combinations of Riemann tensor components, the Riemann tensor itself is *nonlinear* in the metric and its first and second derivatives. Therefore, the Einstein equation represents (up to) 10 nonlinear, (usually) coupled, second-order differential equations to solve for the metric. Finding solutions is *not* an easy task!

Indeed, the most straightforward way to solve the Einstein equation in a given situation is to make a guess about the form of the solution (based on the situation's symmetries) with as few unknown components as possible, use the Einstein equation to generate simpler differential equations for the few unknowns, and solve for those unknowns. We will see several examples of this process in future chapters.

However, there are many situations that simply aren't symmetric enough to reduce the complexity much. In such cases, solving the equations numerically is the only realistic option. Even this can be extraordinarily difficult, and there is an extensive literature in the journals about methods for solving the Einstein equation in various contexts. Still, the advent of powerful computing resources in the past few decades has made it possible to apply the Einstein equation to many more situations than were accessible in the past. This has partly contributed to the blossoming of research opportunities in general relativity in recent times.

Even so, the difficulty of solving the Einstein equation analytically means that one can pose simple and yet important questions about the theory that are extremely difficult to resolve. One of the most famous is the "Cosmic Censorship Hypothesis," which in simple form states that singularities (such as the one at $r = 0$ in Schwarzschild spacetime, where the mass-energy density becomes infinite and the geometry becomes undefined) are always "clothed" by event horizons (like the Schwarzschild event horizon at $r = 2GM$), making the singularity unobservable at infinity. The hypothesis was proposed by Roger Penrose in 1969 (*Rivista Nuovo Cimento*, **1**, 252–276) and is very important (because otherwise it can be argued that general relativity does not lead to useful predictions about the future), but it remains unproven. (See Wald, *General Relativity*, University of Chicago Press, 1984, pp. 302–305 for a detailed but fairly advanced discussion.)

BOX 21.1 The Divergence of the Ricci Tensor

At the origin of a locally inertial reference frame (LIF), the absolute gradient is the same as the ordinary gradient, the first derivatives of the metric are all zero, and the Riemann tensor reduces to

$$R_{\alpha\beta\mu\nu} = \tfrac{1}{2}(\partial_\beta\partial_\mu g_{\alpha\nu} + \partial_\alpha\partial_\nu g_{\beta\mu} - \partial_\beta\partial_\nu g_{\alpha\mu} - \partial_\alpha\partial_\mu g_{\beta\nu}) \tag{21.24}$$

The Ricci tensor is defined to be

$$R^{\gamma\sigma} = g^{\gamma\beta}g^{\sigma\nu}R^{\alpha}{}_{\beta\alpha\nu} = g^{\gamma\beta}g^{\sigma\nu}g^{\alpha\mu}R_{\alpha\beta\mu\nu} \tag{21.25}$$

Note also that because the metric has zero absolute gradient

$$\nabla_\delta R^{\gamma\sigma} = g^{\gamma\beta}g^{\sigma\nu}g^{\alpha\mu}\nabla_\delta R_{\alpha\beta\mu\nu} \tag{21.26}$$

From here, you can pretty easily show that

$$\nabla_\sigma R^{\gamma\sigma} = \tfrac{1}{2}g^{\gamma\beta}g^{\sigma\nu}g^{\alpha\mu}(\partial_\sigma\partial_\beta\partial_\mu g_{\alpha\nu} - \partial_\sigma\partial_\beta\partial_\nu g_{\alpha\mu}) \neq 0 \tag{21.27}$$

Exercise 21.1.1. Verify this. (*Hint:* Applying the partial derivative to equation 21.24 will yield four terms. But you can rename indices $\alpha \leftrightarrow \sigma$ and $\mu \leftrightarrow \nu$ to show that the second and fourth terms cancel. You can't do this with the first and third terms, though, and the triple derivatives of the metric, like the double derivatives, do not all vanish at the origin of a LIF.)

BOX 21.2 Finding the Value of b

Start with the Bianchi Identity:

$$\nabla_\sigma R_{\alpha\beta\mu\nu} + \nabla_\nu R_{\alpha\beta\sigma\mu} + \nabla_\mu R_{\alpha\beta\nu\sigma} = 0 \qquad (21.28)$$

Let's now multiply all terms by $g^{\gamma\sigma} g^{\alpha\mu} g^{\beta\nu}$ and sum over the repeated indices:

$$\nabla_\sigma g^{\gamma\sigma} g^{\alpha\mu} g^{\beta\nu} R_{\alpha\beta\mu\nu} + \nabla_\nu g^{\gamma\sigma} g^{\alpha\mu} g^{\beta\nu} R_{\alpha\beta\sigma\mu} + \nabla_\mu g^{\gamma\sigma} g^{\alpha\mu} g^{\beta\nu} R_{\alpha\beta\nu\sigma} = 0 \qquad (21.29)$$

(We can pull the metric factors inside the absolute derivative because the absolute gradient of the metric is zero, meaning that the metric behaves like a constant with respect to absolute gradients.) If you check the first term against equation 21.6, you will see that the final two metric factors summed with the Riemann tensor as shown yield the curvature scalar R:

$$\nabla_\sigma g^{\gamma\sigma} R + \nabla_\nu g^{\gamma\sigma} g^{\alpha\mu} g^{\beta\nu} R_{\alpha\beta\sigma\mu} + \nabla_\mu g^{\gamma\sigma} g^{\alpha\mu} g^{\beta\nu} R_{\alpha\beta\nu\sigma} = 0 \qquad (21.30)$$

Now, you can show that by the symmetries of the Riemann tensor, we can rearrange indices in the second two terms as follows:

$$\nabla_\sigma g^{\gamma\sigma} R - \nabla_\nu g^{\gamma\sigma} g^{\alpha\mu} g^{\beta\nu} R_{\mu\sigma\alpha\beta} - \nabla_\mu g^{\gamma\sigma} g^{\alpha\mu} g^{\beta\nu} R_{\nu\sigma\beta\alpha} = 0 \qquad (21.31)$$

Exercise 21.2.1. Identify the symmetries that I have used in the last step.

By using the definition of the Ricci tensor and renaming indices, you can show that

$$\tfrac{1}{2}\nabla_\sigma g^{\gamma\sigma} R - \nabla_\sigma R^{\gamma\sigma} = 0 \qquad (21.32)$$

This is the negative of equation 21.9 with $b = -1/2$.

Exercise 21.2.2. Verify equation 21.32.

BOX 21.3 Showing that $-R + 4\Lambda = \kappa T$

If we multiply both sides of equation 21.10 by $g_{\mu\nu}$, we get

$$g_{\mu\nu}R^{\mu\nu} - \tfrac{1}{2}g_{\mu\nu}g^{\mu\nu}R + \Lambda g_{\mu\nu}g^{\mu\nu} = \kappa g_{\mu\nu}T^{\mu\nu} \tag{21.33}$$

Notice that since the metric with upstairs indices is the matrix inverse of the version with downstairs indices, $g_{\mu\nu}g^{\mu\nu} = \delta^{\mu}{}_{\mu} = 4$.

Exercise 21.3.1. Why is this equal to 4?

You can use this and the definitions of R and T to show that $-R + 4\Lambda = \kappa T$.

Exercise 21.3.2. Verify this.

(This was a crucial step in converting the Einstein equation to the more practically useful form $R^{\mu\nu} = \kappa(T^{\mu\nu} - \tfrac{1}{2}g^{\mu\nu}T) + \Lambda g^{\mu\nu}$.)

HOMEWORK PROBLEMS

P21.1 Pick an arbitrary origin in empty space. Assume that the gravitational potential Φ is zero at the origin and is spherically symmetric around that origin. The expression for the Laplacian in such a case of spherical symmetry is

$$\nabla^2 \Phi = \frac{1}{r^2} \frac{d}{dr}\left(r^2 \frac{d\Phi}{dr}\right) \quad (21.34)$$

a. Show that the solution to the empty-space Einstein equation in the Newtonian limit implies that

$$-\frac{d\Phi}{dr} = \frac{\Lambda}{3} r \quad (21.35)$$

Since the left side is the negative gradient of the potential, which is the same as the local acceleration of gravity, we see that any particle displaced from the origin will experience an effective gravitational force *away* from the origin (if Λ is positive) that increases linearly with r. Note also that the acceleration increases with distance, meaning that Λ is more significant in large systems.

b. The observational fact that the Newtonian potential with $\Lambda = 0$ seems to accurately describe planetary motion in our solar system limits the maximum value that Λ can have on scales where $r \sim 10^{12}$ m and for $GM_\odot \approx 1500$ m. Show that requiring that vacuum energy's effect be much smaller than the ordinary gravitational acceleration ($\frac{1}{3}\Lambda r \ll GM_\odot/r^2$) implies that

$$\frac{\Lambda}{8\pi G} \ll 10^{-7} \frac{\text{kg}}{\text{m}^3} \quad (21.36)$$

P21.2 Prove that $G^{\mu\nu} = 0$ if and only if $R^{\mu\nu} = 0$.

P21.3 a. Calculate $T^{\mu\nu} - \frac{1}{2} g^{\mu\nu} T$ for a perfect fluid in an arbitrary coordinate system.

b. Find the components (in terms of ρ_0 and p_0) of this tensor sum in a LOF where the fluid is at rest.

P21.4 Will observers in all LIFs observe the same vacuum stress-energy $T^{\mu\nu}_{\text{vac}}$, irrespective of their motion relative to each other? Explain your reasoning. (*Hint:* No calculation necessary!)

P21.5 Imagine a universe having only two dimensions instead of four and no vacuum energy. Argue that in such a universe, the equation $R_{\mu\nu} = 0$ (taking equation 21.14 as the Einstein equation in this case) implies that spacetime must be flat in a vacuum, and therefore that there can be no gravitational interactions between two objects separated by a vacuum. (*Hint:* Remember that the 2D Riemann tensor has only one possibly nonzero independent component. Explicitly write out the sum for the Ricci tensor components in terms of this component. Be sure to consider the possibility that the metric is not diagonal.)

P21.6 Imagine a universe having only three dimensions instead of four and no vacuum energy. Argue that in such a universe, the vacuum Einstein equation $R_{\mu\nu} = 0$ implies that spacetime must be flat in a vacuum, and therefore that there can be no gravitational interactions between two objects separated by a vacuum. (*Hint:* Use the method of box 19.3 to discover the six independent Riemann tensor components in a 3D spacetime. Then take advantage of the fact that we can always define a LIF at any point in the 3D spacetime and show that $R_{\mu\nu} = 0$ implies that all six independent Riemann components must be zero in this LIF. But if the Riemann tensor is zero in any coordinate system, it must be zero in all, so this proves that the spacetime is flat.)

P21.7 Use the results of box 19.6 to show that the Einstein equation in two dimensions implies that the stress-energy that should act as the source for the curvature of a spherical surface in fact must be zero. (The fact that zero stress-energy is connected to a curved space in this case is another indication that the Einstein equation does not make sense in two dimensions.)

P21.8 Consider the metric

$$ds^2 = -e^{2gx} dt^2 + dx^2 + dy^2 + dz^2 \quad (21.37)$$

This is a candidate metric for describing a uniform gravitational field for the reasons discussed in part a.

a. Use the geodesic equation to argue that at every point in this spacetime, a free particle initially at rest will experience an x-acceleration of $d^2x/d\tau^2 = -g$. (*Hint:* What is $dt/d\tau$ for a particle at rest in this metric?)

b. Argue that the only nonzero Christoffel symbols for this metric are $\Gamma^t_{tx} = \Gamma^t_{xt} = g$ and $\Gamma^x_{tt} = g e^{2gx}$.

c. Argue that the only nonzero Riemann tensor components for this metric are R_{txtx} and its permutations. Calculate the value of R_{txtx}.

d. Argue that only R_{tt} and R_{xx} are nonzero for this metric and calculate their values.

e. Show that the *vacuum* Einstein equation (with no vacuum energy) requires that $g = 0$, meaning that there is no gravitational field.

f. Could g be nonzero inside some kind of fluid? Show that the Einstein equation implies that the stress-energy tensor for a fluid that could possibly create such a field must have all components being equal to zero except for $T^{yy} = T^{zz} = g^2/8\pi G$. Explain why this is absurd.

P21.9 The dominant energy condition (DEC) requires that for *causal* four-vector **a** (meaning that $\mathbf{a} \cdot \mathbf{a} \leq 0$ and $a^t > 0$ in any locally orthonormal frame) then **b** defined by

$$b^\beta \equiv -T^{\beta\nu} g_{\nu\alpha} a^\alpha \quad (21.38)$$

is also causal. This is a proposed constraint on the stress-energy tensor of fluids we might consider to be "physically reasonable": if the DEC is satisfied, then observers in any LOF will observe the fluid to move at less than the speed of light (you don't have to show this: see problem P20.10). Argue that the vacuum stress-energy satisfies the DEC.

22. INTERPRETING THE EQUATION

INTRODUCTION

FLAT SPACETIME
- Review of Special Relativity
- Four-Vectors
- Index Notation

TENSORS
- Arbitrary Coordinates
- Tensor Equations
- Maxwell's Equations
- Geodesics

SCHWARZSCHILD BLACK HOLES
- The Schwarzschild Metric
- Particle Orbits
- Precession of the Perihelion
- Photon Orbits
- Deflection of Light
- Event Horizon
- Alternative Coordinates
- Black Hole Thermodynamics

THE CALCULUS OF CURVATURE
- The Absolute Gradient
- Geodesic Deviation
- The Riemann Tensor

THE EINSTEIN EQUATION
- The Stress-Energy Tensor
- The Einstein Equation
- **Interpreting the Equation**
- The Schwarzschild Solution

COSMOLOGY
- The Universe Observed
- A Metric for the Cosmos
- Evolution of the Universe
- Cosmic Implications
- The Early Universe
- CMB Fluctuations & Inflation

GRAVITATIONAL WAVES
- Gauge Freedom
- Detecting Gravitational Waves
- Gravitational Wave Energy
- Generating Gravitational Waves
- Gravitational Wave Astronomy

SPINNING BLACK HOLES
- Gravitomagnetism
- The Kerr Metric
- Kerr Particle Orbits
- Ergoregion and Horizon
- Negative-Energy Orbits

this depends on this

Introduction. In this chapter, we will explore some of the implications of the Einstein equation, some general and some that apply only to nearly Newtonian contexts. The focus in this chapter will be mostly conceptual, but we will lay some groundwork for a later (more comprehensive) discussion of the weak-field limit.

Energy Conservation as a Geometric Necessity. In the last chapter, we derived the specific form of the Einstein tensor $G^{\mu\nu}$ by requiring that it be consistent with the local law of conservation of energy $\nabla_\mu T^{\mu\nu} = 0$, which we took as given. However, we can look at the Einstein equation in a different way in which energy conservation emerges as a *necessary consequence* of the theory.

Consider how we might develop the Einstein equation $G^{\mu\nu} = 8\pi G T^{\mu\nu}$ without imposing the conservation-law constraint. $G^{\mu\nu}$ must still be symmetric, like $T^{\mu\nu}$, so the Einstein equation represents 10 (second-order, nonlinear, differential) equations in the 10 independent components of the symmetric metric tensor $g_{\mu\nu}$. This should be enough information to determine *completely* those components.

But we *should not be able* to completely determine the components of $g_{\mu\nu}$! The components of the metric depend partly on our *arbitrary* choice of coordinate system. Given four arbitrary coordinate transformation equations $x'^\mu = f^\mu(x^\nu)$, we should be able to transform the components $g_{\mu\nu}$ into $g'_{\mu\nu}$ for the new coordinates and still have a *completely valid solution* to the Einstein equation in the same physical context. This means that the Einstein equation can at most place *six* equations-worth of constraint on $g_{\mu\nu}$, so that it takes four more arbitrary coordinate-choice equations to determine completely the ten components of $g_{\mu\nu}$.

But how can the *ten* equations implicit in $G^{\mu\nu} = 8\pi G T^{\mu\nu}$ really represent only *six* equations-worth of constraint on $g_{\mu\nu}$? This can be true only if $G^{\mu\nu}$ is constructed so that it *automatically* satisfies four internal equations of constraint, no matter *what* $g_{\mu\nu}$ might be. If this were true, then $G^{\mu\nu} = 8\pi G T^{\mu\nu}$ would really represent only *six* independent equations, as four of the ten equations implicit in $G^{\mu\nu} = 8\pi G T^{\mu\nu}$ could be generated from the other six using the four internal equations.

Now, if we define $G^{\mu\nu} \equiv R^{\mu\nu} - \frac{1}{2} g^{\mu\nu} R$, then (because of the Bianchi identity) $G^{\mu\nu}$ *automatically* satisfies the four internal constraint equations $\nabla_\nu G^{\mu\nu} = 0$ no matter what $g_{\mu\nu}$ is. $G^{\mu\nu}$ so defined turns out to be the *only* symmetric, second-rank tensor we can construct out of the Riemann tensor and the metric that (1) is linear in second derivatives of the metric, (2) is free of higher-order derivatives of the metric, (3) is zero in flat space, and (4) satisfies such a four-equation internal constraint (Misner, Thorne, and Wheeler, *Gravitation*, Freeman, 1973, p. 417).

But if $\nabla_\nu G^{\mu\nu} = 0$ is necessary to preserve the arbitrary nature of coordinates, then the Einstein equation $G^{\mu\nu} = 8\pi G T^{\mu\nu}$ itself *requires* that $\nabla_\nu T^{\mu\nu} = 0$. Therefore, we see that conservation of energy and momentum emerges as a *consequence* of requiring that we should be able to choose coordinates arbitrarily.

So, there are at least two ways to look at the gravitational field. In one view (presented in the last chapter), the *source* is primary. It is constrained *a priori* by conservation laws, which then constrain the form that the fields generated by those sources can take. In the other more modern view (discussed here), the *field* is primary and is constrained by a more fundamental principle (the arbitrary nature of coordinate systems). The field then tells the *source* how to behave (i.e., to conserve energy and momentum): without the field, there would be no constraints on the behavior of the source! (For more about this, see the reference noted above.)

The conservation of energy and momentum is also an example of Noether's theorem, which states quite generally that symmetries in the laws of physics imply conservation laws. In this case, the Einstein equation is symmetric with respect to coordinate changes; as a consequence, energy and momentum are conserved.

The Connection to the Geodesic Hypothesis. It turns out that even the geodesic hypothesis is involved in this discussion. One can prove quite generally that the geodesic motion of objects emerges from $\nabla_\nu T^{\mu\nu} = 0$, which in turn follows from the coordinate-independence of the Einstein equation (as we have just seen).

In what follows, I will (for the sake of simplicity) show only that $\nabla_\nu T^{\mu\nu} = 0$ implies that particles of "dust" follow geodesics. Since a normal object (e.g., a satellite) consists of particles that for the most part move together (the pressure being relativistically negligible), this will be a good approximation for many realistic objects. The general proof for all non-rotating objects is similar (but involves more math).

The "dust" stress-energy tensor is $T^{\mu\nu} = \rho_0 u^\mu u^\nu$, where ρ_0 is the density of the dust in its rest frame and u^μ is its four-velocity. By the product rule,

$$0 = \nabla_\nu T^{\mu\nu} = \nabla_\nu(\rho_0 u^\nu u^\mu) = u^\mu \nabla_\nu(\rho_0 u^\nu) + \rho_0 u^\nu \nabla_\nu u^\mu \qquad (22.1)$$

Remember that $-1 = \boldsymbol{u} \cdot \boldsymbol{u} \equiv u^\mu g_{\mu\alpha} u^\alpha$ which also implies that $\nabla_\nu(u^\mu g_{\mu\alpha} u^\alpha) = 0$. Using both of these results, you can show (see box 22.1) that $\nabla_\nu(\rho_0 u^\nu) = 0$. If we substitute this back into equation 22.1, we find that

$$0 = u^\nu \nabla_\nu u^\mu \equiv u^\nu \left(\frac{\partial u^\mu}{\partial x^\nu} + \Gamma^\mu_{\beta\nu} u^\beta \right) = \frac{\partial u^\mu}{\partial x^\nu} \frac{dx^\nu}{d\tau} + \Gamma^\mu_{\beta\nu} u^\beta u^\nu$$

$$\Rightarrow \quad 0 = \frac{du^\mu}{d\tau} + \Gamma^\mu_{\beta\nu} u^\beta u^\nu = \frac{d^2 x^\mu}{d\tau^2} + \Gamma^\mu_{\beta\nu} \frac{dx^\beta}{d\tau} \frac{dx^\nu}{d\tau} \qquad (22.2)$$

since $u^\mu \equiv dx^\mu/d\tau$. This is precisely the geodesic equation in the form given in equation 17.12. Therefore, we see that $\nabla_\nu T^{\mu\nu} = 0$ implies that an object consisting of particles that move together (with negligible pressure) must follow a geodesic.

The Weak-Field Limit. In order to better understand the physical implications of the Einstein equation, it helps to look at the equation's weak-field limit, where we can more easily draw connections and contrasts with Newtonian gravity. If the field in a given region of spacetime is "weak," then spacetime is only slightly curved, and throughout that region, we ought to be able to use a metric such that

$$g_{\mu\nu} = \eta_{\mu\nu} + h_{\mu\nu} \quad \text{where } h_{\mu\nu} = h_{\nu\mu} \text{ and } |h_{\mu\nu}| \ll 1 \qquad (22.3)$$

where $\eta_{\mu\nu}$ is the flat-space metric (note that $h_{\mu\nu}$ must be symmetric because both $g_{\mu\nu}$ and $\eta_{\mu\nu}$ are symmetric). The implied coordinates are *quasi*-cartesian, and in what follows we will call them t, x, y, and z, even though they are not *exactly* equal to the corresponding cartesian coordinates. The metric perturbations $h_{\mu\nu}$ are (generally) functions of both position and time.

Taking the "weak-field limit" means that we will drop terms of order $|h_{\mu\nu}|^2$ and higher in all the equations that follow. Given this assumption, you can show (see box 22.2) that the definition of inverse metric ($g_{\alpha\sigma} g^{\sigma\nu} \equiv \delta^\nu_\alpha$) implies that

$$g^{\mu\nu} = \eta^{\mu\nu} - h^{\mu\nu} \quad \text{where} \quad h^{\mu\nu} = \eta^{\mu\alpha} \eta^{\nu\beta} h_{\alpha\beta} \qquad (22.4)$$

This means that we can raise or lower the indices of a quantity that is of the order of magnitude of $h_{\mu\nu}$ using the *flat-space* inverse metric $\eta^{\mu\nu}$ or metric $\eta_{\mu\nu}$, respectively, instead of having to use the full metric $g^{\mu\nu}$ or $g_{\mu\nu}$. For example,

$$h^\mu_{\ \nu} \equiv g^{\mu\alpha} h_{\alpha\nu} = (\eta^{\mu\alpha} - h^{\mu\alpha}) h_{\alpha\nu} = \eta^{\mu\alpha} h_{\alpha\nu} - h^{\mu\alpha} h_{\alpha\nu} \approx \eta^{\mu\alpha} h_{\alpha\nu} \qquad (22.5)$$

to first order in $h_{\mu\nu}$ (the term omitted in the last step is of order $|h_{\mu\nu}|^2$). This will be *very* handy to keep in mind for what follows.

The Einstein equation with both indices lowered says that

$$R_{\beta\nu} = 8\pi G(T_{\beta\nu} - \tfrac{1}{2} g_{\beta\nu} T) \quad \text{where} \quad T \equiv g_{\mu\nu} T^{\mu\nu} \qquad (22.6)$$

(This is true in arbitrary coordinates, not just the weak-field limit.) You can show (see box 22.3) that, to first order in $h_{\mu\nu}$, the Riemann tensor becomes

$$R_{\alpha\beta\mu\nu} = \tfrac{1}{2}(\partial_\beta \partial_\mu h_{\alpha\nu} + \partial_\alpha \partial_\nu h_{\beta\mu} - \partial_\alpha \partial_\mu h_{\beta\nu} - \partial_\beta \partial_\nu h_{\alpha\mu}) \qquad (22.7)$$

(The mnemonic for the indices is that "inner togetherness is positive," just as in the similar expression for $R_{\alpha\beta\mu\nu}$ in a locally inertial frame.) Note that the entire tensor is of the order of $h_{\mu\nu}$. Using this, you can also show (see box 22.4) that the Ricci tensor $R_{\beta\nu} \equiv g^{\alpha\mu} R_{\alpha\beta\mu\nu}$ to first order in $h_{\mu\nu}$ becomes

$$R_{\beta\nu} = \tfrac{1}{2}\bigl(-\eta^{\alpha\mu}\partial_\alpha\partial_\mu h_{\beta\nu} + \partial_\beta H_\nu + \partial_\nu H_\beta\bigr) \tag{22.8a}$$

$$\text{where} \quad H_\nu \equiv \eta^{\alpha\mu}\bigl(\partial_\mu h_{\alpha\nu} - \tfrac{1}{2}\partial_\nu h_{\alpha\mu}\bigr) \tag{22.8b}$$

A Constraint on the Coordinates. Now, as discussed in the beginning of this chapter, the Einstein equation cannot actually determine the metric completely: because of its internal constraints, it imposes only six equations-worth of constraint on the components of $h_{\mu\nu}$. Even in the weak field limit (where our coordinates are always quasi-cartesian), we can introduce four transformation equations $x'^\mu = f^\mu(x^\nu)$ that tweak the coordinates slightly so that they become another quasi-cartesian coordinate system whose metric will *still* satisfy the Einstein equation.

We will show in a later chapter that we can in fact *always* use our four-equations-worth of freedom to find quasi-cartesian coordinates where the four-component quantity H_ν defined above evaluates to *zero*. I am not going to prove this now, but note that at least we have exactly the right number of equations-worth of freedom to do so. The six equations-worth of information in the Einstein equation coupled with the four-equations-worth of constraint $H_\nu = 0$ provide us with precisely the 10 equations we need to completely determine the 10 components of $h_{\mu\nu}$.

If we apply this coordinate condition, the last two terms in equation 22.8a for the Ricci tensor are zero, so $R_{\beta\nu} = -\tfrac{1}{2}\eta^{\alpha\mu}\partial_\alpha\partial_\mu h_{\beta\nu}$. Note also that since the Ricci tensor is of order $h_{\mu\nu}$, $8\pi G T_{\beta\nu}$ is of the same order, and thus $g_{\beta\nu}T = \eta_{\beta\nu}T$ to our level of approximation. The Einstein equation therefore becomes

$$-\tfrac{1}{2}\eta^{\alpha\mu}\partial_\alpha\partial_\mu h_{\beta\nu} \equiv -\tfrac{1}{2}\Box^2 h_{\beta\nu} = 8\pi G(T_{\beta\nu} - \tfrac{1}{2}\eta_{\beta\nu}T) \tag{22.9}$$

where $\Box^2$ (the four-dimensional del operator) $\equiv \eta^{\alpha\mu}\partial_\alpha\partial_\mu = -\partial^2/\partial t^2 + \nabla^2$."

The Stationary-Source Limit. So far, we have assumed only that the gravitational field is weak. Now let's additionally assume that the source of the field is "stationary" (meaning that its stress-energy is independent of time, even if the source's fluid is moving, as in the case of a constantly rotating star). If the source is stationary, then the field it creates will also be time-independent. If this is true, then $\eta^{\alpha\mu}\partial_\alpha\partial_\mu = \partial_x\partial_x + \partial_y\partial_y + \partial_z\partial_z = \nabla^2$, and the Einstein equation becomes

$$\nabla^2 h_{\beta\nu} = -16\pi G(T_{\beta\nu} - \tfrac{1}{2}\eta_{\beta\nu}T) \tag{22.10}$$

This is a very simple set of differential equations that we can actually solve!

To see how, consider the analogous equation $\nabla^2 \Phi_N = 4\pi G\rho$ for the Newtonian gravitational potential Φ_N. Mathematically, this has exactly the same form as each component of equation 22.10. We know that for a static Newtonian field, we can find the potential $\Phi_N(\vec{r})$ that satisfies $\nabla^2 \Phi_N = 4\pi G\rho$ everywhere by adding up the potentials created at $\vec{r}$ by the mass at each point $\vec{r}_s$ in the source's mass distribution:

$$\Phi_N(\vec{r}) = \int_{\text{src}} \frac{-G\rho(\vec{r}_s)\,dV}{|\vec{r}-\vec{r}_s|} \tag{22.11}$$

The solutions to equation 22.10 are therefore (by analogy):

$$h_{\beta\nu}(\vec{r}) = 4\int_{\text{src}} \frac{G(T_{\beta\nu}-\tfrac{1}{2}\eta_{\beta\nu}T)\,dV}{|\vec{r}-\vec{r}_s|} = 2\int_{\text{src}} \frac{G(2T_{\beta\nu}-\eta_{\beta\nu}T)\,dV}{|\vec{r}-\vec{r}_s|} \tag{22.12}$$

where the stress-energy tensor is evaluated at $\vec{r}_s$.

Perfect-Fluid Sources. The stress-energy tensor for a perfect fluid in arbitrary coordinates is $T_{\beta\nu} = (\rho_0 + p_0)u_\beta u_\nu + g_{\beta\nu}p_0$ (see equation 20.16), where ρ_0 and p_0 are the fluid's density and pressure in its rest frame, and u_ν is the fluid's four-velocity in covector form. You can show (see box 22.5) that to first order in $h_{\mu\nu}$ and for low-velocity fluids (e.g., "slow rotation"), we have

$$2T_{tt} - \eta_{tt}T \approx \rho_0 + 3p_0 \equiv \textbf{gravitoelectric energy density} \equiv \rho_g \tag{22.13a}$$

$$-T_{ti} + \tfrac{1}{2}\eta_{ti}T \approx (\rho_0 + p_0)u_i \equiv \textbf{gravitomagnetic current density} \equiv \Pi_i \tag{22.13b}$$

$$2T_{ii} - \eta_{ii}T \approx \rho_0 - p_0 \equiv \textbf{curvature energy density} \equiv \rho_c \tag{22.13c}$$

and $T_{ij} \approx 0$ if $i \neq j$ (i and j range only over the spatial indices). The quantity $\rho_0 - p_0$ is called **curvature energy density** because (according to equation 22.12) it is the source of deviations h_{ij} from flat space in the spatial part of the metric.

How Particles Respond. To understand the other names, consider a particle that moves non-relativistically through such a weak and stationary field. You can show (see box 22.6) that to first order in both the metric perturbation $h_{\mu\nu}$ and the particle's speed v, the geodesic equation for this particle reduces to

$$\frac{d^2x^i}{dt^2} \approx \tfrac{1}{2}\eta^{ik}\partial_k h_{tt} + \eta^{ik}(\partial_k h_{tj} - \partial_j h_{tk})v^j \equiv \eta^{ik}(-\partial_k \Phi_G + F_{kj}v^j) \qquad (22.14)$$

where i, j and k range only over the spatial coordinates and v^j refers to the particle's spatial coordinate velocity components dx^j/dt.

Note that the term $-\partial_k \Phi_G = \partial_k(\tfrac{1}{2}h_{tt})$ causes the particle to accelerate relative to our quasi-cartesian coordinates even if it is initially at rest. This is analogous to what an electric field does in electrodynamics, so we call this term the **gravitoelectric** aspect of the gravitational field. Equations 22.12 and 22.13 imply that the source of the **gravitoelectric potential** Φ_G is the integral of $-G\rho_g(\vec{r}_s)\,dV/|\vec{r}-\vec{r}_s|$, where ρ_g is the *gravitoelectric energy density* (hence the name).

For most sources, the random motion of fluid particles is not relativistic, so the pressure is negligible compared to the energy density, so the energy density is essentially the same as the mass density and the gravitoelectric potential reduces to the Newtonian potential. However, the vacuum stress-energy $T^{\mu\nu}_{\text{vac}} = -(\Lambda/8\pi G)g^{\mu\nu}$ behaves in the weak-field limit as if it were the stress-energy for a perfect fluid at rest with $p_0 = -\rho_0 = -\Lambda/8\pi G$. In such a case, the gravitoelectric energy density is negative, and the gravitoelectric field it creates is actually repulsive!

The term $F_{kj} = \partial_k h_{tj} - \partial_j h_{tk}$ in equation 22.14 makes an additional contribution to the particle's acceleration that is proportional to its coordinate velocity. This is analogous to what a magnetic field does in electrodynamics, so we call this the **gravitomagnetic** aspect of the gravitational field. Equations 22.12 and 22.13 imply that the source of this aspect is the *gravitomagnetic current density* Π_i.

Spherical Objects. Finally, consider the external field of a non-rotating object. Equation 22.12 applies only if the "weak-field" approximation is true everywhere inside and outside the object. If particles in such an object are to be confined by its "weak" field, their random motions must be non-relativistic, implying that the fluid pressure is negligible compared to its energy density and $\rho_c \approx \rho_g \approx$ Newtonian mass density ρ. For a non-rotating object, the spatial part of the fluid four-velocity is zero throughout the object, so the gravitomagnetic current density $\Pi_i = 0$.

At points outside a spherically symmetric object centered at the origin, we know that the integral in the Newtonian equation 22.11 simply evaluates to $\Phi(r) = -GM/r$, where $M \equiv \int \rho(r)\,dV$ and $r \equiv [x^2+y^2+z^2]^{1/2}$. In this limit, equation 22.12 yields

$$h_{tt} = h_{xx} = h_{yy} = h_{zz} = \frac{2GM}{r}; \quad \text{all other } h_{\mu\nu} = 0. \qquad (22.15)$$

The metric in the empty space around such an object is therefore

$$ds^2 = -\left(1-\frac{2GM}{r}\right)dt^2 + \left(1+\frac{2GM}{r}\right)(dx^2+dy^2+dz^2) \qquad (22.16)$$

This is the same as the Schwarzschild metric evaluated to first order in GM/r (see problem P22.1). Note also that equation 22.14 implies that a particle outside such a star will obey the Newtonian gravitational acceleration equation

$$\frac{d^2\vec{x}}{dt^2} = -\vec{\nabla}\Phi_G = -\vec{\nabla}\Phi_N = -\vec{\nabla}\left(-\frac{GM}{r}\right) = -\frac{GM}{r^2}\hat{r} \qquad (22.17)$$

Concluding Note. The nice distinctions between the various field aspects and types of mass density become blurred and blended when fields are strong and/or sources are relativistic. Even so, the weak-field stationary limit provides a good introduction to the qualitative behavior of solutions of the Einstein equation.

BOX 22.1 Conservation of Four-Momentum Implies $0 = \nabla_\nu(\rho_0 u^\nu)$

The absolute gradient of $-1 = \mathbf{u} \cdot \mathbf{u} \equiv u^\mu g_{\mu\alpha} u^\alpha$ is $\nabla_\nu(u^\mu g_{\mu\alpha} u^\alpha) = 0$.

Exercise 22.1.1. Argue that the latter implies $u^\mu g_{\mu\alpha} \nabla_\nu u^\alpha = 0$. (*Hint:* Remember that the gradient of the metric is zero, and that you can rename bound indices).

Equation 22.1 states that $0 = u^\mu \nabla_\nu(\rho_0 u^\nu) + \rho_0 u^\nu \nabla_\nu u^\mu$. If you multiply both sides of this equation by $g_{\mu\alpha} u^\alpha$ and use $-1 = u^\mu g_{\mu\alpha} u^\alpha$ and the result of the previous exercise, you should be able to show that $0 = \nabla_\nu(\rho_0 u^\nu)$.

Exercise 22.1.2. Verify that this is correct.

BOX 22.2 The Inverse Metric in the Weak-Field Limit

The definition of the inverse metric is the matrix $g^{\mu\nu}$ such that $g^{\mu\nu} g_{\nu\alpha} = \delta^\mu_\alpha$. Let's define $g^{\mu\nu} \equiv \eta^{\mu\nu} + b^{\mu\nu}$, where $|b^{\mu\nu}| \ll 1$. Note also that $\eta^{\mu\nu} \eta_{\nu\alpha} = \delta^\mu_\alpha$.

Exercise 22.2.1. Use these facts to show that to first order in the metric perturbation, the definition of the inverse metric implies that $b^{\mu\nu} = -h^{\mu\nu} \equiv -\eta^{\mu\alpha} \eta^{\nu\beta} h_{\alpha\beta}$.

BOX 22.3 The Riemann Tensor in the Weak-Field Limit

The definition of the Riemann tensor is

$$R_{\alpha\beta\mu\nu} = g_{\alpha\gamma}R^{\gamma}{}_{\beta\mu\nu} = g_{\alpha\gamma}(\partial_\mu \Gamma^{\gamma}_{\beta\nu} - \partial_\nu \Gamma^{\gamma}_{\beta\mu} + \Gamma^{\gamma}_{\mu\sigma}\Gamma^{\sigma}_{\beta\nu} - \Gamma^{\gamma}_{\nu\sigma}\Gamma^{\sigma}_{\beta\mu}) \quad (22.18)$$

The definition of the Christoffel symbols is

$$\Gamma^{\gamma}_{\mu\nu} = \tfrac{1}{2}g^{\gamma\sigma}[\partial_\mu g_{\nu\sigma} + \partial_\nu g_{\sigma\mu} - \partial_\sigma g_{\mu\nu}] \quad (22.19)$$

Our goal in this box is to evaluate both in the weak-field limit. The Christoffel symbols evaluated to first order in the metric perturbation are

$$\Gamma^{\gamma}_{\mu\nu} = \tfrac{1}{2}\eta^{\gamma\sigma}[\partial_\mu h_{\nu\sigma} + \partial_\nu h_{\sigma\mu} - \partial_\sigma h_{\mu\nu}] \quad (22.20)$$

Exercise 22.3.1. Explain this result. Why did I replace $g_{\mu\nu}$ by $h_{\mu\nu}$ inside the brackets but $g^{\gamma\sigma}$ by $\eta^{\gamma\sigma}$ outside the brackets?

Note that in the weak-field limit, the Christoffel symbols are of order $h_{\mu\nu}$, so the Christoffel-symbol products in the definition of the Riemann tensor will be of order $|h_{\mu\nu}|^2$ and so can be ignored. If you substitute equation 22.20 into equation 22.18 and evaluate the result to first order in $h_{\mu\nu}$, you should get equation 22.7:

$$R_{\alpha\beta\mu\nu} = \tfrac{1}{2}(\partial_\beta \partial_\mu h_{\alpha\nu} + \partial_\alpha \partial_\nu h_{\beta\mu} - \partial_\alpha \partial_\mu h_{\beta\nu} - \partial_\beta \partial_\nu h_{\alpha\mu}) \quad (22.7r)$$

Exercise 22.3.2. Verify this last result.

BOX 22.4 The Ricci Tensor in the Weak-Field Limit

In the weak-field limit, the definition of the Ricci tensor implies that

$$R_{\beta\nu} \equiv g^{\alpha\mu} R_{\alpha\beta\mu\nu} = \tfrac{1}{2}\eta^{\alpha\mu}(\partial_\beta \partial_\mu h_{\alpha\nu} + \partial_\alpha \partial_\nu h_{\beta\mu} - \partial_\alpha \partial_\mu h_{\beta\nu} - \partial_\beta \partial_\nu h_{\alpha\mu}) \quad (22.21)$$

where I have used equation 22.7 for the weak-field Riemann tensor. Since the Riemann tensor is of order $h_{\mu\nu}$ in this limit, I can change the $g^{\alpha\mu}$ in front to $\eta^{\alpha\mu}$ (including the correction term would lead only to terms in the whole expression of order $|h_{\mu\nu}|^2$). Split the last term into two, appropriately rename some indices, and take advantage of the fact that the order of partial derivatives is irrelevant to show that this can be rewritten as given by equation 22.8:

$$R_{\beta\nu} = \tfrac{1}{2}\left(-\eta^{\alpha\mu}\partial_\alpha \partial_\mu h_{\beta\nu} + \partial_\beta H_\nu + \partial_\nu H_\beta\right) \quad (22.8ar)$$

$$\text{where} \quad H_\nu \equiv \eta^{\alpha\mu}\left(\partial_\mu h_{\alpha\nu} - \tfrac{1}{2}\partial_\nu h_{\alpha\mu}\right) \quad (22.8br)$$

Exercise 22.4.1. Verify this. (*Hint:* Both $\eta^{\mu\alpha}$ and $h_{\mu\nu}$ are symmetric.)

BOX 22.5 The Stress-Energy Sources of the Metric Perturbation

The stress-energy tensor for a perfect fluid in arbitrary coordinates (written with lower indices) is $T_{\mu\nu} = (\rho_0 + p_0)u_\mu u_\nu + p_0 g_{\mu\nu}$, where u_μ is the fluid's bulk four-velocity (written in covector form) and ρ_0 and p_0 are its rest density and pressure.

Exercise 22.5.1. Argue that in *all* coordinate systems, $T = g^{\mu\nu}T_{\mu\nu} = -\rho_0 + 3p_0$. (*Hints:* Remember the definition of the inverse metric and that $-1 = \boldsymbol{u} \cdot \boldsymbol{u}$.)

Now let's apply our approximations. Remember that the stress-energy tensor is of order $h_{\mu\nu}$, so to first order in $h_{\mu\nu}$, $T_{\beta\nu} - \frac{1}{2}g_{\beta\nu}T \approx T_{\beta\nu} - \frac{1}{2}\eta_{\beta\nu}T$. If we also assume that the bulk fluid velocity is small, so that $u_t \approx -1$ and $(u_i)^2$ is negligible, you can easily show that $2T_{tt} - \eta_{tt}T \approx \rho_0 + 3p_0$, $2T_{ti} - \eta_{ti}T \approx -2(\rho_0 + p_0)u_i$, and $2T_{ij} - \eta_{ij}T \approx \rho_0 - p_0$ if $i = j$ and zero otherwise.

Exercise 22.5.2. Verify these three results.

BOX 22.6 The Geodesic Equation for a Slow Particle in a Weak Field

In otherwise arbitrary coordinates where t is a time coordinate, we can write

$$\frac{d^2 x^i}{d\tau^2} = \frac{dt}{d\tau}\frac{d}{dt}\left(\frac{dt}{d\tau}\frac{dx^i}{dt}\right) = u^t u^t \frac{d^2 x^i}{dt^2} + \frac{du^t}{d\tau} v^i \qquad (22.22)$$

since $u^t \equiv dt/d\tau$ and $v^i \equiv dx^i/dt$ (i ranges over only spatial indices). The geodesic equation implies that $0 = du^t/d\tau + \Gamma^t_{\mu\nu} u^\mu u^\nu$.

Exercise 22.6.1. Argue that if the particle moves slowly enough so that $(u^i)^2 = (u^t v^i)^2$ is negligible, then $v^i(du^t/d\tau) \approx -v^i \Gamma^t_{tt}$. Then argue that for a stationary field, $\Gamma^t_{tt} = 0$ to first order in the metric perturbation.

Using the last result, we can write the geodesic equation in the form

$$0 \approx u^t u^t \frac{d^2 x^i}{dt^2} + \Gamma^i_{\mu\nu} u^\mu u^\nu \;\Rightarrow\; \frac{d^2 x^i}{dt^2} \approx -\Gamma^i_{tt} - \Gamma^i_{tj} v^j - \Gamma^i_{kt} v^k \qquad (22.23)$$

using $v^j = u^j/u^t$ and dropping terms of order $v^j v^k$. Note also that $\Gamma^i_{tj} = \Gamma^i_{jt}$, so the last two terms are really equivalent. If you combine the last terms (renaming the sum over k in the last term to a sum over j) and use equation 22.20, you can show that

$$\frac{d^2 x^i}{dt^2} \approx \tfrac{1}{2}\eta^{ik}\partial_k h_{tt} + \eta^{ik}(\partial_k h_{tj} - \partial_j h_{tk}) v^j \qquad (22.24)$$

Exercise 22.6.2. Verify this. (*Hint:* Recall that $\partial_t h_{\mu\nu} = 0$ for a stationary field.)

HOMEWORK PROBLEMS

P22.1 Consider the weak-field metric for a non-rotating spherical star given by equation 22.16. Using the coordinate transformations $r \equiv [x^2 + y^2 + z^2]^{1/2}$, $\phi \equiv \tan^{-1}(y/x)$, and $\theta \equiv \tan^{-1}([x^2 + y^2]^{1/2}/z)$ that we would use in flat space, we can rewrite the metric's *spatial* part as follows:

$$ds^2 = \left(1 + \frac{2GM}{r}\right)(dr^2 + r^2 d\theta^2 + r^2 \sin^2\theta \, d\phi^2) \quad (22.25)$$

As you can see from the metric, the r coordinate here is *not* a circumferential coordinate, as the distance around the equator is $2\pi(1 + 2GM/r)^{1/2}r$.

a. If we define $r_c \equiv (1 + 2GM/r)^{1/2}r$, then r_c will be a true circumferential radial coordinate according to the metric. Show that $dr_c = dr$ through first order in the metric perturbation $2GM/r$ ($= h_{rr}$ in this context), meaning that the leading correction terms to $dr_c = dr$ are of order $(2GM/r)^2$. (*Hint:* Take the differential of both sides and then use the binomial approximation.)

b. In terms of r_c, the spatial part of the metric is thus

$$ds^2 = \left(1 + \frac{2GM}{r}\right)dr_c^2 + r_c^2(d\theta^2 + \sin^2\theta \, d\phi^2) \quad (22.26)$$

Show that this is the same as the spatial part of the Schwarzschild metric through first order in $2GM/r$. (Note that the circumferential radial coordinate in the Schwarzschild metric corresponds to r_c here, not r.)

P22.2 Express the gravitoelectric energy density ρ_g, the gravitomagnetic current density Π_i, and the curvature energy density ρ_c in terms of Λ and G if the stress-energy tensor is the vacuum stress-energy. (*Hint:* Use the definitions of these quantities in terms of $T_{\mu\nu}$ that appear on the left sides of equations 22.13.)

P22.3 Show that the gravitomagnetic contribution to a particle's acceleration (as given by equation 22.14) is always perpendicular to the particle's spatial velocity $\vec{v}$. (*Hint:* Show that the component parallel to the particle's spatial velocity is zero.)

P22.4 Show that the wave $h_{\mu\nu} = A_{\mu\nu}\cos(\omega t - kx)$, where $A_{\mu\nu}$ is a constant matrix and ω and k are positive constants, satisfies the weak-field Einstein equation 22.9 for empty space if $\omega = k$. What does $\omega = k$ tell us about the speed of this traveling wave? What constraints does the requirement that $H_\nu = 0$ put on the matrix $A_{\mu\nu}$?

P22.5 We will see in chapter 36 that in the space around a rotating spherical star, we have (in addition to the diagonal metric perturbations $h_{tt} = h_{xx} = h_{yy} = h_{zz} = 2GM/r$ listed in equation 22.15) $h_{tx} = h_{xt} = 2GSy/r^3$ and $h_{ty} = h_{yt} = -2GSx/r^3$, where S is the star's angular momentum.

a. Calculate the gravitomagnetic F_{kj} matrix at an arbitrary point in the space around this star.

b. Imagine that a particle is moving in the $+x$ direction with speed $v \ll 1$ at a point along the x axis a distance r from the star's center. What are the magnitude and direction of the gravitomagnetic acceleration it experiences?

P22.6 Consider an infinite thin wire along the z axis that moves with coordinate speed $V \ll 1$ in the $+z$ direction. Assume that the wire has energy density per unit length of λ and assume that its internal pressure is negligible in comparison to the energy density.

a. Using the stationary weak-field approximation, write down the integrals that deliver the components of $h_{\mu\nu}$ at a point a distance $r \equiv \sqrt{x^2 + y^2}$ from the nearest point on the wire. (*Hint:* Note that to first order in V, $u_i = V$.)

b. Calculate both $\vec{\nabla}\Phi_G$ and F_{kj} at such a point. (*Hint:* The integrals you found in the first part are infinite for an infinite wire. However, both $\vec{\nabla}\Phi_G$ and F_{kj} depend on derivatives of these integrals with respect to variables not involved in the integration. First, use symmetry to argue that the value of $h_{\mu\nu}$ cannot depend on z. Take the derivatives of the integrands with respect to the other variables first and *then* integrate to get finite results for the other derivatives of $h_{\mu\nu}$. Feel free to look up any integrals you need.)

c. Calculate the components of the gravitoelectric and gravitomagnetic contributions ($\eta^{ik}\partial_k\Phi_G$ and $\eta^{ik}F_{kj}v^j$, respectively) to the acceleration of a test particle moving with coordinate velocity $\vec{v}$ (with $v \ll 1$) near the wire. Express your answer in terms of r.

d. Compare with what classical electrostatics and magnetostatics would predict for a positively charged particle moving near a moving positively charged wire.

P22.7 Imagine two infinite thin plates of matter parallel to the xy plane, one at $z = b$ and one at $z = -b$, where b is a constant. Assume that the energy per unit area on each plate is σ, that the pressure in each plate is negligible compared to its energy density, and that the lower plate moves with coordinate speed V in the $-x$ direction while the upper the same speed in the $+x$ direction, where $V \ll 1$.

a. Using the stationary, weak-field approximation, write down the integrals that deliver the components of $h_{\mu\nu}$ at an arbitrary point at vertical position z between the plates. (*Hint:* Note that to first order in V, $u_i = V$.)

P22.7 (continued)

b. Calculate both $\vec{\nabla}\Phi_G$ and F_{kj} at such a point. (*Hint:* The integrals you found in the first part are infinite for an infinite plate. However, both $\vec{\nabla}\Phi_G$ and F_{kj} depend on derivatives of these integrals with respect to variables not involved in the integration. First, argue from symmetry that the values of $h_{\mu\nu}$ cannot depend on x or y. Then take the derivatives of the integrands with respect to z first and then integrate to get a finite result to find $\partial_z h_{\mu\nu}$. Feel free to look up any integrals that you need.)

c. Calculate the components of the gravitoelectric and gravitomagnetic contributions ($\eta^{ik}\partial_k\Phi_G$ and $\eta^{ik}F_{kj}v^j$, respectively) to the acceleration of a test particle moving with coordinate velocity $\vec{v}$ (with $v \ll 1$) in the region between the plates.

d. Compare with what classical electrostatics and magnetostatics would predict for a positively charged particle moving between two positively charged moving plates.

23. THE SCHWARZSCHILD SOLUTION

INTRODUCTION

FLAT SPACETIME
- Review of Special Relativity
- Four-Vectors
- Index Notation

TENSORS
- Arbitrary Coordinates
- Tensor Equations
- Maxwell's Equations
- Geodesics

SCHWARZSCHILD BLACK HOLES
- The Schwarzschild Metric
- Particle Orbits
- Precession of the Perihelion
- Photon Orbits
- Deflection of Light
- Event Horizon
- Alternative Coordinates
- Black Hole Thermodynamics

THE CALCULUS OF CURVATURE
- The Absolute Gradient
- Geodesic Deviation
- The Riemann Tensor

THE EINSTEIN EQUATION
- The Stress-Energy Tensor
- The Einstein Equation
- Interpreting the Equation
- **The Schwarzschild Solution**

COSMOLOGY
- The Universe Observed
- A Metric for the Cosmos
- Evolution of the Universe
- Cosmic Implications
- The Early Universe
- CMB Fluctuations & Inflation

GRAVITATIONAL WAVES
- Gauge Freedom
- Detecting Gravitational Waves
- Gravitational Wave Energy
- Generating Gravitational Waves
- Gravitational Wave Astronomy

SPINNING BLACK HOLES
- Gravitomagnetism
- The Kerr Metric
- Kerr Particle Orbits
- Ergoregion and Horizon
- Negative-Energy Orbits

this ▸ depends on ▸ this

23. THE SCHWARZSCHILD SOLUTION

Introduction. Now that we have derived the Einstein equation and discussed its implications in the weak field approximation, we will solve this equation *exactly* in several important cases. The first case that we will consider is the derivation of the Schwarzschild metric, which will exhibit the basic steps one follows to solve the Einstein equation.

How To Solve the Einstein Equation. In the last chapter, we were reminded that even though the Einstein equation determines the *geometry* of spacetime surrounding a given mass distribution, we still are completely free to define the *coordinates* used to describe that geometry. Therefore, the first step in solving the Einstein equation is to define the coordinates. We usually do this by coming up with a coordinate system that expresses and takes advantage of any symmetries that might pertain to the problem at hand. This will allow us to develop a trial metric that contains as few unknown metric components as is possible.

The next step is to substitute the trial metric into the Einstein equation to generate a set of coupled differential equations for the unknown metric components. If we can solve these equations for the metric components, then we have found a coordinate system that describes the geometry determined by the Einstein equation. If we are unable to solve the Einstein equation or get self-contradictory results, we may have overdetermined the trial metric and need to make it more general. If we have undetermined metric components left over, we probably have yet more coordinate choices to make.

So, in summary, the steps are as follows:

1. Use symmetry to define a coordinate system as completely as possible.
2. Set up a trial metric with as few undetermined coefficients as possible.
3. Substitute the trial metric into the Einstein equation.
4. Solve the resulting differential equations for the unknown metric components.

Consequences of Spherical Symmetry. Consider the spacetime around a spherically symmetric source (a source that nonetheless may be expanding or contracting). "Spherical symmetry" means that we can define a nested set of two-dimensional surfaces in the spacetime surrounding the source that have the same intrinsic geometry as an ordinary two-dimensional sphere. Let's define angular coordinates θ and ϕ in the usual way on each such spherical surface, and define a radial coordinate r to be the circumference of such a sphere divided by 2π (i.e., r is a circumferential radial coordinate). Then the metric for each spherical surface is

$$ds^2 = r^2(d\theta^2 + \sin^2\theta \, d\phi^2) \tag{23.1}$$

This part of the metric applies only to a single nested sphere: there is nothing in what we have done so far that precludes giving each sphere its own set of θ and ϕ coordinates (i.e., having the spheres' polar axes point in different directions). But we can require these coordinate systems to line up by requiring that the line defined by $\theta =$ constant and $\phi =$ constant be perpendicular to each sphere. This amounts to saying that the basis vectors $\mathbf{e}_\theta$ and $\mathbf{e}_\phi$ are perpendicular to the $\mathbf{e}_r$ basis vector. This in turn requires that $g_{r\theta} = \mathbf{e}_r \cdot \mathbf{e}_\theta = 0$ and $g_{r\phi} = \mathbf{e}_r \cdot \mathbf{e}_\phi = 0$. With this choice of coordinates, the purely spatial part of our metric reads

$$ds^2 = g_{rr} dr^2 + r^2(d\theta^2 + \sin^2\theta \, d\phi^2) \tag{23.2}$$

Now, if we have a metric term like $g_{t\phi} dt \, d\phi$, then it would imply that the geometry of spacetime treats displacements where $d\phi > 0$ differently from displacements where $d\phi < 0$. This would give a preferred orientation to spacetime, contrary to the assumption of spherical symmetry. Therefore, spherical symmetry should allow us to be able to choose coordinates so that $g_{t\phi} = 0$. A similar argument applies to $g_{t\theta}$. Therefore our metric now has the form

$$ds^2 = g_{tt} dt^2 + 2g_{rt} dr \, dt + g_{rr} dr^2 + r^2(d\theta^2 + \sin^2\theta \, d\phi^2) \tag{23.3}$$

Choosing a Time Coordinate. At this point, our time coordinate t is completely arbitrary. We could use our coordinate freedom to define the value of either g_{tt} or g_{rt}. There are some constraints in our choice, however: g_{tt} must be nonzero (or we don't have a spacetime) and should be negative wherever the other metric components are positive and $g_{rt} = 0$ (or we don't have a time coordinate). The constraints on the choice of g_{rt} are a little less clear.

However, imagine that we make some arbitrary choice of time coordinate that defines g_{tt} and then use the Einstein equation to find g_{rt}. Consider then making a coordinate transformation of the form

$$t' = t + f(r, t) \tag{23.4}$$

where $f(r,t)$ is some function of r and the original coordinate t. You can show (see box 23.1) that an appropriate choice of $f(r,t)$ sets $g'_{rt} = 0$ in our transformed coordinate system. But if we can choose the time coordinate so that $g'_{rt} = 0$ after the fact, no matter what our original choice of time coordinate might have been, then our coordinate freedom allows us to choose to set $g_{rt} = 0$ *now*, *before* solving the Einstein equation, and thus make the metric diagonal. There are real advantages to eliminating any off-diagonal components, so let's use our coordinate freedom this way.

The General Spherically Symmetrical Metric. Therefore, one possibility for a general spherically symmetric metric, where we have used our coordinate freedom to choose r to be a circumferential coordinate and t so that $g_{rt} = 0$, is

$$ds^2 = -A(r,t)\,dt^2 + B(r,t)\,dr^2 + r^2(d\theta^2 + \sin^2\theta\,d\phi^2) \tag{23.5}$$

This will be our trial metric for the Einstein equation.

Solving the Einstein Equation. In empty space and assuming a negligible cosmological constant Λ, the Einstein equation (in the form given by equation 21.23) implies that $R^{\mu\nu} = 0$. This means that $R_{\mu\nu} = g_{\mu\alpha}g_{\nu\beta}R^{\alpha\beta} = 0$ also. We will find it convenient to use the latter form.

The "Diagonal Metric Worksheet" (available as an appendix and online) is a powerful tool for evaluating Christoffel symbols and Ricci tensor components for any diagonal metric. You can use this worksheet (see box 23.2) to show that

$$R_{tt} = \frac{1}{2B}\left[+\frac{\partial^2 A}{\partial r^2} - \frac{1}{2A}\left(\frac{\partial A}{\partial r}\right)^2 - \frac{1}{2B}\frac{\partial A}{\partial r}\frac{\partial B}{\partial r} + \frac{2}{r}\frac{\partial A}{\partial r} - \frac{\partial^2 B}{\partial t^2} + \frac{1}{2B}\left(\frac{\partial B}{\partial t}\right)^2 + \frac{1}{2A}\frac{\partial A}{\partial t}\frac{\partial B}{\partial t}\right] \tag{23.6a}$$

$$R_{rr} = \frac{1}{2A}\left[-\frac{\partial^2 A}{\partial r^2} + \frac{1}{2A}\left(\frac{\partial A}{\partial r}\right)^2 + \frac{1}{2B}\frac{\partial A}{\partial r}\frac{\partial B}{\partial r} + \frac{2A}{Br}\frac{\partial B}{\partial r} + \frac{\partial^2 B}{\partial t^2} - \frac{1}{2B}\left(\frac{\partial B}{\partial t}\right)^2 - \frac{1}{2A}\frac{\partial A}{\partial t}\frac{\partial B}{\partial t}\right] \tag{23.6b}$$

$$R_{\theta\theta} = -\frac{r}{2AB}\frac{\partial A}{\partial r} + \frac{r}{2B^2}\frac{\partial B}{\partial r} + 1 - \frac{1}{B} \tag{23.6c}$$

$$R_{\phi\phi} = \sin^2\theta\, R_{\theta\theta} \tag{23.6d}$$

$$R_{tr} = +\frac{1}{Br}\frac{\partial B}{\partial t} \tag{23.6e}$$

The other components of the Ricci tensor are identically zero. Note that R_{tt} and R_{rr} mostly involve the same terms in the brackets, but with opposite signs and a single term (the fourth in each case) that is different. Therefore,

$$\frac{B}{A}R_{tt} + R_{rr} = \frac{1}{r}\left(\frac{1}{A}\frac{\partial A}{\partial r} + \frac{1}{B}\frac{\partial B}{\partial r}\right) \tag{23.7}$$

The Empty-Space Spherical Solution. Requiring that the component $R_{tr} = 0$ immediately implies that

$$R_{tr} = +\frac{1}{Br}\frac{\partial B}{\partial t} = 0 \quad \Rightarrow \quad \frac{\partial B}{\partial t} = 0 \tag{23.8}$$

23. THE SCHWARZSCHILD SOLUTION

If we plug $R_{tt} = R_{rr} = 0$ into equation 23.7, we find that

$$0 = \frac{B}{A}R_{tt} + R_{rr} = \frac{1}{r}\left(\frac{1}{A}\frac{\partial A}{\partial r} + \frac{1}{B}\frac{\partial B}{\partial r}\right) \Rightarrow \frac{1}{A}\frac{\partial A}{\partial r} = -\frac{1}{B}\frac{\partial B}{\partial r} \quad (23.9)$$

You can then plug this result into equation 23.6c to eliminate the references to A, solve the resulting equation for $\partial B/\partial r$, and integrate. Since B is independent of t, the result (see box 23.3) is

$$\frac{1}{B} = 1 + \frac{C}{r} \quad (23.10)$$

where C is a constant of integration.

Since $\partial B/\partial t = 0$ by equation 23.8, the right side of 23.9 is independent of time. The left side must therefore be as well. This will be true only if any time dependence in A has the form $A(t,r) = f(t)a(r)$, because in that case,

$$-\frac{1}{B}\frac{dB}{dr} = \frac{1}{A}\frac{\partial A}{\partial r} = \frac{1}{f(t)a(r)}f(t)\frac{da}{dr} = \frac{1}{a}\frac{da}{dr} \quad (23.11)$$

(note that I am using dB/dr and da/dr instead of $\partial B/\partial r$ and $\partial a/\partial r$ because I know that B and a depend only on r). Since we know what $B(r)$ is, we can solve equation 23.11 for $a(r)$: the result (see box 23.4) is

$$a = \frac{K}{B} = K\left(1 + \frac{C}{r}\right) \quad (23.12)$$

where K is another constant of integration. Therefore, the empty-space Einstein equation implies that the metric has the form

$$ds^2 = -Kf(t)\left(1 + \frac{C}{r}\right)dt^2 + \frac{dr^2}{1 + C/r} + r^2(d\theta^2 + \sin^2\theta\, d\phi^2) \quad (23.13)$$

with K, $f(t)$, and C unspecified. However, note that $Kf(t)$ must be positive, because if it were negative, we would either have no time coordinate (if $1 + C/r$ is positive) or two time coordinates (if $1 + C/r$ is negative), neither of which is acceptable for a spacetime. But this means that we can redefine the t-coordinate so that

$$dt_{\text{new}} = dt_{\text{old}}\sqrt{Kf(t)} \quad (23.14)$$

Therefore, the empty-space Einstein equation and our remaining freedom to choose the time coordinate implies that we can always put the metric in the vacuum spacetime outside a spherically symmetric object in the *time-independent* form

$$ds^2 = -\left(1 + \frac{C}{r}\right)dt^2 + \frac{dr^2}{1 + C/r} + r^2(d\theta^2 + \sin^2\theta\, d\phi^2) \quad (23.15)$$

even if the source of this gravitational field is *not* time-independent (e.g., even if it is expanding or contracting violently). The important statement that the spacetime surrounding a spherically symmetric time-*dependent* object is time-*independent* is called **Birkhoff's theorem**.

Note that the choice of time coordinate specified by equation 23.14 implies that t corresponds to the time measured by an observer at infinity.

The Schwarzschild Solution. To complete the derivation of the Schwarzschild metric, we need to determine the value of C. We can do this by considering the acceleration of an object that is initially at rest in this spacetime and making this consistent at large r with the expected Newtonian result.

The spatial components of the four-momentum $\boldsymbol{u}$ of an object at rest are zero. The condition $-1 = \boldsymbol{u} \cdot \boldsymbol{u}$ then requires that $u^t = (-g_{tt})^{-1/2}$. The geodesic equation then implies that the acceleration of such an object is

$$0 = \frac{d^2x^\alpha}{d\tau^2} + \Gamma^\alpha_{\mu\nu}u^\mu u^\nu \Rightarrow \frac{d^2x^\alpha}{d\tau^2} = -\Gamma^\alpha_{tt}(u^t)^2 = +\frac{\Gamma^\alpha_{tt}}{g_{tt}} = -\frac{1}{A}\Gamma^\alpha_{tt} \quad (23.16)$$

where $A = -g_{tt} = (1 + C/r)$. You can use the Christoffel-symbol entries on the Diagonal Metric Worksheet to show that of all the Γ^{α}_{tt} symbols, only

$$\Gamma^r_{tt} = +\frac{1}{2B}\frac{\partial A}{\partial r} \qquad (23.17)$$

is nonzero (see box 23.5). Therefore, the r component of the geodesic equation implies that

$$\frac{d^2r}{d\tau^2} = -\frac{1}{A}\Gamma^r_{tt} = -\frac{1}{2AB}\frac{\partial A}{\partial r} = -\frac{1}{2}\frac{\partial A}{\partial r} = +\frac{C}{2r^2} \qquad (23.18)$$

As $r \to \infty$, r becomes equivalent to ordinary radial distance and $d\tau$ becomes the same as dt. We therefore recover the Newtonian result that $d^2r/dt^2 = -GM/r^2$ if and only if $C \equiv -2GM$. Substituting this result back into equation 23.15 yields the **Schwarzschild solution**:

$$ds^2 = -\left(1 - \frac{2GM}{r}\right)dt^2 + \left(1 - \frac{2GM}{r}\right)^{-1}dr^2 + r^2(d\theta^2 + \sin^2\theta \, d\phi^2) \qquad (23.19)$$

The Meaning of the Constant M. Note that the constant M in this case corresponds to the mass of the gravitational source as defined by observation of a falling object in the Newtonian limit, so we might call it the source's **Newtonian mass**. For ordinary stars or planets, this will correspond very closely with the integral of the object's rest-mass density over its volume, but for sources involving relativistic fluids and/or where spacetime is significantly curved in the source's interior, M does *not* correspond to the source's total mass-energy. For example, we have seen that in the weak-field limit the pressure in the source can contribute significantly to the gravitational field if the source's fluid is relativistic (so that its pressure p is comparable to its mass-density ρ), so it would figure (along with the source's mass-energy density) in the value of M determined by a distant observer.

However, note that Birkhoff's theorem does imply that a star's Newtonian mass M (as measured by the distant observer) will remain constant as the star collapses or explodes (as long as the star remains spherical). One might therefore interpret Birkhoff's theorem as a conservation law for the Newtonian mass M.

Beyond the Schwarzschild Solution. The Schwarzschild solution provides a useful springboard for studying other spacetimes. Some of the homework problems in this chapter explore extensions of the Schwarzschild solution when there is also a vacuum energy present (problem P23.3), the Reissner-Nordström solution for a spherical star with electrical charge Q (problem P23.4), and in the interior of a spherical, non-rotating star (problem P23.5). One can also apply analogous methods to other highly symmetrical spacetimes, such as spacetimes with planar symmetry (problem P23.1) or cylindrical symmetry (problem P23.6). The planar case is not physically plausible for a variety of reasons, but the cylindrical case may have an application to cosmic strings, which are hypothetical defects in spacetime left over from the Big Bang that may appear to be nearly infinite, incredibly thin lines of mass-energy. The solution studied in problem P23.6 implies that cosmic strings will (strangely) exert no gravitational forces on even very close objects but can bend initially parallel beams of light passing on opposite sides of the string, opening up possibilities for their detection.

If you are interested in these any of these issues, I invite you to do the problems and explore the references at the end of each problem.

BOX 23.1 Diagonalizing the Spherically Symmetric Metric

Let's take the differential of equation 23.4:

$$dt' = dt + \frac{\partial f}{\partial t} dt + \frac{\partial f}{\partial r} dr = dt\left(1 + \frac{\partial f}{\partial t}\right) + \frac{\partial f}{\partial r} dr$$

$$\Rightarrow \quad dt = \left(1 + \frac{\partial f}{\partial t}\right)^{-1} \left(dt' - \frac{\partial f}{\partial r} dr\right) \tag{23.20}$$

You can show that substituting this into the metric equation 23.3 converts

$$ds^2 = g_{tt} dt^2 + 2g_{tr} dt\, dr + g_{rr} dr^2 + r^2(d\theta^2 + \sin^2\theta\, d\phi^2) \tag{23.3r}$$

to

$$ds^2 = g'_{tt} dt'^2 + 2g'_{tr} dt'\, dr + g'_{rr} dr^2 + r^2(d\theta^2 + \sin^2\theta\, d\phi^2) \tag{23.21a}$$

where

$$g'_{tt} = \left(1 + \frac{\partial f}{\partial t}\right)^{-2} g_{tt} \tag{23.21b}$$

$$g'_{rt} = \left(1 + \frac{\partial f}{\partial t}\right)^{-2} \left[g_{rt}\left(1 + \frac{\partial f}{\partial t}\right) - g_{tt} \frac{\partial f}{\partial r}\right] \tag{23.21c}$$

$$g'_{rr} = g_{rr} + \left(1 + \frac{\partial f}{\partial t}\right)^{-2} \left[g_{tt}\left(\frac{\partial f}{\partial r}\right)^2 - 2g_{rt}\left(1 + \frac{\partial f}{\partial t}\right)\frac{\partial f}{\partial r}\right] \tag{23.21d}$$

Exercise 23.1.1. Verify that equations 23.21 are correct.

To set $g'_{rt} = 0$, we need to have

$$\frac{\partial f}{\partial r} = \frac{g_{rt}}{g_{tt}}\left(1 + \frac{\partial f}{\partial t}\right) \tag{23.22}$$

But since we know g_{tt} and g_{rt} by hypothesis, we ought to be able to solve this differential equation for $f(r,t)$, and thus eliminate g'_{rt} from the metric. Defining $g'_{tt} \equiv -A$, $g'_{rr} \equiv B$, and dropping the prime from the t yields equation 23.5.

BOX 23.2 The Components of the Ricci Tensor

The Diagonal Metric Worksheet can be downloaded and printed from this text's website or copied from the appendix. The notes at the worksheet's beginning say the following:

"Consider the general diagonal metric

$$ds^2 = -A(dx^0)^2 + B(dx^1)^2 + C(dx^2)^2 + D(dx^3)^2$$

where dx^0, dx^1, dx^2, and dx^3 are completely arbitrary coordinates and A, B, C, and D are arbitrary functions of any or all of the coordinates. [... Note that:]

$$A_0 \equiv \frac{\partial A}{\partial x^0}, \quad B_{12} \equiv \frac{\partial^2 B}{\partial x^1 \partial x^2}, \text{ and so on.}"$$

In our case, A on the worksheet = A for our trial metric, B on the worksheet = B for our trial metric, $C = r^2$, and $D = r^2 \sin^2\theta$. Also, $x^0 \equiv t$, $x^1 \equiv r$, $x^2 \equiv \theta$, $x^3 \equiv \phi$. Our metric does not depend on ϕ, so any term involving a 3 subscript is zero.

When I use the worksheet, I write above each term listed in the worksheet the equivalent term for the case in question as well as crossing out any term that is zero. I then assemble all nonzero terms at the bottom. The calculation below illustrates.

$$R_{00} = 0 \quad + \frac{1}{2B}\frac{\partial^2 A}{\partial r^2} \atop + \frac{1}{2B}A_{11} \quad + \frac{1}{2C}\cancel{A_{22}} \quad + \frac{1}{2D}\cancel{A_{33}}$$

$$+ 0 \quad -\frac{1}{2B}\frac{\partial^2 B}{\partial t^2} \atop -\frac{1}{2B}B_{00} \quad -\frac{1}{2C}\cancel{C_{00}} \quad -\frac{1}{2D}\cancel{D_{00}}$$

$$+ 0 \quad +\frac{1}{4B^2}\left(\frac{\partial B}{\partial t}\right)^2 \atop +\frac{1}{4B^2}B_0^2 \quad +\frac{1}{4C^2}\cancel{C_0^2} \quad +\frac{1}{4D^2}\cancel{D_0^2}$$

$$+ 0 \quad +\frac{1}{4AB}\frac{\partial A}{\partial t}\frac{\partial B}{\partial t} \atop +\frac{1}{4AB}A_0 B_0 \quad +\frac{1}{4AC}A_0\cancel{C_0} \quad +\frac{1}{4AD}A_0\cancel{D_0}$$

$$-\frac{1}{4BA}\left(\frac{\partial A}{\partial r}\right)^2 \quad -\frac{1}{4B^2}\frac{\partial A}{\partial r}\frac{\partial B}{\partial r} \quad +\frac{1}{4Br^2}\frac{\partial A}{\partial r}\cancel{2r} \quad +\frac{1}{4Br^2\sin^2\theta}\frac{\partial A}{\partial r}2r\sin^2\theta$$
$$-\frac{1}{4BA}A_1 A_1 \quad -\frac{1}{4B^2}A_1 B_1 \quad +\frac{1}{4BC}A_1 C_1 \quad +\frac{1}{4BD}A_1 D_1$$

$$-\frac{1}{4CA}\cancel{A_2}A_2 \quad +\frac{1}{4CB}\cancel{A_2}B_2 \quad -\frac{1}{4C^2}\cancel{A_2}C_2 \quad +\frac{1}{4CD}\cancel{A_2}D_2$$

$$-\frac{1}{4DA}\cancel{A_3}A_3 \quad +\frac{1}{4DB}\cancel{A_3}B_3 \quad +\frac{1}{4DC}\cancel{A_3}C_3 \quad -\frac{1}{4D^2}\cancel{A_3}D_3$$

$$\boxed{R_{00} = \frac{1}{2B}\left[\frac{\partial^2 A}{\partial r^2} - \frac{\partial^2 B}{\partial t^2} + \frac{1}{2B}\left(\frac{\partial B}{\partial t}\right)^2 + \frac{1}{2A}\frac{\partial A}{\partial t}\frac{\partial B}{\partial t} - \frac{1}{2A}\left(\frac{\partial A}{\partial r}\right)^2 - \frac{1}{2B}\frac{\partial A}{\partial r}\frac{\partial B}{\partial r} + \frac{2}{r}\frac{\partial A}{\partial r}\right]}$$

If you compare the handwritten result above with equation 23.6a, you will see that they are the same (though the terms have been rearranged).

Exercise 23.2.1. Using a similar approach, show that 23.6b–e are correct. Do your work directly on the worksheet. Also check that $R_{t\theta} = 0$ identically in this case. I have provided the worksheet sections you need on the next two pages.

23. THE SCHWARZSCHILD SOLUTION

Diagonal Metric Worksheet: Consider the general diagonal metric

$$ds^2 = -A(dx^0)^2 + B(dx^1)^2 + C(dx^2)^2 + D(dx^3)^2$$

where dx^0, dx^1, dx^2, and dx^3 are completely arbitrary coordinates and A, B, C, and D are arbitrary functions of any or all of the coordinates. ... In this worksheet, I use the following shorthand notation:

$$A_0 \equiv \frac{\partial A}{\partial x^0}, \quad B_{12} \equiv \frac{\partial^2 B}{\partial x^1 \partial x^2}, \text{ and so on.}$$

$R_{11} = \frac{1}{2A}B_{00} \quad + \quad 0 \quad - \quad \frac{1}{2C}B_{22} \quad - \quad \frac{1}{2D}B_{33}$

$\quad - \quad \frac{1}{2A}A_{11} \quad + \quad 0 \quad - \quad \frac{1}{2C}C_{11} \quad - \quad \frac{1}{2D}D_{11}$

$\quad + \quad \frac{1}{4A^2}A_1^2 \quad + \quad 0 \quad + \quad \frac{1}{4C^2}C_1^2 \quad + \quad \frac{1}{4D^2}D_1^2$

$\quad - \quad \frac{1}{4A^2}B_0 A_0 \quad - \quad \frac{1}{4AB}B_0 B_0 \quad + \quad \frac{1}{4AC}B_0 C_0 \quad + \quad \frac{1}{4AD}B_0 D_0$

$\quad + \quad \frac{1}{4BA}B_1 A_1 \quad + \quad 0 \quad + \quad \frac{1}{4BC}B_1 C_1 \quad + \quad \frac{1}{4BD}B_1 D_1$

$\quad - \quad \frac{1}{4CA}B_2 A_2 \quad + \quad \frac{1}{4CB}B_2 B_2 \quad + \quad \frac{1}{4C^2}B_2 C_2 \quad - \quad \frac{1}{4CD}B_2 D_2$

$\quad - \quad \frac{1}{4DA}B_3 A_3 \quad + \quad \frac{1}{4DB}B_3 B_3 \quad - \quad \frac{1}{4DC}B_3 C_3 \quad + \quad \frac{1}{4D^2}B_3 D_3$

$R_{11} = $

$R_{01} = -\frac{1}{2C}C_{01} \quad - \quad \frac{1}{2D}D_{01} \quad + \quad \frac{1}{4C^2}C_0 C_1 \quad + \quad \frac{1}{4D^2}D_0 D_1$

$\quad + \quad \frac{1}{4AC}A_1 C_0 \quad + \quad \frac{1}{4AD}A_1 D_0 \quad + \quad \frac{1}{4BC}B_0 C_1 \quad + \quad \frac{1}{4BD}B_0 D_1$

$R_{01} = $

$R_{02} = -\frac{1}{2B}B_{02} \quad - \quad \frac{1}{2D}D_{02} \quad + \quad \frac{1}{4B^2}B_0 B_2 \quad + \quad \frac{1}{4D^2}D_0 D_2$

$\quad + \quad \frac{1}{4AB}A_2 B_0 \quad + \quad \frac{1}{4AD}A_2 D_0 \quad + \quad \frac{1}{4CB}C_0 B_2 \quad + \quad \frac{1}{4CD}C_0 D_2$

$R_{02} = $

THE DETAILS

$R_{22} = \frac{1}{2A}C_{00} \quad - \frac{1}{2B}C_{11} \quad + 0 \quad - \frac{1}{2D}C_{33}$

$\quad - \frac{1}{2A}A_{22} \quad - \frac{1}{2B}B_{22} \quad + 0 \quad - \frac{1}{2D}D_{22}$

$\quad + \frac{1}{4A^2}A_2^2 \quad + \frac{1}{4B^2}B_2^2 \quad + 0 \quad + \frac{1}{4D^2}D_2^2$

$\quad - \frac{1}{4A^2}C_0A_0 \quad + \frac{1}{4AB}C_0B_0 \quad - \frac{1}{4AC}C_0C_0 \quad + \frac{1}{4AD}C_0D_0$

$\quad - \frac{1}{4BA}C_1A_1 \quad + \frac{1}{4B^2}C_1B_1 \quad + \frac{1}{4BC}C_1C_1 \quad - \frac{1}{4BD}C_1D_1$

$\quad + \frac{1}{4CA}C_2A_2 \quad + \frac{1}{4CB}C_2B_2 \quad + 0 \quad + \frac{1}{4CD}C_2D_2$

$\quad - \frac{1}{4DA}C_3A_3 \quad - \frac{1}{4DB}C_3B_3 \quad + \frac{1}{4DC}C_3C_3 \quad + \frac{1}{4D^2}C_3D_3$

$R_{22} = $

$R_{33} = \frac{1}{2A}D_{00} \quad - \frac{1}{2B}D_{11} \quad - \frac{1}{2C}D_{22} \quad + 0$

$\quad - \frac{1}{2A}A_{33} \quad - \frac{1}{2B}B_{33} \quad - \frac{1}{2C}C_{33} \quad + 0$

$\quad + \frac{1}{4A^2}A_3^2 \quad + \frac{1}{4B^2}B_3^2 \quad + \frac{1}{4C^2}C_3^2 \quad + 0$

$\quad - \frac{1}{4A^2}D_0A_0 \quad + \frac{1}{4AB}D_0B_0 \quad + \frac{1}{4AC}D_0C_0 \quad - \frac{1}{4AD}D_0D_0$

$\quad - \frac{1}{4BA}D_1A_1 \quad + \frac{1}{4B^2}D_1B_1 \quad - \frac{1}{4BC}D_1C_1 \quad + \frac{1}{4BD}D_1D_1$

$\quad - \frac{1}{4CA}D_2A_2 \quad - \frac{1}{4CB}D_2B_2 \quad + \frac{1}{4C^2}D_2C_2 \quad + \frac{1}{4CD}D_2D_2$

$\quad + \frac{1}{4DA}D_3A_3 \quad + \frac{1}{4DB}D_3B_3 \quad + \frac{1}{4DC}D_3C_3 \quad + 0$

$R_{33} = $

BOX 23.3 Solving for B

Equation 23.9 (repeated here) tells us that

$$\frac{1}{A}\frac{\partial A}{\partial r} = -\frac{1}{B}\frac{\partial B}{\partial r} \qquad (23.9r)$$

According to equation 23.6c, the condition that $R_{\theta\theta} = 0$ implies that

$$0 = -\frac{r}{2AB}\frac{\partial A}{\partial r} + \frac{r}{2B^2}\frac{\partial B}{\partial r} + 1 - \frac{1}{B} \qquad (23.23)$$

If you use equation 23.9 to eliminate the factors of A from equation 23.23, you should find that the latter reduces to

$$1 = -\frac{r}{B^2}\frac{\partial B}{\partial r} + \frac{1}{B} = \frac{\partial}{\partial r}\left(\frac{r}{B}\right) \qquad (23.24)$$

Exercise 23.3.1. Show that equation 23.24 is correct.

You can easily integrate this to show that

$$\frac{1}{B} = 1 + \frac{C}{r} \qquad (23.25)$$

where C is an unknown constant of integration.

Exercise 23.3.2. Show the steps that lead from equation 23.24 to equation 23.25. Remember that we have seen earlier that B is time-independent, so the partial derivative in equation 23.24 is the same as an ordinary r-derivative.

BOX 23.4 Solving for $a(r)$

Equation 23.11 (duplicated here for convenience) says that

$$-\frac{1}{B}\frac{dB}{dr} = \frac{1}{a}\frac{da}{dr} \qquad (23.11r)$$

You can easily integrate both sides of this expression and then take the exponential of both sides to show that

$$a(r) = \frac{K}{B} \qquad (23.26)$$

Exercise 23.4.1. Verify equation 23.26.

BOX 23.5 The Christoffel Symbols with t-t as Subscripts

Keeping in mind that we now know that both A and B depend only on r, you can easily use the Christoffel-symbol part of the "Diagonal Metric" worksheet to show

$$\Gamma^r_{tt} = \frac{1}{2B}\frac{\partial A}{\partial r} \quad \text{and} \quad \Gamma^t_{tt} = \Gamma^\theta_{tt} = \Gamma^\phi_{tt} = 0 \qquad (23.27)$$

Exercise 23.5.1. Verify equation 23.27. I have copied the relevant part of the worksheet below.

$$\Gamma^0_{00} = \tfrac{1}{2A}A_0 \qquad \Gamma^1_{00} = \tfrac{1}{2B}A_1 \qquad \Gamma^2_{00} = \tfrac{1}{2C}A_2 \qquad \Gamma^3_{00} = \tfrac{1}{2D}A_3$$

23. THE SCHWARZSCHILD SOLUTION

HOMEWORK PROBLEMS

P23.1 Consider the case of a static, plane-symmetric spacetime. A plane-symmetric spacetime is one where the successive planes labeled by some coordinate x have the geometry of a flat space. The metric on any such plane can be labeled by cartesian coordinates y and z such that the metric on that plane is

$$ds^2 = dy^2 + dz^2 \qquad (23.28)$$

We can choose an x coordinate so that its basis vector is everywhere perpendicular to the successive planes (so that $g_{xy} = \mathbf{e}_x \cdot \mathbf{e}_y = 0$ and $g_{xz} = \mathbf{e}_x \cdot \mathbf{e}_z = 0$), and define that coordinate to be the actual physical distance between successive planes. Our metric for the spacetime must therefore have the form

$$ds^2 = -g_{tt}\, dt^2 + 2g_{tx}\, dt\, dx + dx^2 + dy^2 + dz^2 \qquad (23.29)$$

Since this metric is (by hypothesis) static, g_{tt} and g_{tx} can depend, at most, on x.

a. Argue that because there should be no distinction between the past and future for this static metric, g_{tx} must be zero, and the metric therefore must have the form

$$ds^2 = -A\, dt^2 + dx^2 + dy^2 + dz^2 \qquad (23.30)$$

b. Use a printout of the Diagonal Metric Worksheet to show that the only nonzero Christoffel symbols for this metric are

$$\Gamma^x_{tt} = \frac{1}{2}\frac{dA}{dx}, \quad \Gamma^t_{tx} = \Gamma^t_{xt} = \frac{1}{2A}\frac{dA}{dx} \qquad (23.31)$$

c. Use the same worksheet to show that the empty-space Einstein equation $R_{\mu\nu} = 0$ is satisfied if

$$\frac{d^2 A}{dx^2} = \frac{1}{2A}\left(\frac{dA}{dx}\right)^2 \qquad (23.32)$$

d. Argue from the definition of the Riemann tensor that the only potentially nonzero components of that tensor are R_{xtxt} and its permutations.

e. Show that if 23.32 is true, then $R_{xtxt} = 0$. This means that the spacetime must be flat (i.e., it contains no actual gravitational field). Therefore, the Einstein equation forbids a non-flat plane symmetric spacetime in empty space.

P23.2 The Schwarzschild solution is still a valid solution to the empty-space Einstein equation if we choose the constant of integration C to be positive. What happens if we make this choice? Carefully justify your responses to each of the questions below.

a. Consider a particle initially at rest at some r coordinate. If released, will it fall toward smaller r or larger r?

b. Are circular orbits possible in this case?

c. Is there an event horizon? If so, where?

d. Is there a singularity at $r = 0$? If so, is it a geometric singularity or a coordinate singularity? (*Hint:* Consider the result of problem P19.8 and set $2GM$ in that problem $-C$.) If the latter, is it hidden from the universe by an event horizon?

P23.3 Let's look at the Schwarzschild solution again, but this time let's *not* assume that the vacuum energy is negligible. In this case, the Einstein equation (in the form given by equation 21.14) implies that in empty space

$$R_{\mu\nu} = \Lambda g_{\mu\nu} \qquad (23.33)$$

Specifically, this implies that $R_{tt} = -A\Lambda$, $R_{rr} = B\Lambda$, and $R_{\theta\theta} = r^2 \Lambda$.

a. Carefully go through the derivation of the Schwarzschild metric given in this chapter, noting any changes required. Show that in this case we have

$$A = \frac{1}{B} = 1 + \frac{C}{r} - \frac{\Lambda}{3}r^2 \qquad (23.34)$$

where C is a constant of integration. (*Hint:* Only one equation changes.)

b. Assume (as is experimentally the case) that Λ is sufficiently small so that there is a large range of r such that $1 \gg |C|/r \gg \Lambda r^2$. Re-evaluate equation 23.18 in this case and show that if we are to agree with Newtonian predictions for r within this range, we must still have $C = -2GM$.

c. However, argue that if r is sufficiently large, the gravitational field becomes repulsive.

d. How would a nonzero Λ change the radial equation of motion given by equation 10.8?

P23.4 In this problem, we will derive the **Reissner-Nordström solution** for a black hole with electrical charge Q. As in the Schwarzschild solution, we will consider a spherically symmetric metric of the form given by equation 23.5, except that we will accept Birkhoff's theorem and assume that both A and B depend only on r. We will also assume that the black hole's charge creates a static electric field in the spacetime around the black hole.

a. In this case, the spacetime around the black hole is not strictly empty: the electric field has a nonzero stress-energy tensor that pervades the black hole's exterior. If you did problem P20.5, you should have found that the electromagnetic stress-energy tensor is

$$T^{\mu\nu} = \frac{1}{4\pi k}\left(F^{\mu\alpha} g_{\alpha\beta} F^{\nu\beta} - \tfrac{1}{4} g^{\mu\nu} F^{\sigma\gamma} F_{\sigma\gamma}\right) \qquad (23.35)$$

where $F^{\mu\nu}$ is the electromagnetic field tensor and k is the Coulomb constant. Show that no matter what the metric might be, $T \equiv T^\mu{}_\mu = 0$ for the electromagnetic stress-energy tensor.

b. Assuming spherical symmetry means that the electric field created by a black hole can have only a radial component and that this component can depend only on the radial coordinate r. This in turn means that we can write the electromagnetic field tensor in the form

$$F_{\mu\nu} = \begin{bmatrix} 0 & -E(r) & 0 & 0 \\ E(r) & 0 & 0 & 0 \\ 0 & 0 & 0 & 0 \\ 0 & 0 & 0 & 0 \end{bmatrix} \qquad (23.36)$$

where $E(r)$ is some unknown function of r (note the subscripted indices!). Show that if we assume that this is true, then the only nonzero values of $F^{\mu\nu}$ are

$$F^{tr} = -F^{rt} = \frac{E}{AB} \qquad (23.37)$$

c. Since $T = 0$ for this field, $T_{\mu\nu} - \frac{1}{2}g_{\mu\nu}T$ (the quantity that appears on the right side of the Einstein equation) is simply $T_{\mu\nu}$ here. Show that in this case,

$$8\pi G T_{tt} = GE^2/kB \qquad (23.38a)$$
$$8\pi G T_{rr} = -GE^2/kA \qquad (23.38b)$$
$$8\pi G T_{\theta\theta} = GE^2 r^2/kAB \qquad (23.38c)$$

d. The t-t and r-r components of the Einstein equation therefore imply that

$$R_{tt} = \frac{GE^2}{kB}, \; R_{rr} = -\frac{GE^2}{kA} \;\Rightarrow\; BR_{tt} + AR_{rr} = 0 \quad (23.39)$$

Use equations 23.6 (setting all time derivatives to zero) to show that the equation above implies that

$$\frac{1}{A}\frac{dA}{dr} + \frac{1}{B}\frac{dB}{dr} = 0 \qquad (23.40)$$

e. Multiply both sides of the above by AB and argue that this equation means that AB = constant. Since the spacetime metric must approach flat space at infinity, both A and B must go to 1 at infinity. This means that AB must be equal to 1, i.e., that $B = 1/A$, everywhere. Note that equation 23.37 then also implies that electric field's r component $E_r \equiv F^{rt}$ is equal to the function $E(r)$ that we defined in equation 23.36.

f. Use the definition of the absolute gradient and Christoffel symbol part of the Diagonal Metric Worksheet to show that the t component of the Maxwell equation $\nabla_\nu F^{\mu\nu} = 4\pi k J^\mu = 0$ for empty space implies that

$$\frac{d}{dr}(r^2 E) = 0 \;\Rightarrow\; E(r) = \frac{b}{r^2} \qquad (23.41)$$

where b is a constant of integration (note that we just proved that $AB = 1$). If this is to reduce to the electric field of a point charge at infinity, then we must have $b = kQ$, where Q is the black hole's total charge.

g. If $AB = 1$, then the θ-θ component of the Einstein equation tells us that $R_{\theta\theta} = GE^2r^2/k$. Use equation 23.6c, equation 23.40, and equation 23.41 to argue that

$$1 - \frac{GkQ^2}{r^2} = \frac{d}{dr}(rA)$$

$$\Rightarrow \; A(r) = 1 - \frac{2GM}{r} + \frac{GkQ^2}{r^2} \qquad (23.42)$$

if we choose the constant of integration so that we get the Schwarzschild solution when $Q = 0$. We also know that $B = 1/A$, so this completely determines the metric.

h. For a spherical star, we have event horizons wherever $g_{tt} = 0$. Show that if $kQ^2 < GM^2$, the Reissner-Nordström spacetime has *two* event horizons, and specify their locations. Also show that if $kQ^2 > GM^2$, there are *no* event horizons (meaning that the singularity at $r = 0$ is exposed). For a variety of reasons, the latter case is considered unphysical.

For more about the Reissner-Nordström spacetime, see Hobson, Efstathiou, and Lasenby, *General Relativity*, Cambridge, 2006, pp. 296–305, from which I have adapted the derivation in this problem. It also turns out that no one has been able to find a physically reasonable solution like this for any kind of external field *other* than an electromagnetic field, including various kinds of scalar, vector, tensor, and spinor fields. This led theorist John A. Wheeler to famously remark that "a black hole has no hair."

P23.5 The Schwarzschild solution applies only in the vacuum outside a spherical star. In this problem, we will explore solving the Einstein equation *inside* a spherical, static, and non-rotating star. We will again consider a spherical metric of the form given by equation 23.5, except that we will assume that everything is completely static, meaning that A and B depend only on r. We will also assume that the star's interior consists of a perfect fluid. We are looking for a solution to the Einstein equation $R_{\mu\nu} = 8\pi G(T_{\mu\nu} - \frac{1}{2}g_{\mu\nu}T)$.

a. Show that for a perfect fluid,

$$T_{\mu\nu} - \frac{1}{2}g_{\mu\nu}T = (\rho + p)u_\mu u_\nu + \frac{1}{2}(\rho - p)g_{\mu\nu} \quad (23.43)$$

where ρ is the density of the fluid in its rest LIF and p is its pressure in that frame. Spherical symmetry implies that ρ and p can at most depend on r.

b. Argue that since the fluid is at rest, the spatial components of u_μ are zero and $u_t = -\sqrt{A}$.

c. Show that the Einstein equation implies that

$$R_{tt} = 4\pi G(\rho + 3p)A \qquad (23.44a)$$
$$R_{rr} = 4\pi G(\rho - p)B \qquad (23.44b)$$
$$R_{\theta\theta} = 4\pi G(\rho - p)r^2 \qquad (23.44c)$$

and therefore that

$$\frac{R_{tt}}{A} + \frac{R_{rr}}{B} + \frac{2R_{\theta\theta}}{r^2} = 16\pi G\rho \qquad (23.45)$$

d. Use equations 23.6 (setting all time derivatives to zero) to show that the equation above reduces to

$$\frac{d}{dr}\left[r\left(1 - \frac{1}{B}\right)\right] = 8\pi G\rho r^2 \qquad (23.46)$$

e. Integrate both sides of this from 0 to r to get

$$B = \left[1 - \frac{2Gm(r)}{r}\right]^{-1} \qquad (23.47a)$$

where $\quad m(r) \equiv \int_0^r 4\pi\rho r^2 dr \qquad (23.47b)$

The quantity $m(r) \equiv \int_0^r 4\pi\rho r^2 dr$ *looks* like the mass enclosed by a radius r, but remember that true radial distance corresponding to a coordinate step dr is a bit larger than dr, so this is less than the actual total mass. Once we reach the star's surface, the Schwarzschild

P23.5 (part e, continued)

solution will take over and matching the two metrics at the surface requires that m(surface) $= M$, the star's gravitational mass. So we can think of $m(r)$ as being the *gravitational* mass inside r, which will be the total mass minus some gravitational binding energy.

f. The fastest way to find the condition on A turns out to be to use the energy conservation equation $\nabla_\nu T^{\mu\nu} = 0$ (which is implicit in the Einstein equation, as we have seen in previous chapters). Show that the $\mu = r$ component of this equation implies that

$$\frac{1}{A}\frac{dA}{dr} = -\frac{2}{\rho + p}\frac{dp}{dr} \qquad (23.48)$$

(*Hint:* Remember that the absolute gradient of the metric is zero. You will have to calculate one Christoffel symbol and argue that others are zero.)

g. Use equations 23.6c, 23.44c, 23.47a, and 23.48 to show that

$$\frac{dp}{dr} = -\frac{\rho+p}{r^2}\left[\frac{4\pi G p r^3 + Gm(r)}{1 - 2Gm(r)/r}\right] \qquad (23.49)$$

This is the **Oppenheimer-Volkoff equation** for stellar structure. This equation, the equation $dm/dr = 4\pi\rho r^2$, and an "equation of state" that specifies $p(\rho)$ provide three equations in the three unknowns $m(r)$, $p(r)$, and $\rho(r)$. Except for the simplest cases, these equations are pretty hard to solve analytically, but they are pretty easy to solve numerically (as I personally did as part of my doctoral thesis) for a specified equation of state and central density $\rho(0)$. If you like, you can also integrate equations 23.48 along the way to find A and use 23.47a to find B at any r. For more information about stellar interiors, see Hobson, Efstathiou, and Lasenby, *General Relativity*, Cambridge, 2006, pp. 288–296, from which I have adapted this problem.

P23.6 Cosmic strings are hypothetical structures left over from the Big Bang. If they exist, they would have diameters much smaller than a nucleus but could have lengths comparable to galaxies or even larger. Consider an infinite, straight, and static string lying along the z axis of our coordinate system. The metric that describes both the interior and the exterior of this string will be axially symmetric around the z axis and independent of time. The simplest metric satisfying these constraints is

$$ds^2 = -dt^2 + dr^2 + [f(r)]^2 d\phi^2 + dz^2 \qquad (23.50)$$

where t is the time measured by any observer at rest, r is the radial distance from z axis, ϕ is an angular coordinate, and z is the distance measured parallel to the string. Let's see if such a metric meaningfully satisfies the Einstein equation.

a. A simplified hypothetical stress-energy for the exotic trapped energy inside the string is $T^t{}_t = T^z{}_z = -\sigma(r)$ and all other $T^\mu{}_\nu = 0$, where $\sigma(r)$ is some function of r with units of energy density. Show that the Einstein equation in this case reduces to

$$R_{rr} = R_{\phi\phi}/f^2 = 8\pi G\sigma; \text{ other } R_{\mu\nu} = 0 \qquad (23.51)$$

b. Use the Diagonal Metric Worksheet to show that all the Ricci tensor components for this metric except for R_{rr} and $R_{\phi\phi}$ are identically zero. Also show that R_{rr} is indeed equal to $R_{\phi\phi}/f^2$. Therefore, *all* the components of the Einstein equation are satisfied if $R_{rr} = 8\pi G\sigma$.

c. Show that $R_{rr} = 8\pi G\sigma$ reduces to

$$\frac{d^2 f}{dr^2} = -8\pi G f\sigma \qquad (23.52)$$

d. In order to be non-singular at the origin, the metric should reduce to the flat space metric there. This will happen if $f(r) \to r$ as $r \to 0$ goes to zero, meaning that $f(0) = 0$, and $f' \equiv df/dr = 1$ at the origin. Note also that the metric implies that the physical area of a ring of arbitrary radius r and width dr is $2\pi f\, dr$. Thus the string's energy per unit length will be

$$\mu \equiv \int_0^{r_s} \sigma 2\pi f\, dr \qquad (23.53)$$

where r_s is the radius beyond which $\sigma(r) = 0$ (remember that r_s will be small compared to an atomic nucleus). Show that if we integrate equation 23.52 from $r = 0$ to any radius beyond r_s, we get

$$f'(\text{for } r > r_s) = f'(r_s) = 1 - 4G\mu \qquad (23.54)$$

For physically reasonable strings, σ is such that $4G\mu$ will be significantly less than 1.

e. So $f(r) = (1 - 4G\mu)r + K$ for $r > r_s$, where K is a constant of integration. Argue that since r_s is very small compared to any macroscopic value of r, we can take $K \approx 0$. (*Hint:* Consider how df/dr compares to 1.)

f. So our metric for $r \gg r_s$ must be

$$ds^2 = -dt^2 + dr^2 + dz^2 + (1 - 4G\mu)^2 r^2 d\phi^2 \qquad (23.55)$$

Argue that if we define a new angular coordinate $\tilde{\phi} \equiv (1 - 4G\mu)\phi$, then the metric becomes mathematically equivalent to a flat space in cylindrical coordinates.

Comments: This means that the Riemann tensor will evaluate to zero everywhere outside the string, meaning that the spacetime is locally flat with no gravitational field. This is still not a normal Euclidean spacetime, though, because the circumferences of circles are smaller than $2\pi r$ (because $f < r$ in the original coordinates, and because $\tilde{\phi} < 2\pi$ for a complete circle in the new coordinates). A surface of constant t and z therefore has the *global* geometry of a cone that we can model by taking a flat circle and cutting out a pie-shaped wedge of angle $2\pi - 2\pi(1 - 4G\mu) = 8\pi G\mu$. As discussed in Helliwell and Konkowski, *American Journal of Physics*, **55**, 5 (1987), this means that cosmic string can bend initially parallel light beams, which opens up possibilities for their detection. The derivation here is adapted from Vilenkin's chapter in *300 Years of Gravitation* (Hawking & Israel, eds., Cambridge, 1987, pp. 499–522), which offers a more advanced discussion of cosmic strings.

24. THE UNIVERSE OBSERVED

INTRODUCTION

FLAT SPACETIME
- Review of Special Relativity
- Four-Vectors
- Index Notation

TENSORS
- Arbitrary Coordinates
- Tensor Equations
- Maxwell's Equations
- Geodesics

SCHWARZSCHILD BLACK HOLES
- The Schwarzschild Metric
- Particle Orbits
- Precession of the Perihelion
- Photon Orbits
- Deflection of Light
- Event Horizon
- Alternative Coordinates
- Black Hole Thermodynamics

THE CALCULUS OF CURVATURE
- The Absolute Gradient
- Geodesic Deviation
- The Riemann Tensor

THE EINSTEIN EQUATION
- The Stress-Energy Tensor
- The Einstein Equation
- Interpreting the Equation
- The Schwarzschild Solution

COSMOLOGY
- **The Universe Observed**
- A Metric for the Cosmos
- Evolution of the Universe
- Cosmic Implications
- The Early Universe
- CMB Fluctuations & Inflation

GRAVITATIONAL WAVES
- Gauge Freedom
- Detecting Gravitational Waves
- Gravitational Wave Energy
- Generating Gravitational Waves
- Gravitational Wave Astronomy

SPINNING BLACK HOLES
- Gravitomagnetism
- The Kerr Metric
- Kerr Particle Orbits
- Ergoregion and Horizon
- Negative-Energy Orbits

this depends on this

24. THE UNIVERSE OBSERVED

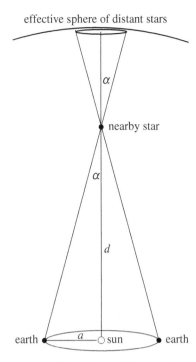

FIG. 24.1 This diagram illustrates how to measure the distance to a nearby star using parallax. By measuring the maximum angular radius α of the star's apparent motion relative to distant stars during the year, one can estimate the distance to the star: $d = a/\tan\alpha \approx a/\alpha$ (if α is small and in radians) where a is the known radius of the earth's orbit. The distance to the star is 1 pc = 1 **parsec** (= 3.26 ly) if α is 1 arc-second. (**Note:** Since all distances in the universe are calibrated from such measurements, the parsec is a natural unit for working astronomers. The light-year is more natural and meaningful for relativity, though, and will be used here consistently.)

Introduction. This chapter opens a five-chapter sequence about applying general relativity to cosmology. Starting in the next chapter, we will solve the Einstein equation for the geometry of the universe. But as discussed in the last chapter, we need to know the symmetries that might apply to in order to choose the appropriate trial metric, and we also need to know about what stress-energy to use. In this chapter, then, we will explore what we can know about the universe by direct observation.

Measuring the Universe. Until the 1900s, astronomers thought that the most distant objects in the universe were stars. However, the distance to the stars was understood only to be "extremely large" until 1838, when Friedrich Bessell won a competition by using the method of parallax to measure the distance to the star 61 Cygni to be roughly 10 ly (the refined modern result is 11.36 ly).

The method of **parallax** (illustrated in figure 24.1) involves observing a star from extreme positions in the earth's orbit. Using the known diameter of the earth's orbit (see box 24.1 for a discussion of how *this* was measured), one can calculate the distance to a nearby star by measuring the shift in its angular position relative to more distant stars. No one was able to do this at all until 1838 because the angular shifts of even the very nearest stars are very small (about 8×10^{-5} degrees in the case of 61 Cygni). Until the 1990s, the parallax distances to only a few hundred of the nearest stars were well known. In 1989, however, the European Space Agency launched the Hipparcos satellite, which was able (by making observations unhindered by the earth's atmosphere) to measure the parallax distances to over 20,800 stars to better than 10% accuracy. This experiment was extremely important because distances to *all* other objects in the universe are based on these results.

Measuring the distances to all more distant objects involves finding "standard candles" of known luminosity. If an object has a known luminosity L (total energy emitted per time), then by measuring the energy flux F (energy per time per area) we receive from that object at the earth, we can calculate its distance ($F = L/4\pi r^2$, so $r = \sqrt{L/4\pi F}$). But since stars come in many sizes and brightnesses, it is difficult to know the luminosity of any given star *a priori*.

Astronomers between 1910 and 1997 used the statistical characteristics of stellar populations to determine the intrinsic luminosity of star *clusters* and thus their distance (see box 24.2 for more details). In 1912, Henrietta Leavitt discovered within such clusters an identifiable subclass of stars known as **Cepheid variables**, whose brightness oscillates with time. She was able to demonstrate that these stars' average luminosities were linked to their oscillation periods in a mathematically well-defined way. Measuring both the average flux from such a star and its oscillation period thus determines its distance. Since these stars are typically 1000 to 10,000 times brighter than the sun, one can observe them from quite far away.

However, using the statistical characteristics of stellar populations to calibrate the Cepheid period-luminosity relation was difficult to do well. In 1997, Michael Feast published results from 26 Cepheids observed by Hipparcos that provided the first *direct* connection between parallax measurements and the Cepheid "standard candle." His re-calibration of the Cepheid standard candle led to a 10% increase in all universal distances not directly determined by parallax.

Before 1925, most astronomers assumed that the entire universe consisted simply of isolated stars, the most distant of which (according to Cepheid measurements) were less than 100,000 ly away. However, as early as the 1750s, some (including the philosopher Immanuel Kant) speculated that certain fuzzy objects known as "nebulae" were "island universes" similar to the local collection of stars known as the Milky Way. This remained an open question in 1920, when astronomers Harlow Shapley and Heber Curtis famously debated the topic. However, in the early 1920s Edwin Hubble, using the new 100-inch telescope on Mount Wilson to closely observe a number of nebulae, found that some contained dim but identifiable Cepheids, and in 1925, he published results that confirmed that these nebulae were in fact **galaxies**, vast collections of stars that were *millions* of light-years away.

At present, we believe that the observable universe (out to about 12 Gly of light-travel distance) contains ~10^{11} galaxies (where "~" means "order of magnitude"). A typical galaxy is ~50,000 ly across and contains ~10^{11} stars. Our own fairly large spiral galaxy contains about 2 to 4×10^{11} stars in a disk 100,000 ly in diameter and about 1000 ly thick, but with a central bulge about 6000 ly thick. Our solar system is about 28,000 ly from the galaxy's center and 20 ly above its central plane.

Galaxies come in many shapes, including irregular, spiral, and elliptical, which evolve through time as galaxies collapse and interact; astronomers believe that elliptical galaxies are the end result of multiple collisions between galaxies. Figure 24.2 is a modern Hubble Space Telescope photograph showing roughly 1000 galaxies within an angular area smaller than 1/30 of the full moon.

Even at present, we can locate individual Cepheid variables only out to about 100 Mly, which is a tiny fraction of the observable universe. In the late 1970s, however, astronomers argued that all "Type Ia" supernovae appeared (on the basis of other measurements and theoretical arguments) to have the same intrinsic luminosity, and therefore could serve as standard candles. Such supernovae occur in a given galaxy about once every 300 y on the average, so one might reasonably expect to see a supernova happening once a year in a reasonably large cluster of galaxies, and each can be nearly as bright as an entire galaxy (see figure 24.3). This class of supernovae is at present the most useful standard candle for measuring cosmological distances.

FIG. 24.2 This is portion of the "Hubble Deep Field" photograph taken by the Hubble Space Telescope (HST). Virtually every spot of light seen here is a galaxy (the field of view is so small that few stars in our own galaxy are in the frame). Credits: R. Williams (STSci), the Hubble Deep Field Team, and NASA.

Though the discussion above is somewhat oversimplified, we see that at the present, three steps (parallax, Cepheid variables, and Type Ia supernovae) can take us from the nearest stars to very distant galaxies.

The Expanding Universe. As early as 1918, astronomers had noticed that the light from some "nebulae" was distinctly shifted to the red, something that could be accurately determined by observing the wavelength of identifiable spectral lines. This could be explained if these nebulae were moving with significant velocities away from the earth, though it seemed unclear why objects in our own galaxy should be all be moving so rapidly away from us. After identifying these "nebulae" as distant galaxies, Georges Lemaître in 1927 and Hubble in 1929 showed that there was a nearly linear relationship between the distance d to a galaxy and its recession speed v inferred from its redshift (see box 24.3 for a discussion of how to make this calculation). We can express this relationship ("**Hubble's law**") in the form

$$v \approx H_0 d \tag{24.1}$$

FIG. 24.3 This HST photograph shows supernova 1994D in the spiral galaxy NGC 4526, which is about 65 Mly away. This Type Ia supernova is the bright dot in the lower left. Credits: NASA, ESA, The Hubble Key Project Team, and the High-Z Supernova Search Team.

where the constant of proportionality H_0 is the **Hubble constant**. Until about 15 years ago, difficulties in calibrating the supernova standard candle meant that H_0 was not known within a factor of 2, but in the past few years, H_0 has been measured (using data from a variety of sources including supernovae)[1] to be 70.4($\pm$1.5) km/s per Mpc (where 1 Mpc $\equiv$ 3.26 Mly). In GR units (see box 24.4),

$$H_0 = 7.61(\pm 0.16) \times 10^{-27} \text{ m}^{-1} = 2.28(\pm 0.05) \times 10^{-18} \text{ s}^{-1} = [13.9(\pm 0.3) \text{ Gy}]^{-1} \tag{24.2}$$

depending on whether we measure distance in distance or time units (either way, v is unitless). The last quantity is actually most useful, because we often express cosmological distances and times in light years and years.

The universe is therefore clearly expanding with time. If all galaxies are moving away from us, doesn't this mean that we are at the center of the universe? No. The linearity of the Hubble relationship means the expansion looks the same from any galaxy. Qualitatively, if you take a picture of the universe and superpose on it a picture that is uniformly enlarged by, say, 25%, you can see that if we align the two pictures so that any given galaxy is motionless, *all* the other galaxies will appear to be moving radially away from it. See box 24.5 for more discussion of this issue.

If we assume that the amount of matter in the universe is conserved, the fact that the universe is expanding means that it was more dense in the past than it is now. Indeed,

---

1. See http://lambda.gsfc.nasa.gov/product/map/dr2/params/lcdm_all.cfm.

Indeed, if any given galaxy has *always* moved with its current speed as given by the Hubble relation $v \approx H_0 d$, then at a time $t_0 = H_0^{-1} = 13.9$ Gy ago, the positions of the galaxies would coincide and the universe would be infinitely dense. We will see in a few chapters that the Hubble "constant" H_0 is actually *not* constant in time, but the conclusion still holds: at $t_0 \approx 13.7$ Gy ago, the universe's density was essentially infinite and the universe has been expanding since. We call t_0 the time of the **Big Bang**.

The Universe Is Isotropic. Indeed, in 1965, Arno Penzias and Robert Wilson (while calibrating a microwave antenna for another purpose) accidently discovered that the universe sends us microwave photons uniformly from all directions (for which they won a Nobel Prize in 1978). We now know that every point in the sky emits these **cosmic microwave background** (CMB) photons as if the entire sky were a black body with a temperature of 2.725($\pm$0.001) K.

This background was a predicted consequence of the Big Bang. As we will see, as the universe expands, it also cools. Until about 380,000 y after the Big Bang, the universe was so dense and hot that protons and electrons could not stably combine to form hydrogen atoms, and (because free charged particles strongly interact with light) the universe was opaque. At about 380,000 y after the Big Bang (which we call the time of **recombination**), the universe became cool enough to form neutral (and thus transparent) hydrogen atoms. Photons emitted by the incandescent hydrogen gas (whose temperature was then about 3000 K) thus became free to travel large distances. The CMB photons that we now see were radiated by parts of the universe that at the time of recombination were about 43 Mly away from the glob of gas that eventually became us. These photons (now highly red-shifted, so that they appear to be radiated by a black body at 2.7 K instead of 3000 K) are just now reaching us.

The CMB is slightly bluer in one direction and redder in the opposite direction. If we interpret this as being a Doppler shift due to our solar system and galaxy moving at 369 km/s relative to the average rest frame of the gas emitting the CMB and subtract the effect of such a motion, we find the CMB to be isotropic (independent of direction) to better than 1 part in 100,000. Since these photons were produced by parts of the universe up to 86 Mly apart even at the time of recombination (parts that are, of course, much more distant at present), we see that at the time of recombination (at least), the density of the universe was extraordinarily isotropic.

The Universe Is Homogeneous. An isotropic universe is not necessarily homogeneous (independent of *position* at a given time): we *could* be near the center of a spherically symmetric expanding universe whose density varies with radius. But this seems very unlikely. To observe the extraordinary isotropy in the CMB, we would have to be quite improbably close to the center (see problem P24.6).

The best independent evidence of the universe's homogeneity comes from recent large-scale surveys of the distribution of galaxies. Figure 24.4 shows a plot of the data from the 2dF survey, displaying redshift data for more than 220,000 galaxies. The survey does show that on small scales (less than about 0.3 Gly) the distribution of galaxies is *not* smooth: we see filaments, voids, walls, and other complex features. It also *seems* that the density of galaxies falls off with radius, but this is a selection problem: as distance increases, fewer and fewer galaxies are bright enough to yield reliable redshifts. Out to about 1.3 Gly (ignoring small-scale variations), where we can observe redshifts for most of the galaxies in the surveyed region, the distribution of galaxies looks pretty homogeneous. There is no evidence that our location in the universe is special.

List of Ingredients. According to our current understanding (combining data from a variety of sources as of 2010: see http://lambda.gsfc.nasa.gov/product/map/current/best_params.cfm), a list of ingredients for the universe looks like this[2]:

---

2. This list assumes a model of the universe that is partly based on general relativity, so it is not a pure list of observational "facts." Nonetheless, it expresses our current best knowledge. Percentages are of the total average energy density, $9.32(\pm 0.20) \times 10^{-27}$ kg/m^3.

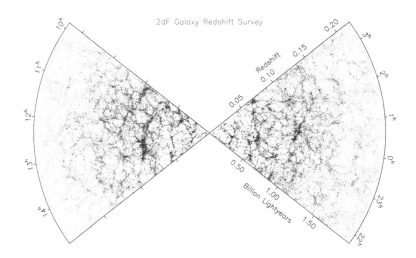

FIG. 24.4 This plot shows the density of galaxies versus distance for a relatively thin slice of the universe (about 10° thick perpendicular to the plot). Credit: The 2dF Galaxy Redshift Survey team (www2.aao.gov.au/2dFGRS), courtesy of Matthew Colless.

Ordinary matter (protons, neutrons, electrons, etc.)	4.56($\pm$0.16)%
Radiation (photons and neutrinos)	$\approx$ 0.0084%
Non-baryonic dark matter	22.7($\pm$1.4)%
Dark energy	72.8 ($\pm$1.6)%

Matter. The visible matter in the universe (i.e., stars and other objects that emit detectable electromagnetic radiation) is mostly concentrated in galaxies. However, the total amount of matter in stars comprises only about 0.5% of the mass of the universe. There is more matter in the form of gas clouds, dust, stars too small to shine, neutron stars, and black holes. The average density of all the ordinary matter in the universe is about 4×10^{-28} kg/m^3, or about one proton per four cubic meters.

Radiation. The photon gas associated with the CMB is not very dense, and the universal photon gas created by starlight is far less dense still. Radiation's total contribution to the universe's stress-energy is thus negligible.

Dark Matter. Since the 1930s, it has been suspected that there is more matter in the universe than we can see, and this has been confirmed in a number of ways in recent decades (box 24.6 discusses several of these methods). However, as we will see in a later chapter, the physics of the early universe puts a sharp upper limit on how much of this matter can be ordinary "baryonic" matter (protons and neutrons), and the amount of dark matter we observe through its gravitational effects is several times larger. We know virtually nothing about what the remainder of the dark matter consists of, though the currently favored proposal is that it consists of WIMPs (Weakly Interacting Massive Particles) that are part of some generalizations of the Standard Model of particle physics. Models of the universe best fit the observed universe (particularly the observed tiny fluctuations in the CMB) if these WIMPs are "cold" (non-relativistic) and comprise about 22% of the universe's energy density. Experiments that might directly observe these WIMPs are currently underway.

Dark Energy. In early 1998, two research teams observing distant Type Ia supernovae presented strong evidence that the expansion of the universe was accelerating. This result has since been confirmed by detailed observations of the CMB, gravitational lensing, and more supernovae, and also resolves a problem where the universe appeared to be younger than its oldest stars.

As we saw in chapter 22, the gravitoelectric field is attractive (and thus should act to *slow down* the universal expansion) if $\rho_0 + 3p_0$ is positive. Vacuum energy is the only known thing for which $\rho_0 + 3p_0$ is negative. Indeed, the accelerated expansion of the universe appears to be entirely consistent with a vacuum energy whose stress-energy is $T_{\text{vac}}^{\mu\nu} = -\Lambda g^{\mu\nu}/8\pi G$ and whose effective energy density $T_{\text{vac}}^{tt} = \Lambda/8\pi G$ comprises about 73% of the universe's density.

BOX 24.1 Measuring Astronomical Distances in the Solar System

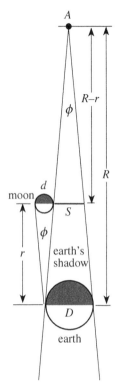

FIG. 24.5 This figure illustrates Aristarchus's method for estimating the distance to the moon using lunar eclipse data. (This diagram is obviously not to scale.)

Measuring the distance to any celestial object using only measurements that can be made on the surface of the earth is extremely difficult. The only practical way to do this is to use triangulation in some form, but the distances to even the closest celestial objects are so large that this presents severe challenges.

The ancient Greeks were able to measure only two astronomical distances to any accuracy: the distance to the moon and the earth's radius. Aristarchus of Samos (c. 310 to 230 BCE) in a work called *On the Sizes and Distances* was the first to publish estimates of the distances to the moon and sun based on measurements. Aristarchus's method for estimating the distance to the moon was based on his observations of the angular diameter of the earth's shadow cast on the moon during a lunar eclipse. The argument (in modern language) goes like this (see figure 24.5): The ancients knew from solar eclipses that the sun and moon had the same angular size when viewed from the earth. Therefore (if the sun is far enough away that it does not matter much whether we view the sun from the earth or from point A behind the earth), the angles marked ϕ on the diagram are the same. By observing a lunar eclipse, one can measure the apparent diameter S of the earth's shadow at the moon's position as a multiple of the moon's diameter d. Then, by similar triangles, $S/d = (R - r)/r = R/r - 1$. The ancients had also measured the actual angular diameter of the moon $\phi \approx d/r$. Finally, it is also clear from the diagram that $\phi \approx D/R$. Using these results, you can eliminate R in the first expression and find the moon's distance in earth diameters (r/D) in terms of known quantities.

Exercise 24.1.1. Find r/D in terms of the known quantities S/d and ϕ. The best ancient values of S/d and ϕ were about 2.5 and 0.5°, respectively. Estimate the distance to the moon in earth diameters: you should get about 33. (The modern result for this ratio is 30.16.)

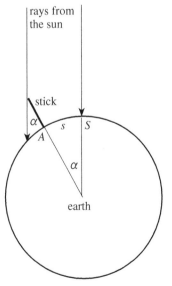

FIG. 24.6 This diagram (also obviously not to scale) illustrates Eratosthenes' method of measuring the earth's circumference).

Aristarchus's values for S/d and ϕ were roughly 2 and 2° respectively, which led to an erroneous estimate of $r/D \approx 10$. The latter estimate is particularly surprising, since Aristarchus was elsewhere quoted as knowing that the moon's angular diameter was 0.5°. In any case, other ancient astronomers quickly revised his result while using the same method. Ptolemy's estimate for r/D in early 2nd century CE was 29.5, very close to the modern value. In spite of his poor estimates, Aristarchus was a visionary astronomer who was the first (not Copernicus!) to propose a heliocentric model of the solar system, possibly partly because the logic of his model makes it clear that the sun is much larger than the earth.

To determine the absolute distance to the moon, one needs to know the diameter of the earth. Eratosthenes of Cyrene (276–194 BCE) provided the first reasonable estimate c. 240 BCE. According to the story, he was told that in the town of Syene, the sun at noon on the summer solstice could be seen reflected in a deep well, meaning that it was directly overhead. He used the shadow of a vertical stick to measure the angle of the sun in Alexandria on the same date to be ≈ 1/50 of a circle (a bit more than 7°) away from vertical. He assumed that Syene was due south of Alexandria (it isn't quite), that the distance between Syene and Alexandria was

BOX 24.1 (continued) Measuring Astronomical Distances in the Solar System

$s \approx 5000$ *stadia*, the earth was spherical, and that the sun was far enough away that light rays from the sun would be essentially parallel at the two locations (a *big* assumption at the time). As shown in figure 24.6, these assumptions imply that s is 1/50th of the earth's circumference, meaning that the latter was 250,000 *stadia*. How good this result was depends on which of the various Greek *stadia* Eratosthenes was talking about, but the best guess was that his *stadion* was about 185 m, leading to a circumference of 46,000 km, about 15% larger than the currently accepted result of 40,000 km, which is not bad considering how crude his estimates were and how difficult it was to actually measure the distance between Syene and Alexandria.

The Greeks also tried but failed to measure the distance to the sun accurately. Aristarchus was again the first to publish a measurement-based estimate: figure 24.7 shows his method. When the moon is exactly at first quarter, the angle between the earth, moon, and light from the sun is 90°. If one can measure the angle marked θ on the figure, one can calculate the distance to the sun by trigonometry. Aristarchus estimated the angle to be 87°, and thus the distance to the sun to be about 20 times the distance between the earth and moon.

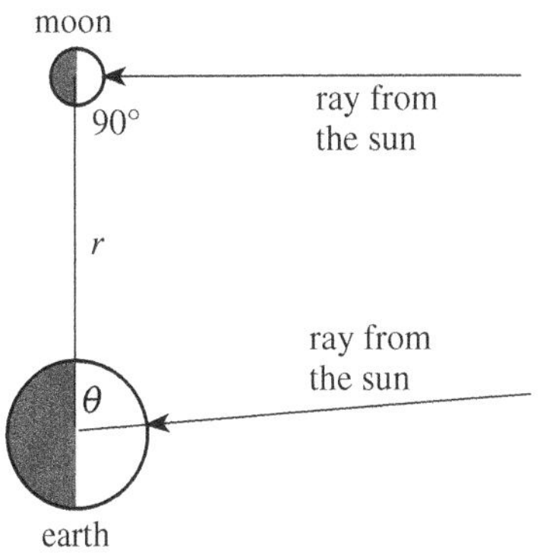

FIG. 24.7 This diagram illustrates Aristarchus's method of measuring the distance to the sun.

Exercise 24.1.2. Verify this result.

It is actually very difficult to use this method to measure the distance to the sun, partly because it is difficult to determine when the moon is exactly at first quarter, and partly because θ is so close to 90° that small errors in θ lead to huge errors in the distance. In any case, Aristarchus's estimate was *way* off: the actual angle is 89° 50′, and the distance to the sun is more like 390 times the moon's orbital radius. However, in this case, no one could do much better than Aristarchus, and his estimate was accepted until the middle 1600s.

Edmond Halley had proposed in 1716 that one might accurately measure the distance to the sun by observing a transit of Venus (i.e., Venus crossing the face of the sun) from different locations on the surface of the earth. Figure 24.8 illustrates the logic of this method. Observers at different points on the earth will see Venus and the sun from slightly different angles and thus will measure different times for the transit. By accurately recording the transit times, one can calculate the distance to Venus very accurately in terms of the distance between viewers, and since the ratio of the earth's orbital radius to Venus's orbital radius is easy to establish by measuring the maximum angle that Venus makes with respect to the sun, one can then calculate the distance to the sun. The next transits were predicted for 1761 and 1769.

In both years, expeditions were mounted to perform the measurement from widely separated (north-south) positions on the earth. The resulting estimate of the distance was 153 million kilometers, which differs from the current accepted value by less than 2%. This result was refined during observations of later transits.

In 1959, a detectable radar pulse was bounced off of Venus for the first time, allowing its distance to be calculated by measuring the pulse's light travel time. Modern methods based on measuring light travel time have reduced the uncertainty in the earth-sun distance to mere tens of meters.

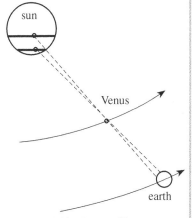

FIG. 24.8 This diagram illustrates how to determine the distance to Venus by observing a transit of Venus across the sun's face. A northern observer on the earth will see Venus cross the sun's disk a bit lower than a southern one. The tracks' angular separation is actually comparable to Venus's angular diameter when viewed from the earth, and so too small to measure accurately. But Halley realized that measuring the transit's *duration* allows one to accurately calculate the tracks' separation (and thus the distance).

BOX 24.2 Determining the Distance to Stellar Clusters

In 1910, Ejnar Hertzsprung and Henry Russell produced the first version of what astrophysicists now call a **Hertzsprung-Russell** (or **H-R**) **diagram**. Such a diagram plots the absolute luminosity of the stars in a given population (expressed as how bright the stars *would* be at a distance of 10 pc = 32.6 ly) versus their **B-V color index** (which expresses star's relative brightness when viewed through blue and yellow filters, a measure of their surface temperature). Figure 24.9 shows such a diagram for 20,853 stars mapped by the Hipparcos satellite. Such H-R diagrams not only were useful for unraveling the secrets of stellar evolution, but they provided the first means for estimating the distances to distant clusters of stars.

To do this, astronomers first prepared an H-R diagram for all stars whose distances could be determined by parallax. For a given cluster (whose stars all have essentially the same but unknown distance), one would then plot a similar diagram, except with *observed* brightness plotted against the color index. Since a star's color index does not depend on its brightness, if we assume that stars are alike everywhere, the graphs should look the same except for being offset vertically by an amount that depends on how far away the cluster is compared to the 10-pc standard distance. For example, if a star with the same color index as the sun and occupying the same position on the diagram is 100 times less bright than the sun would be at 10 pc, the cluster must have a distance of 10×10 pc = 100 pc = 326 ly.

Large distant clusters would often contain Cepheid variables, so this method of determining cluster distances helped astronomers calibrate the Cepheid period-luminosity relationship. However, this calibration is very challenging to do accurately. In the early 1900s, less than 100 stars had well-measured parallax distances, so the base diagram for all comparisons was pretty fuzzy. Moreover, clusters contain stars that all formed at the same time, which is not true of all the nearest stars, undermining the assumption that the two populations are similar. (However, comparing H-R diagrams for clusters of different ages did prove very important for helping astronomers understand the processes of stellar evolution.) This is why using the Hipparcos data to calibrate the Cepheid period-luminosity relation *directly* was so valuable (and why it led to a 10% readjustment of that relation).

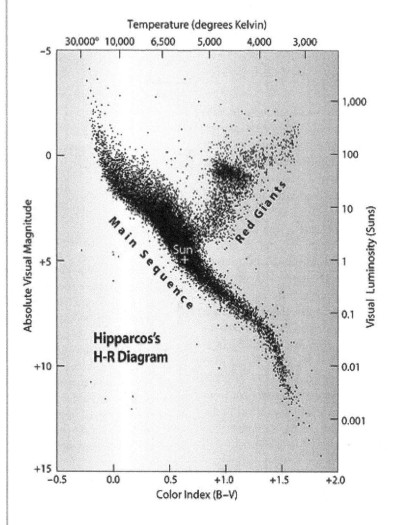

FIG. 24.9 This diagram shows an H-R diagram for the 20,853 stars in the Hipparcos catalog whose distances and color indexes were both reasonably well known. This omits a *huge* number of dim, low-temperature stars, whose colors and parallaxes are hard to measure accurately. A complete diagram showing *all* nearby stars would have many more stars on the lower right of the main sequence. (Note, though, that one would not see such stars in a distant cluster either.) Courtesy of Michael Perryman.

BOX 24.3 How the Doppler Shift Is Connected to Radial Speed

Consider a light-emitting object that moves directly away from an observer with speed v. *We can use the spacetime diagram shown in figure* 24.10 *to calculate the Doppler shift seen by the observer.* Let the emission of two sequential crests of a light wave be events E and F along the object's worldline. These crests travel to the observer along the dashed light worldlines shown and arrive at the observer at events G and H, respectively. The time between emission events in the *object's* frame is the proper time $d\tau_E$; the time between those events in the *observer's* frame is dt_E. The time between reception events in the observer's frame is dt_R (we consider all these intervals to be infinitesimal).

If the object is moving *away* from the observer, then the observer's time dt_R between reception events is longer than the object's time $d\tau_E$ between emission events for two reasons: (1) $dt_E > d\tau_E$ because the proper time between two events is always shorter than the coordinate time, and (2) $dt_R > dt_E$ because the light from event F has to go dx_E farther than that from event E to get back to the observer. Specifically,

$$d\tau_E^2 = dt_E^2 - dx_E^2 \quad \text{and} \quad dt_R = dt_E + dx_E \qquad (24.3)$$

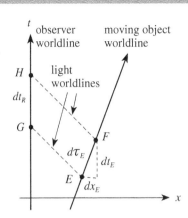

FIG. 24.10 A spacetime diagram to help us calculate the relativistic Doppler shift of light waves emitted by a moving object.

(in GR units). The object's radial speed is $v \equiv dx_E/dt_E$: you can use this to eliminate dx_E in the equations above. Finally, note that the wavelength λ of a light wave is equal to its period T in GR units: $\lambda = cT = T$. You can use these bits of information to show that the ratio between the received and emitted wavelengths is therefore

$$\frac{\lambda_R}{\lambda_E} = \sqrt{\frac{1+v}{1-v}} \qquad (24.4)$$

Exercise 24.3.1. Verify that this is correct.

The fractional *change* in the wavelength is therefore

$$\frac{\Delta\lambda}{\lambda_E} \equiv \frac{\lambda_R - \lambda_E}{\lambda_E} = \sqrt{\frac{1+v}{1-v}} - 1 \approx v \quad \text{if } v \ll 1 \qquad (24.5)$$

Exercise 24.3.2. Verify the last step using the binomial approximation.

The approximation $\Delta\lambda/\lambda_E \approx v$ is sound until v becomes significant compared to 1; beyond that, we should use the full expression $\Delta\lambda/\lambda_E = (\sqrt{1+v}/\sqrt{1-v}) - 1$.

BOX 24.4 Values of the Hubble Constant

Exercise 24.4.1. Using $H_0 = 70.4(\pm 1.5)$ km/s per Mpc and 1 Mpc = 3.26 Mly, verify the values and uncertainties given in equation 24.2. Remember that all the values must have the same *fractional* uncertainty.

BOX 24.5 Every Point Is the Expansion's "Center"

We can prove mathematically that a linear expansion looks the same from various viewpoints (at least in flat space) as follows. Consider two observers A and B that both look at a third galaxy G, as shown in figure 24.11. Observer A sees both B and G receding as specified by the Hubble law, here expressed in vector form:

$$\vec{v}_B = H_0 \vec{r}_B \quad \text{and} \quad \vec{v}_G = H_0 \vec{r}_G \tag{24.6}$$

If we subtract the first expression from the second, we get

$$\vec{v}_G - \vec{v}_B = H_0(\vec{r}_G - \vec{r}_B) \tag{24.7}$$

But as figure 24.11 shows, $\vec{r}_G - \vec{r}_B$ is simply the displacement between observer B and galaxy G, and $\vec{v}_G - \vec{v}_B$ (since it is parallel to $\vec{r}_G - \vec{r}_B$) is the radial velocity of G relative to B. Therefore, we will see that observer B will see (the arbitrary) galaxy G receding from it as specified by the Hubble law. The expansion of the universe thus looks the same to both observers A and B.

Figure 24.12 illustrates this more intuitively. The array of black dots (representing galaxies now) is the same as the array of white dots (representing the same set of galaxies some time ago) except that the latter is 20% smaller. No matter which white dot we take to be our origin, all of the other dots (galaxies) appear to be moving radially away from it as the universe expands (with more distant galaxies moving more rapidly). Again, the expansion looks the same from every point.

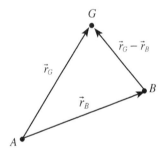

FIG. 24.11 The vector positions of observer B and galaxy G relative to observer A, and the position of G relative to B.

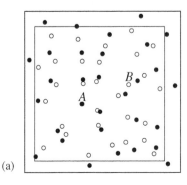
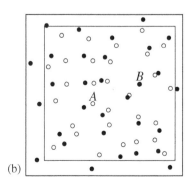

FIG. 24.12 Diagram (a) shows the expansion as viewed by an observer in galaxy A, who considers herself to be at rest; diagram (b) shows the same as viewed by the observer in galaxy B, who considers himself to be at rest. In each case, the observer sees all *other* galaxies to be moving away with speeds proportional to distance.

BOX 24.6 The Evidence for Dark Matter

The most vivid evidence for dark matter comes from galaxy rotation curves. Since 1970, astrophysicists have been able to measure the Doppler shifts of both stellar light and the 21-cm radio emissions from hydrogen gas with enough accuracy to measure the rotational speed of spiral galaxies. Figure 24.13 shows data from measurements of the spiral galaxy NGC 2403. We also can also calculate what we expect this rotational speed *should* be using Newtonian physics (since the fields are weak here, this should be fine). If the galaxy were spherically symmetric, we would know from Gauss's law that the mass $M(r)$ within a given radius r attracts as if it were located at the center and the mass outside has no effect. Therefore, the orbital speed for a star of mass m in a circular orbit at a given radius r is given by

$$\frac{GM(r)m}{r^2} = ma = m\frac{v^2}{r} \quad \Rightarrow \quad v = \sqrt{\frac{GM(r)}{r}} \qquad (24.8)$$

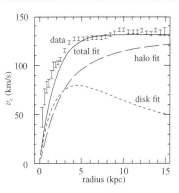

FIG. 24.13 This diagram shows that a large halo of dark matter is required to match the rotation curve of the spiral galaxy NGC 2403. Credit: Chris Mihos (adapted).

Now, a spiral galaxy is *not* spherically symmetric, so the actual prediction will be somewhat more complicated, but it can easily be done numerically in a way that actually fits the distribution of stars in the galaxy disk. The "disk fit" curve in figure 24.13 represents such a prediction. You can see that for radii beyond about 5 kpc (16 kly), the velocity curve begins to fall off as something like $r^{-1/2}$, because virtually all of the galaxy's visible mass is within this radius. Of course, there may be mass in the galaxy in the form of dust, gas, dim stars, giant planets, etc. that we can't see, but if it is distributed in the same way as the stars are, the curve's *shape* will not change, only its amplitude. (The "disk fit" curve shown actually assumes that there is twice as much matter in the spiral disk as can be seen.)

However, you can see that this *nowhere near* fits the actual data for NGC 2403. The stars in the disk of this galaxy orbit much more rapidly than expected. Moreover, no adjustment of the "disk fit" curve *amplitude* will make it fit, because the velocity data are actually flat for large radii, not decreasing as $r^{-1/2}$. The only way to fit these data is to assume that in addition to the visible matter in the disk (and any unseen matter distributed in the same way) the galaxy is surrounded by a huge halo of dark matter. It is hard to estimate the total amount of matter in the halo (because the amount continues to increase beyond radii for which we can see anything), but the total galactic mass implied by these data is at least 10 times as much as that in the form of stars. (See problem P24.7 for more discussion of these data.)

The statistical analysis of velocities of galaxies in clusters can also be used to estimate a cluster's total gravitational mass. Specifically, the **virial theorem** of mechanics (see problem P24.8) implies that $\langle v^2 \rangle = \frac{1}{2} GM \langle 1/r \rangle$, where $\langle v^2 \rangle$ is the averaged squared speed of galaxies in the cluster, M is the cluster's total mass, and $\langle 1/r \rangle$ is the average inverse galactic separation. Since $\langle v^2 \rangle$ and $\langle 1/r \rangle$ can be estimated from observations, we can compute the cluster's mass M. The result is typically roughly 50 to 80 times the mass one would estimate from the stars in the galaxies.

Other evidence comes from gravitational lensing: the degree of lensing observed in photographs such as figure 13.5 implies masses for the foreground galaxy cluster much larger than associated with the visible stars. The evidence that puts the tightest constraints on the total dark-matter density at present comes from examining the tiny fluctuations in the cosmic microwave background. We will discuss this in more detail in a later chapter on the early universe (chapter 28).

Collectively, these observations imply that the total density of cold (nonrelativistic) dark matter comprises about 27% of the total energy density of the universe. However, considerations regarding nucleosynthesis in the early universe

BOX 24.6 (continued) The Evidence for Dark Matter

(which we will also discuss in Chapter 28) strictly limit the amount that can be ordinary matter (i.e. the quarks and leptons of the Standard Model of particle physics) to about 4.6% of the total energy density. The remainder must be weakly-interacting (i.e., interacting with ordinary matter only through gravity) massive (so that they are non-relativistic) particles, or WIMPs. We know virtually nothing else about the physical nature of these particles at present.

HOMEWORK PROBLEMS

P24.1 Near noon one day in late March, you are observing the sun from a solar observatory in Lewiston, Maine (44.1° north latitude) and observe a gigantic alien mothership crossing the face of the sun. (Because the spaceship was very black, no one had spotted it before.) You immediately call a friend at the La Silla observatory in Chile (29.25° south latitude), and the two of you are able to record the ship's position relative to the sun's face simultaneously. You discover that your friend sees the ship to be displaced about 3.2 (±0.1)% of the sun's apparent diameter away from where you see it relative to the sun. How far is the alien ship from earth? (*Hints*: Lewiston and La Silla have nearly the same longitude. The spring equinox is in late March, so at noon local time, the sun will be very nearly directly overhead at 0° north latitude. Draw careful pictures on the scale of the earth and on the scale of the earth's orbit, and do *not* assume that the sun is infinitely far away. Work algebraically to determine the ship's distance r in terms of the sun's angular diameter $\alpha = 0.525°$ when viewed from earth, the distance to the sun $R = 1.50 \times 10^8$ km, and an appropriate projection r_o of the separation between the observatories. The distance between the earth's center and each observatory is about $r_e = 6370$ km. You may make approximations that yield errors smaller than the basic measurement's uncertainty of about 3%).

P24.2 In this problem, we will explore the practical process of determining the distance to a Cepheid star.

a. Astronomers express the relative brightness of stars using a system of **magnitudes**. The difference between two stars' **observed magnitudes** m_1 and m_2 is related to the ratio of their observed fluxes f_1 and f_2 as follows:

$$m_1 - m_2 \equiv -\frac{5}{2}\log\left(\frac{f_1}{f_2}\right) \quad (24.9)$$

where "log" represents a base-10 logarithm. A star's **absolute magnitude** M is the observed magnitude that it *would* have if it were 10 pc from earth. Show that the relation between a star's observed magnitude m and its absolute magnitude M is

$$m - M = -5 + 5\log\left(\frac{d}{1\text{ pc}}\right) \quad (24.10)$$

where d is the star's distance from the earth.

b. Also argue that the difference between a star's absolute magnitudes M and the sun's absolute magnitude $M_\odot$ is related to the star's luminosity L as follows:

$$M - M_\odot = -\frac{5}{2}\log\left(\frac{L}{L_\odot}\right) \quad (24.11)$$

where $L_\odot$ is the luminosity of the sun.

c. A recent article (Turner, *Astrophysics and Space Science*, **326**, 2009, pp. 219–231) claims the following up-to-date formula for the relationship between a Cepheid's luminosity L and its period P after analyzing the latest parallax information from Hipparcos and the Hubble Space Telescope:

$$\log\left(\frac{L}{L_\odot}\right) = 2.415(\pm 0.035) + 1.148(\pm 0.044)\log\left(\frac{P}{1\text{ day}}\right) \quad (24.12)$$

Let's consider a specific Cepheid mentioned in the article: ζ Gem. This star has an observed magnitude of 3.80 and a period of 10.151 days. The sun's absolute magnitude is 4.83. From this information, determine the distance to ζ Gem in parsecs and estimate the uncertainty in that distance.

d. The measured parallax of ζ Gem is 2.79 ± 0.81 mas (milliarcseconds). Calculate its distance from its parallax. Is the result consistent with your result for part c?

P24.3 Olbers's paradox is a famous statement of a logical problem with an infinite, eternal, and static model of the universe (the default model before the 20th century). The argument goes like this: If you look in any direction in an infinite universe, your line of sight will eventually run into some star's surface. Therefore the night sky should be as bright as the face of the sun! As this is clearly not the case, the model must be wrong. Though the paradox is linked to H. W. Olbers (1758–1840), astronomers were aware of the issue back into the early 1600s. The paradox remained unresolved before the 20th century.

a. The most important reason that the sky is not as bright as the face of the sun is that the universe is not eternal but has a finite age of about 13.7 Gy. To see what a difference this makes, imagine that the universe is completely Newtonian and static except that it was born 13.7 Gy ago. Imagine that essentially immediately after its birth, the universe was populated by stars with the currently observed luminosity density of about $0.2 \times 10^9 \ L_\odot/(\text{Mpc})^3$ (in other words, about 200 million stars like the sun in every cubic megaparsec) and has remained unchanged since. Determine the total flux (in W/m²) of starlight arriving at the earth from all the stars in the universe incident on a flat surface facing the zenith, and compare to the solar flux that such a surface would receive from the sun at the zenith (≈ 1000 W/m²). (*Hints:* Out to what distance would one be able to see stars at the current time in this model? Also note that the flux that a surface facing the zenith will receive from a star at an angle of θ from the zenith will be $\cos\theta$ times the flux that a surface directly facing the star would receive. Integrate over all angles, assuming that the universe is isotropic. You should find that the total flux that a surface facing the zenith should receive from the half of the sky visible to it will be half of the flux it would receive if it could directly face each star simultaneously.)

b. How much older would the universe have to be before the flux of starlight on such a surface would be comparable to the sun at zenith?

P24.4 Consider a distant galaxy whose light has an observed fractional redshift $z \equiv (\lambda_R - \lambda_E)/\lambda_E$ of 0.30.

a. Use the exact relativistic Doppler-shift formula to calculate the galaxy's corresponding radial speed v. Is the approximation that $z \approx v$ very accurate?

b. Assuming that the relationship $v \approx H_0 d$ still holds for such large speeds, what is the galaxy's approximate distance from the earth according to this law?

P24.5 Could observed galactic redshifts be entirely or even significantly due to gravitational redshifts? Consider the light from a large, dense, and solitary elliptical galaxy with a mass of 10^{12} solar masses and a radius of 50,000 ly. Assume that the spacetime surrounding the galaxy is approximately Schwarzschild spacetime. What is the approximate fractional gravitational redshift $z \equiv (\lambda_R - \lambda_E)/\lambda_E$ of light emitted by stars on the surface of such a galaxy? Considering that galaxies typically have random velocities roughly 100 km/s on top of their cosmological expansion velocities, is the gravitational redshift likely to be significant?

P24.6 Imagine that the universe is *not* homogeneous, but rather spherically symmetric (i.e., its density depends on radius) and expanding according to the Hubble-like law $v = H_0 r$, where r is now the distance of a given galaxy *from the universe's center* (not from us). The hot gas that radiates the cosmic microwave background would in this model be a shell at the universe's outer edge that is moving away from the center at nearly the speed of light. This gas must be moving near the speed of light because hydrogen atoms will form (and thus become transparent) at a temperature of about 3000 K. Since the peak wavelength of light emitted by a black body is inversely proportional to the temperature, for an observer at the center to see this as radiation from a black body at 2.7 K, we must have

$$\frac{\lambda_R}{\lambda_E} = \sqrt{\frac{1+v}{1-v}} \approx \frac{3000 \text{ K}}{2.7 \text{ K}} = 1100 \quad (24.13)$$

a. Define $u \equiv \lambda_R/\lambda_E = \sqrt{1+v}/\sqrt{1-v}$. Show in fact that for large u ($1/u^2 \ll 1$), we have

$$v \approx 1 - \frac{2}{u^2} \quad (24.14)$$

The gas speed in our case will therefore differ from 1 by less than two parts in a million. (*Hint:* Solve for $v(u)$, and use the binomial approximation.)

b. If we are offset from the center by a distance dr, we will have a velocity dv toward the near part of the gas and away from the far part. As noted in the text, we do in fact seem to be moving relative to the gas at a speed of 369 km/s. Taking our offset as being the explanation of this relative speed, how far are we from the universe's center?

c. Assuming that all positions throughout the universe are equally likely, estimate the probability that our speed relative to CMB would be this small or smaller.

P24.7 On the Internet at http://burro.astr.cwru.edu/JavaLab/RotcurveWeb/main.html, you will find a Java applet that lets you try to fit an actually observed galactic rotation curve with simple models of the galactic disk and dark matter halo. Read the "Background" and "Applet Physics" sections, and then run the applet. We will analyze the spiral galaxy NGC 2403.

a. Using the measured rotation curve and equation 24.8 estimate the mass (in solar masses) enclosed by the most distant point on the rotation curve.

b. By adjusting the free parameters, fit the NGC 2403 curve with disk and halo models that minimize the chi-squared error estimate but keep the disk's mass-to-luminosity ratio below 2 (larger results are not really credible). Note that you can click on the scroll-bar ends for fine adjustments. Take a screen shot of your best-fit graph and print it out.

c. Again applying equation 24.8 to the most distant data point, estimate the ratio of dark matter to visible matter for NGC 2403.

P24.8 One can prove the virial theorem of Newtonian mechanics as follows. Consider a cluster of N galaxies, which for the sake of simplicity we will assume to have the same mass m. We will assume that the basic physical properties of this cluster are independent of time; i.e., the overall size of the cluster and its galaxies' average velocities will be independent of time.

a. Consider the **virial** for this system, defined to be

$$Q \equiv \sum_i \vec{p}_i \cdot \vec{r}_i \qquad (24.15)$$

For a cluster that is fixed in size, we'd expect the value of Q to be zero, since at any given radius we would expect to see as many galaxies moving inward as moving outward. Therefore, let's assume that $dQ/dt \approx 0$ as well for this system. Show that this assumption implies that

$$0 \approx \sum_i m v_i^2 + \sum_i \vec{F}_i \cdot \vec{r}_i \qquad (24.16)$$

where $\vec{F}_i$ is the total force acting on the ith galaxy.

b. The gravitational forces between galaxies obey Newton's third law. Show that this means that

$$\sum_i \vec{F}_i \cdot \vec{r}_i = \sum_i \sum_j \vec{F}_{ij} \cdot \vec{r}_i = \tfrac{1}{2} \sum_i \sum_j \vec{F}_{ij} \cdot \vec{r}_{ij} \qquad (24.17)$$

where $\vec{F}_{ij}$ is the force on the ith galaxy due to the jth galaxy ($\vec{F}_{ij} \equiv 0$ if $i = j$), and $\vec{r}_{ij} \equiv \vec{r}_i - \vec{r}_j$ is the position of the ith galaxy relative to the jth galaxy. (*Hint:* Divide the double sum into two equal parts and rename indices on the second part.)

c. The gravitational force on the ith galaxy due to the jth points toward the jth galaxy, and thus points in the same direction as $-\vec{r}_{ij}$. Use this and the result of part b to show that

$$\sum_i \vec{F}_i \cdot \vec{r}_i = -Gm^2 \tfrac{1}{2} \sum_i \sum_{j \neq i} \frac{1}{r_{ij}} \qquad (24.18)$$

d. Now, note that the cluster average of v^2 is

$$\langle v^2 \rangle \equiv \frac{1}{N} \sum_i v_i^2 = \frac{1}{Nm} \sum_i m v_i^2 \qquad (24.19)$$

Use this and previous results to show that

$$\langle v^2 \rangle \approx \tfrac{1}{2} GM \left\langle \frac{1}{r} \right\rangle \qquad (24.20)$$

where $M = Nm$ is the cluster's total mass and the average $\langle 1/r \rangle$ is over all galaxy *pairs*. You may assume that $N \gg 1$. (Note that these are *cluster* averages, not *time* averages of cluster *totals* as you will often see in other treatments of the theorem.) Therefore, if one can measure $\langle v^2 \rangle$ and $\langle 1/r \rangle$ for a galactic cluster, then one can then solve for the cluster's total mass M.

To be more accurate, we would have to generalize the final result for unequal galactic masses and re-express it in terms of the measurable Doppler-shift velocities and angular separations. Even uncorrected, though, it gives the right order of magnitude. Fritz Zwicky first noted in the 1930s that this approach yielded a mass for the Coma galactic cluster that was about 100 times larger than one would expect from the luminous matter.

25. A METRIC FOR THE COSMOS

INTRODUCTION

FLAT SPACETIME
Review of Special Relativity
Four-Vectors
Index Notation

TENSORS
Arbitrary Coordinates
Tensor Equations
Maxwell's Equations
Geodesics

SCHWARZSCHILD BLACK HOLES
The Schwarzschild Metric
Particle Orbits
Precession of the Perihelion
Photon Orbits
Deflection of Light
Event Horizon
Alternative Coordinates
Black Hole Thermodynamics

THE CALCULUS OF CURVATURE
The Absolute Gradient
Geodesic Deviation
The Riemann Tensor

THE EINSTEIN EQUATION
The Stress-Energy Tensor
The Einstein Equation
Interpreting the Equation
The Schwarzschild Solution

COSMOLOGY
The Universe Observed
A Metric for the Cosmos
Evolution of the Universe
Cosmic Implications
The Early Universe
CMB Fluctuations & Inflation

GRAVITATIONAL WAVES
Gauge Freedom
Detecting Gravitational Waves
Gravitational Wave Energy
Generating Gravitational Waves
Gravitational Wave Astronomy

SPINNING BLACK HOLES
Gravitomagnetism
The Kerr Metric
Kerr Particle Orbits
Ergoregion and Horizon
Negative-Energy Orbits

this depends on this

Introduction. Our goal in this chapter is to develop a trial metric that expresses the symmetries that we seem to see in the universe, and then solve for undetermined aspects of the metric using the Einstein equation. This will ultimately allow us to determine "equations of motion" for the expansion of the universe.

The Basic Cosmological Principle. The fundamental assumption that underlies virtually all modern cosmologies, as discussed in the previous chapter, is the idea that the universe is *homogeneous* and *isotropic* (i.e., the universe looks the same at every point and in every direction). Observations seem consistent with this model on the very largest distance scales (billions of light years) in spite of the universe's complex structure at smaller distance scales. This essentially expresses the belief of the physics community that there should be nothing special about the earth's place in the universe (the radical antithesis of the geocentric hypothesis).

If the universe is truly and completely homogeneous and isotropic, the universe can have no boundary (since the universe would look very different to an observer at the boundary than it would to an observer well inside). If we imagine the universe to have only two spatial dimensions instead of three, we can more easily see that there are two distinctly different ways that we could construct an "edge-less" universe. On the one hand, the universe might be infinite in all directions, with galaxies peppered with constant density over the surface of what amounts to an infinite sheet of paper. On the other hand, the universe could be finite, but shaped like the surface of a globe, wrapping back on itself so that there are no edges. In either case, you can see that the universe would look the same to all observers.

We also know from observation that the universe is expanding with time and seems to have begun with a Big Bang at a certain specific moment in the past.

A Trial Metric for the Universe. In outline, our procedure in what follows is to (1) find a trial metric that manifests the symmetries implied by homogeneity and isotropy and allows for expansion and the Big Bang, (2) plug it into the Einstein equation, and (3) solve for any undetermined parts.

Imagine that we have a huge number of observers spread like buoys throughout the universe, each at rest with respect to the universal "gas" whose particles are randomly moving galaxies (or, equivalently, at rest with regard to the cosmic microwave background). Imagine that we start each observer's clock at the Big Bang and define the coordinate time t at any event in spacetime to be the clock-time registered by the observer located at that event. Since the universe is homogeneous, all of our observers should be completely equivalent, so the universe should look the same at every place and in every direction at a given instant of coordinate time.

Since the universe is isotropic around any origin we choose, the spatial metric at a given instant of coordinate time t should be spherically symmetric in space. This means that spheres of constant radius around the origin should have the metric of simple spherical surfaces: $ds^2 = r_c^2(d\theta^2 + \sin^2\theta\, d\phi^2)$, where r_c is the surface's circumferential radius. In the past, we have labeled these spheres using r_c, but in this case it is more convenient to choose our radial coordinate r to be the actual radial distance from the origin. With that choice, a spherically symmetric metric for three-dimensional space at a given instant of time becomes

$$ds^2 = dr^2 + [f(r,t)]^2(d\theta^2 + \sin^2\theta\, d\phi^2) \qquad (25.1)$$

where the function $f(r,t)$ expresses the as-yet-unknown (and possibly time-dependent) distinction between the radial coordinate r and the circumferential radius that arises if the space happens to be curved.

Now, if the universe is expanding homogeneously, then as time passes, the spatial separation ds between *any* pair of floating observers must scale as some position-independent factor $a(t)$. The most convenient way to express this is to assign each floating observer a *fixed* radial coordinate $\bar{r}$ (the bar signifying "steady") such that its distance from the origin at any coordinate time is $r \equiv a(t)\bar{r}$. Expressed in terms of $\bar{r}$, the spatial metric at any instant of coordinate time becomes

$$ds^2 = [a(t)]^2\{d\bar{r}^2 + [q(\bar{r})]^2(d\theta^2 + \sin^2\theta\, d\phi^2)\} \tag{25.2}$$

where $a(t)q(\bar{r}) = f(r,t)$. We describe the new $\bar{r}$ coordinate as **comoving** with our floating observers and thus with the expanding "gas" of galaxies.

Authors often make the comoving radial coordinate $\bar{r}$ unitless and assign distance units to $a(t)$. However, I think that this makes things unnecessarily abstract. Let's instead define the comoving radial coordinate $\bar{r}$ of an observer or galaxy to be its actual radial distance at *the present time*. The scale factor $a(t)$ then becomes unitless, expressing the universe's fractional size at time t compared to the present.

Finally, noting that $dt = d\tau = \sqrt{-ds^2}$ by design for an observer at rest in our comoving coordinates, a complete and plausible trial metric for the universe is

$$ds^2 = -dt^2 + [a(t)]^2 d\bar{r}^2 + [a(t)q(\bar{r})]^2 d\theta^2 + [a(t)q(\bar{r})]^2 \sin^2\theta\, d\phi^2 \tag{25.3}$$

A term of the form $2g_{\bar{r}t}\, dt\, d\bar{r}$ would be inconsistent with an isotropic universe, since such a term would give a different spacetime interval for a differential step inward than a differential step outward. This plausibility argument gives us some hope that the *diagonal* trial metric above might yield a meaningful solution.

Ricci Tensor Components for This Metric. You can use the Diagonal Metric Worksheet (see box 25.1) to show that for this metric

$$R_{tt} = -\frac{3\ddot{a}}{a} \quad \text{(where } \ddot{a} \equiv \frac{d^2 a}{dt^2}\text{)} \tag{25.4a}$$

$$R_{\bar{r}\bar{r}} = a\ddot{a} + 2\dot{a}^2 - \frac{2q''}{q} \quad \text{(where } \dot{a} \equiv \frac{da}{dt},\ q'' \equiv \frac{d^2 q}{d\bar{r}^2}\text{)} \tag{25.4b}$$

$$R_{\theta\theta} = 2q^2\dot{a}^2 + q^2 a\ddot{a} - (q')^2 - qq'' + 1 \quad \text{(where } q' \equiv \frac{dq}{d\bar{r}}\text{)} \tag{25.4c}$$

$$R_{\phi\phi} = \sin^2\theta\, R_{\theta\theta} \tag{25.4d}$$

and all off-diagonal Ricci components are zero. It is actually most convenient to express these Ricci components in the form where the first index is raised. You can show (see box 25.2) that these are

$$R^t{}_t = +\frac{3\ddot{a}}{a} \tag{25.5a}$$

$$R^{\bar{r}}{}_{\bar{r}} = \frac{\ddot{a}}{a} + 2\frac{\dot{a}^2}{a^2} - \frac{2q''}{qa^2} \tag{25.5b}$$

$$R^\theta{}_\theta = R^\phi{}_\phi = 2\frac{\dot{a}^2}{a^2} + \frac{\ddot{a}}{a} + \frac{1}{a^2 q^2}[1 - (q')^2 - qq''] \tag{25.5c}$$

The Stress-Energy Tensor for the Galactic "Gas." We will treat the galaxies in the universe as the randomly moving molecules of an ideal gas. The general equation for the stress-energy tensor (including the vacuum energy) is

$$T^{\mu\nu} = (\rho_0 + p_0) u^\mu u^\nu + p_0 g^{\mu\nu} - \frac{\Lambda}{8\pi G} g^{\mu\nu} \tag{25.6}$$

where ρ_0 is the energy density of the galactic "gas" in its rest frame and p_0 is its pressure at any given instant of time (see equations 20.16 and 21.21). Again, we will find it most convenient to express the stress-energy tensor in a form with the first index raised and the second one lowered:

$$T^\mu{}_\nu = (\rho_0 + p_0) u^\mu g_{\nu\alpha} u^\alpha + p_0 \delta^\mu{}_\nu - \frac{\Lambda}{8\pi G} \delta^\mu{}_\nu \tag{25.7}$$

The galactic "gas" as a whole is at rest in the comoving coordinates that we are using (by definition!), so $u^{\bar{r}} = u^\theta = u^\phi = 0$. Note also that

$$-1 = u^\mu g_{\mu\nu} u^\nu = g_{tt}(u^t)^2 = -(u^t)^2 \quad \Rightarrow \quad u^t = 1 \tag{25.8}$$

25. A METRIC FOR THE COSMOS

Thus (see box 25.3) the nonzero components of the stress energy tensor are

$$T^t{}_t = -\rho_0 - \frac{\Lambda}{8\pi G}, \qquad T^{\bar r}{}_{\bar r} = T^\theta{}_\theta = T^\phi{}_\phi = p_0 - \frac{\Lambda}{8\pi G} \qquad (25.9)$$

(Note how simple these components are!) The stress-energy scalar is thus

$$T \equiv T^\mu{}_\mu = T^t{}_t + T^{\bar r}{}_{\bar r} + T^\theta{}_\theta + T^\phi{}_\phi = -(\rho_0 - 3p_0) - \frac{4\Lambda}{8\pi G} \qquad (25.10)$$

Note that homogeneity requires that ρ_0 and p_0 are independent of $\bar r$, θ, and ϕ, and Λ is by definition a constant.

The Einstein Field Equation for the Universe. We saw in chapter 21 that we could write the Einstein equation in the form

$$R^{\mu\nu} = 8\pi G\left(T^{\mu\nu} - \tfrac{1}{2}g^{\mu\nu}T\right) \qquad (25.11)$$

(see equation 21.23 and note that I am absorbing the cosmological term into the stress-energy tensor). If we lower the second index of this equation, it becomes $R^\mu{}_\nu = 8\pi G(T^\mu{}_\nu - \tfrac{1}{2}\delta^\mu{}_\nu T)$. You can show (see box 25.4) that this equation has the following nonzero components:

$$R^t{}_t = -4\pi G(\rho_0 + 3p_0) + \Lambda \qquad (25.12a)$$

$$R^{\bar r}{}_{\bar r} = R^\theta{}_\theta = R^\phi{}_\phi = 4\pi G(\rho_0 - p_0) + \Lambda \qquad (25.12b)$$

These equations specify everything we need to know about the gravitational behavior of the universal galactic "gas."

Solving for q. Note that equation 25.12b requires that $R^{\bar r}{}_{\bar r} = R^\theta{}_\theta$. According to equations 25.5, this means that

$$0 = R^{\bar r}{}_{\bar r} - R^\theta{}_\theta = 2\frac{\cancel{\dot a^2}}{a^2} + \cancel{\frac{\ddot a}{a}} - \frac{2q''}{a^2 q} - 2\frac{\cancel{\dot a^2}}{a^2} - \cancel{\frac{\ddot a}{a}} - \frac{1}{a^2 q^2}[1 - (q')^2 - qq'']$$

$$0 = -\frac{2qq''}{a^2 q^2} - \frac{1}{a^2 q^2} + \frac{(q')^2}{a^2 q^2} + \frac{\cancel{qq''}}{\cancel{a^2 q^2}} \;\Rightarrow\; (q')^2 - qq'' = 1 \qquad (25.13)$$

This deceptively simple-looking differential equation is hard to solve for q (though it *is* possible with a simple trick: see problem P25.2). However, you can easily show (see box 25.5) that the following solutions do satisfy this equation:

$$q(\bar r) = \begin{Bmatrix} R\sin(\bar r/R) \\ \bar r \\ R\sinh(\bar r/R) \end{Bmatrix} \qquad (25.14)$$

where R is a constant. These solutions look different, but you can think of them as being a *single* solution with three different manifestations depending on the sign of a single constant of integration $K \equiv \pm 1/R^2$: note that $\sinh(K^{1/2}\bar r) = \sinh(i\bar r/R) = i\sin(\bar r/R)$ when $K < 0$ but $\sinh(K^{1/2}\bar r) = \sinh(\bar r/R)$ when $K > 0$, and both reduce to the center solution as $K \to 0$ ($R \to \infty$). These three solutions also exhaust the possibilities for solutions that go to zero as $\bar r \to 0$. Therefore, we see that the metric of a homogeneous, isotropic universe expressed in our coordinates must be

$$ds^2 = -dt^2 + a^2\left[d\bar r^2 + \begin{Bmatrix} R^2\sin^2(\bar r/R) \\ \bar r^2 \\ R^2\sinh^2(\bar r/R) \end{Bmatrix}(d\theta^2 + \sin^2\theta\, d\phi^2)\right] \qquad (25.15)$$

The Meaning of R. What do these three possibilities mean physically? In the center case, the spatial metric at any given instant of time is simply

$$ds^2 = a^2 d\bar r^2 + (a\bar r)^2(d\theta^2 + \sin^2\theta\, d\phi^2) = dr^2 + r^2(d\theta^2 + \sin^2\theta\, d\phi^2) \qquad (25.16)$$

since the actual distance from the origin r at any given instant of time is $r \equiv a\bar r$. This is the metric of *flat space* in spherical coordinates. Therefore, this solution corresponds to a universe whose spatial geometry is always flat.

The other two cases correspond to universes whose spatial geometry is curved. We can more easily visualize just *how* space is curved in each case by considering the geometry of the $\theta = \pi/2$ equatorial plane at an instant of coordinate time. In the first case, the metric of this plane is

$$ds^2 = a^2 d\bar{r}^2 + a^2 R^2 \sin^2(\bar{r}/R) d\phi^2 = (aR)^2 (d\chi^2 + \sin^2\chi \, d\phi^2) \tag{25.17}$$

where $\chi \equiv \bar{r}/R$. As illustrated in figure 25.1, this is precisely the metric of a sphere with radius aR. So in this case, the geometry of the equatorial plane (and by spherical symmetry, any plane through the origin) is the same as that of a spherical surface.

In the third case, the metric of the equatorial plane is

$$ds^2 = (aR)^2 (d\chi^2 + \sinh^2\chi \, d\phi^2) \tag{25.18}$$

This is similar to the first case, except that the circumference $2\pi aR \sinh\chi$ of a circle is *greater* than 2π times its radius $aR\chi$ instead of being smaller, and there is no way to construct an accurate embedding diagram for the equatorial plane's geometry analogous to that shown in figure 25.1 (see problem P25.5). However, this geometry is *qualitatively* like that of a saddle in that the circumferences of circles on the saddle are also larger than 2π times their radii, as illustrated in figure 25.2.

In both cases, the value of the constant R specifies how large $\bar{r}$ (the radial distance in today's universe) must be so that the spatial curvature becomes evident. Note that when $q(\bar{r}) = R\sin(\bar{r}/R)$, $\bar{r}/R = \chi$ specifies the angle in figure 25.1 in radians. When $\chi \ll 1$, $R\sin(\bar{r}/R) \approx R(\bar{r}/R) = \bar{r}$, so the geometry appears to be flat. But as χ becomes significant compared to 1 (i.e., as $\bar{r}$ becomes significant compared to R), the deviation from flat geometry becomes evident. The same qualitative argument applies when $q(\bar{r}) = R\sinh(\bar{r}/R)$.

So, in summary,

$q(\bar{r}) = R\sin(\bar{r}/R) \Rightarrow$ spatial geometry: sphere (25.19a)
$q(\bar{r}) = \bar{r} \Rightarrow$ spatial geometry: flat (25.19b)
$q(\bar{r}) = R\sinh(\bar{r}/R) \Rightarrow$ spatial geometry: saddle-like (25.19c)

We will see in the next chapter that the value of the scale constant R and which of these solutions applies depends on the the density of energy in the universe.

Cosmologists speak of the first solution as describing a **closed universe**, because the equatorial plane, if it is *topologically* as well as *geometrically* like a spherical surface, would have finite area, and thus the total volume of the universe would be finite. Similarly, cosmologists speak of the other two solutions as describing an **open universe**, because if the equatorial plane is *topologically* as well as *geometrically* like a flat plane or saddle surface, then it must be infinite to be homogeneous.

However, one should be careful *not* to confuse topology with geometry. The Einstein equation specifies the local *geometry* at every point in space and time, but not the *topology*. For example, a sheet of paper rolled into a cylinder everywhere has the *geometry* of a flat plane (because we do not have to rip or crumple the paper to roll it into a cylinder, the way we would have to crumple or rip it to make it conform with a sphere or saddle) but not the same *topology* as an infinite flat plane (if we go far enough around the cylinder, we come back to our starting point, which is not true on an ordinary plane). Similarly, the saddle-like surface of the third solution *could* be connected to itself like the inner surface of a donut.

Conversely, the topology of the equatorial plane for the first solution *could* be like an infinite set of spheres like beads on a string: when one reaches the opposite side of the sphere from the origin, instead of just returning on the same sphere, one could move on to the next sphere on the string, meaning that increasing χ could label new positions on the equatorial plane (on successive spheres) all the way to infinity, instead of returning to points previously labeled.

The topology of the universe does have observable consequences. While studies done so far are not conclusive, no evidence has been found as yet to suggest that the radial coordinate repeats itself in any direction.[1]

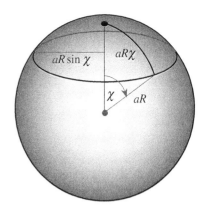

FIG. 25.1 This embedding diagram shows the spatial geometry of the universe's equatorial plane (at a given coordinate time) for the solution where $q(\bar{r}) = R\sin(\bar{r}/R)$. Note that $\chi \equiv \bar{r}/R$, and that for this geometry, the circumference of a circle of radius $r = aR\chi$ is smaller than $2\pi aR\chi = 2\pi r$.

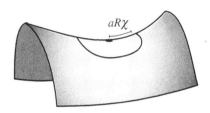

FIG. 25.2 This diagram illustrates a two-dimensional surface whose geometry is *similar* to the spatial geometry of the universe's equatorial plane at a given coordinate time for the solution where $q(\bar{r}) = R\sinh(\bar{r}/R)$. Note that for this geometry, the circumference of a circle of constant radius $r = aR\chi$ is *longer* than $2\pi aR\chi = 2\pi r$.

---

1. See Luminet, et al., "Is Space Finite?" *Scientific American*, **280**, 4 (1999) for a discussion of the topic.

BOX 25.1 The Universal Ricci Tensor

Exercise 25.1.1. Use the Diagonal Metric Worksheet (see the next two pages for a copy of the relevant portion) to show that equations 25.4 are correct. Also check that at least $R_{t\theta}$ of the off-diagonal components is zero. (This is an important exercise: please try at least a few components!) Remember that $A = 1$, $B = a^2$, $C = a^2 q^2$, and $D = a^2 q^2 \sin^2\theta$ here, and that $A_1 \equiv \partial A/\partial \bar{r}$, $B_{01} = \partial^2 B/\partial t\, \partial \bar{r}$, etc.

BOX 25.2 Raising One Index of the Universal Ricci Tensor

Exercise 25.2.1. Verify that equations 25.5 follow from equations 25.4 by raising the index of each Ricci component.

BOX 25.3 The Stress-Energy Tensor with One Index Lowered

Exercise 25.3.1. Show that equation 25.9 follows from equations 25.7 and 25.8.

$R_{00} = 0 \qquad + \tfrac{1}{2B}A_{11} \qquad + \tfrac{1}{2C}A_{22} \qquad + \tfrac{1}{2D}A_{33}$

$+ \quad 0 \qquad - \tfrac{1}{2B}B_{00} \qquad - \tfrac{1}{2C}C_{00} \qquad - \tfrac{1}{2D}D_{00}$

$+ \quad 0 \qquad + \tfrac{1}{4B^2}B_0^2 \qquad + \tfrac{1}{4C^2}C_0^2 \qquad + \tfrac{1}{4D^2}D_0^2$

$+ \quad 0 \qquad + \tfrac{1}{4AB}A_0 B_0 \qquad + \tfrac{1}{4AC}A_0 C_0 \qquad + \tfrac{1}{4AD}A_0 D_0$

$- \tfrac{1}{4BA}A_1 A_1 \qquad - \tfrac{1}{4B^2}A_1 B_1 \qquad + \tfrac{1}{4BC}A_1 C_1 \qquad + \tfrac{1}{4BD}A_1 D_1$

$- \tfrac{1}{4CA}A_2 A_2 \qquad + \tfrac{1}{4CB}A_2 B_2 \qquad - \tfrac{1}{4C^2}A_2 C_2 \qquad + \tfrac{1}{4CD}A_2 D_2$

$- \tfrac{1}{4DA}A_3 A_3 \qquad + \tfrac{1}{4DB}A_3 B_3 \qquad + \tfrac{1}{4DC}A_3 C_3 \qquad - \tfrac{1}{4D^2}A_3 D_3$

$\boxed{R_{00} = }$

$R_{11} = \tfrac{1}{2A}B_{00} \qquad + 0 \qquad - \tfrac{1}{2C}B_{22} \qquad - \tfrac{1}{2D}B_{33}$

$- \tfrac{1}{2A}A_{11} \qquad + 0 \qquad - \tfrac{1}{2C}C_{11} \qquad - \tfrac{1}{2D}D_{11}$

$+ \tfrac{1}{4A^2}A_1^2 \qquad + 0 \qquad + \tfrac{1}{4C^2}C_1^2 \qquad + \tfrac{1}{4D^2}D_1^2$

$- \tfrac{1}{4A^2}B_0 A_0 \qquad - \tfrac{1}{4AB}B_0 B_0 \qquad + \tfrac{1}{4AC}B_0 C_0 \qquad + \tfrac{1}{4AD}B_0 D_0$

$+ \tfrac{1}{4BA}B_1 A_1 \qquad + 0 \qquad + \tfrac{1}{4BC}B_1 C_1 \qquad + \tfrac{1}{4BD}B_1 D_1$

$- \tfrac{1}{4CA}B_2 A_2 \qquad + \tfrac{1}{4CB}B_2 B_2 \qquad + \tfrac{1}{4C^2}B_2 C_2 \qquad - \tfrac{1}{4CD}B_2 D_2$

$- \tfrac{1}{4DA}B_3 A_3 \qquad + \tfrac{1}{4DB}B_3 B_3 \qquad - \tfrac{1}{4DC}B_3 C_3 \qquad + \tfrac{1}{4D^2}B_3 D_3$

$\boxed{R_{11} = }$

25. A METRIC FOR THE COSMOS

$R_{22} = \frac{1}{2A}C_{00} \quad - \quad \frac{1}{2B}C_{11} \quad + \quad 0 \quad - \quad \frac{1}{2D}C_{33}$

$\quad - \quad \frac{1}{2A}A_{22} \quad - \quad \frac{1}{2B}B_{22} \quad + \quad 0 \quad - \quad \frac{1}{2D}D_{22}$

$\quad + \quad \frac{1}{4A^2}A_2^2 \quad + \quad \frac{1}{4B^2}B_2^2 \quad + \quad 0 \quad + \quad \frac{1}{4D^2}D_2^2$

$\quad - \quad \frac{1}{4A^2}C_0 A_0 \quad + \quad \frac{1}{4AB}C_0 B_0 \quad - \quad \frac{1}{4AC}C_0 C_0 \quad + \quad \frac{1}{4AD}D_0 C_0$

$\quad - \quad \frac{1}{4BA}C_1 A_1 \quad + \quad \frac{1}{4B^2}C_1 B_1 \quad + \quad \frac{1}{4BC}C_1 C_1 \quad - \quad \frac{1}{4BD}C_1 D_1$

$\quad + \quad \frac{1}{4CA}C_2 A_2 \quad + \quad \frac{1}{4CB}C_2 B_2 \quad + \quad 0 \quad + \quad \frac{1}{4CD}C_2 D_2$

$\quad - \quad \frac{1}{4DA}C_3 A_3 \quad - \quad \frac{1}{4DB}C_3 B_3 \quad + \quad \frac{1}{4DC}C_3 C_3 \quad + \quad \frac{1}{4D^2}C_3 D_3$

$\boxed{R_{22} = }$

$R_{33} = \frac{1}{2A}D_{00} \quad - \quad \frac{1}{2B}D_{11} \quad - \quad \frac{1}{2C}D_{22} \quad + \quad 0$

$\quad - \quad \frac{1}{2A}A_{33} \quad - \quad \frac{1}{2B}B_{33} \quad - \quad \frac{1}{2C}C_{33} \quad + \quad 0$

$\quad + \quad \frac{1}{4A^2}A_3^2 \quad + \quad \frac{1}{4B^2}B_3^2 \quad + \quad \frac{1}{4C^2}C_3^2 \quad + \quad 0$

$\quad - \quad \frac{1}{4A^2}D_0 A_0 \quad + \quad \frac{1}{4AB}D_0 B_0 \quad + \quad \frac{1}{4AC}D_0 C_0 \quad - \quad \frac{1}{4AD}D_0 D_0$

$\quad - \quad \frac{1}{4BA}D_1 A_1 \quad + \quad \frac{1}{4B^2}D_1 B_1 \quad - \quad \frac{1}{4BC}D_1 C_1 \quad + \quad \frac{1}{4BD}D_1 D_1$

$\quad - \quad \frac{1}{4CA}D_2 A_2 \quad - \quad \frac{1}{4CB}D_2 B_2 \quad + \quad \frac{1}{4C^2}D_2 C_2 \quad + \quad \frac{1}{4CD}D_2 D_2$

$\quad + \quad \frac{1}{4DA}D_3 A_3 \quad + \quad \frac{1}{4DB}D_3 B_3 \quad + \quad \frac{1}{4DC}D_3 C_3 \quad + \quad 0$

$\boxed{R_{33} = }$

$$R_{02} = -\tfrac{1}{2B}B_{02} \quad -\tfrac{1}{2D}D_{02} \quad +\tfrac{1}{4B^2}B_0B_2 \quad +\tfrac{1}{4D^2}D_0D_2$$

$$+\tfrac{1}{4AB}A_2B_0 \quad +\tfrac{1}{4AD}A_2D_0 \quad +\tfrac{1}{4CB}C_0B_2 \quad +\tfrac{1}{4CD}C_0D_2$$

$R_{02} =$

BOX 25.4 The Einstein Equation with One Index Lowered

Exercise 25.4.1. Show that equations 25.12 follow from equations 25.9 through 25.11.

BOX 25.5 Verifying the Solutions for q

Exercise 25.5.1. Show that $q(\bar{r}) = R\sin(\bar{r}/R)$, $q(\bar{r}) = \bar{r}$, and $q(\bar{r}) = R\sinh(\bar{r}/R)$ all solve the differential equation $(q')^2 - qq'' = 1$.

HOMEWORK PROBLEMS

P25.1 Use the Diagonal Metric Worksheet to verify equation 25.4c for $R_{\theta\theta}$. Also check that $R_{\bar{r}\theta} = 0$.

P25.2 In this problem, we will find all the functions $q(\bar{r})$ that solve the differential equation $(q')^2 - qq'' = 1$. Since this is a second-order differential equation, we will have to integrate twice, so the general solution will involve two arbitrary constants of integration.

a. First, let's define $p \equiv q' \equiv dq/d\bar{r}$. Note that

$$q'' = \frac{dp}{d\bar{r}} = \frac{dp}{dq}\frac{dq}{d\bar{r}} = \frac{dp}{dq}p \quad (25.20)$$

Show that if you substitute this into the differential equation $(q')^2 - qq'' = 1$, we find that

$$2\frac{dq}{q} = \frac{2p\,dp}{p^2 - 1} \quad (25.21)$$

b. Integrate both sides of this expression to show that

$$\left(\frac{dq}{d\bar{r}}\right)^2 = 1 + Kq^2 \quad (25.22)$$

where K is a constant of integration that can be positive, negative, or zero. (*Hint:* Note that the integral of dx/x is $\ln|x|$, not $\ln x$. We can assume that q is positive, since it represents a circumferential radius, but we cannot assume the same for $p^2 - 1$.)

c. Finally, integrate equation 25.22 to find $q(\bar{r})$ for all possible values of K (feel free to use a table of integrals). Fix the constant of integration by requiring that $q = 0$ when $\bar{r} = 0$.

P25.3 Use the geodesic equation to show that in a universe whose metric is given by equation 25.15, the equations of motion for a free particle with nonzero mass passing through the origin are given by

$$\theta = \text{constant}, \phi = \text{constant},$$

$$a^2\frac{d\bar{r}}{d\tau} = \text{constant}, \left(\frac{dt}{d\tau}\right)^2 = 1 + a^2\left(\frac{d\bar{r}}{d\tau}\right)^2 \quad (25.23)$$

(*Hint:* Find the last using $\mathbf{u} \cdot \mathbf{u} = -1$.)

P25.4 Show that if we take our radial coordinate to be q instead of $\bar{r}$, we can write the universal metric as

$$ds^2 = -dt^2 + a^2\left[\frac{dq^2}{1 - kq^2} + q^2 d\Omega^2\right] \quad (25.24)$$

(where $d\Omega^2 \equiv d\theta^2 + \sin^2\theta\,d\phi^2$) for *all three* cases of $q(\bar{r})$. This is the **Friedman-Walker-Robertson (FRW) metric** for the universe. How is the constant k related to the constant R in each of the three cases?

P25.5 Consider a curved 2D space with the metric

$$ds^2 = (aR)^2 d\chi^2 + (aR)^2 \sinh^2\chi\,d\phi^2 \quad (25.25)$$

(This is the metric of the equatorial plane of the cosmos at a given instant when $q(\bar{r}) = R\sinh(\bar{r}/R)$ and $\chi \equiv \bar{r}/R$.)

Show that (unlike the geometry of a spherical surface or the equatorial plane of Schwarzschild spacetime) we *cannot* use the methods of chapter 11 to construct a 2D axisymmetric surface (i.e., a surface whose vertical coordinate z does not depend on ϕ) that embeds this geometry in a 3D Euclidean space.

P25.6 The spatial metric of a t = constant slice of a universe where $q = R\sinh(\bar{r}/R)$ is

$$ds^2 = (aR)^2[d\chi^2 + \sinh^2\chi(d\theta^2 + \sin^2\theta\,d\phi^2)] \quad (25.26)$$

where $\chi \equiv \bar{r}/R$ and a and R are constants on this surface. Even though we can't create a 2D embedding diagram of this in a 3D Euclidean space (see problem P25.5), we can create an "embedding diagram" of this three-dimensional surface in a four-dimensional *flat spacetime* with spherical coordinates t, r, θ, ϕ and the usual flat spacetime spherical metric $ds^2 = -dt^2 + dr^2 + r^2(d\theta^2 + \sin^2\theta\,d\phi^2)$. Show that metric on the surface defined by $t^2 - r^2 = (aR)^2$ (a "hyper-hyperboloid") can be put into the same form as that given in equation 25.26 if one makes a suitable definition for $r(\chi)$. Sadly, this successful embedding does not help us visualize the geometry! (*Hint:* Express dt as a function of dr on this surface. Note that r is a coordinate in the *flat* spacetime: $r \neq \bar{r}$ in this situation. Remember also the hyperbolic "trig" identity $\cosh^2 x = 1 + \sinh^2 x$.)

P25.7 Use a computer graphing tool to accurately plot the area of a sphere of radius $\bar{r}$ as a function of $\bar{r}$ from 0 to $\bar{r} = \pi R$ for the three possible spatial geometries for a homogeneous universe at some specific instant of time (implying that a = constant). (*Hints:* It is easier and more meaningful if you actually plot $\alpha \equiv \text{area}/4\pi R^2$ versus $\chi = \bar{r}/R$. To find the area of a sphere, use the metric to find the *physical* area of a patch with coordinate area $d\theta\,d\phi$ and integrate over all θ and ϕ.)

P25.8 Let's *assume* that the topology of the equatorial plane in a universe with $q(\bar{r}) = R\sin(\bar{r}/R)$ is the same as the topology of a finite sphere. Find the total volume of a such a closed universe in terms of aR. (*Hints:* Use the metric to find the volume of a differential volume element in terms of $d\chi, d\theta$, and $d\phi$. Then integrate over all possible values of the coordinates. What are the limits on these values and why?)

P25.9 The Milne universe is a model of the universe that was briefly popular in the 1930s and provided an alternative and more intuitive picture of the universe than general relativity's model of a genuinely expanding and curved spacetime. Imagine that spacetime is *not* affected by the matter in it, but rather is everywhere and always the flat spacetime of special relativity with the metric

$$ds^2 = -dt^2 + dr^2 + r^2(d\theta^2 + \sin^2\theta\,d\phi^2) \quad (25.27)$$

in spherical coordinates. Imagine further that there was a creation event O that we take to define $t = 0$ and $r = 0$ in our

P25.9 (continued)

coordinate system, and at this creation event, an infinite number of galaxies were created with all possible radial speeds up to the speed of light. Subsequently, these galaxies move away from the spatial origin, at every instant of time t comprising a great puffball of galaxies whose outer edge expands at the speed of light.

Such a universe violates the cosmological principles of isotropy and homogeneity, in that at any instant of time t, the density of galaxies is *not* the same at all points in space. But will it look that way to an observer riding on one of the galaxies? The edge of the puffball moves away from the origin at the speed of light, like a spherical shell of light emitted from the origin event. Since every observer was at the origin at $t = 0$, and since the principle of relativity ensures that each observer will see the speed of light to be constant, each observer will consider him- or herself to be always at the center of that expanding puffball. Moreover, for the reasons discussed in box 24.5, each observer will also see every other galaxy to be moving away from his or her position with a speed proportional to their distance, consistent with Hubble's law.

So is it possible for such an observer to *think* that he is in an isotropic and homogeneous universe, even though it isn't really so from a global perspective? Consider a set of observers who have all measured the same proper time τ from the origin event. The universe will appear isotropic and homogeneous to such observers if the global density of galaxies is such that the *local* density of galaxies around each observer is the same for all. Consider a small spherical ball around one observer. Since all other galaxies are moving radially away from our observer, this ball (if it is to contain the same number of galaxies) must expand so that its radius is proportional to the proper time τ that our observer has measured since the beginning. Let's say that its radius is $R = q\tau$, where q is a constant but small number (so that the ball is small). The local density of galaxies for this observer is therefore $n_0 = N/(\frac{4}{3}\pi R^3) = N/(\frac{4}{3}\pi q^3 \tau^3) = K/\tau^3$, where N is the fixed number of galaxies in the small but expanding ball and K is a constant $= N/(\frac{4}{3}\pi q^3)$. What we want to do is ensure that this is the same for every observer at the same observer proper time τ from the origin event.

a. Argue that this will be so if, at an instant of time t in the global flat-space coordinate system, the number density of galaxies as a function of r is given by

$$n(t,r) = \frac{K}{t^3[1 - (r/t)^2]^2} \quad (25.28)$$

(*Hints:* Note that the proper time from the origin event for an observer at r and t is $\tau = (t^2 - r^2)^{1/2}$ and such an observer's speed is $v = r/t$. Note also that in the global frame, the observer's little ball of N galaxies is Lorentz-contracted and thus has a smaller volume than the observer thinks it has.)

b. So, this works in principle, if the creation event happened to create galaxies with this particular special density distribution! What else do you see that may be somewhat implausible about this density distribution?

c. Show that if we label events in our flat spacetime by τ (the proper time since the creation event measured by an observer passing by that event) and ρ (the "rapidity" $\equiv \tanh^{-1} v$ of that observer) instead of t and r, the flat spacetime metric in these new coordinates is

$$ds^2 = -d\tau^2 + \tau^2(d\rho^2 + \sinh^2\rho \, d\Omega^2) \quad (25.29)$$

where $d\Omega^2 \equiv d\theta^2 + \sin^2\theta \, d\phi^2$. (*Hint:* First, argue that $t = \tau \cosh \rho$ and $r = \tau \sinh \rho$ satisfy the defining equations for both τ and ρ in this case.)

d. Does this metric look somewhat familiar? Compare and contrast with the metric given in equation 25.15. Is it physically equivalent to the latter? Discuss. (There is no "right" answer here: just be as thoughtful and complete as you can in your comparison.)

The Milne model is inconsistent with general relativity (a huge problem considering the success of GR in other contexts) and has an implausible density function. There also appears to be an ongoing lively discussion about what cosmological observations it violates (if any). In spite of its problems, it is useful for exploring exactly which aspects of cosmology *really* depend on general relativity. For more about the Milne universe model, see Rindler, *Relativity*, 2nd ed., Oxford, 2006, pp. 360–363, from which I have adapted this problem.

26. EVOLUTION OF THE UNIVERSE

INTRODUCTION

FLAT SPACETIME
- Review of Special Relativity
- Four-Vectors
- Index Notation

TENSORS
- Arbitrary Coordinates
- Tensor Equations
- Maxwell's Equations
- Geodesics

SCHWARZSCHILD BLACK HOLES
- The Schwarzschild Metric
- Particle Orbits
- Precession of the Perihelion
- Photon Orbits
- Deflection of Light
- Event Horizon
- Alternative Coordinates
- Black Hole Thermodynamics

THE CALCULUS OF CURVATURE
- The Absolute Gradient
- Geodesic Deviation
- The Riemann Tensor

THE EINSTEIN EQUATION
- The Stress-Energy Tensor
- The Einstein Equation
- Interpreting the Equation
- The Schwarzschild Solution

COSMOLOGY
- The Universe Observed
- A Metric for the Cosmos
- **Evolution of the Universe**
- Cosmic Implications
- The Early Universe
- CMB Fluctuations & Inflation

GRAVITATIONAL WAVES
- Gauge Freedom
- Detecting Gravitational Waves
- Gravitational Wave Energy
- Generating Gravitational Waves
- Gravitational Wave Astronomy

SPINNING BLACK HOLES
- Gravitomagnetism
- The Kerr Metric
- Kerr Particle Orbits
- Ergoregion and Horizon
- Negative-Energy Orbits

this depends on this

26. EVOLUTION OF THE UNIVERSE

Introduction. In the previous chapter, we developed a trial metric for an homogeneous and isotropic universe that had the form

$$ds^2 = -dt^2 + a^2[d\bar{r}^2 + q^2(d\theta^2 + \sin^2\theta\, d\phi^2)] \quad (26.1)$$

where $\bar{r}$ is a radial coordinate that is comoving with the galactic "gas" as the universe expands, $a(t)$ is a unitless time-dependent quantity that specifies the scale of the universe, and $q(\bar{r})$ is either $R\sin(\bar{r}/R), \bar{r},$ or $R\sinh(\bar{r}/R)$. We have (arbitrarily) defined the fixed $\bar{r}$ coordinate of an object at rest with respect to the galactic "gas" to be equal to its radial distance from the origin at the present time, implying that the scale factor a has the value 1 at the present time. In the last chapter, we used the Einstein equation to determine the possible solutions for $q(\bar{r})$: our goal in this chapter is to use the Einstein equation to link these solutions to the total energy density of the universe and to solve for the time-dependent scale factor $a(t)$.

The Einstein Equation Revisited. If we substitute equations 25.5 into the Einstein equation components given by equations 25.12, we get

$$\frac{3\ddot{a}}{a} = R^t{}_t = 8\pi G(T^t{}_t - \tfrac{1}{2}\delta^t{}_t T) = -4\pi G(\rho_0 + 3p_0) + \Lambda \quad (26.2a)$$

$$\frac{\ddot{a}}{a} + 2\frac{\dot{a}^2}{a^2} - \frac{2q''}{qa^2} = R^{\bar{r}}{}_{\bar{r}} = 8\pi G(T^{\bar{r}}{}_{\bar{r}} - \tfrac{1}{2}\delta^{\bar{r}}{}_{\bar{r}} T) = 4\pi G(\rho_0 - p_0) + \Lambda \quad (26.2b)$$

where $\dot{a} \equiv da/dt, \ddot{a} \equiv d^2a/dt^2, q'' \equiv d^2q/d\bar{r}^2$, and ρ_0 and p_0 are the energy density and pressure of the galactic "gas" in its own rest frame. But since we know from our previous work that $q(\bar{r}) = R\sin(\bar{r}/R), \bar{r},$ or $R\sinh(\bar{r}/R)$, we know that

$$\frac{2q''}{qa^2} = \begin{cases} -2/R^2a^2 & \text{for } q = R\sin(\bar{r}/R) \\ 0 & \text{for } q = \bar{r} \\ +2/R^2a^2 & \text{for } q = R\sinh(\bar{r}/R) \end{cases} \equiv \frac{2K}{a^2}, \text{ where } K \equiv \mp\frac{1}{R^2} \quad (26.3)$$

for the three cases, respectively. K is the constant of integration we encountered before (note that $K = 0$ corresponds to the curvature radius R going to infinity). If we substitute this back into equation 26.2b, we see that equations 26.2 become

$$\frac{3\ddot{a}}{a} = -4\pi G(\rho_0 + 3p_0) + \Lambda \quad (26.4a)$$

$$\frac{\ddot{a}}{a} + 2\frac{\dot{a}^2}{a^2} - \frac{2K}{a^2} = 4\pi G(\rho_0 - p_0) + \Lambda \quad (26.4b)$$

(The θ-θ and ϕ-ϕ components of the Einstein equation are the same as equation 26.4b: see box 26.1). Note that the energy density ρ_0 and the pressure p_0 of the galactic "gas" depend on time, so equations 26.4 represent two equations in three unknown time-dependent quantities, so we do not yet have enough information to solve these equations.

Local Conservation of Energy. We can get some useful information that we need from the local conservation-of-energy law

$$0 = \nabla_\mu T^{t\mu} = \nabla_\mu T^{\mu t} \quad (26.5)$$

where the last step follows because the stress-energy tensor is symmetric. We can lower the t index and expand the absolute divergence to get

$$0 = \partial_\mu T^\mu{}_t + \Gamma^\mu{}_{\alpha\mu} T^\alpha{}_t - \Gamma^\beta{}_{t\mu} T^\mu{}_\beta \quad (26.6)$$

You can use the Diagonal Metric Worksheet to compute the necessary Christoffel symbols (or even compute them by hand; they are not that difficult). The number of Christoffel symbols you need is greatly reduced by the fact that $T^\mu{}_\nu$ is diagonal. The result (see box 26.2) is the simple relationship

$$\frac{d}{dt}(\rho_0 a^3) = -p_0 \frac{d}{dt}(a^3) \quad (26.7)$$

This result has a simple physical interpretation. Remember that the coordinate $\bar{r}$ is comoving with the galactic "gas" in the expanding universe. The number of galaxies enclosed by a given coordinate $\bar{r}$ is therefore constant. But the metric implies that the physical radius of the volume enclosed by a given value of $\bar{r}$ is proportional to a, so the physical volume enclosed is proportional to a^3. Let's say that $V = Ba^3$, where B is the constant of proportionality. The total energy U in this volume is thus $U = \rho_0 B a^3$. Equation 26.7 can therefore be written

$$\frac{dU}{dt} = -p_0 B \frac{d}{dt} a^3 \quad \Rightarrow \quad dU = -p_0 dV \tag{26.8}$$

This is the first law of thermodynamics for the universal "gas" inside the volume V: the change in the total energy inside the volume has to be equal to the work energy flowing into the volume. (Note that there can be no heat flow across the boundary because the universe is homogeneous, so all points in space at a given instant of cosmic time t have the same temperature.)

However, it should be noted that equation 26.5 is a consequence of the definition of the Einstein equation, so it does not actually tell us anything that is not already implicit in equations 26.4 (which also specify the implications of the Einstein equation). So while we will find equation 26.7 helpful in solving the field equations in a moment, it does not provide the missing information we need.

Equations of State. What we really need to complete the solution of equations 26.4 is how the pressure p_0 of the galactic "gas" depends on its density ρ_0. An equation that specifies $p(\rho)$ is called an **equation of state**.

In general, the "stuff" in the universe has three important components, non-relativistic *matter*, relativistic *radiation*, and *vacuum energy*. The **matter** component is (now) represented by galaxies and the dark matter that accompanies them. The measured random velocities of galaxies with respect to each other are on the order of 100 km/s, which (though large by human standards) is very small compared to the speed of light. The pressure of the universal "gas" whose "molecules" are galaxies will thus be negligible compared to its energy density, and we can accurately model this component as if it were pressureless dust. Equation 26.7 implies that

$$\frac{d}{dt}(\rho_m a^3) = 0 \quad \Rightarrow \quad \rho_m a^3 = \text{const.} = \rho_{m0} \quad \text{(for \textbf{matter})} \tag{26.9a}$$

where ρ_m is the portion of the total energy density ρ_0 that is matter and ρ_{m0} is that density at the present time (note that $a = 1$ at the present time).

However, some of the "stuff" of the universe consists of photons, neutrinos, and other highly relativistic particles. As discussed in problem P20.4, the pressure p_r of a photon gas is related to its energy density ρ_r as follows: $p_r = \frac{1}{3}\rho_r$. If you plug this relationship back into equation 26.7, you can show (see box 26.3) that

$$\rho_r a^4 = \text{const.} = \rho_{r0} \quad \text{(for \textbf{radiation})} \tag{26.9b}$$

where ρ_{r0} is the present density of radiation.

Now, a photon gas in thermal equilibrium with its surroundings at an absolute temperature T has an energy density that is proportional to T^4 (this is a consequence of the Stefan-Boltzmann law). This means that the effective temperature of the photon gas (and whatever is in thermal equilibrium with it) varies as follows:

$$Ta = \text{const.} \tag{26.10}$$

In other words, the temperature of any "radiation" component of the universe varies in *inverse* proportion to the universe's scale a.

Finally, there is the **vacuum energy**. As discussed in chapter 21, we can treat the cosmological constant term as if it were a type of energy that we can include on the stress-energy side of the Einstein equation. In what follows, it will help us to treat the density of this energy like that of other sources. According to equation 21.21, the effective stress-energy tensor for this vacuum energy term is

$$T^{\mu\nu} = -g^{\mu\nu}\frac{\Lambda}{8\pi G} \quad \Rightarrow \quad \rho_v \equiv T^{tt} = -g^{tt}\frac{\Lambda}{8\pi G} = +\frac{\Lambda}{8\pi G} \qquad (26.11)$$

We see that the vacuum energy density is constant and so does *not* vary with a.

Note also that if we consider the *pressure* of the vacuum to be

$$p_v = (T^{\bar r}{}_{\bar r})_{\text{vac}} = -\frac{\Lambda}{8\pi G}\delta^{\bar r}{}_{\bar r} = -\frac{\Lambda}{8\pi G} \qquad (26.12)$$

(see equation 25.9), then $\rho_v - p_v = \Lambda/4\pi G$, so the right side of equation 26.4b can be written as $4\pi G(\rho_{0,\text{tot}} - p_{0,\text{tot}})$, where $\rho_{0,\text{tot}}$ and $p_{0,\text{tot}}$ include contributions from matter, radiation, *and* vacuum. We see from that equation that the *difference* $\rho_{0,\text{tot}} - p_{0,\text{tot}}$ (by specifying K) uniquely determines the type and magnitude of spatial curvature here, just as that difference does in the weak-field limit.

The Friedman Equation. Now we are finally ready to finish our solution of the Einstein equation. If you add the negative of equation 26.4a to 3 times equation 26.4b (as discussed in box 26.4), you will find that the terms involving $\ddot{a}/a$ cancel on the right and the pressure terms cancel on the left, leaving the simpler equation

$$\dot{a}^2 - \frac{8\pi G}{3}(\rho_m + \rho_r + \rho_v)a^2 = K \qquad (26.13)$$

where I have written $\rho_0 = \rho_m + \rho_r$ and used equation 26.11 to express the vacuum energy term as an energy density. This is the **Friedman equation** for the time-evolution of the universe.

There is some hidden time dependence in this equation, because the densities of matter and radiation depend on a and thus on time. We can use equations 26.9 to make this dependence explicit. Remember that we have defined the value of a to be unity at the present time. Using this notation and equations 26.9, we can write equation 26.13 in the form

$$\dot{a}^2 - \frac{8\pi G}{3}\left(\frac{\rho_{m0}}{a^3} + \frac{\rho_{r0}}{a^4} + \rho_v\right)a^2 = K \qquad (26.14)$$

where ρ_{m0} and ρ_{r0} are the matter and energy densities at the present (ρ_v is constant in time, so $\rho_{v0} = \rho_v$). If we know the values of K, ρ_{m0}, ρ_{r0}, and ρ_v, we can in principle solve this differential equation for $a(t)$.

The Hubble Parameter. While we might be able to measure the densities of matter, radiation, and vacuum in the present universe, it is hard to see how we might determine K and a. There is a clever way to address this problem, though.

We saw earlier that the distance from the origin to a particular galaxy at a given fixed value of $\bar r$ is given by $d = a\bar r$. The rate at which the distance to that galaxy increases due to the expansion of the universe (as reflected by the increase in the value of the universal scale factor a) is thus $\dot{a}\bar r$. We can interpret this rate of increase of distance as a recessional velocity v. So at any instant of cosmological time t,

$$v \equiv \dot{a}\bar r = \frac{\dot{a}}{a}(a\bar r) = \frac{\dot{a}}{a}d = Hd \quad \text{where} \quad H \equiv \frac{\dot{a}}{a} \qquad (26.15)$$

H is therefore the Hubble "constant" at that t. Note that H is *not* generally constant with time. It only *appears* constant if we limit ourselves to observing the motion of relatively nearby galaxies, so that the difference in the cosmological t between light's departure from a galaxy and its detection on earth is tiny compared to the age of the universe. Therefore, I will call H the **Hubble parameter**.

However, we can measure H at the *present* time by examining the distances and apparent recessional velocity of relatively nearby galaxies. The present value of H is therefore $H_0 \equiv \dot{a}_0/a_0 = \dot{a}_0$ (since $a_0 \equiv 1$). I will call H_0 (the present value of the Hubble parameter H) the **Hubble constant**. If we divide both sides of equation 26.14 by $\dot{a}^2$ and evaluate it at the present time, we get (see box 26.5)

$$1 - \frac{8\pi G}{3H_0^2}(\rho_{m0} + \rho_{r0} + \rho_v) = \frac{K}{H_0^2} \qquad (26.16)$$

The Critical Density. Now, $|K| = 1/R^2$, where R is the scale of the universe's spatial curvature and the sign of K determines the *type* of that curvature. We see here that both the value and sign depends on how the present total energy density of the universe compares to the value of $3H_0^2/8\pi G$. If $\rho_{tot} \equiv \rho_{m0} + \rho_{r0} + \rho_v > 3H_0^2/8\pi G$, then K is negative, meaning that the geometry of the spatial part of the universe is spherical, with radius $aR = a/|K|^{1/2}$. If $\rho_{tot} < 3H_0^2/8\pi G$, then K is positive, meaning that the universe's spatial geometry is like that of a saddle surface, with aR again specifying roughly the radial scale where this curvature becomes important. If $\rho_{tot} = 3H_0^2/8\pi G$, then K is zero, and the universe has a flat spatial geometry. We therefore define the **critical density** ρ_c for the universe at the present to be

$$\rho_c \equiv \frac{3H_0^2}{8\pi G} \tag{26.17}$$

and compare the present energy densities of matter, radiation, and the vacuum to this critical density by defining the unitless ratios

$$\Omega_m \equiv \frac{\rho_{m0}}{\rho_c}, \quad \Omega_r \equiv \frac{\rho_{r0}}{\rho_c}, \quad \Omega_v \equiv \frac{\rho_v}{\rho_c} \tag{26.18}$$

Then we can rewrite equation 26.16 in the form

$$1 - (\Omega_m + \Omega_r + \Omega_v) = \frac{K}{H_0^2} \equiv \Omega_k \tag{26.19}$$

Therefore, if we can measure the present density of matter, radiation, and vacuum energy, and we know the Hubble constant H_0, then we can determine the present value of the **curvature parameter** Ω_k. The sign of Ω_k determines the spatial curvature of the universe just as the sign of K does: if Ω_k is positive, the universe's spatial geometry is saddle-shaped. If Ω_k is negative, then the universe's spatial geometry is spherical. If $\Omega_k = 0$, then $K = 0$, and the universe's spatial geometry is flat.

An Equation of Motion for the Universe. If you divide both sides of equation 26.14 by $\dot{a}_0^2 = H_0^2$ and use equations 26.17 through 26.19, you can show (see box 26.6) that we can express the Friedman equation in the form

$$\left(\frac{1}{H_0}\frac{da}{dt}\right)^2 = \Omega_k + \frac{\Omega_m}{a} + \frac{\Omega_r}{a^2} + \Omega_v a^2 \tag{26.20}$$

We can in principle solve this equation for $a(t)$ at all times, thereby comparing the scale of the universe at any time as a fraction or multiple of the present scale. Taking the absolute value of equation 26.19 and using equation 26.3 allow us to determine the spatial comoving curvature scale R given H_0, Ω_m, Ω_r, and Ω_v:

$$R = \frac{1}{H_0\sqrt{|\Omega_k|}} = \frac{1}{H_0\sqrt{|1 - \Omega_m - \Omega_r - \Omega_v|}} \tag{26.21}$$

($K = 0$ corresponds to infinite curvature scale, meaning that space is flat.) From these last two equations we see that *the four parameters H_0, Ω_m, Ω_r, and Ω_v completely determine the evolution and spatial geometry of the universe.*

Equation 26.20 has the form of a one-dimensional conservation of energy equation where the $(1/H_0)^2(da/dt)^2$ term is the kinetic energy, the curvature parameter plays the role of the conserved total energy, and the remaining terms (when negated) play the role of an a-dependent potential energy. Interpreted this way, we see that in determining how a evolves with time t, matter density acts like a simple attractive gravitational force, radiation acts like the potential for an attractive $1/a^3$-dependent force, and the vacuum energy acts like a repulsive spring-like force. This means that you can predict the dynamical behavior of the universe by drawing an effective potential energy graph that expresses these ideas. You can practice this by working through box 26.7, where you will consider the possible behaviors for a matter-dominated universe ($\Omega_r \approx 0$, $\Omega_v \approx 0$) for values of Ω_m both greater than and less than 1.

BOX 26.1 The Other Components of the Einstein Equation

According to equation 25.5c,

$$R^\theta{}_\theta = R^\phi{}_\phi = 2\frac{\dot{a}^2}{a^2} + \frac{\ddot{a}}{a} + \frac{1}{a^2 q^2}[1 - (q')^2 - qq''] \tag{26.22}$$

Equation 25.12b tells us that the Einstein equation implies that

$$R^\theta{}_\theta = R^\phi{}_\phi = 4\pi G(\rho_0 - p_0) + \Lambda \tag{26.23}$$

One can see that the θ-θ and ϕ-ϕ components of the Einstein equation yield only one distinct differential equation connecting the metric functions a and q with ρ_0, p_0 and Λ. Moreover, you can show that if $q = R\sin(\tilde{r}/R), \tilde{r},$ or $R\sinh(\tilde{r}/R)$, then this differential equation becomes equivalent to equation 26.4b, repeated here for convenience:

$$\frac{\ddot{a}}{a} + 2\frac{\dot{a}^2}{a^2} - \frac{2K}{a^2} = 4\pi G(\rho_0 - p_0) + \Lambda \tag{26.4br}$$

Exercise 26.1.1. Verify that the last statement is true.

BOX 26.2 Consequences of Local Energy/Momentum Conservation

According to equation 25.9, the universal stress-energy tensor is diagonal and has components $T^t{}_t = -\rho_0 - \Lambda/8\pi G$, $T^r{}_r = T^\theta{}_\theta = T^\phi{}_\phi = p_0 - \Lambda/8\pi G$. Using this and the Diagonal Metric Worksheet, you can prove that $d(\rho_0 a^3)/dt = -p_0 d(a^3)/dt$ (equation 26.7) follows from $0 = \partial_\mu T^\mu{}_t + \Gamma^\mu{}_{\alpha\mu} T^\alpha{}_t - \Gamma^\beta{}_{t\mu} T^\mu{}_\beta$ (equation 26.6).

Exercise 26.2.1. Verify this. (*Hint:* You should find that you need to calculate only Christoffel symbols of the form $\Gamma^i{}_{ij}$, where i and j are spatial indices. These are very easy to calculate from the definition of the Christoffel symbols if you don't want to bother with the Diagonal Metric Worksheet.)

BOX 26.3 Deriving the Density/Scale Relationship for Radiation

Exercise 26.3.1. Show that $d(\rho_0 a^3)/dt = -p_0 d(a^3)/dt$ (equation 26.7) and the relationship $p_r = \frac{1}{3}\rho_r$ implies $\rho_r a^4 =$ constant (equation 26.9b).

BOX 26.4 Deriving the Friedman Equation

In equations 26.4 (repeated here), we saw that the Einstein equation becomes

$$\frac{3\ddot{a}}{a} = -4\pi G(\rho_0 + 3p_0) + \Lambda \qquad (26.4ar)$$

$$\frac{\ddot{a}}{a} + 2\frac{\dot{a}^2}{a^2} - \frac{2K}{a^2} = 4\pi G(\rho_0 - p_0) + \Lambda \qquad (26.4br)$$

Exercise 26.4.1. Add the negative of equation 26.4a to three times equation 26.4b to derive the Friedman equation $\dot{a}^2 - \frac{8}{3}\pi G(\rho_0 + \rho_v)a^2 = K$ (equation 26.13).

BOX 26.5 The Friedman Equation for the Present Time

Equation 26.14 (repeated here for convenience) says that

$$\dot{a}^2 - \frac{8\pi G}{3}\left(\frac{\rho_{m0}}{a^3} + \frac{\rho_{r0}}{a^4} + \rho_v\right)a^2 = K \qquad (26.14r)$$

If we divide both sides of this by $\dot{a}^2$ and evaluate at the present time using using $H_0 \equiv \dot{a}_0$ and $a_0 = 1$, we get equation 26.16 (repeated below).

$$1 - \frac{8\pi G}{3H_0^2}(\rho_{m0} + \rho_{r0} + \rho_v) = \frac{K}{H_0^2} \qquad (26.16r)$$

Exercise 26.5.1. Verify this.

BOX 26.6 Deriving the Friedman Equation in Terms of the Omegas

Equation 26.20 (repeated here for convenience) claims that

$$\left(\frac{1}{H_0}\frac{da}{dt}\right)^2 = \Omega_k + \frac{\Omega_m}{a} + \frac{\Omega_r}{a^2} + \Omega_v a^2 \qquad (26.20r)$$

Exercise 26.6.1. Show that if we divide both sides of equation 26.14 (see above) by $\dot{a}_0^2 = H_0^2$ and use $\rho_c \equiv 3H_0^2/8\pi G$, $\Omega_m \equiv \rho_{m0}/\rho_c$, $\Omega_r \equiv \rho_{r0}/\rho_c$, $\Omega_v \equiv \rho_v/\rho_c$, and $K/H_0^2 \equiv \Omega_k$ (equations 26.17–26.19), we get the above.

BOX 26.7 The Behavior of a Matter-Dominated Universe

Assume that the universe always has been dominated by matter ($\Omega_r \approx \Omega_v \approx 0$) throughout its history. Equation 26.20 then becomes

$$\left(\frac{1}{H_0}\frac{da}{dt}\right)^2 - \frac{\Omega_m}{a} = \Omega_k \qquad (26.24)$$

Exercise 26.7.1. Interpreting this as a one-dimensional "conservation of energy" equation, plot a potential energy graph, and use the graph to qualitatively describe the evolution of a in the case where $\Omega_m > 1$ and when $\Omega_m < 1$. Also describe how the time evolution of the universe is connected with its spatial curvature in this case. (*Hint:* What does $\Omega_m > 1$ mean for the value of Ω_k? See equation 26.19.)

HOMEWORK PROBLEMS

P26.1 Use a "potential energy graph" approach to discuss the qualitative behavior of a radiation-dominated universe where $\Omega_m \approx \Omega_v \approx 0$. Describe the evolution of the universal scale a in the case where $\Omega_r > 1$ and when $\Omega_r < 1$.

P26.2 Use a "potential energy graph" approach to discuss the qualitative behavior of an empty vacuum-dominated universe where $\Omega_m \approx \Omega_r \approx 0$. Qualitatively describe the evolution of the universal scale a in the cases where $\Omega_v > 1$ and $\Omega_v < 1$. For what such universes will there be a Big Bang? (Empty vacuum-dominated universes are called **Lemaître universes**.)

P26.3 When Einstein first applied general relativity to the problem of cosmology in 1917, it was reasonable to consider the universe to be a homogeneous, isotropic, and *static* collection of stars similar to the stars near to the earth. Einstein could find a static solution to the Einstein equation $R_{\mu\nu} - \tfrac{1}{2} g_{\mu\nu} R = 8\pi G T_{\mu\nu}$ if and only if he added the "cosmological constant" term $\Lambda g_{\mu\nu}$ to the left side of the equation. As we saw in chapter 21, we now consider this "cosmological constant" term to be instead a "vacuum energy" term that we add to the equation's *right* side.

a. Use a "potential energy graph" approach to argue that one can indeed find a static solution to the Friedman equation in the form given in 26.14 if ρ_{m0} and ρ_v are both nonzero, $\rho_{r0} \approx 0$, and K has exactly the right value.

b. Assuming that $a = 1$ at the present, express ρ_v and K in terms of ρ_{m0} assuming that the universe is static at present and ρ_{r0} is negligibly small.

c. A plausible mass density for the universe in Einstein's time might have been the approximate local density of stars near the earth, which is very roughly 0.05 solar masses per cubic parsec. Determine the value of ρ_{m0} corresponding to this value, and use your result from part b to determine the values of ρ_v and K.

d. Is your value for ρ_v comfortably smaller than the upper limit established by solar system measurements (see equation 21.36 in problem P21.1)?

e. You should have found K to be negative for this hypothetical static universe, implying that its spatial geometry is spherical. Assuming that it is also *topologically* spherical, find its radius in light-years, its total volume in cubic light-years, and the total mass of matter in solar masses. (*Hint:* Be sure to use the metric to find the volume, which you should find to be $V = 2\pi^2 R^3$.)

f. However, use the "potential energy diagram" to argue that this static universe is *unstable* (something that wasn't initially clear to Einstein).

As we saw in chapter 24, Lemaître and Hubble established in 1927 and 1929, respectively, that the universe was in fact expanding. In 1931, Einstein formally abandoned the cosmological constant term, later calling it "the biggest blunder of his life" according to George Gamow in his autobiography *My Worldline* (Viking Press, 1970, p. 44).

P26.4 Consider a model of the universe where there is no vacuum energy, only matter and radiation. Argue that the age of such a universe must be less than $H_0^{-1} = 13.7$ Gy. [*Hint:* Use a "potential energy graph" approach to determine qualitatively how a depends on time, and sketch a qualitative graph of $a(t)$. Note that H_0 is related to the present slope of such a graph.]

P26.5 Consider the case of an empty, vacuum-dominated universe where $\Omega_m \approx \Omega_r \approx 0$ and $\Omega_v = 1$. (Such a universe is called a **De Sitter universe**.)

a. Show that this universe expands exponentially:
$$a(t) = e^{+H_0(t-t_0)} \quad \text{or} \quad a(t) = e^{-H_0(t-t_0)} \quad (26.25)$$

b. Argue that for this to be consistent with an expanding universe, we must choose the first solution.

c. What is the age of the universe in this case?

d. Show that the Hubble parameter $H \equiv \dot{a}/a$ in such a universe happens to actually be constant in time.

P26.6 (Important!) Argue that the universe's present age is
$$t_0 = \frac{1}{H_0} \int_0^1 \frac{da}{\sqrt{\Omega_k + \Omega_r a^{-2} + \Omega_m a^{-1} + \Omega_v a^2}} \quad (26.26)$$

P26.7 Consider a universe whose metric is
$$ds^2 = -dt^2 + \left(\frac{t}{t_0}\right)(dx^2 + dy^2 + dz^2) \quad (26.27)$$
where t_0 is some constant.

a. Explain how we can interpret this metric as a special case of the general universal metric given by equation 26.1. What is $a(t)$ in this case? What is the age of the universe when $a = 1$?

b. Is the spatial geometry of this universe saddle-like, flat, or spherical? Explain your reasoning.

c. Is this universe radiation, matter, or vacuum dominated? Explain your reasoning.

P26.8 When I first learned cosmology in the 1970s, both the average matter density of the universe and the value of the Hubble constant H_0 were so poorly known that it was possible (though improbable) that $\Omega_m > 1$. This was also during the time that most physicists believed that $\Omega_v = 0$ (and, more correctly, that Ω_r is negligible). Let's consider the evolution of such a super-critical "matter-only" universe.

a. Argue that the spatial geometry of such a universe is spherical ("closed").

P26.8 (continued)

b. Equation 26.20 for the time evolution of the universe in this case (where $\Omega_r = \Omega_v = 0$) is still a nonlinear differential equation. To solve such an equation, one must use intelligent guessing, trickery, or both. Usually people use intelligent guessing to get the solution the first time and then invent clever tricks to find the solution more elegantly after it is known. We will use such a trick in this case. Let's define a "time parameter" ψ such that

$$(\Omega_m - 1)^{1/2} H_0 \, dt = a \, d\psi \quad (26.28)$$

and $\psi = 0$ at $t = 0$. Show that if we consider a to be a function of of the new parameter ψ, we can re-express equation 26.20 in the form

$$\left(\frac{da}{d\psi}\right)^2 + a^2 = \frac{\Omega_m}{\Omega_m - 1} a \quad (26.29)$$

c. Take the ψ-derivative of both sides of this expression to show that

$$\frac{d^2 a}{d\psi^2} + a = \frac{1}{2} \frac{\Omega_m}{\Omega_m - 1} \quad (26.30)$$

d. This is the harmonic oscillator equation with a constant driving term. In a differential equations or mechanics course, you may have learned that the general solution to such an equation is the most general solution to the *homogeneous* equation

$$\frac{d^2 a}{d\psi^2} + a = 0 \quad (26.31)$$

plus any *particular* solution to the *full* equation. Argue that $a = \frac{1}{2}\Omega_m/(\Omega_m - 1) = \text{constant} \equiv A$ is a solution to the full equation.

e. The solution to the homogeneous harmonic oscillator equation is $a = B \sin \psi + C \cos \psi$, where B and C are constants determined by initial conditions. Therefore, the general solution to equation 26.30 is

$$a(\psi) = A + B \sin \psi + C \cos \psi \quad (26.32)$$

Argue that requiring that $a \to 0$ as $\psi \to 0$ puts no constraints on B but requires that $C = -A$.

f. Argue, however, that this solution will not satisfy the *original* relation in equation 26.29 as $\psi \to 0$ unless we also have $B = 0$.

g. Use equation 26.28 to determine t in terms of H_0, Ω_m and ψ. Your answer to this part and equation 26.32 with $B = 0$ and $A = -C = \frac{1}{2}\Omega_m/(\Omega_m - 1)$ provide a parametric solution for $a(t)$ in terms of the parameter ψ.

h. Argue that such a universe expands, reaches a maximum scale a, and then contracts to a Big Crunch. Is this consistent with the results of box 26.7?

i. If $H_0^{-1} = 13.9$ Gy, and $\Omega_m = 1.10$, how long after the Big Bang does the universe reach its maximum spatial size, and what is the radius of its spherical geometry at that point? How long does the universe last between the Big Bang and the Big Crunch? (*Hint:* See equation 26.21.)

j. Argue that a graph of $Ra(t)$ has the shape of a cycloid, i.e., the path of a point on the rim of a rolling wheel. (*Hint:* Look up "cycloid" online.)

P26.9 Consider a universe where $\Omega_v > 1$ and matter and energy densities are negligible ($\Omega_m \approx 0$ and $\Omega_r \approx 0$). It turns out that such a universe will never have a Big Bang singularity, but will have an instant of maximal (finite) density. Define that instant to be $t = 0$. Assume that observers in this universe at some time t_0 measure the Hubble constant to be $H_0 = (15 \text{ Gy})^{-1}$.

a. Show that for such a universe, $a = b \cosh(\omega t)$, where $b = \sqrt{(\Omega_v - 1)/\Omega_v}$ and $\omega = H_0 \sqrt{\Omega_v}$. If this universe is expanding at time t_0, will it ever cease expanding? If so, at what time t?

b. Imagine that observers in this universe determine from observations of their cosmic microwave background that $\Omega_v = 2$. How old is their universe at time t_0?

c. Is the spatial geometry of this universe spherical, flat, or saddle-like?

d. What is the curvature scale R of this universe (which is the scale over which the spatial curvature of the universe becomes evident)?

P26.10 We have expressed our equations of motion for the universal scale factor a in terms of the *constants* Ω_m, Ω_r, and Ω_v, which are ratios of the *current* mass, radiation, and vacuum energy densities (respectively) to the *current* critical density. However, a hypothetical observer at a different cosmic time t would determine these constants to have different values. Define

$$\Omega_m(t) = \frac{\rho_m(t)}{\rho_c(t)} = \frac{8\pi G}{3[H(t)]^2} \rho_m(t) \quad (26.33)$$

where $\rho_m(t)$ is the density of matter at time t, $\rho_c(t)$ is the critical density at time t, and $H(t) \equiv \dot{a}/a$ is the Hubble parameter at time t. We can define Ω_r and Ω_v similarly.

a. Show that we can write

$$\Omega_m(t) = \frac{\Omega_m}{a^3}\left(\frac{H_0}{H}\right)^2 \quad \text{and} \quad \Omega_v(t) = \Omega_v\left(\frac{H_0}{H}\right)^2 \quad (26.34)$$

where Ω_m and Ω_v are the values we would measure.

b. Imagine a vacuum-dominated universe where the current value of $\Omega_v \approx 1$ and Ω_m and Ω_r are both $\ll \Omega_v$. Argue that if the value of Ω_m is not *strictly* zero, then observers in the distant past would determine the value of $\Omega_m(t)$ to be greater than Ω_m, while $\Omega_v(t) = \Omega_v$ always. Argue therefore that the approximation that the universe is vacuum dominated must break down at some point sufficiently far in the past. (*Hint:* You can use the results of problem P26.5.)

27. COSMIC IMPLICATIONS

INTRODUCTION

FLAT SPACETIME
- Review of Special Relativity
- Four-Vectors
- Index Notation

TENSORS
- Arbitrary Coordinates
- Tensor Equations
- Maxwell's Equations
- Geodesics

SCHWARZSCHILD BLACK HOLES
- The Schwarzschild Metric
- Particle Orbits
- Precession of the Perihelion
- Photon Orbits
- Deflection of Light
- Event Horizon
- Alternative Coordinates
- Black Hole Thermodynamics

THE CALCULUS OF CURVATURE
- The Absolute Gradient
- Geodesic Deviation
- The Riemann Tensor

THE EINSTEIN EQUATION
- The Stress-Energy Tensor
- The Einstein Equation
- Interpreting the Equation
- The Schwarzschild Solution

COSMOLOGY
- The Universe Observed
- A Metric for the Cosmos
- Evolution of the Universe
- **Cosmic Implications**
- The Early Universe
- CMB Fluctuations & Inflation

GRAVITATIONAL WAVES
- Gauge Freedom
- Detecting Gravitational Waves
- Gravitational Wave Energy
- Generating Gravitational Waves
- Gravitational Wave Astronomy

SPINNING BLACK HOLES
- Gravitomagnetism
- The Kerr Metric
- Kerr Particle Orbits
- Ergoregion and Horizon
- Negative-Energy Orbits

this depends on this

317

27. COSMIC IMPLICATIONS

Introduction. In the last chapter, we determined how the scale factor $a(t)$ of the universe behaves with time and how its behavior is linked with the curvature of space. In this chapter, we will explore some of the consequences of these results and see something about why we currently believe that the universe is flat ($K = 0$).

A Conformal Time Coordinate. In chapter 25, we developed the following metric for the cosmos:

$$ds^2 = -dt^2 + a^2\left[d\bar{r}^2 + q^2(d\theta^2 + \sin^2\theta\, d\phi^2)\right] \tag{27.1}$$

where $q(\bar{r}) = R\sin(\bar{r}/R), \bar{r},$ or $R\sinh(\bar{r}/R)$, depending on how the density of the universe compares to the critical density. The coordinate $\bar{r}$ is comoving with the expanding universe, in that an object at rest with respect to the "gas" of galaxies has a fixed $\bar{r}$ coordinate (galaxies themselves remain at essentially fixed $\bar{r}$ if we ignore their small random motions). At a given instant of cosmic time t, the physical radial distance from the origin to any object at a coordinate $\bar{r}$ is

$$d = \int_0^d ds = a(t)\int_0^{\bar{r}} d\bar{r} = a\bar{r} \tag{27.2}$$

When describing the radial motion of light, it is convenient to define a "conformal" time coordinate $\bar{t}$ ("t-bar") such that

$$\bar{t} \equiv \int_0^t \frac{dt}{a(t)} \quad \Rightarrow \quad dt = a\, d\bar{t} \tag{27.3}$$

Equation 27.1 for the cosmic metric then becomes

$$ds^2 = a^2\left[-d\bar{t}^2 + d\bar{r}^2 + q^2(d\theta^2 + \sin^2\theta\, d\phi^2)\right] \tag{27.4}$$

The advantage of this coordinate system is that for radially moving light,

$$0 = ds^2 = a^2(-d\bar{t}^2 + d\bar{r}^2) \quad \Rightarrow \quad d\bar{r} = \pm d\bar{t} \tag{27.5}$$

where the positive sign applies to light that moves radially *outward* as time moves forward and the minus sign to light that moves radially *inward*.

In a spacetime diagram in $\bar{t}, \bar{r}$ coordinates, radially moving light therefore follows straight lines at 45° angles. Figure 27.1 shows such a spacetime diagram where we are at the spatial origin. Note that as $\bar{t}$ progresses, the triangle moves up the $\bar{t}$ axis, and the range of $\bar{r}$ that we can see grows. Since individual galaxies are approximately at rest with respect to the comoving $\bar{r}$ coordinate, this means that as time passes, we will be able to see more galaxies and thus more of the universe (what we could actually see if we watched for millions of years would be the formation of new stars and galaxies just after the Big Bang).

The present physical distance to the edge of the visible universe (i.e., to those places where we are now receiving light from very shortly after the Big Bang) is

$$d_{\text{horiz}} = a\bar{r}_{\text{horiz}} = \bar{r}_{\text{horiz}} = \bar{t}_0 \tag{27.6}$$

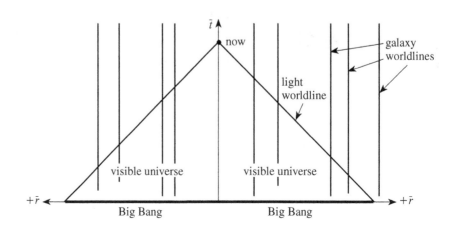

FIG. 27.1 This is a spacetime diagram of the universe. Light from the Big Bang* is currently reaching us from a radial coordinate $\bar{r}$ equal to our present conformal time coordinate $\bar{t}$. As time passes, we will be able to receive light from a greater proportion of the universe; for example, we will begin to see the galaxy whose worldline is on the far right begin to form.

*Technically, we can receive light only from when the universe became transparent, which was about 380,000 y after the Big Bang.

in all directions, where $\bar{t}_0$ is the present value of the $\bar{t}$ coordinate. (Note that d_{horiz} is a radius, not a diameter.) Of course, the distance that the light currently reaching us from the horizon has actually traveled to get to us is

$$t_0 = \int_0^{\bar{t}_0} a(\bar{t})\,d\bar{t} \qquad (27.7)$$

which is smaller, because $a(\bar{t}) \leq a(\bar{t}_0)$ over the range of integration.

The Cosmic Redshift. In principle, one could determine both the recession velocity v and its physical distance d by attaching one end of a tape measure to the center of that galaxy, stretching it between the earth and that galaxy and then watching the tape move past the earth as the distance between the earth and that galaxy increases due to the cosmic expansion. Unfortunately, we can't actually get to a distant galaxy to do this, and no one seems to have left tape measures lying around the universe. So we have to come up with another way to measure these quantities.

The only practical way to determine the rate at which the distance between us and a distant galaxy is growing is to measure the redshift of the light from that galaxy. Let's define the redshift z to be the fractional shift in wavelength:

$$z \equiv \frac{\lambda_r - \lambda_e}{\lambda_e} \qquad (27.8)$$

where λ_e is the wavelength of an arbitrary spectral line in the light emitted by the galaxy and λ_r is its wavelength when it is received by us on the earth. Define our position to be $\bar{r} = 0$. Light arriving at our location will be following an ingoing radial worldline by definition, so $d\bar{t} = -d\bar{r}$ for every step along that worldline. Integrating this, we find that the equation of motion for a given photon of this light is

$$\bar{t} - \bar{t}_e = -(\bar{r} - \bar{r}_e) \qquad (27.9)$$

So for a photon reaching earth at the present time $\bar{t}_0$, we have

$$\bar{t}_0 - \bar{t}_e = -(0 - \bar{r}_e) \quad \Rightarrow \quad \bar{t}_e = \bar{t}_0 - \bar{r}_e \qquad (27.10)$$

Now consider two wavecrests that are emitted by the source galaxy a very short $d\bar{t}$ apart. Since equation 27.9 is linear, they must arrive at the earth with the *same* coordinate separation $d\bar{t}$. The metric implies, however, that the actual physical time between the emission of the wavecrests (assuming that the source is at a fixed $\bar{r}$ coordinate) is $dt = \sqrt{-ds^2} = a_e d\bar{t}$, where a_e is the scale of the universe at the time of emission, and the physical time between the reception of the wavecrests at the earth is given by $a_0 d\bar{t}$, where $a_0 = 1$ is the present scale of the universe. Since the wavelength of light is proportional to the time between adjacent crests, we have

$$z \equiv \frac{\lambda_r - \lambda_e}{\lambda_e} = \frac{a_0 d\bar{t} - a_e d\bar{t}}{a_e d\bar{t}} = \frac{a_0}{a_e} - 1 = \frac{1}{a_e} - 1 \qquad (27.11)$$

This is an exact expression for the redshift. Since $\bar{t}_e = \bar{t}_0 - \bar{r}_e$ (see equation 27.10), if we know how a depends on $\bar{t}$, we can calculate $a_e = a(\bar{t}_e) = a(\bar{t}_0 - \bar{r}_e)$.

We can see how this connects to the Hubble constant H_0 as follows. Note that if $\bar{r}_e$ is small compared to the time scale over which a changes significantly, then we can expand $a_e = a(\bar{t}_0 - \bar{r}_e)$ in a Taylor series around $a(\bar{t}_0)$:

$$a_e = a(\bar{t}_0 - \bar{r}_e) = a(\bar{t}_0) - \bar{r}_e \left(\frac{da}{d\bar{t}}\right)_0 + \cdots \approx 1 - \bar{r}_e \left(\frac{da}{d\bar{t}}\right)_0 \qquad (27.12)$$

If you rearrange equation 27.11 to make one side equal to $a(\bar{t}_0 - \bar{r}_e)$ and use the definitions of H_0 and the present distance d to a galaxy, you can show that

$$\frac{1}{1+z} \approx 1 - H_0 d \qquad (27.13)$$

(see box 27.1). If $H_0 d$ is small (and it better be, because it represents the second term in the Taylor expansion in equation 27.12), then you can invert both sides again and use the binomial approximation to get

$$z \approx H_0 d \qquad (27.14)$$

to first order in $H_0 d$ (see box 27.2).

In special relativity, the redshift z for a source receding at speed v is

$$z + 1 = \frac{\lambda_r}{\lambda_e} = \sqrt{\frac{1+v}{1-v}} \approx (1 + \tfrac{1}{2}v)(1 + \tfrac{1}{2}v) \approx 1 + v \quad \Rightarrow \quad z \approx v \qquad (27.15)$$

as long as $v \ll 1$. Therefore, a galaxy's redshift z due to cosmic expansion will look the same as if the galaxy were moving away from us at speed $v = z$ (even though the galaxy is not really moving so much as the space between us and it is expanding). This means that for galaxies whose distances are such that $H_0 d \ll 1$, equation 27.14 is equivalent to the simple Hubble relation $v \approx H_0 d$. However, I want you to recognize that this relation is an *approximation* valid for small z. For large z, we have to use the exact equation 27.11 to calculate a given galaxy's z, which means that we have to know the function $a(\bar{t})$ and the galaxy's coordinate $\bar{r}_e$.

Luminosity Distance. This presents another problem, as we cannot *directly* measure a galaxy's coordinate $\bar{r}_e$. What we *can* do is measure how bright a distant object is compared to how bright we think it *should* be. Let's define an object's **luminosity distance** d_L to be such that the observed flux f (energy per unit time per unit area) at the earth is equal to $f = L/4\pi d_L^2$, where L is the object's presumed luminosity. This would be equal to the object's actual physical distance if (1) the universe's spatial geometry is flat, and (2) the universe were not expanding. How is it related to observed quantities in an actual universe?

Imagine that a source at a radial coordinate $\bar{r}_e$ emits N photons with energy h/λ_e (in GR units) during a certain tiny time interval Δt_e. The source's luminosity is thus

$$L_s = \frac{\text{energy}}{\text{time}} = \frac{Nh}{\lambda_e \Delta t_e} \qquad (27.16)$$

As time passes, these photons spread out and the universe continues to expand. When an earth observer finally receives some of these photons at the present time, they have traveled a coordinate distance of $\bar{r}_e$ away from the source in all directions and a has attained its present value of 1, so (according to the metric), the photons have been spread out over a physical area of $4\pi q_e^2$, where $q_e \equiv q(\bar{r}_e) = R\sin(\bar{r}_e/R)$ or $\bar{r}_e$ or $R\sinh(\bar{r}_e/R)$ (depending on the value of Ω_k). Each photon's wavelength will also be red-shifted to λ_r such that $z = \lambda_r/\lambda_e - 1 \Rightarrow \lambda_r = \lambda_e(1+z)$. Moreover, the time Δt_r during which photons arrive at the earth has also been red-shifted compared to the time of emission Δt_e by the same factor: $\Delta t_r = \Delta t_e(1+z)$. Therefore (see box 27.3), the luminosity distance to the source is given by

$$d_L = \frac{q_e}{a_e} = \frac{q(\bar{r}_e)}{a(\bar{t}_0 - \bar{r}_e)} \qquad (27.17)$$

If we know the functions $a(\bar{t})$ and $q(\bar{r})$, we can in principle use equation 27.17 to calculate the source's $\bar{r}_e$ coordinate from its measured flux at the earth.

The Universe Is Flat. Conversely, we can measure the luminosity distances and redshifts to distant objects and use equations 27.17 and 27.11 to find the best fit to the functions $a(\bar{t})$ and $q(\bar{r})$. The cosmology community was shaken in 1999 when two separate groups published analyses of very distant supernovae arguing that the data were fit best by the functions $a(\bar{t})$ and $q(\bar{r})$ produced by a universe model where $\Omega_m \approx 0.3$ and $\Omega_v \approx 0.7$ (see figure 27.2). This was subsequently and independently corroborated by analysis of the CMB. Most recently, CMB data from the Wilkinson Microwave Anisotropy Probe (WMAP) has been able to establish cosmic parameters to within a few percent (we will see how in the next chapter). In particular, the most WMAP data in combination with the best data from other sources implies that Ω_k is $-0.0023(\pm 0.0056)$, which is consistent with zero, the age of the universe is $t_0 = 13.75(\pm 0.11)$ Gy, and Ω_m and Ω_v are $0.272(\pm 0.016)$ and $0.728(\pm 0.016)$ respectively (Jarosik et al., *Astrophysical Journal Supplement*, **192**:14, 2011). Various other lines of evidence suggest that $\Omega_r \approx 0.000084$.

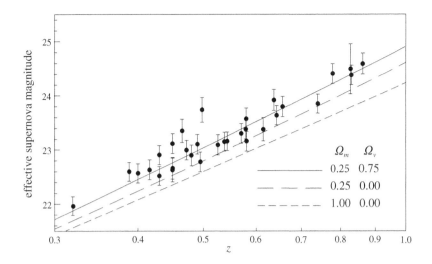

FIG 27.2 A graph of fairly recent supernova data illustrating how those data constrain possible models of the universe. This graph was adapted from R. A. Knop, et al., *Astrophysical Journal*, **598**, 2003, 102.

Calculating the Evolution of a Flat Universe. We are finally in a position to calculate the evolution of the universe. Note that equation 27.3 implies that $d\bar{t}/dt = 1/a$. Let's define a unitless time parameter $\eta \equiv \bar{t}H_0$. If you set $\Omega_k = 0$ in equation 26.20 and use the results just discussed, you can show (see box 27.4) that we get the completely unitless evolution equation

$$\frac{da}{d\eta} = \sqrt{\Omega_r + \Omega_m a + \Omega_v a^4} \tag{27.18}$$

While we cannot easily solve this equation *analytically* for $a(\eta)$, we can easily solve it numerically (see box 27.5). Solving this equation numerically is easier than solving equation 26.20 because it is both completely unitless and non-singular at $a = 0$. Once you have generated a table of $a_i = a(\eta_i)$, one can calculate $\bar{t}_i = \eta_i/H_0$ and t_i by integrating $dt = a(\bar{t})d\bar{t} = a(\eta)d\eta/H_0$, and thereby assemble graphs of $a(\bar{t})$ and $a(t)$. Such graphs are shown in figure 27.3 for the current best-fit values of Ω_m and Ω_v. You can also use equations 27.11 and 27.17 and figure 27.3a to calculate the redshift z or the luminosity distance for any $\bar{r}$ (or vice versa).

FIG. 27.3 These graphs display the scale factor a as a function of (a) conformal time $\bar{t}$ and (b) coordinate time t for a model of the universe where $\Omega_m = 0.272$, $\Omega_v = 0.728$ and $\Omega_r = 0.000084$. Graph (a) is useful for evaluating equations 27.17 and 27.11.

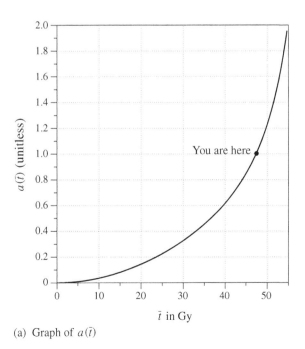

(a) Graph of $a(\bar{t})$

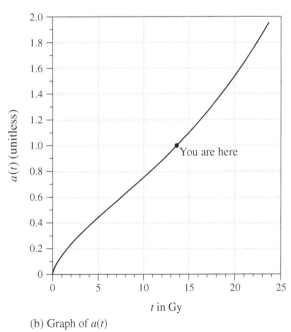

(b) Graph of $a(t)$

BOX 27.1 Connecting the Redshift z to the Hubble Constant

Exercise 27.1.1. Verify that $(1 + z)^{-1} \approx 1 - H_0 d$ (equation 27.13) follows from $z = (1/a_e) - 1$ (equation 27.11) and $a_e \approx 1 - \bar{r}_e (da/d\bar{t})_0$ (equation 27.12) in the limit that H is changing only slowly.

BOX 27.2 Deriving the Hubble Relation in Terms of Redshift z

Exercise 27.2.1. Verify that $z \approx H_0 d$ (equation 27.14) follows from equation 27.13 (see above) in the limit that $H_0 d \ll 1$.

BOX 27.3 The Luminosity Distance

Exercise 27.3.1. Show that $d_L = q_e/a(\bar{t}_0 - \bar{r}_e)$ (equation 27.17) follows from $z = (1/a_e) - 1$ (equation 27.11) and 27.16 and the information following the latter.

BOX 27.4 The Differential Equation for $a(\eta)$

According to equation 26.20, the Friedman equation for a reads

$$\left(\frac{1}{H_0}\frac{da}{dt}\right)^2 + \left(-\frac{\Omega_m}{a} - \frac{\Omega_r}{a^2} - \Omega_v a^2\right) = \Omega_k \qquad (27.19)$$

Exercise 27.4.1. Show that this, along with $\Omega_k = 0$, $\eta \equiv \bar{t}H_0$, and $d\bar{t}/dt = 1/a$, yields $da/d\eta = [\Omega_r + \Omega_m a + \Omega_v a^4]^{1/2}$ (equation 27.18).

BOX 27.5 How to Generate a Numerical Solution for Equation 27.18

To find a numerical solution for $da/d\eta = [\Omega_r + \Omega_m a + \Omega_v a^4]^{1/2}$ (equation 27.18), first divide up the η number line into equal steps of small $\Delta\eta$ starting at $\eta = 0$ and set up a table to store a value of a after each step. Then write equation 27.18 using a centered difference as an approximation for the derivative:

$$\frac{a_{i+1} - a_{i-1}}{2\Delta\eta} = \sqrt{\Omega_r + \Omega_m a_i + \Omega_v a_i^4}$$

$$\Rightarrow \quad a_{i+1} = a_{i-1} + 2\Delta\eta\sqrt{\Omega_r + \Omega_m a_i + \Omega_v a_i^4} \quad (27.20)$$

where a_i is the value of a at time η_i. So, given the values of a_i and a_{i-1}, one can use this to calculate a_{i+1}. We need only two values $a(0)$ and $a_1 \equiv a(\Delta\eta)$ to start the calculation. Since the universe starts at infinite density, we will assume that $a(0) = 0$. At very small values of a, the radiation term Ω_r in equation 27.20 dominates (indeed, with the values of Ω_m and Ω_r that we think actually apply to our universe, the radiation term is very important until η is larger than about 0.2). Therefore, you can calculate a_1 by setting up the difference equation 27.20 to span the step between $a(0)$ and a_1 (this step is only $\Delta\eta$ wide, not $2\Delta\eta$) and solving for a_1 to get $a_1 \approx \sqrt{\Omega_r}\,\Delta\eta$.

Exercise 27.5.1. Verify this last result.

Exercise 27.5.2. Let $\Delta\eta = 0.01$, $\Omega_r = 0.0001$, $\Omega_m = 0.3$, and $\Omega_v = 0.7$. Use the method described above to calculate a_0 through a_4.

HOMEWORK PROBLEMS

P27.1 What is the present physical distance d_{horiz} to the matter that created CMB we see now, according to the present best-fit model of the universe? Compare your result to $t_0 = 13.75$ Gy, the distance the light has actually traveled. (*Hint:* Use figure 27.3a.)

P27.2 Suppose that we discover that a certain kind of galaxy has a fixed linear size D. One could then compute its effective distance d_A by dividing its size D by its measured angular size as viewed from earth. Find a formula for d_A in terms of $\bar{r}_e$ that is analogous to equation 27.17 for the luminosity distance. Do not assume that the universe is flat.

P27.3 Using the method described in box 27.5, set up a spreadsheet, a Mathematica program, or some other kind of computer program to create a graph of $a(t)$ for our universe. You will have to develop a differencing scheme for calculating $t(\eta)$ analogous to the scheme we developed to find $a(\eta)$. Submit both a graph and a printout or description of your program, and compare your graph to figure 27.3b.

P27.4 Using the spreadsheet or program you created for problem P27.3 or a spreadsheet supplied by your instructor, create a graph of d_L versus z for our universe. Submit a printout or description of your modifications to the program or spreadsheet as well as a printout of the graph.

P27.5 A galaxy's measured redshift is $z = 1$. What is that galaxy's luminosity distance according to our current model of the universe? (*Hint:* Use figure 27.3a.)

P27.6 Imagine a model universe that (like ours) has negligible radiation energy ($\Omega_r \approx 0$) and $\Omega_m + \Omega_v + \Omega_r = 1$ but (unlike ours) has no vacuum energy, so that $\Omega_m \approx 1$. Assume also that $H_0^{-1} = 13.9$ Gy (as in our universe).
a. Is the spatial geometry of this universe spherical, flat, or saddle-like? How do you know?
b. Show that $a(t) = \left(\tfrac{3}{2} H_0 t\right)^{2/3}$ for this model universe.
c. Find the age of this model universe, and compare with 13.2 Gy, which is the best independent estimate for the age of the oldest known star in our galaxy.
d. The universe became transparent when $z = 1090$. How many years was this after the Big Bang in this model universe? (Calculate this, don't just quote a result from some source.)
e. Show that $\bar{t} = (12 H_0^{-2} t)^{1/3}$ in this model.
f. Imagine that we observe a galaxy with a redshift $z = 1$. How long ago did light leave the galaxy? How far away is the galaxy at the present time (if we could lay a tape measure between that galaxy and ours)? What is the galaxy's luminosity distance d_L? Express all of your answers in terms of both H_0^{-1} and Gy.

P27.7 If the only kind of matter in the universe were ordinary baryonic matter, then $\Omega_m \approx 0.04$, which is pretty small. An approximation to such a universe would be a completely empty universe model where $\Omega_m = \Omega_r = \Omega_v = 0$. Assume that $H_0^{-1} = 13.9$ Gy, as in our universe.
a. Show that in such a universe $a(t) = H_0 t$.
b. Show that $R = H_0^{-1}$, and write down the metric for this model universe.
c. Find the age of this model universe, and compare with 13.2 Gy, which is the best independent estimate for the age of the oldest known star in our galaxy.
d. The universe became transparent when $z = 1090$. How many years was this after the Big Bang in this model universe? (Calculate this, don't just quote a result from some source.)
e. Show that $\bar{t} = H_0^{-1} \ln(t/T)$, where T is a nonzero coordinate time where we arbitrarily set $\bar{t} = 0$. (You should find that in this model we *can't* set $\bar{t} = 0$ when $t = 0$. However, ultimately this does not matter.)
f. Imagine that we observe a galaxy with a redshift $z = 1$. How long ago did light leave the galaxy? How far away is the galaxy at the present time (if we could lay a tape measure between that galaxy and ours)? What is the galaxy's luminosity distance d_L? Express all of your answers in terms of both H_0^{-1} and Gy. [*Hint:* Note that $\sinh x \equiv \tfrac{1}{2}(e^x - e^{-x})$.]

P27.8 Imagine a model universe that has $\Omega_m + \Omega_v + \Omega_r = 1$ (like ours) but has no dark energy or dark matter, and (like ours) has a smidgen of baryonic matter ($\Omega_m \approx 0$) and (unlike ours) has lots of radiation (perhaps in the form of exotic massless neutrinos) so that $\Omega_r \approx 1$. Assume that in this universe, $H_0^{-1} = 13.9$ Gy (as in our universe).
a. Show that in such a universe $a(t) = \sqrt{2 H_0 t}$.
b. Is the spatial geometry of this universe spherical, flat, or saddle-like? How do you know?
c. Find the age of this model universe, and compare with 13.2 Gy, which is the best independent estimate for the age of the oldest known star in our galaxy.
d. The universe became transparent when $z = 1090$. How many years was this after the Big Bang in this model universe? (Calculate this, don't just quote a result from some source.)
e. Show that $\bar{t} = \sqrt{2 H_0^{-1} t}$.
f. Imagine that we observe a galaxy with a redshift $z = 1$. How long ago did light leave the galaxy? How far away is the galaxy at the present time (if we could lay a tape measure between that galaxy and ours)? What is the galaxy's luminosity distance d_L? Express all of your answers in terms of both H_0^{-1} and Gy.

P27.9 Assume that we now observe galaxies to have redshift z if they emitted light at a certain cosmic time t. Show that the temperature T of the CMB at that t is

$$T = T_0(1 + z) \qquad (27.21)$$

where T_0 is the temperature of the cosmic microwave background now. (*Hint:* Consider equation 26.10.)

P27.10 Quantum field theory provides a nice potential explanation for the vacuum energy density. The vacuum could very well (indeed, really should) have a zero point energy, much as a quantum harmonic oscillator or particle-in-a-box has nonzero energy even in the ground state. We can estimate what the energy density should be using dimensional analysis. Since this energy density should involve quantum mechanics, gravity, and relativity, it makes sense that it should be within an order of magnitude or so of whatever combination of $\hbar$, G, and c has units of mass density.

a. Show that the Planck mass density $\rho_{Pl} \equiv c^5/\hbar G^2$ is the only combination of these three constants that has the correct (SI) units.

b. Evaluate the Planck mass density and compare to the value of ρ_v that follows from the observed value of Ω_v and the definition of the same. Do you see a problem?

Comments: This mystery remains unsolved, and it is very troubling to thoughtful physicists. Adler et al., *American Journal of Physics,* **63**, 7 (1995) presents a reasonably accessible discussion of the quantum-mechanical aspects of this issue in much more depth, though it was written at a time before the discovery that the cosmological vacuum energy is indeed nonzero. See Shaw and Barrow, *Physical Review D*, **83**, 2011, 043518 for a review of the cosmological aspects of the problem and one testable example of a proposed solution.

28. THE EARLY UNIVERSE

INTRODUCTION

FLAT SPACETIME
Review of Special Relativity
Four-Vectors
Index Notation

TENSORS
Arbitrary Coordinates
Tensor Equations
Maxwell's Equations
Geodesics

SCHWARZSCHILD BLACK HOLES
The Schwarzschild Metric
Particle Orbits
Precession of the Perihelion
Photon Orbits
Deflection of Light
Event Horizon
Alternative Coordinates
Black Hole Thermodynamics

THE CALCULUS OF CURVATURE
The Absolute Gradient
Geodesic Deviation
The Riemann Tensor

THE EINSTEIN EQUATION
The Stress-Energy Tensor
The Einstein Equation
Interpreting the Equation
The Schwarzschild Solution

COSMOLOGY
The Universe Observed
A Metric for the Cosmos
Evolution of the Universe
Cosmic Implications
The Early Universe
CMB Fluctuations & Inflation

GRAVITATIONAL WAVES
Gauge Freedom
Detecting Gravitational Waves
Gravitational Wave Energy
Generating Gravitational Waves
Gravitational Wave Astronomy

SPINNING BLACK HOLES
Gravitomagnetism
The Kerr Metric
Kerr Particle Orbits
Ergoregion and Horizon
Negative-Energy Orbits

this depends on this

Introduction. In the last chapter, we talked about how we could model the evolution of the universe given the parameters $\Omega_r, \Omega_m, \Omega_v,$ and H_0 at the present time. I also said that detailed observations of the cosmic microwave background (CMB) have helped determine the present values of these parameters. In the next two chapters, we will discuss the physics of the universe for the first 380,000 years after the Big Bang to see how studying relics of the early universe (including the CMB) can give us insight into the values of $\Omega_r, \Omega_m, \Omega_v$ and H_0 at the present.

Review of the Universal Evolution Equations. According to equation 26.20, the evolution of the universal scale a is governed by the equation

$$\left(\frac{1}{H_0}\frac{da}{dt}\right)^2 = \Omega_k + \frac{\Omega_m}{a} + \frac{\Omega_r}{a^2} + \Omega_v a^2 \tag{28.1}$$

where $\Omega_k \equiv 1 - \Omega_m - \Omega_r - \Omega_v$ and $\Omega_{(m,r,v)} \equiv 8\pi G \rho_{0(m,r,v)}/3H_0^2$. We also saw (see equations 26.9 and 26.10) that the density ρ_r of any relativistic "radiation," the absolute temperature T of that radiation, and the density ρ_m of any non-relativistic "matter" are related to the universe's scale according to

$$\rho_r \propto a^{-4}, \quad T \propto a^{-1}, \quad \text{and} \quad \rho_m \propto a^{-3} \tag{28.2}$$

Single-Component Universes. As we will see, at various times in the early universe, different single components dominated the right side of equation 28.1. While it is difficult to solve equation 28.1 analytically in general, we can easily find analytical solutions when one term dominates. You can show (see box 28.1) that

$$[a(t)]^2 - [a(t_s)]^2 = 2\sqrt{\Omega_r} H_0(t - t_s) \quad \text{(when radiation dominates)} \tag{28.3}$$

$$[a(t)]^{3/2} - [a(t_s)]^{3/2} = \tfrac{3}{2}\sqrt{\Omega_m} H_0(t - t_s) \quad \text{(when matter dominates)} \tag{28.4}$$

$$a(t) = a(t_s) e^{(t-t_s)H_0} \quad \text{(when vacuum energy dominates and } \Omega_k = 0\text{)} \tag{28.5}$$

where t_s is the time when the component in question starts to dominate.

The Radiation Era. For a short while after the Big Bang, the universe was very small ($a \ll 1$) and, according to equation 28.2, very hot. For sufficiently small a, equation 28.1 implies that no matter what the other terms on the right of that equation might be, the radiation term will dominate in its effect on the universe's evolution. Moreover, when the temperature is sufficiently high, all particles that are in thermal equilibrium with photons will be relativistic, so components of the universe's total energy density that are later considered "matter" will at earlier times be considered "radiation." Let's call the interval after the Big Bang during which radiation dominates in determining the evolution of the universe the "radiation era."

We can roughly estimate the duration of this era as follows. The estimated value of Ω_r is 0.000084 (if we include neutrinos) and that of Ω_m is 0.272. The radiation and matter terms contribute equally in equation 28.1 when $\Omega_m/a \approx \Omega_r/a^2 \Rightarrow a = \Omega_r/\Omega_m \approx 3 \times 10^{-4}$. Assuming that $a(t_s) \approx 0$ and $t_s = 0$ at the beginning of the radiation era, equation 28.3 implies that $t \approx 72{,}000$ y (see box 28.2).

This is only a crude estimate, primarily because matter does *not* have strictly zero effect before this time, so estimating the precise time of equality would require a numerical model. The radiation era does not suddenly end, but rather smoothly transitions to a matter-dominated universe during a period centered on that time.

Time-Temperature Relation. Equation 28.2 tells us that $T \propto a^{-1}$, and equation 28.3 says that $a \propto t^{1/2}$ during the radiation era, so $t \propto 1/T^2$ during that era. With more work (see box 28.3) we can evaluate the constants of proportionality:

$$t \approx \frac{10^{20}\,\text{s}\cdot\text{K}^2}{T^2} \approx \frac{3 \times 10^{12}\,\text{eV}^2\text{s}}{(k_B T)^2} \tag{28.6}$$

within a factor of 2 or so (the exact number depends on how many particle species count as "radiation" at a given time).

CONCEPT SUMMARY

Equilibrium Stages. Just after the Big Bang, all possible particles and antiparticles strongly interacted with each other and were therefore in thermal equilibrium. But as the universe expanded and cooled, certain particle interactions became impossible and certain classes of particles became non-relativistic. One can divide the history of the early universe into stages corresponding to the types of interactions that were possible and/or the types of particles that qualify as "radiation."

A given reaction's rate Γ per particle can be calculated at a given time using

$$\Gamma = n\langle\sigma v\rangle \qquad (28.7)$$

where n is number density of reactants, σ is the reaction's energy-dependent cross section, and v is the relative particle speed (with σv averaged over the thermal distribution of speeds and particle energies). We can understand this formula as follows: A particle with cross-sectional area σ moving with speed v through a motionless sea of reactants would sweep out a volume $\sigma v\,dt$ in a time dt. If the reactant number density is n, then the probability that the particle will hit something during dt is $n\sigma v\,dt$. Dividing by dt gives the rate above. Note that since n has units of m^{-3} and the cross section has units of m^2, Γ has units of s^{-1} (reactions/particle/s).

The key to careful calculations of what interactions are possible at a given time in the rapidly expanding early universe is the **Gamow relation**, which states that a reaction is possible at a given time t only if the number of reactions per unit time per particle Γ exceeds 1 reaction per Hubble time $= 1/t_H = H(t) = \dot{a}/a$ at that time:

$$n\langle\sigma v\rangle \equiv \Gamma \geq H \qquad (28.8)$$

As the universe expands, thermal energies fall, and since reaction cross sections typically decrease sharply with decreasing energy, there will typically come a time when the reaction rate falls below H, i.e., below one reaction per particle in the universe's (approximate) age at that point. Sometimes a reaction will cease simply because energy would not be conserved at the typical thermal energies roughly $k_B T$ (where k_B is Boltzmann's constant) that are available to reactants at the time. However, the value of $k_B T$ by itself is not always a good guide. If the number density n of reactants is very high, the reaction may go even if $k_B T$ is quite a bit smaller than what would be required to conserve energy, as in a thermal distribution there will still be a few particles having enough energy. Conversely, if n is low, the reaction may cease (even if there is ample energy) simply because particles rarely find each other.

During a given transition, the densities of what qualifies as "matter" and "radiation" or even "vacuum" might change. To the extent that we can model these transitions as sudden, we can use the adjustable constant $a(t_s)$ in equations 28.3 through 28.5 to match the value of $a(t)$ at the beginning of the new stage to that at the end of the previous stage, so that $a(t)$ remains continuous. However, it is often the case that the universe grows so much during a given stage that $a(t) \approx 0$ at the stage's beginning compared to its magnitude at the end, so matching becomes moot.

Figure 28.1 is a logarithmic time-temperature number line that shows (very approximately) some important stages in the early universe. In what follows, I will focus on three particular stages whose relics help us fix the present values of Ω_r, Ω_m, Ω_v and H_0.

Neutrino Decoupling. By the time that the universe was a few milliseconds old, the normal matter in the universe consisted of a plasma of protons, neutrons, electrons, positrons, neutrinos, and photons in thermal equilibrium. The equilibrium between protons, neutrons, and neutrinos was maintained by weak nuclear interactions such as $e^+ + n \longleftrightarrow p + \bar{\nu}$ and $\nu + n \longleftrightarrow p + e^-$. The cross section (in SI units) for such interactions turns out to be $\sigma \sim G_F^2 E^2/(\hbar c)^4$, where E is the particle energy and G_F is the **Fermi constant** for the weak interaction $= 8.8 \times 10^{-17}$ eV·nm^3. You can show (see Box 28.4) that these reactions shut down (i.e., fail to satisfy the Gamow relation) when $k_B T$ falls below about 0.7 MeV, a few seconds after the Big Bang. At this time neutrinos become decoupled from normal matter and the ratio of protons and neutrons becomes

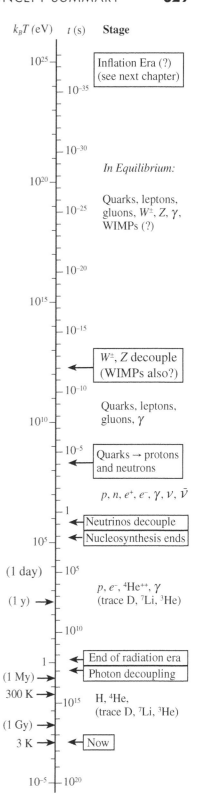

FIG. 28.1 This diagram shows a logarithmic timeline of equilibrium stages in the early universe. The items in boxes indicate important transitions; the lists on the right without boxes indicate the particles in thermal equilibrium with each other at each stage.

"frozen" at the value fixed by thermal equilibrium just before that time. While in thermal equilibrium, protons and neutrons can be considered to be two quantum states separated by 1.29 MeV (the difference in the rest energies of the two particles). Because protons and neutrons are *not* relativistic at this time, the relative probabilities of the two states (and thus the relative numbers $[p]$ and $[n]$ of protons and neutrons) is given by a simple Boltzmann factor:

$$\frac{[n]}{[p]} \approx e^{-\Delta E/k_B T} \approx e^{-1.29/0.7} \approx \frac{1}{6.3} \qquad (28.9)$$

Only a bit later, the reaction $2\gamma \longleftrightarrow e^+ + e^-$ that kept positrons and electrons in thermal equilibrium with photons became impossible in the forward direction (since the forward reaction requires gamma photons with energies of at least 0.511 MeV). Shortly thereafter, essentially all positrons are annihilated, leaving a small excess of electrons and many photons.

Primordial Nucleosynthesis. At this time, protons and neutrons often collided to form deuterium and a photon: $p + n \to D + \gamma$. Initially, there were enough photons around with energies above 1.1 MeV to drive this reaction in reverse as well, so protons, neutrons, deuterium nuclei and photons were in thermal equilibrium. But as the universe continued to cool, the reverse reaction became impossible and deuterium began to accumulate, enabling further reactions such as $n + D \to {}^3H + \gamma$, $p + D \to {}^3He + \gamma$, $p + {}^3H \to {}^4He + \gamma$, and $n + {}^3He \to {}^4He + \gamma$. Indeed, formation of ${}^4He$ was strongly favored thermodynamically, because it is extraordinarily tightly bound (and thus hard to break up) compared to other small nuclei.

However, nucleosynthesis beyond helium was *not* practical. There are no stable nuclei with $A = 5$ or $A = 8$, meaning that one could not create larger nuclei either by adding protons or neutrons one at a time or by banging together two ${}^4He$ nuclei. Creating larger nuclei instead required reactions such as $D + {}^4He \to {}^6Li + \gamma$ and ${}^3H + {}^4He \to {}^7Li + \gamma$, which were rare because D and ${}^3H$ were *much* less common than protons or neutrons. The result was that during the next 200 s or so (roughly 3 minutes), virtually all neutrons ended up in ${}^4He$ nuclei, with only a few ending up in deuterium and lithium nuclei. After 200 s, the number densities of all reactants became too small for further nuclear reactions.

We can therefore roughly predict how much helium should have been formed during this period. Taking the decay of free neutrons ($n \to p + e^- + \bar{\nu}$) into account, the ratio of neutrons to protons at the end of the 200-s period falls to something more like 1/7. If *all* of these neutrons end up being bound into ${}^4He$ nuclei, and approximating the mass of a ${}^4He$ nucleus to be two times the mass of its neutrons, the fraction of the total mass of ordinary matter that is ${}^4He$ would be about

$$\frac{\text{mass}({}^4He)}{\text{mass}(\text{all baryons})} \approx \frac{2[n]}{[p]+[n]} = \frac{2([n]/[p])}{1+([n]/[p])} \approx \frac{2/7}{1+1/7} = \frac{2/7}{8/7} = 0.25 \qquad (28.10)$$

This is very close to the observed ratio, suggesting that we are on the right track.

Now, the precise value of this ratio is sensitive to the temperature of neutrino decoupling, which in turn depends on the number density of reactants. Therefore, we can work backwards using the experimentally observed ratio to infer the number density of reactants (specifically, protons and neutrons) at the transition. We can also do similar calculations working backward from the observed ratios of deuterium, ${}^3He$, and ${}^7Li$ to the same initial density of protons and neutrons. Since the total number of protons and neutrons in the universe has been conserved since well before this transition, we can from these data determine Ω_b, the ratio of the density of ordinary (baryonic) matter at present to the critical density.

Figure 28.2 shows the results of such calculations. The thick gray curves on this graph represent how the ratios of ${}^4He$, D, ${}^3He$, and ${}^7Li$ theoretically depend on the total density of baryonic matter (i.e., protons and neutrons) at the present time. The vertical side of the box associated with each curve represents the uncertainty in the actually measured ratio for that nucleus. The range of values of Ω_b that are consistent with *all* of the observed ratios is therefore represented by the vertical gray bar. The result is that these observations imply that Ω_b must lie in a pretty restricted range centered on about 0.044. Yet we know from the arguments involving galaxy rotation and cluster statistics discussed in chapter 24 that the universe must contain much more matter than this. Thus most of the matter in the universe *cannot* be baryons (i.e., protons and neutrons), but rather something else. Models of the universe best reproduce the universe's current structure if this matter is "cold" (i.e., non-relativistic), so this matter must consist of as-yet-unknown particles with nonzero rest mass which interact only weakly with ordinary matter (the WIMPs discussed in chapter 24).

Photon Decoupling. Excluding dark matter and neutrinos (both of which were thermally decoupled from everything else at this time), the expanding universe then consisted of photons and non-relativistic matter, which was comprised mostly of protons and electrons (the *number* density of helium nuclei was only about 6% of those for protons and electrons). During this time, scattering interactions such as $e^- + \gamma \longleftrightarrow e^- + \gamma$ and $p + \gamma \longleftrightarrow p + \gamma$ and the reversible interaction $p + e^- \longleftrightarrow H + \gamma$ kept hydrogen atoms, protons, electrons, and photons in thermal equilibrium. Because the number density of photons was so much larger than that of protons during this period (see box 28.5), there remained enough high-energy photons around to keep the last of these reactions going in both directions at temperatures even where $k_B T$ was significantly smaller than hydrogen's ionization energy (13.6 eV). Indeed, a careful calculation involving the Gamow relation shows that the reaction remained possible until $k_B T \approx 0.26$ eV, or $T \approx 3000$ K.

When the universe cooled below this temperature, protons and electrons could combine to form hydrogen and photons, but not the reverse. Before long, essentially all free protons and electrons combined to form neutral hydrogen. Because there were no free charged particles left to scatter photons, and because neutral hydrogen has a very low cross section for photons whose energies are below 1 eV (such photons do not have enough energy to excite the hydrogen atom and an electrically neutral atom does not even scatter low-energy photons well), photons became essentially decoupled from matter at this point (i.e., the universe became transparent). Since the universe has remained transparent since, we are able today to observe the remaining photons as the cosmic microwave background (CMB). This transition is often confusingly called the time of "recombination," even though electrons and protons had never been combined before this point.

We can estimate the time of photon decoupling as follows. Note that the observed blackbody temperature T of the photons is now 2.7 K (not 3000 K), T is inversely dependent on the most probable photon wavelength, and Ta is constant. According to equation 27.11, this implies that the universal scale a_{PD} at the decoupling time is given by

$$z = \frac{1}{a_{PD}} - 1 \quad \Rightarrow \quad a_{PD} = \frac{1}{z+1} \approx \frac{1}{1100} \tag{28.11}$$

This scale factor is roughly three times the universe's scale factor during the transition from radiation dominance to matter dominance, so we can get a *very* rough estimate of the age of the universe at the time of photon decoupling by assuming that it was matter-dominated the whole time. Using equation 28.4 with $a(t_s) = 0$ and $t_s = 0$, and using $\Omega_m = 0.272$ and $H_0 = (13.9 \text{ Gy})^{-1}$, we find that

$$t_{PD} \approx \frac{2 a_{PD}^{3/2}}{3 \sqrt{\Omega_m} H_0} = \frac{2(1/1100)^{3/2}}{3\sqrt{0.272}} (13.9 \text{ Gy}) = 490{,}000 \text{ y} \tag{28.12}$$

This is the right order of magnitude, but because the universe is *not* really matter-dominated the whole time, this estimate is high: a numerical model that takes radiation into account yields $t_{PD} \approx 380{,}000$ y (see problem P28.1).

In the next chapter, we will see how observing the detailed characteristics of the CMB can help us determine the universe's spatial geometry and cement the case for the existence of vacuum energy.

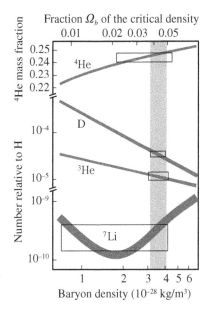

FIG. 28.2 This diagram shows the theoretical abundances of various nuclei as a function of the current density of baryonic matter. The vertical side of the box associated with each curve corresponds to the uncertainty in the measured value of that nucleus's abundance: the horizontal side thus corresponds to the range of baryon densities consistent with that abundance. The vertical gray bar specifies the range of baryon densities consistent with all measured abundances. This figure is adapted from Charbonnel, *Nature*, **415**, 2002, 27–29.

BOX 28.1 Single-Component Universes

When the universal scale a is sufficiently small, the Ω_r/a^2 term will dominate over all other terms on the right side of equation 28.1. If we ignore all these other terms, the universal evolution equation becomes

$$\left(\frac{1}{H_0}\frac{da}{dt}\right)^2 \approx \frac{\Omega_r}{a^2} \quad \Rightarrow \quad \left(a\frac{da}{dt}\right)^2 = H_0^2 \Omega_r \quad \Rightarrow \quad a\,da = H_0\sqrt{\Omega_r}\,dt \quad (28.13)$$

Taking the definite integral of both sides yields

$$\int_{a(t_s)}^{a(t)} a\,da = H_0\sqrt{\Omega_r}\int_{t_s}^{t} dt \quad \Rightarrow \quad \tfrac{1}{2}[a(t)]^2 - \tfrac{1}{2}[a(t_s)]^2 = H_0\sqrt{\Omega_r}\,(t - t_s) \quad (28.14)$$

Multiplying both sides of the last expression by 2 yields equation 28.3. Note that if we can consider $a(t_s) \approx 0$ and $t_s \approx 0$ compared to later times of interest, this expression reduces to the simple statement that $a(t) \propto t^{1/2}$.

Exercise 28.1.1. Follow the same steps to show that equations 28.4 and 28.5 are also correct. In the case of matter dominance, what simpler expression results if we can assume that $a(t_s) \approx 0$ and $t_s \approx 0$ compared to later times of interest?

BOX 28.2 The Transition to Matter Dominance

We will discuss the cosmic neutrino background in box 28.4 and problem P28.2 but for now, let's *assume* that the combination of the neutrino and photon backgrounds have a current fractional energy density of $\Omega_r = 0.000084$. We will also assume that the radiation era goes essentially all the way back to the Big Bang, so that $a(t_s) \approx 0$ and $t_s \approx 0$. The current measured value of the Hubble constant is $H_0 = (13.9 \times 10^9 \text{ y})^{-1}$. Note that we are also assuming that the universe remains completely radiation dominated up until the instant that $a(t) = \Omega_r/\Omega_m \approx 3.09 \times 10^{-4}$.

Exercise 28.2.1. Show that these assumptions imply that the transition from radiation dominance to matter dominance (the time when $a(t) = \Omega_r/\Omega_m \approx 3.09 \times 10^{-4}$) happens at about $t \approx 72{,}000$ y. (**Note:** A more accurate calculation that accounts for the rising influence of matter puts this moment at about $t = 53{,}000$ y.)

BOX 28.3 The Time-Temperature Relation

The Stefan-Boltzmann relation implies that the light power radiated by a flat blackbody surface whose area is A and absolute temperature is T is given by

$$\frac{dE}{dt} = \sigma_{SB} A T^4 \tag{28.15}$$

where σ_{SB} is the Stefan-Boltzmann constant $= 5.6703 \times 10^{-8}$ W·m^{-2}·K^{-4} in SI units. Now this light will move at speed c away from the surface, and so in a time interval dt will fill a box of volume $Ac\,dt$. The density of photon energy in the box is therefore $\rho_r = dE/(Ac\,dt) = (\sigma_{SB}/c)T^4 = \sigma_{SB} T^4$ in GR units.

Relativistic quantum field theory implies that a slightly modified version of this equation applies to all particles in the relativistic limit where particles have energies that are huge compared to their masses (photons have strictly zero mass). The modified equation is

$$\rho_r = \frac{g^*}{2} \sigma_{SB} T^4 \quad \text{(GR units)} \quad \text{where} \quad g^* \equiv \sum_i g_{ib} + \tfrac{7}{8} \sum_i g_{if} \tag{28.16}$$

where g_{ib} is the number of spin states available to the ith type of boson in the plasma and g_{if} is the same for the ith type of fermion (the factor of 7/8 comes from the fact that the thermal distribution of fermions is somewhat different than that for bosons because of the Pauli exclusion principle). Photons have $g^* = 2$ in spite of being spin-1 particles (the $S_z = 0$ state is suppressed in the relativistic limit). At the earliest times, when all quarks, leptons, gluons, WIMPs, vector bosons, photons, and neutrinos were in equilibrium, $g^* \sim 100$ but decreased with time as more and more particles decoupled and/or became non-relativistic. The value of g^* during most of the stages of interest to us was below 11.

BOX 28.3 (continued) The Time-Temperature Relation

Local conservation of energy (see equation 26.9b) and the definitions of Ω_r and ρ_c in equations 26.17 and 26.18 together imply that

$$\rho_r a^4 = \rho_{r0} = \frac{\rho_{r0}}{\rho_c} \rho_c = \Omega_r \frac{3H_0^2}{8\pi G} \quad \Rightarrow \quad \Omega_r = \frac{8\pi G \rho_r a^4}{3H_0^2} \qquad (28.17)$$

If you plug this into $[a(t)]^2 = 2\sqrt{\Omega_r} H_0 t$ (equation 28.3, assuming that $a(t_s) \approx 0$ and $t_s \approx 0$ compared to times of interest) and then use equation 28.16 above to eliminate the density ρ_r, you can show that $t = B\sqrt{2/g^*}/(k_B T)^2$ with $B = 3.43 \times 10^{12}$ eV²·s.

Exercise 28.3.1. Verify that this is true. (Remember that your resulting equation will be in GR units: to evaluate the constant, I recommend simply multiplying your expression for the constant by enough powers of c to convert it to the appropriate SI units, and use the SI values of G, σ_{SB}, and k_B. Finally, convert J to eV.)

Note that $\sqrt{2/g^*}$ ranges from 1 for a pure photon gas to about 0.44 when neutrinos, electrons, and positrons were all still relativistic and interacting. So 3×10^{12} eV²·s represents $B\sqrt{2/g^*}$ within a factor of 2 or so. Obtaining more precision is not really worthwhile, because even equation 28.3 technically does not apply during transitions: during a transition, the mix of relativistic particles changes, implying that Ω_r (the fractional density that particular mix would have if projected to the present) is not constant. So equation 28.6 for $t(k_B T)$ should be taken only as an approximation.

BOX 28.4 Neutrino Decoupling

According to equation 28.16, the energy density of neutrinos (when they are in thermal equilibrium with their surroundings at temperature T) is proportional to T^4. Now the average energy of *photons* emitted by a black body with temperature T is proportional to T, so it is plausible (and quantum statistical mechanics verifies) that this is true of neutrinos as well. This means that the *number* density of neutrinos should be proportional to the total energy density divided by the average energy per neutrino, which will be proportional to $T^4/T = T^3$. The actual number (since it involves quantum mechanics, relativity, and thermal physics) should involve T, k_B, c, and $\hbar$, and the only combination with the right SI units that depends on T^3 is $n \sim (k_B T/\hbar c)^3$.

Exercise 28.4.1. Verify that this is indeed the *only* arrangement of these constants that yields something with the right units and is also proportional to T^3.

By the usual rules of dimensional analysis, we expect the actual number density to be equal to $(k_B T/\hbar c)^3$ to within a dimensionless constant of order of magnitude 1. As stated in the main text, the total cross section for the reactions that keep neutrinos connected to protons and neutrons is $\sigma \sim G_F^2 E^2/(\hbar c)^4$. The thermal average of the neutrino energy will be of order of magnitude $k_B T$, and for relativistic particles, $v \approx c$. Therefore, the reactions' rate should be

$$\Gamma = n \langle \sigma v \rangle \sim \left(\frac{k_B T}{\hbar c}\right)^3 \frac{G_F^2 (k_B T)^2 c}{(\hbar c)^4} = \frac{G_F^2 (k_B T)^5 c}{(\hbar c)^7} \qquad (28.18)$$

Since the universe at the time of neutrino decoupling had been radiation-dominated since very shortly after the Big Bang, you can easily show using box 28.1 that

$$H \equiv \frac{\dot{a}}{a} = \frac{1}{2t} \qquad (28.19)$$

Exercise 28.4.2. Verify that this is correct.

BOX 28.4 (continued) Neutrino Decoupling

The Gamow relationship tells us that the reactions will cease when Γ falls below $H = 1/(2t)$. You can show that this will happen when $k_B T$ falls below 1.14 MeV.

Exercise 28.4.3. Combine $\Gamma = 1/(2t)$ with $t = B\sqrt{2/g^*}/(k_B T)^2$ to verify the last statement. Note that $\hbar c = 197.3$ eV·nm, and the speed of light is 3×10^{17} nm/s.

The actual energy when all dimensionless constants are correctly included is about 0.7 MeV, the number quoted in the text (it is astonishing that the result above is this close, considering all the approximations we have made).

Neutrino decoupling will produce a neutrino background similar to the CMB. For all forms of radiation, we have seen that that $T \propto a^{-1}$, so since $Ta = $ constant $= T_{\text{now}}$ (because $a = 1$ now), we would expect that no matter when it was formed, the temperature of the neutrino background (if we could measure it or even detect the neutrinos at all) would be the same as that of the CMB. However, this is not quite true. Remember that shortly after neutrino decoupling, the reaction $e^+ + e^- \to \gamma + \gamma$ stopped being possible in the reverse direction. The resulting annihilation of all positrons in the universe produced many new high-energy photons, which had the effect of increasing the temperature of the photon gas.

We can estimate the temperature increase as follows. Consider a volume of plasma enclosed by a comoving boundary that expands with the plasma as the universe expands. Since the universe is homogeneous, there can be no net heat flow across the boundary, so the expansion process is adiabatic. If we also assume that the reactions that maintain thermal equilibrium are faster than the annihilation process, then we can consider the expansion process to be quasistatic as well.

BOX 28.4 (continued) Neutrino Decoupling

The entropy of a gas does not change during an adiabatic quasistatic process. Now, it turns out that the entropy *density* of a relativistic gas (which can be found by counting quantum states) is proportional to g^*T^3 (see problem P28.5). Therefore, if the total entropy inside our comoving boundary is to remain constant, then technically it is $g^*(Ta)^3$ that remains constant, not Ta (though the latter is constant *between* transitions because g^* is constant between transitions).

Before electron-positron annihilation, g^* for the relativistic plasma consisting of electrons, positrons, and photons in equilibrium was 11/2. Afterward, when there were only photons left, $g^* = 2$. Therefore the photon gas after this transition was $(11/4)^{1/3}$ times hotter than before the transition.

Exercise 28.4.4. Verify that $g^* = 11/2$ in the first case. (*Hint:* Photons, positrons, and electrons all have two spin states, but positrons and electrons are fermions.)

Since the neutrinos decoupled before the annihilation of the positrons, the cosmic neutrino background will reflect the plasma temperature before this heating, while the CMB will reflect the temperature afterward. Since the latter has a temperature of 2.73 K at present, the cosmic neutrino background must have a temperature of $(2.73 \text{ K})(4/11)^{1/3} = 1.95$ K.

BOX 28.5 The Number Density of Photons

As discussed in the previous box, the number density of relativistic particles in thermal equilibrium at temperature T is $n \sim (k_B T/\hbar c)^3$: the expression for photons including the unitless constants turns out to be $n_\gamma = 0.244(k_B T/\hbar c)^3$. After photon decoupling, the number of CMB photons remains constant, so the number *density* varies as $1/a^3$ (consistent with $T \propto a^{-1}$). The density of matter also varies as $1/a^3$ (see equation 26.9a), so the ratio of photon number density to matter number density should be the same at the time of photon decoupling as it is today. The present number density of nucleons is $n_p \approx \rho_{b0}/m_p = \Omega_b \rho_c/m_p$, where ρ_c is the critical density (see equations 26.17 and 26.18), ρ_{b0} is the normal-matter portion of the current total matter density ρ_{m0}, and Ω_b is the present baryon density fraction ≈ 0.046. You can show from this that $n_\gamma/n_p = 1.6 \times 10^9$. Thus, there are billions photons per proton at the time of decoupling: this is why photon decoupling is delayed until $k_B T$ falls far below the hydrogen ionization energy 13.6 eV.

Exercise 28.5.1. Verify that $n_\gamma/n_p = 1.6 \times 10^9$.

28. THE EARLY UNIVERSE

HOMEWORK PROBLEMS

P28.1 Use your numerical model from problem P27.3 or a computer program provided by your instructor to verify that the age of the universe at the time of photon decoupling is $t_{PD} \approx 380{,}000$ y when the influence of both matter and radiation are taken into account. (*Hint:* If you use your own computer model, you might need to take smaller steps in η to get good results. Remember to use $\Omega_r = 0.000084$ if we include neutrinos.)

P28.2 In this problem, you will show that the neutrino background density fraction is $\Omega_{rn} = 0.000034$. We saw in box 28.4 that the neutrino background temperature T_n is given by $T_n/T_\gamma = (4/11)^{1/3}$, where T_γ is the CMB temperature = 2.73 K. Use this and equation 28.16 to argue that the ratio of the neutrino and photon energy densities is 0.68. Since $\Omega_{r\gamma} = 0.000050$, this means that $\Omega_{rn} = 0.000034$, and thus that the total radiation density fraction is $\Omega_r = 0.000084$, as claimed. (*Hint:* When calculating g^* for neutrinos, remember that neutrinos are spin-1/2 fermions and that there are three types: electron, muon, and tau neutrinos.)

P28.3 What if instead of three neutrino families (electron neutrinos, muon neutrinos, and tau neutrinos) there were *four* families (with one remaining undiscovered)?

a. Review the calculations in box 28.4 to find the fraction by which this would raise the value of $k_B T$ at neutrino decoupling, and so find the new value of $k_B T$ at decoupling *assuming* that with three families it is 0.7 MeV.

b. Determine by what fraction this temperature change will change the ratio $[n]/[p]$ of the number of protons compared to the number of neutrons at the time of neutrino decoupling. (*Hint:* Assume that the same fraction of neutrons decays in both cases.)

c. Determine how this affects the mass fraction of ^{4}He in the universe. Would this significantly contradict the experimental evidence, or is this possible? (*Hint:* Consider the uncertainty box in figure 28.2.)

P28.4 Reactions such as $p + \bar{p} \longleftrightarrow 2\gamma$ and $n + \bar{n} \longleftrightarrow 2\gamma$ kept protons p, neutrons n, and their antiparticles $\bar{p}$ and $\bar{n}$ in thermal equilibrium with photons when $k_B T$ is much larger than the proton mass-energy of about 1 GeV. However, once $k_B T$ fell below this value, protons and neutrons could annihilate with their antiparticles but could no longer be created. Some (still not completely understood) imbalance between these particles meant that after this grand annihilation, there were a few protons and neutrons left. Estimate the time at which the total number of protons and neutrons in the universe became fixed. (Don't worry about the details of the Gamow relation, just do a crude estimate.)

P28.5 In this problem, we will calculate the entropy of a plasma of relativistic particles. Consider a "box" of relativistic plasma with volume V held at a constant temperature T. We know that the plasma's internal energy U and pressure p are

$$U(T, V) = \rho_r V = \tfrac{1}{2} g^* \sigma_{SB} T^4 V \qquad (28.20a)$$

$$p(T) = \tfrac{1}{3} \rho_r = \tfrac{1}{6} g^* \sigma_{SB} T^4 \qquad (28.20b)$$

Now imagine that we slowly expand the box while holding its temperature fixed. By the first law of thermodynamics, $dU = dQ + dW = dQ - p\,dV$, where W is the work done *on* the gas. Note that since T is constant, we can easily calculate ΔU and ΔW for finite volume changes. For a quasistatic isothermal process, the entropy change is $\Delta S = Q/T$. Show that if we assume that $S = 0$ at zero volume, then

$$S = \tfrac{2}{3} g^* \sigma_{SB} T^3 V \qquad (28.21)$$

(This implies, as claimed in box 28.4, that the entropy density $s = S/V = \tfrac{2}{3} g^* \sigma_{SB} T^3 \propto g^* T^3$.)

P28.6 Problem P28.2 argues that if neutrinos remain a relativistic gas from neutrino decoupling to the present, then their present fractional contribution to the total energy density of the universe would be small ($\Omega_{rn} \approx 0.000034$). However, this assumes that neutrinos have zero rest mass, so that their energy density is simply their present thermal "kinetic energy" (roughly $k_B T$) times the present neutrino number density n. However, we now suspect that neutrinos have *some* rest mass. Thus, neutrinos *could* contribute more significantly to the universe's total energy density, because then their energy density would be $\approx mn$ (where m is the neutrino mass averaged over the three types) and m could be $\gg k_B T$. In such a case neutrinos would *not* be relativistic at present, and they would be counted not as radiation but as exotic (non-baryonic) dark matter. We know from cosmological observations that the fractional exotic matter contribution to the universe's total energy density is $\Omega_{ex} = \Omega_m - \Omega_b \approx 0.272 - 0.046 \approx 0.226$. Use this cosmological information to put an estimated upper limit on the value of the average neutrino rest mass m. (*Hint:* This problem is very easy if you think about it. Remember that the actual number of neutrinos has been fixed since the time of neutrino decoupling, so their number density will be essentially the same whether they have rest mass or not.)

Comments: Models of galactic evolution and other observations put even more severe limits on neutrino masses, making it improbable that neutrinos contribute all that much to the universe's exotic dark matter. See Goobar et al., *Journal of Cosmology and Astroparticle Physics*, 0606:019, 2006 (also available at http://arxiv.org/abs/astro-ph/0602155).

29. CMB FLUCTUATIONS AND INFLATION

INTRODUCTION

FLAT SPACETIME
Review of Special Relativity
Four-Vectors
Index Notation

TENSORS
Arbitrary Coordinates
Tensor Equations
Maxwell's Equations
Geodesics

SCHWARZSCHILD BLACK HOLES
The Schwarzschild Metric
Particle Orbits
Precession of the Perihelion
Photon Orbits
Deflection of Light
Event Horizon
Alternative Coordinates
Black Hole Thermodynamics

THE CALCULUS OF CURVATURE
The Absolute Gradient
Geodesic Deviation
The Riemann Tensor

THE EINSTEIN EQUATION
The Stress-Energy Tensor
The Einstein Equation
Interpreting the Equation
The Schwarzschild Solution

COSMOLOGY
The Universe Observed
A Metric for the Cosmos
Evolution of the Universe
Cosmic Implications
The Early Universe
CMB Fluctuations & Inflation

GRAVITATIONAL WAVES
Gauge Freedom
Detecting Gravitational Waves
Gravitational Wave Energy
Generating Gravitational Waves
Gravitational Wave Astronomy

SPINNING BLACK HOLES
Gravitomagnetism
The Kerr Metric
Kerr Particle Orbits
Ergoregion and Horizon
Negative-Energy Orbits

this depends on this

29. CMB FLUCTUATIONS AND INFLATION

Introduction. In this final chapter on applications of general relativity to the universe, we will see how we can "read" information about the shape and evolution of the universe from the cosmic microwave background.

CMB Fluctuations and Cosmic Flatness. In the last chapter, we saw how photon decoupling produces the cosmic microwave background (CMB) about 380,000 y after the Big Bang. It turns out that the angular size of the largest fluctuations in the CMB provides a way of measuring the curvature of space in the universe. Fluctuations in the CMB come from mechanical (sound) oscillations in the baryon-photon plasma at the time of photon decoupling. The largest possible physical size of such an oscillation would be one where a single oscillation wavelength fits within the maximum distance that sound could travel in the plasma during the time before decoupling. Now, the speed of sound in an ideal gas is given by

$$v_s = \sqrt{\frac{p_0}{\rho_0}} \tag{29.1}$$

where p_0 and ρ_0 are the gas's pressure and density in its rest frame. Before photon decoupling, virtually all of the plasma's pressure and most of its density come from the photon-gas component. If we ignore the baryon contribution to the density,[1] then $p_0 = \frac{1}{3}\rho_0$ and the sound speed will be a constant $v_s = \sqrt{1/3}$. (Including the baryon density makes the speed only a bit smaller for times after the radiation era, but greatly complicates the analysis.) According to equation 27.5, by the time t_{PD} of photon decoupling, light could travel a comoving coordinate distance of $\Delta\bar{r} = \bar{t}_{PD}$ away from any given origin, so sound should be able to travel a coordinate separation of $\Delta\bar{r}_s = v_s\bar{t}_{PD}$ away from the origin in that time. According to the universal metric

$$ds^2 = -dt^2 + a^2\{d\bar{r}^2 + [q(\bar{r})]^2(d\theta^2 + \sin^2\theta\, d\phi^2)\} \tag{29.2}$$

a radial coordinate separation of $\Delta\bar{r}_s = v_s\bar{t}_{PD}$ corresponds to a *physical* distance at the decoupling time of $\Delta s = a_{PD}\Delta\bar{r}_s$, or

$$\Delta s = a_{PD}v_s\bar{t}_{PD} \approx \frac{a_{PD}\bar{t}_{PD}}{\sqrt{3}} = \frac{a_{PD}}{\sqrt{3}}\int_0^{t_{PD}} \frac{dt}{a(t)} \tag{29.3}$$

where the last step follows from the definition of the *t*-bar coordinate given in equation 27.3. So this Δs should be the physical size of the largest CMB fluctuations at the time of decoupling.

If we assume that Δs is small and lies on or near the equatorial plane $\theta = \pi/2$, the universal metric tells us that the fluctuation's angular size $\Delta\phi$ will be such that $\Delta s \approx a_{PD}q(\bar{r}_{PD})\Delta\phi$ or

$$\Delta\phi \approx \frac{\Delta s}{a_{PD}q(\bar{r}_{PD})} \approx \frac{1}{\sqrt{3}\, q(\bar{r}_{PD})}\int_0^{t_{PD}} \frac{dt}{a(t)} \tag{29.4}$$

where $q(\bar{r}_{PD}) = R\sin(\bar{r}_{PD}/R), \bar{r}_{PD}$, or $R\sinh(\bar{r}_{PD}/R)$ depending on whether Ω_k is negative, zero, or positive, respectively, and $\bar{r}_{PD}$ is the comoving radial coordinate (i.e., the present radial distance) of the matter whose CMB photons we are receiving now. Since we are receiving the light at $\bar{r} = 0$ and at time t_0, we have

$$\bar{r}_{PD} = \bar{t}_0 - \bar{t}_{PD} = \int_{t_{PD}}^{t_0} \frac{dt}{a(t)} \tag{29.5}$$

We can roughly evaluate the integrals in equations 29.4 and 29.5 by assuming that the universe's evolution has been matter-dominated since photon decoupling. While the universe has actually been vacuum dominated for the last few billion years, note that the integrand is large when a is small, so the integral strongly emphasizes early times when the approximation is excellent. You can show (see box 29.1) that this approximation and $R = H_0^{-1}|\Omega_k|^{-1/2}$ (equation 26.21) imply that

---

[1]. While the density of *all* matter is greater than the density of radiation at the time of photon decoupling, the density of *baryons* is still significantly smaller. Because dark matter is uncoupled from baryons and photons at this time, it has no effect on the speed of sound.

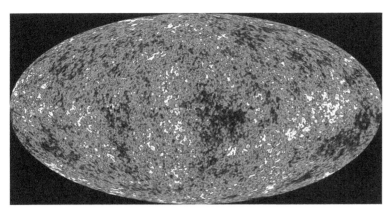

FIG. 29.1 This is a map of the CMB (corrected for the earth's motion and contributions from the galaxy). The fluctuations represent temperature changes on the order of tens of μK. **Credit:** NASA and the Wilkinson Microwave Anisotropy Probe (WMAP) team.

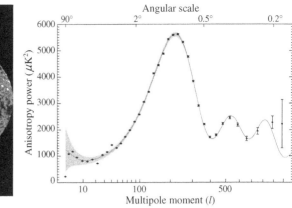

FIG. 29.2 This is the 2D analogue to a Fourier transform of the fluctuations showing that the biggest fluctuations have an angular size of about 1°. **Credit:** NASA and the WMAP team.

$$\Delta\phi \approx \frac{a_{PD}^{1/2} u}{\sqrt{3}\sin u}, \frac{a_{PD}^{1/2}}{\sqrt{3}}, \frac{a_{PD}^{1/2} u}{\sqrt{3}\sinh u} \quad \text{where} \quad u \equiv 2\sqrt{\frac{|\Omega_k|}{\Omega_m}} \qquad (29.6)$$

where the choices correspond to Ω_k negative, zero, and positive, respectively. This means that if we know a_{PD} and Ω_m, measuring the angular width of these largest fluctuations can help us determine both the sign and magnitude of Ω_k: $\Delta\phi$ in the middle case is about $(1100)^{-1/2}/\sqrt{3} \approx 0.0174$ rad $\approx 1°$, and will be bigger than this if Ω_k is negative and smaller if Ω_k is positive.

Figure 29.1 shows fluctuations in the observed CMB (with the Doppler shift from the earth's motion and galactic microwaves removed). You can see that it is credible that there would be roughly 360 speckles along the equator. Figure 29.2 shows a graph that represents what amounts to a two-dimensional version of a Fourier transform. The first peak shows that the largest speckles have an angular width of about 1°, consistent with the idea that $\Omega_k = 0$. A sophisticated statistical analysis of these data in fact specifies that the CMB data combined with certain other astronomical data yield a best-fit value for Ω_k of $-0.0023(+0.0054, -0.0056)$: see Jarosik et al., *Astrophysical Journal Supplement*, **192**:14, 2011.

The detailed shape of the CMB spectrum depends on H_0, Ω_b, and Ω_m (as well as a few other parameters), so fitting the detailed shape of the curve can determine these quantities (though the analysis is not as straightforward as that considered above). The determination of Ω_k described above then fixes Ω_v. So CMB data alone allow us to determine both the evolution and dynamics of the universe. The independent and highly precise information about these parameters that became available in 2003 from the Wilkinson Microwave Anisotropy Probe's measurements of the CMB has revolutionized the study of cosmology.

The Horizon and Flatness Problems. The CMB data present two significant problems with our model of the universe that we have not yet exposed and addressed. The first of these is the **horizon problem**. We have seen that the CMB is extremely isotropic, down to a few parts in 10^5. This can be easily explained only if every part of the plasma that eventually produced the CMB was in thermal equilibrium with every other part at some time in the past. However, this seems impossible! Consider two patches of plasma that are 180° apart in our sky at present. Photons emitted by one patch have barely had time to make it to *our* location (midway between the patches) during the lifetime of the universe, and thus have a long way yet to go before reaching the other patch. Therefore, these patches could *not* have been in thermal equilibrium in the past: they are too far apart to exchange photons within even the present age of the universe. So why is the CMB so uniform?

The second issue is the **flatness problem**. The CMB evidence discussed above argues quite convincingly that the universe is very nearly flat at the present. Indeed, we might say that it is *too* flat to be believable! You can show (see box 29.2) that the original Friedman equation for the evolution of the universe implies that an observer at cosmic time t would measure the curvature parameter to be

$$\Omega_k(t) \equiv 1 - \Omega_m(t) - \Omega_r(t) - \Omega_v(t) = \frac{K}{\dot{a}^2} \qquad (29.7)$$

where

$$\Omega_m(t) \equiv \frac{8\pi G}{3H^2}\rho_m(t) \equiv \frac{\rho_m(t)}{\rho_c(t)}, \quad \Omega_r(t) \equiv \frac{\rho_r(t)}{\rho_c(t)}, \quad \text{and} \quad \Omega_v(t) \equiv \frac{\rho_v}{\rho_c(t)} \qquad (29.8)$$

are the mass, radiation, and vacuum density fraction that that observer would measure at time t (compare with equation 26.18). For a radiation-dominated universe, $a \propto t^{1/2}$, so $\Omega_k(t) = K/\dot{a}^2 \propto [t^{-1/2}]^{-2} = t \propto a^2$. For a matter-dominated universe, $a \propto t^{2/3}$, so $\Omega_k(t) = K/\dot{a}^2 \propto [t^{-1/3}]^{-2} = t^{2/3} \propto a$. Since the universe was initially radiation-dominated and then matter-dominated, we see that all known earlier times, an observer would measure the universe to be *flatter* than at present. Indeed, you can show (see box 29.3) that $\Omega_k(t)$ at $t \sim 1$ μs (roughly the earliest time at which we thoroughly understand the physics of the universal plasma) was about 22 orders of magnitude smaller than Ω_k at present! Therefore, we see that for the universe to be even *close* to flat at present, the universe must have been *unbelievably* flat just after the Big Bang. Since there are no obvious *a priori* constraints on the amount of energy in this universe, it seems implausible that the universe's initial conditions were so finely tuned as to deliver a nearly flat universe at the present by accident. So why is the universe so flat?

Inflation to the Rescue. The most widely accepted solution to both of these problems is to postulate that the early universe went through an era of rapid exponential expansion (called **inflation**) early in its history. Such an expansion could occur if the physics of the early universe allowed vacuum energy to strongly dominate over other forms of energy for a suitable period of time. Under such circumstances, as we have seen (see equation 28.5), the scale of the universe in a vacuum-dominated universe increases exponentially. Indeed, you can show (see box 29.4) that

$$a(t) = a(t_s)e^{\sqrt{\frac{8}{3}\pi G \rho_v}(t-t_s)} \qquad (29.9)$$

in a flat, vacuum-dominated universe where ρ_v is the vacuum energy density at the time of inflation and t_s is the time at which inflation starts.

There is even a plausible physical explanation for how this might happen. Potential Grand Unified Theories (GUTs) that unify the strong nuclear interaction with the weak and electromagnetic interactions assume that at high enough temperatures (typically assumed to be above where $k_B T \sim 10^{15}$ GeV $\sim 10^5$ J), these three interactions would be identical. As temperatures fall past this point, the universe would go through a phase transition where the symmetry would be broken and differences between the strong and electroweak interactions would appear. This same scheme works well for the electroweak interaction, which at $k_B T$ below roughly 100 GeV breaks apart into separate weak and electromagnetic interactions because the vector bosons (W^+, W^-, and Z) that carry the weak interaction have masses in the 90 GeV range, while the photon is strictly massless. Therefore the essential symmetry of these interactions is manifest only at very high energies where the vector bosons can be created and destroyed as easily as photons.

In both the electroweak theory and the (proposed) GUTs, the symmetry is broken by (different) spin-zero fields $\phi(x^\alpha)$ whose corresponding quantum particles are called **Higgs bosons**. When this field has a nonzero value, it has the effect of making particles that interact with it (e.g., the vector bosons) behave as if they were massive, even if they are intrinsically massless like the photon. The Higgs boson that breaks the symmetry for the electroweak interaction was recently discovered to have a mass of about 125 GeV, while that of the GUT Higgs boson is expected to be roughly 10^{15} GeV.

Like most fields (e.g., the electromagnetic field), the Higgs fields carry energy. However, unlike most other fields, the lowest-energy state of the Higgs field is not *necessarily* the field where $\phi(x^\alpha)$ has a zero value. Rather, it is postulated that as the temperature falls, the energy associated with values of ϕ slowly changes from the shape shown in figure 29.3a (where the lowest energy corresponds to $\phi = 0$) to a shape more like that shown in figure 29.3b, where the lowest energy corresponds to $\phi \neq 0$. Since the actual value of $\phi(x^\alpha)$ will tend to hover around its lowest-energy value, initially we have $\phi \approx 0$ (meaning that all particles affected by the Higgs field are massless and the interactions in question are symmetrical) but ϕ will eventually transition to the lower-energy state $\phi \neq 0$ where the symmetries are broken.

However, there will be a time in between these states where the energy graph for the Higgs field has evolved to that shown in figure 29.3b, but the value of ϕ has not yet moved away from $\phi \approx 0$. In such a case, the field will have a nonzero energy relative to the true vacuum (the lowest energy state). This field energy behaves as if it were a vacuum energy. If this energy is large enough, then during the time that it takes ϕ to move from zero to its final nonzero "true vacuum" value, the universe will be vacuum dominated and will expand exponentially.

If we assume that the inflation arises from a transition involving the GUT Higgs field, then the transition temperature is such that $k_B T \sim 10^{15}$ GeV. According to equation 28.6 (adjusted downward a bit because g^* is very large at this time), the corresponding time is roughly 10^{-36} s after the Big Bang. The energy density of the relativistic plasma at this time would have been on the order of magnitude of $\rho_r \approx k_B T n \sim k_B T (k_B T / \hbar c)^3$ $\sim 10^{80}$ kg/m³ (see box 29.5). Since the vacuum energy must be significantly larger than this to dominate, the time required for the size of the universe to expand by a factor of e must be $t_e \sim [\tfrac{8}{3}\pi G \rho_v]^{-1/2} \sim [10 \cdot \tfrac{8}{3}\pi G \rho_r]^{-1/2}$ 10^{-36} s (again, see box 29.5). If the transition were to take as long as $70 t_e$ (less than 10^{-34} s) then the universe would grow by a factor of about $e^{70} \sim 10^{30}$ in this very brief time. One can design equations for how the field energy depends on temperature for which it would take this much time.

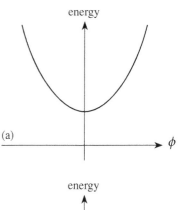

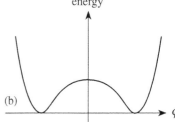

FIG. 29.3 These figures show the energy as a function of the value of the scalar Higgs field (a) before the GUT transition, and (b) after that transition. (The function evolves smoothly between these shapes during the transition). Note that the field's lowest-energy state (which qualifies as being the "true vacuum state") corresponds to $\phi = 0$ at the beginning, but not later. Indeed, if the state of the field remains $\phi = 0$ after the transition, the field will have an energy relative to the true vacuum state which behaves like a vacuum energy density.

Solving the Horizon Problem. How does this solve the horizon problem? Just before inflation, the size of a causally-connected region of the universe is on the order of magnitude of the distance that light can travel in the time since the Big Bang, which is roughly $c(10^{-36}$ s$) \sim 10^{-28}$ m. The universe expands by about 10^{30} during inflation, expanded by a further factor of roughly $(t_{RM}/t_{\inf})^{1/2} \sim (53{,}000$ y $/ 10^{-34}$ s$)^{1/2} \sim 10^{23}$ from the end of the inflation to the time of radiation-matter equality (remember that $a \sim t^{1/2}$ in a radiation dominated universe) and then a further factor of $(3 \times 10^{-4})^{-1} \sim 10^4$ (see chapter 28) since then. Therefore, the size of our initially causally-connected region has now expanded to a radius of roughly $(10^{-28}$ m$) \cdot 10^{30} \cdot 10^{23} \cdot 10^4 \sim 10^{29}$ m now, which is about 10,000 Gly, much farther than the edge of the visible universe now (see problem P29.4). Therefore, we and everything we can currently see would have been part of a single thermally-interacting mass before inflation.

Solving the Flatness Problem. Note that during inflation, $\dot{a} \propto e^{t/t_e}$, so

$$\frac{\Omega_k(\text{after inflation})}{\Omega_k(\text{before inflation})} = e^{-2t_{\inf}/t_e} \sim 10^{-60} \tag{29.10}$$

Therefore, no matter what Ω_k was before inflation, it will be essentially zero shortly after it starts to inflate (see problem P29.7 for more details), and even if it has increased by a factor of $\sim (t_{\text{now}}/t_{\inf}) \sim 10^{51}$ times after inflation was complete (note that $\Omega_k \propto t$ in a radiation-dominated universe), it will *still* be essentially zero.

To put it more intuitively, if you take a highly curved marble and expand it by a factor of 10^{30}, it would have a radius of about 1 Tly. If you were to stand on the surface of this expanded marble it would look completely flat, and no measurement you could practically make would contradict this assumption.

Other predictions of inflation have been tested and seem sound, so though it is a pretty speculative hypothesis, it seems to explain some crucial mysteries!

The Search for an Electroweak Higgs Boson. The inflation model assumes the validity of the Higgs mechanism. While detection of the GUT Higgs boson is not practical at present, detection of the analogous electroweak Higgs boson *is* practical. Indeed, just before this book went to press, physicists at the Large Hadron Collider (currently the world's largest particle accelerator) have just confirmed the existence of the electroweak Higgs boson. Such confirmation enhances the plausibility of the inflation scenario. On the other hand, if physicists find that this boson's properties are significantly different than expected, then the whole inflation proposal could suffer a significant blow.

Problems with the Inflation Model. While it is probably fair to say that the inflation scenario is the currently favored solution to the horizon and flatness problems, it does have some serious problems that are yet to be resolved. Recently, Paul Steinhardt (one of the original creators of the inflation model) wrote an article detailing problems that theoreticians have discovered with the inflation model in the 30 years since its invention. Steinhardt summarizes these problems as follows:

1. **The right kind of inflation may be very improbable.** The inflation solution requires that the potential energy curve shown in figure 29.3*b* have just the right shape for inflation to last long enough to properly inflate the universe. There is no known reason why this curve *must* be shaped this way. Indeed, some plausible methods of estimating the probability that the curve has the right shape lead to the conclusion that universes quite unlike ours are vastly more probable, and indeed that it is much more likely that the universe simply had the right flatness and homogeneity *without* inflation!

2. **Eternal inflation.** Quantum fluctuations mean that the inflation period ends in certain regions of space sooner than in some others. This seems good, because it explains the observed fluctuations in density that ultimately give rise to galaxies. However, since regions that haven't quite stopped inflating grow exponentially, it turns out mathematically that they should quickly occupy vastly more volume than the isolated regions (such as the visible portion of the universe we see) where inflation has stopped. Moreover, fluctuations in these huge inflating regions seem to inevitably give rise to new inflating regions. So inflation never stops, and this eternal inflation creates an infinite number of new non-inflating regions, some with very different characteristics than our own. If this is so, then inflation no longer provides the explanation of why our particular non-inflating region has the properties it does. No one yet knows yet how to construct a plausible model of inflation where this doesn't happen.

Most physicists think that these problems simply show that the inflation model needs further development, and that we will eventually discover *why* the potential energy curve must have the required shape and also how inflation can be turned off in an orderly way. However, these problems have led some physicists (such as Steinhardt) to give up on the inflation model and look for other solutions to the flatness and horizon problems. For example, people have noted that if quantum gravity allows a contracting universe to "bounce" instead of reaching the mathematical singularity predicted by general relativity, then it is possible that regions of space vastly separated *after* such a bounce would have had plenty of time to come into thermal equilibrium during the contraction phase *before* the bounce. (A successful "bounce" model would separately have to resolve the "flatness" problem.)

For more information about the problems with the inflation model, see Steinhardt, "The Inflation Debate," *Scientific American,* **304**, 4, 2011, and references therein. Guth, *The Inflationary Universe,* Basic Books, 1998, provides an accessible introduction to the inflation model written by one of its creators. Chapter 16 of Hobson, Efstathiou, and Lasenby, *General Relativity,* Cambridge, 2006, provides a detailed discussion of inflation at a higher level than this text.

BOX 29.1 The Angular Width of the Largest CMB Fluctuations

To evaluate the integral, start with the fundamental universal evolution equation in the matter-dominated limit:

$$\left(\frac{1}{H_0}\frac{da}{dt}\right)^2 \approx \frac{\Omega_m}{a} \quad \Rightarrow \quad \frac{1}{H_0}\frac{da}{dt} \approx \frac{\Omega_m^{1/2}}{a^{1/2}} \tag{29.11}$$

If you solve this for dt and substitute the result into the integrals appearing in equations 29.4 and 29.5, you can show that

$$\int_0^{t_{PD}} \frac{dt}{a(t)} \approx \frac{2a_{PD}^{1/2}}{H_0\sqrt{\Omega_m}}, \quad \int_{t_{PD}}^{t_0} \frac{dt}{a(t)} \approx \frac{2(1-a_{PD}^{1/2})}{H_0\sqrt{\Omega_m}} \approx \frac{2}{H_0\sqrt{\Omega_m}} \tag{29.12}$$

Exercise 29.1.1. Verify these results. (See problem P29.2 for a discussion of the soundness of the matter-domination approximation.)

Exercise 29.1.2. Substitute this result into equation 29.5 and use $R = H_0^{-1}|\Omega_k|^{-1/2}$ to verify that $\bar{r}_{PD}/R \approx 2[|\Omega_k|/\Omega_m]^{1/2} \equiv u$.

BOX 29.1 (continued) The Angular Width of the Largest CMB Fluctuations

Exercise 29.1.3. Use the results from the previous exercises to verify that $\Delta\phi \approx (\frac{1}{3}a_{PD})^{1/2}(u/\sin u)$, $(\frac{1}{3}a_{PD})^{1/2}$, or $(\frac{1}{3}a_{PD})^{1/2}(u/\sinh u)$, depending on the sign of Ω_k.

BOX 29.2 The Equation for $\Omega_k(t)$

The basic Friedman equation (equation 26.13) states that

$$\dot{a}^2 - \frac{8\pi G}{3}(\rho_m + \rho_r + \rho_v)a^2 = K \tag{29.13}$$

If you divide both sides by $\dot{a}^2$ and use $H \equiv \dot{a}/a$ (see equation 26.15) and the definitions in equation 29.8, you can easily show that $\Omega_k(t) \equiv 1 - \Omega_m(t) - \Omega_r(t) - \Omega_v(t)$ is indeed equal to $K/\dot{a}^2$, as claimed in equation 29.7.

Exercise 29.2.1. Verify that this is true.

BOX 29.3 Cosmic Flatness at the End of Nucleosynthesis

We know that $a(t_{RM}) \approx 3 \times 10^{-4}$ at the time of radiation-matter equality. Since the universe has been (approximately) matter dominated since that time, then $\Omega_k = K/\dot{a}^2 \propto a$, so $\Omega_k(t_{RM})/\Omega_k(\text{now}) \approx a(t_{RM}) \approx 3 \times 10^{-4}$. Earlier than that time, the universe was radiation dominated, so $\Omega_k = K/\dot{a}^2 \propto t$. Since $t_{RM} \approx 53{,}000$ y, you can use this to find $\Omega_k(1\,\mu s)/\Omega_k(t_{RM})$ and thus $\Omega_k(1\,\mu s)/\Omega_k(\text{now})$.

Exercise 29.3.1. Use the approach described to show that $\Omega_k(1\,\mu s)/\Omega_k(\text{now}) \approx 1.8 \times 10^{-22}$.

BOX 29.4 The Exponential Inflation Formula

The difference between a vacuum energy density ρ_v and ρ_m or ρ_r is that ρ_v does *not* change as the universe expands (though it may change for other physical reasons).

Exercise 29.4.1. Use this fact, the basic Friedman equation (equation 29.13) with $K = 0$, and the assumption of vacuum dominance ($\rho_m = \rho_r \approx 0$) to show that $a(t) = a(t_s)e^{\sqrt{\frac{8}{3}\pi G \rho_v}(t - t_s)}$ (see equation 29.9). This form of the equation is valuable because it no longer refers to the present state of the universe.

BOX 29.5 Inflation Calculations

Exercise 29.5.1. Verify that the energy density $\rho \approx k_B T (k_B T/\hbar c)^3 \sim 10^{80}$ kg/m³ at the GUT transition where $k_B T \sim 10^{15}$ GeV. (*Hint:* Be very careful with units, and remember that we are working in GR units here.)

Exercise 29.5.2. Use equation 28.6 to estimate when this transition occurs. (Note that $g^* \sim 100$ at this time: see box 28.3. Does this make much difference?)

> **BOX 29.5 (continued) Inflation Calculations**
>
> **Exercise 29.5.3.** Verify that $t_e \sim [10 \cdot \frac{8}{3}\pi G\rho]^{-1/2} \approx 10^{-36}$ s.

HOMEWORK PROBLEMS

P29.1 What if the value of $\Delta\phi$ measured by WMAP turned out to be 1.5° instead of 1°? What would the value of Ω_k then be (assuming a_{PD} and Ω_m remain the same)?

P29.2 Use a numerical model (either the model you created for problem P27.3 or a program your instructor provides) to determine better estimates for the values of the integrals in equations 29.3 and 29.5. Express your results as a multiple of $a_{PD}^{1/2}$, where $a_{PD} = 1/1100$. How much does including these better results change the result for $\Delta\phi$ in flat space (see below equation 29.6)?

P29.3 Suppose that a certain identifiable type of galaxy had a standard diameter D. Derive an expression for the angular size of such a galaxy as a function of redshift z, and show that beyond a certain z (which you should identify), the angular size actually gets *larger* as z gets larger. Assume that the universe is flat and matter-dominated from the time of galaxy formation to the present (a pretty good approximation). [*Hints:* First find the angular size as a function of a, then use $a = (1 + z)^{-1}$. You cannot use equation 29.6 directly, because some approximations were used in that equation that do not apply here. Do *not* assume that $\Omega_m = 1$.] (This problem was adapted from Hartle, *Gravity*, Addison-Wesley, 2003, problem 19.6.)

P29.4 It has taken roughly 13.75 Gy for the light from the CMB to reach us. But how far is the matter emitting the CMB away from us *at present*? *This* is the distance that we want to compare with the 10,000-Gly distance mentioned in the section "Solving the Horizon Problem." (*Hint:* The result is *not* 13.75 Gly: there is a big difference between the cosmic time t and the cosmic time parameter $\bar{t}$.)

P29.5 What is the minimum duration for inflation (in terms of t_e and in terms of seconds) required to ensure that the entire universe visible at present was in thermal equilibrium before inflation? (*Hints:* You will first have to resolve the question raised in problem P29.4. You can assume that t at the start of inflation roughly 3×10^{-36} s and $t_e \approx 0.9 \times 10^{-36}$ s. Redo the calculation in the "Solving the Horizon Problem" section to an accuracy of about one significant digit, and note that t_{inf}, the time that inflation ends, is the time at the start of inflation plus the duration of inflation, and the duration is what you are trying to find.)

P29.6 Suppose that future experimental work leads us to conclude that the GUT transition should occur when $k_B T \approx 10^{14}$ GeV instead of 10^{15} GeV. By what factors does this change the time of the onset of inflation, the density of energy during the transition, and the time required for the universe to grow by a factor of e during inflation?

P29.7 Perhaps you noted a certain circularity between equations 29.9 and 29.10: in proving the former, we had to assume that the universe *was* flat, but then we used that solution in the second to argue that the universe *becomes* flat. The goal of this problem is to resolve this puzzle.

a. As in box 29.4, let's start with the basic Friedman equation (equation 29.13) with $\rho_r \approx \rho_m \approx 0$ compared to ρ_v but this time let's assume that K is *not* zero. Show that if we also assume the universe is expanding, we can put the Friedman equation in the form

$$\frac{da}{\sqrt{a^2 + c}} = b\, dt \qquad (29.14)$$

where $b \equiv \sqrt{8\pi G \rho_v / 3}$ and $c = K/b^2$ are constants.

b. Integrate this from t_s to t to show that

$$a + \sqrt{a^2 + c} = k e^{b(t-t_s)} \equiv q \qquad (29.15)$$

where $k \equiv a_s + \sqrt{a_s^2 + c}$ and $a_s \equiv a(t_s)$. (You may look up the integral: just be sure to specify your source.)

c. Show that the above implies that

$$a = \frac{1}{2}\left(q - \frac{c}{q}\right) \qquad (29.16)$$

d. Argue that once the expansion has endured for a time equal to a relatively small multiple of $t_e = 1/b$, this will become an exponential expansion (even if $K \neq 0$).

P29.8 Find the physical distance between two points in an exponentially expanding universe separated by a fixed comoving coordinate separation $\Delta \bar{r}$ and show that this physical distance grows faster than the speed of light if it is greater than H^{-1} at any instant. (This implies that a causally connected patch of the universe must have physical radius smaller than H^{-1}, and therefore that the inflation process inside such a patch cannot be affected by whatever goes on outside of it. This problem was adapted from Hobson, Efstathiou, and Lasenby, *General Relativity*, Cambridge, 2006, problem 16.10.)

P29.9 In this problem, we will discuss a possible mathematical theory for the Higgs field. The Lagrangian for a particle moving in one dimension under the influence of a potential energy function is

$$L = T - V = \tfrac{1}{2} m \dot{x}^2 - V(x) \qquad (29.17)$$

and the particle's equation of motion is thus

$$m\ddot{x} + \frac{dV}{dx} = 0 \qquad (29.18)$$

By analogy, the simplest Lagrangian for a scalar field in a locally inertial frame (LIF) can be written

$$L = -\tfrac{1}{2}\eta^{\mu\nu}(\partial_\mu \phi)(\partial_\nu \phi) - V(\phi) \qquad (29.19)$$

and the corresponding field equation is

$$-\partial^\mu \partial_\mu \phi + \frac{\partial V}{\partial \phi} = 0 \qquad (29.20)$$

Note how the field value at a given point corresponds to a particle coordinate, and the derivatives of the field to a particle velocity. (The signs in these equations make the time derivatives positive, to correspond more closely with equations 29.17 and 29.18.)

If the field's "potential energy" $V(\phi)$ has the simple-harmonic-oscillator form $V(\phi) = \tfrac{1}{2}\mu^2 \phi^2$, then the field equation becomes $\partial^\mu \partial_\mu \phi + \mu^2 \phi = 0$, which is the well-known Klein-Gordon equation for a free quantum particle of mass μ, the typical case that one meets when beginning a study of relativistic quantum field theory. Our field with $V(\phi)$ unspecified is a simple generalization of a field very familiar to particle theorists.

We can find the stress-energy tensor $T^{\mu\nu}$ for a field of this type by requiring simply that (1) $T^{\mu\nu}$ be symmetric, (2) $T^{\mu\nu}$ be quadratic in the derivatives of ϕ (i.e., involve linear combinations of the kinetic-energy-like terms for the field), (3) terms involving the field's "kinetic energy" $\dot{\phi}^2$ and V contribute positively to the energy density T^{tt}, and (4) that $T^{\mu\nu}$ satisfies the law of local energy conservation $\partial_\mu T^{\mu\nu} = 0$. The possible terms satisfying the first two constraints are $a(\partial^\mu \phi)(\partial^\nu \phi), b\eta^{\mu\nu}(\partial_\alpha \phi)(\partial^\alpha \phi)$, and $c\eta^{\mu\nu}V$, where a, b, and c are constants. The third constraint requires that a be positive and c negative. We can absorb the absolute values of a and c into the definitions of ϕ and V. Therefore, the most general form of a stress-energy tensor for the field that satisfies criteria 1–3 is

$$T^{\mu\nu} = (\partial^\mu \phi)(\partial^\nu \phi) + b\eta^{\mu\nu}(\partial_\alpha \phi)(\partial^\alpha \phi) - \eta^{\mu\nu} V \quad (29.21)$$

a. Show that criterion (4) requires that $b = -\tfrac{1}{2}$. (*Hints:* Use equation 29.20 and feel free to rename summed indices. Note also that $M^{\mu\nu} B_\nu = M^\mu{}_\nu B^\nu$ for arbitrary tensors **M** and **B**, as you can show by writing out what it means to raise and lower indices.)

b. Argue that if the spatial derivatives of ϕ are zero (expressing homogeneity) and the time derivatives of ϕ are small enough so that $\dot{\phi}^2 \ll V$ (meaning that the transition to a new value takes some time) then the stress-energy has the form of a vacuum energy (expressed in a LIF) whose density is proportional to $V(\phi)$. This gives some credibility to the idea that a simple scalar field could cause inflation.

30. GAUGE FREEDOM

INTRODUCTION

FLAT SPACETIME
- Review of Special Relativity
- Four-Vectors
- Index Notation

TENSORS
- Arbitrary Coordinates
- Tensor Equations
- Maxwell's Equations
- Geodesics

SCHWARZSCHILD BLACK HOLES
- The Schwarzschild Metric
- Particle Orbits
- Precession of the Perihelion
- Photon Orbits
- Deflection of Light
- Event Horizon
- Alternative Coordinates
- Black Hole Thermodynamics

THE CALCULUS OF CURVATURE
- The Absolute Gradient
- Geodesic Deviation
- The Riemann Tensor

THE EINSTEIN EQUATION
- The Stress-Energy Tensor
- The Einstein Equation
- Interpreting the Equation
- The Schwarzschild Solution

COSMOLOGY
- The Universe Observed
- A Metric for the Cosmos
- Evolution of the Universe
- Cosmic Implications
- The Early Universe
- CMB Fluctuations & Inflation

GRAVITATIONAL WAVES
- **Gauge Freedom**
- Detecting Gravitational Waves
- Gravitational Wave Energy
- Generating Gravitational Waves
- Gravitational Wave Astronomy

SPINNING BLACK HOLES
- Gravitomagnetism
- The Kerr Metric
- Kerr Particle Orbits
- Ergoregion and Horizon
- Negative-Energy Orbits

this depends on this

Introduction. This chapter starts a new section of the course, which focuses on the generation and detection of gravitational waves. Gravitational waves are required by *any* relativistic theory of gravitation because information about changes in a dynamic source's gravitational field can move away from the source no faster than the speed of light. General relativity certainly does predict the existence of such waves and places other specific constraints on their character, how they can be detected, and how they are connected to dynamic sources. We will be exploring all of these topics in the next few chapters.

Review of the Weak-Field Limit. The gravitational waves produced by virtually any realistic astrophysical source will (far from the source) be only tiny perturbations of flat spacetime. Therefore, we can use the weak-field approximation discussed in chapter 22 to express the Einstein equation in linear form. This approximation makes the mathematics much easier.

In that chapter we saw that when fields are weak, we can use "nearly cartesian" coordinates where the metric of spacetime is

$$g_{\mu\nu} = \eta_{\mu\nu} + h_{\mu\nu} \quad \text{where} \quad h_{\mu\nu} = h_{\nu\mu} \text{ and } |h_{\mu\nu}| \ll 1 \tag{30.1}$$

(see equation 22.3). We call $h_{\mu\nu}$ the **metric perturbation**. We also saw (see equation 22.4) that to first order in $h_{\mu\nu}$, the inverse metric is given by

$$g^{\mu\nu} = \eta^{\mu\nu} - h^{\mu\nu} \quad \text{where} \quad h^{\mu\nu} \equiv \eta^{\mu\alpha}\eta^{\nu\beta}h_{\alpha\beta} \tag{30.2}$$

and (according to equation 22.7) the Riemann tensor is

$$R_{\alpha\beta\mu\nu} = \tfrac{1}{2}[\partial_\beta\partial_\mu h_{\alpha\nu} + \partial_\alpha\partial_\nu h_{\beta\mu} - \partial_\alpha\partial_\mu h_{\beta\nu} - \partial_\beta\partial_\nu h_{\alpha\mu}] \tag{30.3}$$

In this weak-field limit (where we are dropping terms that are second order or higher in the metric perturbation $h_{\mu\nu}$), we can in fact raise and lower the indices of $h_{\mu\nu}$ using the flat-space metric (for example, $h^\mu{}_\nu \equiv \eta^{\mu\alpha}h_{\alpha\nu}$). We can also raise or lower the indices of *gradients* of $h_{\mu\nu}$ using the flat-space metric, because (for example) $\partial^\alpha h_{\mu\nu} \equiv g^{\alpha\beta}\partial_\beta h_{\mu\nu} \approx (\eta^{\alpha\beta} - h^{\alpha\beta})\partial_\beta h_{\mu\nu} \approx \eta^{\alpha\beta}\partial_\beta h_{\mu\nu}$ to first order in the perturbation. Therefore, we can generally treat equations involving the metric perturbation *as if* $h_{\mu\nu}$ were simply a tensor field in a flat spacetime.

The Trace-Reversed Perturbation. In chapter 22, we found it convenient to work with the Einstein equation in the form $R_{\mu\nu} = 8\pi G(T_{\mu\nu} - \tfrac{1}{2}g_{\mu\nu}T)$. It turns out that when working with gravitational waves, it will be more convenient to use the equation in its more original form:

$$G^{\gamma\sigma} \equiv R^{\gamma\sigma} - \tfrac{1}{2}g^{\gamma\sigma}R = 8\pi G T^{\gamma\sigma} \tag{30.4}$$

Using equation 30.3, the definition $R^{\gamma\sigma} \equiv g^{\gamma\beta}g^{\sigma\nu}R_{\beta\nu} = g^{\gamma\beta}g^{\sigma\nu}g^{\alpha\mu}R_{\alpha\beta\mu\nu}$ of the Ricci tensor with upper indices, and the definition $R \equiv g^{\alpha\mu}g^{\beta\nu}R_{\alpha\beta\mu\nu}$ of the curvature scalar, you can show (see box 30.1) that to first order in the metric perturbation, the Einstein equation becomes

$$\tfrac{1}{2}\big(\partial^\gamma\partial_\mu h^{\mu\sigma} + \partial^\sigma\partial_\mu h^{\mu\gamma} - \partial^\gamma\partial^\sigma h - \partial^\mu\partial_\mu h^{\gamma\sigma}$$
$$- \eta^{\gamma\sigma}\partial_\beta\partial_\mu h^{\mu\beta} + \eta^{\gamma\sigma}\partial^\mu\partial_\mu h\big) = 8\pi G T^{\gamma\sigma} \quad \text{where} \quad h \equiv \eta^{\mu\nu}h_{\mu\nu} \tag{30.5}$$

We can achieve a modest simplification of this equation if we define the **trace-reversed metric perturbation**

$$H_{\mu\nu} \equiv h_{\mu\nu} - \tfrac{1}{2}\eta_{\mu\nu}h \tag{30.6}$$

You can easily show from this definition (see box 30.2) that $H \equiv \eta^{\mu\nu}H_{\mu\nu} = -h$ (which is where the "trace-reversed" name comes from) and that

$$h_{\mu\nu} = H_{\mu\nu} - \tfrac{1}{2}\eta_{\mu\nu}H \tag{30.7}$$

This equation implies that if we can determine the trace-reversed perturbation, we can easily calculate the actual metric perturbation $h_{\mu\nu}$.

You can show (see box 30.3) that if we substitute equation 30.7 into equation 30.5 and use $h = -H$, we find that the weak-field Einstein equation becomes

$$\Box^2 H^{\gamma\sigma} - \partial^\gamma \partial_\mu H^{\mu\sigma} - \partial^\sigma \partial_\mu H^{\mu\gamma} + \eta^{\gamma\sigma} \partial_\beta \partial_\mu H^{\mu\beta} = -16\pi G T^{\gamma\sigma} \quad (30.8)$$

where $\Box^2 \equiv \eta^{\alpha\beta} \partial_\alpha \partial_\beta \equiv \partial^\beta \partial_\beta = -(\partial^2/\partial t^2) + \nabla^2$ (the "box" notation is deliberate and conventional, not a computer's indication of a missing character). This is simpler, but still looks pretty difficult to solve for $H_{\mu\nu}$. However, we have not yet taken advantage of our ability to choose coordinates.

Gauge Transformation Basics. Imagine that we make a small coordinate transformation of our "nearly cartesian" coordinates x^α to new "nearly cartesian" coordinates x'^α using the transformation equation

$$x'^\alpha \equiv x^\alpha + \xi^\alpha, \quad \text{where } \xi^\alpha = \xi^\alpha(t, x, y, z) \text{ and } |\xi^\alpha| \ll 1 \quad (30.9)$$

Note that this means that $x^\alpha = x'^\alpha - \xi^\alpha$. The coordinate transformation partials that we would use to transform from one coordinate system to the other are

$$\frac{\partial x'^\alpha}{\partial x^\beta} = \frac{\partial}{\partial x^\beta}[x^\alpha + \xi^\alpha] = \delta^\alpha_\beta + \partial_\beta \xi^\alpha \quad (30.10a)$$

$$\frac{\partial x^\beta}{\partial x'^\alpha} = \frac{\partial}{\partial x'^\alpha}[x'^\beta - \xi^\beta] = \delta^\beta_\alpha - \frac{\partial \xi^\beta}{\partial x'^\alpha} = \delta^\beta_\alpha - \frac{\partial x^\mu}{\partial x'^\alpha} \frac{\partial \xi^\beta}{\partial x^\mu}$$

$$= \delta^\beta_\alpha - (\delta^\mu_\alpha - \partial'_\alpha \xi^\mu) \partial_\mu \xi^\beta \approx \delta^\beta_\alpha - \delta^\mu_\alpha \partial_\mu \xi^\beta = \delta^\beta_\alpha - \partial_\alpha \xi^\beta \quad (30.10b)$$

where in the next-to-last step I have dropped a term of order $|\xi^\alpha|^2$.

If you apply the tensor transformation to the metric tensor itself in this case, you can show (see box 30.4) that the metric perturbations transform as

$$h'_{\mu\nu} = h_{\mu\nu} - \partial_\mu \xi_\nu - \partial_\nu \xi_\mu \quad (30.11a)$$

$$H'_{\mu\nu} = H_{\mu\nu} - \partial_\mu \xi_\nu - \partial_\nu \xi_\mu + \eta_{\mu\nu} \partial_\alpha \xi^\alpha \quad (30.11b)$$

You can also show (see box 30.5) that such a transformation does not affect the Riemann tensor to our level of approximation:

$$R'_{\alpha\beta\mu\nu} = \tfrac{1}{2}[\partial_\beta \partial_\mu h'_{\alpha\nu} + \partial_\alpha \partial_\nu h'_{\beta\mu} - \partial_\alpha \partial_\mu h'_{\beta\nu} - \partial_\beta \partial_\nu h'_{\alpha\mu}]$$

$$\approx \tfrac{1}{2}[\partial_\beta \partial_\mu h_{\alpha\nu} + \partial_\alpha \partial_\nu h_{\beta\mu} - \partial_\alpha \partial_\mu h_{\beta\nu} - \partial_\beta \partial_\nu h_{\alpha\mu}] = R_{\alpha\beta\mu\nu} \quad (30.12)$$

Therefore, such a transformation will have *no effect* on the Einstein equation in the weak-field limit: if $h_{\mu\nu}$ (or $H_{\mu\nu}$) satisfies the equation, then $h'_{\mu\nu}$ (or $H'_{\mu\nu}$) will as well. Therefore, given any solution to the weak-field Einstein equation we can generate an infinite *family* of solutions by using different (arbitrary but small) transformations of the type given by equation 30.9.

The Electromagnetic Analogy. This indeterminacy in the solution to the weak-field Einstein equation has its fundamental root in our complete freedom to choose what coordinate system we want to use to describe spacetime. As noted earlier, though, in the weak-field limit, we can *pretend* that either $h_{\mu\nu}$ or $H_{\mu\nu}$ is a tensor field that exists in *flat* spacetime in the same way that the electromagnetic field tensor $F^{\mu\nu}$ exists in a flat spacetime. We can then look on this freedom as being analogous to the freedom we have in choosing electromagnetic potentials. For example, in static situations, the electric potential ϕ is defined only up to a scalar function of time $g(t)$, since adding $g(t)$ to the potential does not change the electric field, which is the physically measurable aspect of the potential:

$$\vec{E}' \equiv -\vec{\nabla}\phi' = -\vec{\nabla}[\phi + g(t)] = -\vec{\nabla}\phi + 0 = -\vec{\nabla}\phi \equiv \vec{E} \quad (30.13)$$

Similarly, adding a function of the form $\vec{\nabla} f$ (where f is any scalar function of position and time) to the vector potential $\vec{A}$ does not affect the physically measurable magnetic field $\vec{B}$ generated by that vector potential:

$$\vec{B}' \equiv \vec{\nabla} \times \vec{A}' = \vec{\nabla} \times (\vec{A} + \vec{\nabla} f) = \vec{\nabla} \times \vec{A} + 0 \equiv \vec{B} \qquad (30.14)$$

where I have used the identity $\vec{\nabla} \times \vec{\nabla} f = 0$. In electromagnetism, we call our ability to add functions to the potentials describing the field without changing any physically observable aspects of the field **gauge freedom** (for now-irrelevant historical reasons). We will use the same term to describe the analogous phenomenon in the weak-field limit of the Einstein equation. We will call a coordinate transformation of the type given in equation 30.9 a **gauge transformation**, and any specific "nearly cartesian" coordinate choice for expressing $h_{\mu\nu}$ or $H_{\mu\nu}$ a **gauge**.

Just as a clever choice of coordinates can make solving a problem easier, so we will find that a clever choice of gauge can make finding solutions easier in both electromagnetic theory and general relativity.

The Lorenz Gauge. Imagine that we know that a given trace-reversed perturbation $H^{\mu\nu}$ solves the weak-field Einstein equation given by equation 30.8. As discussed in box 30.6, we can *always* find a gauge transformation that converts it to a new solution $H'^{\mu\nu}$ that satisfies the Lorenz gauge condition

$$\partial_\mu H'^{\mu\nu} = 0 \qquad (30.15)$$

Since we can always do this, rather than imposing this condition *after* solving the weak-field Einstein equation, we can require that we are looking for solutions of this type *before* we solve that equation. Since the last three terms in the Einstein equation as given by equation 30.8 involve factors of $\partial_\mu H^{\mu\nu}$, *requiring* that $\partial_\mu H'^{\mu\nu} = 0$ at the start eliminates all but the first term in that equation. So this requirement effectively replaces the complicated equation 30.8 by two simpler equations

$$\Box^2 H^{\mu\nu} = -16\pi G T^{\mu\nu} \qquad (30.16a)$$

$$\partial_\mu H^{\mu\nu} = 0 \qquad (30.16b)$$

These equations are especially nice because not only are they linear in the trace-reversed perturbation, but the trace-reversed perturbation component equations are also completely *uncoupled* in the weak-field Einstein equation, a tremendous advantage for finding solutions. This pair of equations will be our starting point for studying gravitational waves.

Additional Gauge Freedom. It turns out that requiring $\partial_\mu H^{\mu\nu} = 0$ does not exhaust our gauge freedom. As discussed in box 30.7, given a $H^{\mu\nu}$ that satisfies the Lorenz gauge condition, we can generate new solutions $H'^{\mu\nu}$ using transformation functions $\xi^\alpha(t, x, y, z)$ such that

$$\Box^2 \xi^\alpha \equiv \partial^\mu \partial_\mu \xi^\alpha = 0 \qquad (30.17)$$

then the new $H'^{\mu\nu}$ will *still* satisfy the Lorenz gauge condition. This allows us to generate new solutions that satisfy both of equations 30.16. In the next chapter, we will find this additional gauge freedom to be very helpful in constraining gravitational wave solutions to the simplest possible form.

BOX 30.1 The Weak-Field Einstein Equation in Terms of $h_{\mu\nu}$

According to equation 30.3 (repeated here), we have

$$R_{\alpha\beta\mu\nu} = \tfrac{1}{2}[\partial_\beta\partial_\mu h_{\alpha\nu} + \partial_\alpha\partial_\nu h_{\beta\mu} - \partial_\alpha\partial_\mu h_{\beta\nu} - \partial_\beta\partial_\nu h_{\alpha\mu}] \qquad (30.3r)$$

to first order in the metric perturbation. Note also that the defining equations for the Ricci tensor and the curvature scalar become (to the same order of approximation)

$$R^{\gamma\sigma} \equiv g^{\gamma\beta}g^{\sigma\nu}R_{\beta\nu} = g^{\gamma\beta}g^{\sigma\nu}g^{\alpha\mu}R_{\alpha\beta\mu\nu} \approx \eta^{\gamma\beta}\eta^{\sigma\nu}\eta^{\alpha\mu}R_{\alpha\beta\mu\nu} \qquad (30.18a)$$

$$R \equiv g^{\alpha\mu}g^{\beta\nu}R_{\alpha\beta\mu\nu} \approx \eta^{\alpha\mu}\eta^{\beta\nu}R_{\alpha\beta\mu\nu} \qquad (30.18b)$$

This is true because the Riemann tensor is already first order in $h_{\mu\nu}$, so any terms beyond the leading term in each metric factor would simply generate terms that are second order or higher in $h_{\mu\nu}$. If you substitute the above into the Einstein equation $G^{\gamma\sigma} \equiv R^{\gamma\sigma} - \tfrac{1}{2}g^{\gamma\sigma}R = 8\pi GT^{\gamma\sigma}$, you should get equation 30.5, repeated here,

$$\tfrac{1}{2}(\partial^\gamma\partial_\mu h^{\mu\sigma} + \partial^\sigma\partial_\mu h^{\mu\gamma} - \partial^\gamma\partial^\sigma h - \partial^\mu\partial_\mu h^{\gamma\sigma}$$
$$- \eta^{\gamma\sigma}\partial_\beta\partial_\mu h^{\mu\beta} + \eta^{\gamma\sigma}\partial^\mu\partial_\mu h) = 8\pi GT^{\gamma\sigma} \qquad (30.5r)$$

to first order in the metric perturbation (note that $h \equiv \eta^{\mu\nu}h_{\mu\nu}$).

Exercise 30.1.1. Verify this result. Note that as discussed in the overview, you can use the inverse flat-space metric $\eta^{\mu\nu}$ to raise the index of any partial derivative operator (as long as it acts on the metric perturbation) or to raise either index of the metric perturbation itself.

BOX 30.2 The Trace-Reverse of $h_{\mu\nu}$

We have defined the trace-reverse of $h_{\mu\nu}$ to be $H_{\mu\nu} \equiv h_{\mu\nu} - \frac{1}{2}\eta_{\mu\nu}h$, where $h =$ the "trace" of $h_{\mu\nu}$ is defined to be $h \equiv \eta^{\alpha\beta}h_{\alpha\beta}$.

Exercise 30.2.1. Show from this that $H \equiv \eta^{\mu\nu}H_{\mu\nu} = -h$.

Exercise 30.2.2. Show, therefore, that $h_{\mu\nu} = H_{\mu\nu} - \frac{1}{2}\eta_{\mu\nu}H$.

BOX 30.3 The Weak-Field Einstein Equation in Terms of $H^{\mu\nu}$

If we raise both indices of $h_{\mu\nu} = H_{\mu\nu} - \frac{1}{2}\eta_{\mu\nu}H$, we get $h^{\mu\nu} = H^{\mu\nu} - \frac{1}{2}\eta^{\mu\nu}H$. If we substitute the latter into equation 30.5 (repeated here),

$$\tfrac{1}{2}\big(\partial^\gamma \partial_\mu h^{\mu\sigma} + \partial^\sigma \partial_\mu h^{\mu\gamma} - \partial^\gamma \partial^\sigma h - \partial^\mu \partial_\mu h^{\gamma\sigma}$$
$$- \eta^{\gamma\sigma} \partial_\beta \partial_\mu h^{\mu\beta} + \eta^{\gamma\sigma} \partial^\mu \partial_\mu h\big) = 8\pi G T^{\gamma\sigma} \qquad (30.5r)$$

and use the fact that $H = -h$, you should be able to derive the weak-field Einstein equation given in equation 30.8 (repeated here):

$$\partial^\mu \partial_\mu H^{\gamma\sigma} - \partial^\gamma \partial_\mu H^{\mu\sigma} - \partial^\sigma \partial_\mu H^{\mu\gamma} + \eta^{\gamma\sigma} \partial_\beta \partial_\mu H^{\mu\beta} = -16\pi G T^{\gamma\sigma} \qquad (30.8r)$$

Exercise 30.3.1. Verify that this is true.

BOX 30.4 Gauge Transformations of the Metric Perturbations

Consider a gauge transformation from nearly cartesian coordinates x^α to nearly cartesian coordinates x'^α given by $x'^\alpha = x^\alpha + \xi^\alpha$ where $|\xi^\alpha(t, x, y, z)| \ll 1$. The tensor transformation rule for the metric implies that

$$g'_{\mu\nu} = \frac{\partial x^\alpha}{\partial x'^\mu} \frac{\partial x^\beta}{\partial x'^\nu} g_{\alpha\beta} \tag{30.19}$$

Note that in the weak-field limit, $g'_{\mu\nu} \equiv \eta_{\mu\nu} + h'_{\mu\nu}$ and $g_{\alpha\beta} \equiv \eta_{\alpha\beta} + h_{\alpha\beta}$. For a gauge transformation, we also know (from equation 30.10b) that

$$\frac{\partial x^\alpha}{\partial x'^\mu} = \delta^\alpha_\mu - \partial_\mu \xi^\alpha \tag{30.20}$$

You can use these results and definitions to show that, to first order in the metric perturbation and $|\xi^\alpha|$, equation 30.11a (repeated here) is true:

$$h'_{\mu\nu} = h_{\mu\nu} - \partial_\mu \xi_\nu - \partial_\nu \xi_\mu \tag{30.11ar}$$

where $\xi_\mu = \eta_{\mu\alpha} \xi^\alpha$ to our level of approximation.

Exercise 30.4.1. Verify equation 30.11a.

If you then substitute the above into $H_{\mu\nu} \equiv h_{\mu\nu} - \frac{1}{2}\eta_{\mu\nu} h$, you should be able to show that equation 30.11b (repeated here) is also true:

$$H'_{\mu\nu} = H_{\mu\nu} - \partial_\mu \xi_\nu - \partial_\nu \xi_\mu + \eta_{\mu\nu} \partial_\alpha \xi^\alpha \tag{30.11br}$$

Exercise 30.4.2. Verify equation 30.11b.

BOX 30.5 A Gauge Transformation Does Not Change $R_{\alpha\beta\mu\nu}$

According to equation 30.12 (repeated here), the Riemann tensor is invariant under gauge transformations of the type considered in the previous box:

$$\begin{aligned} R'_{\alpha\beta\mu\nu} &= \tfrac{1}{2}[\partial_\beta\partial_\mu h'_{\alpha\nu} + \partial_\alpha\partial_\nu h'_{\beta\mu} - \partial_\alpha\partial_\mu h'_{\beta\nu} - \partial_\beta\partial_\nu h'_{\alpha\mu}] \\ &\approx \tfrac{1}{2}[\partial_\beta\partial_\mu h_{\alpha\nu} + \partial_\alpha\partial_\nu h_{\beta\mu} - \partial_\alpha\partial_\mu h_{\beta\nu} - \partial_\beta\partial_\nu h_{\alpha\mu}] = R_{\alpha\beta\mu\nu} \quad (30.12r) \end{aligned}$$

Exercise 30.5.1. Use equation 30.11a (see the previous box) to verify that this is true. (You should find that all the terms involving ξ^α cancel.)

BOX 30.6 Lorenz Gauge

The gauge transformation for the trace-reversed perturbation $H_{\mu\nu}$ is given by equation 30.11b, which is repeated here:

$$H'_{\mu\nu} = H_{\mu\nu} - \partial_\mu \xi_\nu - \partial_\nu \xi_\mu + \eta_{\mu\nu} \partial_\alpha \xi^\alpha \qquad (30.11br)$$

Raising both free indices of this equation yields simply

$$H'^{\mu\nu} = H^{\mu\nu} - \partial^\mu \xi^\nu - \partial^\nu \xi^\mu + \eta^{\mu\nu} \partial_\alpha \xi^\alpha \qquad (30.11cr)$$

You can show from this equation that given any initial $H^{\mu\nu}$, we can make $\partial_\mu H'^{\mu\nu} = 0$ as long as we choose our transformation functions ξ^α so that

$$\partial^\mu \partial_\mu \xi^\nu = \partial_\mu H^{\mu\nu} \qquad (30.21)$$

Exercise 30.6.1. Verify that this is correct.

This is a set of four simple second-order differential equations for $\xi^\nu(t, x, y, z)$. The differential operator $\partial^\mu \partial_\mu \equiv \Box^2 = -(\partial^2/\partial t^2) + \nabla^2$ has been well studied, and one can prove mathematically that solutions to equations of the form $\Box^2 f = g$ always exist for well-defined driving functions g. Therefore we can indeed always (in principle) find suitable transformation functions ξ^ν to make $\partial_\mu H'^{\mu\nu} = 0$ true.

(Indeed, the solutions f to the inhomogeneous differential equation $\Box^2 f = g$ actually represent *families* of solutions, because if any given function f solves the equation, then so does $f + bf_0$, where b is a constant and f_0 is a solution to the homogeneous differential equation $\Box^2 f_0 = 0$. This is the foundation for the additional gauge freedom discussed in the next box.)

BOX 30.7 Additional Gauge Freedom

Imagine that we already have a solution $H^{\mu\nu}$ to the weak-field Einstein equation (equation 30.16) that satisfies the Lorenz gauge condition $\partial_\mu H^{\mu\nu} = 0$. Imagine that we then do a subsequent gauge transformation where the transformation functions ξ^μ are any solutions to the differential equation $0 = \Box^2 \xi^\nu \equiv \partial^\mu \partial_\mu \xi^\nu$. The resulting transformed solution $H'^{\mu\nu}$ will still solve the Einstein equation because that equation is invariant under gauge transformations (see box 30.5).

Exercise 30.7.1. Show that the transformed $H'^{\mu\nu}$ components also still satisfy the Lorenz gauge condition $\partial_\mu H'^{\mu\nu} = 0$. Therefore the transformed $H'^{\mu\nu}$ matrix thus generated is a new solution to both of equations 30.16.

HOMEWORK PROBLEMS

P30.1 In chapter 22 we took a different approach to simplifying the weak-field Einstein equation. There, we saw that we could write the weak-field Ricci tensor in the form given by equation 22.8:

$$R_{\beta\nu} = \tfrac{1}{2}\left(-\eta^{\alpha\mu}\partial_\alpha \partial_\mu h_{\beta\nu} + \partial_\beta H_\nu + \partial_\nu H_\beta\right)$$

where $H_\nu \equiv \eta^{\alpha\mu}\left(\partial_\mu h_{\alpha\nu} - \tfrac{1}{2}\partial_\nu h_{\alpha\mu}\right)$ (30.22)

(Note that this H_ν is not the same as $H_{\mu\nu}$, which has two indices instead of one.) We also claimed in that chapter that we could always use our freedom to choose coordinates to find solutions such that $H_\nu = 0$. Prove that this is equivalent to requiring that the solution satisfy the Lorenz gauge condition. (That is, show that $H_\nu = 0$ if and only if $\partial_\mu H^{\mu\nu} = 0$.)

P30.2 Consider a stationary, spherical, non-rotating, and non-relativistic object whose internal pressure is negligible compared to its rest energy density ρ. The weak-field Einstein equation for such an object reduces to

$$\nabla^2 H^{tt} = -16\pi G\rho \qquad (30.23a)$$

other $\nabla^2 H^{\mu\nu} \approx 0 \qquad (30.23b)$

The corresponding equation $\nabla^2 \phi = -4\pi k\rho_c$ in electrostatics (where ϕ is the electrostatic potential and ρ_c is the charge density) has the solution $\phi(r) = kQ/r$ outside a spherically symmetric source at $r = 0$, where

$$Q = \int \rho_c\, dV = \text{total charge} \qquad (30.24)$$

and $\phi(r)$ is the potential due to that charge. (One can always add a solution satisfying $\nabla^2 \phi = 0$ to this solution, but such a solution will represent fields from external objects, not the central source in question.)

a. Use this analogy to determine H^{tt} in the exterior of this stationary, spherical, non-rotating, and non-relativistic object, and to argue that in the absence of fields from external objects, the other components of $H^{\mu\nu} \approx 0$.

b. Show that *any* time-independent solution for H^{tt} will satisfy the Lorenz gauge condition as long as the other components of $H^{\mu\nu}$ are zero.

c. Show that the metric of spacetime around this object (to first order in the metric perturbation) is

$$ds^2 = -\left(1 - \frac{2GM}{r}\right)dt^2 + \left(1 + \frac{2GM}{r}\right)(dx^2 + dy^2 + dz^2) \qquad (30.25)$$

P30.3 Imagine that in a certain region of spacetime we have $H_{\mu\nu} = 0$ except for $H_{tt} = Br^2 + C$, where B and C are constants (B has units of m^{-2}) and $r^2 \equiv x^2 + y^2 + z^2$.

a. Show that $H_{\mu\nu}$ satisfies the Lorenz gauge condition.

b. Use equation 30.16 to argue that this could be a solution to the weak-field Einstein equation *inside* a spherical star centered on the origin whose fluid has *uniform* and *constant* energy density ρ and negligible pressure and is not moving around the star's center of mass. Also determine how the constant B must be related to ρ.

c. Define $m(r) = \frac{4}{3}\pi r^3 \rho$ = the Newtonian mass-energy enclosed by the radius r. Calculate all of the metric components $g_{\mu\nu}$ in terms of $m(r)$ and C.

d. At the star's surface $r = R$, the metric must match the exterior solution given by equation 30.25 in problem P30.2 [note that M in that problem = $m(R)$ here]. Use this to determine the value of C and write out the final metric equation for this spacetime in these coordinates.

e. Use a computer graphing tool to plot a quantitatively accurate graph of h_{tt} (in terms of GM/R) as a function of r from $r = 0$ to $4R$. Submit a printout of this graph.

f. What condition must be satisfied at all points inside such a star if we are to satisfy the weak-field condition $|h_{\mu\nu}| \ll 1$? Is this plausible for normal matter?

P30.4 Say that in a certain region of spacetime, $H_{tt} = H_{xx} = H_{yy} = H_{zz} = A$ (where A is a constant) and all other $H_{\mu\nu} = 0$.

a. Argue that this constant field trivially satisfies both the Lorenz gauge condition and the weak-field Einstein equation for empty space.

b. Find $h_{\mu\nu}$ for this field, and argue that the field has no physical consequences.

P30.5 Imagine that in a certain region of spacetime we have $H_{\mu\nu} = 0$ except for $H_{xx} = -H_{yy} = A\cos(\omega t - \omega z)$, where ω is a constant with units of m^{-1}.

a. Argue that this represents a plane wave whose crests move in the $+z$ direction at the speed of light.

b. Show that this wave obeys the Lorenz gauge condition.

c. Show that this wave is a solution to the weak-field Einstein equation for empty space.

d. Determine the metric for the spacetime through which this gravitational wave moves.

e. What condition must be satisfied at all points if we are to satisfy the weak-field condition $|h_{\mu\nu}| \ll 1$?

P30.6 In this problem, we will consider conservation of energy in the weak-field limit.

a. Show from the general weak-field Einstein equation for the trace-reversed metric perturbation (equation 30.8) that $\partial_\mu T^{\mu\nu} = 0$, i.e., that (non-gravitational!) energy and momentum are conserved in the implied nearly cartesian coordinate system.

b. Show that this also follows from the form of the weak-field Einstein equation given in equations 30.16 as long as the solution $H^{\mu\nu}$ satisfies *both* of those equations. (We see that the extra terms in equation 30.8 are required to maintain energy conservation in absence of the Lorenz gauge condition.)

P30.7 Consider a plane-wave metric perturbation having the form $h_{\mu\nu} = A_{\mu\nu}\sin(k_\sigma x^\sigma)$, where $A^{\mu\nu}$ is a constant matrix and k_σ is a constant wave-number covector. We might consider such a perturbation as representing a hypothetical gravitational wave in empty space.

a. Show that for such a metric perturbation

$$\partial_\alpha h_{\mu\nu} = k_\alpha A_{\mu\nu} \cos(k_\sigma x^\sigma) \quad (30.26)$$

b. Show that for such a metric perturbation,

$$R_{\alpha\beta\mu\nu} = -\tfrac{1}{2}\big[k_\beta k_\mu A_{\alpha\nu} + k_\alpha k_\nu A_{\beta\mu} - k_\alpha k_\mu A_{\beta\nu} - k_\beta k_\nu A_{\alpha\mu}\big]\sin(k_\sigma x^\sigma) \quad (30.27)$$

c. Show that

$$R_{\beta\nu} = -\tfrac{1}{2}\big(k_\beta A_\nu + k_\nu A_\beta - k^2 A_{\beta\nu}\big)\sin(k_\sigma x^\sigma) \quad (30.28a)$$

$$\text{where } A_\nu = k_\mu \eta^{\mu\alpha} A_{\alpha\nu} - \tfrac{1}{2} k_\nu A, \quad (30.28b)$$

$k^2 \equiv k^\mu k_\mu \equiv \eta^{\alpha\mu} k_\alpha k_\mu$, and $A \equiv \eta^{\mu\nu} A_{\mu\nu}$. Note that A_ν is some constant four-vector for this perturbation.

d. In empty space, the Einstein equation requires that $R_{\beta\nu} = 0$. Show that whenever $k^2 \neq 0$, our metric perturbation will satisfy $R_{\beta\nu} = 0$ only if

$$A_{\beta\nu} = \frac{1}{k^2}(k_\beta A_\nu + k_\nu A_\beta) \quad (30.29)$$

e. Substitute this result into equation 30.27 to show that in such a case, the Riemann tensor is identically zero. This implies that unless $k^2 = 0$, our hypothetical wave *cannot* represent a physical gravitational wave (which would curve spacetime) but rather simply a periodic oscillation of our nearly cartesian coordinates in flat space.

f. If $k^2 = 0$, however, our perturbation *could* represent a physical gravitational wave. If so, show that such a wave must move at the speed of light. (*Hint:* One crest of the wave will be where $k_\sigma x^\sigma = \pi/2$. Write out the sum and take the t-derivative. For the sake of simplicity, assume that the wave moves in the $+z$ direction, where $k_\alpha = [-\omega, 0, 0, \omega]$.)

31. DETECTING GRAVITATIONAL WAVES

INTRODUCTION

FLAT SPACETIME
- Review of Special Relativity
- Four-Vectors
- Index Notation

TENSORS
- Arbitrary Coordinates
- Tensor Equations
- Maxwell's Equations
- Geodesics

SCHWARZSCHILD BLACK HOLES
- The Schwarzschild Metric
- Particle Orbits
- Precession of the Perihelion
- Photon Orbits
- Deflection of Light
- Event Horizon
- Alternative Coordinates
- Black Hole Thermodynamics

THE CALCULUS OF CURVATURE
- The Absolute Gradient
- Geodesic Deviation
- The Riemann Tensor

THE EINSTEIN EQUATION
- The Stress-Energy Tensor
- The Einstein Equation
- Interpreting the Equation
- The Schwarzschild Solution

COSMOLOGY
- The Universe Observed
- A Metric for the Cosmos
- Evolution of the Universe
- Cosmic Implications
- The Early Universe
- CMB Fluctuations & Inflation

GRAVITATIONAL WAVES
- Gauge Freedom
- **Detecting Gravitational Waves**
- Gravitational Wave Energy
- Generating Gravitational Waves
- Gravitational Wave Astronomy

SPINNING BLACK HOLES
- Gravitomagnetism
- The Kerr Metric
- Kerr Particle Orbits
- Ergoregion and Horizon
- Negative-Energy Orbits

this depends on this

363

31. DETECTING GRAVITATIONAL WAVES

Introduction. Gravitational waves are ripples in the curvature of spacetime that carry information about the changes in an object's gravitational field away from that object through empty space. Now that we have developed a reasonably simple form for the linearized field equations, we can explore gravitational wave solutions of the empty-space field equations (in the weak-field limit, at least).

In the last chapter, we found that by choosing to work in Lorenz gauge (where $\partial_\mu H^{\mu\nu} = 0$), the weak-field Einstein equation reduces to the simple form given in equation 30.16a. In empty space, the relevant equations for exploring gravitational wave solutions are therefore

$$\partial^\alpha \partial_\alpha H^{\mu\nu} = 0, \text{ subject to the restriction } \partial_\mu H^{\mu\nu} = 0 \quad (31.1)$$

Recall also that even if we confine ourselves to working in Lorenz gauge, we still have some freedom left to choose coordinates: if $H^{\mu\nu}$ solves equations 31.1, then

$$H'^{\mu\nu} = H^{\mu\nu} - \partial^\mu \xi^\nu - \partial^\nu \xi^\mu + \eta^{\mu\nu} \partial_\alpha \xi^\alpha \text{ with } \partial^\alpha \partial_\alpha \xi^\mu = 0 \quad (31.2)$$

does as well. This will be useful to us later.

A Plane-Wave Solution. Since $\partial^\alpha \partial_\alpha H^{\mu\nu} = [-(\partial^2/\partial t^2) + \nabla^2]H^{\mu\nu} = 0$ is the basic wave equation, we expect that we should be able to find wave-like solutions. Let's attempt a plane-wave solution of the form

$$H^{\mu\nu}(t,x,y,z) = A^{\mu\nu} \cos(k_\sigma x^\sigma) = A^{\mu\nu} \cos(\vec{k} \cdot \vec{r} - \omega t) \quad (31.3)$$

where $A^{\mu\nu}$ is a constant matrix and k_σ is a constant covector $= [-\omega, k_x, k_y, k_z]$. Such a wave is a plane wave with crests perpendicular to the $\vec{k}$ direction that move in the $\vec{k}$ direction with speed $v = \omega/k$. The Einstein equation, the Lorenz gauge condition, and the symmetry of $H^{\mu\nu}$ (which follows from the symmetry of $h^{\mu\nu}$) require that

$$\text{Einstein Equation} \Rightarrow k^\alpha k_\alpha = 0 \quad (31.4a)$$

$$\text{Lorenz Gauge} \Rightarrow k_\mu A^{\mu\nu} = 0 \quad (31.4b)$$

$$\text{Symmetry} \Rightarrow A^{\mu\nu} = A^{\nu\mu} \quad (31.4c)$$

(see box 31.1). The first equation implies that the wave moves with speed $v = 1$:

$$0 = k^\alpha k_\alpha = k_\alpha k_\beta \eta^{\alpha\beta} = (-\omega)^2 \eta^{tt} + (k_x)^2 \eta^{xx} + (k_y)^2 \eta^{yy} + (k_z)^2 \eta^{zz}$$

$$\Rightarrow 0 = -\omega^2 + k^2 \Rightarrow \omega = k \Rightarrow v = \frac{\omega}{k} = 1 \quad (31.5)$$

Therefore, gravitational waves (like electromagnetic waves) move at the maximum speed allowed by the theory of relativity. The other two equations put limitations on the matrix $A^{\mu\nu}$ that we can use to simplify that matrix.

Transverse-Traceless Gauge. Consider now an additional gauge transformation of the form

$$\xi^\mu = B^\mu \sin(k_\sigma x^\sigma) \quad (31.6)$$

where k_σ is the same as in the trial solution and the components of the vector B^μ are constants. Note that

$$\partial^\alpha \partial_\alpha \xi^\mu = \partial^\alpha \partial_\alpha B^\mu \sin(k_\sigma x^\sigma) = -k^\alpha k_\alpha B^\mu \sin(k_\sigma x^\sigma) = 0 \quad (31.7)$$

so this gauge transformation satisfies the restriction described in equation 31.2. Now, it turns out (see box 31.2) that if we appropriately choose the components of B^μ, we can convert any original choice of $A^{\mu\nu}$ satisfying equations 31.4b and 31.4c to a new $A^{\mu\nu}$ such that

$$A^{t\mu} = A^{\mu t} = 0 \quad \text{and} \quad A^\mu{}_\mu = 0 \quad (31.8)$$

We say that a gravitational plane wave solution fulfilling the conditions specified in equation 31.8 is in **transverse-traceless gauge** and denote it with the symbol $H^{\mu\nu}_{TT}$. Note that according to equation 30.7,

$$h_{\mu\nu}^{TT} = H_{\mu\nu}^{TT} - \tfrac{1}{2}\eta_{\mu\nu}H^{TT} = H_{\mu\nu}^{TT} \qquad (31.9)$$

since $H^{TT} \equiv (H^{TT})^{\mu}{}_{\mu} = A^{\mu}{}_{\mu}\cos(k_{\sigma}x^{\sigma}) = 0$. Therefore, there is no distinction in transverse-traceless gauge between the basic metric perturbation $h_{\mu\nu}$ and the trace-reversed perturbation $H_{\mu\nu}$. This is, therefore, a very nice gauge to work in (it is too bad that we can pull this off only for wave solutions in empty space!).

A Wave Propagating in the $+z$ Direction. We can make things more concrete by considering a plane wave propagating in a certain specific direction. The components for a wave-vector k_{σ} that satisfies the condition in equation 31.4a and describes a wave propagating in the $+z$ direction are

$$k_{\sigma} = [-\omega, 0, 0, \omega] \qquad (31.10)$$

where ω is the angular frequency of the wave. In this situation, the Lorenz condition reduces to $0 = k_{\mu}A^{\mu\nu} = -\omega A^{t\nu} + \omega A^{z\nu} \Rightarrow A^{t\nu} = A^{z\nu}$. This, together with the transverse-traceless gauge conditions and the symmetry of $A^{\mu\nu}$, implies that

TT Gauge and symmetry: $\quad A^{t\nu} = A^{\nu t} = 0 \qquad (31.11a)$

Lorenz Gauge and symmetry: $\quad A^{\nu z} = A^{z\nu} = A^{t\nu} = 0 \qquad (31.11b)$

TT Gauge: $\quad A^{xx} + A^{yy} = 0 \qquad (31.11c)$

This means that only $A^{xy} = A^{yx} \equiv A_{\times}$ and $A^{xx} = -A^{yy} = A_{+}$ are nonzero in $A^{\mu\nu}$.

This means that the $A_{\mu\nu}$ matrix for the most general possible transverse-traceless gravitational wave moving in the $+z$ direction can be written in the form

$$A^{\mu\nu} = A_{+}\begin{bmatrix} 0 & 0 & 0 & 0 \\ 0 & 1 & 0 & 0 \\ 0 & 0 & -1 & 0 \\ 0 & 0 & 0 & 0 \end{bmatrix} + A_{\times}\begin{bmatrix} 0 & 0 & 0 & 0 \\ 0 & 0 & 1 & 0 \\ 0 & 1 & 0 & 0 \\ 0 & 0 & 0 & 0 \end{bmatrix} \qquad (31.12)$$

Therefore, such a wave will be in general some linear combination of two distinct kinds of solutions, which we call the **polarizations** of the gravitational wave (in analogy to the two possible linear polarizations of electromagnetic waves). We call these two polarization states "upright" and "diagonal" or "plus" and "cross" (for reasons that will become clear shortly).

The Physical Effects of Such a Wave. When considering a wave's physical effects, we will find it useful to express the metric perturbation in the form $h_{\mu\nu}^{TT} = H_{\mu\nu}^{TT} = A_{\mu\nu}\cos(k_{\sigma}x^{\sigma})$ (with lowered indices) because this allows us to calculate the metric $g_{\mu\nu} = \eta_{\mu\nu} + h_{\mu\nu}$. To our level of approximation, $A_{\mu\nu} = \eta_{\mu\alpha}\eta_{\nu\beta}A^{\alpha\beta}$, and since $\eta_{\mu\nu}$ is diagonal and only η_{tt} is negative, we have $A_{ij} = +A^{ij}$, $A_{ti} = -A^{ti}$, $A_{it} = -A^{it}$, and $A_{tt} = +A^{tt}$, where i and j are spatial indices. In transverse-traceless gauge, $A^{t\nu} = 0$, so $A_{\mu\nu} = +A^{\mu\nu}$ for all elements. Therefore, equation 31.12 also describes the components of $A_{\mu\nu}$ with *lowered* indices.

Now, then, consider a particle initially at rest. Its four-velocity to leading order is $u^{\alpha} = [(-g_{tt})^{-1/2}, 0, 0, 0] \approx [1, 0, 0, 0]$. As a gravitational wave of the type described above moves by the particle, you can show (see box 31.3) that the particle's acceleration according to the geodesic equation is

$$\frac{d^2x^{\alpha}}{d\tau^2} = -\Gamma^{\alpha}_{\mu\nu}u^{\mu}u^{\nu} = 0 \quad (!) \qquad (31.13)$$

This looks like it says that the wave has no effect on the particle at all! However, this equation doesn't really say this. What it *is* saying is that the coordinate system that we have adopted by using the transverse-traceless gauge is a comoving coordinate system where the coordinates of a free particle remain fixed.

To find out about the real effects of the wave, therefore, we have to use the metric to determine the physical separation of nearby particles. Consider a set of particles in the xy plane that (before the wave comes by) are arranged in a ring of radius R. Now imagine

366 31. DETECTING GRAVITATIONAL WAVES

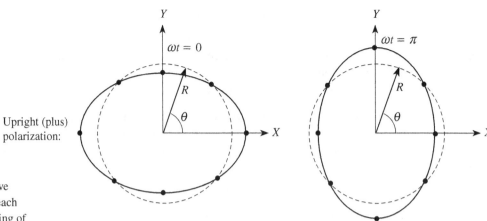

Upright (plus) polarization:

FIG. 31.1 How a gravitational wave moving in the +z direction with each possible polarization deforms a ring of particles. The X and Y coordinates are actual cartesian coordinates based on distances from the origin, not the co-moving coordinates that we are using in our metric.

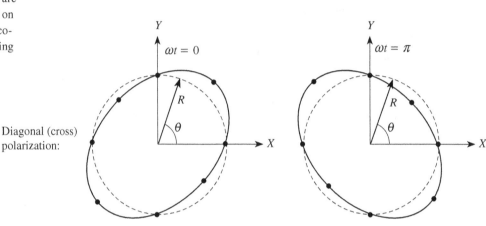

Diagonal (cross) polarization:

that an "uprightly" or "plus" polarized plane gravitational wave ($A_\times = 0$) moves in the +z direction through this ring. The displacements $\Delta x = R\cos\theta$ and $\Delta y = R\sin\theta$ of any given particle relative to the circle's center will be fixed in the transverse-traceless coordinate system, as we have just found. You can show, however, that the metric implies that the physical separation of this particle from the circle's center is actually

$$\Delta s = R[1 + \tfrac{1}{2}A_+ \cos\omega t \cos 2\theta] \qquad (31.14a)$$

Similarly, for a "diagonally" or "cross" polarized wave, we have

$$\Delta s = R[1 + \tfrac{1}{2}A_\times \cos\omega t \sin 2\theta] \qquad (31.14b)$$

(see box 31.4). Each wave polarization deforms the ring as shown in figure 31.1. This is why the international symbol for dangerous gravitational radiation is

(OK, this is a geeky insider GR joke.)

Typical Amplitudes of Astrophysical Sources. Actually, as we will see, a wide variety of astrophysical sources will produce gravitational waves with an amplitude A_+ or $A_\times \approx 10^{-22}$ to 10^{-18} measured at the earth. This means that free particles separated by $R = 10^6$ m (= 1000 km) will oscillate back and forth with amplitudes on the order of 10^{-14} m (about the size of a large atomic nucleus) if struck by a gravitational wave with an amplitude of $A_{+/\times} \approx 10^{-20}$. The implication is that gravitational waves from astrophysical sources are going to be hard to detect (and the deformations shown in the diagrams above are grossly over-exaggerated)! Though such a gravitational wave can carry a relatively high energy density (a 1-kHz wave with such an amplitude has an intensity of about 30 W/m^2, which is about 1/30 the intensity of direct sunlight), the tiny effect that such a wave has on particles means that the energy in a gravitational wave moving through matter is not very effectively absorbed by that matter. The sun could be swallowed by a solar-mass black hole (something that would generate very high-amplitude gravitational waves at the earth's location, maybe as large as $A_{+/\times} \approx 10^{-8}$) and you would never notice the waves passing through your body.

Gravitational Wave Detectors. In spite of the difficulties, scientists are currently planning and building gravitational wave detectors that can detect waves of this magnitude. The basic design for almost any such detector involves a set of floating masses that are extraordinarily well isolated from external effects, and some kind of laser-ranging system that accurately measures the distance between the objects using interferometry. (You might think that this would be impossible with visible-light lasers, since the displacement of each mass when the wave comes by will be such a tiny fraction of the wavelength of light. But if the laser is bright enough, one can detect a shift in the interference pattern that is an extraordinarily tiny fraction of the wavelength.)

An experiment called LIGO (Laser Interferometer Gravitational-wave Observatory) is currently operating at sites in Washington State (see figure 31.2) and Louisiana (having a pair of detectors enables one to discard spurious signals that do not affect both at the same time). The "floating masses" for these detectors consist of mirrors suspended at the ends of a 4-km L-shaped interferometer in such a way as to make them extraordinarily isolated, both seismically and thermally, from their surroundings. To reduce the effects of gas molecules hitting the mirrors and to prevent scattering of the laser light by molecules, the whole interferometer system is put in a vacuum. LIGO is able to detect waves with amplitudes below 10^{-22} over a frequency range from 60 Hz to 800 Hz, and has already detected a number of black hole and neutron-star merger events. Leaders of the LIGO experiment won the 2017 Nobel Prize in physics for this triumph. For more information, visit www.ligo.caltech.edu.

FIG. 31.2 An aerial photograph of the LIGO facility at Hanford, WA. One complete 4 km leg of the interferometer "L" is shown receding in the distance. Courtesy of the LIGO Laboratory.

I personally have been working to prepare the way for LISA (Laser Interferometer Space Antenna), which is currently projected to launch in 2034 as an ESA (European Space Agency) mission. The floating masses in the case of LISA would be "proof masses" isolated at the centers of three satellites arranged in an equilateral triangle. The array would orbit the sun about 20° behind the earth's position, and would be roughly 2.5 million km on a side (the vacuum system comes for free in this case!). In spite of the longer interferometer arms, LISA will have a somewhat lower sensitivity than LIGO, but will be most sensitive to waves with frequencies between 10^{-2} Hz and 10^{-4} Hz. While LIGO has to wait for high-frequency events that happen rarely, LISA will be able to detect waves produced by known sources (binary star systems), and thus would be certain to detect gravitational waves if GR is at all correct. For more information, visit elisascience.org.

BOX 31.1 Constraints on Our Trial Solution

The Einstein equation and Lorenz gauge condition require that trace-reversed metric perturbation solutions obey the constraints $\partial^\alpha \partial_\alpha H^{\mu\nu} = 0$ and $\partial_\mu H^{\mu\nu} = 0$. For our trial plane-wave solution $H^{\mu\nu} = A^{\mu\nu} \cos(k_\sigma x^\sigma)$, you can show that these equations imply that $k^\alpha k_\alpha = 0$ and $k_\mu A^{\mu\nu} = 0$.

Exercise 31.1.1. Verify both of these results.

BOX 31.2 The Transformation to Transverse-Traceless Gauge

According to equation 31.6, the gauge transformation $\xi^\mu = B^\mu \sin(k_\sigma x^\sigma)$ will transform our original wave $H^{\mu\nu} = A^{\mu\nu}\cos(k_\sigma x^\sigma)$ to a new wave such that

$$H'^{\mu\nu} = H^{\mu\nu} - \partial^\mu \xi^\nu - \partial^\nu \xi^\mu + \eta^{\mu\nu}\partial_\alpha \xi^\alpha$$
$$= H^{\mu\nu} - \eta^{\mu\gamma}\partial_\gamma \xi^\nu - \eta^{\nu\beta}\partial_\beta \xi^\mu + \eta^{\mu\nu}\partial_\alpha \xi^\alpha \quad (31.15)$$

Now, for the given transformation,

$$\partial_\mu \xi^\nu = B^\nu \frac{\partial}{\partial x^\mu}[\sin(k_\sigma x^\sigma)] = B^\nu \cos(k_\sigma x^\sigma) k_\beta \frac{\partial x^\beta}{\partial x^\mu}$$
$$= B^\nu \cos(k_\sigma x^\sigma) k_\beta \delta^\beta{}_\mu = k_\mu B^\nu \cos(k_\sigma x^\sigma) \quad (31.16)$$

Using this and the definition $H'^{\mu\nu} \equiv A'^{\mu\nu}\cos(k_\sigma x^\sigma)$, you can easily show that

$$A'^{\mu\nu} = A^{\mu\nu} - k^\mu B^\nu - k^\nu B^\mu + \eta^{\mu\nu} k_\alpha B^\alpha \quad (31.17)$$

Exercise 31.2.1. Verify equation 31.17.

The "transverse" condition therefore requires that

$$0 = A'^{\mu t} = A^{\mu t} - k^\mu B^t - k^t B^\mu + \eta^{\mu t} k_\alpha B^\alpha \quad (31.18)$$

You can easily show that the "traceless" condition implies that

$$0 = A'^\mu{}_\mu = A^\mu{}_\mu + 2k_\mu B^\mu \quad (31.19)$$

Exercise 31.2.2. Verify equation 31.19.

BOX 31.2 (continued) The Transformation to Transverse-Traceless Gauge

If we solve equation 31.19 ($0 = A'^\mu{}_\mu = A^\mu{}_\mu + 2k_\alpha B^\alpha$) for $k_\alpha B^\alpha$ and plug the result into equation 31.18 ($0 = A'^{\mu t} = A^{\mu t} - k^\mu B^t - k^t B^\mu + \eta^{\mu t} k_\alpha B^\alpha$), we get

$$0 = A^{\mu t} - k^\mu B^t - k^t B^\mu - \tfrac{1}{2}\eta^{\mu t} A^\sigma{}_\sigma \qquad (31.20)$$

(Note that I have renamed the index in the last term to avoid confusion with the free μ index.) To see that we can really find values of B^μ that solve this equation, it helps to take advantage of our freedom to orient our reference frame so that the $+z$ axis coincides with the direction of the wave's motion. Then $k_\sigma = [-\omega, 0, 0, \omega]$, and we can divide equation 31.20 into individual component equations as follows:

$$0 = A^{tt} - k^t B^t - k^t B^t - \tfrac{1}{2}\eta^{tt} A^\sigma{}_\sigma = A^{tt} - 2\omega B^t + \tfrac{1}{2}A^\sigma{}_\sigma \qquad (31.21a)$$

$$0 = A^{tx} - k^t B^x - k^x B^t - \tfrac{1}{2}\eta^{tx} A^\sigma{}_\sigma = A^{tx} - \omega B^x \qquad (31.21b)$$

$$0 = A^{ty} - k^t B^y - k^y B^t - \tfrac{1}{2}\eta^{ty} A^\sigma{}_\sigma = A^{ty} - \omega B^y \qquad (31.21c)$$

$$0 = A^{tz} - k^t B^z - k^z B^t - \tfrac{1}{2}\eta^{tz} A^\sigma{}_\sigma = A^{tz} - \omega B^z - \omega B^t \qquad (31.21d)$$

The first three equations we can solve for B^t, B^x, and B^y directly. We can then plug our solution into the last equation to solve for B^z.

Exercise 31.2.3. Carry out this plan to express the components of B^μ entirely in terms of the original wave amplitude matrix $A^{\mu\nu}$.

Since we can find a solution for waves moving in the $+z$ direction, and since there should be nothing special about any particular direction, we ought to be able to find a solution for B^μ that puts *any* plane wave into transverse-traceless gauge.

BOX 31.3 A Particle at Rest Remains at Rest in TT Coordinates

Equation 31.13 (repeated here) claims that when we describe a gravitational wave in a transverse-traceless (TT) coordinate system, its effect on a particle initially at rest (whose four-velocity is thus $u^\alpha \approx [1,0,0,0]$ in our nearly cartesian TT coordinate system) according to the geodesic equation is given by

$$\frac{d^2 x^\alpha}{d\tau^2} = -\Gamma^\alpha_{\mu\nu} u^\mu u^\nu = 0 \qquad (31.13r)$$

Exercise 31.3.1. Check that equation 31.13 is true for a wave expressed in transverse-traceless gauge. (*Hint:* Write out the relevant Christoffel symbols in terms of the metric, and note the conditions imposed by the Lorenz and transverse-traceless gauge conditions. You won't need the full space below.)

BOX 31.4 The Effect of a Gravitational Wave on a Ring of Particles

Consider a given ring particle that has (small) transverse-traceless coordinate displacements $\Delta t = 0$, $\Delta x = R\cos\theta$, $\Delta y = R\sin\theta$, and $\Delta z = 0$ from the origin as a gravitational wave moves through. We have seen that the transverse-traceless coordinates of such a particle will remain fixed. However, such a particle's actual squared displacement from the origin is (according to the metric equation)

$$\Delta s^2 = g_{\mu\nu}\Delta x^\mu \Delta x^\nu = (\eta_{\mu\nu} + h^{TT}_{\mu\nu})\Delta x^\mu \Delta x^\nu = (\eta_{\mu\nu} + H^{TT}_{\mu\nu})\Delta x^\mu \Delta x^\nu$$
$$= [\eta_{\mu\nu} + A_{\mu\nu}\cos(k_\sigma x^\sigma)]\Delta x^\mu \Delta x^\nu = [\eta_{\mu\nu} + A_{\mu\nu}\cos\omega t]\Delta x^\mu \Delta x^\nu \quad (31.22)$$

since $\cos(k_\sigma x^\sigma) = \cos(-\omega t + k_z \cdot 0) = \cos(-\omega t) = \cos\omega t$ if all the particles are at $z = 0$. Since only Δx and Δy are nonzero and $\eta_{xy} = 0$, this reduces to

$$\Delta s^2 = [\eta_{xx} + A_{xx}\cos\omega t]\Delta x^2 + [\eta_{yy} + A_{yy}\cos\omega t]\Delta y^2 + 2A_{xy}\cos\omega t\, \Delta x \Delta y \quad (31.23)$$

If the wave is upright polarized ($A_{xx} = -A_{yy} \equiv A_+$, $A_{xy} = 0$), we have

$$\Delta s^2 = R^2[1 + A_+ \cos\omega t]\cos^2\theta + R^2[1 - A_+ \cos\omega t]\sin^2\theta$$
$$= R^2[1 + A_+ \cos\omega t(\cos^2\theta - \sin^2\theta)] = R^2[1 + A_+ \cos\omega t \cos 2\theta]$$
$$\Rightarrow \quad \Delta s = R[1 + A_+ \cos\omega t \cos 2\theta]^{1/2} \approx R[1 + \tfrac{1}{2}A_+ \cos\omega t \cos 2\theta] \quad (31.24)$$

where in the last step I used the binomial approximation.

Similarly, you can show that for a cross-polarized wave, the separation is

$$\Delta s = R[1 + \tfrac{1}{2}A_\times \cos\omega t \sin 2\theta] \quad (31.25)$$

Exercise 31.4.1. Verify the last result.

HOMEWORK PROBLEMS

P31.1 (Important! Be sure to work either this problem or the next.) Consider a gravitational wave having the form $H^{\mu\nu} = A^{\mu\nu}\cos(k_\alpha x^\alpha)$ where $k_t = -\omega$, $k_x = k_y = 0$, $k_z = +\omega$ and

$$A^{\mu\nu} = \begin{bmatrix} a & 0 & 0 & a \\ 0 & b & 0 & 0 \\ 0 & 0 & c & 0 \\ a & 0 & 0 & a \end{bmatrix} \quad (31.26)$$

where a, b, and c are different constants.

a. Show that this wave satisfies all the conditions implied by equation 31.4.
b. Use the results of exercise 31.2.3 to calculate the component of B^μ that will put this matrix into transverse-traceless gauge.
c. Use equation 31.17 to then calculate the $A'^{\mu\nu}$ matrix.
d. Show that the $A'^{\mu\nu}$ does in fact satisfy the transverse-traceless conditions specified in equation 31.11, and calculate A_+ and $A_\times$ in terms of a, b, and c.
e. For waves moving in a coordinate axis direction, you can find the components of $A'^{\mu\nu}$ by simply (1) setting to zero all the components of $A^{\mu\nu}$ that have a t index or an index corresponding to the axis direction of the wave's motion, and (2) subtracting half the trace of the remaining components of $A^{\mu\nu}$ from each diagonal element. Verify that this simple two-step scheme does in fact yield the result you found in part c.

P31.2 (Important! Be sure to work either this problem or the previous.) Consider a gravitational wave having the form $H^{\mu\nu} = A^{\mu\nu}\cos(k_\alpha x^\alpha)$ where $k_t = -\omega$, $k_x = k_y = 0$, $k_z = +\omega$ and

$$A^{\mu\nu} = \begin{bmatrix} 0 & a & 0 & 0 \\ a & c & b & a \\ 0 & b & -c & 0 \\ 0 & a & 0 & 0 \end{bmatrix} \quad (31.27)$$

where a, b, and c are different constants.

a. Show that this wave satisfies all the conditions implied by equation 31.4.
b. Use the results of exercise 31.2.3 to calculate the component of B^μ that will put this matrix into transverse-traceless gauge.
c. Use equation 31.17 to then calculate the $A'^{\mu\nu}$ matrix.
d. Show that $A'^{\mu\nu}$ does in fact satisfy the transverse-traceless conditions specified in equation 31.11, and calculate A_+ and $A_\times$ in terms of a, b, and c.
e. For waves moving in a coordinate axis direction, you can find the components of $A'^{\mu\nu}$ by simply (1) setting to zero all the components of $A^{\mu\nu}$ that have a t index or an index corresponding to the axis direction of the wave's motion, and (2) subtracting half the trace of the remaining components of $A^{\mu\nu}$ from each diagonal element. Verify that this simple two-step scheme does in fact yield the result you found in part c.

P31.3 Check that values of B^μ that you calculated in exercise 31.2.3 really do produce a *traceless* wave; that is, verify that $A'^\mu{}_\mu = 0$ no matter what $A^{\mu\nu}$ is. (*Hint:* Remember that we are using Lorenz gauge.)

P31.4 Show that if we define the X and Y coordinates of a particle responding to a plus-polarized gravitational wave using the equations

$$X = x[1 + \tfrac{1}{2}A_+\cos\omega t] \quad (31.28a)$$
$$Y = y[1 - \tfrac{1}{2}A_+\cos\omega t] \quad (31.28b)$$

then $\Delta s^2 = X^2 + Y^2$ for all particles in the ring, where Δs^2 is given by 31.14a. (*Hint:* Remember that the particle's transverse-traceless coordinates x and y remain fixed as the wave passes, and that $x^2 + y^2 = R^2$).

P31.5 Use the results of problem P31.4 to show that the deformed shape of the ring is an ellipse (at least to lowest order in A_+). (*Hint:* The equation of an ellipse is

$$\frac{X^2}{a^2} + \frac{Y^2}{b^2} = \text{const.} \quad (31.29)$$

where a and b are constants.)

P31.6 Consider the following gravitational wave

$$h^{TT}_{\mu\nu} = \begin{bmatrix} 0 & 0 & 0 & 0 \\ 0 & A\cos(k_\alpha x^\alpha) & A\sin(k_\alpha x^\alpha) & 0 \\ 0 & A\sin(k_\alpha x^\alpha) & -A\cos(k_\alpha x^\alpha) & 0 \\ 0 & 0 & 0 & 0 \end{bmatrix} \quad (31.30)$$

where $k_t = -\omega$, $k_x = k_y = 0$, $k_z = +\omega$ and A is a suitably small constant. This is a superposition of a plus-polarized wave with a cross-polarized wave that is $\pi/2$ out of phase. We call this a circularly polarized gravitational wave.

a. Argue that such a wave perturbs a ring of particles in the xy plane in such a way that their shape becomes an ellipse (see problem P31.5) that rotates in the that plane. (*Hint:* Look up the trigonometric identity for the cosine of a difference between two angles.)
b. What is the rotation rate of the ellipse in terms of ω? Does it rotate clockwise or counterclockwise when viewed with the z axis facing us?

P31.7 Imagine that a rigid rod separates two small spheres of mass m so that they are a distance L apart along the x axis. Define the origin of that axis so that the centers of the spheres are initially located at $x = \pm\tfrac{1}{2}L$. Now imagine that a plus-polarized plane gravitational wave with amplitude A_+ and angular frequency ω passes through this object. Determine the x component of the force that the rod must exert on each sphere to keep them a fixed distance apart. (*Hint:* We know that the distance between freely floating objects would change as the gravitational wave goes by, so objects that are constrained to have a fixed separation cannot be following geodesics in our spacetime. Remember also that in transverse-traceless coordinates, freely floating objects would remain at a fixed coordinate x. There are several reasonable ways to determine the force. Choose one that makes sense to you and explain it carefully.)

P31.8 We do not *have* to use the weak-field limit to find gravitational wave solutions to the Einstein equation. From a simple analogy to electromagnetic waves, we might expect gravitational waves to be transverse and move with the speed of light. This suggests a trial metric of the form

$$ds^2 = -dt^2 + [p(u)]^2 dx^2 + [q(u)]^2 dy^2 + dz^2 \quad (31.31)$$

where $p(u)$ and $q(u)$ are arbitrary functions of $u \equiv t - z$. This is a plane wave in the metric (because it is independent of x and y) where the disturbance in the metric moves in the $+z$ direction and is "transverse" because only components of the metric transverse to the z direction (the wave's direction of travel) are affected.

a. Use the Diagonal Metric Worksheet to prove that this metric satisfies the empty-space Einstein equation if and only if

$$\frac{\ddot{p}}{p} + \frac{\ddot{q}}{q} = 0 \quad (31.32)$$

where $\dot{p} = dp/dt = -dp/dz$, $\ddot{p} = d^2p/dt^2 = d^2p/dz^2$, and similarly for q.

b. One of the problems with our metric is that for some choices of p and q, the metric is really flat space in disguise. Use the Diagonal Metric Worksheet to calculate all the Christoffel symbols for this metric. Then calculate R_{xtxt} and R_{ytyt} to show a *necessary* condition for our metric to describe flat spacetime is that $\ddot{p} = \ddot{q} = 0$. (This turns out to be the sufficient condition as well.)

c. Show that $p = \cos\omega u$, $q = \cosh\omega u$ is a non-flat exact plane wave solution (though not a very realistic one.)

d. Show that $p^2 = 1 + A\cos\omega u$ and $q^2 = 1 - A\cos\omega u$ (the solutions to the linearized Einstein equation for plus-polarized gravitational waves) satisfy the equation $\ddot{p}/p + \ddot{q}/q = 0$ only if A is small enough that we can ignore terms of order A^n for $n > 1$. Also argue that this is a real wave (meaning that it describes a non-flat spacetime) even to just first order in A.

e. (Optional). Argue that $\ddot{p} = \ddot{q} = 0$ is a *sufficient* condition for our metric to describe flat space. (*Hint:* You will have to determine all of the possibly nonzero components of the Riemann tensor.)

32. GRAVITATIONAL WAVE ENERGY

INTRODUCTION

FLAT SPACETIME
- Review of Special Relativity
- Four-Vectors
- Index Notation

TENSORS
- Arbitrary Coordinates
- Tensor Equations
- Maxwell's Equations
- Geodesics

SCHWARZSCHILD BLACK HOLES
- The Schwarzschild Metric
- Particle Orbits
- Precession of the Perihelion
- Photon Orbits
- Deflection of Light
- Event Horizon
- Alternative Coordinates
- Black Hole Thermodynamics

THE CALCULUS OF CURVATURE
- The Absolute Gradient
- Geodesic Deviation
- The Riemann Tensor

THE EINSTEIN EQUATION
- The Stress-Energy Tensor
- The Einstein Equation
- Interpreting the Equation
- The Schwarzschild Solution

COSMOLOGY
- The Universe Observed
- A Metric for the Cosmos
- Evolution of the Universe
- Cosmic Implications
- The Early Universe
- CMB Fluctuations & Inflation

GRAVITATIONAL WAVES
- Gauge Freedom
- Detecting Gravitational Waves
- **Gravitational Wave Energy**
- Generating Gravitational Waves
- Gravitational Wave Astronomy

SPINNING BLACK HOLES
- Gravitomagnetism
- The Kerr Metric
- Kerr Particle Orbits
- Ergoregion and Horizon
- Negative-Energy Orbits

this depends on this

32. GRAVITATIONAL WAVE ENERGY

Introduction. In chapter 31, we saw that a passing gravitational wave will distort a ring of floating particles and cause them to oscillate relative to each other. This means that the particles gain kinetic energy in a reference frame based on the center of the ring, which implies that the gravitational wave must carry energy from its source to the particles. Our goal in this chapter is to try to find out how to calculate how much energy a gravitational wave of a given amplitude carries.

Gravitational Energy in General Relativity. This is not a trivial task. The energy stored in a gravitational field is a very subtle and sticky subject in general relativity. In the Einstein equation $G^{\mu\nu} = 8\pi G T^{\mu\nu}$, the stress-energy tensor $T^{\mu\nu}$ describes the density of matter and energy *excluding* whatever energy might be associated with the gravitational field. Everything having to do with the gravitational field is embedded in the Einstein tensor $G^{\mu\nu}$: the way that the gravitational field energy acts as a source for the field is expressed through the nonlinear aspects of $G^{\mu\nu}$.

We have talked before about how the equation $\nabla_\mu T^{\mu\nu} = 0$ expresses the principle of conservation of energy and momentum. Actually, if you look back at the derivation, you will see that I demonstrated this only in the context of a locally cartesian coordinate system: it is $\partial_\mu T^{\mu\nu} = 0$ that really implies conservation of energy and momentum. It turns out that in a general curved spacetime, the equation $\nabla_\mu T^{\mu\nu} = 0$ cannot be integrated over a finite region of space (using Gauss's law) to yield conserved quantities the way that it can in a flat spacetime. Therefore, conservation of non-gravitational energy and momentum makes sense only in a flat space. In fact, it is the symmetry of flat spacetime with respect to displacements in time and space that give rise to the conserved quantities that we interpret as conservation of energy and momentum: an arbitrary curved spacetime does not have these symmetries.

Indeed, at any given point in a curved spacetime, we can define a locally inertial reference frame, and sufficiently near the center of that coordinate system, the equation $\partial_\mu T^{\mu\nu} = 0$ implies that *non-gravitational* energy and momentum are conserved, but it says nothing about how gravitational energy fits in. So if a gravitational field is going to exchange "energy" with matter, it must do so in a manner that can be seen only if we integrate over a sufficiently large area to "see" the non-local curvature of spacetime. This is essentially what we did with the ring of particles in figure 31.1: we cannot observe the effects of a gravitational wave on a single particle! This all means that we will *never* be able to find a tensor quantity that expresses the density of gravitational field energy at a given event in spacetime. If such a tensor quantity were to exist, it would have to be zero at the center of a locally cartesian coordinate system that we could set up at that event, and if it is zero in that coordinate system then (by the tensor transformation rule) it will be zero in all coordinate systems (!).

But if we cannot define a tensor that describes the local energy density of gravitational energy, then it becomes very difficult to decide what this concept even means. One can simply give up and say that energy is simply not conserved in a curved spacetime, so particles can pick up energy from a wave that has no energy to give (see for example, J. L. Martin, *General Relativity*, Halsted Press, 1988, p. 164). On the other hand, one can work to find an effective stress-energy pseudotensor $T^{\mu\nu}_{\text{eff}}$ that includes something about gravity and that satisfies $\partial_\mu T^{\mu\nu}_{\text{eff}} = 0$ even in a curved space and can thus be integrated to find conserved quantities (see Ciufolini and Wheeler, *Gravitation and Inertia*, Princeton, 1995, pp. 45–48). This approach does yield a rather strange mathematical object that has meaning only when integrated over a finite region of spacetime, and even then, the conserved quantities do not transform like tensors (and thus do not have a coordinate-independent meaning) except under certain circumstances.

The bottom line is that energy conservation is a very slippery concept in general relativity that must be approached with great care and delicacy. (Thomas Helliwell, a theorist at Harvey Mudd College, remembers a quite heated argument over this very issue at Caltech some years ago.)

In Almost Flat Spacetime, We Can *Pretend*. Happily, in the weak-field limit, spacetime is almost flat and therefore energy and momentum are almost conserved. In this limit, it turns out that we can employ a generally-accepted trick that satisfies our intuition about how gravitational waves should conserve energy and allows us to calculate an effective energy carried by the waves in this limit.

The Einstein equation to first order in the metric perturbation $h_{\mu\nu}$ (and in the coordinates defined by the Lorenz gauge condition) says that

$$-2G^{(1)}_{\mu\nu} = \partial_\alpha \partial^\alpha H_{\mu\nu} = -16\pi G T_{\mu\nu} \qquad (32.1)$$

where $G^{(1)}_{\mu\nu}$ is the Einstein tensor $G_{\mu\nu}$ evaluated to first order in $h_{\mu\nu}$ and $H_{\mu\nu}$ is the trace-reversed perturbation $H_{\mu\nu} = h_{\mu\nu} - \frac{1}{2}\eta_{\mu\nu}h$. This equation describes how a gravitational field is generated by the density of matter and non-gravitational energy expressed by the stress-energy tensor $T_{\mu\nu}$. This equation does not express anything about how the energy of the gravitational field feeds back on itself to create a stronger gravitational field because this issue is expressed by the nonlinearities in $G_{\mu\nu}$ that we are expressly ignoring in the weak-field limit.

In order to look at this gravitational energy, we have to expand the field equations to second order in the metric perturbation:

$$-2G^{(1)}_{\mu\nu} - 2G^{(2)}_{\mu\nu} = \partial_\alpha \partial^\alpha H_{\mu\nu} - 2G^{(2)}_{\mu\nu} = -16\pi G T_{\mu\nu} \qquad (32.2)$$

This will capture at least the leading terms in the self-feedback effect. Now, remember that in our gravitational wave model, we are pretending that spacetime is flat and that gravity is completely described by the tensor field $H_{\mu\nu}$ that sits on top of that flat spacetime. If we move the $G^{(2)}_{\mu\nu}$ term to the other side, we get

$$\partial_\alpha \partial^\alpha H_{\mu\nu} = -16\pi G T_{\mu\nu} + 2G^{(2)}_{\mu\nu} = -16\pi G(T_{\mu\nu} + T^{GW}_{\mu\nu}) \qquad (32.3)$$

$$\text{where } T^{GW}_{\mu\nu} \equiv -\frac{G^{(2)}_{\mu\nu}}{8\pi G} \qquad (32.4)$$

In this way of interpreting the field equation, $T^{GW}_{\mu\nu}$ is acting along with the non-gravitational stress-energy $T_{\mu\nu}$ to create the gravitational field, so it is acting like a stress-energy of the gravitational field. Moreover, $\partial_\mu H^{\mu\nu} = 0$ in the Lorenz gauge we are working in, so if we take the divergence of both sides of equation 32.3 (and raise indices appropriately), we find that

$$\partial_\mu (T^{\mu\nu} + T^{\mu\nu}_{GW}) = 0 \qquad (32.5)$$

which expresses conservation of the sum of matter-energy and gravitational energy (as well as corresponding momentum conservation laws), at least in the context of the assumed flat-space background. We can therefore interpret $T^{GW}_{\mu\nu}$ as being the *effective* stress-energy of the gravitational field in this limit. As discussed on the previous page, this quantity has no meaning at a point but becomes a meaningful tensor-like quantity only after it has been averaged over several wavelengths of the gravitational wave, so technically, we need to modify the definition of $T^{GW}_{\mu\nu}$ given above to read

$$T^{GW}_{\mu\nu} \equiv -\frac{\langle G^{(2)}_{\mu\nu} \rangle}{8\pi G} \quad \text{(where the average is over several wavelengths)} \qquad (32.6)$$

This is, therefore, our final expression for the effective stress-energy carried by a gravitational wave in the weak-field limit.

Evaluating the Stress-Energy for a Gravitational Wave. Our goal in this section is to evaluate this tensor for a given gravitational wave of the type discussed in the last chapter. Consider a plus-polarized gravitational wave moving in the $+z$ direction. Let's define for such a wave the following quantities as convenient short-hand expressions:

$$h_+(t,z) \equiv A_+ \cos(\omega t - \omega z) = h_{xx}^{TT} = -h_{yy}^{TT} \tag{32.7a}$$

$$\dot{h}_+ \equiv \partial_t h_+ = -\partial_z h_+ = -A_+ \omega \sin(\omega t - \omega z) \tag{32.7b}$$

$$\ddot{h}_+ \equiv \partial_t \partial_t h_+ = \partial_z \partial_z h_+ = -\partial_z \partial_t h_+ = -\omega^2 A_+ \cos(\omega t - \omega z) = -\omega^2 h_+ \tag{32.7c}$$

The metric is completely diagonal for this perturbation, so we can use the Diagonal Metric Worksheet to evaluate the Ricci tensor, with

$$A = 1, \ B = 1 + h_+, \ C = 1 - h_+, \ D = 1 \tag{32.8a}$$

$$B_0 = C_3 = \dot{h}_+, \ B_3 = C_0 = -\dot{h}_+, \ B_{00} = -B_{03} = -C_{00} = B_{33} = C_{03} = -C_{33} = \ddot{h}_+ \tag{32.8b}$$

You can use the worksheet (see box 32.1) to show that, to second order in h_+, the diagonal elements of the Ricci tensor are

$$R_{tt} = R_{zz} = h_+ \ddot{h}_+ + \tfrac{1}{2}\dot{h}_+ \dot{h}_+, \quad R_{xx} = R_{yy} = 0 \tag{32.9}$$

Now, note that if we average over several gravitational wave wavelengths,

$$\langle \dot{h}_+ \dot{h}_+ + h_+ \ddot{h}_+ \rangle = \langle [-\omega A_+ \sin\theta]^2 + A_+ \cos\theta[-\omega^2 A_+ \cos\theta] \rangle$$
$$= A_+^2 \omega^2 \langle \sin^2\theta - \cos^2\theta \rangle = -A_+^2 \omega^2 \langle \cos 2\theta \rangle = 0 \tag{32.10}$$

where $\theta = \omega t - \omega z$. This means that

$$\langle h_+ \ddot{h}_+ + \tfrac{1}{2}\dot{h}_+ \dot{h}_+ \rangle = \langle h_+ \ddot{h}_+ + \dot{h}_+ \dot{h}_+ \rangle - \tfrac{1}{2}\langle \dot{h}_+ \dot{h}_+ \rangle = 0 - \tfrac{1}{2}\langle \dot{h}_+ \dot{h}_+ \rangle \tag{32.11}$$

So the nonzero second-order diagonal components of the Ricci tensor are

$$\langle R_{tt}^{(2)} \rangle = \langle R_{zz}^{(2)} \rangle = -\tfrac{1}{2}\langle \dot{h}_+ \dot{h}_+ \rangle \tag{32.12}$$

when averaged over several wavelengths. You can show (see box 32.2) that the averaged second-order portion of the curvature scalar is zero:

$$\langle R^{(2)} \rangle = 0 \tag{32.13}$$

The effective energy density of an uprightly polarized gravitational wave is, then,

$$T_{tt}^{GW} = -\frac{\langle G_{tt}^{(2)} \rangle}{8\pi G} = -\frac{\langle R_{tt}^{(2)} \rangle}{8\pi G} = +\frac{\langle \dot{h}_+ \dot{h}_+ \rangle}{16\pi G} \tag{32.14}$$

The energy contributed by a diagonally polarized wave is trickier to calculate because we cannot use the Diagonal Metric Worksheet but rather must calculate the Christoffel symbols, Riemann tensor components, and Ricci tensor components by hand. However, the result cannot be any different than equation 32.14, because we can convert an uprightly polarized wave to a diagonally polarized wave simply by rotating our coordinate system by 45° around the z axis. Therefore, the formula for the total energy density of an arbitrary gravitational wave must be

$$T_{tt}^{GW} = \frac{1}{16\pi G}\langle \dot{h}_+ \dot{h}_+ + \dot{h}_\times \dot{h}_\times \rangle \quad \text{where} \quad h_\times \equiv h_{xy}^{TT} = h_{yx}^{TT} \tag{32.15}$$

You can easily show (see box 32.3) that this expression can be written

$$T_{tt}^{GW} = \frac{1}{32\pi G}\langle \dot{h}_{jk}^{TT} \dot{h}_{TT}^{jk} \rangle \tag{32.16}$$

This last expression, which involves a sum over the purely spatial indices j and k, is general and applies to waves moving in any spatial direction. This is (finally) the sought-after general formula for a gravitational waves' effective energy density.

The gravitational wave flux (energy transported per unit time per unit area by the wave in the direction of its motion) is given by

$$\text{flux} = \frac{1}{32\pi G}\langle \dot{h}_{jk}^{TT} \dot{h}_{TT}^{jk} \rangle = T_{tt}^{GW} \ (!) \tag{32.17}$$

This result is discussed in problems P32.1 and P32.2.

BOX 32.1 The Ricci Tensor

Exercise 32.1.1. Use the Diagonal Metric Worksheet to verify equations 32.9. I have provided a copy of the relevant pages of the worksheet in the next two pages for your convenience.

BOX 32.2 The Averaged Curvature Scalar

Exercise 32.2.1. Show that equation 32.9 and 32.12 imply equation 32.13. Don't forget that we are keeping only terms that are second order in h_+.

BOX 32.3 The General Energy Density of a Gravitational Wave

Exercise 32.3.1. Argue that equation 32.15 can be written in the form given in equation 32.16.

380 32. GRAVITATIONAL WAVE ENERGY

$R_{00} = 0 \qquad + \tfrac{1}{2B}A_{11} \qquad + \tfrac{1}{2C}A_{22} \qquad + \tfrac{1}{2D}A_{33}$

$+ \quad 0 \qquad - \tfrac{1}{2B}B_{00} \qquad - \tfrac{1}{2C}C_{00} \qquad - \tfrac{1}{2D}D_{00}$

$+ \quad 0 \qquad + \tfrac{1}{4B^2}B_0^2 \qquad + \tfrac{1}{4C^2}C_0^2 \qquad + \tfrac{1}{4D^2}D_0^2$

$+ \quad 0 \qquad + \tfrac{1}{4AB}A_0 B_0 \qquad + \tfrac{1}{4AC}A_0 C_0 \qquad + \tfrac{1}{4AD}A_0 D_0$

$- \tfrac{1}{4BA}A_1 A_1 \qquad - \tfrac{1}{4B^2}A_1 B_1 \qquad + \tfrac{1}{4BC}A_1 C_1 \qquad + \tfrac{1}{4BD}A_1 D_1$

$- \tfrac{1}{4CA}A_2 A_2 \qquad + \tfrac{1}{4CB}A_2 B_2 \qquad - \tfrac{1}{4C^2}A_2 C_2 \qquad + \tfrac{1}{4CD}A_2 D_2$

$- \tfrac{1}{4DA}A_3 A_3 \qquad + \tfrac{1}{4DB}A_3 B_3 \qquad + \tfrac{1}{4DC}A_3 C_3 \qquad - \tfrac{1}{4D^2}A_3 D_3$

$\boxed{R_{00} = }$

$R_{11} = \tfrac{1}{2A}B_{00} \qquad + \quad 0 \qquad - \tfrac{1}{2C}B_{22} \qquad - \tfrac{1}{2D}B_{33}$

$- \tfrac{1}{2A}A_{11} \qquad + \quad 0 \qquad - \tfrac{1}{2C}C_{11} \qquad - \tfrac{1}{2D}D_{11}$

$+ \tfrac{1}{4A^2}A_1^2 \qquad + \quad 0 \qquad + \tfrac{1}{4C^2}C_1^2 \qquad + \tfrac{1}{4D^2}D_1^2$

$- \tfrac{1}{4A^2}B_0 A_0 \qquad - \tfrac{1}{4AB}B_0 B_0 \qquad + \tfrac{1}{4AC}B_0 C_0 \qquad + \tfrac{1}{4AD}B_0 D_0$

$+ \tfrac{1}{4BA}B_1 A_1 \qquad + \quad 0 \qquad + \tfrac{1}{4BC}B_1 C_1 \qquad + \tfrac{1}{4BD}B_1 D_1$

$- \tfrac{1}{4CA}B_2 A_2 \qquad + \tfrac{1}{4CB}B_2 B_2 \qquad + \tfrac{1}{4C^2}B_2 C_2 \qquad - \tfrac{1}{4CD}B_2 D_2$

$- \tfrac{1}{4DA}B_3 A_3 \qquad + \tfrac{1}{4DB}B_3 B_3 \qquad - \tfrac{1}{4DC}B_3 C_3 \qquad + \tfrac{1}{4D^2}B_3 D_3$

$\boxed{R_{11} = }$

$$R_{22} = \tfrac{1}{2A}C_{00} \quad - \tfrac{1}{2B}C_{11} \quad + 0 \quad - \tfrac{1}{2D}C_{33}$$

$$- \tfrac{1}{2A}A_{22} \quad - \tfrac{1}{2B}B_{22} \quad + 0 \quad - \tfrac{1}{2D}D_{22}$$

$$+ \tfrac{1}{4A^2}A_2^2 \quad + \tfrac{1}{4B^2}B_2^2 \quad + 0 \quad + \tfrac{1}{4D^2}D_2^2$$

$$- \tfrac{1}{4A^2}C_0 A_0 \quad + \tfrac{1}{4AB}C_0 B_0 \quad - \tfrac{1}{4AC}C_0 C_0 \quad + \tfrac{1}{4AD}C_0 D_0$$

$$- \tfrac{1}{4BA}C_1 A_1 \quad + \tfrac{1}{4B^2}C_1 B_1 \quad + \tfrac{1}{4BC}C_1 C_1 \quad - \tfrac{1}{4BD}C_1 D_1$$

$$+ \tfrac{1}{4CA}C_2 A_2 \quad + \tfrac{1}{4CB}C_2 B_2 \quad + 0 \quad + \tfrac{1}{4CD}C_2 D_2$$

$$- \tfrac{1}{4DA}C_3 A_3 \quad - \tfrac{1}{4DB}C_3 B_3 \quad + \tfrac{1}{4DC}C_3 C_3 \quad + \tfrac{1}{4D^2}C_3 D_3$$

$$\boxed{R_{22} = }$$

$$R_{33} = \tfrac{1}{2A}D_{00} \quad - \tfrac{1}{2B}D_{11} \quad - \tfrac{1}{2C}D_{22} \quad + 0$$

$$- \tfrac{1}{2A}A_{33} \quad - \tfrac{1}{2B}B_{33} \quad - \tfrac{1}{2C}C_{33} \quad + 0$$

$$+ \tfrac{1}{4A^2}A_3^2 \quad + \tfrac{1}{4B^2}B_3^2 \quad + \tfrac{1}{4C^2}C_3^2 \quad + 0$$

$$- \tfrac{1}{4A^2}D_0 A_0 \quad + \tfrac{1}{4AB}D_0 B_0 \quad + \tfrac{1}{4AC}D_0 C_0 \quad - \tfrac{1}{4AD}D_0 D_0$$

$$- \tfrac{1}{4BA}D_1 A_1 \quad + \tfrac{1}{4B^2}D_1 B_1 \quad - \tfrac{1}{4BC}D_1 C_1 \quad + \tfrac{1}{4BD}D_1 D_1$$

$$- \tfrac{1}{4CA}D_2 A_2 \quad - \tfrac{1}{4CB}D_2 B_2 \quad + \tfrac{1}{4C^2}D_2 C_2 \quad + \tfrac{1}{4CD}D_2 D_2$$

$$+ \tfrac{1}{4DA}D_3 A_3 \quad + \tfrac{1}{4DB}D_3 B_3 \quad + \tfrac{1}{4DC}D_3 C_3 \quad + 0$$

$$\boxed{R_{33} = }$$

32. GRAVITATIONAL WAVE ENERGY

HOMEWORK PROBLEMS

P32.1 Explain physically why the gravitational wave flux should (in GR units) have the same magnitude as the gravitational wave energy density. (*Hint:* Consider a surface area A perpendicular to the wave direction. All of the energy in a rectangular volume with face area A and length $v\,\Delta t$ parallel to the wave motion will hit the surface area in time Δt. Draw a picture of this, and then compute the flux in terms of the energy density.)

P32.2 If the gravitational wave is moving in the $+z$ direction, then the flux of energy transported in that direction should be T^{tz}_{GW} by definition (see figure 20.2). Calculate the value of this component in the same way that we calculated T^{GW}_{tt} and show that we get the same result as given by equation 32.14. (*Hint:* Use the Diagonal Metric Worksheet to compute the Ricci tensor component that you need.)

P32.3 Consider a gravitational wave $h^{TT}_{\mu\nu} = A_{\mu\nu}\cos(k_\alpha x^\alpha)$ such that $k_t = \omega$, $k_x = k_y = 0$, and $k_z = -\omega$ and

$$A_{\mu\nu} = \begin{bmatrix} 0 & 0 & 0 & 0 \\ 0 & A_+ & A_\times & 0 \\ 0 & A_\times & -A_+ & 0 \\ 0 & 0 & 0 & 0 \end{bmatrix} \quad (32.18)$$

a. Show that the effective stress-energy tensor for this gravitational wave is

$$T^{GW}_{\mu\nu} = \frac{\omega^2(A_+^2 + A_\times^2)}{32\pi G}\begin{bmatrix} 1 & 0 & 0 & -1 \\ 0 & 0 & 0 & 0 \\ 0 & 0 & 0 & 0 \\ -1 & 0 & 0 & 1 \end{bmatrix} \quad (32.19)$$

(*Hint:* Use the Diagonal Metric Worksheet to determine the off-diagonal terms of $R^{(2)}_{\mu\nu}$ when $A_\times = 0$ and then generalize the result as we did before.)

b. Use this result and the definition of the components of the stress-energy tensor to show that this gravitational wave transports energy *only* in the $+z$ direction and that the flux of energy in that direction is

$$\text{flux} = \frac{\omega^2(A_+^2 + A_\times^2)}{32\pi G} \quad (32.20)$$

(Be sure to carefully discuss any sign issues.)

P32.4 Consider a gravitational wave of the type described in problem P32.3. Imagine that we observe this wave in an inertial frame S' moving with speed β in the $+z$ direction relative to our initial frame S. (Note that S' is moving in the $+z$ direction, not the $+x$ direction.)

a. What are the values of A'_+ and $A'_\times$ in frame S' compared to those of A_+ and $A_\times$ in frame S? Does this make intuitive sense, based on your understanding of special relativity?

b. Transform the stress-energy tensor given in equation 32.19 to the frame S' and so the values of the wave's energy density and flux in the frame S' in terms of their values in S. Explain why these results make physical sense considering the Doppler shift we should observe in that frame. (*Hint:* You may want to raise the indices in the stress-energy tensor before transforming.)

P32.5 Calculate the flux (in W/m²) of an uprightly polarized gravitational wave with amplitude $A_+ = 10^{-20}$ and frequency 1000 Hz. (We might receive such waves from a supernova in our galaxy.)

P32.6 The binary star system ι Boötis is about 42 ly away from earth. The stars in this system orbit each other with a period of about 6.5 h. This system generates gravitational waves having an amplitude of roughly 10^{-21} at the earth. Estimate the flux (in W/m²) of a plus-polarized gravitational wave from this source, and compare with the flux of electromagnetic energy from the sun at the earth's surface (which is about 1000 W/m²). **Note:** We will see in chapter 34 that the frequency of gravitational waves emitted by a binary system is twice the frequency of the stars' orbit.

P32.7 The binary star system μ Scorpii is about 355 ly away from earth. The stars in this system orbit each other with a period of about 125,000 s. This system generates gravitational waves having an amplitude of about 3×10^{-20} at the earth. Estimate the flux (in W/m²) of a plus-polarized gravitational wave from this source, and compare with the flux of electromagnetic energy from the sun at the earth's surface (which is about 1000 W/m²). **Note:** We will see in chapter 34 that the frequency of gravitational waves emitted by a binary system is twice the frequency of the stars' orbit.

P32.8 Imagine that some cataclysm in our galaxy suddenly illuminates the earth with 800-Hz gravitational waves with the same energy flux (about 1000 W/m²) as light from the sun. (This might be possible if a supernova explodes in a near portion of the galaxy.) Estimate the fractional change in the distance between two suspended mirrors in the LIGO detector. Will LIGO be able to detect this event?

33. GENERATING GRAVITATIONAL WAVES

INTRODUCTION

FLAT SPACETIME
| Review of Special Relativity |
| Four-Vectors |
| Index Notation |

TENSORS
| Arbitrary Coordinates |
| Tensor Equations |
| Maxwell's Equations |
| Geodesics |

SCHWARZSCHILD BLACK HOLES
| The Schwarzschild Metric |
| Particle Orbits |
| Precession of the Perihelion |
| Photon Orbits |
| Deflection of Light |
| Event Horizon |
| Alternative Coordinates |
| Black Hole Thermodynamics |

THE CALCULUS OF CURVATURE
| The Absolute Gradient |
| Geodesic Deviation |
| The Riemann Tensor |

THE EINSTEIN EQUATION
| The Stress-Energy Tensor |
| The Einstein Equation |
| Interpreting the Equation |
| The Schwarzschild Solution |

COSMOLOGY
| The Universe Observed |
| A Metric for the Cosmos |
| Evolution of the Universe |
| Cosmic Implications |
| The Early Universe |
| CMB Fluctuations & Inflation |

GRAVITATIONAL WAVES
| Gauge Freedom |
| Detecting Gravitational Waves |
| Gravitational Wave Energy |
| **Generating Gravitational Waves** |
| Gravitational Wave Astronomy |

SPINNING BLACK HOLES
| Gravitomagnetism |
| The Kerr Metric |
| Kerr Particle Orbits |
| Ergoregion and Horizon |
| Negative-Energy Orbits |

this depends on this

33. GENERATING GRAVITATIONAL WAVES

Introduction. In this chapter, we will explore how we can calculate the gravitational waves emitted by a dynamic source.

Crude Estimates. A crude estimate of the *maximum* strength of a gravitational wave would be where some components of the metric perturbation have an absolute value of order of magnitude 1. This is the order of magnitude of the "perturbation" away from flat space that one would experience near the event horizon of a black hole, where $g_{tt} = -(1 - 2GM/r) \approx -(1-1)$.

A wave of such an amplitude might be produced near the event horizon of two merging black holes (which is about the most violent process for producing a dynamically changing spacetime one can imagine). If the black holes each have the order of magnitude of a few solar masses, then $2GM$ for the final product will be on the order of magnitude of 10 km = 10^4 m. If $h = [A_+^2 + A_\times^2]^{1/2} \approx 1$ at a radius of $2GM$, then (since amplitudes for gravitational waves, like EM waves, fall off as $1/r$ as r increases) $h \approx 10^{-20}$ at a radius of 10^{24} m $\approx 10^8$ ly, which is a radius that encloses many millions of galaxies. Considering that such events will be extremely rare, and that most binary coalescence events will be weaker producers of gravitational radiation than this, a gravitational wave detector will need to have a sensitivity of at least $h \approx 10^{-21}$ to have a good chance of seeing such events. LIGO is currently operating at a sensitivity of about $h \approx 10^{-22}$, but has seen nothing to date.

The Small-Source Approximation. It is difficult to make more accurate guesses about the gravitational waves produced by coalescing stars short of creating detailed computer models of such events. It is also difficult to predict with any precision how often such events will occur.

Simple binary star systems, however, are *known* and *steady* sources of gravitational waves. To estimate the flux of gravitational radiation from such sources, we need to be able to calculate more accurately the gravitational radiation produced by a source involving moving massive parts.

In what follows, I will describe how to do this using the *small-weak-slow* source approximation, where we assume that

1. The source is *small* compared to both a wavelength of the wave and the distance to the observer.

2. The source is *weak* in that $|h_{\mu\nu}| \ll 1$ even very near the source. (This will be true for most sources of gravitational waves except coalescing black holes.)

3. The source is *slow* in that the parts of the source move at speeds $\ll 1$. (This will also be true for most wave sources except for coalescing neutron stars.)

Let's see how this approximation can help. The weak-field Einstein equation in a coordinate system satisfying the Lorenz gauge condition tells us the following:

$$\partial^\alpha \partial_\alpha H^{\mu\nu} = \left[-\frac{\partial^2}{\partial t^2} + \nabla^2\right] H^{\mu\nu} = -16\pi G T^{\mu\nu} \tag{33.1}$$

We can compare this equation to the analogous time-dependent equation for the electric potential $[-(\partial/\partial t)^2 + \nabla^2]\phi = -4\pi k \rho_c$, where ρ_c is the charge density. You may be aware that the way to calculate the potential $\phi(t, \vec{R})$ in *static* situations is to divide the source into a set of cells each having a volume dV small enough to be considered a point charge with charge $q_i \equiv \rho_c(\vec{r}_i) dV$, and then sum the point charge potentials kq_i/u_i over all cells in the source (where u_i is the distance $u_i \equiv |\vec{R} - \vec{r}_i|$ between the field position $\vec{R}$ and the cell's position $\vec{r}_i$). It turns out that the only additional thing we need to do in *dynamic* situations is to account for the light travel time required for information to get from the source to the field point $\vec{R}$ by evaluating each cell's potential at time $t - u_i$. By analogy, the solution to equation 33.1 is

$$H^{\mu\nu}(t, \vec{R}) = 4G \int_{\text{src}} \frac{T^{\mu\nu}(t-u, \vec{r}) dV}{u} \quad \text{where } u \equiv |\vec{R} - \vec{r}| \tag{33.2}$$

If the source is *small* compared to the distance from the observer, then $R \approx u$. If in addition, the source is small compared to a wavelength of the wave, then $t - u \approx t - R$ for all points in the source. In such a case, equation 33.2 becomes

$$H^{\mu\nu}(t,\vec{R}) = \frac{4G}{R}\left[\int_{src} T^{\mu\nu} dV\right]_{\text{at } t-R} \quad (33.3)$$

From now on, let's *assume* that all integrals over the source are calculated at the "retarded" time $t - R$ (so that we don't continue to write this over and over).

You can show (see box 33.1) that if the source's center of mass is at rest in our coordinate system, then $H^{tt} = 4GM/R$ = const. and $H^{ti} = H^{it} = 0$. Therefore, the only potentially "waving" components of the trace-reversed metric perturbation are the spatial components H^{ik}. We will turn our attention to these components now.

A Useful Identity. It turns out that the divergence theorem in conjunction with conservation of energy and momentum in the source (neglecting the energy and momentum of the gravitational waves, which will be very small in comparison) imply the following very useful identity (see box 33.2):

$$\int_{src} T^{jk} dV = \frac{1}{2}\frac{d^2}{dt^2}\int_{src} T^{tt} x^j x^k dV = \frac{1}{2}\frac{d^2}{dt^2}\int_{src} \rho x^j x^k dV \quad (33.4)$$

If we plug this result into equation 33.3, we get

$$H^{jk}(\text{at time } t) = \frac{2G}{R}\ddot{I}^{jk}(\text{at time } t - R), \text{ where } I^{jk} \equiv \int_{src} \rho x^j x^k dV \quad (33.5)$$

where j and k are spatial indices, I^{jk} is the source's **quadrupole moment 3-tensor**, and the double-dot over the I is a compact way of indicating a double time derivative: $\ddot{I}^{jk} \equiv d^2 I^{jk}/dt^2$. Note that this tensor is symmetric: $I^{jk} = I^{kj}$.

The Reduced Quadrupole Moment Tensor. It turns out to be useful (for a variety of reasons) to express our equations for gravitational radiation in terms of the source's **reduced quadrupole moment** 3-tensor $\mathcal{I}^{jk}$ defined as follows:

$$\mathcal{I}^{jk} \equiv \int_{src} \rho(x^j x^k - \tfrac{1}{3}\eta^{jk} r^2) dV \quad (33.6)$$

This is partly because $\mathcal{I}^{jk}$ has zero trace $\eta_{jk}\mathcal{I}^{jk}$ (because $\eta_{jk}x^j x^k - \tfrac{1}{3}\eta_{jk}\eta^{jk} r^2 = r^2 - \tfrac{1}{3}3r^2 = 0$), and partly because if we expand the Newtonian gravitational potential Φ at some large distance R from a compact and static but asymmetrical source whose center of mass is at the origin, we get

$$\Phi = -\frac{GM}{R} - \left(\frac{3G\mathcal{I}_{jk}}{2R^3}\right)\left(\frac{X^j}{R}\right)\left(\frac{X^k}{R}\right) + \text{ higher-order terms} \quad (33.7)$$

where X^j is a component of the radius vector $\vec{R}$ from the source to the observer. Thus the reduced quadrupole moment tensor expresses the leading component of the asphericity of the source's gravitational field for large R. We will see that $\mathcal{I}^{jk}$ expresses the aspect of the source that creates the physically significant (i.e., coordinate-independent) aspects of the source's gravitational radiation.

Point-Like Objects. In many practical applications, gravitational waves are created by systems (e.g., binary star systems) that we can model as interacting point-like objects. In such a case, the source's mass-energy density ρ in equation 33.6 is nonzero only in a tiny region around each object's center. In such a case, the position coordinates for all points *within* the ith object are essentially equal to its center coordinates x_i^j, so we can pull these coordinates out in front of the integral, yielding

$$\mathcal{I}^{jk} \approx \sum_{\text{objects } i} (x_i^j x_i^k - \tfrac{1}{3}\eta^{jk} r_i^2)\int_{src\,i} \rho\, dV = \sum_{\text{objects } i} m_i(x_i^j x_i^k - \tfrac{1}{3}\eta^{jk} r_i^2) \quad (33.8)$$

The Transverse-Traceless Part. Now, the only physically significant part of a gravitational wave once it has reached the observer is the wave's *transverse-traceless* part. If we are far enough from the source that its gravitational waves are essentially plane waves, then we can write the trace-reversed metric perturbation as a sum of Fourier component plane waves of the form $H^{\mu\nu} = A^{\mu\nu}\cos(k_\sigma x^\sigma + \theta_0)$, where $A^{\mu\nu}$ is a constant matrix. We saw in chapter 31 that a gauge transformation of the form $\xi^\mu = B^\mu \sin(k_\sigma x^\sigma + \theta_0)$ will put this given Fourier component into transverse-traceless gauge if we choose the values of B^μ correctly. Specifically (see box 33.3), for gravitational waves moving in the $+z$ direction, you can show that

$$A^{xx}_{TT} = -A^{yy}_{TT} = \tfrac{1}{2}(A^{xx} - A^{yy}), \quad A^{xy}_{TT} = A^{yx}_{TT} = A^{yx}, \quad \text{all other } A^{\mu\nu} = 0 \quad (33.9)$$

Note that all reference to the frequency ω and the phase θ_0 has disappeared. Therefore if we multiply both sides of equations 33.9 by $\cos(k_\sigma x^\sigma + \theta_0)$ and reassemble the Fourier sum, we can say that for waves moving in the $+z$ direction

$$H^{xx}_{TT} = \tfrac{1}{2}(H^{xx} - H^{yy}) = \frac{2G}{R}\tfrac{1}{2}(\ddot{I}^{xx} - \ddot{I}^{yy}) = \frac{2G}{R}\tfrac{1}{2}(\ddot{\mathcal{I}}^{xx} - \ddot{\mathcal{I}}^{yy}) \equiv \frac{2G}{R}\ddot{\mathcal{I}}^{xx}_{TT} \quad (33.10a)$$

$$H^{yy}_{TT} = \tfrac{1}{2}(H^{yy} - H^{xx}) = \frac{2G}{R}\tfrac{1}{2}(\ddot{I}^{yy} - \ddot{I}^{xx}) = \frac{2G}{R}\tfrac{1}{2}(\ddot{\mathcal{I}}^{yy} - \ddot{\mathcal{I}}^{xx}) \equiv \frac{2G}{R}\ddot{\mathcal{I}}^{yy}_{TT} \quad (33.10b)$$

$$H^{xy}_{TT} = H^{xy} = \frac{2G}{R}\ddot{I}^{xy} = \frac{2G}{R}\ddot{\mathcal{I}}^{xy} \equiv \frac{2G}{R}\ddot{\mathcal{I}}^{xy}_{TT} \quad (33.10c)$$

where the quadrupole moment tensors are evaluated at time $t - R$. (Note that $\ddot{I}^{xx} - \ddot{I}^{yy} = \ddot{\mathcal{I}}^{xx} - \ddot{\mathcal{I}}^{yy}$ because the extra term $-\tfrac{1}{3}\eta^{ij}r^2$ that appears in the *reduced* quadrupole moment tensor is the same in both $\ddot{\mathcal{I}}^{xx}$ and $\ddot{\mathcal{I}}^{yy}$ and so cancels out in the difference. This extra term is also zero in all off-diagonal elements, so $\ddot{\mathcal{I}}^{jk} = \ddot{I}^{jk}$ if $j \neq k$.) Equations 33.10 therefore define what we mean by the "transverse-traceless" parts of the reduced quadrupole moment tensor for waves moving in the $+z$ direction. In many cases, we will be able to orient our coordinate system so waves for a given observer are moving in this direction, so we will find these results quite useful.

If the waves are moving in a direction parallel to either the $+x$ or $+y$ axes, one can adapt the method shown in equations 33.10. What one needs to do can be summarized verbally by a simple two-step process: (1) set to zero all components of $H^{\mu\nu}$ that have a t index or an index corresponding to the wave's direction (to make it *transverse*), and (2) subtract half the trace of the remaining components from each of the remaining two diagonal elements (to make it *traceless*).

However, we sometimes need to calculate the transverse-traceless components of a wave moving in an arbitrary direction indicated by a unit vector $\vec{n}$. We can do this essentially by projecting the wave onto the plane perpendicular to $\vec{n}$ (making it *transverse*) and then subtracting symmetric fractions of the trace from each remaining nonzero diagonal element (to make the result *traceless*). The result (as you will prove in box 33.4) is

$$h^{jk}_{TT} = H^{jk}_{TT} = \frac{2G}{R}\ddot{I}^{jk}_{TT} = \frac{2G}{R}\ddot{\mathcal{I}}^{jk}_{TT} \quad (33.11)$$

where $\quad \ddot{\mathcal{I}}^{jk}_{TT} \equiv \left(P^j_m P^k_n - \tfrac{1}{2}P^{jk}P_{mn}\right)\ddot{\mathcal{I}}^{mn} \quad$ with $\quad P^j_m \equiv \delta^j_m - n^j n_m \quad (33.12)$

where $\ddot{\mathcal{I}}^{jk}_{TT}$ is the quadrupole moment tensor's "transverse-traceless part" (evaluated at time $t - R$, of course). You will also see in box 33.4 that this general formula reduces to the results given above when $\vec{n} = [0, 0, 1]$.

The Flux of Gravitational Wave Energy. In chapter 32 we found that the effective energy density of a gravitational plane wave is given by

$$T^{tt}_{GW} = \frac{1}{32\pi G}\langle \dot{h}^{jk}_{TT}\dot{h}^{TT}_{jk}\rangle \quad (33.13)$$

where the average is over several wavelengths of the wave (see equation 32.16). We also saw (see equation 32.17) that this is the flux (energy per unit area per unit time) carried by that wave in the direction it is moving. If we substitute equation 33.11 into this expression, we find that this flux is

$$\text{Flux of GW energy } = \frac{G}{8\pi R^2}\langle \dddot{\mathcal{I}}^{jk}_{TT}\dddot{\mathcal{I}}^{TT}_{jk}\rangle \quad (33.14)$$

It is more practical in applications to calculate $\dddot{\mathcal{I}}^{jk}$ (which depends on the source alone and not also on the direction the waves are moving when observed) than $\dddot{\mathcal{I}}^{jk}_{TT}$. You can use equation 33.12 in conjunction with the above to show that

$$\text{Flux} = \frac{G}{16\pi R^2}\left\langle 2\dddot{\mathcal{I}}_{jk}\dddot{\mathcal{I}}^{jk} - 4n^j n^k \dddot{\mathcal{I}}_{jm}\dddot{\mathcal{I}}^m{}_k + n^i n^j n^k n^m \dddot{\mathcal{I}}_{ij}\dddot{\mathcal{I}}_{km}\right\rangle \tag{33.15}$$

correctly expresses flux in terms of $\dddot{\mathcal{I}}^{jk}$ (see box 33.5).

Integrating the Flux. We can find the total energy radiated by the source by computing the energy radiated per unit time through a differential area element $dA = R^2 \sin\theta\, d\theta\, d\phi$ on the surface of a sphere surrounding the source in the direction specified by a unit 3-vector $\vec{n} = [\sin\theta\cos\phi, \sin\theta\sin\phi, \cos\theta]$, and then integrating that result over the entire sphere. The energy radiated per unit time through this area is simply Flux·dA, so the total energy radiated per unit time (which results in a decrease in the source's total energy E by the same amount) is given by

$$-\frac{dE}{dt} = \int_{\text{sphere}} \text{Flux} \cdot dA = \frac{GR^2}{16\pi R^2}\int_0^\pi d\theta \int_0^{2\pi} \left\langle \dddot{\mathcal{I}}^{jk}_{TT}\dddot{\mathcal{I}}^{TT}_{jk}\right\rangle \sin\theta\, d\phi$$

$$= \frac{G}{16\pi}\int_0^\pi d\theta \int_0^{2\pi} \left\langle 2\dddot{\mathcal{I}}_{jk}\dddot{\mathcal{I}}^{jk} - 4n^j n^k \dddot{\mathcal{I}}_{ji}\dddot{\mathcal{I}}^i{}_k + n^i n^j n^k n^m \dddot{\mathcal{I}}_{ij}\dddot{\mathcal{I}}_{km}\right\rangle \sin\theta\, d\phi \tag{33.16}$$

Note that the integrand in the last expression depends on θ and ϕ only through the components of the unit vector $\vec{n}$: the components $\dddot{\mathcal{I}}^{jk}$ depend only on the orientation of the source in our coordinate system and not on the direction $\vec{n}$ of the observer relative to the source. Therefore, the integral becomes

$$-\frac{dE}{dt} = \frac{2G}{16\pi}\left\langle \dddot{\mathcal{I}}_{jk}\dddot{\mathcal{I}}^{jk}\right\rangle \int_0^\pi d\theta \int_0^{2\pi}\sin\theta\, d\phi - \frac{4G}{16\pi}\left\langle \dddot{\mathcal{I}}_{ji}\dddot{\mathcal{I}}^i{}_k\right\rangle \int_0^\pi d\theta \int_0^{2\pi} n^j n^k \sin\theta\, d\phi$$

$$+ \frac{G}{16\pi}\left\langle \dddot{\mathcal{I}}_{ij}\dddot{\mathcal{I}}_{km}\right\rangle \int_0^\pi d\theta \int_0^{2\pi} n^i n^j n^k n^m \sin\theta\, d\phi \tag{33.17}$$

The first integral is well known:

$$\int_0^\pi d\theta \int_0^{2\pi} \sin\theta\, d\phi = 4\pi \tag{33.18}$$

You can show (see box 33.6) that the other integrals are

$$\int_0^\pi d\theta \int_0^{2\pi} n^j n^k \sin\theta\, d\phi = +\frac{4\pi}{3}\eta^{jk} \tag{33.19}$$

$$\int_0^\pi d\theta \int_0^{2\pi} n^i n^j n^k n^m \sin\theta\, d\phi = +\frac{4\pi}{15}(\eta^{ij}\eta^{km} + \eta^{jk}\eta^{im} + \eta^{ik}\eta^{jm}) \tag{33.20}$$

If we substitute these results into equation 33.17, we get

$$-\frac{dE}{dt} = \frac{G}{4}\left\langle 2\dddot{\mathcal{I}}_{jk}\dddot{\mathcal{I}}^{jk} - \tfrac{4}{3}\eta^{jk}\dddot{\mathcal{I}}_{ji}\dddot{\mathcal{I}}^i{}_k + \tfrac{1}{15}(\eta^{ij}\eta^{km} + \eta^{jk}\eta^{im} + \eta^{ik}\eta^{jm})\dddot{\mathcal{I}}_{ij}\dddot{\mathcal{I}}_{km}\right\rangle$$

$$= \frac{G}{4\cdot 15}\left\langle 30\dddot{\mathcal{I}}_{jk}\dddot{\mathcal{I}}^{jk} - 20\dddot{\mathcal{I}}_{ji}\dddot{\mathcal{I}}^{ij} + \dddot{\mathcal{I}}^i{}_i\dddot{\mathcal{I}}^k{}_k + \dddot{\mathcal{I}}_{ij}\dddot{\mathcal{I}}^{ji} + \dddot{\mathcal{I}}_{ij}\dddot{\mathcal{I}}^{ij}\right\rangle$$

$$= \frac{G}{4\cdot 15}\left\langle 30\dddot{\mathcal{I}}_{jk}\dddot{\mathcal{I}}^{jk} - 20\dddot{\mathcal{I}}_{jk}\dddot{\mathcal{I}}^{jk} + 0 + \dddot{\mathcal{I}}_{jk}\dddot{\mathcal{I}}^{jk} + \dddot{\mathcal{I}}_{jk}\dddot{\mathcal{I}}^{jk}\right\rangle$$

$$= \frac{12G}{60}\left\langle \dddot{\mathcal{I}}_{jk}\dddot{\mathcal{I}}^{jk}\right\rangle = \frac{G}{5}\left\langle \dddot{\mathcal{I}}_{jk}\dddot{\mathcal{I}}^{jk}\right\rangle \tag{33.21}$$

where, in going from the second line to the third, I used the symmetry of $\mathcal{I}^{jk}$ and the fact that it is traceless (see just below equation 33.6) and renamed some indices. So to summarize, a source's gravitational wave "luminosity" (the rate at which it emits gravitational wave energy at the expense of its internal energy E) is

$$L_{GW} = -\frac{dE}{dt} = \frac{G}{5}\left\langle \dddot{\mathcal{I}}_{jk}\dddot{\mathcal{I}}^{jk}\right\rangle \tag{33.22}$$

This is a very important and useful result.

BOX 33.1 $H^{t\mu}$ for a Compact Source Whose CM is at Rest

According to the definition of the stress-energy tensor (see equation 20.7 and figure 20.2), the $T^{t\mu}$ components of the stress-energy tensor in a LIF are

$$T^{tt} = \text{density of energy} = \rho \tag{33.23a}$$

$$T^{ti} = T^{it} = \text{density of } i\text{-momentum} \tag{33.23b}$$

This should also be true (to the level of our approximations) in the "nearly cartesian" coordinates we use in the weak-field limit. Therefore, we have

$$\int_{\text{src}} T^{tt} dV = \int_{\text{src}} \rho \, dV \equiv M, \text{ and } \int_{\text{src}} T^{ti} dV = \int_{\text{src}} T^{it} dV = P^i \tag{33.24}$$

where M is the total mass-energy inside the source and P^i is its net i-momentum. But if we anchor our coordinates to the source's center of mass, then $P^i = 0$, because the source (considered as a unit) is not moving in our coordinate system. If the source is also "small" and "slow," then equation 33.3 implies that

$$H^{tt} = \frac{4G}{R} \int_{\text{src}} T^{tt} dV = \frac{4GM}{R}, \quad H^{ti} = \frac{4G}{R} \int_{\text{src}} T^{ti} dV = \frac{4G}{R} P^i = 0 \tag{33.25}$$

and $H^{it} = 0$ similarly. Therefore, even if the source is highly dynamic, these components of the metric perturbation contribute nothing to any gravitational waves that the source might emit. Q. E. D.

BOX 33.2 A Useful Identity

In a LIF, conservation of energy implies that $\partial_\mu T^{\mu\nu} = 0$. If we break the sum into time and space parts, we get

$$0 = \partial_\mu T^{\mu\nu} = \partial_t T^{t\nu} + \partial_i T^{i\nu} \quad \Rightarrow \quad \partial_t T^{t\nu} = -\partial_i T^{i\nu} \tag{33.26}$$

Again, to our level of approximation, this will still be true in the "nearly cartesian" coordinates we are using in the weak-field approximation. Now note that in the expression $T^{tt} x^j x^k dV$ that appears in equation 33.4, x^j and x^k represent the coordinates of a fixed "cell" of volume dV inside the source. The energy density T^{tt} inside the cell may vary with time, but its position will not. Therefore,

$$\partial_t \partial_t (T^{tt} x^j x^k) = (\partial_t \partial_t T^{tt}) x^j x^k \quad \text{since } x^j \text{ and } x^k \text{ are fixed "cell" coordinates}$$

$$= -(\partial_t \partial_i T^{it}) x^j x^k \quad \text{because of equation 33.26}$$

$$= -(\partial_i \partial_t T^{ti}) x^j x^k \quad \text{because } \partial_t \partial_i = \partial_i \partial_t \text{ and } T^{it} = T^{ti}$$

$$= +(\partial_i \partial_m T^{mi}) x^j x^k \quad \text{because of equation 33.26}$$

$$= (\partial_i \partial_m T^{im}) x^j x^k \quad \text{because } T^{im} = T^{mi}$$

$$= \partial_i \partial_m (T^{im} x^j x^k) - 2\partial_m (T^{mj} x^k + T^{mk} x^j) + 2T^{jk} \tag{33.27}$$

Exercise 33.2.1. Verify that the last step follows from the product rule. Note the extra room provided on the next page.

BOX 33.2 (continued) A Useful Identity

Now, the divergence theorem says that for an arbitrary 3-vector field $\vec{F}(x,y,z)$,

$$\int_V \vec{\nabla} \cdot \vec{F}\, dV = \oint_S \vec{F} \cdot d\vec{A} \quad \text{or} \quad \int_V \partial_i F^i\, dV = \oint_S F^i\, dA_i \quad \text{in index notation} \qquad (33.28)$$

The same is true for the integral of an arbitrary 3-tensor field $F^{ij}(x,y,z)$:

$$\int_V \partial_i F^{ij}\, dV = \oint_S F^{ij}\, dA_i \qquad (33.29)$$

(To see this, you can think of the jth column of the matrix F^{ij} as being an independent 3-vector field.) Now, imagine integrating both sides of equation 33.27 over a volume large enough to completely enclose the source, so that $T^{jk} = 0$ on the surface of the volume. You can use the divergence theorem to show that the first two terms on the right of equation 33.27 integrate to zero, leaving

$$\int_V \partial_t \partial_t (T^{tt} x^j x^k)\, dV = 2 \int_V T^{jk}\, dV \qquad (33.30)$$

Exercise 33.2.2. Verify that equation 33.30 is correct.

One can then pull the double partial time derivative outside of the left integral, where it becomes an ordinary double time derivative (since the integral does not depend on position) and divide both sides by 2 to get equation 33.4.

BOX 33.3 The Transverse-Traceless Components of $A^{\mu\nu}$

We saw in box 31.2 that a gauge transformation of the form $\xi^\mu = B^\mu \sin(k_\sigma x^\sigma)$ acting on the plane wave $H^{\mu\nu} = A^{\mu\nu} \cos(k_\sigma x^\sigma)$ transforms the components of the constant matrix $A^{\mu\nu}$ as follows (see equation 31.17):

$$A_{TT}^{\mu\nu} = A^{\mu\nu} - k^\mu B^\nu - k^\nu B^\mu + \eta^{\mu\nu} k_\alpha B^\alpha \tag{33.31}$$

In the same box, you showed that for waves moving in the $+z$ direction at the speed of light (for which $k_\mu = [-\omega, 0, 0, \omega]$ and $k^\mu = [\omega, 0, 0, \omega]$) the values for B^μ that you need to get to the transverse-traceless gauge are

$$B^t = \frac{[A^{tt} + \tfrac{1}{2} A^\mu{}_\mu]}{2\omega}, \quad B^x = \frac{A^{xt}}{\omega}, \quad B^y = \frac{A^{yt}}{\omega}, \quad B^z = \frac{[2A^{zt} - A^{tt} - \tfrac{1}{2} A^\mu{}_\mu]}{2\omega} \tag{33.32}$$

Now, the Lorenz gauge conditions imply that $k_\alpha A^{\alpha\nu} = 0$. This implies that

$$0 = k_\alpha A^{\alpha t} = k_t A^{tt} + k_z A^{zt} = -\omega A^{tt} + \omega A^{zt} \Rightarrow A^{tt} = A^{zt} \tag{33.33a}$$

Similarly,

$$A^{tx} = A^{zx}, \quad A^{ty} = A^{zy}, \text{ and } A^{tz} = A^{zz} \tag{33.33b}$$

Remember also that $A^{\mu\nu} = A^{\nu\mu}$, because the metric (and thus the metric perturbation) has to be symmetric. If you substitute these results back into equation 33.32, you can express B^t and B^z entirely in terms of spatial components of $A^{\mu\nu}$:

$$B^t = \frac{[A^{zz} + \tfrac{1}{2} A^{xx} + \tfrac{1}{2} A^{yy}]}{2\omega} \quad \text{and} \quad B^z = \frac{[A^{zz} - \tfrac{1}{2} A^{xx} - \tfrac{1}{2} A^{yy}]}{2\omega} \tag{33.34}$$

Exercise 33.3.1. Verify that equation 33.34 is correct.

If you substitute the results from equation 33.34 back into equation 33.31, you can show that $A_{TT}^{xx} = -A_{TT}^{yy} = \tfrac{1}{2}(A^{xx} - A^{yy})$ and $A_{TT}^{xy} = A_{TT}^{yx} = A^{xy}$.

Exercise 33.3.2. Check these results. (Problem P33.2 shows that the other components of $A_{TT}^{\mu\nu} = 0$.)

BOX 33.4 How to Find $\bar{I}_{TT}^{jk}$ for Waves Moving in the $\vec{n}$ Direction

Our goal in this box is to calculate the transverse-traceless components of the matrix $A^{\mu\nu}$ for a gravitational wave $H^{\mu\nu} = A^{\mu\nu}\cos(k_\sigma x^\sigma + \theta_0)$ traveling in the direction indicated by the unit vector $\vec{n}$. Note that for all possible directions of wave travel, the transverse-traceless gauge conditions require that $A^{t\mu} = A^{\mu t} = 0$ (see equation 31.8), so we need worry only about the spatial components A^{jk}.

Secondly, in arriving at equation 33.9 for waves moving in the $+z$ direction, we set to zero all components with a z index in the original arbitrary A^{jk} matrix. This amounts to projecting the original A^{jk} matrix onto the xy plane. Therefore, it follows that for waves in traveling in the $\vec{n}$ direction, our first task should be to project the general A^{jk} matrix onto the plane perpendicular to the $\vec{n}$ direction.

It turns out that the linear operator $P_m^j \equiv \delta_m^j - n^j n_m$ converts an arbitrary vector $\vec{v}$ to a vector $v_T^j \equiv P_m^j v^m$ that is the projection of $\vec{v}$ on the plane perpendicular (transverse) to $\vec{n}$. To see this, note that $\vec{n} \cdot \vec{v}_T = 0$ no matter what $\vec{v}$ is:

$$\vec{n} \cdot \vec{v}_T = \eta_{ij} n^i (P_m^j v^m) = n_j(\delta_m^j - n^j n_m)v^m = (n_m - n_j n^j n_m)v^m = 0 \quad (33.35)$$

since $n_j n^j \equiv \vec{n} \cdot \vec{n} = 1$, because $\vec{n}$ is a unit vector by hypothesis. Now, if $\vec{v}_T$ is truly a *projection* onto the plane, then projecting it *again* should not change the vector, that is, we should have $v_T^k = P_j^k v_T^j = P_j^k (P_m^j v^m)$. You can easily show that the definition of the projection operator given implies that no matter what it acts upon,

$$P_j^k P_m^j = P_m^k \quad \text{(also } P_{kj} P_m^j = P_{km}, \ P_j^k P^{jm} = P^{km}\text{)} \quad (33.36)$$

Exercise 33.4.1. Verify that $P_j^k P_m^j = P_m^k$. (The others follow simply by raising or lowering one free index. Note that in all cases, one index of one projection operator must be summed with one index of the other. However, since P_{jk} is symmetric, i.e., $P_{jk} = \eta_{jk} - n_j n_k = P_{kj}$, it doesn't matter *which* index we sum.)

One last check may convince you that this is correct. Note that if $\vec{n} = [0, 0, 1]$, then $n^j n_m = 1$ if $j = m = z$, and zero otherwise. In this particular case, $P_m^j \equiv \delta_m^j - n^j n_m$ is zero either if $j \neq m$ or if $j = m = z$ (the two terms cancel in the last case), so the only two nonzero elements are $P_x^x = P_y^y = 1$. You can see that such a matrix is precisely what is needed to project a vector on the xy plane:

$$\begin{bmatrix} 1 & 0 & 0 \\ 0 & 1 & 0 \\ 0 & 0 & 0 \end{bmatrix} \begin{bmatrix} v^x \\ v^y \\ v^z \end{bmatrix} = \begin{bmatrix} v^x \\ v^y \\ 0 \end{bmatrix} \quad (33.37)$$

So now how do we project a *matrix* like A^{mn} on the plane transverse to $\vec{n}$? Each upper index of A^{mn} transforms like a vector index, so a plausible guess would be $A_T^{jk} = P_m^j A^{mn} P_n^k$ (one projection operator for each index of A^{mn}). We can easily see that in the case where $\vec{n} = [0, 0, 1]$, this equation (in matrix notation) implies that

$$\begin{bmatrix} 1 & 0 & 0 \\ 0 & 1 & 0 \\ 0 & 0 & 0 \end{bmatrix} \left(\begin{bmatrix} A^{xx} & A^{xy} & A^{xz} \\ A^{yx} & A^{yy} & A^{yz} \\ A^{zx} & A^{zy} & A^{zz} \end{bmatrix} \begin{bmatrix} 1 & 0 & 0 \\ 0 & 1 & 0 \\ 0 & 0 & 0 \end{bmatrix} \right) = \begin{bmatrix} 1 & 0 & 0 \\ 0 & 1 & 0 \\ 0 & 0 & 0 \end{bmatrix} \begin{bmatrix} A^{xx} & A^{xy} & 0 \\ A^{yx} & A^{yy} & 0 \\ A^{zx} & A^{zy} & 0 \end{bmatrix} = \begin{bmatrix} A^{xx} & A^{xy} & 0 \\ A^{yx} & A^{yy} & 0 \\ 0 & 0 & 0 \end{bmatrix} \quad (33.38)$$

which is exactly what we want. Note that *both* factors of the projection operator are required to set both A^{jz} and A^{zk} to zero.

BOX 33.4 (continued) How to Find $\ddot{I}^{jk}_{TT}$ for Waves Moving in the $\vec{n}$ Direction

So we now know how to make a matrix *transverse*. To make it also *traceless* in the manner shown in equation 33.10, we need to subtract the transverse matrix's trace from the transverse matrix's diagonal elements in such a way as to leave the result still transverse. The trace of the transverse matrix is

$$\eta_{ab} A_T^{ab} = \eta_{ab} P_m^a P_n^b A^{mn} = P_{am} P_n^a A^{mn} = P_{ma} P_n^a A^{mn} = P_{mn} A^{mn} \quad (33.39)$$

where I have used equation 33.36 to simplify the result. Again, for the special case where $\vec{n} = [0, 0, 1]$, this yields

$$P_{mn} A^{mn} = \eta_{mn} A^{mn} - n_m n_n A^{mn} = A^{xx} + A^{yy} + A^{zz} - A^{zz} = A^{xx} + A^{yy} \quad (33.40)$$

which is what we would want in that case. To subtract this from the diagonal of A_T^{jk} shown in equation 33.38 in a way that treats each nonzero diagonal element equally, and yet does not do anything to already zero elements, we can multiply the scalar trace by $\frac{1}{2} P^{jk}$ and subtract it from A_T^{jk}. In our special case, this yields

$$\begin{bmatrix} A^{xx} & A^{xy} & 0 \\ A^{yx} & A^{yy} & 0 \\ 0 & 0 & 0 \end{bmatrix} - \frac{1}{2} \begin{bmatrix} 1 & 0 & 0 \\ 0 & 1 & 0 \\ 0 & 0 & 0 \end{bmatrix} (A^{xx} + A^{yy}) = \begin{bmatrix} \frac{1}{2}(A^{xx} - A^{yy}) & A^{xy} & 0 \\ A^{yx} & \frac{1}{2}(A^{yy} - A^{xx}) & 0 \\ 0 & 0 & 0 \end{bmatrix} \quad (33.41)$$

which is precisely what we want. So our abstract expression for A_{TT}^{jk} is thus

$$A_{TT}^{jk} = A_T^{jk} - \tfrac{1}{2} P^{jk} P_{mn} A^{mn} = (P_m^j P_n^k - \tfrac{1}{2} P^{jk} P_{mn}) A^{mn} \quad (33.42)$$

Since this is a tensor equation (in flat 3-space), if it yields correct results in the coordinate system where $\vec{n} = [0, 0, 1]$, it should work in arbitrary rotated coordinates systems where $\vec{n}$ points in an arbitrary direction. Moreover, since $\ddot{I}^{jk} \propto H^{jk} \propto A^{jk}$, we can use this same linear operator to find H_{TT}^{jk} and $\ddot{I}_{TT}^{jk}$ as well. This proves the result given in equation 33.12.

Exercise 33.4.2. Using abstract notation and the definition of A_{TT}^{jk} above, show that the trace of A_{TT}^{jk} is zero (i.e., $\eta_{jk} A_{TT}^{jk} = 0$) no matter what the components of $\vec{n}$ and the original matrix A^{mn} might be (assuming that $\vec{n}$ is a *unit* vector, though).

BOX 33.5 Flux in Terms of $\mathcal{I}^{jk}$

According to equation 33.15, the flux of gravitational wave energy emitted by a source in the particular direction indicated by a unit vector $\vec{n}$ is $(G/8\pi R^2)\langle \dddot{\mathcal{I}}_{TT}^{jk} \dddot{\mathcal{I}}_{jk}^{TT}\rangle$, where the transverse-traceless components of $\dddot{\mathcal{I}}_{TT}^{jk}$ depend on $\vec{n}$ according to

$$\dddot{\mathcal{I}}_{TT}^{jk} \equiv (P_m^j P_n^k - \tfrac{1}{2} P^{jk} P_{mn}) \dddot{\mathcal{I}}^{mn}, \qquad P_m^j \equiv \delta_m^j - n^j n_m \tag{33.43}$$

as we have seen. Using this equation, we see that

$$\dddot{\mathcal{I}}_{TT}^{jk} \dddot{\mathcal{I}}_{jk}^{TT} \equiv (P_m^j P_n^k - \tfrac{1}{2} P^{jk} P_{mn}) \dddot{\mathcal{I}}^{mn} (P_j^a P_k^b - \tfrac{1}{2} P_{jk} P^{ab}) \dddot{\mathcal{I}}_{ab} \tag{33.44}$$

By multiplying out the binomials, using $P_j^k P_m^j = P_m^k$, $P_{kj} P_m^j = P_{km}$, $P_j^k P^{jm} = P^{km}$, etc., to simplify things, then using the definition of P_m^j, the fact that $\vec{n}$ is a unit vector ($n_j n^j = 1$), and the fact that the *reduced* quadrupole moment tensor is traceless ($\eta_{jk} \mathcal{I}^{jk} = 0$) and symmetric ($\mathcal{I}^{jk} = \mathcal{I}^{kj}$), you can show that

$$\dddot{\mathcal{I}}_{TT}^{jk} \dddot{\mathcal{I}}_{jk}^{TT} = \dddot{\mathcal{I}}^{ab} \dddot{\mathcal{I}}_{ab} - 2 n^b n^m \dddot{\mathcal{I}}_m^a \dddot{\mathcal{I}}_{ab} + \tfrac{1}{2} n^a n^b n^m n^n \dddot{\mathcal{I}}_{ab} \dddot{\mathcal{I}}_{mn} \tag{33.45}$$

Equation 33.15 follows from this.

Exercise 33.5.1. Verify equation 33.45. Remember that $A^{\mu\nu} A_{\nu\alpha} = A^\mu{}_\nu A^\nu{}_\alpha$, etc., i.e., we can raise one summed index in a term as long as we lower the other. This exercise is not so bad if you use $P_j^k P_m^j = P_m^k$, etc. to simplify things *before* you start using the definition of P_j^k. You should find, for example, that equation 33.44 simplifies readily to $\dddot{\mathcal{I}}_{TT}^{jk} \dddot{\mathcal{I}}_{jk}^{TT} \equiv (P_m^a P_n^b - \tfrac{1}{2} P^{ab} P_{mn}) \dddot{\mathcal{I}}^{mn} \dddot{\mathcal{I}}_{ab}$. Also remember that since P_j^k and $\mathcal{I}^{jk}$ are symmetric, it is irrelevant which index is first and which is second.

BOX 33.6 Evaluating the Integrals in the Power Calculation

In this box, we want to prove equations 33.19 and 33.20, repeated here:

$$\int_0^\pi d\theta \int_0^{2\pi} n^j n^k \sin\theta \, d\phi = +\frac{4\pi}{3} \eta^{jk} \quad (33.19r)$$

$$\int_0^\pi d\theta \int_0^{2\pi} n^i n^j n^k n^m \sin\theta \, d\phi = +\frac{4\pi}{15}(\eta^{ij}\eta^{km} + \eta^{ik}\eta^{jm} + \eta^{ik}\eta^{jm}) \quad (33.20r)$$

where $\vec{n} = [\sin\theta\cos\phi, \sin\theta\sin\phi, \cos\theta]$ in spherical coordinates. Consider the first of these integrals. First let's examine the cases where $j \neq k$:

$$\int_0^\pi d\theta \int_0^{2\pi} n^x n^y \sin\theta \, d\phi = \int_0^\pi \sin^3\theta \, d\theta \int_0^{2\pi} \sin\phi \cos\phi \, d\phi$$

$$= \int_0^\pi \sin^3\theta \, d\theta \int_0^{2\pi} \tfrac{1}{2}\sin 2\phi \, d\phi = 0 \quad (33.46a)$$

since $\sin 2\phi$ is as often positive as negative in the integration range of ϕ. Similarly,

$$\int_0^\pi d\theta \int_0^{2\pi} n^x n^z \sin\theta \, d\phi = \int_0^\pi \sin^2\theta \cos\theta \, d\theta \int_0^{2\pi} \cos\phi \, d\phi = 0 \quad (33.46b)$$

$$\int_0^\pi d\theta \int_0^{2\pi} n^y n^z \sin\theta \, d\phi = \int_0^\pi \sin^2\theta \cos\theta \, d\theta \int_0^{2\pi} \sin\phi \, d\phi = 0 \quad (33.46c)$$

We see that the integrals where $j \neq k$ are all zero. The remaining integrals involve $n^x n^x$, $n^y n^y$, and $n^z n^z$. We would expect these to integrate to the *same* value, since there is no intrinsic difference between these coordinate directions. The integral for $n^z n^z$ is the easiest to evaluate:

$$\int_0^\pi d\theta \int_0^{2\pi} n^z n^z \sin\theta \, d\phi = \int_0^\pi \cos^2\theta \sin\theta \, d\theta \int_0^{2\pi} d\phi$$

$$= -2\pi\left[\tfrac{1}{3}\cos^3\theta\right]_0^\pi = -\tfrac{2}{3}\pi[-1-1] = +\tfrac{4}{3}\pi \quad (33.47)$$

Since the same applies to the $n^x n^x$ and $n^y n^y$ integrals, and since the off-diagonal integrals are zero, equation 33.19 nicely summarizes the results.

For the case described in equation 33.20, it is also plausible that the integrals will be zero unless the i, j, k, and m are equal in pairs (any remaining *odd* power of a component of $\vec{n}$ will yield an integrand that is as often positive as negative over the integration range). Since the orientation of our coordinate system is arbitrary, we would expect the integrals for $n^x n^x n^y n^y$, $n^x n^x n^z n^z$, and $n^y n^y n^z n^z$ (and permutations of these) to be equivalent. We would also expect that the integrals for $n^x n^x n^x n^x$, $n^y n^y n^y n^y$, and $n^z n^z n^z n^z$ are equivalent, though they may not be the same as the first three (and their permutations).

Exercise 33.6.1. As in equations 33.46 and 33.47, calculate the integrals for the $n^x n^x n^z n^z$ and $n^z n^z n^z n^z$ cases (you should get $4\pi/15$ and $4\pi/5$, respectively).

BOX 33.6 (continued) Evaluating the Integrals in the Power Calculation

Now, examine the expression given by equation 33.20 again. If the four indices are equal in distinct pairs, then either $i = j$ (and thus $k = m$) or $i = k$ (and thus $j = m$) or $i = m$ (and thus $j = k$). In each of these cases, *one* of the terms in the parentheses on the right side of equation 33.20 will be nonzero and the others zero, so the result will be $4\pi/15$. If all of the indices are the same, then *all three* terms in the parentheses will be nonzero, and the result will be $4\pi/5$. If the indices are not equal in pairs or are not all equal, then none of the terms in the parentheses will be nonzero, and the result will be zero. Therefore, equation 33.20 compactly expresses the result for all possible cases!

HOMEWORK PROBLEMS

P33.1 Use the crude estimation technique discussed in the first section to estimate the distance at which a detector sensitive to waves above $h \sim 10^{-21}$ could detect the coalescence of two supermassive black holes, each with $M \sim 10^6 \, M_\odot$.

P33.2 In exercise 33.3.2, you found expressions for the transverse-traceless matrix elements $A_{TT}^{xx}, A_{TT}^{yy}, A_{TT}^{xy}$, and A_{TT}^{yx} in terms of the arbitrary original matrix $A^{\mu\nu}$ for a gravitational wave moving in the $+z$ direction. Show that all other components of $A_{TT}^{\mu\nu}$ are zero, no matter what $A^{\mu\nu}$ is.

P33.3 Show that if $\vec{n} = [0, 0, 1]$, equation 33.12 reduces to the specific results for $\ddot{\cancel{I}}_{TT}^{jk}$ given in equations 33.10.

P33.4 Show abstractly that for an arbitrary symmetric A^{mn},

$$n_j A_{TT}^{jk} = n_j (P_m^j P_n^k - \tfrac{1}{2} P^{jk} P_{mn}) A^{mn} = 0 \quad \text{and}$$

$$n_k A_{TT}^{jk} = n_k (P_m^j P_n^k - \tfrac{1}{2} P^{jk} P_{mn}) A^{mn} = 0 \quad (33.48)$$

meaning that the transverse-traceless projection of A^{mn} is fully transverse. (*Hint:* You can use a symmetry argument for the second case.)

P33.5 Consider a dumbbell consisting of two very small objects with equal masses M connected by a rigid massless rod of length $2L$. Assume that the dumbbell's center of mass is at the origin, the dumbbell is free to rotate in the xy plane, and the rod makes an angle of $\theta(t)$ with the x axis.

a. Calculate $\cancel{I}^{jk}$ for this system as a function of $\theta(t)$.
b. Calculate $h_{\mu\nu}^{TT}$ at a position in the $+z$ direction that is a distance $R \gg L$ away from the origin if the dumbbell rotates with a constant angular frequency ω. [*Hint:* Note that $\cos^2\theta = \tfrac{1}{2}(1 + \cos 2\theta)$, $\sin^2\theta = \tfrac{1}{2}(1 - \cos 2\theta)$, and $2\sin\theta\cos\theta = \sin 2\theta$.]
c. At what rate will this rotating dumbbell radiate gravitational wave energy?

P33.6 Consider a crossed *pair* of dumbbells, each consisting of two very small objects with equal masses M connected by a rigid massless rod of length $2L$. Let both dumbbells lie in the xy plane, let the center of mass of each coincide with the origin, and let the angle between their rods be $90°$. Let one designated rod make an angle of $\theta(t)$ with the x axis: this will define the cross's orientation.

a. Calculate $\mathcal{I}^{jk}$ for this crossed pair of dumbbells.

b. Argue that if this crossed pair rotates around the z axis, it will radiate *no* gravitational waves in any direction.

P33.7 Consider a dumbbell consisting of two very small objects with equal masses M both the same distance $L(t)$ from the origin, which is the system's center of mass. Assume that the dumbbell lies along the x axis and that the masses oscillate along that line so that the distance $L(t)$ varies sinusoidally with time: $L(t) = L_0 + A \sin \omega t$, with $A < L_0$.

a. Calculate $\mathcal{I}^{jk}$ for this system as a function of time.

b. Calculate $h_{\mu\nu}^{TT}$ at a position in the $+z$ direction that is a distance $R \gg L$ away from the origin. [*Hint:* Note that $\sin^2\theta = \tfrac{1}{2}(1 - \cos 2\theta)$.]

c. At what rate will this vibrating dumbbell radiate gravitational wave energy?

P33.8 Imagine that two small identical balls of mass m move in opposite directions along the z axis at a constant velocity toward an ideal massless spring of length L centered at the origin. The balls hit the ends of the spring simultaneously at $t = 0$ and compress the spring to half its length before rebounding back out to infinity. The duration of the collision (from the moment when the balls first touch the spring to the time they leave the spring) is t_c. Remember from classical mechanics that a mass on a spring oscillates sinusoidally with angular frequency ω, so each ball's motion during the collision in this case will be exactly half an oscillation of a sine function away from each ball's position when it first touches the spring, and $\omega \equiv \pi/t_c$.

a. Argue that this system's reduced quadrupole moment tensor during the collision is given by

$$\mathcal{I}^{jk} = A(t) \begin{bmatrix} -1 & 0 & 0 \\ 0 & -1 & 0 \\ 0 & 0 & 2 \end{bmatrix}, \quad \text{where}$$

$$A(t) = \frac{mL^2}{48}(9 - 8\sin\omega t - \cos 2\omega t) \quad (33.49)$$

b. Calculate $h_{\mu\nu}^{TT}$ at a position in the $+z$ direction that is a distance $R \gg L$ away from the origin. What can you conclude about the gravitational wave energy flux in this direction? (*Hint:* You may be surprised by the result, but think about it.)

c. Calculate $h_{\mu\nu}^{TT}$ at a position in the $+x$ direction a distance $R \gg L$ away from the origin. Express your result in terms of L, t_c, G, m, and R. Is the wave cross-polarized, upright-polarized, or a mixture of both?

P33.9 Consider a long, thin rod of mass M and length L rotating in deep space around the z axis (in the xy plane) with an angular frequency of ω.

a. Argue that the quadrupole moment tensor for such a rod when it makes an angle of ωt with the x axis is

$$I^{ij} = \tfrac{1}{12}ML^2 \begin{bmatrix} \cos^2\omega t & \cos\omega t \sin\omega t & 0 \\ \cos\omega t \sin\omega t & \sin^2\omega t & 0 \\ 0 & 0 & 0 \end{bmatrix} \quad (33.50)$$

(*Hints:* Note that I^{ij} here is the quadrupole moment tensor not the reduced quadrupole moment. Express the position of every point on the rod in terms of ℓ and ωt, where ℓ ranges from $-\tfrac{1}{2}L$ to $+\tfrac{1}{2}L$ along the rod.)

b. Find h_{TT}^{ij} for waves emitted in the $+y$ direction. [*Hint:* $\cos^2\theta = \tfrac{1}{2}(1 + \cos 2\theta)$, $\sin^2\theta = \tfrac{1}{2}(1 - \cos 2\theta)$, and $\sin\theta\cos\theta = \tfrac{1}{2}\sin 2\theta$.]

c. This rod has a rotational kinetic energy of $K^{\text{rot}} = \tfrac{1}{2}I\omega^2 = \tfrac{1}{24}ML^2\omega^2$ (where $I = \tfrac{1}{12}ML^2$ is the moment of inertia of a thin rod). If the rod has a length of 10 m, a mass of 1000 kg, and rotates at 10 turns per second (a huge rate for something this big), about what fraction of its rotational kinetic energy will be emitted in the form of gravitational wave energy per year?

34. GRAVITATIONAL WAVE ASTRONOMY

INTRODUCTION

FLAT SPACETIME
Review of Special Relativity
Four-Vectors
Index Notation

TENSORS
Arbitrary Coordinates
Tensor Equations
Maxwell's Equations
Geodesics

SCHWARZSCHILD BLACK HOLES
The Schwarzschild Metric
Particle Orbits
Precession of the Perihelion
Photon Orbits
Deflection of Light
Event Horizon
Alternative Coordinates
Black Hole Thermodynamics

THE CALCULUS OF CURVATURE
The Absolute Gradient
Geodesic Deviation
The Riemann Tensor

THE EINSTEIN EQUATION
The Stress-Energy Tensor
The Einstein Equation
Interpreting the Equation
The Schwarzschild Solution

COSMOLOGY
The Universe Observed
A Metric for the Cosmos
Evolution of the Universe
Cosmic Implications
The Early Universe
CMB Fluctuations & Inflation

GRAVITATIONAL WAVES
Gauge Freedom
Detecting Gravitational Waves
Gravitational Wave Energy
Generating Gravitational Waves
Gravitational Wave Astronomy

SPINNING BLACK HOLES
Gravitomagnetism
The Kerr Metric
Kerr Particle Orbits
Ergoregion and Horizon
Negative-Energy Orbits

this depends on this

Introduction. In this chapter, we will be looking at some astrophysical applications of the gravitational wave mathematics that we have developed.

Gravitational Radiation from a Rotating Dumbbell. The most *common* cosmic sources of (unfortunately quite weak) gravitational waves are binary star systems. Let's approximate a binary star system as a pair of point masses m_1 and m_2 separated by a fixed distance D, so they orbit each other as if they were the ends of a dumbbell rotating around its center of mass.

Let's set up our coordinate system so that its origin is at the dumbbell's center of mass and so that the plane of the dumbbell's rotation coincides with the xy plane. The orbital radii of the two masses are then

$$r_1 = \left(\frac{m_2}{m_1 + m_2}\right)D \quad \text{and} \quad r_2 = \left(\frac{m_1}{m_1 + m_2}\right)D \tag{34.1}$$

respectively. (Note that $r_1/r_2 = m_2/m_1$, which is appropriate if these are distances from the center of mass at the origin, and that $r_1 + r_2 = D$.) Let's also define $t = 0$ to be the instant when mass 1 crosses the $+x$ axis. The coordinates x_1, y_1 and x_2, y_2 of the masses at a given instant of time t are then

$$x_1 = r_1 \cos \omega t = \frac{m_2 D}{m_1 + m_2} \cos \omega t \quad y_1 = r_1 \sin \omega t = \frac{m_2 D}{m_1 + m_2} \sin \omega t \tag{34.2a}$$

$$x_2 = -r_2 \cos \omega t = \frac{-m_1 D}{m_1 + m_2} \cos \omega t \quad y_2 = -r_2 \sin \omega t = \frac{-m_1 D}{m_1 + m_2} \sin \omega t \tag{34.2b}$$

where ω is the angular frequency of the orbit. If we treat the ends of the dumbbell as point masses, you can show (see box 34.1) that the reduced quadrupole moment tensor for this system is

$$\not{I}^{jk} = \frac{m_1 m_2 D^2}{m_1 + m_2} \begin{bmatrix} \cos^2 \omega t - \frac{1}{3} & \cos \omega t \sin \omega t & 0 \\ \sin \omega t \cos \omega t & \sin^2 \omega t - \frac{1}{3} & 0 \\ 0 & 0 & -\frac{1}{3} \end{bmatrix} \tag{34.3}$$

We can simplify this by defining the quantities

$$\eta \equiv \frac{m_1 m_2}{(m_1 + m_2)^2} \quad \text{and} \quad M \equiv m_1 + m_2 \tag{34.4}$$

M is just the dumbbell's total mass and η is a unitless quantity that ranges from $\frac{1}{4}$ (if $m_1 = m_2$) to roughly zero (if $m_1 \gg m_2$ or vice versa). We can also use the trigonometric identities $\cos\theta \sin\theta = \frac{1}{2} \sin 2\theta$, $\cos^2 \theta = \frac{1}{2} + \frac{1}{2}\cos 2\theta$, and $\sin^2 \theta = \frac{1}{2} - \frac{1}{2}\cos 2\theta$ to simplify the angular terms in this matrix so that they no longer involve products.

$$\not{I}^{jk} = \frac{1}{2} M \eta D^2 \begin{bmatrix} \frac{1}{3} + \cos 2\omega t & \sin 2\omega t & 0 \\ \sin 2\omega t & \frac{1}{3} - \cos 2\omega t & 0 \\ 0 & 0 & -\frac{2}{3} \end{bmatrix} \tag{34.5}$$

(Note that $\frac{1}{2} - \frac{1}{3} = \frac{1}{6} = \frac{1}{2} \cdot \frac{1}{3}$.) Getting rid of the trigonometric products makes it much easier to evaluate the double time derivative:

$$\ddot{\not{I}}^{jk} = 2M\eta D^2 \omega^2 \begin{bmatrix} -\cos 2\omega t & -\sin 2\omega t & 0 \\ -\sin 2\omega t & \cos 2\omega t & 0 \\ 0 & 0 & 0 \end{bmatrix} \tag{34.6}$$

This tensor happens to be already in transverse-traceless gauge for radiation in the $+z$ direction, so for an observer in the $+z$ direction and a distance R_0 from the system's center of mass, the metric perturbation is $h_{TT}^{jk} = 2G\ddot{\not{I}}_{TT}^{jk}/R_0$, which here is

$$h_{TT}^{jk} = -\frac{4GM}{R_0} \eta D^2 \omega^2 \begin{bmatrix} \cos[2\omega(t - R_0)] & \sin[2\omega(t - R_0)] & 0 \\ \sin[2\omega(t - R_0)] & -\cos[2\omega(t - R_0)] & 0 \\ 0 & 0 & 0 \end{bmatrix} \tag{34.7}$$

Note that $A_+ = A_\times = -4GM\eta D^2 \omega^2 / R_0$, so the wave has equal amounts of plus and cross polarization. Also, note that the plus and cross polarizations are 90° out of phase, implying that the wave is *circularly* polarized (the ring-distortion ellipse will rotate counterclockwise instead of oscillating in and out). Finally, note that the wave has a frequency equal to *twice* the rotational frequency of the system.

To find the gravitational waves radiated in another direction, we can calculate $\ddot{I}^{jk}_{TT}$ using the more general equation 33.12:

$$\ddot{I}^{jk}_{TT} \equiv \left(P^j_m P^k_n - \tfrac{1}{2} P^{jk} P_{mn}\right)\ddot{I}^{mn} \quad \text{with} \quad P^j_m \equiv \delta^j_m - n^j n_m \quad (34.8)$$

where $\vec{n}$ is a unit vector in the desired direction, and then use $h^{jk}_{TT} = 2G\ddot{I}^{jk}_{TT}/R_0$. For example, consider waves moving in the $+x$ direction. You can show (see problem P34.1) that the expression in transverse-traceless gauge for waves moving in the $+x$ direction from this source yields

$$h^{jk}_{TT} = \frac{2GM\eta D^2 \omega^2}{R_0} \begin{bmatrix} 0 & 0 & 0 \\ 0 & \cos[2\omega(t-R_0)] & 0 \\ 0 & 0 & -\cos[2\omega(t-R_0)] \end{bmatrix} \quad (34.9)$$

This wave is purely plus-polarized ($A_\times = 0$), and A_+ has half the magnitude compared to waves moving in the $+z$ direction. This result is important, because almost all the close binaries we know about are eclipsing binaries, so we are looking at the orbital plane (the xy plane by assumption) edge on, that is, from a direction we can take to be the $+x$ direction.

You can show (see box 34.2) that the total power carried away in *all* directions by the gravitational waves generated by this dumbbell system is

$$-\frac{dE}{dt} = \frac{G}{5}\langle \dddot{I}^{jk}\dddot{I}_{jk}\rangle = \frac{32(GM)^2\eta^2 D^4\omega^6}{5G} \quad (34.10)$$

Note the astonishing 6th-power dependence of this power on rotational frequency!

Application to a Binary Star System. If this rotating dumbbell is actually two stars orbiting each other, then the orbital frequency ω is related to the distance D between the stars, so we can eliminate one in favor of the other. Since the more easily observed quantity is the orbital frequency, let's eliminate D.

Assume that the stars move slowly enough that their velocities are non-relativistic and that they are far enough apart so that Newtonian gravitational theory is adequate to predict their motion. Newton's second law applied to m_1 tells us that

$$\frac{Gm_1m_2}{D^2} = \frac{m_1v_1^2}{r_1} \Rightarrow \frac{Gm_2}{D^2} = r_1\left(\frac{v_1}{r_1}\right)^2 = \frac{m_2 D}{m_1+m_2}\omega^2 \Rightarrow D^3 = \frac{GM}{\omega^2} \quad (34.11)$$

Plugging this into equation 34.7, we find that the polarization amplitudes for a wave traveling in the $+z$ direction (perpendicular to the plane of the orbit) are

$$A_+ = A_\times = \frac{4GM\eta}{R_0}\omega^2 \frac{(GM)^{2/3}}{\omega^{4/3}} = 4\frac{GM}{R_0}\eta(GM\omega)^{2/3} \quad (34.12)$$

The radiated power is

$$-\frac{dE}{dt} = \frac{32(GM)^2\eta^2\omega^6}{5G}\left(\frac{GM}{\omega^2}\right)^{4/3} = \frac{32\eta^2}{5G}(GM\omega)^{10/3} \quad (34.13)$$

This shows that rate of energy loss increases dramatically as the system's total mass M increases or the orbital frequency ω increases.

Now, this energy must come at the expense of the system's orbital energy, which you can show (see box 34.3) is given by

$$E = -\frac{Gm_1m_2}{2D} = -\frac{G(\eta M^2)\omega^{2/3}}{2(GM)^{1/3}} = -\tfrac{1}{2}M(GM\omega)^{2/3}\eta \quad (34.14)$$

Thus the stars in the binary pair will maintain neither a fixed separation D nor a constant angular frequency ω, as assumed in the derivation so far: rather ω will increase with time and D will decrease with time as the binary's orbital energy is radiated away. This means that the calculations we have made are not quite right: for example, equation 34.6 for $\ddot{I}^{jk}$ is not exact because we are ignoring the time dependence of both D and ω. However, as long as the energy leaks away only very slowly, we are justified in ignoring these time derivatives.

One way to quantitatively express how "slowly" the energy is radiated is to calculate the rate of change of the orbit's period $T = 2\pi/\omega$. You can show (see box 34.4) that this rate is given by

$$\frac{dT}{dt} = \frac{dT}{d\omega}\frac{d\omega}{dE}\frac{dE}{dt} = -\frac{192\pi}{5}\eta(GM\omega)^{5/3} \tag{34.15}$$

Note that the orbit's period $T = 2\pi/\omega$ in meters of light travel time will be very large compared to GM for the pair, so $GM\omega \propto GM/T$ will be extremely tiny. This justifies our approximation that D and ω are approximately constant.

A Realistic Binary System. A strongly radiating binary pair close to the earth is the ι Boötis system, which consists of two stars with masses of $1.0\,M_\odot$ and $0.5\,M_\odot$ orbiting with an approximate period of $23{,}300$ s $= 7.0 \times 10^{12}$ m. The distance to this system is about 42 ly $= 4.0 \times 10^{17}$ m. For this system (see box 34.5),

$$A_+ = 7.7 \times 10^{-21} \tag{34.16a}$$

$$-\frac{dE}{dt} = 1.1 \times 10^{23}\text{ W} \tag{34.16b}$$

$$\frac{dT}{dt} = -8.4 \times 10^{-14} \tag{34.16c}$$

For comparison, note that the sun radiates electromagnetic energy at a rate of about 3.9×10^{26} W. Even though this system radiates an appreciable power, the rate of change of its period is not measurable. To be able to detect waves from this system, a gravitational wave detector would have to be able to detect fractional distance changes between two floating masses that are smaller than 10^{-20} at a frequency of $2/23{,}300$ s $= 86$ μHz. (The LIGO detector has the appropriate sensitivity only in a frequency range of a few hundred Hz, not in this very low frequency range.)

The Evidence for Gravitational Waves. In 1974, Joseph Taylor and Russell Hulse discovered a pulsar (designated PSR B1913+16) whose radio signals were periodically Doppler shifted in such a way as to indicate that it was a part of a binary pair (they won the 1993 Nobel Prize in physics for this discovery). A pulsar is a rapidly rotating neutron star that emits radio pulses synchronized with its rotation rate. Since these pulses are extremely steady, it is possible to measure the Doppler shift with extraordinary accuracy. This allows one to determine many characteristics of the binary system to extraordinary accuracy as well. For example, we know that for this binary system:

Pulsar period:	0.059029995271(2) s
Pulsar mass:	1.4414(2) solar masses
Companion mass:	1.3867(2) solar masses
Period of orbit:	27906.98163(2) s
Projected semimajor axis:	2.341775(8) s ($\perp$ to the line of sight)
Orbital eccentricity:	0.6171338(4)
Precession of periastron:	4.226595(5)°/y
dT/dt (measured):	$-2.4184(9) \times 10^{-12}$

(These data are from Taylor and Weisberg, "Relativistic Binary Pulsar B1913+16: Thirty Years of Observations and Analysis," in *Binary Radio Pulsars*, ASP Conference Series, 2004, Rasio & Stairs, eds., and Ohanian and Ruffini, *Gravitation and Spacetime*, 2nd ed., Norton, 1994, p. 266. Numbers in parentheses indicate the uncertainty in the final digit.) The last quantity is particularly interesting, because we can predict it.

Equation 34.15 does not work in this case, because the orbit is not even remotely circular (the eccentricity is more than 0.617) and the emission of gravitational radiation is greatly enhanced during the part of the orbit where the partners are close together. But one can make corrections for this, and one finds that, when one does, the ratio of the measured rate to the predicted rate is 1.0013 ± 0.0021. This is very strong evidence that gravitational waves are being emitted by this system at the rate predicted by general relativity. So while we have not yet detected gravitational waves directly, this measurement provides very solid evidence that gravitational waves exist.

A graph of the observed orbital phase shift as a function of time is shown in figure 34.1.

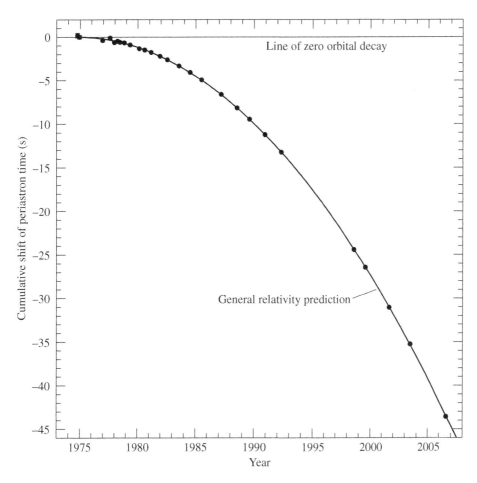

FIG. 34.1 This graph shows PSR B1913+16's orbital decay since 1975. The datapoints reflect the observed *change* in the periastron time (relative to when it should have occurred if the orbital period was constant) plotted versus date. The smooth curve illustrates the change that general relativity predicts *should* occur due to orbital energy carried away by gravitational waves. From Weisberg, Nice, and Taylor. "Timing Measurements of the Relativistic Binary Pulsar PSR B1913+16," *Astrophysical Journal*, **722**, 2010, 1030. Courtesy of Joel Weisberg,

BOX 34.1 The Dumbbell I^{jk}

The expression for the reduced quadrupole moment tensor for the dumbbell is

$$I^{jk} \equiv \int \rho(x^j x^k - \tfrac{1}{3}\eta^{jk} r^2)\,dV = m_1(x_1^j x_1^k - \tfrac{1}{3}\eta^{jk} r_1^2) + m_2(x_2^j x_2^k - \tfrac{1}{3}\eta^{jk} r_2^2) \quad (34.17)$$

Show that plugging in the results from equation 34.2 yields

$$I^{jk} = [m_1 r_1^2 + m_2 r_2^2] \begin{bmatrix} \cos^2 \omega t - \tfrac{1}{3} & \cos \omega t \sin \omega t & 0 \\ \cos \omega t \sin \omega t & \sin^2 \omega t - \tfrac{1}{3} & 0 \\ 0 & 0 & -\tfrac{1}{3} \end{bmatrix} \quad (34.18)$$

Exercise 34.1.1. Check equation 34.18.

You can then eliminate r_1 and r_2 in favor of D in the $m_1 r_1^2 + m_2 r_2^2$ factor to yield the result given by equation 34.3, duplicated here for convenience:

$$I^{jk} = \frac{m_1 m_2 D^2}{m_1 + m_2} \begin{bmatrix} \cos^2 \omega t - \tfrac{1}{3} & \cos \omega t \sin \omega t & 0 \\ \sin \omega t \cos \omega t & \sin^2 \omega t - \tfrac{1}{3} & 0 \\ 0 & 0 & -\tfrac{1}{3} \end{bmatrix} \quad (34.3r)$$

Exercise 34.1.2. Show that we can indeed convert the quantity in brackets in equation 34.18 to that in front of the matrix here in equation 34.3.

BOX 34.2 The Power Radiated by a Rotating Dumbbell

The triple time derivative of the reduced quadrupole moment tensor is

$$\dddot{\mathcal{F}}^{jk} = 4M\eta D^2 \omega^3 \begin{bmatrix} \sin 2\omega t & -\cos 2\omega t & 0 \\ -\cos 2\omega t & -\sin 2\omega t & 0 \\ 0 & 0 & 0 \end{bmatrix} \quad (34.19)$$

You can show that after doing the sums implied by $\dddot{\mathcal{F}}^{jk}\dddot{\mathcal{F}}_{jk}$ and taking the average over several wavelengths of the wave, the result is equation 34.10, which is repeated here for convenience:

$$-\frac{dE}{dt} = \frac{G}{5}\langle \dddot{\mathcal{F}}^{jk}\dddot{\mathcal{F}}_{jk}\rangle = \frac{32(GM)^2 \eta^2 D^4 \omega^6}{5G} \quad (34.10r)$$

Exercise 34.2.1. Verify equation 34.10.

BOX 34.3 The Total Energy of an Orbiting Binary Pair

Assuming that Newtonian approximations are accurate, a binary system's total energy is given by

$$E = \tfrac{1}{2}m_1 v_1^2 + \tfrac{1}{2}m_2 v_2^2 - \frac{Gm_1 m_2}{D} = \tfrac{1}{2}m_1 r_1^2 \omega^2 + \tfrac{1}{2}m_2 r_2^2 \omega^2 - \frac{Gm_1 m_2}{D} \quad (34.20)$$

You can use the values of r_1 and r_2 from equation 34.1 (repeated here),

$$r_1 = \left(\frac{m_2}{m_1 + m_2}\right)D \quad \text{and} \quad r_2 = \left(\frac{m_1}{m_1 + m_2}\right)D \quad (34.1r)$$

and equation 34.11, which asserts that $D^3 = G(m_1 + m_2)/\omega^2$, to show that

$$E = -\frac{Gm_1 m_2}{2D} \quad (34.21)$$

Exercise 34.3.1. Verify the last equation, and then fill in the remaining steps leading to $E = -\tfrac{1}{2}M(GM\omega)^{2/3}\eta$.

BOX 34.4 The Time-Rate-of-Change of the Orbital Period

You can use equations 34.14 and 34.13 (repeated here for convenience)

$$E = -\tfrac{1}{2}M(GM\omega)^{2/3}\eta \quad \text{and} \quad -\frac{dE}{dt} = \frac{32\eta^2}{5G}(GM\omega)^{10/3} \quad (34.22)$$

and $T = 2\pi/\omega$ to prove equation 34.15, repeated here for convenience:

$$\frac{dT}{dt} = \frac{dT}{d\omega}\frac{d\omega}{dE}\frac{dE}{dt} = -\frac{192\pi}{5}\eta(GM\omega)^{5/3} \quad (34.15r)$$

Exercise 34.4.1. Fill in the missing steps.

BOX 34.4 (continued) The Time-Rate-of-Change of the Orbital Period

BOX 34.5 Characteristics of ι Boötis

Exercise 34.5.1. Verify the values of $A_+ = 4(GM/R_0)\eta(GM\omega)^{2/3}$, dE/dt, and dT/dt for the system ι Boötis (*Hint:* Calculate the unitless quantities $GM\omega$ and $\eta = m_1 m_2/(m_1 + m_2)^2$ first. Note that GM for the sun is 1477 m.)

HOMEWORK PROBLEMS

P34.1 Use equation 34.8 to find the transverse-traceless form of the metric perturbation

$$h_{TT}^{ik} = \frac{2G}{R_0} \ddot{I}_{TT}^{ik} \qquad (34.23)$$

for gravitational waves radiated in the $+x$ direction from a binary star system orbiting in the xy plane.

P34.2 The binary μ Scorpii has masses $m_1 = m_2 = 12 M_\odot$, a period of 125,000 s, and a distance of 355 ly. Find the gravitational wave amplitude A_+ at the earth, the total gravitational wave energy it radiates (in W), and the rate dT/dt at which its period is changing. Assume that the stars' orbits are circular and that we are viewing the orbital plane edge-on from a direction we can take to be the $+x$ direction relative to the system.

P34.3 The star known as Algol (β Persei) is actually a close binary consisting of two stars, one with mass $3.70 M_\odot$ and one with mass $0.80 M_\odot$. These stars orbit each other with a period of 2.87 d. The system is 93 ly from the earth. Find the gravitational wave amplitude A_+ at the earth, the total gravitational wave energy it radiates (in W), and the rate dT/dt at which its period is changing. Assume that the stars' orbits are circular and that we are viewing the orbital plane edge-on from a direction we can take to be the $+x$ direction relative to the system.

P34.4 PSR J0737-3039 is a recently-discovered binary pulsar system consisting of two neutron stars, one with mass $1.337 M_\odot$ and one with mass $1.250 M_\odot$, orbiting each other with a period of 2.4 h. The orbit is only mildly eccentric, and we are viewing the orbital plane almost edge on from a direction we can take to be the $+x$ direction relative to the system. This system is roughly 1800 ly from the earth. Find the gravitational wave amplitude A_+ at the earth, the total gravitational wave energy it radiates (in W), and the rate dT/dt at which its period is changing.

P34.5 Use equation 34.15 to calculate dT/dt for the binary pulsar PSR 1913+16 and compare with the measured result. (Remember that equation 34.15 is not really appropriate for this system because the stars' orbits are not even close to being circular.)

P34.6 Estimate the rate at which the orbiting earth radiates energy in the form of gravitational waves. Is it likely that the earth will spiral into the sun any time soon as a result of this energy loss?

P34.7 Rank the eight planets in our solar system in terms of their gravitational wave luminosity due to their orbital motion around the sun.

P34.8 The orbiting earth will slowly spiral into the sun as it radiates gravitational wave energy. Estimate how long it will take for the radius of the earth's orbit to decrease by 1 km due to this effect. (*Hint:* Note that according to Kepler's third law, the period T and the radius r of an orbiting object are related by $T^2 = (4\pi^2/GM)r^3$, where M is the mass of the primary.)

P34.9 Consider a pair of particles with equal mass m that are connected by a spring, so that each particle's position oscillates along the x axis as follows:

$$x_1 = A \cos \omega t, \quad x_2 = -A \cos \omega t \qquad (34.24)$$

Note that these particles exchange positions every half-cycle of the oscillation.

a. Find the reduced quadrupole moment tensor for this system.
b. Find the polarization and amplitude of waves emitted in the $+z$ direction.
c. Calculate the rate at which this system loses energy to gravitational waves.
d. Calculate the rate dA/dt at which the system's oscillation amplitude changes, assuming that the system's total energy can be described by Newtonian mechanics.

P34.10 Consider two cars with equal mass m that approach each other along the x axis with equal speeds v_0. Let $x = 0$ to be midway between the cars. At time $t = -T$, they contact each other while their centers of mass are both a distance $2d$ from the origin. At time $t = 0$, they have both come to rest, with their centers of mass a distance d from the origin (they have crumpled about to half their length).

a. Show that if we assume that the cars' acceleration during the collision is constant, then $T = 2d/v_0$ and the magnitude of the cars' acceleration is $a = v_0^2/2d$.
b. Find the reduced quadrupole moment tensor for this system as a function of $x(t)$.
c. Argue that the cars radiate no meaningful gravitational waves until they actually begin to collide.
d. Find a formula for the total energy radiated in the form of gravitational waves during the collision as a fraction of the total initial kinetic energy of the cars. Express your result in terms of G, m, v_0, and d. (*Hint:* We can essentially take the average "over several wave cycles" by integrating over the entire pulse.)
e. Calculate this fraction if $m = 1000$ kg, $v_0 = 30$ m/s, and $d = 1$ m. (*Hint:* Remember to use GR units.)

(**Note:** You might consider this calculation a bit bogus, because the gravitational wave in this case is not a simple cosinusoidal wave, but rather a pulse wave. So how do we know that we can find the right gauge transformation to put it into TT gauge? It turns out that we can define a transverse-traceless gauge for *any* arbitrary gravitational wave, and as long as the wave is a plane wave, we can use the projection operator 34.8 to do this. See Misner, Thorne, and Wheeler, *Gravitation*, Freeman, 1973, pp. 946–949.)

35. GRAVITOMAGNETISM

INTRODUCTION

FLAT SPACETIME
- Review of Special Relativity
- Four-Vectors
- Index Notation

TENSORS
- Arbitrary Coordinates
- Tensor Equations
- Maxwell's Equations
- Geodesics

SCHWARZSCHILD BLACK HOLES
- The Schwarzschild Metric
- Particle Orbits
- Precession of the Perihelion
- Photon Orbits
- Deflection of Light
- Event Horizon
- Alternative Coordinates
- Black Hole Thermodynamics

THE CALCULUS OF CURVATURE
- The Absolute Gradient
- Geodesic Deviation
- The Riemann Tensor

THE EINSTEIN EQUATION
- The Stress-Energy Tensor
- The Einstein Equation
- Interpreting the Equation
- The Schwarzschild Solution

COSMOLOGY
- The Universe Observed
- A Metric for the Cosmos
- Evolution of the Universe
- Cosmic Implications
- The Early Universe
- CMB Fluctuations & Inflation

GRAVITATIONAL WAVES
- Gauge Freedom
- Detecting Gravitational Waves
- Gravitational Wave Energy
- Generating Gravitational Waves
- Gravitational Wave Astronomy

SPINNING BLACK HOLES
- **Gravitomagnetism**
- The Kerr Metric
- Kerr Particle Orbits
- Ergoregion and Horizon
- Negative-Energy Orbits

this depends on this

35. GRAVITOMAGNETISM

Introduction. This chapter discusses the phenomenon of gravitomagnetism in the weak-field limit, drawing useful analogies with electricity and magnetism and discussing experimental consequences. This chapter will also prepare the way for a discussion of rotating black holes in subsequent chapters.

Review of the Weak-Field Limit. Consider again the weak-field limit discussed in chapter 22, where we assume that the metric is given by $g_{\mu\nu} = \eta_{\mu\nu} + h_{\mu\nu}$ where $h_{\mu\nu} = h_{\nu\mu}$ and $|h_{\mu\nu}| \ll 1$. Given the coordinate condition

$$0 = \eta^{\alpha\mu}\left(\partial_\mu h_{\alpha\nu} - \tfrac{1}{2}\partial_\nu h_{\alpha\mu}\right) \quad \text{(see equation 22.8b)} \tag{35.1}$$

(which we saw in problem P30.1 was equivalent to the Lorenz gauge condition), the Einstein equation becomes simply

$$\partial^\alpha \partial_\alpha h^{\mu\nu} = \left(-\frac{\partial^2}{\partial t^2} + \nabla^2\right) h^{\mu\nu} = -16\pi G\left(T^{\mu\nu} - \tfrac{1}{2}\eta^{\mu\nu}T\right) \tag{35.2}$$

(see equation 22.9). We have seen that once we have chosen our coordinate condition (in this case, equation 35.1), we can treat the metric perturbation exactly as if it were a tensor field in flat spacetime.

Solving the Einstein Equation. We can compare this equation to the analogous time-dependent equation for the electric potential $[-(\partial/\partial t)^2 + \nabla^2]\phi = -4\pi k \rho_c$, where ρ_c is the charge density. You may be aware that in *static* situations the way to calculate the potential $\phi(t, \vec{R})$ at position $\vec{R}$ is to divide the source into a set of cells each having a volume dV small enough to be considered a point charge with charge $q_i \equiv \rho_c(\vec{r}_i) dV$, and then sum the point charge potentials kq_i/s_i over all cells in the source (where s_i is the separation $s_i \equiv |\vec{R} - \vec{r}_i|$ between the field position $\vec{R}$ and the cell's position $\vec{r}_i$ and $k = 1/4\pi\varepsilon_0$ is the Coulomb constant). It turns out that the only additional thing we need to do in *dynamic* situations is to account for the light travel time required for information to get from the source to the field point $\vec{R}$ by evaluating kq_i/s_i for each cell at time $t - s_i$.

By analogy, the solution to equation 35.2 is therefore (quite generally!)

$$h^{\mu\nu}(t, \vec{R}) = 4G \int_{\text{src}} \frac{1}{s}\left[T^{\mu\nu}(t-s, \vec{r}) - \tfrac{1}{2}\eta^{\mu\nu}T(t-s, \vec{r})\right]dV \tag{35.3}$$

The Slow-Source Approximation. Let's now make the approximation that the source consists of a non-relativistic perfect fluid. In such a case, the pressure p_0 in the source will be negligible compared to its energy density ρ_0, and the fluid four-velocity u^α at every point will be such that $u^t \approx 1$ and $u^i \approx v^i \ll 1$, where i is a spatial index. In this approximation, we will keep terms only to first order in $u^i \approx v^i$. The general form of the perfect-fluid stress-energy is $T^{\mu\nu} = (\rho_0 + p_0)u^\mu u^\nu + p_0 g^{\mu\nu}$, but in this slow-source approximation, the stress-energy components will be

$$T^{tt} \approx \rho_0 u^t u^t \approx \rho_0 \tag{35.4a}$$

$$T^{ti} = T^{it} \approx \rho_0 u^t u^i \approx \rho_0 v^i \tag{35.4b}$$

$$T^{ij} \approx \rho_0 u^i u^j \approx 0 \quad \text{(since this is second order in } u^i\text{)} \tag{35.4c}$$

This means that $T \equiv \eta_{\alpha\beta}T^{\alpha\beta} \approx \eta_{tt}T^{tt} + 0 + 0 + 0 = -\rho_0$, so, we have

$$T^{tt} - \tfrac{1}{2}\eta^{tt}T \approx \rho_0 - \tfrac{1}{2}(-1)(-\rho_0) = \tfrac{1}{2}\rho_0 \tag{35.5a}$$

$$T^{ti} - \tfrac{1}{2}\eta^{ti}T \approx \rho_0 u^t u^i + 0 \approx \rho_0 u^i \approx \rho_0 v^i \tag{35.5b}$$

$$T^{ij} - \tfrac{1}{2}\eta^{ij}T \approx 0 - \tfrac{1}{2}\eta^{ij}(-\rho_0) \approx +\tfrac{1}{2}\eta^{ij}\rho_0 \tag{35.5c}$$

In this case, the components of equation 35.3 become

$$h^{tt}(t, \vec{R})(= h^{xx} = h^{yy} = h^{zz}) = 2\int_{\text{src}} \frac{G\rho_0(t-s, \vec{r})}{s} dV \tag{35.6a}$$

$$h^{ti}(t, \vec{R}) = h^{it}(t, \vec{R}) = 4\int_{\text{src}} \frac{GJ^i(t-s, \vec{r})}{s} dV \tag{35.6b}$$

where $\vec{J} \equiv \rho_0 \vec{v}$ is the energy current density. The other components of the metric perturbation are approximately zero.

Define the gravitational scalar and vector potentials to be

$$\Phi_G \equiv -\tfrac{1}{2} h^{tt} = -\int_{\text{src}} \frac{G\rho_0}{s} dV \tag{35.7a}$$

$$A_G^i \equiv -\tfrac{1}{4} h^{ti} = -\tfrac{1}{4} h^{it} = -\int_{\text{src}} \frac{GJ^i}{s} dV \tag{35.7b}$$

You can show (see box 35.1) that the coordinate condition in equation 35.1 requires that these potentials obey the relationship

$$\vec{\nabla} \cdot \vec{A}_G = -\frac{\partial \Phi_G}{\partial t} \tag{35.8}$$

The Gravitational Maxwell Equations. Let's also define the gravitoelectric and gravitomagnetic fields such that

$$\vec{E}_G \equiv -\vec{\nabla} \Phi_G - \frac{\partial \vec{A}_G}{\partial t}, \quad \vec{B}_G \equiv \vec{\nabla} \times \vec{A}_G \tag{35.9}$$

These equations are exactly the same as the definitions of the electric and magnetic fields in terms of potentials.

You can show (see box 35.2) that given these definitions, the Einstein equation implies that the gravitoelectric and gravitomagnetic fields obey

$$\vec{\nabla} \cdot \vec{E}_G = -4\pi G \rho_0 \tag{35.10a}$$

$$\vec{\nabla} \times \vec{B}_G - \frac{\partial \vec{E}_G}{\partial t} = -4\pi G \vec{J} \tag{35.10b}$$

$$\vec{\nabla} \cdot \vec{B}_G = 0 \tag{35.10c}$$

$$\vec{\nabla} \times \vec{E}_G + \frac{\partial \vec{B}_G}{\partial t} = 0 \tag{35.10d}$$

These equations are *identical* to Maxwell's equations (in GR units) except that the Coulomb constant $k = 1/4\pi\varepsilon_0 = \mu_0/4\pi$ (in GR units) gets replaced by G and the signs of the right sides of the first two equations are negative in the gravity case instead of being positive (as in the electromagnetic case). The sign change is because the gravitational force between positive masses is attractive, but the electromagnetic force between positive charges is repulsive.

If the field is static (so that the time derivatives of all field quantities are zero), then, as we saw in chapter 22, the geodesic equation for a particle moving with a non-relativistic velocity $\vec{V}$ becomes

$$\frac{d^2 x^i}{dt^2} \approx \tfrac{1}{2} \eta^{ik} \partial_k h_{tt} + \eta^{ik} (\partial_k h_{tj} - \partial_j h_{tk}) V^j \tag{35.11}$$

(see equation 22.14). From this, you can show (see box 35.3) that the gravitational force that we would infer from Newton's second law as acting on the particle is

$$\vec{F}_G = m \frac{d^2 \vec{x}}{dt^2} = m(\vec{E}_G + \vec{V} \times 4\vec{B}_G) \tag{35.12}$$

where m is the mass of the particle. Again, this is the same as for the Lorentz equation $\vec{F}_{EM} = q(\vec{E} + \vec{V} \times \vec{B})$ for the electromagnetic field, except that (1) the mass m replaces its charge q on the right side and (2) the gravitomagnetic field exerts four times the force that we would expect from the electromagnetic analogy. (Unlike the Lorentz equation, though, equation 35.12 does not apply to dynamic fields.)

The point is that gravity does exhibit gravitomagnetic effects that are almost exactly like magnetic effects except that (1) the sign of $\vec{B}_G$ is reversed (so we should use left-hand rules instead of right-hand rules) and (2) the effect of the gravitomagnetic field is four times larger than we would expect from the analogy.

This is very useful, because we can use electromagnetic analogies to anticipate, understand, and even calculate some gravitomagnetic effects that would be very tricky to explore otherwise. The next few sections will illustrate this.

Gravitomagnetic Effects on a Gyroscope. We know from electromagnetic theory that a simple current loop of area A carrying current i has a magnetic moment $\vec{\mu}$ with magnitude $\mu = iA$ pointing perpendicular to the loop's plane in the sense indicated by your right thumb when your right fingers curl in the direction of the current flow. In a magnetic field $\vec{B}$, such a current loop experiences a torque $\vec{\tau} = \vec{\mu} \times \vec{B}$ that seeks to align the loop's magnetic moment with the field.

You can show (see box 35.4) that a spinning object (a gyroscope) has an analogous gravitomagnetic moment $\vec{\mu}_G = \frac{1}{2}\vec{s}$, where $\vec{s}$ is the gyroscope's total spin angular momentum. By analogy, in a gravitomagnetic field $\vec{B}_G$, such a gyroscope should experience a torque

$$\vec{\tau} = \vec{\mu}_G \times 4\vec{B}_G = \vec{s} \times 2\vec{B}_G \qquad (35.13)$$

(remember that a gravitomagnetic field exerts 4 times more force on a moving mass than the corresponding magnetic field would exert on a moving charge). As discussed in box 35.5, exerting such a torque on a gyroscope causes it to precess around the field direction with an angular velocity of

$$\vec{\Omega}_{LT} = -2\vec{B}_G \qquad (35.14)$$

if we define $\vec{\Omega}_{LT}$ to point as your right thumb does when your fingers curl in the direction of the precession. Observing this so-called **Lense-Thirring precession** of a gyroscope at a point in empty space provides a practical way to measure both the magnitude and direction of any gravitomagnetic field present at that location.

Lense-Thirring Precession Near a Spinning Object. Another established result from electromagnetic theory is that any steadily spinning charged object with spherical symmetry produces a dipole magnetic field in its exterior:

$$\vec{B}(\vec{r}) = \frac{\mu_0}{4\pi r^3}[3(\vec{\mu}\cdot\hat{r})\hat{r} - \vec{\mu}] \qquad (35.15)$$

where $\vec{\mu}$ is the object's total magnetic moment, $\mu_0 = 4\pi k$ (in GR units), $\vec{r}$ is the displacement from the object's center to the point where the field is being evaluated, and $\hat{r}$ is a unit vector pointing in the $\vec{r}$ direction (see the "Dipole" entry in Wikipedia). By analogy, then, the gravitomagnetic field produced by a spherical star or planet with total spin angular momentum $\vec{S}$ is

$$\vec{B}_G(\vec{r}) = -\frac{G}{r^3}[3(\vec{\mu}_G\cdot\hat{r})\hat{r} - \vec{\mu}_G] = \frac{G}{2r^3}[\vec{S} - 3(\vec{S}\cdot\hat{r})\hat{r}] \qquad (35.16)$$

where the factor of 2 comes from $\vec{\mu}_G = \frac{1}{2}\vec{S}$ (as discussed above) and the minus sign comes from the reversal of the gravitomagnetic field compared to the analogous magnetic field). We can use this to estimate both the magnitude and direction of the Lense-Thirring effect near a rotating body of interest.

For example, consider a gyroscope in an equatorial orbit around the earth. Since the earth's spin angular momentum $\vec{S}$ points perpendicular to the earth's equatorial plane from south to north (check with your right hand), for a point on the earth's equatorial plane, $\vec{S}\cdot\hat{r} = 0$, so on that plane, $\vec{B}_G = G\vec{S}/2r^3$ oriented parallel to the earth's spin $\vec{S}$. The orbiting gyroscope's precession relative to distant stars will be easiest to observe if its spin $\vec{s}$ is perpendicular to this direction (i.e, it lies *in* the equatorial plane): let's assume this. The angular speed of precession will then be

$$\Omega_{LT} = 2B_G = \frac{GS}{r^3} = \frac{GI\omega}{r^3} \qquad (35.17)$$

where I is the earth's moment of inertia and ω is its spin angular speed $= 2\pi$/day.

To go further, we need to estimate the earth's moment of inertia. We can quite generally express an axially symmetric object's moment of inertia as

$$I = \alpha M R^2 \tag{35.18}$$

where M is the object's mass, R is its radius, and α is a constant ($0 < \alpha \leq 1$) that depends on the distribution of mass in the object: α larger if the mass is concentrated near the object's rim and smaller if it is concentrated in the center. For a uniform sphere, $\alpha = 2/5$, but since the earth is denser near its center, we would expect α for the earth to be somewhat smaller: detailed estimates based on the earth's measured density profile imply that $\alpha = 0.33$. Therefore, a good estimate of the Lense-Thirring precession rate for our orbiting gyroscope would be

$$\Omega_{LT} = \frac{G(\alpha M R^2)\omega}{r^3} = 0.33 \frac{GM}{R}\left(\frac{R}{r}\right)^3 \omega \tag{35.19}$$

where R is the earth's radius ≈ 6380 km. Note that GM for the earth is 4.45 mm. Therefore, for a gyroscope in low-earth orbit where $r \approx R$, we have

$$\Omega_{LT} \approx \frac{0.33(4.5 \times 10^{-3} \text{ m})}{6,380,000 \text{ m}}\left(\frac{2\pi \text{ rad}}{\text{day}}\right)\left(\frac{365 \text{ day}}{\text{y}}\right) = \frac{5.4 \times 10^{-7} \text{ rad}}{\text{y}} \tag{35.20}$$

This corresponds to about 0.11 arcseconds per year, which is obviously a very small number (and therefore is very difficult to measure).

Geodetic Precession. There is a second effect that will also cause our hypothetical orbiting gyroscope to precess. Because of the curvature of spacetime, a gyroscope orbiting even a non-spinning object will precess: this phenomenon is called *geodetic precession*. Since this is not a gravitomagnetic effect, I will not discuss it here, but problem P35.7 will guide you through the derivation if you are interested. Again assuming that the gyroscope orbits in the equatorial plane and has its spin lying in that plane, the angle through which the gyroscope precesses is

$$\Delta\phi_{gd} \approx \frac{3\pi GM}{R} \quad \text{per orbit} \tag{35.21}$$

Since a near-earth orbit takes about 85 minutes, the precession rate is

$$\Omega_{gd} = \frac{\Delta\phi_{gd}}{T} = \frac{3\pi(4.45 \times 10^{-3} \text{ m})}{(85 \text{ min})(6,380,000 \text{ m})}\left(\frac{1 \text{ min}}{60 \text{ s}}\right)\left(\frac{3.16 \times 10^7 \text{ s}}{1 \text{ y}}\right)$$

$$= 4.1 \times 10^{-5} \text{ rad/y} \tag{35.22}$$

This is almost two orders of magnitude larger than the Lense-Thirring effect for a gyroscope in low earth orbit.

Gravity Probe B. On April 20, 2004, NASA launched the *Gravity Probe B* mission to measure both of these effects. The probe carried four precision gyroscopes, each consisting of a nearly perfect quartz sphere electrostatically suspended in a chamber. (The gyroscopes are listed in the *Guinness World Records* as being the most perfect spherical objects ever made.) All the gyroscopes' spin axes were carefully monitored relative to the direction of a distant guide star. The satellite was launched into a polar orbit with a radius of almost exactly 7000 km (rather than an equatorial orbit) to separate the geodetic and gravitomagnetic effects. (Unfortunately, since $\vec{B}_G$ is not constant for such an orbit, the calculation of the predicted precession is much harder than what we did above.)

To ensure that the solar wind and other external forces did not push the spacecraft off a geodesic path, one of the spheres was used as a "proof mass:" it was allowed to float freely in its chamber and the spacecraft's thrusters were fired as necessary to keep that chamber centered on the sphere. This ensured that the spacecraft and the other three gyroscopes followed the purely geodesic path blazed by the protected proof mass. (The thrusters cleverly used helium gas that had boiled out of the Dewar flask of liquid helium that kept the apparatus supercooled to minimize thermal effects.)

When data from the spacecraft were first analyzed, noise appeared to be much larger than expected, swamping even the geodetic result. Subsequent investigation showed that the problem had to do with unexpected electromagnetic torques on the gyroscope spheres due to patchy areas of stray electrostatic charge on the gyroscopes. Fortunately, the problem could be modeled and corrected using calibration data taken at the beginning of the mission. Unfortunately, the problem also led to measurement uncertainties about 20 to 30 times larger than hoped.

The final analysis of the data was finally published in May of 2011 (Everitt et al., *Physical Review Letters* **106**, 221101). The combined results from all four gyroscopes imply a measured geodetic precession of 6602 ± 18 mas/y and a Lense-Thirring precession of 37.2 ± 7.2 mas/y, to be compared with predictions from general relativity of 6606 mas/y and 39.2 mas/y, respectively. The direct measurement of the geodetic precession is currently the most precise (at 0.3% uncertainty) on record. Because of the stray charge problems, the Lense-Thirring measurement is good only to about ±19%. In 2004, this effect had been measured using the LAGEOS satellites to an accuracy of about 10%, but that result had to assume that general relativity correctly predicts the value of the geodetic precession. The *Gravity Probe B* results therefore represent the best completely independent measurements of these effects, as well as an important verification of Einstein's theory. (For an accessible discussion of these findings, see Blau, "Gravity Probe B concludes its 50-year quest," *Physics Today,* **64,** 7, July 2011, pp. 14–16.)

BOX 35.1 The Lorenz Condition for the Potentials

The coordinate condition specified in equation 35.1 tells us that our metric perturbation must satisfy $0 = \eta^{\alpha\mu}(\partial_\mu h_{\alpha\nu} - \frac{1}{2}\partial_\nu h_{\alpha\mu})$. If we raise the ν index and perform the summations, this condition becomes

$$0 = (\partial_\mu h^{\mu\sigma} - \tfrac{1}{2}\eta^{\sigma\nu}\partial_\nu h) \tag{35.23}$$

Exercise 35.1.1. Substitute $h^{tt} = h^{xx} = h^{yy} = h^{zz} = -2\Phi_G$ and $h^{ti} = -4A_G^i$ into this expression to show that it implies that $\vec{\nabla}\cdot\vec{A}_G = -\partial\Phi_G/\partial t$ (equation 35.8).

BOX 35.2 The Maxwell Equations for the Gravitational Field

The definitions of the gravitoelectric and gravitomagnetic fields are given by equation 35.9: $\vec{E}_G \equiv -\vec{\nabla}\Phi_G - \partial\vec{A}_G/\partial t$, $\vec{B}_G \equiv \vec{\nabla}\times\vec{A}_G$. With the help of these definitions, the weak-field Einstein equations $\partial^\alpha \partial_\alpha h^{\mu\nu} = -16\pi G(T^{\mu\nu} - \frac{1}{2}\eta^{\mu\nu}T)$, the values $T^{tt} - \frac{1}{2}\eta^{tt}T = \frac{1}{2}\rho_0$ and $T^{ti} - \frac{1}{2}\eta^{ti}T = J^i$, the results of the previous box, and the following identities from vector calculus,

$$\vec{\nabla}\times(\vec{\nabla}\times\vec{F}) = \vec{\nabla}(\vec{\nabla}\cdot\vec{F}) - \nabla^2\vec{F} \qquad (35.24a)$$

$$\vec{\nabla}\cdot(\vec{\nabla}\times\vec{F}) = 0 \quad \text{(basically because } \vec{\nabla} \text{ is } \perp \text{ to } \vec{\nabla}\times\vec{F}) \qquad (35.24b)$$

$$\vec{\nabla}\times\vec{\nabla}f = 0 \quad \text{(for similar reasons)} \qquad (35.24c)$$

you should be able to derive the Maxwell equations for the gravitational field.

Exercise 35.2.1. Verify the Maxwell equations (equations 35.10).

BOX 35.3 The Gravitational Lorentz Equation

If you start with the geodesic equation for a stationary source given by equation 35.11 (repeated here for your convenience),

$$\frac{d^2 x^i}{dt^2} \approx \tfrac{1}{2}\eta^{ik}\partial_k h_{tt} + \eta^{ik}(\partial_k h_{tj} - \partial_j h_{tk})V^j \tag{35.11r}$$

you should be able to derive the gravitational version of the Lorentz force equation $d^2\vec{x}/dt^2 = \vec{E}_G + \vec{V} \times 4\vec{B}_G$.

Exercise 35.3.1. Do this. (*Hint:* I think that you will find the vector calculus identity $\vec{A} \times (\vec{B} \times \vec{C}) = (\vec{A} \cdot \vec{C})\vec{B} - (\vec{A} \cdot \vec{B})\vec{C}$ helpful.)

BOX 35.4 The "Gravitomagnetic Moment" of a Spinning Object

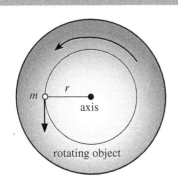

FIG. 35.1 This diagram shows a top view of a rotating object (gray). The white circle represents a small "bit" of the object's mass that is being carried around the the axis (which points toward us) by the object's rotation.

In this box, we want to find for a spinning object the gravitomagnetic moment $\vec{\mu}_G$ that corresponds to the magnetic moment $\mu = iA$ for a simple loop of area A conducting a current i. Consider an axially symmetric object rotating around its axis of symmetry, and specifically consider a small bit of mass m carried by the object's rotation in a circular path with radius r (see figure 35.1) around the object's axis.

Current is defined to be the charge per unit time passing a certain point. By analogy, the quantity that should replace i in the gravitomagnetic moment should be the *mass* per unit time that passes a certain point. For our hypothetical bit of mass, this is m/T, where $T = 2\pi r/v$ is the time required for a complete turn, and v is the particle's speed around the axis. The quantity we should substitute for A is the area enclosed by the mass's circular orbit.

Exercise 35.4.1. Show, then, that $\mu = iA$ becomes $\mu_G = \tfrac{1}{2}mvr$ (for the bit).

BOX 35.4 (continued) The "Gravitomagnetic Moment" of a Spinning Object

Note that mvr is the *magnitude* of the mass bit's angular momentum $\vec{r} \times m\vec{v}$ around the axis. The direction of this angular momentum (as you can check using the cross product right-hand rule) is perpendicular to the plane of the orbit in the direction indicated by your right thumb when your right fingers curl in the direction the mass is orbiting.

Since this analysis is the same for every bit of mass in the spinning object, we can just sum over all such small bits to find the relationship between the object's *total* gravitomagnetic moment and its *total* spin angular momentum:

$$\vec{\mu}_G = \hat{s} \sum \tfrac{1}{2} m_i v_i r_i = \tfrac{1}{2}\vec{s} \qquad (35.25)$$

BOX 35.5 Angular Speed of Gyroscope Precession

According to equation 35.13 (repeated here in somewhat modified form for convenience), a spinning object in a gravitomagnetic field experiences a torque

$$\vec{\tau} = \vec{s} \times 2\vec{B}_G = \frac{d\vec{s}}{dt} \qquad (35.26)$$

where the last step applies (from the definition of torque) if no other torques act on the object. The differential change $d\vec{s}$ in the spinning object's angular momentum is perpendicular to both $\vec{B}_G$ and the object's spin vector $\vec{s}$ itself, so it will not change that spin vector's magnitude, only its direction. Figure 35.2 shows the situation. As $\vec{s}$ changes direction, $d\vec{s}$ remains perpendicular to it and so drags the tip of $\vec{s}$ around in a circle whose plane is perpendicular to $\vec{B}_G$.

Now, if we take the magnitude of both sides of equation 35.26, we get

$$\mathrm{mag}\!\left(\frac{d\vec{s}}{dt}\right) = \mathrm{mag}(\vec{s} \times 2\vec{B}_G) = 2sB_G \sin\theta \qquad (35.27)$$

where θ is the angle between $\vec{s}$ and $\vec{B}_G$. Now note from the diagram that the differential angle $d\phi$ through which the gyroscope's spin precesses in time dt is given by $ds/(s\sin\theta)$. From this, you can show pretty easily that

$$\Omega_{LT} = \frac{d\phi}{dt} = 2B_G \qquad (35.28)$$

independent of θ (!).

Exercise 35.5.1. Verify this result.

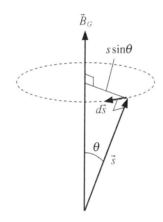

FIG. 35.2a This diagram shows how $d\vec{s}$ is related to $\vec{s}$ and $\vec{B}_G$. Note that no matter where the tip of $\vec{s}$ is along the circle, $d\vec{s}$ points tangent to the circle.

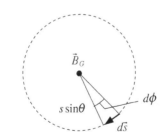

FIG. 35.2b This is a top view of the same situation, showing how $d\phi$ is related to $d\vec{s}$ and the projection of $\vec{s}$ on the plane of the circle.

Note from the diagram that the direction of the precession is clockwise in figure 35.2 (check this using the right-hand rule for the cross product). If we define the direction of $\vec{\Omega}_{LT}$ to point in the direction of your right thumb when your right fingers curl in the precession direction, then we see that $\vec{\Omega}_{LT} = -2\vec{B}_G$, which is equation 35.14.

HOMEWORK PROBLEMS

P35.1 Go through the derivation of equation 35.11 in box 22.6 and continue through the calculation in box 35.3 to determine what new terms come into play in the gravitational version of the Lorentz force equation if we do *not* assume that the fields are static.

P35.2 Imagine a wire along the *x* axis whose linear mass density is λ and which moves at a non-relativistic speed $\vec{v}$ in the +*x* direction. Calculate the magnitude and direction of the gravitomagnetic acceleration experienced by a particle a distance R from the wire that is moving with a velocity $\vec{V}$ that is (a) *parallel* to the wire's motion (b) *opposite* to the wire's motion, and (c) perpendicularly *toward* the wire. [*Hint:* You may use the introductory physics result that the magnitude of the magnetic field created by an infinite wire carrying current I is $B = (\mu_0/2\pi)I/r = 2kI/rc^2$ where $k = \mu_0 c^2/4\pi$ is the Coulomb constant.]

P35.3 What are the units of the gravitoelectric and gravitomagnetic fields in GR units? In SI units, the electric and magnetic fields have units of force/charge and force/(charge·velocity), respectively. What are the corresponding statements for the gravitoelectric and gravitomagnetic fields in GR units? Does it make sense that $\vec{B}_G$ has the same units as an angular velocity (see equation 35.14)?

P35.4 Argue that (ignoring geodetic precession) a gyroscope falling toward either pole of the earth (with its axis perpendicular to the earth's axis) precesses in the same direction as the earth rotates, while a gyroscope in an equatorial orbit precesses opposite to the earth's rotation (like a small gear in contact with a big gear). (Physicists sometimes conceptualize this as being the result of "frame dragging," as if inertial frames near a spinning object are "dragged" with the object's rotation, closer ones more emphatically than farther ones. This image can sometimes lead to incorrect results, as we will see in chapter 37.)

P35.5 Consider a gyroscope in orbit with $r = 1200$ km around a pulsar (neutron star) with a radius of 12 km, whose mass is such that $GM = 1.2$ km and which has a rotational period of 1.0 s. Assume the gyroscope's axis lies in the plane of its orbit. Estimate both the Lense-Thirring and geodetic precession rates for this gyroscope. (Use Newtonian mechanics to estimate the gyroscope's orbital period).

P35.6 Consider an object falling radially inward (from rest at nearly infinity) in a rotating planet's equatorial plane.
  a. Will the object be deflected from its radial path? Explain using an electromagnetic analogy.
  b. For an earth-like planet, estimate the object's gravitomagnetic acceleration a distance r from the planet's center in terms of G, the planet's mass M, its radius R, and its rotational period T.

P35.7 This problem will guide you through the derivation of the geodetic precession for a gyroscope in a circular equatorial orbit of with radial coordinate R around a *non-spinning* star or planet whose exterior spacetime is described by Schwarzschild coordinates. In an inertial frame orbiting with the gyroscope, the gyroscope will *not* precess. If we think of the gyroscope's spin four-vector **s** as a vector field defined at points along the worldline, this amounts to saying that field's absolute gradient is zero along the gyroscope's worldline: $\nabla_\mu s^\alpha = 0$. Multiplying both sides by the gyroscope's four-velocity u^μ and summing over μ yield

$$0 = u^\mu(\nabla_\mu s^\alpha) = u^\mu\left(\frac{\partial s^\alpha}{\partial x^\mu} + \Gamma^\alpha_{\mu\nu} s^\nu\right)$$

$$= \frac{\partial s^\alpha}{\partial x^\mu}\frac{dx^\mu}{d\tau} + \Gamma^\alpha_{\mu\nu} u^\mu s^\nu = \frac{ds^\alpha}{d\tau} + \Gamma^\alpha_{\mu\nu} u^\mu s^\nu \quad (35.29)$$

in arbitrary coordinates (note that this is $ds^\alpha/d\tau = 0$ in a LIF). In its own LIF the gyroscope's spin vector is purely spatial and its four-velocity has no spatial components, so $\mathbf{s} \cdot \mathbf{u} = 0$. In arbitrary coordinates, this becomes

$$0 = \mathbf{s} \cdot \mathbf{u} = g_{\mu\nu} s^\mu u^\nu \quad (35.30)$$

a. Show that in Schwarzschild coordinates

$$\mathbf{s} \cdot \mathbf{u} = 0 \;\Rightarrow\; s^t = R^2\left(1 - \frac{2GM}{R}\right)^{-1}\Omega s^\phi \quad (35.31)$$

where $\Omega = d\phi/dt = (d\phi/d\tau)(d\tau/dt) = u^\phi/u^t$ is the gyroscope's constant *orbital* angular speed.

b. Use the Christoffel symbol section of the Diagonal Metric Worksheet (or a direct calculation) to show that in the equatorial plane,

$$\Gamma^r_{\phi\phi} = -r\left(1 - \frac{2GM}{r}\right), \quad \Gamma^\phi_{r\phi} = \Gamma^\phi_{\phi r} = \frac{1}{r}$$

$$\Gamma^r_{tt} = \frac{GM}{r^2}\left(1 - \frac{2GM}{r}\right) \quad (35.32)$$

Also determine which other Christoffel symbols are nonzero (though you don't have to calculate them).

c. If $s^\theta = 0$ initially (i.e., **s** is lying in the equatorial plane), then by symmetry, s^θ will remain zero, since there is no difference between above the equatorial plane and below it. Assuming this, and using $u^t = dt/d\tau$ to convert τ-derivatives to t-derivatives, use the results above to show that

$$\frac{ds^r}{dt} = (R - 3GM)\Omega s^\phi, \quad \frac{ds^\phi}{dt} = -\frac{\Omega}{R} s^r \quad (35.33)$$

d. Eliminate s^r from these two equations to arrive at

$$\frac{d^2 s^\phi}{dt^2} + \left(1 - \frac{3GM}{R}\right)\Omega^2 s^\phi = 0 \quad (35.34)$$

Argue, therefore, that the value of s^ϕ sinusoidally oscillates with a frequency of $\Omega' = (1 - 3GM/R)^{1/2}\Omega$.

e. If **s** were to maintain a truly fixed orientation in space, the value of s^ϕ would nonetheless oscillate with frequency Ω. Explain why. The geodetic precession frequency, therefore, should be $\Omega - \Omega'$. From this, verify equation 35.21 if $GM/R \ll 1$.

36. THE KERR METRIC

INTRODUCTION

FLAT SPACETIME
- Review of Special Relativity
- Four-Vectors
- Index Notation

TENSORS
- Arbitrary Coordinates
- Tensor Equations
- Maxwell's Equations
- Geodesics

SCHWARZSCHILD BLACK HOLES
- The Schwarzschild Metric
- Particle Orbits
- Precession of the Perihelion
- Photon Orbits
- Deflection of Light
- Event Horizon
- Alternative Coordinates
- Black Hole Thermodynamics

THE CALCULUS OF CURVATURE
- The Absolute Gradient
- Geodesic Deviation
- The Riemann Tensor

THE EINSTEIN EQUATION
- The Stress-Energy Tensor
- The Einstein Equation
- Interpreting the Equation
- The Schwarzschild Solution

COSMOLOGY
- The Universe Observed
- A Metric for the Cosmos
- Evolution of the Universe
- Cosmic Implications
- The Early Universe
- CMB Fluctuations & Inflation

GRAVITATIONAL WAVES
- Gauge Freedom
- Detecting Gravitational Waves
- Gravitational Wave Energy
- Generating Gravitational Waves
- Gravitational Wave Astronomy

SPINNING BLACK HOLES
- Gravitomagnetism
- **The Kerr Metric**
- Kerr Particle Orbits
- Ergoregion and Horizon
- Negative-Energy Orbits

this depends on this

36. THE KERR METRIC

Overview. In this chapter, we will look first at the weak-field solution for a spherically symmetric rotating object to expose the core characteristics of the metric of spacetime surrounding such an object. This will provide some motivation for the second part of the chapter, where I will present an exact vacuum solution, called the Kerr metric, for the spacetime surrounding a rotating black hole.

Weak-Field Solution for a Rotating Sphere. Consider a slowly rotating spherically symmetric object. If the stress-energy tensor is also time-independent inside the object (indicating that the object's mass and fluid velocity distributions are unchanging in spite of its rotation), we say that the object is *stationary*. In such a case, the metric perturbations will be time-independent and we can also ignore the retardation due to light travel time. In such circumstances, equations 35.6 from the last chapter imply that the metric perturbations will be simply

$$h^{tt}(\vec{R}) = h^{xx}(\vec{R}) = h^{yy}(\vec{R}) = h^{zz}(\vec{R}) = 2G \int_{\text{src}} \frac{\rho_0(r)}{|\vec{R} - \vec{r}|} dV \qquad (36.1a)$$

$$h^{ti}(\vec{R}) = h^{it}(\vec{R}) = 4G \int_{\text{src}} \frac{\rho_0(r) v^i(\vec{r})}{|\vec{R} - \vec{r}|} dV, \text{ other } h^{\mu\nu} \approx 0 \qquad (36.1b)$$

where $\vec{R}$ is the position where the field is being evaluated and $\vec{r}$ represents the varying position of the differential volume element in the integration.

The integral in equation 36.1a is no different than for a non-rotating spherical object, and no different (except for the constants) from the integral for the electrostatic potential created by a spherical charge distribution. Therefore we can conclude without further effort that the result of this integral must be

$$h^{tt}(\vec{R}) = h^{xx}(\vec{R}) = h^{yy}(\vec{R}) = h^{zz}(\vec{R}) = \frac{2GM}{R} \qquad (36.2)$$

where M is the object's total mass.

The integral in equation 36.1b is more difficult. To simplify things, let's assume that our object rotates rigidly with angular velocity $\vec{\Omega}$: then $\vec{v}(\vec{r}) = \vec{\Omega} \times \vec{r}$ (check this with your right hand). Moreover, for the sake of concreteness, let's choose our coordinates so that the object's center of mass is at rest and orient our coordinate system so that $\vec{\Omega}$ points along the +z axis. In such a case, we have

$$v^x = -\Omega y, \; v^y = \Omega x, \; v^z = 0 \qquad (36.3)$$

Finally, for future reference, note that the object's total spin angular momentum $\vec{S}$ is given by the integral

$$\vec{S} = I\vec{\Omega} = \vec{\Omega} \int_{\text{src}} \rho_0 r_\perp^2 \, dV = \vec{\Omega} \int_{\text{src}} \rho_0 (x^2 + y^2) \, dV \qquad (36.4)$$

where I is the object's moment of inertia and $r_\perp$ is the distance from the axis to the volume element.

The most straightforward way to work out the integral is to expand $|\vec{R} - \vec{r}|^{-1}$ as a power series in r/R. The expansion turns out to be

$$\frac{1}{|\vec{R} - \vec{r}|} = \frac{1}{R} + \frac{\vec{R} \cdot \vec{r}}{R^3} + \frac{3(\vec{R} \cdot \vec{r})^2 - R^2 r^2}{2R^5} + \cdots \qquad (36.5)$$

You can show that at least the first two terms are correct using the binomial approximation (see box 36.1). Then the integral for h^{tx}, for example, becomes

$$h^{tx}(\vec{R}) = -\frac{4G\Omega}{R} \int_{\text{src}} \rho_0(r) y \, dV \;-\; \frac{4G\Omega}{R^3} \int_{\text{src}} \rho_0(r) [\vec{R} \cdot \vec{r}] y \, dV + \cdots \qquad (36.6)$$

Now, the *first* of these integrals is zero, because the density is the same for a volume element at any given x, y, z as it is at $x, -y, z$, so the contributions from the two volume elements will cancel in the sum. Since all volume elements in the source can be paired with its opposite this way, the terms in the integral's sum all cancel in pairs. (Alternatively, you can note that the first integral gives you the y-position of the object's center of mass, which we have defined to be the origin.) You can show (see box 36.2) that the second of these integrals evaluates to $-2GSY/R^3$, where $\vec{R} = [X, Y, Z]$. It turns out that the higher-order terms in the expansion all integrate to zero (see box 36.3). The integral for h^{ty} is completely analogous, so we have

$$h^{tx}(\vec{R}) = h^{xt}(\vec{R}) = -\frac{2GS}{R^3}Y \quad \text{and} \quad h^{ty}(\vec{R}) = h^{yt}(\vec{R}) = +\frac{2GS}{R^3}X \qquad (36.7)$$

Since $g_{\mu\nu} = \eta_{\mu\nu} + h_{\mu\nu}$ and since $h_{tx} = \eta_{t\alpha}\eta_{x\beta}h^{\alpha\beta} = \eta_{tt}\eta_{xx}h^{tx} = -h^{tx}$ (and similarly $h_{ty} = -h^{ty}$) the metric of the spacetime surrounding a slowly rotating object is

$$ds^2 = -\left(1 - \frac{2GM}{R}\right)dt^2 + \left(1 + \frac{2GM}{R}\right)(dX^2 + dY^2 + dZ^2)$$
$$- \frac{4GS}{R^3}(X\,dY - Y\,dX)\,dt \qquad (36.8)$$

If we re-express this in terms of polar coordinates R, θ, ϕ such that

$$X = R\sin\theta\cos\phi, \quad Y = R\sin\theta\sin\phi, \quad Z = R\cos\theta \qquad (36.9)$$

then (as discussed in box 36.4) this metric becomes

$$ds^2 = -\left(1 - \frac{2GM}{R}\right)dt^2 + \left(1 + \frac{2GM}{R}\right)(dR^2 + R^2 d\theta^2 + R^2\sin^2\theta\,d\phi^2)$$
$$- \frac{4GMa}{R}\sin^2\theta\,d\phi\,dt \qquad (36.10)$$

where $a \equiv S/M$ is the object's *angular momentum per unit mass*. This is essentially the Schwarzschild metric to first order in $2GM/R$ plus an off-diagonal term that depends on the object's spin angular momentum.

Introduction to the Kerr Solution. The metric given in equation 36.10 is useful for many calculations regarding objects near to rotating planets and ordinary stars, but it does not apply to objects with strong fields (such as black holes or neutron stars). To explore the behavior of objects near black holes or neutron stars, we need an exact vacuum solution to the Einstein equation that is more general than the Schwarzschild solution.

What should such a solution look like? First of all, we will narrow our consideration to stationary compact objects that are completely surrounded by empty space. We will then look for a vacuum solution with the following characteristics:

1. The metric should reduce to that for flat space spherical coordinates as $r \to \infty$.
2. The metric should be time-independent.
3. The metric should be axially symmetric. If we take the axis of symmetry to be perpendicular to the $\theta = \pi/2$ plane and through $r = 0$, then axial symmetry implies that the metric will be independent of ϕ.

We can begin by looking for solutions of the form

$$ds^2 = g_{tt}dt^2 + g_{rr}dr^2 + g_{\theta\theta}d\theta^2 + g_{\phi\phi}d\phi^2 + 2g_{t\phi}dt\,d\phi \qquad (36.11)$$

where $g_{tt}, g_{rr}, g_{\theta\theta}, g_{\phi\phi}$, and $g_{t\phi}$ are functions of r and θ alone. This trial metric differs from the trial metric that we used in the Schwarzschild case in two important ways: (1) we do not presume to identify $g_{\theta\theta} = r^2$ and $g_{\phi\phi} = r^2\sin^2\theta$, because we cannot assume that constant r slices will actually be spherically symmetric, and (2) we have allowed for the possibility of an off-diagonal $g_{t\phi} = g_{\phi t}$ term in the metric. This is partly because we already know from equation 36.10 that such a term appears in the weak-field limit. Even ignoring this, such a term cannot be excluded *a priori*, since it is possible in this axisymmetric case that the spacetime interval between two events might depend on whether the $d\phi$ between events goes with the object's rotation (which would be $d\phi > 0$ if we take the object's rotation to be in the $+\phi$ direction) or opposite to its rotation ($d\phi < 0$). On the other hand, we would expect the metric to be independent of a change of sign in both dt and $d\phi$ since if we run a movie of the object backward, we would see it rotating in the opposite sense, so $d\phi < 0$ would now correspond to rotating "with" the star if $dt < 0$ also. So having a $d\phi\, dt$ term in the metric makes perfect sense. On the other hand, other off-diagonal terms don't make the same kind of sense (for example, if there were a $dr\,dt$ term, why would it make sense for a step outward to correspond to a different distance than a step inward, or for a step outward and forward in time to be equivalent to a step inward and backward in time?). So we can at least attempt a solution using the form given in equation 36.10: we can always add in more terms if we have to.

The Kerr Solution. However, the task of solving for the unknown metric coefficients is much more difficult in this case than in either the Schwarzschild or cosmological cases, where the spherical symmetry required solving for only two unknown metric functions, not five. Moreover, because the trial metric is not diagonal, we cannot use the Diagonal Metric Worksheet. Instead, we have to go back to the basic definitions of the Ricci tensor and solve the (at least) five coupled nonlinear differential equations that result.

I am not going to make you do this. Suffice it to say that (after considerable effort) one can show that a solution satisfying our criteria is

$$ds^2 = -\left(1 - \frac{2GMr}{r^2 + a^2\cos^2\theta}\right)dt^2 + \left(\frac{r^2 + a^2\cos^2\theta}{r^2 - 2GMr + a^2}\right)dr^2 + (r^2 + a^2\cos^2\theta)\,d\theta^2$$

$$+ \left(r^2 + a^2 + \frac{2GMra^2\sin^2\theta}{r^2 + a^2\cos^2\theta}\right)\sin^2\theta\,d\phi^2 - \left(\frac{4GMra\sin^2\theta}{r^2 + a^2\cos^2\theta}\right)d\phi\,dt \qquad (36.12)$$

This is the **Kerr geometry** described in **Boyer-Lindquist coordinates**. For simplicity in what follows, I will refer to this as the **Kerr metric**, and to $t, r, \theta,$ and ϕ in this metric as the **Kerr coordinate system**, even though technically Roy Kerr himself used different coordinates when he found this geometry in 1963.

The parameters GM and a are constants of integration (both with units of distance) that can be measured operationally by observing the motion of particles in such a spacetime. Note the following:

1. If $a = 0$, the solution reduces to the Schwarzschild solution (check this!).
2. We can therefore interpret M as corresponding to the object's mass, as we did in the Schwarzschild case.
3. The parameter a, on the other hand, must have something to do with the object's rotation, since that is what is different here.
4. In fact, in the limit that $r \gg a$, you can see that to first order in $2GM/r$, the metric above is the same as the weak-field metric given by equation 36.10 (see box 36.5 for some of the details). Therefore, we can interpret a as being equal to the object's angular momentum per unit mass: $a = S/M$.

The Importance of the Kerr Solution. The Kerr solution is very important for astrophysics for the following reasons.

1. This is an exact solution for the empty-space Einstein equation and is therefore valid for the empty space outside an arbitrarily massive uncharged black hole.
2. In particular, this is the *only* possible exterior geometry for a black hole formed by gravitational collapse of an uncharged mass distribution. It has been shown (Price, *Physical Review* D, **5**, 1972, 2419ff) that as any non-spherical mass distribution collapses to a black hole, the non-spherical complexities in the gravitational field will radiate away in the form of gravitational waves, leaving a gravitational field with a Kerr geometry behind. (If the angular momentum of the mass distribution is exactly zero, then the resulting field has the Schwarzschild geometry instead. But as we have seen, the Schwarzschild geometry is just a special case of the Kerr geometry with $a = 0$.)

3. In fact, astrophysical objects that might collapse to black holes will almost inevitably have nonzero angular momenta, indeed enough so that black holes formed by almost any conceivable process will have $a \approx GM$ (Bardeen, *Nature*, **226**, 1970, 64–65), though recent evidence adds some doubt to this.

4. Physicists also strongly believe that a black hole will not form with $a \geq GM$: the high angular momentum of the collapsing object will lead it to spin off some material. This means that astrophysical black holes formed by collapse will almost certainly be Kerr black holes with $a < GM$.

5. Astrophysicists are now nearly certain that there are massive black holes at the centers of all galaxies, and that accretion disks around these black holes in certain galaxies supply the energy source for quasars.

Therefore, we have good astrophysical reasons to study the Kerr geometry as thoroughly as we can!

BOX 36.1 Expanding $|\vec{R} - \vec{r}|^{-1}$ to First Order in r/R

Note that

$$\frac{1}{|\vec{R} - \vec{r}|} = \frac{1}{\sqrt{(\vec{R} - \vec{r})^2}} = \frac{1}{\sqrt{(\vec{R} - \vec{r}) \cdot (\vec{R} - \vec{r})}} = \frac{1}{\sqrt{R^2 - 2\vec{R} \cdot \vec{r} + r^2}} \quad (36.13)$$

The expansion in equation 36.5 is an expansion of this quantity in powers of r, assuming that $r < R$. Using the binomial approximation, you can argue that to first order in the ratio r/R, the expression above becomes

$$\frac{1}{|\vec{R} - \vec{r}|} \approx \frac{1}{R} + \frac{\vec{R} \cdot \vec{r}}{R^3} \quad (36.14)$$

Exercise 36.1.1. Verify equation 36.14.

BOX 36.2 The Integral for h^{tx}

If we write out the integral in question here in terms of the coordinates of $\vec{R} = [X, Y, Z]$ and $\vec{r} = [x, y, z]$, we find that

$$-\frac{4G\Omega}{R^3} \int_{src} \rho_0(r)(\vec{R} \cdot \vec{r}) y \, dV = -\frac{4G\Omega}{R^3} \int_{src} \rho_0(r)(Xx + Yy + Zz) y \, dV$$

$$= -\frac{4G\Omega X}{R^3} \int_{src} \rho_0(r) xy \, dV - \frac{4G\Omega Y}{R^3} \int_{src} \rho_0(r) y^2 \, dV - \frac{4G\Omega Z}{R^3} \int_{src} \rho_0(r) zy \, dV \quad (36.15)$$

Exercise 36.2.1. Use logic similar to that used to show that the first term in equation 36.6 is zero to argue that the first and third terms above must be zero. Explain why this logic does *not* apply to the second term.

So we see that

$$h^{tx}(\vec{R}) = -\frac{4G\Omega Y}{R^3} \int_{src} \rho_0(r) y^2 \, dV \quad (36.16)$$

But we also know from equation 36.4 that

$$S = \Omega \int_{src} \rho_0(r)[x^2 + y^2] dV \quad (36.17)$$

Exercise 36.2.2. Explain why this means that $h^{tx} = -2GSY/R^3$.

BOX 36.3 Why the Other Terms in the Expansion Integrate to Zero

According to equation 36.5, the third term in the expansion for $h^{tx}(\vec{R})$ started in equation 36.6 would be

$$-\frac{6G\Omega}{R^5}\int_{\text{src}} \rho_0(r)\left[(\vec{R}\cdot\vec{r})^2 - \tfrac{1}{3}R^2 r^2\right] y\, dV =$$

$$-\frac{6G\Omega}{R^5}\int_{\text{src}} \rho_0(r)\left[(Xx + Yy + Zz)^2 - \tfrac{1}{3}R^2 r^2\right] y\, dV =$$

$$-\frac{6G\Omega}{R^5}\int_{\text{src}} \rho_0(r)\left[X^2 x^2 + Y^2 y^2 + Z^2 z^2 + 2XYxy + 2XZxz + 2YZyz - \tfrac{1}{3}R^2 r^2\right] y\, dV =$$

$$-\frac{6G\Omega}{R^5}\int_{\text{src}} \rho_0(r)\left[X^2 x^2 y + Y^2 y^3 + Z^2 z^2 y + 2XYxy^2 + 2XZxzy + 2YZy^2 z - \tfrac{1}{3}R^2 r^2 y\right] dV \tag{36.18}$$

Exercise 36.3.1. Using the same kind of logic used in the last box, verify that each of the terms in the integral on the last line integrates to zero. (The same is true for all the higher-order terms in the expansion for $|\vec{R} - \vec{r}|^{-1}$).

BOX 36.4 Transforming the Weak-Field Solution to Polar Coordinates

We already know that in polar coordinates such that
$$X = R\sin\theta\cos\phi, \quad Y = R\sin\theta\sin\phi, \quad Z = R\cos\theta \qquad (36.19)$$
the quantity $dX^2 + dY^2 + dZ^2 = dR^2 + R^2 d\theta^2 + R^2 \sin^2\theta\, d\phi^2$. You can also easily show that
$$X dY - Y dX = R^2 \sin^2\theta\, d\phi \qquad (36.20)$$

Exercise 36.4.1. Verify equation 36.20.

This means that the metric given in equation 36.8 (repeated here),
$$ds^2 = -\left(1 - \frac{2GM}{R}\right)dt^2 + \left(1 + \frac{2GM}{R}\right)(dX^2 + dY^2 + dZ^2)$$
$$-\frac{4GS}{R^3}(X\, dY - Y\, dX)\, dt \qquad (36.8r)$$

becomes the metric given in equation 36.10 (repeated here),
$$ds^2 = -\left(1 - \frac{2GM}{R}\right)dt^2 + \left(1 + \frac{2GM}{R}\right)(dR^2 + R^2 d\theta^2 + R^2 \sin^2\theta\, d\phi^2)$$
$$-\frac{4GMa}{R}\sin^2\theta\, d\phi\, dt \qquad (36.10r)$$

with $a \equiv S/M$.

BOX 36.5 The Weak-Field Limit of the Kerr Metric

If we assume that $a \sim GM \ll r$, then you can easily show that to first order in a/r and GM/r, the Kerr metric given in equation 36.12 (repeated below),

$$ds^2 = -\left(1 - \frac{2GMr}{r^2 + a^2\cos^2\theta}\right)dt^2 + \left(\frac{r^2 + a^2\cos^2\theta}{r^2 - 2GMr + a^2}\right)dr^2 + (r^2 + a^2\cos^2\theta)d\theta^2$$
$$+ \left(r^2 + a^2 + \frac{2GMra^2\sin^2\theta}{r^2 + a^2\cos^2\theta}\right)\sin^2\theta\, d\phi^2 - \left(\frac{4GMra\sin^2\theta}{r^2 + a^2\cos^2\theta}\right)d\phi\, dt \quad (36.12r)$$

becomes

$$ds^2 = -\left(1 - \frac{2GM}{r}\right)dt^2 + \left(1 + \frac{2GM}{r}\right)dr^2 + r^2 d\theta^2 + r^2 \sin^2\theta\, d\phi^2$$
$$- \frac{4GMa}{r}\sin^2\theta\, d\phi\, dt \quad (36.21)$$

Exercise 36.5.1. Verify equation 36.21. (*Hint:* Divide top and bottom of most of the metric terms by r^2 and then throw away any terms involving a^2/r^2. Then use the binomial approximation on the g_{rr} term.)

This is *almost* the same as the weak-field solution given in equation 36.10 (if we replace r with R) except that in the latter the $(1 + 2GM/R)$ factor multiplies *all* of the spatial terms, not just the first. However, if you look back at problem P22.1, you will find that if we actually define $r \equiv (1 + 2GM/R)^{1/2}R$ instead of simply equating R with r, then you get the result given above to first order in GM/r. Therefore, we see that the Kerr metric does reduce to the weak-field limit if $r \gg a \sim GM$. (It is not obvious that this should be so, because the weak-field limit was derived assuming a non-relativistic source, but here we see that this limit applies to the far field of black holes or neutron stars.)

36. THE KERR METRIC

HOMEWORK PROBLEMS

P36.1 According to equation 36.10, the spatial part of the weak-field metric for a rotating object is

$$ds^2 = \left(1 + \frac{2GM}{R}\right)(dR^2 + R^2 d\theta^2 + R^2 \sin^2\theta \, d\phi^2) \quad (36.22)$$

a. If we define $r \equiv (1 + 2GM/R)^{1/2} R$, then r will be a true circumferential coordinate. Show that to first order in $2GM/R$, $dr = dR$.

b. In terms of r, then, the metric becomes

$$ds^2 = \left(1 + \frac{2GM}{R}\right) dr^2 + r^2 d\theta^2 + r^2 \sin^2\theta \, d\phi^2 \quad (36.23)$$

Argue that this is the same as

$$ds^2 = \left(1 + \frac{2GM}{r}\right) dr^2 + r^2 d\theta^2 + r^2 \sin^2\theta \, d\phi^2 \quad (36.24)$$

to the order of approximation we are considering.

P36.2 Estimate the value of a/GM for the earth and the sun. Note that the sun rotates about once a month, though it does not actually rotate rigidly. Its radius is about 700,000 km. (*Hints:* $I \approx \frac{2}{5} MR^2$ for a uniform sphere. A Newtonian calculation is acceptable here.)

P36.3 Estimate the value of a/GM for a neutron star with mass $1.4 \, M_\odot$ and a radius of 12 km that is rotating about 1000 times a second (the fastest known pulsars rotate about this quickly). Also, what is the speed of a point on the star's equator? (*Hints:* $I \approx \frac{2}{5} MR^2$ for a uniform sphere. A Newtonian calculation is acceptable for an estimation.)

P36.4 Let's see what the Kerr metric tells us about physical meaning of its coordinates.

a. Does the Kerr metric coordinate t correspond to the time measured at infinity (as in the Schwarzschild case)? Defend your response.

b. Does the Kerr metric coordinate r correspond to radial distance measured from $r = 0$, the circumference of a circle divided by 2π (like the Schwarzschild r coordinate), or neither? Defend your response.

c. What is the circumference of a circle of constant r in the equatorial plane divided by 2π?

d. Consider a purely radial line (along which t, θ, and ϕ are constant), a line purely in the θ direction (along which t, r, and ϕ are constant), and a line purely in the ϕ direction (along which t, r, and θ are constant). Are such lines always orthogonal? Defend your responses.

P36.5 Calculate the physical distance between $r = 3GM$ and $r = 4GM$ in the equatorial plane for an extreme Kerr black hole ($a = GM$).

P36.6 a. Show that in the limit that $M \to 0$, the Kerr metric reduces to

$$ds^2 = -dt^2 + \frac{\rho^2 dr^2}{r^2 + a^2} + \rho^2 d\theta^2 + (r^2 + a^2)\sin^2\theta \, d\phi^2 \quad (36.25)$$

where $\rho^2 \equiv r^2 + a^2 \cos^2\theta$.

b. The spacetime geometry around a zero-mass object *should* be flat spacetime. Show that if you calculate the differentials of

$$x = \sqrt{r^2 + a^2} \sin\theta \cos\phi \quad (36.26a)$$
$$y = \sqrt{r^2 + a^2} \sin\theta \sin\phi \quad (36.26b)$$
$$z = r \cos\theta \quad (36.26c)$$

and substitute them into the flat-space metric, you get the metric given in equation 36.25. Therefore, that metric *is* simply flat spacetime in funny coordinates.

c. Since this metric is axially symmetric, we can explore this coordinate system in the xz plane without loss of generality. Show that r = constant curves in the xz plane are ellipses (i.e., show that these curves satisfy the equation $x^2/b^2 + z^2/c^2 = 1$ for some constants b and c).

d. Show that θ = constant curves are hyperbolas (i.e., show that these curves satisfy $x^2/b^2 - z^2/c^2 = 1$ for some constants b and c).

e. Sketch (or make a computer to draw) curves of constant r and θ in the xz plane for $a = GM$.

P36.7 Show that one can write the Kerr metric in the form

$$ds^2 = -\frac{\rho^2 R^2}{\Sigma^2} dt^2 + \frac{\rho^2}{R^2} dr^2 + \rho^2 d\theta^2 + \frac{\Sigma^2 \sin^2\theta}{\rho^2}(d\phi - \omega \, dt)^2 \quad (36.27)$$

where $\rho^2 \equiv r^2 + a^2 \cos^2\theta$, $R^2 \equiv r^2 - 2GMr + a^2$, $\Sigma^2 \equiv (r^2 + a^2)^2 - a^2 R^2 \sin^2\theta$, and $\omega \equiv 2GMra/\Sigma^2$. This form is interesting because it is more evocative of an object rotating with angular speed ω. [*Hint:* First show that

$$\Sigma^2(\rho^2 - 2GMr) = \rho^4 R^2 - (2GMra \sin\theta)^2 \quad (36.28)$$

by expanding both sides and comparing terms.]

P36.8 The Kerr-Newman metric for a spinning black hole with electrical charge Q is as given in equation 36.27 except that $R^2 = r^2 - 2GMr + a^2 + GkQ^2$. This clearly reduces to the Kerr metric when $Q = 0$.

a. Show that this metric reduces to the Reisner-Nordström metric for a charged, non-rotating black hole (see problem P23.4) if $a = 0$.

b. Show that this metric reduces to the Schwarzschild metric if $a = 0$ and $Q = 0$.

Note: Problem P36.6 implies that this metric also reduces to flat space if $M = 0$, $a = 0$, and $Q = 0$. Physicists strongly believe that there are *no* other solutions to the empty-space Einstein equations for compact objects (even though this is not absolutely certain): every black hole can be characterized by the three numbers M, a, and Q alone. In particular, this means that no matter what originally formed the black hole, it cannot have any other kind of external field other than the gravitational and electromagnetic fields characterized by these three numbers: in John Wheeler's memorable phrasing, "a black hole has no hair."

37. PARTICLE ORBITS IN KERR SPACETIME

INTRODUCTION

FLAT SPACETIME
- Review of Special Relativity
- Four-Vectors
- Index Notation

TENSORS
- Arbitrary Coordinates
- Tensor Equations
- Maxwell's Equations
- Geodesics

SCHWARZSCHILD BLACK HOLES
- The Schwarzschild Metric
- Particle Orbits
- Precession of the Perihelion
- Photon Orbits
- Deflection of Light
- Event Horizon
- Alternative Coordinates
- Black Hole Thermodynamics

THE CALCULUS OF CURVATURE
- The Absolute Gradient
- Geodesic Deviation
- The Riemann Tensor

THE EINSTEIN EQUATION
- The Stress-Energy Tensor
- The Einstein Equation
- Interpreting the Equation
- The Schwarzschild Solution

COSMOLOGY
- The Universe Observed
- A Metric for the Cosmos
- Evolution of the Universe
- Cosmic Implications
- The Early Universe
- CMB Fluctuations & Inflation

GRAVITATIONAL WAVES
- Gauge Freedom
- Detecting Gravitational Waves
- Gravitational Wave Energy
- Generating Gravitational Waves
- Gravitational Wave Astronomy

SPINNING BLACK HOLES
- Gravitomagnetism
- The Kerr Metric
- **Kerr Particle Orbits**
- Ergoregion and Horizon
- Negative-Energy Orbits

this ➡ depends on ➡ this

427

37. PARTICLE ORBITS IN KERR SPACETIME

Overview. In the last chapter, I introduced the full Kerr metric for the spacetime surrounding a black hole or rotating object. This chapter will explore geodesics for massive particles orbiting in such an object's equatorial plane.

Conserved Quantities. The geodesic equations of motion are

$$0 = \frac{d}{d\tau}\left(g_{\mu\nu}\frac{dx^\nu}{d\tau}\right) - \tfrac{1}{2}\partial_\mu g_{\alpha\beta}\frac{dx^\alpha}{d\tau}\frac{dx^\beta}{d\tau} \tag{37.1}$$

Because the metric components are independent of t and ϕ, the geodesic equation implies that two quantities are conserved for geodesic motion, just as in the Schwarzschild case. We can find these conserved quantities e and ℓ by setting the μ index in equation 37.1 to t and ϕ, respectively:

If $\mu = t$: $\quad 0 = \dfrac{d}{d\tau}\left(g_{tt}\dfrac{dt}{d\tau} + g_{t\phi}\dfrac{d\phi}{d\tau}\right) + 0 \;\Rightarrow\; e \equiv -g_{tt}\dfrac{dt}{d\tau} - g_{t\phi}\dfrac{d\phi}{d\tau}$ \hfill (37.2a)

If $\mu = \phi$: $\quad 0 = \dfrac{d}{d\tau}\left(g_{\phi t}\dfrac{dt}{d\tau} + g_{\phi\phi}\dfrac{d\phi}{d\tau}\right) + 0 \;\Rightarrow\; \ell \equiv g_{\phi t}\dfrac{dt}{d\tau} + g_{\phi\phi}\dfrac{d\phi}{d\tau}$ \hfill (37.2b)

As $r \to \infty$, $g_{t\phi}$ becomes small compared to g_{tt} (which approaches -1) and $g_{\phi\phi}$ (which approaches $r^2\sin^2\theta$), so physically, $e = (dt/d\tau)_\infty$ = relativistic energy per unit mass at infinity and $\ell = \left(r^2\sin^2\theta[d\phi/d\tau]\right)_\infty$ = z-component of angular momentum per unit mass at infinity.

You can solve equations 37.2 for $dt/d\tau$ and $d\phi/d\tau$: the results are

$$\frac{dt}{d\tau} = \frac{g_{\phi\phi}e + g_{t\phi}\ell}{[g_{t\phi}]^2 - g_{\phi\phi}g_{tt}}, \quad \frac{d\phi}{d\tau} = \frac{-g_{\phi t}e - g_{tt}\ell}{[g_{t\phi}]^2 - g_{\phi\phi}g_{tt}} \tag{37.3}$$

(see box 37.1). The quantity $[g_{t\phi}]^2 - g_{\phi\phi}g_{tt}$ comes up often in Kerr-metric calculations: with a bit of work (see box 37.2), you can show that

$$[g_{t\phi}]^2 - g_{tt}g_{\phi\phi} = [r^2 + a^2 - 2GMr]\sin^2\theta \equiv R^2\sin^2\theta \tag{37.4}$$

(Note that this is a new definition for R unrelated to its use in the last chapter.)

The Metric Components in the Equatorial Plane. The motion of objects in Kerr spacetime is extremely complicated. We can greatly simplify things by considering only orbits in the equatorial plane. The equations below display the metric components in terms of $\rho^2 \equiv r^2 + a^2\cos^2\theta$ and $R^2 \equiv r^2 + a^2 - 2GMr$, in general and on the equatorial plane (where $\theta = \pi/2$):

Everywhere: \hfill In the equatorial plane:

$$g_{tt} = -\left(1 - \frac{2GMr}{\rho^2}\right) \qquad g_{tt} = -\left(1 - \frac{2GM}{r}\right) \tag{37.5a}$$

$$g_{t\phi} = g_{\phi t} = -\frac{2GMra\sin^2\theta}{\rho^2} \qquad g_{t\phi} = g_{\phi t} = -\frac{2GMa}{r} \tag{37.5b}$$

$$g_{rr} = \frac{\rho^2}{R^2} \qquad g_{rr} = \frac{r^2}{R^2} \tag{37.5c}$$

$$g_{\theta\theta} = \rho^2 \qquad g_{\theta\theta} = r^2 \tag{37.5d}$$

$$g_{\phi\phi} = \left(r^2 + a^2 + \frac{2GMra^2\sin^2\theta}{\rho^2}\right)\sin^2\theta \qquad g_{\phi\phi} = r^2 + a^2 + \frac{2GMa^2}{r} \tag{37.5e}$$

It is reasonable to focus on the equatorial plane, because one can show that $\theta =$ constant is a solution to the geodesic equation for any particle whose initial conditions include $\theta = \pi/2$, and $d\theta/d\tau = 0$, so a particle initially orbiting in the equatorial plane will continue to orbit in that plane. Indeed, we can arrive at the same result more easily using a simple symmetry argument: since the equatorial plane is a plane of symmetry for the Kerr geometry, a particle originally moving in that plane cannot move off it (which side, top or bottom, would it choose, as both are equivalent?). However, note that unlike the Schwarzschild case, studying motion in the equatorial plane does *not* give a complete picture of orbital motion in this geometry: the Kerr geometry is *not* spherically symmetric, so the Kerr equatorial plane is qualitatively different than other planes through the origin.

Equatorial Equations of Motion. For equatorial orbits in Kerr spacetime, we can derive "conservation of energy-like" and "force-like" equations for the particle's radial motion. For a particle with nonzero rest mass in an equatorial orbit, the fact that $u \cdot u = -1$ implies that

$$-1 = g_{tt}\left(\frac{dt}{d\tau}\right)^2 + g_{rr}\left(\frac{dr}{d\tau}\right)^2 + g_{\phi\phi}\left(\frac{d\phi}{d\tau}\right)^2 + 2g_{t\phi}\frac{dt}{d\tau}\frac{d\phi}{d\tau} \quad (37.6)$$

If you substitute in the results of equation 37.5 and do some algebra (see box 37.3), you should find that

$$\tilde{E} \equiv \tfrac{1}{2}(e^2 - 1) = \tfrac{1}{2}\left(\frac{dr}{d\tau}\right)^2 - \frac{GM}{r} + \frac{\ell^2 + a^2(1-e^2)}{2r^2} - \frac{GM(\ell - ea)^2}{r^3} \quad (37.7)$$

Both this and the Schwarzschild "conservation-of-energy-like" equation

$$\tilde{E} = \tfrac{1}{2}\left(\frac{dr}{d\tau}\right)^2 - \frac{GM}{r} + \frac{\ell^2}{2r^2} - \frac{GM\ell^2}{r^3} \quad (37.8)$$

(to which the Kerr equation reduces if $a = 0$) can be interpreted as conservation-of-energy equations of the form $\tilde{E} = K + V_{\text{eff}}(r)$, where the effective potential energy function $V_{\text{eff}}(r)$ corresponds to the last three terms in each equation. In the Kerr case, though, $V_{\text{eff}}(r)$ depends on $e^2 = 2\tilde{E} - 1$ and thus on $\tilde{E}$, meaning that we cannot draw a single r-dependent potential energy curve that applies for all energies. This makes equation 37.7 less useful for determining the radial motion of an orbiting particle in Kerr spacetime than equation 37.8 is for Schwarzschild spacetime. However, note that in both cases $V_{\text{eff}}(r)$ has the form $-A/r + B/r^2 - C/r^3$, so qualitatively, the graphs and types of possible motions will be similar.

You can find a "force-like" equation by taking the τ derivative of equation 37.7 and dividing through by $dr/d\tau$. The result is

$$\frac{d^2 r}{d\tau^2} = -\frac{GM}{r^2} + \frac{\ell^2 + a^2(1-e^2)}{r^3} - \frac{3GM(\ell - ae)^2}{r^4} \quad (37.9)$$

which reduces to the analogous Schwarzschild "force-like" equation if $a = 0$. Equation 37.3 in turn yields a differential equation for $\phi(\tau)$: in the equatorial plane, this equation evaluates to

$$\frac{d\phi}{d\tau} = \frac{-g_{t\phi} e - g_{tt}\ell}{R^2} = \frac{(2GMa/r)e + (1 - 2GM/r)\ell}{r^2 - 2GMr + a^2} \quad (37.10)$$

If we convert equations 37.9 and 37.10 to difference equations, they can be integrated numerically to find $r(\tau)$ and $\phi(\tau)$ for any equatorial orbit given M, a, ℓ, and e.

Zero-Angular-Momentum Trajectories. In the Schwarzschild case, a particle with zero angular-momentum-per-unit-mass-at-infinity ℓ will fall inward along a radial line. Equation 37.10 implies that this is *not* the case for zero-angular-momentum particles in Kerr spacetime:

$$\frac{d\phi}{d\tau} = \frac{2GMae}{rR^2} = \frac{2GMae}{r(r^2 - 2GMr + a^2)} \quad \text{when } \ell = 0 \quad (37.11)$$

Note that $e > 0$ and as long as $r > 2GM$, the denominator is unambiguously positive for all a, so $d\phi/d\tau$ will have the same sign as a (which is positive if the field source is rotating counterclockwise when viewed from the $\theta = 0$ direction). Thus an infalling particle will be swept in the same direction as the source's rotation (see figure 37.1). This deflection can be interpreted as a gravitomagnetic effect: using the left-hand rule, the source's gravitomagnetic field in the particle's location in figure 37.1 comes up out of the plane of the drawing, so the gravitomagnetic "force" $\vec{F}_{GM} = 4\vec{V} \times \vec{B}_G$ acts in to the left, "along with" the source's rotation.

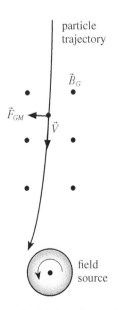

FIG. 37.1 Deflection of a particle falling inward with $\ell = 0$. Note that the particle is swept in the direction of the source's rotation.

This phenomenon is called the "dragging of inertial frames." Freely falling frames *are* inertial frames in general relativity, and we can see from the calculation above that falling reference frames will be "dragged" in the direction of the field source's rotation as they fall inward. However, one should *not* view this as if the "fabric of spacetime" were being swept around the source in the direction of its rotation, and objects moving in that spacetime are frictionally dragged along with that fabric. Such an image works in this particular case, but not in others. For example, consider a free particle that happens to be moving radially outward at a certain instant of time. The "fabric drag" image would suggest that such a particle would *still* be deflected in the direction of the source's rotation, but the gravitomagnetic analogy suggests (correctly) that the particle will be deflected *against* the source's rotation (see problem P37.2 for a derivation from equation 37.3). In this case, the "fabric drag" model is misleading.

Kepler's Third Law. Another important case that we can discuss in some detail is the case of circular orbits. We can easily find the period T of equatorial circular orbits for a particle with nonzero rest mass as follows. The geodesic equation (equation 37.1) with $\mu = r$ and assuming an equatorial orbit with $d\theta/d\tau = 0$ yield

$$0 = \frac{d}{d\tau}\left(g_{rr}\frac{dr}{d\tau}\right) - \frac{1}{2}\left[\frac{\partial g_{tt}}{\partial r}\left(\frac{dt}{d\tau}\right)^2 + \frac{\partial g_{rr}}{\partial r}\left(\frac{dr}{d\tau}\right)^2 + \frac{\partial g_{\phi\phi}}{\partial r}\left(\frac{d\phi}{d\tau}\right)^2 + 2\frac{\partial g_{t\phi}}{\partial r}\frac{dt}{d\tau}\frac{d\phi}{d\tau}\right] \quad (37.12)$$

For a circular orbit, $dr/d\tau = 0$, so the first and third terms will be zero. Define the particle's angular velocity $\Omega \equiv d\phi/dt$: then $d\phi/d\tau = (d\phi/dt)(dt/d\tau) = \Omega(dt/d\tau)$. If we plug this into the remaining nonzero terms, we get

$$0 = \left(\frac{dt}{d\tau}\right)^2\left(\frac{\partial g_{tt}}{\partial r} + 2\frac{\partial g_{t\phi}}{\partial r}\Omega + \frac{\partial g_{\phi\phi}}{\partial r}\Omega^2\right) \quad (37.13)$$

If you divide through by $(dt/d\tau)^2$, plug in the equatorial-plane values of the metric components, and solve the quadratic equation for Ω, you will find (see box 37.4)

$$\Omega = \frac{\sqrt{GM}}{\sqrt{GM}\,a \pm r^{3/2}} \quad \Rightarrow \quad T = \left|\frac{2\pi}{\Omega}\right| = \sqrt{\frac{4\pi^2}{GM}r^{3/2} \pm 2\pi a} \quad (37.14)$$

Note that in the limit $a \to 0$, this becomes Kepler's third law.

Inspection of these equations show that $|\Omega|$ is smaller and T is larger when Ω has the same sign as a, that is, the orbit is in the same direction as the spin of the source of the gravitational field. We see that for a given radius, there are two different orbital periods, a larger period if the particle orbits "with" and a shorter period if it orbits "against" the source's rotation. This may seem counter-intuitive if you are using the "dragging fabric" image of Kerr spacetime, but it is fully consistent with the gravitomagnetic model. As shown in figure 37.2, the left-hand rule implies that the spinning field source will create a gravitomagnetic field that is directed upward through the plane of the drawing. A particle orbiting "with" the rotation of the source experiences an *inward* gravitoelectric force but (according to the right-hand rule) an *outward* gravitomagnetic force. On the other hand, a particle orbiting "against" the source's rotation will experience an *inward* gravitomagnetic force. Therefore, the net inward gravitational force is larger in the "against" case than in the "with" case. Since the radii of both orbits are the same by assumption, the greater acceleration caused by the greater force implies that the particle's orbital speed must be larger in the "against" case, so its period is smaller.

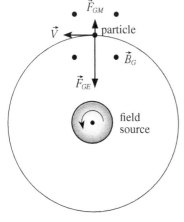

FIG. 37.2 Gravitomagnetism explains why a particle orbits more slowly when it orbits with the source's rotation than against it. Note that (by the right-hand rule) the gravitomagnetic force on the particle is outward in the "with" case but inward in the "against" case, so there is less net inward force, thus less net inward acceleration, and thus less speed at a given radius in the "with" case.

The Innermost Stable Circular Orbit. For Schwarzschild black holes, we found that the innermost stable circular orbit (ISCO) was $r = 6GM$. We can determine the radii of stable circular orbits in the Kerr case by solving simultaneously the three equations $\frac{1}{2}(e^2 - 1) = V_{\text{eff}}(r)$ (since $dr/d\tau = 0$ in a circular orbit), $dV_{\text{eff}}/dr = 0$, and $d^2V_{\text{eff}}/dr^2 \geq 0$ (the conditions for a local minimum in the effective potential energy) for the three unknowns r_c, ℓ, and e for a stable circular orbit (given M and a). The algebra is so difficult and tedious that I will not make you go through it, but the final result is not so gruesome: the innermost stable orbits are such that

$$r^2 - 6GMr - 3a^2 \pm 8a\sqrt{GMr} = 0 \qquad (37.15)$$

where the upper sign corresponds to an orbit "with" the source's rotation and the lower sign to an orbit "against" the source's rotation. You can see that this becomes the Schwarzschild result $r = 6GM$ when $a = 0$. You can show (see box 37.5) that when a has its extreme value GM, the ISCO radii are $r = GM$ for a co-rotating orbit and $r = 9GM$ for a counter-rotating orbit.

BOX 37.1 Calculating Expressions for $dt/d\tau$ and $d\phi/d\tau$

According to equations 37.2, the definitions of e and ℓ are

$$e \equiv -g_{tt}\frac{dt}{d\tau} - g_{t\phi}\frac{d\phi}{d\tau} \quad \text{and} \quad \ell \equiv g_{\phi t}\frac{dt}{d\tau} + g_{\phi\phi}\frac{d\phi}{d\tau} \qquad (37.16)$$

Exercise 37.1.1. Solve these equations for $dt/d\tau$ and $d\phi/d\tau$, and therefore verify equations 37.3.

BOX 37.2 Verify the Value of $[g_{t\phi}]^2 - g_{tt}g_{\phi\phi}$

In the full Kerr metric, the metric components g_{tt}, $g_{t\phi}$, and $g_{\phi\phi}$ are

$$g_{tt} = -\left(1 - \frac{2GMr}{r^2 + a^2\cos^2\theta}\right), \quad g_{t\phi} = g_{\phi t} = -\frac{2GMra\sin^2\theta}{r^2 + a^2\cos^2\theta}$$

$$\text{and} \quad g_{\phi\phi} = \left(r^2 + a^2 + \frac{2GMra^2\sin^2\theta}{r^2 + a^2\cos^2\theta}\right)\sin^2\theta \qquad (37.17)$$

From these definitions, it is simple algebra to prove equation 37.4, repeated here:

$$[g_{t\phi}]^2 - g_{tt}g_{\phi\phi} = [r^2 + a^2 - 2GMr]\sin^2\theta \equiv R^2\sin^2\theta \qquad (37.4r)$$

Exercise 37.2.1. Verify this result.

BOX 37.3 The "Energy-Conservation-Like" Equation of Motion

If we explicitly write out the sum in $-1 = \boldsymbol{u} \cdot \boldsymbol{u} = g_{\mu\nu}u^\mu u^\nu$ for equatorial motion in Kerr spacetime, we get

$$-1 = g_{tt}\left(\frac{dt}{d\tau}\right)^2 + g_{rr}\left(\frac{dr}{d\tau}\right)^2 + g_{\phi\phi}\left(\frac{d\phi}{d\tau}\right)^2 + 2g_{t\phi}\frac{dt}{d\tau}\frac{d\phi}{d\tau} \qquad (37.18)$$

If you substitute the values of $g_{tt}, g_{rr}, g_{\phi\phi}$, and $g_{t\phi}$ from equations 37.5 into the above and use equation 37.3, you should be able to derive equation 37.7, repeated here for the sake of convenience:

$$\tfrac{1}{2}(e^2 - 1) = \tfrac{1}{2}\left(\frac{dr}{d\tau}\right)^2 - \frac{GM}{r} + \frac{\ell^2 + a^2(1 - e^2)}{2r^2} - \frac{GM(\ell - ea)^2}{r^3} \qquad (37.7r)$$

Exercise 37.3.1. Verify this equation.

BOX 37.4 Kepler's Third Law

Equation 37.13 implies that a particle in a circular orbit in Kerr spacetime must satisfy the following relationship

$$0 = \frac{\partial g_{tt}}{\partial r} + 2\frac{\partial g_{t\phi}}{\partial r}\Omega + \frac{\partial g_{\phi\phi}}{\partial r}\Omega^2 \quad \text{where} \quad \Omega \equiv \frac{d\phi}{dt} \qquad (37.19)$$

If you substitute into this equation the equatorial values of g_{tt}, $g_{\phi\phi}$, and $g_{t\phi}$, you should be able to show that

$$0 = GM - 2GMa\Omega + (GMa^2 - r^3)\Omega^2 \qquad (37.20)$$

Exercise 37.4.1. Verify equation 37.20.

If you use the quadratic equation to solve equation 37.20 for Ω, you can show with a bit of extra algebra (remember that $a^2 - b^2 = [a+b][a-b]$) that

$$\Omega = \frac{\sqrt{GM}}{\sqrt{GM}\,a \pm r^{3/2}} \qquad (37.21)$$

Exercise 37.4.2. Verify equation 37.21.

BOX 37.5 The Radii of ISCOs When $a = GM$

If we divide equation 37.15 by $(GM)^2$, we find that the radii for the innermost stable circular orbits in Kerr spacetime are such that

$$\rho^2 - 6\rho - 3\alpha^2 \pm 8\alpha\sqrt{\rho} = 0 \quad \text{where } \rho \equiv \frac{r}{GM}, \ \alpha \equiv \frac{a}{GM} \qquad (37.22)$$

Exercise 37.5.1. Verify that when $\alpha = 1$, then $\rho = 1$ is a solution when the sign of the last term is positive and $\rho = 9$ is a solution when the sign of the last term is negative. (Optional: Use a computer algebra tool to find a general solution for arbitrary α. Do not attempt this by hand.)

HOMEWORK PROBLEMS

P37.1 Find expressions for e and ℓ for a particle that is dropped from rest at $r = r_0$ in the equatorial plane. Note that $\ell \neq 0$ for such a particle if $r_0 < \infty$.

P37.2 Assume that a particle has $d\phi/d\tau = 0$ but $dr/d\tau > 0$ at a given initial radius $r_0 > 2GM$ in the equatorial plane. Show that as the particle moves outward, its angular velocity becomes negative. (*Hint*: Let $dt/d\tau$ at $r = r_0$ be u_0. Calculate e and ℓ for such a particle in terms of u_0.)

P37.3 Consider a freely falling particle in the equatorial plane of an extreme Kerr black hole ($a = GM$) that is moving inward from extremely large r with $e = \frac{5}{4}$, $\ell = -\frac{5}{4}GM$.
 a. Argue that at large r, the particle's proper angular velocity $d\phi/d\tau$ is negative (i.e., *opposite* to the star's rotation).
 b. Find the value of r where the particle's proper angular velocity "turns around" and becomes positive (going *with* the star's rotation) in spite of ℓ being negative.

P37.4 Assume that a freely falling particle initially has zero radial velocity but a negative but finite angular-momentum-per-unit-mass ℓ at an essentially infinite radial coordinate r in the equatorial plane. This means that the particle's angular velocity $d\phi/d\tau$ is initially very small but negative. Find an expression (in terms of GM, a, and $|\ell|$) for the radius r where the particle's angular velocity $d\phi/d\tau$ "turns around" and becomes positive. Can this radius ever be smaller than $r = 2GM$? (*Hint*: First determine the value for this particle's energy-per-unit-mass e.)

P37.5 Assume that a freely falling particle initially has zero radial velocity but $d\phi/d\tau = -\omega_0 < 0$ at $r = r_0 > 2GM$ in the equatorial plane. This particle will drop to lower radii as time passes. Find (in terms of ω_0 and r_0) the radius r where the particle's angular velocity $d\phi/d\tau$ "turns around" and becomes positive. [*Hints*: This problem is *much* harder than problem P37.4 but is still doable. To save writing, define $A \equiv -g_{tt}$, $B \equiv -g_{\phi t}$, and $C \equiv g_{\phi\phi}$ (I have chosen the signs so that A, B and C are all positive) and let A_0, B_0, and C_0 be the values of these metric components at $r = r_0$. Note that according to equation 37.4, $B^2 + AC = R^2$. Start by using the metric equation to determine $dt/d\tau$ at r_0, and use that to find e and ℓ for this particle. Don't plug in the expressions for A, B, and C until the very end. Be sure to be attentive to the implicit GR units of every expression along the way, and make sure that your final result makes physical sense in the limits that a goes to zero, ω_0 goes to zero, etc.]

P37.6 A robot probe is orbiting in the equatorial plane of a neutron star for which $GM = 2.1$ km. The radial coordinate of the orbit is $r = 20$ km. Assume that the neutron star has an angular momentum per unit mass of $a = 2.1$ km $= GM$.
 a. Find the probe's orbital period T (as measured at infinity) for both co-rotating and counter-rotating orbits.
 b. What time interval $\Delta\tau$ will the probe's clock register for one complete orbit of the star in both cases? (*Hint*: $d\tau^2 = -ds^2 = -g_{\mu\nu}dx^\mu dx^\nu$ for any infinitesimal segment of the orbit. Note that r is constant for a circular orbit. Note also that $|d\phi/dt| \equiv |\Omega| \equiv |2\pi/T|$ is also constant for this orbit.)

P37.7 a. Show that $dV_{\text{eff}}/dr = 0$ and $d^2V_{\text{eff}}/dr^2 = 0$ (which must apply for the innermost circular orbit) imply, respectively, that

$$0 = GMr^2 - A^2r + 3GMB^2 \quad (37.23a)$$
$$0 = 2GMr^2 - 3A^2r + 12GMB^2 \quad (37.23b)$$

where $A^2 \equiv \ell^2 + a^2(1-e^2)$ and $B^2 \equiv (\ell - ae)^2$.
 b. Consider the extreme case where $a = GM$. The text claims that $r = GM$ is the innermost stable co-rotating circular orbit in such a case. Use equations 37.23 to show that for such an orbit

$$e = \sqrt{\tfrac{1}{3}}, \quad \ell = \sqrt{\tfrac{4}{3}}\,GM \quad (37.24)$$

This last result implies that in an accretion disk, a particle spiralling in from rest at infinity (where it has $e = 1$) can release as much as $(1 - \sqrt{\tfrac{1}{3}}) = 42\%$ of its rest mass in the form of thermal energy.

P37.8 In this problem we will consider circular photon orbits in the equatorial plane of a Kerr black hole.
 a. Argue that photons in a circular orbit (r = constant) in the equatorial plane must satisfy the following equation in Kerr spacetime (where $\Omega \equiv d\phi/dt$):

$$0 = g_{tt} + 2g_{t\phi}\Omega + g_{\phi\phi}\Omega^2 \quad (37.25)$$

 b. Equation 37.14 follows from the geodesic equation for massive particles (see equation 37.12) but 37.14 does not refer to the particle's proper time, so it should be valid for photons as well. Show that this equation implies that for an equatorial circular orbit around an extreme Kerr black hole

$$\frac{1}{GM\Omega} = 1 \pm \left(\frac{r}{GM}\right)^{3/2} \quad (37.26)$$

where the + and − solutions correspond co-rotating and counter-rotating orbits respectively.
 c. Photons orbiting in circular orbits in the equatorial plane around a Kerr black hole must therefore satisfy both equations 37.25 and 37.26. We can write equation 37.25 more conveniently as

$$0 = \frac{g_{tt}}{\Omega^2} + \frac{2g_{t\phi}}{\Omega} + g_{\phi\phi}$$

$$\Rightarrow \quad 0 = \frac{g_{tt}}{(GM\Omega)^2} + \frac{2g_{t\phi}}{GM(GM\Omega)} + \frac{g_{\phi\phi}}{(GM)^2} \quad (37.27)$$

Verify that equations 37.26 and 37.27 are simultaneously satisfied if $r = GM$ in the co-rotating case and $r = 4GM$ in the counter-rotating case. You do not have to show that these are the *only* solutions (though they are): simply verify that they work. This shows that there are possible circular orbits for light at these two radii.

38. ERGOREGION AND HORIZON

INTRODUCTION

FLAT SPACETIME
| Review of Special Relativity |
| Four-Vectors |
| Index Notation |

TENSORS
| Arbitrary Coordinates |
| Tensor Equations |
| Maxwell's Equations |
| Geodesics |

SCHWARZSCHILD BLACK HOLES
| The Schwarzschild Metric |
| Particle Orbits |
| Precession of the Perihelion |
| Photon Orbits |
| Deflection of Light |
| Event Horizon |
| Alternative Coordinates |
| Black Hole Thermodynamics |

THE CALCULUS OF CURVATURE
| The Absolute Gradient |
| Geodesic Deviation |
| The Riemann Tensor |

THE EINSTEIN EQUATION
| The Stress-Energy Tensor |
| The Einstein Equation |
| Interpreting the Equation |
| The Schwarzschild Solution |

COSMOLOGY
| The Universe Observed |
| A Metric for the Cosmos |
| Evolution of the Universe |
| Cosmic Implications |
| The Early Universe |
| CMB Fluctuations & Inflation |

GRAVITATIONAL WAVES
| Gauge Freedom |
| Detecting Gravitational Waves |
| Gravitational Wave Energy |
| Generating Gravitational Waves |
| Gravitational Wave Astronomy |

SPINNING BLACK HOLES
| Gravitomagnetism |
| The Kerr Metric |
| Kerr Particle Orbits |
| **Ergoregion and Horizon** |
| Negative-Energy Orbits |

this depends on this

Overview. The surface corresponding to radius $r = 2GM$ around a Schwarzschild black hole has *two* interesting features. First of all, it is an **infinite-redshift surface** where $g_{tt} = 0$, which means that clocks at rest on that surface measure zero proper time relative to clocks at infinity and radially outgoing photons ($d\theta = d\phi = 0$) are at rest ($dr = 0$). As we approach this surface, the time Δt measured by observers at infinity between events separated by a specific proper time $\Delta\tau$ as measured by local observers at rest goes to infinity: hence the name "infinite-redshift surface."

The same surface in Schwarzschild spacetime is also an *event horizon*, i.e., a surface beyond which no ingoing particle can return. This is signaled by g_{rr} and g_{tt} switching signs at that surface, which implies (in the case of a diagonal metric) that the r and t coordinates switch roles and decreasing r becomes the future.

In Kerr spacetime, the infinite redshift surface (where $g_{tt} = 0$) does *not* coincide with the event horizon. We will see instead that the infinite redshift surface *encloses* the event horizon, meaning that objects can cross the infinite redshift surface and yet return to large r. The region of spacetime between these surfaces (the **ergoregion**) has some fascinating properties. Our task in this chapter is to locate the infinite redshift surface and the event horizon in Kerr spacetime.

The Infinite-Redshift Surface. A clock at rest ($dr = d\theta = d\phi = 0$) in Kerr spacetime registers a proper time $d\tau = \sqrt{-g_{tt}}\, dt$ between events separated by coordinate time dt. Such a clock will register zero time (relative to clocks at infinity, which measure coordinate time t) anywhere that $g_{tt} = 0$. If you substitute in the full Kerr expression for g_{tt} from equation 37.5a and solve $g_{tt} = 0$ for r, you will find (see box 38.1) that Kerr spacetime has *two* infinite-redshift surfaces at

$$r = GM \pm \sqrt{(GM)^2 - a^2 \cos^2\theta} \tag{38.1}$$

as long as $a < GM$. It turns out that the inner surface is completely enclosed by the event horizon and so has no possible physical connection to the exterior of a Kerr black hole. We will therefore focus our attention on the *outer* surface in what follows, which defines the outer boundary of the ergoregion:

$$r_e = GM + \sqrt{(GM)^2 - a^2 \cos^2\theta} \tag{38.2}$$

This surface has an r coordinate that ranges from $GM + \sqrt{(GM)^2 - a^2}$ at the poles to $2GM$ on the equator. As $a \to 0$, $r_e \to 2GM$, as one would expect.

The Static Limit. Consider an observer in a spaceship whose engines keep the ship at a constant r and θ but do not constrain the ship's angular coordinate ϕ. Since such an observer must follow a timelike worldline, we must have

$$0 \leq d\tau^2 = -ds^2 = -(g_{tt}dt^2 + 2g_{t\phi}dt\,d\phi + g_{\phi\phi}d\phi^2)$$
$$\Rightarrow \quad 0 \geq g_{tt} + 2g_{t\phi}\Omega + g_{\phi\phi}\Omega^2 \tag{38.3}$$

where $\Omega \equiv d\phi/dt$. Applying the quadratic equation to the lower limit, we see that the observer's angular velocity Ω must lie in the range $\Omega_{\min} \leq \Omega \leq \Omega_{\max}$, where

$$\Omega_{\min} = -\frac{g_{t\phi}}{g_{\phi\phi}} - \sqrt{\frac{[g_{t\phi}]^2 - g_{tt}g_{\phi\phi}}{[g_{\phi\phi}]^2}}, \quad \Omega_{\max} = -\frac{g_{t\phi}}{g_{\phi\phi}} + \sqrt{\frac{[g_{t\phi}]^2 - g_{tt}g_{\phi\phi}}{[g_{\phi\phi}]^2}} \tag{38.4}$$

(Note that equations 37.5 imply that $g_{t\phi}$ is negative and $g_{\phi\phi}$ is positive everywhere in Kerr spacetime, so the first term in each of these expressions is positive.) Having nonzero $dr/d\tau$ and/or $d\theta/d\tau$ only reduces the range of Ω (see box 38.2), so these limits represent the extremes of possible angular velocities Ω at constant r. In the equatorial plane ($\theta = \pi/2$), these angular velocities turn out to be

$$\Omega_{\max,\min} = +\frac{2GMa}{r^3 + a^2 r + 2GMa^2} \pm \sqrt{\frac{r^2 + a^2 - 2GMr}{(r^2 + a^2 + 2GMa^2/r)^2}} \tag{38.5}$$

(see box 38.3). In the limit that $r \to \infty$, note that $r\Omega_{\min} \to -1$ and $r\Omega_{\max} \to +1$, which correspond to the limits imposed by the speed of light, as we would expect.

On the infinite-redshift surface (the outer surface of the ergoregion), $g_{tt} = 0$, so equation 38.4 implies that $\Omega_{\min} = 0$, which in turn implies that only photons can be at rest on that surface. On that surface, all massive particles must have angular velocities in the range $0 < \Omega < \Omega_{\max} = -2g_{t\phi}/g_{\phi\phi} = 4GMa/(r_e^3 + a^2 r_e + 2GMa^2)$. For r inside the infinite-redshift surface, $g_{tt} > 0$ and the quantity in the square root of equation 38.4 is smaller than $|g_{t\phi}/g_{\phi\phi}|$, so all objects must have positive angular velocities (i.e., they must move *with* the rotation of the black hole):

$$0 < \Omega_{\min} \leq \Omega \leq \Omega_{\max} \quad \text{(inside the infinite redshift surface)} \quad (38.6)$$

The infinite-redshift surface in Kerr spacetime is therefore also called the **static limit**: no particle can remain at rest inside this surface.

The Event Horizon. As r decreases inside the infinite redshift surface, g_{tt} becomes more and more positive. Solutions to equation 38.4 cease to exist at all for when g_{tt} becomes large enough that $[g_{t\phi}]^2 - g_{tt}g_{\phi\phi}$ becomes negative: this implies that even a powered spaceship is unable to maintain a constant r coordinate where this is true. If all objects must have nonzero dr inside the surface where $[g_{t\phi}]^2 - g_{tt}g_{\phi\phi} = 0$, it is plausible that this is because all objects must move *inward* inside this surface. If this is true, then the surface where $[g_{t\phi}]^2 - g_{tt}g_{\phi\phi} = 0$ corresponds to the *event horizon* for a Kerr black hole.

Equation 37.4 tells us that $[g_{t\phi}]^2 - g_{tt}g_{\phi\phi} = (r^2 + a^2 - 2GMr)\sin^2\theta \equiv R^2 \sin^2\theta$, so the radial coordinate of this surface is such that

$$0 = R^2 = r^2 + a^2 - 2GMr \quad \Rightarrow \quad r_\pm = GM \pm \sqrt{(GM)^2 - a^2} \quad (38.7)$$

where I used the quadratic equation in the last step. Again we find two solutions, one inside the other. However, assuming these surfaces really are event horizons, then whatever is going on inside the outer horizon is inaccessible to observers in the outer universe. Therefore we will focus our attention on describing the outer horizon (the surface where $r_+ = GM + \sqrt{(GM)^2 - a^2}$) in what follows. Note that in the limit that $a \to 0$, the r-coordinate of this surface becomes $r = 2GM$, as we would expect.

Unlike the infinite-redshift surface, our hypothetical horizon has a constant radial coordinate even when $a \neq 0$, and it is always inside the infinite-redshift surface except at the poles, where the two surfaces coincide. Figure 38.1 shows the two surfaces plotted in r-θ coordinates for an extreme Kerr black hole where $a = GM$.

Please note that this figure displays (a cross section of) the surfaces in r and θ coordinates, which have no immediately obvious geometrical interpretation. For example, the fact that the event horizon surface has a constant r coordinate does *not* imply that the surface is a sphere. Indeed, you can show (see box 38.4) that the metric on the event horizon's surface is

$$ds^2 = \rho_+^2 d\theta^2 + \left(\frac{2GMr_+}{\rho_+}\right)^2 \sin^2\theta\, d\phi^2 \quad \text{where} \quad \rho_+^2 \equiv r_+^2 + a^2 \cos^2\theta \quad (38.8)$$

Since both $g_{\theta\theta}$ and $g_{\phi\phi}$ in this metric depend on θ, the geometry of this surface is *not* the geometry of a sphere (unless $a = 0$). You can show, for example (see problem P38.1), that for an extreme Kerr black hole ($a = GM$), the distance around the equator of the event horizon is $4\pi GM = 12.56 GM$, but the distance around a curve of constant longitude through the poles is $7.6GM$. This is more like the geometry of an oblate spheroid. Indeed (as discussed in problem P38.2), if $a/GM < \sqrt{3/4} \approx 0.86$, one can construct an embedding diagram for the event horizon's surface that visually displays its intrinsic geometry. Such embedding diagrams for different values of a indeed look like oblate spheroids that become flatter and flatter as a increases.

From the metric given by equation 38.8, however, one can fairly easily show (see box 38.5) that the physical area of the event horizon is

$$A = 8\pi GMr_+ \quad (38.9)$$

This result will be useful to us in the next chapter.

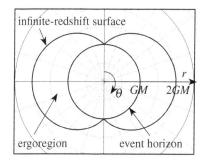

FIG. 38.1 A polar $r(\theta)$ plot of the infinite-redshift and event horizon surfaces for an extreme Kerr black hole ($a = GM$). The r-coordinate of the event horizon is $r_+ = GM$: the equatorial r-coordinate of the infinite-redshift surface is $r_e = 2GM$.

Locating the Event Horizon. How can we be sure that the surface $r = r_+$ really is an event horizon? In the Schwarzschild case, I argued that the surface defined by $r = 2GM$ was an event horizon because inside that radius, $g_{tt} > 0$ and $g_{rr} < 0$, so t became a *spatial* coordinate and r a *time* coordinate. Where r is a time coordinate, an object must move inward with the same inevitability as time moving forward: a decreasing r coordinate becomes the future of every object. Since an event horizon is precisely the surface inside which outward movement becomes impossible, we see that (for compact objects where we can define a useful radial coordinate, anyway) the surface where r switches from being a spatial coordinate to a time coordinate *is* the event horizon.

In the Kerr case it is true that inside $r = r_+$, we have $g_{tt} > 0$ (because we are also inside the infinite-redshift surface) and $g_{rr} = (r^2 + a^2 \cos^2\theta)/(r^2 + a^2 - 2GMr) < 0$, so it indeed looks as if t is a spatial coordinate and r a time coordinate inside $r = r_+$. However, identifying coordinate types by merely inspecting the sign of the corresponding diagonal metric component works only with *diagonal* metrics. To see the problem, note that in the Kerr ergoregion, we have both $g_{tt} > 0$ and $g_{rr} > 0$. We might naively conclude, therefore, that there is no time coordinate at all in this region!

We can resolve this puzzle as follows. The presence of the off-diagonal $g_{t\phi} = g_{\phi t}$ terms in the Kerr metric implies that the t and ϕ coordinates are physically connected in a way that would not be true in a diagonal metric. While neither coordinate seems unambiguously a time coordinate in the ergoregion (because g_{tt} and $g_{\phi\phi}$ are both positive), there can still be a time coordinate lurking behind the combination of the two. To see this, consider the $t\phi$ submatrix of the Kerr metric

$$g_{\mu\nu} = \begin{bmatrix} g_{tt} & g_{t\phi} \\ g_{\phi t} & g_{\phi\phi} \end{bmatrix} \quad (t\phi \text{ submatrix}) \tag{38.10}$$

A coordinate transformation from the Kerr coordinates t and ϕ to some new coordinates $T(t, \phi)$ and $\Phi(t, \phi)$ at a given event will affect only the components of this submatrix, because for such a transformation, the partial derivatives in the general transformation expression

$$g'_{\alpha\beta} = \frac{\partial x^\mu}{\partial x'^\alpha} \frac{\partial x^\nu}{\partial x'^\beta} g_{\mu\nu} \tag{38.11}$$

will be zero when either μ or ν are equal to r or θ. Now, *if* there is a time coordinate lurking behind t and ϕ, then we should be able to find coordinate transformations $T(t, \phi)$ and $\Phi(t, \phi)$ that transform the submatrix to the locally flat form

$$g'_{\mu\nu} = \begin{bmatrix} -1 & 0 \\ 0 & 1 \end{bmatrix} \quad (T\Phi \text{ submatrix, time coordinate lurking}) \tag{38.12a}$$

where the T and Φ coordinates are unambiguously time and space coordinates, respectively. On the other hand, if t and ϕ are both physically spatial coordinates (no time coordinate lurks behind them), then the best we can do is transform them to two new locally flat *spatial* coordinates, in which case the submatrix will be

$$g'_{\mu\nu} = \begin{bmatrix} 1 & 0 \\ 0 & 1 \end{bmatrix} \quad (T\Phi \text{ submatrix, no time coordinate lurking}) \tag{38.12b}$$

Now, box 38.6 argues that a coordinate transformation at an event cannot change the *sign* of a matrix's determinant at that event. The determinant of the $g'_{\mu\nu}$ submatrix is negative if there is a time coordinate lurking and positive if not. Therefore, there is a time coordinate lurking behind the Kerr coordinates t and ϕ if and only if

$$\det \begin{bmatrix} g_{tt} & g_{t\phi} \\ g_{\phi t} & g_{\phi\phi} \end{bmatrix} = g_{tt}g_{\phi\phi} - [g_{t\phi}]^2 \tag{38.13}$$

is negative, i.e., $[g_{t\phi}]^2 - g_{tt}g_{\phi\phi} > 0$. This is true everywhere outside of $r = r_+$, so we see that even in the ergoregion, there must be a physical time coordinate lurking behind the Kerr t and ϕ coordinates.

However, inside $r = r_+$ (or more technically, within the region between $r = r_+$ and $r = r_-$), we have $[g_{t\phi}]^2 - g_{tt}g_{\phi\phi} = (r^2 + a^2 - 2GMr)\sin^2\theta < 0$, so by the argument above, t and ϕ are unambiguously spatial coordinates in this region. Moreover, the Kerr metric component $g_{rr} = (r^2 + a^2\cos^2\theta)/(r^2 + a^2 - 2GMr)$ is negative everywhere in this region, and since there is no possibly competing time coordinate lurking behind the $t\phi$ coordinates, we can conclude that the r coordinate is unambiguously the Kerr metric's true time coordinate in this region, with decreasing r corresponding to the future. Therefore, the surface $r = r_+$ is indeed an event horizon analogous to the Schwarzschild horizon.

Cosmic Censorship. Note that solutions to equation 38.7 for the event horizon radii fail to exist if $a > GM$, meaning that there are no event horizons in such a case. Now Kerr spacetime has a true geometric singularity (a place where the curvature of spacetime becomes infinite, analogous to $r = 0$ in Schwarzschild spacetime) where $r = 0$. If $a > GM$, there would be no event horizons surrounding this singularity, meaning that observers could visit this physical absurdity and send information back. This would raise a host of deep theoretical problems having to do with such issues as causality and self-consistency of the theory.

The **Cosmic Censorship Hypothesis** (first proposed by Roger Penrose in 1969) asserts that the gravitational collapse of a physically reasonable mass distribution can never produce such a "naked" singularity "unclothed" by an event horizon. As of this writing, this hypothesis remains unproven, but there is evidence to suggest that a Kerr spacetime with $a > GM$ would be unstable, meaning that it would spontaneously radiate gravitational waves until $a < GM$. Moreover, there are published arguments that suggest that any collapsing physical object with $a > GM$ would fragment into pieces before forming a black hole. We will therefore assume that the Cosmic Censorship Hypothesis is true and that all astrophysical black holes have $a < GM$ and thus have their singularities discreetly clothed by event horizons.

BOX 38.1 The Radii Where $g_{tt} = 0$

Equation 37.5a tells us that for the full Kerr metric

$$g_{tt} = -\left(1 - \frac{2GMr}{r^2 + a^2\cos^2\theta}\right) = -\left(\frac{r^2 + a^2\cos^2\theta - 2GMr}{r^2 + a^2\cos^2\theta}\right) \quad (38.14)$$

Exercise 38.1.1. Use the quadratic equation and the equation above to argue that $g_{tt} = 0$ where $r = GM \pm \sqrt{(GM)^2 - a^2\cos^2\theta}$. (Note that the denominator in the expression above is always positive.)

BOX 38.2 The Angular Speed Range When dr and/or $d\theta \neq 0$

Equation 38.4 was derived assuming that $dr = d\theta = 0$. If this is not the case, the quadratic equation implies that the limits on Ω become

$$\Omega_{\text{max,min}} = -\frac{g_{t\phi}}{g_{\phi\phi}} \pm \sqrt{\frac{[g_{t\phi}]^2 - [g_{tt} + g_{rr}v_r^2 + g_{\theta\theta}v_\theta^2]g_{\phi\phi}}{[g_{\phi\phi}]^2}} \qquad (38.15)$$

where $v_r \equiv dr/dt$ and $v_\theta \equiv d\theta/dt$.

Exercise 38.2.1. Verify that equation 38.15 is correct.

Exercise 38.2.2. Argue that v_r and/or v_θ nonzero does indeed make the range of possible angular velocities smaller. In your argument, consider carefully the signs of all the metric components.

BOX 38.3 Angular-Speed Limits in the Equatorial Plane

The expressions found in equations 37.5 for the metric components evaluated in the Kerr equatorial plane imply that the limits on an object's angular speed on that plane (when its radial velocity is also zero) are given by equation 38.5 (repeated here for convenience):

$$\Omega_{max,min} = +\frac{2GMa}{r^3 + a^2 r + 2GMa^2} \pm \sqrt{\frac{r^2 + a^2 - 2GMr}{(r^2 + a^2 + 2GMa^2/r)^2}} \qquad (38.16)$$

Exercise 38.3.1. Verify that equation 38.16 is correct.

Exercise 38.3.2. Verify that in the limit that $r \to \infty$, we indeed have $r\Omega_{min} \to -1$ and $r\Omega_{max} \to +1$.

BOX 38.4 The Metric of the Event Horizon's Surface

The outer Kerr event horizon at a given instant of time is the surface such that $dt = 0$ and $r = r_+$ (meaning that $dr = 0$). From the expressions for the full Kerr metric components found in equations 37.5, you can verify that the metric on the event horizon's surface is given by equation 38.8, repeated here for convenience:

$$ds^2 = \rho_+^2 \, d\theta^2 + \left(\frac{2GMr_+}{\rho_+}\right)^2 \sin^2\theta \, d\phi^2 \qquad (38.17)$$

where $\rho_+^2 \equiv r_+^2 + a^2 \cos^2\theta$.

Exercise 38.4.1. Verify that equation 38.17 is correct. Note that r_+ is defined to satisfy the relation $r_+^2 + a^2 - 2GMr_+ = 0$ (see equation 38.7).

BOX 38.5 The Area of the Outer Kerr Event Horizon

According to equation 38.17, the physical length of an infinitesimal step purely in the θ direction along the surface of the outer Kerr event horizon is $ds = \rho_+ d\theta$. Similarly, the physical length of an infinitesimal step purely in the ϕ direction is $ds = (2GMr_+/\rho_+)\sin\theta\, d\phi$. Since these directions are orthogonal ($g_{\theta\phi} = 0$), this means that the physical area corresponding to a patch of the event horizon surface with the differential coordinate area $d\theta\, d\phi$ is

$$dA = \rho_+ d\theta \cdot \left(\frac{2GMr_+}{\rho_+}\right)\sin\theta\, d\phi = 2GMr_+ \sin\theta\, d\theta\, d\phi \tag{38.18}$$

Exercise 38.5.1. Integrate this over all θ and ϕ to verify that the outer Kerr horizon's total area is $A = 8\pi GMr_+$, as claimed in equation 38.9. (*Hint:* Note that r_+, unlike ρ_+, is a constant on the event horizon's surface.)

BOX 38.6 Transformations Preserve the Metric Determinant's Sign

The general transformation equation for the components of the metric is

$$g'_{\alpha\beta} = \frac{\partial x^\mu}{\partial x'^\alpha}\frac{\partial x^\nu}{\partial x'^\beta} g_{\mu\nu} \tag{38.19}$$

When evaluated at a particular event in spacetime, the partial derivatives become simply numbers, so the metric transformation at any particular event is simply a linear transformation, which we can write in matrix language as

$$[\,\boldsymbol{g}'\,] = [\,\boldsymbol{B}\,]^T [\,\boldsymbol{g}\,][\,\boldsymbol{B}\,] \quad \text{where} \quad B^\nu_\beta \equiv \left[\frac{\partial x^\nu}{\partial x'^\beta}\right]_{\text{event}} \tag{38.20}$$

Now, it is a property of the matrix determinant (as you can verify by checking a text on linear algebra or by searching online) that the determinant of a matrix product is the product of the matrix determinants and that the determinant of a transposed matrix is the same as that of the original matrix. Therefore,

$$\det[\,\boldsymbol{g}'\,] = \det\!\left([\,\boldsymbol{B}\,]^T[\,\boldsymbol{g}\,][\,\boldsymbol{B}\,]\right) = \det[\,\boldsymbol{B}\,]^T \cdot \det[\,\boldsymbol{g}\,] \cdot \det[\,\boldsymbol{B}\,]$$

$$= (\det[\,\boldsymbol{B}\,])^2 \det[\,\boldsymbol{g}\,] \tag{38.21}$$

Therefore, if $\det[\boldsymbol{B}]$ is nonzero, a transformation will not change the sign of the metric's determinant.

BOX 38.6 (continued) Transformations Preserve the Metric Determinant's Sign

Can we be confident that det[B] is nonzero? Note that a coordinate transformation is quite generally invertible, because the transformation partials satisfy the basic identity

$$\frac{\partial x^\mu}{\partial x'^\alpha} \frac{\partial x'^\alpha}{\partial x^\nu} = \delta^\mu_\nu \quad \Rightarrow \quad [\,B\,][\,B\,]^{-1} = [\,I\,] \tag{38.22}$$

A linear transformation B cannot have an inverse unless its determinant is nonzero. Therefore, as long as the transformation is non-singular at the event (i.e., the partial derivatives of the transformation are defined), det[B] must be nonzero.

Note also that a matrix's determinant is unchanged if we rearrange its rows and/or columns (e.g., to bring the rows and columns of the t and ϕ submatrix together). Finally, note that the argument also applies to any *submatrix* of the metric as long as the transformation B does not involve coordinates outside of that submatrix. Therefore, since the argument earlier in this chapter regarding the $t\phi$ submatrix of the Kerr metric involved a hypothetical transformation $T(t,\phi)$ and $\Phi(t,\phi)$ to new coordinates T and Φ that did *not* involve the coordinates r and θ outside of the $t\phi$ submatrix, the claims made there should hold.

Exercise 38.6.1. According to equation 15.5, the metric for "global rain" coordinates in Schwarzschild spacetime is $ds^2 = -(1 - 2GM/r)d\mathring{t}^2 + 2\sqrt{2GM/r}\,d\mathring{t}\,dr + dr^2 + r^2 d\theta^2 + r^2\sin^2\theta\,d\phi^2$. By evaluating the determinant of the $\mathring{t}r$ submatrix of $\boldsymbol{g}$, prove that there is a time coordinate lurking behind this metric at all r. (This is a good thing, because we specifically *chose* $\mathring{t}$ so that it was a meaningful time coordinate at all radii.)

HOMEWORK PROBLEMS

P38.1 Show that the outer horizon of an extreme Kerr black hole ($a = GM$) has an equatorial circumference of $4\pi GM$, while the circumference of a closed path of fixed longitude through the poles is $7.6 GM$. (*Hint:* Consult box 38.5 for a discussion of the physical lengths of infinitesimal steps in the θ and ϕ directions. Computing the polar circumference requires a numerical integration.)

P38.2 We can construct an embedding diagram for a Kerr black hole's event horizon using the techniques discussed in box 11.5.

a. The metric of a three-dimensional flat space in cylindrical coordinates is given by

$$ds^2 = dr_E^2 + r_E^2 d\phi^2 + dz^2 \qquad (38.23)$$

where r_E is the radial coordinate in the 3D flat space. We can define a surface in this space by specifying $z(\theta)$. Prove that this surface will have the same metric as given in equation 38.8 as long as

$$r_E(\theta) = \frac{2GMr_+}{\rho_+(\theta)} \sin\theta \quad \text{and}$$

$$\frac{dz}{d\theta} = \pm\sqrt{\rho_+^2 - \left(\frac{dr_E}{d\theta}\right)^2} \qquad (38.24)$$

where $\rho_+^2 \equiv r_+^2 + a^2 \cos^2\theta$.

b. Show that

$$\frac{dr_E}{d\theta} = \frac{(2GMr_+)^2 \cos\theta}{\rho_+^3} \qquad (38.25)$$

(*Hint:* Remember that the event horizon's radius r_+ is defined such that $r_+^2 + a^2 - 2GMr_+ = 0$.)

c. The expression for $dz/d\theta$ above can in principle be integrated to determine $z(\theta)$. While it is very difficult to do this analytically, we can fairly easily do this numerically. Use an appropriate computer tool to find $z(\theta)$ and plot it versus the cylindrical radial coordinate $r_E(\theta)$ for $a = 0.85GM$. [*Hints:* If you want, you can even do this with a spreadsheet. First divide the angular range from 0 to $\pi/2$ into a set of discrete angles θ_i separated by some fixed $\Delta\theta$. Evaluate $r_E(\theta_i)$ at each discrete angle. Then approximate the differential equation for $dz/d\theta$ given above by the centered difference equation

$$\frac{z(\theta_{i+1}) - z(\theta_i)}{\Delta\theta} = \left[\sqrt{\rho_+^2 + \left(\frac{dr_E}{d\theta}\right)^2}\right]_{\theta_{i+1/2}} \qquad (38.26)$$

evaluating the right side at $\theta_{i+1/2} = \theta_i + \frac{1}{2}\Delta\theta$, the angle halfway between θ_{i+1} and θ_i. If you solve this for $z(\theta_{i+1})$ and iterate, you can calculate $z(\theta_i)$ at all angles given its value at some initial angle. It is actually easiest to set $z(\pi/2) = 0$ (so the surface has zero height at the equator) and work backward to $z(0)$. Once you have a table of $z(\theta_i)$ and $r_E(\theta_i)$, you can easily plot the curve. Even just 10 discrete angles gives reasonable results. You should find that the surface resembles a flattened spheroid.]

P38.3 A powered space probe is holding itself at a constant radial coordinate $r = 1.5GM$ in the equatorial plane of an extreme ($a = GM$) Kerr black hole. What range of ordinary rotational speeds $v = \sqrt{g_{\phi\phi}}\,\Omega$ (as measured by an observer at infinity) can this probe have?

P38.4 In chapter 14, we saw that for an object falling in Schwarzschild spacetime the proper time required to reach $r = 0$ from any finite $r > 2GM$ was finite and well defined: this was one of the pieces of evidence that the Schwarzschild event horizon was not a geometric singularity. Consider a doomed observer freely falling in the equatorial plane of Kerr spacetime from rest at a very large r.

a. Use equation 37.7 to argue that the proper time the observer measures between passing some radius r_0 and reaching $r = 0$ is

$$\tau = \frac{1}{\sqrt{2GM}} \int_0^{r_0} \sqrt{\frac{r^3}{r^2 + a^2}}\, dr \qquad (38.27)$$

(*Hint:* First determine e and ℓ for the observer in question.) This integral is clearly finite and well defined, so this is evidence that neither of the two Kerr event horizons are geometric singularities, but rather some kind of coordinate problem.

b. This deceptively simple integral has no solution that can be written in terms of simple functions. However, we can evaluate it numerically. Re-express the integral in terms of the unitless variable $x = r/GM$. Then use a computer tool (e.g., WolframAlpha, Mathematica, Mathcad, a personally written computer program, or a spreadsheet) to numerically evaluate this integral for $a = GM$ and $r_0 = 3GM$. Compare your result to the analogous result for Schwarzschild spacetime. (*Hint:* To evaluate the integral using a computer program or spreadsheet, first divide the range from 0 to x_0 into discrete steps: a Δx of about 0.1 or smaller is adequate. To find τ_i at the ith step, calculate $x_{i-1/2} \approx \frac{1}{2}(x_i + x_{i-1})$, and then use this value to evaluate the integrand at the halfway point between the $(i-1)$th step and the ith step. Multiply the result by Δx and add to τ_{i-1} to get τ_i. This "centered-difference" approach is straightforward but more accurate than other simple methods. To get still more accurate results, you might optionally use a 4th-order Runge-Kutta method: look it up.)

P38.5 Consider a doomed observer freely falling in the equatorial plane of Kerr spacetime from rest and $\ell = 0$ at a very large r, starting with an angular coordinate of $\phi = 0$.

a. Use equations 37.7 and 37.10 to argue that the observer's ϕ coordinate as the observer passes r is

$$\phi(r) = \sqrt{2GM}\, a \int_r^\infty \frac{1}{R^2} \sqrt{\frac{r}{r^2 + a^2}}\, dr \qquad (38.28)$$

where $R^2 = r^2 - 2GMr + a^2$. (*Hint:* First determine e for the observer in question.) Note that since $R^2 \to 0$ at the event horizon (see equation 38.7), the integrand is singular at the event horizon.

P38.5 (continued)

b. The integral in equation 38.28 cannot be expressed in terms of simple functions. Rewrite it in terms of $x = r/GM$ with $a = GM$ and use a computer tool to numerically integrate it from some fairly large upper limit to a lower limit near the event horizon at $x = 1$ and argue that $\phi(r)$ diverges as the lower limit gets closer to the event horizon. (*Hint:* See problem P38.4 for some suggestions about how you might perform a numerical integration using a computer program or spreadsheet.)

Comment: If the problems with the Kerr coordinate system at the event horizon are simply coordinate problems rather than geometric singularities, then the ϕ coordinate is part of the problem. Any coordinate system for the Kerr spacetime that is non-singular down to the origin (analogous to the "global rain" system for Schwarzschild spacetime) will have to somehow "unwind" this infinite wrapping of ϕ around the event horizon.

P38.6 It turns out that (just as in the case of a Schwarzschild black hole) the radial velocity dr/dt (as measured by an observer at infinity) of any object dropped into a Kerr black hole approaches zero as the object approaches the outer event horizon. Show, however, that such an object's angular velocity (as observed at infinity) approaches

$$\left(\frac{d\phi}{dt}\right)_H \equiv \Omega_H = \frac{a}{2GMr_+} \quad (38.29)$$

An observer at infinity will thus see objects that have been dropped into the black hole in the past to be stuck at the event horizon radius but swirling around it with this angular velocity. We can therefore take this in some sense to be the "angular velocity" of the black hole.

P38.7 As problem P38.6 makes clear, we can take equation 38.29 as defining the rate at which a Kerr black hole's event horizon spins (as viewed from infinity). Calculate this angular speed Ω_H in rotations per second, its rotation period T in seconds, and the horizon's ordinary equatorial rotation speed $v \equiv \sqrt{g_{\phi\phi}\Omega_H^2}$ as a fraction of the speed of light for an extreme Kerr black hole with $M = 2M_\odot$.

P38.8 Use the methods discussed in this chapter to argue that if $a > GM$, then g_{rr} is positive everywhere and the t-ϕ submatrix always has a time coordinate lying behind it, so there is no event horizon in this case.

P38.9 In this problem, we will consider observers at a fixed radial coordinate r in the equatorial plane of a Kerr black hole that have zero ℓ: we call such observers **zero-angular-momentum observers**, or **ZAMOs**. We can imagine such observers as riding on rings around a Kerr black that rotate at just the correct rate so that $\ell = 0$.

a. Show that the rings must rotate at an angular speed Ω_Z (as measured at infinity) such that

$$\Omega_Z \equiv \left(\frac{d\phi}{dt}\right)_Z = \frac{2GMa}{r(r^2 + a^2 + 2GMa^2/r)} \quad (38.30)$$

b. Does Ω_Z increase or decrease as we move to smaller r?

c. Show that as we approach the event horizon, the rings' angular speed Ω_Z approaches the horizon rotation rate given in equation 38.29.

d. Calculate the ordinary rotation speed $v \equiv \sqrt{g_{\phi\phi}\Omega_Z^2}$ of the rings for observers in the equatorial plane of an extreme Kerr black hole at $r = GM$, $2GM$, $5GM$, $10GM$, $25GM$, and $100GM$.

e. Consider a freely falling object with $\ell = 0$ moving inward at an arbitrary speed in the equatorial plane. Argue that a ZAMO will always see this object to move past his or her ring in a purely radial direction, even if the object is a photon. (*Hint:* In the Kerr coordinate system, what is the falling object's angular velocity at a given r? What is the ZAMO's angular velocity at that same r?)

P38.10 To better determine exactly what a ZAMO (see problem P38.9) would observe, we can define an orthonormal coordinate system for such an observer, as we did in chapter 12 for stationary observers in Schwarzschild spacetime.

a. Use equation 37.3 and the requirement that $-1 = \mathbf{u} \cdot \mathbf{u}$ to argue that the components of a ZAMO's four-velocity in the Kerr coordinate system is

$$u_Z^t = \frac{\sqrt{g_{\phi\phi}}}{R}, \quad u_Z^r = u_Z^\theta = 0, \quad u_Z^\phi = \frac{-g_{t\phi}}{R\sqrt{g_{\phi\phi}}} \quad (38.31)$$

(*Hint:* Remember the metric's off-diagonal term!)

b. This, then, is the observer's time basis vector $\mathbf{o}_t$. The observer's $\mathbf{o}_\phi$ basis vector must be perpendicular to $\mathbf{o}_t$. Since $\mathbf{o}_t$ has both a t and ϕ component in Kerr coordinates, let's assume that $\mathbf{o}_\phi$ does as well: let these components be a and b, respectively. Show that the requirement that $\mathbf{o}_t \cdot \mathbf{o}_\phi = 0$ (somewhat counter-intuitively) requires that $a = 0$. (*Hint:* Again, remember the metric's off-diagonal term, and note also that with an off-diagonal metric, $\mathbf{u} \cdot \mathbf{w} = g_{tt}u^t w^t + g_{t\phi}u^t w^\phi + g_{\phi t}u^\phi w^t + \cdots$)

c. Show that the requirement that $1 = \mathbf{o}_\phi \cdot \mathbf{o}_\phi$ implies that $b = 1/\sqrt{g_{\phi\phi}}$.

d. Find basis vectors $\mathbf{o}_r$ and $\mathbf{o}_\theta$ satisfying the requirements that they be perpendicular to each other and the vectors we have already found and that they have unit magnitude. Express them in terms of the metric components.

e. Find the Kerr components of the four-velocity $\mathbf{u}_{obj}$ of an arbitrary object in the equatorial plane that is freely falling with $\ell = 0$, and, by evaluating $\mathbf{u}_{obj} \cdot \mathbf{o}_\phi$ and $\mathbf{u}_{obj} \cdot \mathbf{o}_\theta$ formally prove that such object has a purely radial velocity in a ZAMO's coordinate system.

39. NEGATIVE-ENERGY ORBITS

INTRODUCTION

FLAT SPACETIME
- Review of Special Relativity
- Four-Vectors
- Index Notation

TENSORS
- Arbitrary Coordinates
- Tensor Equations
- Maxwell's Equations
- Geodesics

SCHWARZSCHILD BLACK HOLES
- The Schwarzschild Metric
- Particle Orbits
- Precession of the Perihelion
- Photon Orbits
- Deflection of Light
- Event Horizon
- Alternative Coordinates
- Black Hole Thermodynamics

THE CALCULUS OF CURVATURE
- The Absolute Gradient
- Geodesic Deviation
- The Riemann Tensor

THE EINSTEIN EQUATION
- The Stress-Energy Tensor
- The Einstein Equation
- Interpreting the Equation
- The Schwarzschild Solution

COSMOLOGY
- The Universe Observed
- A Metric for the Cosmos
- Evolution of the Universe
- Cosmic Implications
- The Early Universe
- CMB Fluctuations & Inflation

GRAVITATIONAL WAVES
- Gauge Freedom
- Detecting Gravitational Waves
- Gravitational Wave Energy
- Generating Gravitational Waves
- Gravitational Wave Astronomy

SPINNING BLACK HOLES
- Gravitomagnetism
- The Kerr Metric
- Kerr Particle Orbits
- Ergoregion and Horizon
- **Negative-Energy Orbits**

this depends on this

39. NEGATIVE-ENERGY ORBITS

Overview. One of the most interesting features of the ergoregion of a Kerr black hole is that particles in that region can exist in trajectories that have negative energy-at-infinity. This allows one to extract energy from a Kerr black hole using a process first described by Roger Penrose in 1969. In this chapter, we will discuss negative-energy trajectories, the Penrose process, and various "thermodynamic" limits applying to Kerr black holes.

"Conservation of Energy" Revisited. In chapter 37, we saw that in the equatorial plane, free particles with nonzero rest mass follow geodesics that obey a "conservation-of-energy-like" equation given by equation 37.7:

$$\tfrac{1}{2}(e^2 - 1) = \tfrac{1}{2}\left(\frac{dr}{d\tau}\right)^2 - \frac{GM}{r} + \frac{\ell^2 + a^2(1 - e^2)}{2r^2} - \frac{GM(\ell - ea)^2}{r^3} \tag{39.1}$$

where e is the particle's conserved "energy per unit mass at infinity," ℓ is the particle's conserved "angular momentum per unit mass at infinity," and M and a are the Kerr source's gravitational mass and spin angular momentum per unit mass, respectively. Note that the left side of this equation is a constant that depends on the particle's energy, while the first term on the right has the form of a radial kinetic energy per unit mass. The remainder can be interpreted as an effective potential function that depends only on the radial coordinate r and constants. However, we noted in chapter 37 that this equation was somewhat less useful than the corresponding Schwarzschild "conservation-of-energy-like" equation because in equation 39.1, the effective potential depends on e, so we cannot draw a single r-dependent potential curve that applies for all particle energies.

However, there is a way around this difficulty. If we multiply both sides of this equation by 2 and gather like powers of e, we get

$$0 = 1 - \frac{2GM}{r} + \frac{\ell^2 + a^2}{r^2} - \frac{2GM\ell^2}{r^3} + \left(\frac{dr}{d\tau}\right)^2$$
$$+ \left(\frac{4GM\ell a}{r^3}\right)e - \left(1 + \frac{a^2}{r^2} + \frac{2GMa^2}{r^3}\right)e^2 \tag{39.2}$$

(see box 39.1). This has the form of a quadratic equation in e:

$$0 = -Ae^2 + Be + C + \left(\frac{dr}{d\tau}\right)^2 \tag{39.3a}$$

where $\quad A \equiv 1 + \dfrac{a^2}{r^2} + \dfrac{2GMa^2}{r^3}, \quad B \equiv \dfrac{4GM\ell a}{r^3}, \quad$ and $\tag{39.3b}$

$$C \equiv 1 - \frac{2GM}{r} + \frac{\ell^2}{r^2} + \frac{a^2}{r^2} - \frac{2GM\ell^2}{r^3} \tag{39.3c}$$

We can easily use the quadratic formula to solve this for e:

$$e = \frac{-B \pm \sqrt{B^2 + 4A[C + (dr/d\tau)^2]}}{-2A} = \frac{\tfrac{1}{2}B \mp \sqrt{\tfrac{1}{4}B^2 + AC + A(dr/d\tau)^2}}{A} \tag{39.4}$$

Since A is positive for all values of a and r, the *positive* solution for e will always be *greater* than an r-dependent effective potential we can define to be

$$V_+(r) \equiv \frac{\tfrac{1}{2}B + \sqrt{\tfrac{1}{4}B^2 + AC}}{A} \tag{39.5a}$$

Similarly, the *negative* solution for e will always be *smaller* than

$$V_-(r) \equiv \frac{\tfrac{1}{2}B - \sqrt{\tfrac{1}{4}B^2 + AC}}{A} \tag{39.5b}$$

Figure 39.1 shows these effective potential functions for an extreme Kerr hole with $a = GM$ and a particle with $\ell = -8GM$.

The solution $e \leq V_-(r)$ is actually physically meaningless: one can see that if such solutions were allowed, it would be possible for particles at infinity to have negative relativistic energy per unit mass, which is absurd. (Technically, these solutions cor-

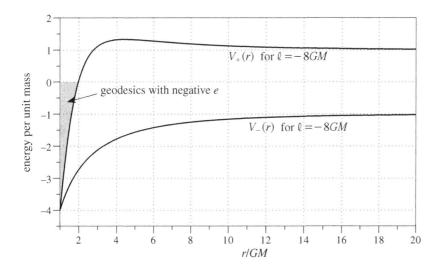

FIG. 39.1 Effective potential curves for a free particle with $\ell = -8GM$ moving in the equatorial plane of an extreme Kerr black hole ($a = GM$). The gray region shows radii where the particle could exist with negative energy at infinity.

respond to particles whose four-momentum points toward the past instead of the future.) Therefore, a real particle must have $e \geq V_+(r)$. This means that we can interpret a graph of $V_+(r)$ just as we would an ordinary potential energy graph: allowed regions for the particle are those radii for which $e > V_+(r)$, forbidden regions are radii for which $e < V_+(r)$, and turning points are where $e = V_+(r)$. The only difference is that $e - V_+(r)$ is no longer equal to the particle's radial kinetic energy per unit mass $\frac{1}{2}(dr/d\tau)^2$, but is related to that quantity only in a complicated way.

The Penrose Process. The interesting thing about the graph of $V_+(r)$ is that you can show that the square root in equation 39.5a is zero at the event horizon (see box 39.2). Thus e can be as small as $\frac{1}{2}B/A = 2GM\ell a/(r^3 + a^2 r + 2GMa^2)$ at the event horizon, which is *negative* if $\ell \leq 0$. Therefore, free particles *can* have trajectories having negative energy-per-unit-mass-at-infinity, though only *inside* the ergoregion (see box 39.3). In the specific case shown in figure 39.1 (which applies to an extreme Kerr hole with $a = GM$ and a particle with $\ell = -8GM$), you can see that negative energy orbits exist in the equatorial plane between the event horizon at $r = GM$ to just inside the ergoregion's outer boundary at $r = 2GM$.

Since such trajectories are confined to the ergoregion, an observer at infinity will never *directly* observe a particle with negative energy (which would be absurd). However, the existence of such trajectories does allow one to extract energy from the Kerr black hole as follows. Imagine that a particle P falls into a Kerr black hole's ergoregion and decays there into two particles Q and R (see figure 39.2). If the decay process happens to put Q into a trajectory with negative energy at infinity, conservation of energy implies that R will end up with *more* total energy at infinity than P originally carried in! If the decay also happens to put particle R into an outbound trajectory, R will carry this higher energy back out to infinity, where it can be captured. However, Q *must* fall into the black hole (any negative-energy trajectory has a inward-facing turning point inside the ergoregion according to figure 39.1): it thus adds its negative energy to the hole, reducing the hole's total energy (as observed at infinity) and thus its mass. The net effect of this process is therefore to extract energy from the Kerr black hole at the expense of its mass.

Since negative-energy geodesics exist only for particles with negative angular momentum, particle Q also adds negative angular momentum to the hole, and thus decreases its spin. Indeed, as we will see shortly, it is helpful to think of the Penrose process as actually mining a Kerr black hole's "spin energy." That is, if we continually use the Penrose process to extract energy from a Kerr black hole, we never can make that black hole vanish entirely: rather, we always eventually end up with a non-spinning Schwarzschild black hole from which we can no longer extract energy. The next few sections will explore this and similar limitations.

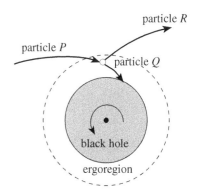

FIG. 39.2 A schematic picture of a Penrose process, as viewed above a Kerr black hole's north pole. A particle P enters the hole's ergoregion and decays into particles Q and R such that Q has negative energy-at-infinity and R is on an escape trajectory. (Q at least must also have negative angular-momentum-at-infinity.) In such a case, R emerges with greater energy than P had originally.

The Core Limitation. We have seen that at the event horizon, the square root in equation 39.5a vanishes, so a particle falling into the black hole through the event horizon must have an energy per unit mass at infinity satisfying the inequality

$$e \geq \frac{B}{2A} = \frac{2GM\ell a}{r_+^3 + a^2 r_+ + 2GMa^2} \tag{39.6}$$

where $r_+ = GM + \sqrt{(GM)^2 - a^2}$ is the radial coordinate of the black hole's event horizon. (Even though equation 39.5 strictly applies only to orbits in the equatorial plane, off-the-plane trajectories only make the inequality worse, so the limit above is quite general.) Now me, where m is the particle's mass energy, is equal to the total change δM in the black hole's mass-energy (as measured at infinity) that results from absorbing the particle. Similarly, $m\ell = \delta S$, the change in the black hole's spin angular momentum $S \equiv Ma$ (as measured at infinity) during the process. So if we multiply both sides of equation 39.6 by m and manipulate the denominator using the relationship $r_+^2 = 2GMr_+ - a^2$, we find (see box 39.4) that any particle absorption process must satisfy the inequality

$$\delta M \geq \frac{a\delta S}{r_+^2 + a^2} = \frac{a\delta S}{2GMr_+} \tag{39.7}$$

This core limitation applies to *any* particle absorption event, not just Penrose decay events that allow us to extract energy.

No Naked Singularities. As discussed in the last chapter, a Kerr black hole must have a spin per unit mass of $a < GM$ if its singularity is to be enclosed by an event horizon. Is it possible for a Penrose process to extract enough mass energy from an originally "clothed" hole to leave it with $a > GM$ and thus expose its singularity? The answer (as one might expect) is "no." You can use equation 39.7 to show (see problem P39.1) that in the case of an already-extreme Kerr black hole, *any* particle absorption process will satisfy the inequality

$$\delta(GM)^2 \geq \delta(a^2) \tag{39.8}$$

implying that $(GM)^2$ will remain safely larger than a^2.

Irreducible Mass. You can also show (see box 39.5) that equation 39.7 implies that no particle-absorption or Penrose process can reduce a black hole's so-called irreducible mass

$$\delta M_{ir} \geq 0 \quad \text{where} \quad M_{ir} \equiv \frac{\sqrt{2GMr_+}}{2G} \tag{39.9}$$

A Schwarzschild black hole's event horizon radius is $r_+ = 2GM$, so $M_{ir} = M$, implying that its mass *is* its irreducible mass and therefore no particle absorption event can decrease that mass. However, you can show (see box 39.6) that a Kerr black hole's mass M is generally *larger* than its irreducible mass:

$$M^2 = M_{ir}^2 + \left(\frac{S}{2GM_{ir}}\right)^2 \tag{39.10}$$

Penrose processes *can* therefore reduce a Kerr black hole's total mass energy, but only until the hole's spin angular momentum S is reduced to zero, whereupon the hole becomes a "dead" Schwarzschild black hole with $M = M_{ir}$. This is what I meant when I said earlier we should think of Penrose processes as mining a Kerr black hole's spin energy.

Entropy of a Black Hole. In chapter 16, we saw that it was useful to think of a Schwarzschild black hole as having an entropy proportional to the area of its event horizon (see equation 16.11). The same idea is useful for Kerr black holes as well. Equation 38.9 tells us that the area of a Kerr black hole's event horizon is

$$A = 8\pi GMr_+ = 4\pi(2GM_{ir})^2 \tag{39.11}$$

Therefore, saying that no particle absorption process can reduce a Kerr black hole's irreducible mass is the same as saying that such a process cannot reduce the area of the hole's event horizon. We see, therefore, that we can think of the irreducibility of a black hole's irreducible mass as being equivalent to saying that particle absorption cannot reduce a black hole's entropy.

Of course, we did see in chapter 16 that a Schwarzschild black hole could radiate energy (and thus lose entropy) through the Hawking radiation process. Though I will not discuss the details here, Kerr black holes also can emit Hawking radiation through basically the same mechanism. This emission does not violate the law of entropy increase, because although the black hole's entropy decreases as its irreducible mass decreases, the emitted radiation increases the entropy of the environment by more than enough to compensate.

Realistic Energy Extraction. It is hard to imagine that the Penrose process would occur frequently enough in any natural setting to liberate significant amounts of spin energy from a Kerr black hole. However, physicists have proposed more plausible natural methods for extracting energy from a Kerr black hole. One such mechanism is the *Blandford-Znajek mechanism* (see Hartle, *Gravity*, Addison-Wesley, 2003, box 15.1 for an overview or Thorne, Price, and MacDonald, *Black Holes: The Membrane Paradigm*, Yale, 1976, pp. 138ff for details). In this situation, a rotating black hole immersed in the magnetic field created by its accretion disk generates a huge voltage difference between its equator and either pole for much the same reasons that such a voltage difference would develop between the equator and pole of a rotating conductor immersed in an external magnetic field. This voltage difference can drive a huge current flow that dissipates electromagnetic energy. Rough estimates (see problem P39.8) show that the power that could be liberated by such a process is comparable to the actual power (10^{35} to 10^{41} W) observed to be radiated by active galactic nuclei, which are presumably powered by black holes. While the Blandford-Znajek mechanism cannot easily be understood or visualized in terms of Penrose processes, it still obeys the law of entropy increase, so equations 39.9 and 39.10 still apply.

Of course, an advanced civilization could easily use the Penrose process to mine energy from a Kerr black hole. The process description I gave earlier is not limited to elementary particles: the initially incoming "particle" P could be a container filled with garbage, which could be oriented by computer control (once the container is in the ergoregion) to explosively eject its garbage Q into a negative-energy trajectory while sending the now-empty container R on an escape trajectory. The empty container R would then emerge from the garbage-dumping "decay event" with more energy than P had originally, and this energy could be extracted in slowing the container down for reuse. Thus the civilization could simultaneously get rid of its garbage and power its industries in a very "green" way using this method!

BOX 39.1 Quadratic Form for Conservation of Energy

We start with equation 39.1, repeated here for convenience:

$$\tfrac{1}{2}(e^2 - 1) = \frac{1}{2}\left(\frac{dr}{d\tau}\right)^2 - \frac{GM}{r} + \frac{\ell^2 + a^2(1 - e^2)}{2r^2} - \frac{GM(\ell - ea)^2}{r^3} \quad (39.1r)$$

If you multiply through by 2 and gather like powers of e, you can show that

$$0 = 1 - \frac{2GM}{r} + \frac{\ell^2 + a^2}{r^2} - \frac{2GM\ell^2}{r^3} + \left(\frac{dr}{d\tau}\right)^2$$

$$+ \left(\frac{4GM\ell a}{r^3}\right)e - \left(1 + \frac{a^2}{r^2} + \frac{2GMa^2}{r^3}\right)e^2 \quad (39.2r)$$

Exercise 39.1.1. Verify equation 39.2r.

BOX 39.2 The Square Root Is Zero at the Event Horizon

According to equations 39.3b and 39.3c, $B \equiv 4GM\ell a/r^3$,

$$A \equiv 1 + \frac{a^2}{r^2} + \frac{2GMa^2}{r^3}, \quad \text{and} \quad C \equiv 1 - \frac{2GM}{r} + \frac{\ell^2}{r^2} + \frac{a^2}{r^2} - \frac{2GM\ell^2}{r^3} \quad (39.12)$$

The square root in equation 39.5a can therefore be written

$$\tfrac{1}{4}B^2 + AC = \frac{1}{r^6}\left[r^6 - 2GMr^5 + (\ell^2 + 2a^2)r^4 - 2GM\ell^2 r^3 \right.$$

$$\left. + a^2(\ell^2 + a^2 - 4G^2M^2)r^2 + 2GMa^4 r\right] \quad (39.13)$$

Exercise 39.2.1. Verify this (use the space on the next page).

BOX 39.2 (continued) The Square Root Is Zero at the Event Horizon

We saw in the last chapter (see equation 38.7) that the event horizon's radial coordinate r_+ satisfies $r_+^2 + a^2 - 2GMr_+ = 0$. This means that

$$r_+^2 = 2GMr_+ - a^2 \qquad (39.14)$$

Exercise 39.2.2. Evaluate the quantity in square brackets in equation 39.13 at $r = r_+$ and use equation 39.14 to reduce the powers of r_+ until everything cancels. The square root is therefore zero at the event horizon.

BOX 39.3 Negative e Is Possible Only in the Ergoregion

According to equation 37.2a, a falling particle's conserved energy per unit mass at infinity e is defined to be

$$e \equiv -g_{tt}\frac{dt}{d\tau} - g_{t\phi}\frac{d\phi}{d\tau} \qquad (39.15)$$

Consider a four-vector field $\mathbf{n}$ such that $\mathbf{n} \equiv (1,0,0,0)$ at every event in spacetime when expressed in the Kerr coordinate system, and let the falling particle's four-velocity be $\mathbf{u} = (dt/d\tau, dr/d\tau, d\theta/d\tau, d\phi/d\tau)$. You can easily show that

$$e = -\mathbf{n} \cdot \mathbf{u} \qquad (39.16)$$

This equation applies everywhere in Kerr coordinates.

Exercise 39.3.1. Verify this last result.

Now imagine a *stationary* observer. Such an observer will have a four-velocity with the form $\mathbf{u}_{obs} = (b,0,0,0) = b\mathbf{n}$, where b is a positive number that depends on the observer's location (found by requiring that $-1 = \mathbf{u} \cdot \mathbf{u}$: see problem P39.4). Imagine further that this observer observes a passing particle with four-momentum and mass m. As we have seen before, the observer will find the particle's *energy* to be $E = -\mathbf{p} \cdot \mathbf{u}_{obs}$. In the case of a freely falling particle,

$$E = -\mathbf{p} \cdot \mathbf{u}_{obs} = -m\mathbf{u} \cdot b\mathbf{n} = -mb\mathbf{u} \cdot \mathbf{n} = -mb\mathbf{n} \cdot \mathbf{u} = +mbe \qquad (39.17)$$

since the dot product is commutative.

Now a stationary observer must measure the energy of any locally passing particle to be positive. Since both m and b are also positive, it follows that wherever we can have put a stationary observer, e must be positive for a falling particle. Since we can post stationary observers anywhere outside the ergoregion, we must have $e > 0$ for any falling particle outside the ergoregion.

On the other hand, we saw in chapter 38 that there are no stationary observers inside the ergoregion, so equation 39.17 does not apply. In fact, since $\mathbf{n} \cdot \mathbf{n} = g_{tt} > 0$ inside the ergoregion, is a *spacelike* vector in this region (and so cannot be proportional to any observer's $_{obs}$, which must be timelike). The projection $e = -\mathbf{n} \cdot \mathbf{u}$ therefore yields a spatial component of the falling particle's four-velocity, which can in principle be either positive or negative.

In conclusion, a falling particle with any part of its trajectory outside the ergoregion (where equation 39.17 applies) *must* have positive e. Only particles with trajectories wholly confined to the ergoregion can possibly have negative e.

BOX 39.4 The Fundamental Limit on δM in Terms of δS

Equation 39.6 (repeated here for convenience) tells us that

$$e \geq \frac{B}{2A} = \frac{2GM\ell a}{r_+^3 + a^2 r_+ + 2GMa^2} \qquad (39.6r)$$

where r_+ is the radial coordinate of the event horizon. If you multiply both sides of this expression by the particle's mass m and use $r_+^2 = 2GMr_+ - a^2$, you should be able to verify equation 39.7, repeated here for convenience:

$$\delta M \geq \frac{a\delta S}{r_+^2 + a^2} = \frac{a\delta S}{2GMr_+} \qquad (39.7r)$$

Exercise 39.4.1. Verify both results on the right side. (This is not a difficult calculation—you will not need all the space provided below.)

BOX 39.5 $\delta M_{ir} \geq 0$

According to equation 39.9, a black hole's irreducible mass is given by

$$M_{ir} \equiv \frac{\sqrt{2GMr_+}}{2G}, \quad \text{so} \quad 2GM_{ir} = \sqrt{2GMr_+} \quad (39.18)$$

where $r_+ = GM + \sqrt{G^2M^2 - a^2}$ is the radial coordinate of the event horizon. If you square both sides, substitute in this value for r_+, and note that the black hole's total spin angular momentum is $S \equiv Ma$, you can show that

$$2G^2 M_{ir}^2 = GMr_+ = G^2M^2 + \sqrt{G^4M^4 - G^2S^2} \quad (39.19)$$

Exercise 39.5.1. Verify both equalities in this equation.

If we take the differential of both sides of this, we end up with

$$4G^2 M_{ir} \delta M_{ir} = 2G^2 M \delta M + \frac{2G^4 M^3 \delta M - G^2 S \delta S}{\sqrt{G^4 M^4 - G^2 S^2}} \quad (39.20)$$

(Check this in your head.) Now note from equation 39.19 that

$$\sqrt{G^4 M^4 - G^2 S^2} = GM(r_+ - GM) \quad (39.21)$$

Multiply equation 39.20 through by this square root, use the substitution, cancel some terms, and pull out a factor of $2GMr_+$ from the right to find

$$2G^2 M(r_+ - GM)(2GM_{ir})\delta M_{ir} = G^2 M(2GMr_+)\delta M - G^2 Ma\, \delta S \quad (39.22)$$

Exercise 39.5.2. Verify this result.

BOX 39.5 (continued) $\delta M_{ir} \geq 0$

Finally, pulling a common factor of $2GMr_+$ from both terms on the right, using equation 39.18, and dividing through by various factors, you can show that

$$\delta M_{ir} = \frac{\sqrt{2GMr_+}}{2(r_+ - GM)}\left[\delta M - \frac{a\delta S}{2GMr_+}\right] \qquad (39.23)$$

Since $r_+ \geq GM$, the quantity in *front* of the square brackets is positive. The quantity *inside* the square brackets is positive according to equation 39.7. Therefore, we are assured that $\delta M_{ir} \geq 0$.

Exercise 39.5.3. Verify equation 39.23.

BOX 39.6 The Spin Energy Contribution to a Black Hole's Mass

Equation 39.19 implies that

$$2G^2 M_{ir}^2 - G^2 M^2 = \sqrt{G^4 M^4 - G^2 S^2} \qquad (39.24)$$

If you square both sides, and do just a bit of algebra, you can prove equation 39.10, repeated here for convenience:

$$M^2 = M_{ir}^2 + \left(\frac{S}{2GM_{ir}}\right)^2 \qquad (39.10r)$$

Exercise 39.6.1. Verify this result.

39. NEGATIVE-ENERGY ORBITS

HOMEWORK PROBLEMS

P39.1 In this problem, you will show that one cannot create a naked singularity by throwing things into an already extreme Kerr black hole. Start with equation 39.7, and note that $\delta S = \delta(Ma) = M\delta a + a\delta M$, since both the hole's mass M and its spin per mass a might change in the process. Also note that for an extreme Kerr hole, $a = GM$ and $r_+ = GM$. Show that under these conditions, equation 39.7 implies that

$$G^2 M\,\delta M \geq a\,\delta a \qquad (39.25)$$

Argue further that this implies that $\delta(GM)^2 \geq \delta(a^2)$, meaning that GM remains safely greater than a after any particle absorption process.

P39.2 Consider an extreme Kerr black hole ($a = GM$).
a. What fraction of such a black hole's mass-energy can in principle be extracted?
b. The black hole at the center of our own galaxy has a mass of about $2.6 \times 10^6\, M_\odot$. Assuming it is an extreme Kerr black hole, how much energy is available for extraction? Compare with the energy radiated by the entire galaxy during one year (roughly 2×10^{44} J).

P39.3 Show that for an extreme Kerr black hole, we can write the effective potential function $V_+(r)$ in unitless form where $\rho \equiv r/GM$ and $h \equiv \ell/GM$ as follows:

$$V_+(\rho) = \frac{2h + \sqrt{g(h,\rho)}}{\rho^3 + \rho + 2}, \quad \text{where}$$

$$g(h,\rho) = \rho(\rho-1)^2[\rho^3 + (h^2+1)\rho + 2] \qquad (39.26)$$

Verify this, plot this function for $h = -8$, and compare the result to figure 39.1.

P39.4 A stationary observer's four-momentum has to have the form $\boldsymbol{u}_{\text{obs}} = (b, 0, 0, 0)$. Use the requirement that $-1 = \boldsymbol{u} \cdot \boldsymbol{u}$ to find b as a function of r outside the ergoregion in Kerr spacetime.

P39.5 Here is another way to show that negative-energy orbits exist, but only in the ergoregion. To simplify the algebra, we will consider a particle moving in the equatorial plane of an extreme Kerr black hole ($a = GM$).
a. Use equations 37.2a and 37.5 to show in such a case that a particle will have negative energy per unit mass at infinity e when its angular velocity $d\phi/dt$ (as measured by an observer at infinity) satisfies the condition

$$1 - \frac{r}{2GM} > GM\frac{d\phi}{dt} \qquad (39.27)$$

b. But a particle's angular velocity $d\phi/dt$ must also be within the range specified by equation 38.5. Adapt that equation for the case where $a = GM$ to show that

$$GM\frac{d\phi}{dt} > \frac{1 - \frac{r}{GM} + \frac{2GM}{r}}{1 + \left(\frac{r}{GM}\right)^2 + \frac{2GM}{r}} \qquad (39.28)$$

c. So $GM(d\phi/dt)$ must be between the two limits specified by equations 39.27 and 39.28 if negative energy solutions are to exist. Define the unitless variable $x = r/GM$ and use some kind of computer application to plot both the upper and lower limits as functions of x between 1 and 3. Show that there are indeed a range of possible negative-energy values for $GM(d\phi/dt)$ that are between these limits, but only within the ergoregion.

d. Calculate the range of possible tangential speeds $v_\phi \equiv \sqrt{g_{\phi\phi}}\,d\phi/dt$ (as measured by an observer at infinity) that a negative-energy particle might have at $r = \frac{3}{2}GM$ in the equatorial plane of an extreme Kerr black hole with $M = 2M_\odot$.

P39.6 Imagine that we have two extreme Kerr black holes of mass m, with their spins aligned in the same direction. Assume that these black holes coalesce in such a way that no angular momentum is lost to their surroundings. Conservation of angular momentum then implies that the final black hole must have a total spin angular momentum of $S = ma + ma = 2m(Gm)$.
a. Show that each initial black hole has an event horizon area of $8\pi(Gm)^2$.
b. Show that if the final black hole has mass M, then the area of its event horizon is

$$A = 8\pi(GM)^2\left[1 + \sqrt{1 - \frac{4m^4}{M^4}}\right] \qquad (39.29)$$

c. Show that if $M = 2m$ (meaning that no energy is radiated either), then the area of the final black hole's event horizon is substantially greater than the sum of the initial holes' horizon areas. Thus the system's entropy has increased in this coalescence process.
d. Calculate the maximum amount of energy (as a multiple of m) that we might in principle extract from this coalescence process consistent with the second law of thermodynamics.

P39.7 We have seen that the area of a Kerr black hole's event horizon cannot be reduced by any process and is therefore plausibly linked to its entropy. In chapter 16, we defined a Schwarzschild black hole's entropy to be

$$\boxed{S} = \frac{k_B}{4G\hbar}A \qquad (39.30)$$

where A is the area of the event horizon (see equation 16.11). Since a Kerr black hole's entropy must reduce to this in the limit that $a \to 0$, let's assume that this is also the definition of a Kerr black hole's entropy. Note that I am using the notation $\boxed{S}$ to distinguish the entropy from the black hole's spin angular momentum S.
a. Use equations 39.29, 39.11, and 39.23 to show that

$$\delta\boxed{S} = \frac{4\pi k_B}{\hbar}\frac{GMr_+}{r_+ - GM}[\delta M - \Omega_H \delta S] \qquad (39.31)$$

where $\Omega_H \equiv a/2GMr_+$ is the event horizon's rotation rate (see problem P38.6).

b. Consider an ordinary non-relativistic object spinning at a rate Ω. Show that if we exert a force on that object's surface that causes its spin angular momentum to increase by δS, that force also does work $\delta W = \Omega\, \delta S$ (according to Newtonian mechanics).

c. We can therefore rewrite equation 39.31 in the form

$$\delta M = \frac{\hbar(r_+ - GM)}{4\pi k_B GM r_+}\delta \boxed{S} + \delta W \qquad (39.32)$$

Since M is the total mass-energy in the black hole, this equation is mathematically equivalent to the first law of thermodynamics $dU = Td\boxed{S} + dW = dQ + dW$, where U is the internal energy of a thermodynamic system, dQ and dW are the heat and work added to the system in an infinitesimal process, and $d\boxed{S} = dQ/T$ is the definition of the thermodynamic entropy. Use this equivalence to identify the absolute temperature T of the Kerr black hole and find an expression for that temperature that eliminates r_+ in favor of GM and a.

d. Show that Kerr black hole's temperature T reduces to the Schwarzschild temperature (see equation 16.7) in the limit that $a \to 0$ and goes to zero when $a \to GM$.

P39.8 In this problem we will make some crude estimates regarding the Blandford-Znajek mechanism for extracting energy from a rotating black hole.

a. Begin by considering a conducting cylinder of radius r_0 rotating in flat spacetime with an angular speed Ω in a uniform external magnetic field $\vec{B}$ pointing along its rotation axis (see figure 39.3). Charges carried around the axis by the object's rotation experience a magnetic force pointing radially away from the axis (check this with your right hand). At a given radius r from the axis, the magnitude of the magnetic force on a charge q is

$$F_B(r) = qvB = qr\Omega B \qquad (39.33)$$

The emf created by this force between the cylinder's "equator" and its pole will be the work per unit charge done by the magnetic force on a particular charge carrier as it moves from the pole to the equator. Argue that this emf is

$$\mathcal{E} = \tfrac{1}{2}\Omega B r_0^2 \quad \text{(in SI units)} \qquad (39.34)$$

b. This emf will drive a current flow $I = \mathcal{E}/R$ through any connection between the equator and pole, where R is the circuit's total resistance. Argue that the electrical power dissipated by this current will be

$$P = \frac{\Omega^2 B^2 r_0^4}{4R} \quad \text{(in SI units)} \qquad (39.35)$$

c. It turns out that in many ways, a black hole's event horizon behaves electromagnetically as if it were a

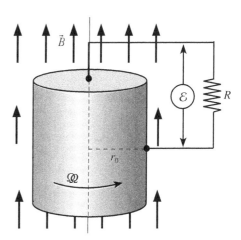

FIG. 39.3 A conducting cylinder rotating in an external magnetic field constitutes a "homopolar generator" that creates an emf between the cylinder's equator and its axis. This emf will drive a current through the schematic circuit shown.

conductive membrane rotating with the angular speed $\Omega_H = a/2GMr_+$ in GR units (see problem P38.6) and a resistance of roughly 100 Ω when conducting a current from pole to equator (see Thorne, Price, and MacDonald, *Black Holes: The Membrane Paradigm*, Yale, 1976, p. 52). While a black hole's event horizon is not a cylinder and the spacetime surrounding the event horizon is not flat, the cylinder analogy above should yield an order-of-magnitude estimate of the power that can be dissipated by a rotating black hole in an external magnetic field. For a numerical estimate, we must estimate the resistance R of the total circuit and the strength of the external field. Power will be maximally dissipated outside the black hole if the resistance of the hot plasma that completes the circuit outside the hole has roughly the same resistance as the hole itself (this is the idea behind impedance matching), so for the sake of argument, let's assume a total circuit resistance of $R \sim 200$ Ω: this will yield an estimate of the *maximum* energy this mechanism might dissipate. It is plausible that the rapidly moving plasma around the hole will create an external magnetic field of the order of magnitude of 1 T near the black hole (maybe even larger). A galaxy's central black hole can have a mass of up to a billion solar masses: let's estimate $GM \sim 10^9$ km $\sim 10^{12}$ m. Given these numbers, very roughly estimate the power dissipated by an extreme Kerr black hole, and compare to the power output of a typical galaxy ($\sim 10^{38}$ W). (This problem is based on the discussion of the Blandford-Znajek process in Hartle, *Gravity*, Addison-Wesley, 2003, pp. 326–327.)

APPENDIX: A DIAGONAL METRIC WORKSHEET

Consider the general diagonal metric

$$ds^2 = -A(dx^0)^2 + B(dx^1)^2 + C(dx^2)^2 + D(dx^3)^2$$

where dx^0, dx^1, dx^2, and dx^3 are completely arbitrary coordinates and A, B, C, and D are arbitrary functions of any or all of the coordinates. This worksheet (adapted from results listed in Rindler, *Essential Relativity*, 2/e, Springer-Verlag, 1977) allows you to quickly calculate the components of $\Gamma^\alpha_{\mu\nu}$ and $R_{\mu\nu} \equiv +R^\alpha{}_{\mu\alpha\nu}$ for any specific special case of such a metric. In this worksheet, I use the following shorthand notation:

$$A_0 \equiv \frac{\partial A}{\partial x^0}, \quad B_{12} \equiv \frac{\partial^2 B}{\partial x^1 \partial x^2}, \text{ and so on.}$$

To use this worksheet, start by crossing out each tabulated term that is zero for the specific metric in question. For the remaining terms, write the term's value in the space above that term. For the Ricci tensor components, you can then gather the terms in the space provided at the bottom. To adapt this worksheet to smaller dimensional spaces or spacetimes, treat the metric components corresponding to any nonexistent coordinates as if they had the value 1 and the remaining metric components as being independent of the nonexistent coordinates.

Christoffel Symbols

$\Gamma^0_{00} = \frac{1}{2A}A_0$ $\quad$ $\Gamma^0_{10} = \Gamma^0_{01} = \frac{1}{2A}A_1$ $\quad$ $\Gamma^0_{20} = \Gamma^0_{02} = \frac{1}{2A}A_2$ $\quad$ $\Gamma^0_{30} = \Gamma^0_{03} = \frac{1}{2A}A_3$

$\Gamma^0_{11} = \frac{1}{2A}B_0$ $\quad$ $\Gamma^0_{22} = \frac{1}{2A}C_0$ $\quad$ $\Gamma^0_{33} = \frac{1}{2A}D_0$ $\quad$ other $\Gamma^0_{\mu\nu} = 0$

$\Gamma^1_{01} = \Gamma^1_{10} = \frac{1}{2B}B_0$ $\quad$ $\Gamma^1_{11} = \frac{1}{2B}B_1$ $\quad$ $\Gamma^1_{12} = \Gamma^1_{21} = \frac{1}{2B}B_2$ $\quad$ $\Gamma^1_{13} = \Gamma^1_{31} = \frac{1}{2B}B_3$

$\Gamma^1_{00} = \frac{1}{2B}A_1$ $\quad$ $\Gamma^1_{22} = -\frac{1}{2B}C_1$ $\quad$ $\Gamma^1_{33} = -\frac{1}{2B}D_1$ $\quad$ other $\Gamma^1_{\mu\nu} = 0$

$\Gamma^2_{02} = \Gamma^2_{20} = \frac{1}{2C}C_0$ $\quad$ $\Gamma^2_{12} = \Gamma^2_{21} = \frac{1}{2C}C_1$ $\quad$ $\Gamma^2_{22} = \frac{1}{2C}C_2$ $\quad$ $\Gamma^2_{32} = \Gamma^2_{23} = \frac{1}{2C}C_3$

$\Gamma^2_{00} = \frac{1}{2C}A_2$ $\quad$ $\Gamma^2_{11} = -\frac{1}{2C}B_2$ $\quad$ $\Gamma^2_{33} = -\frac{1}{2C}D_2$ $\quad$ other $\Gamma^2_{\mu\nu} = 0$

$\Gamma^3_{03} = \Gamma^3_{30} = \frac{1}{2D}D_0$ $\quad$ $\Gamma^3_{13} = \Gamma^3_{31} = \frac{1}{2D}D_1$ $\quad$ $\Gamma^3_{23} = \Gamma^3_{32} = \frac{1}{2D}D_2$ $\quad$ $\Gamma^3_{33} = \frac{1}{2D}D_3$

$\Gamma^3_{00} = \frac{1}{2D}A_3$ $\quad$ $\Gamma^3_{11} = -\frac{1}{2D}B_3$ $\quad$ $\Gamma^3_{22} = -\frac{1}{2D}C_3$ $\quad$ other $\Gamma^3_{\mu\nu} = 0$

Ricci Tensor Components (following three pages)

APPENDIX: A DIAGONAL METRIC WORKSHEET

$R_{00} = 0$ $\qquad + \frac{1}{2B}A_{11} \qquad + \frac{1}{2C}A_{22} \qquad + \frac{1}{2D}A_{33}$

$+ \quad 0 \qquad - \frac{1}{2B}B_{00} \qquad - \frac{1}{2C}C_{00} \qquad - \frac{1}{2D}D_{00}$

$+ \quad 0 \qquad + \frac{1}{4B^2}B_0^2 \qquad + \frac{1}{4C^2}C_0^2 \qquad + \frac{1}{4D^2}D_0^2$

$+ \quad 0 \qquad + \frac{1}{4AB}A_0B_0 \qquad + \frac{1}{4AC}A_0C_0 \qquad + \frac{1}{4AD}A_0D_0$

$- \frac{1}{4BA}A_1A_1 \qquad - \frac{1}{4B^2}A_1B_1 \qquad + \frac{1}{4BC}A_1C_1 \qquad + \frac{1}{4BD}A_1D_1$

$- \frac{1}{4CA}A_2A_2 \qquad + \frac{1}{4CB}A_2B_2 \qquad - \frac{1}{4C^2}A_2C_2 \qquad + \frac{1}{4CD}A_2D_2$

$- \frac{1}{4DA}A_3A_3 \qquad + \frac{1}{4DB}A_3B_3 \qquad + \frac{1}{4DC}A_3C_3 \qquad - \frac{1}{4D^2}A_3D_3$

$$\boxed{R_{00} = }$$

$R_{11} = \frac{1}{2A}B_{00} \qquad + \quad 0 \qquad - \frac{1}{2C}B_{22} \qquad - \frac{1}{2D}B_{33}$

$- \frac{1}{2A}A_{11} \qquad + \quad 0 \qquad - \frac{1}{2C}C_{11} \qquad - \frac{1}{2D}D_{11}$

$+ \frac{1}{4A^2}A_1^2 \qquad + \quad 0 \qquad + \frac{1}{4C^2}C_1^2 \qquad + \frac{1}{4D^2}D_1^2$

$- \frac{1}{4A^2}B_0A_0 \qquad - \frac{1}{4AB}B_0B_0 \qquad + \frac{1}{4AC}B_0C_0 \qquad + \frac{1}{4AD}B_0D_0$

$+ \frac{1}{4BA}B_1A_1 \qquad + \quad 0 \qquad + \frac{1}{4BC}B_1C_1 \qquad + \frac{1}{4BD}B_1D_1$

$- \frac{1}{4CA}B_2A_2 \qquad + \frac{1}{4CB}B_2B_2 \qquad + \frac{1}{4C^2}B_2C_2 \qquad - \frac{1}{4CD}B_2D_2$

$- \frac{1}{4DA}B_3A_3 \qquad + \frac{1}{4DB}B_3B_3 \qquad - \frac{1}{4DC}B_3C_3 \qquad + \frac{1}{4D^2}B_3D_3$

$$\boxed{R_{11} = }$$

APPENDIX: A DIAGONAL METRIC WORKSHEET

$R_{22} = \frac{1}{2A} C_{00} \qquad - \frac{1}{2B} C_{11} \qquad + 0 \qquad - \frac{1}{2D} C_{33}$

$\phantom{R_{22} =} - \frac{1}{2A} A_{22} \qquad - \frac{1}{2B} B_{22} \qquad + 0 \qquad - \frac{1}{2D} D_{22}$

$\phantom{R_{22} =} + \frac{1}{4A^2} A_2^2 \qquad + \frac{1}{4B^2} B_2^2 \qquad + 0 \qquad + \frac{1}{4D^2} D_2^2$

$\phantom{R_{22} =} - \frac{1}{4A^2} C_0 A_0 \qquad + \frac{1}{4AB} C_0 B_0 \qquad - \frac{1}{4AC} C_0 C_0 \qquad + \frac{1}{4AD} C_0 D_0$

$\phantom{R_{22} =} - \frac{1}{4BA} C_1 A_1 \qquad + \frac{1}{4B^2} C_1 B_1 \qquad + \frac{1}{4BC} C_1 C_1 \qquad - \frac{1}{4BD} C_1 D_1$

$\phantom{R_{22} =} + \frac{1}{4CA} C_2 A_2 \qquad + \frac{1}{4CB} C_2 B_2 \qquad + 0 \qquad + \frac{1}{4CD} C_2 D_2$

$\phantom{R_{22} =} - \frac{1}{4DA} C_3 A_3 \qquad - \frac{1}{4DB} C_3 B_3 \qquad + \frac{1}{4DC} C_3 C_3 \qquad + \frac{1}{4D^2} C_3 D_3$

$$\boxed{R_{22} = }$$

$R_{33} = \frac{1}{2A} D_{00} \qquad - \frac{1}{2B} D_{11} \qquad - \frac{1}{2C} D_{22} \qquad + 0$

$\phantom{R_{33} =} - \frac{1}{2A} A_{33} \qquad - \frac{1}{2B} B_{33} \qquad - \frac{1}{2C} C_{33} \qquad + 0$

$\phantom{R_{33} =} + \frac{1}{4A^2} A_3^2 \qquad + \frac{1}{4B^2} B_3^2 \qquad + \frac{1}{4C^2} C_3^2 \qquad + 0$

$\phantom{R_{33} =} - \frac{1}{4A^2} D_0 A_0 \qquad + \frac{1}{4AB} D_0 B_0 \qquad + \frac{1}{4AC} D_0 C_0 \qquad - \frac{1}{4AD} D_0 D_0$

$\phantom{R_{33} =} - \frac{1}{4BA} D_1 A_1 \qquad + \frac{1}{4B^2} D_1 B_1 \qquad - \frac{1}{4BC} D_1 C_1 \qquad + \frac{1}{4BD} D_1 D_1$

$\phantom{R_{33} =} - \frac{1}{4CA} D_2 A_2 \qquad - \frac{1}{4CB} D_2 B_2 \qquad + \frac{1}{4C^2} D_2 C_2 \qquad + \frac{1}{4CD} D_2 D_2$

$\phantom{R_{33} =} + \frac{1}{4DA} D_3 A_3 \qquad + \frac{1}{4DB} D_3 B_3 \qquad + \frac{1}{4DC} D_3 C_3 \qquad + 0$

$$\boxed{R_{33} = }$$

APPENDIX: A DIAGONAL METRIC WORKSHEET

$$R_{01} = -\frac{1}{2C}C_{01} - \frac{1}{2D}D_{01} + \frac{1}{4C^2}C_0 C_1 + \frac{1}{4D^2}D_0 D_1$$

$$+ \frac{1}{4AC}A_1 C_0 + \frac{1}{4AD}A_1 D_0 + \frac{1}{4BC}B_0 C_1 + \frac{1}{4BD}B_0 D_1$$

$$R_{01} =$$

$$R_{02} = -\frac{1}{2B}B_{02} - \frac{1}{2D}D_{02} + \frac{1}{4B^2}B_0 B_2 + \frac{1}{4D^2}D_0 D_2$$

$$+ \frac{1}{4AB}A_2 B_0 + \frac{1}{4AD}A_2 D_0 + \frac{1}{4CB}C_0 B_2 + \frac{1}{4CD}C_0 D_2$$

$$R_{02} =$$

$$R_{03} = -\frac{1}{2B}B_{03} - \frac{1}{2C}C_{03} + \frac{1}{4B^2}B_0 B_3 + \frac{1}{4C^2}C_0 C_3$$

$$+ \frac{1}{4AB}A_3 B_0 + \frac{1}{4AC}A_3 C_0 + \frac{1}{4DB}D_0 B_3 + \frac{1}{4DC}D_0 C_3$$

$$R_{03} =$$

$$R_{12} = -\frac{1}{2A}A_{12} - \frac{1}{2D}D_{12} + \frac{1}{4A^2}A_1 A_2 + \frac{1}{4D^2}D_1 D_2$$

$$+ \frac{1}{4BA}B_2 A_1 + \frac{1}{4BD}B_2 D_1 + \frac{1}{4CA}C_1 A_2 + \frac{1}{4CD}C_1 D_2$$

$$R_{12} =$$

$$R_{13} = -\frac{1}{2A}A_{13} - \frac{1}{2C}C_{13} + \frac{1}{4A^2}A_1 A_3 + \frac{1}{4C^2}C_1 C_3$$

$$+ \frac{1}{4BA}B_3 A_1 + \frac{1}{4BC}B_3 C_1 + \frac{1}{4DA}D_1 A_3 + \frac{1}{4DC}D_1 C_3$$

$$R_{13} =$$

$$R_{23} = -\frac{1}{2A}A_{23} - \frac{1}{2B}B_{23} + \frac{1}{4A^2}A_2 A_3 + \frac{1}{4B^2}B_2 B_3$$

$$+ \frac{1}{4CA}C_3 A_2 + \frac{1}{4CB}C_3 B_2 + \frac{1}{4DA}D_2 A_3 + \frac{1}{4DB}D_2 B_3$$

$$R_{23} =$$

INDEX

2dF Galaxy Redshift Survey, 283

A

Absolute derivative, 217, 249
Absolute divergence, 245, 306
Absolute gradient
    of a covector, 201, 204
    defined, 201
    of the metric, 203, 209, 210, 242, 245, 249, 278
    of a tensor, 200, 201, 227, 242
    of a vector, 201, 204
Absolute temperature, 198, 307, 328, 333, 461
Absolute zero, 192
Abstract-index notation, 44, 45, 47, 51, 75, 90
Accelerated expansion of the universe, 283
Accretion disk, 118, 119, 421, 436, 453
Adler, R. J., 326
Alcock, C., 159
Algol (close binary), 406
Ampere-Maxwell relation, 46, 49, 81
Amplitude matrix (for a gravitational wave), 362, 364, 365, 370, 373, 386, 391, 395
Ancient astronomers, 284
Angular momentum (of a star). *See* Spin angular momentum
Angular momentum per unit mass (of a particle), 116–122, 127–128, 184, 428, 450
Antimatter, 42, 236
Antiparticle(s), 191, 192, 329, 338
Arbitrary coordinates, 53–55, 57, 67, 68, 75, 90, 116, 170, 203, 210, 214, 223, 232, 234, 235, 242, 251, 254–256, 261, 262, 271, 272, 416, 463
Aristarchus, 284, 285

B

Background temperature of the universe, 192, 193
Barrow, J. D., 326
Baryon(s), 283, 290, 325, 330, 331, 337, 340
Barstow, M., 114
Basis vector, 54, 55, 58, 60, 64, 145, 146, 150, 152, 188, 200–202, 204, 214, 217, 230, 242, 266, 276, 448
Bending of light, xvi, 11, 153, 156
Bessell, Friedrich, 280
Beta Persei. *See* Algol
Bianchi identity, 223, 227, 245, 249, 254
Big Bang, 40, 198, 269, 278, 282, 294, 315, 316, 318, 325, 328, 329, 333, 335, 340, 342, 343
Binary coalescence event(s), 384
Binary star system, 141, 367, 382, 384, 385, 398–400, 404, 406

Binding energy, 3, 278
Binomial approximation, 9, 10, 102, 109, 110, 112, 113, 132, 141, 164–166, 192, 194, 263, 287, 291, 319, 372, 418, 421, 425
Binomial expansion, 37
Birkhoff's theorem, 268, 269, 276
Black hole
    area, 119, 190, 192, 193, 197, 439, 445, 452, 453, 460
    entropy, 190, 193, 198, 452, 453, 460
    ergoregion, 437–440, 450, 451, 453, 456, 460
    event horizon, 167, 168, 171, 175–178, 180–183, 185, 188, 190–192, 194, 197, 198, 247, 276, 384, 438–441, 444, 445, 447, 448, 451–455, 457, 458, 460, 461
    extreme Kerr, 426, 436, 439, 447, 448, 450, 451, 460, 461
    Hawking radiation, 190–193, 453
    Kerr, 418, 426, 428, 436, 438, 439, 447, 448, 450–453, 460, 461
    lifetime, 192, 197
    mass, 198
    Schwarzschild, 118, 127, 128, 132, 147, 152, 169, 180, 188, 191, 210, 276, 277, 438, 448, 451–453, 460, 461
    thermodynamics, xvi, 189, 190, 192, 193, 197, 452
    time to reach $r = 0$, 169, 174, 447
Black hole(s), xvi, xviii, 108, 118, 119, 127, 128, 130, 132, 147, 152, 157, 168–170, 175, 176, 178, 180, 182, 183, 188–193, 196–198, 210, 230, 276, 277, 283, 367, 384, 395, 408, 418–421, 425, 426, 428, 430, 436, 438, 439, 441, 447, 448, 450–453, 458–461
Black Holes: The Membrane Paradigm (Thorne, Price, and Macdonald), 453, 461
Blackbody, 40, 41, 119, 190, 192, 193, 197, 198, 282, 291, 331, 333, 335
Blandford-Znajek mechanism, 453, 461
Boltzmann factor, 330
Boltzmann's constant, 192, 198, 329
Bound index, indices, 46, 47, 50, 202, 204, 215, 258
Boyer-Lindquist coordinates (for Kerr spacetime), 420
Brightening effect (for gravitational lensing), 157, 163

C

Cartesian coordinates, 54, 55, 60, 63, 64, 67, 69–71, 75, 76, 78, 80, 82, 88, 96, 101, 102, 104, 145, 147, 177, 200, 210, 220, 242, 245, 255, 276, 352–354, 358, 362, 366, 371, 376, 388
Cepheid variables, 280, 281, 286, 290
Christoffel symbol(s), 200–203, 205–207, 209, 210, 214, 215, 217–220, 222–224, 227–230, 234, 251, 259, 267, 269, 275–278, 306, 311, 371, 374, 378, 416, 463
Christoffel symbols, trick for calculating, 206, 207

Circular orbit, 117–119, 122–125, 127, 128, 130, 131, 138, 140–142, 144, 147, 152, 289, 414, 430, 434, 436
Circularly polarized (gravitational wave), 373, 399
Circumferential radial coordinate, 107, 117, 132, 142, 180, 263, 266, 294, 303
Clock lattice, 14, 54, 145
Clock synchronization, 15, 16
Closed universe, 297, 303
Cluster (galactic). *See* Galactic cluster
Cluster (stellar). *See* Stellar cluster
Colless, Matthew, 283
Computer model(s), 118, 134, 138–140, 142, 325, 338, 384, 447, 448
Computer models of orbits, 134
Conformal time, 318, 321
Conservation
    of angular momentum, 116, 168, 376, 428
    of charge, 81, 84, 85, 234
    of energy, 35, 116, 121, 144, 168, 234, 236, 242, 254, 278, 306, 309, 314, 334, 350, 362, 376, 385, 388, 428, 450, 451, 454
    of four-momentum, 34, 38–40, 42, 234, 239, 244, 258
    of kinetic energy, 35
    law, 234, 235, 269
    of Newtonian energy, 117, 122
    of Newtonian momentum, 34, 38, 39
Constant vector field, 200
Contraction (of a tensor's indices), 18, 27, 69, 73, 75, 223, 246, 344
Contravariant vectors, 66
Coordinate basis, 54–58, 60, 63, 64, 67, 114, 145, 210, 227
Coordinate independence, 54, 67, 68, 200, 210, 223, 254, 376, 385
Coordinate pathology, 169
Coordinate transformation(s), 56, 57, 69, 75, 96, 190, 207, 210, 223, 254, 263, 267, 353, 354, 440, 446
Copernicus, Nicolaus, 284
Cosmic
    Censorship Hypothesis, 247, 441
    flatness, 340, 342–344, 347
    horizon problem, 341, 343, 349
    metric, 318
    microwave background (CMB), 40, 41, 282, 283, 289, 291, 294, 316, 320, 325, 326, 328, 331, 336–342, 345, 346, 349
    neutrino background, 333, 336–338
    redshift, 281, 283, 287, 291, 319–322
    strings, 269, 278
    time coordinate, 307, 308, 316, 318, 326, 342, 349
    time parameter, 316, 321, 349
Cosmological constant, 242, 246, 247, 251, 267, 269, 276, 283, 295, 307–309, 315, 316, 325, 326, 328, 331, 342, 343, 347, 350
Coulomb's law, 2, 11, 78, 232
Covariant
    equation, 78, 79
    vector, 66
Covector, 66–70, 75, 76, 80, 200, 201, 204, 230, 242, 256, 261, 362, 364
Critical angle, 146, 147, 151, 152

Critical density (universe), 309, 316, 318, 330, 331, 337
Crommelin, Andrew, 155
Cross polarization (of gravitational wave), 366, 398
Current density, 49, 79, 84, 256, 257, 263, 331, 409
Curvature energy (mass) density, 256, 257, 263
Curvature parameter (for universe), 309, 342
Curvature scalar, 223, 229, 230, 244, 249, 352, 355, 378, 379
Curvilinear coordinates, 54, 57, 64, 92, 200
Cylindrical coordinates, 142, 278, 447

D

Dark energy, 283, 325
Dark matter, 283, 289–291, 307, 325, 331, 338, 340
Darmour, T., 141
De Sitter universe, 315
Decay of free neutrons, 330
Deflection of light, 10, 153, 161
Density/Scale relationship, 312
Deruelle, N., 141
Deuterium, 330
Diagonal metric, 170, 182, 263, 267, 269, 271, 272, 275–278, 295, 298, 303, 306, 311, 374, 378, 379, 382, 416, 420, 438, 440, 463
Diagonal Metric worksheet, 267, 269, 271, 272, 275–278, 295, 298, 303, 306, 311, 374, 378, 379, 382, 416, 420, 463
Diagonally polarized (gravitational waves), 366, 378
Differential form (of Maxwell's equations), 49, 78, 82
Dimensional analysis, 191, 326, 335
Dipole, 410
Divergence theorem, 385, 389
Dominant Energy Condition (DEC), 242, 251
Doppler shift, 9, 141, 282, 287, 289, 341, 382, 400
Double quasar. *See* Q0957+561
Dot product (of four-vectors), 33, 36, 45, 82, 128, 147, 188, 191, 195, 201, 456
Dragging of inertial frames, 430
Dual vectors, 66
Dumbbell (as a gravitational wave source), 395, 396, 398, 399, 402, 403
Dummy index, 46
Dust (as a type of fluid), 232–235, 237, 255, 283, 289, 307
Dyson, Frank, 155

E

Eclipse expedition, 10, 155
Eddington, Arthur, 155
Effective potential, 117, 125, 130, 144, 145, 149, 309, 429, 430, 450, 451, 460
Effective stress-energy pseudotensor, 307, 376, 377, 382
Efstathiou, G., 277, 278, 344, 350
Einstein
    equation, 8, 11, 106, 200, 222, 223, 243–247, 250, 251, 254–257, 263, 266–268, 276–278, 280, 294, 296, 297, 301, 306–308, 310, 312, 315, 352–355, 357, 361, 362, 364, 368, 374, 376, 377, 384, 408, 409, 419, 420
    ring, 156, 157, 162, 164, 166
    summation convention, 44
    velocity transformation, 18, 28, 34, 38, 42
Einstein, Albert, 18, 28, 34, 35, 38, 42, 44, 80, 130, 136, 212, 213

Electric potential, 81, 213, 353, 384, 408
Electromagnetic
   analogy, 353, 409, 416
   energy, 382, 400, 453
   field equations (see Maxwell's equations)
   field tensor, 46, 47, 51, 52, 68, 75, 76, 79–81, 86, 88, 241, 276, 353
   potential, 81, 88, 213, 353, 384, 408
   stress-energy, 241, 276
   vector potential, 81, 88, 354
Electroweak interaction(s), 342
Electroweak theory, 342
Embedding diagram(s), 133, 137, 142, 297, 303, 439, 447
Empty-space Einstein equation, 251, 268, 276, 374, 420
Empty vacuum-dominated universes, 315
Energy
   at infinity, 176, 192, 450, 451
   conservation, 35, 121, 144, 234, 236, 242, 254, 278, 306, 309, 314, 334, 350, 362, 376, 385, 388, 450, 451, 454
   conservation as a geometric necessity, 254
   current density, 409
   density, 88, 232–235, 238, 241, 242, 246, 256, 257, 263, 278, 282, 283, 289, 290, 295, 306–309, 316, 326, 328, 333, 335, 338, 342, 343, 347, 348, 350, 361, 362, 367, 376, 378, 379, 382, 386, 388, 408
   flux, 233, 280, 382, 396
   in gravitational waves, 375, 382, 386, 387, 393, 395, 396, 406
   per unit mass, 116, 117, 119, 127, 174, 184, 190, 196, 198, 428, 436, 450–452, 456, 460
   stress-energy tensor, 68, 88, 231–235, 237, 238, 241, 242, 244, 247, 251, 255–257, 261, 263, 276, 278, 280, 283, 295, 296, 298, 306, 307, 311, 350, 376, 377, 382, 388, 408, 418
   vacuum, 251, 257, 263, 308, 326, 342, 343, 347
Entropy, 190, 193, 198, 337, 338, 452, 453, 460, 461
Eötvös, Loránd, 3
Equations of motion, 144, 154, 168, 194, 235, 294, 303, 316, 428, 429
Eratosthenes, 284, 285
Ergoregion, xviii, 437–440, 450, 451, 453, 456, 460
Escape trajectory, 451, 453
Euclid's axiom, 8, 212
Euler-Lagrange equation(s), 90, 94, 95
European Space Agency (ESA), 281, 367
Event (definition), 14, 177, 440
Event horizon, 167, 168, 171, 175–178, 180–183, 185, 188, 190–192, 194, 197, 198, 247, 276, 384, 438–441, 444, 445, 447, 448, 451–455, 457, 458, 460, 461
Exact gravitational wave solution, 374

F

Fabric drag model, 430
Falling-room experiment, 7
Faraday's law, 76, 88
Fermi constant (for the weak interaction), 329
Feynman, Richard, 102
First law of thermodynamics, 307, 338, 461
Flamm's paraboloid, 132–134, 142

Flat space
   inverse metric, 255
   metric, 75, 106, 202, 203, 209, 213, 216, 233, 245, 246, 255, 278, 304, 352, 355, 426
   spacetime of special relativity, 54, 75, 76, 90, 303
Flatness problem, 341–344
Fluctuations, 190, 192, 283, 289, 339–341, 344–346
Flux, 82, 233, 280, 291, 320, 378, 382, 384, 386, 387, 393, 396
Flux (of gravitational wave energy), 384, 386, 393
Four-acceleration, 52, 210
Four-current, 46, 78, 79
Four-displacement, 32, 33, 188
Four-force, 241
Four-momentum, 34, 35, 38–40, 42, 46, 49, 52, 68, 75, 79, 88, 113, 114, 128, 145–147, 191, 195, 234, 237, 239–242, 244, 258, 268, 451, 456, 460
Four-potential, 81, 88
Four-vector, 32–34, 39, 40, 44, 45, 47, 52, 57, 69, 74, 75, 78–81, 83, 86, 88, 145, 146, 150, 201, 202, 214, 215, 217, 220, 222, 230, 232, 242, 251, 362, 416, 456
Four-velocity, 32, 33, 35–37, 47, 49, 68, 75, 80, 103, 111, 113, 116, 128, 145, 150, 152, 176, 191, 195, 202, 210, 214, 222, 233–235, 237, 239, 241, 242, 255–257, 261, 365, 371, 408, 416, 448, 456
Frame-independence
   of the scalar product, 35, 36
   of the spacetime interval, 26, 47
Free index (indices), 46, 47, 50, 51, 66, 215, 360, 391
Free object, 4, 5, 11, 14, 19, 50, 90, 92, 93, 104, 120, 251, 303, 365, 430, 451
Freely falling reference frame(s), 5–7, 10–12, 15, 188, 191, 195, 203, 210, 212, 213, 216, 220, 230, 430
Friedman-Walker-Robertson metric, 303
FRW metric. See Friedman-Walker-Robertson metric
Fundamental identity, 56, 72, 73
Fundamental rule, 47
Fusion reaction, 119

G

Galactic cluster, 157, 158, 281, 292
Galaxy, xv, 41, 118, 119, 157, 158, 164, 166, 168, 176, 246, 247, 278, 280–283, 288, 289, 291, 292, 294–296, 304, 306–308, 318–320, 325, 326, 331, 338, 341, 344, 349, 382, 384, 421, 453, 460, 461
Galileo, Galilei, 3
Gamow, G., 315, 329, 331, 336, 338
Gamow relation, 329, 331, 336, 338
Gauge
   definition, 88, 353, 354, 358–361, 364, 369, 386, 390, 406
   freedom, 351, 354, 360, 361
   Lorenz, 88, 354, 360–362, 364, 365, 368, 373, 377, 384, 390, 408
   transformation, 88, 353, 354, 358–361, 364, 369, 386, 390, 406
   transverse-traceless, 364–366, 369–373, 385, 386, 390, 391, 393, 395, 398, 399, 406
Gauss's law, 46, 49, 75, 76, 78, 79, 81, 82, 88, 232, 244, 289, 376

Gauss's law for the magnetic field, 76, 88
Geeky insider GR joke, 367
General Relativity (Hobson, et al.), 277, 278, 344, 350
General relativity in a nutshell, 8, 11
General spherically symmetrical metric, 267
Geodesic
    deviation, 211–215, 220, 245, 246
    equation, xvi, 8, 91, 92, 95–98, 100–104, 107, 111, 116, 120, 124, 130, 134, 141, 142, 144, 171, 177, 202, 203, 206, 214, 215, 217, 218, 228, 251, 255, 257, 262, 268, 269, 303, 365, 371, 409, 414, 428, 430, 436
    hypothesis, 3–5, 8, 11, 90, 254
    as locally straight, 11, 202
    as a path or worldline, 4, 7, 8, 11, 89, 90, 92, 96–99, 101, 104, 142, 171, 188, 203, 212, 214, 215, 222, 230, 255, 373, 428, 450, 451
Geodetic precession, 411, 412, 416
Geometric pathology, 169
Geometry
    of a flat space, 276
    of a sphere, 106, 171, 303, 439
    of spacetime, 8, 190, 266, 420
    of the spatial part (of a spacetime), 309
    of the universe, 64, 280, 297, 309
Global rain coordinates, 180–183, 185, 190, 446, 448
Global rain metric, 181, 185
Goobar, Ariel, 338
GR unit system, 15, 20, 29, 42, 46, 48, 49, 76, 81, 88, 102, 108, 109, 141, 152, 166, 178, 192, 197, 198, 235, 241, 281, 287, 320, 333, 334, 348, 382, 406, 409, 410, 416, 436, 461
Gradient (simple), 66, 68–70, 76, 78, 81
Grand Unified Theory (GUT), 342–344, 348, 349
Gravitation (Misner, Thorne, and Wheeler), 4, 12, 254, 406
Gravitation and Spacetime (Ohanian and Ruffini), 198, 400
Gravitational
    blue-shift/redshift, 6, 11, 109, 112, 113, 291
    energy, 376, 377
    field vector, 212, 232
    lensing, 156–158, 162, 164, 166, 283, 289
    mass, 2, 3, 5, 11, 278, 289, 450
Gravitational wave
    amplitude(s), 367, 370, 373, 376, 382, 384, 406
    dangerous, 367
    detector(s), 367, 382, 384, 400
    exact solution, 374
    polarizations, 365, 366, 373, 398, 399, 406
    wave energy, xvii, 375, 382, 386, 387, 393, 395, 396, 406
Gravitational waves, xv, xviii, 76, 352, 354, 362–367, 371–379, 382–388, 391, 393, 395–401, 406, 421, 441
Gravitoelectric energy density, 256, 257, 263
Gravitoelectric potential, 257
Gravitomagnetic current density, 256, 257, 263
Gravitomagnetism, xvii, 256, 257, 263, 264, 283, 407–411, 413–416, 429, 430
Gravity
    acceleration of, 9, 102, 251
    cannot exist in two or three dimensional spacetimes, 247, 251
    as fictitious, 5, 6, 11
    force of, 6, 134
    Maxwell-like equations for, 409
    newtonian, 106, 108, 118, 128, 255
    quantum theory of, 190, 193, 344
    reality of, 6, 11
    relativistic theory of, 76
    source of, 232, 236, 241
    tidal effects of, 7, 10, 11, 15, 212, 213
Gravity (Hartle), 40, 127, 152, 188, 349, 453, 461
Gravity Probe B, 411, 412
Great circle, 63, 99, 102, 103
Gyroscope(s), 410–412, 415, 416
GZK (Greisen, Zatsepin and Kuzmin) cutoff / paradox, 40, 41

H

H-R (Hertzsprung-Russell) diagram, 286
Halley, Edmond, 285
Harmonic oscillator equation, 131, 154, 316
Hartle, James, 40, 127, 152, 188, 349, 453, 461
Hawking, Stephen, 190, 192, 278, 453
Hawking radiation, 190–193, 453
Headlight effect, 30
Heat engine, 198
Helliwell, T. M., xx, 278, 376
Higgs boson / fields, 342–344, 350
High-Z Supernova Search Team, 281
Hipparcos satellite, 280, 286, 290
Hobson, M. P., 277, 278, 344, 350
Homogeneous (universe), 282, 291, 294, 296, 303, 304, 306, 307, 315, 336
Homopolar generator, 461
Horizon problem, 341, 343, 349
Hubble constant, 281, 282, 288, 308, 309, 315, 316, 319, 322, 333
Hubble relation, 282, 288, 320, 322
Hubble Space Telescope (HST), 158, 281, 290
Hulse, Russell, 141, 400

I

Ideal gas, 235, 241, 295, 340
Impact parameter, 144, 147, 148, 155, 156
Index notation, 43, 47, 49–52, 54, 56, 209, 213, 216, 389
Inertial mass, 2, 3, 5, 11
Inertial reference frame(s), 5–7, 10–12, 14–19, 21, 25–30, 32–36, 38, 42, 52, 54, 57, 68, 69, 75, 76, 78, 79, 83, 88, 93, 113, 175, 177, 202, 203, 210, 212, 220, 222, 224, 232–234, 236, 248, 255, 350, 376, 382, 416, 430
Inertial worldline, 90
Infinite-redshift surface, 438–440
Inflation (of universe), 329, 339, 342–344, 347–350
Innermost stable circular orbit (ISCO), 118, 119, 125, 126, 430, 431, 435, 436
Invariance of charge, 232
Invariant magnitude of the four-velocity, 36
Invariant scalar, 68
Inverse flat-space metric, 355
Inverse Lorentz transformation, 16, 22, 23, 44, 45, 79
Inverse metric, 67, 68, 72, 73, 202, 216, 229, 255, 258, 261, 352
Inverse transformation, 18, 47, 59, 60

Iota Boötis (close binary), 382, 400, 405
Irreducible mass, 452, 453, 458
Isotropic universe, 282, 291, 294–296, 304, 306, 315

J

Jefferson Physical Laboratory (Harvard), 6
Joke, geeky, 367

K

Kepler's third law, 118, 124, 406, 430, 434
Kerr
    black hole, 421, 426, 436, 438, 439, 447, 448, 450–453, 460, 461
    Boyer-Lindquist coordinates, 420
    geometry, 420, 421, 428, 429
    metric, 417–420, 425, 426, 428, 432, 440, 441, 444, 446
    spacetime, 427–430, 433–436, 438, 439, 441, 447, 448, 460
    spin parameter a defined, 420
Kerr, Roy, 420
Kerr-Newman solution, 426
Klein-Gordon equation, 350
Knop, R. A., 321
Konkowski, D. A., 278
Kronecker delta, 45, 46, 48, 68, 72, 73, 216
Kruskal-Szekeres coordinates, 180, 182, 183, 186–188, 190

L

La Silla Observatory, 290
Lagrangian, 90, 91, 102, 114, 350
Laplacian, 251
Lasenby, A. N., 277, 278, 344, 350
Laser Interferometer Gravitational-wave Observatory (LIGO), 367, 382, 384, 400
Laser Interferometer Space Antenna (LISA), 367
Left-hand rule (for gravitomagnetism), 429, 430
Lemaître, Georges, 281, 315
Lemaître universe(s), 315
Lens equation, 162
Lense-Thirring effect, 410–412, 416
Leptons, 283, 290, 329, 333
Lewiston, Maine, 290
Lifetime of a black hole, 192, 197
Light clock, 22
Light cone, 17, 32, 176
Lightlike, 17, 30, 92, 170, 171, 242
Linear approximation. *See* Weak-field limit
Linear transformation law, 18, 32, 38, 39, 445, 446
Linearity of the Lorentz transformation, 38
Linearized field equation. *See* Weak-field Einstein equation
Local acceleration of gravity, 9, 251
Local conservation of energy, 254, 306, 334
Local field equation, 232
Local flatness theorem, 207–209
Locally inertial frame (LIF), 202, 203, 205, 207, 209, 210, 214, 220, 222–225, 227, 232–239, 241, 242, 244, 248, 251, 255, 277, 350, 376, 388, 416
Lorentz contraction, 18, 27, 78, 233, 304
Lorentz transformation (LTE), 16, 18, 21–28, 30, 32, 34, 38, 42, 44–47, 51, 52, 57, 62, 67, 69, 79, 88, 250

Lorenz gauge, 88, 354, 360–362, 364, 365, 368, 373, 377, 384, 390, 408, 412
Lowering indices, 66, 67, 69, 71, 74, 80
Luminosity distance, 320, 321, 323, 325
Lunar eclipse, 284

M

Macdonald, Alan, 21
Macdonald, D. A., 453, 461
MACHO(s), 157, 159, 164, 165
Magnetic moment, 410, 414
Magnetic potential, 86
Magnitudes (for stars)
    absolute, 290
    observed, 290
Manifestly covariant, 68, 78, 79
Mars, 29, 141
Mass
    density, 76, 232, 241, 244, 245, 247, 257, 269, 315, 326, 385, 416
    gravitational, 2, 3, 5, 11, 278, 289, 450
    inertial, 2, 3, 5, 11
Mass spectrometer, 3
Mass-energy, 11, 35, 119, 190, 191, 197, 245, 247, 269, 338, 362, 385, 388, 452, 460, 461
Massless (particles), 114, 168, 325, 342, 343, 395, 396
Master clock, 16, 108, 170
Matrix
    equation, 16, 25, 45, 46
    form, 44, 76
    notation, 38, 44, 391
    product, 445
    trace, 75, 329, 356, 373, 385, 386, 392
Matter number density, 337
Matter-dominated (universe), 309, 314, 328, 331–333, 340, 342, 345, 349
Maximum angle of light deflection, 161
Maxwell's equations, 15, 77, 78, 81, 87, 88, 200, 203, 409, 413
Mean free path, 41
Metric
    absolute gradient, 203, 209, 210, 242, 245, 249, 278
    component, 463, 63, 64, 96, 116, 121, 133, 147, 150, 169, 170, 181, 182, 228–230, 266, 267, 362, 428, 430, 432, 436, 440–444, 448
    determinant, 445, 446
    equation, 17, 18, 29, 30, 45, 58, 60, 63, 90, 101, 106–109, 114, 116, 128, 142, 144, 176, 177, 185, 186, 210, 270, 362, 372, 436
    flat space, 75, 202, 209, 213, 216, 233, 246, 255, 296, 352, 355, 426
    inverse, 67, 68, 72, 75, 111, 202, 216, 229, 255, 258, 261, 352
    Kerr, 417, 418, 420, 425, 426, 428, 432, 440, 441, 444, 446
    perturbation, 255, 257–259, 261–263, 352, 353, 355, 358, 361, 362, 365, 368, 377, 384–386, 388, 390, 398, 406, 408, 409, 412, 418
    Schwarzschild, 105–109, 114, 116, 121, 124, 128, 132, 137, 141, 142, 144, 147, 150, 168–170, 172, 176, 181, 182, 185, 186, 206, 257, 263, 266, 268, 276, 419, 426

Metric (*continued*)
   tensor, 45, 55–58, 60, 66–68, 75, 106, 114, 116, 121, 145, 228, 242, 254, 353
   transformation law, 59, 62
   universe, 294, 295, 303, 315, 340
Mihos, Chris, 289
Milne universe, 303, 304
Misner, Thorne, and Wheeler, *Gravitation*, 4, 12, 254, 406
Model(s) of the universe, 246, 282, 283, 290, 303, 315, 321, 325, 331, 341
Moment of inertia, 92, 385, 386, 396, 410, 411, 418
Moon, 7, 212, 281, 284, 285
Mössbauer effect, 6
Mount Wilson, 280
Mu Scorpii (close binary), 382, 406

N
Naked singularity, 441, 452, 460
NASA, 158, 281, 282, 341, 367, 411
Negative energy, 190, 191, 246, 450, 451, 460
Negative-energy orbits, 190, 191, 449–451, 453, 456, 460
Neutrino
   background, 333, 336–338
   decoupling, 329, 330, 335–338
   mass, 338
   number density, 338
Neutrino(s), 42, 192, 283, 307, 325, 328–331, 333–338
Neutron(s), 3, 40, 283, 329–331, 335, 338
Neutron star(s), xv, 9, 10, 108, 114, 118, 127, 141, 157, 283, 384, 400, 401, 406, 416, 419, 425, 426, 436
New Gravitational wave Observatory (NGO), 367
New York Times, 155
Newton, Isaac, xv, 34, 35, 38, 39, 49, 106–109, 112, 114, 116–118, 122, 126–128, 130, 131, 135, 139, 141, 155, 165, 168, 176, 244–246, 251, 254–257, 268, 269, 276, 362, 385, 426, 461
Newton's first law, 15, 19, 203
Newton's law of universal gravitation, 107
Newton's second law, 2, 5, 75, 78, 213, 214, 235, 399, 409
Newton's third law, 292
Newtonian
   analysis of tidal effects, 212
   approximation, 108, 246, 404
   energy, 117, 122, 130
   equation of tidal deviation, 213, 216, 220, 245
   field equation, 246
   gravitational acceleration, 257
   gravitational potential, 75, 256, 385
   gravity, 106, 108, 118, 255
   law of continuity, 235
   limit, 232, 244–246, 251, 269
   mass, 257, 269
   mass density, 257
   mechanics, xv, 5, 114, 116, 127, 130, 139, 155, 176, 212, 220, 292, 406, 416, 461
   momentum, 34, 38, 39
   orbit, 118
   orbital equation, 135
No-hair theorem, 277, 426
Non-baryonic dark matter, 283, 338

Non-Euclidean geometry, 8, 11, 107
Noninertial reference frame, 5, 15, 16
Nonlinearities (in the Einstein equation), 376, 377
Nordström, Gunnar, 76
Nordström theory of gravity, 76
Nuclear fission, 119
Nuclear reactions, 330
Nucleosynthesis, 289, 329, 330, 347
Number density, 233, 237, 304, 329–331, 335, 337, 338
Numerical model, 138, 324

O
Observable universe, 281
Observer's
   basis vectors, 145, 146, 230, 242
   four-velocity, 35, 150, 152, 242
   frame, 15, 18, 19, 29, 30, 35, 42, 93, 102, 128, 145–147, 150, 152, 176, 178, 212, 230, 242, 287, 304, ix
Occam's Razor, 244
Ohanian, Hans, 198, 400
Olbers, H. W., 290
Olbers's paradox, 290
Open universe, 297
Oppenheimer-Volkoff equation, 278
Orbit, innermost stable circular (ISCO), 118, 119, 125, 126, 430, 431, 436
Orbital
   eccentricity, 400
   energy, 399, 401
   equation, 135, 142, 154
   frequency, 399
   period, 124, 127, 141, 400, 401, 404, 405, 416, 436
   plane, 133, 399, 406
   radius, 136, 285
   speed, 164, 289, 430
Orthonormal basis, 145, 150–152, 202–203, 230, 242

P
Parabolic coordinates, 60, 61, 96–98
Paraboloid, 64, 104, 132–134, 142
Parallax, 280, 281, 286, 290
Particle-antiparticle pair(s), 190, 192, 195
Particle-in-a-box, 326
Particle physics, 283, 290
Path of shortest distance, 90
Pathlength, 3, 92, 96, 97, 99, 102, 103
Pathology (geometric vs. metric), 169
Pauli exclusion principle, 333
Penrose, Roger, 247, 441, 450
Penrose process, 450–453
Perfect fluid, 235, 237, 238, 241, 242, 251, 256, 257, 261, 277, 408
Periastron shift, 141, 400, 401
Perigee, 141
Perihelion shift (of Mercury), xvi, 129, 130, 132–134, 136, 141, 142
Period of orbit (see orbital period)
Perryman, Michael, 286
Perturbation approach, 131, 154
Perturbed worldline, 94

Phase transition, 342
Photon
   decoupling, 329, 331, 337, 338, 340
   equations of motion, 154
   four-momentum, 35, 40
   gas, 283, 307, 334, 336, 337, 340
   number density, 337
   orbits, xvi, 143, 154, 436
   torpedoes, 188
   worldline, 35, 92, 128, 144, 154, 182, 183, 188
Physical meaning of T components. See Stress-energy
Pion(s), 40, 42
Planck mass density, 326
Planck's constant, 113, 152, 192
Plane-symmetric spacetime, 276
Plane-wave
   metric perturbation, 362
   solution, 364, 368
Plus-polarized (gravitational wave), 373, 374, 377, 382, 399
Point-like objects (reduced quadrupole moment), 118, 385
Poisson equation, 75
Polar coordinate basis, 58
Polar coordinates, 58, 63, 64, 70, 71, 75, 103, 132, 137, 152, 210, 220, 419, 424
Polar orbit, 411
Polarization amplitudes (for gravitational wave), 365, 398, 399
Positrons, 329, 330, 334, 336, 337
Potential
   barrier, 118
   electromagnetic, 81, 86, 88
   energy, 117, 119, 125, 130, 144, 145, 149, 191, 213, 309, 314, 315, 344, 350, 429, 430, 451
   energy graph(s), 117, 119, 144, 309, 314, 315, 451
   magnetic, 86, 354
Precession of the perihelion. See Perihelion shift
Pressure, 234, 235, 238, 241, 242, 255–257, 261, 263, 269, 277, 295, 306–308, 338, 340, 361, 362, 408
Price, R., 453, 461
Primordial nucleosynthesis, 330
Princeton University, 158
Principle of Relativity, 15, 16, 21, 34, 38, 39, 54, 68, 78, 304
Project Icarus, 166
Proper time, 18, 27, 32, 33, 35, 36, 42, 69, 75, 76, 79, 90, 93–96, 102, 104, 107–109, 114, 127, 134, 138, 140, 144, 169–171, 174–177, 184, 191, 194, 202, 214, 222, 241, 245, 287, 304, 436, 438, 447
Proton cutoff energy. See GZK cutoff
PSR B1913+16, 141, 401
PSR J0737-3039, 406
Ptolemy, 284
Pulsar(s), xv, 141, 400, 401, 406, 416, 426
Pythagorean theorem, 17, 22, 55, 63, 67, 165

Q

Q0957+561, 164, 166
Quadrupole moment tensor, 385, 386, 393, 396, 398, 402, 403, 406
Quantum
   field theory, , xv, 190, 191, 326, 333, 350
   fluctuation, 190–192, 344
   harmonic oscillator, 326
   mechanics, 190, 326, 335
   statistical mechanics, 335
   theories of gravity, 190, 193
Quark(s), 283, 290, 329, 333
Quasars, xv, 421
Quasi-cartesian coordinates, 255–257

R

Radial-Longitude coordinates, 64
Radially falling observer, 152
Radiation-dominated universe, 315, 331, 333, 335, 342, 343
Radiation era, 328, 329, 333, 340
Raising indices, 69, 81, 83, 86, 298, 360, 391
Reaction cross section, 329
Rebka, G. A., 6
Redshift, 6, 109, 112–114, 168, 178, 281–283, 291, 319–322, 325, 326, 349, 438, 439
Redshift formula, 109, 168
Redshift z, 291, 319–322, 325, 326, 349
Reduced quadrupole moment tensor, 385, 386, 393, 396, 398, 402, 403, 406
Reissner-Nordström metric / solution, 269, 276, 277
Relative acceleration, 7, 8, 212–214, 220, 222
Relativistic concepts
   Doppler, 9, 287
   energy, 34, 35, 52, 116, 117, 119, 146, 198, 232, 241, 428, 450
   energy per unit mass, 116, 117, 428, 450
   fluid dynamics, 235, 239, 240
   gas, 337, 338
   generalization, 34, 78, 80, 81, 232, 244
   limit, 88, 333
   momentum, 34, 88, 233
   quantum field theory, 190, 333, 350
   scalar, 76, 78, 223, 232
   theory of gravitation, 352
   units, 15
Repulsive gravitational field, 246
Reservoir (thermal), 193, 198
Rest-mass density, 269
Ricci tensor, 68, 223, 228–230, 244, 248, 249, 251, 255, 256, 260, 267, 271, 278, 295, 298, 352, 355, 361, 378, 379, 382, 420, 463
Riemann tensor, 68, 209, 212, 215, 218–230, 232, 244–249, 251, 254, 255, 259, 260, 276, 278, 352, 353, 355, 359, 362, 374, 378
   contractions of, 223, 244
   mnemonic, 215
Right-hand rule, 415, 430
Robot observer, 180–182, 184, 188, 436
Rotating rod (as gravitational wave source), 396
Rotation curve (for galaxy), 289, 291
Ruffini, Remo, 198, 400
Runge-Kutta (numerical integration technique), 447
Russell, Henry, 286

S

Saddle-like spatial geometry, 297, 309, 315, 316, 325
Scalar product, 33, 35, 36, 40, 41, 55, 67, 69, 204, 205

Scale factor (in universal metric), 295, 306, 308, 309, 316, 318, 319, 321, 331, 342
Schwarzschild
    alternative coordinates, 179–183, 185, 186, 188, 190, 446, 448
    Christoffel symbols, 207, 210, 220
    circumferential radius, 107, 128, 134, 426
    coordinates (defined), 106, 108
    embedding diagram, 137
    equations of motion, 117, 121, 144
    global-rain coordinates, 180–183, 185, 190, 446, 448
    Kruskal-Szekeres coordinates, 180, 182, 183, 186–188, 190
    mass parameter, 257, 269
    metric, 105–109, 114, 116, 121, 124, 128, 132, 137, 141, 142, 144, 147, 150, 168–170, 172, 176, 181, 182, 185, 186, 206, 257, 263, 266, 268, 276, 419, 426
    orbits, 120, 138–140
    solution, 108, 265, 268, 269, 276, 277, 419, 420
    time, 108, 144, 145, 150, 180, 181, 184, 188, 190
Second law of thermodynamics, 190, 193, 198, 460
Second-order-accurate (difference equations), 138, 139
Self-consistency of the theory, 441
Semilog coordinates, 64, 210, 230
Separation
    four-vector, 214, 217, 222
    vector (Newtonian), 213
Shapiro, Irwin, 165, 166
Shapiro delay, 165, 166
Shaw, D. J., 326
SI units, 15, 20, 35, 48, 196, 326, 329, 333–335, 408, 416, 461
Sign convention, 222, 223
Single-component universes, 328, 332
Singularity, 190, 230, 247, 276, 277, 316, 344, 441, 447, 452, 460
Sinusoidal coordinates, 64
Sirius, 114
Slow-source approximation, 408
Small-angle approximation, 133, 155, 157, 162
Small-source approximation, 384
Small-weak-slow source approximation, 384
Smartphone (as accelerometer), 10
Spacelike
    displacements, 171
    interval, 17
    vector, 456
Spacetime
    diagram, 9, 17, 22, 29, 30, 32, 93, 109, 176, 177, 180, 182, 183, 188, 212, 287, 318
    geometry, 14, 426
    interval, 17, 26, 27, 30, 33, 47, 51, 106, 295, 419
    interval squared, 17, 27, 33
Spacetime Physics (Taylor and Wheeler), 18, 21
Spatial curvature, 132, 133, 297, 308, 309, 314, 316
Spatial geometry, 64, 296, 297, 309, 315, 316, 320, 325, 331
Spatial index or indices, 70, 233, 234, 236, 246, 257, 262, 311, 365, 378, 385, 408
Special relativity, xv, xviii, 13, 18, 33, 34, 46, 54, 57, 67, 75, 76, 90, 93, 114, 145, 234, 303, 320, 382
Spherical charge distribution, 232, 418
Spherical coordinates, 103, 106, 230, 296, 303, 394, 419

Spherical symmetry, 57, 106, 114, 120, 158, 180, 182, 190, 251, 257, 266–268, 270, 276, 277, 282, 289, 291, 294, 297, 361, 410, 418–420, 429
Spherically symmetric metric, 267, 276, 294
Spin angular momentum, 263. 410, 415, 418–421, 436, 450, 452, 458, 460–461
Spin energy contribution, 459
Spin per unit mass (Kerr parameter a), 410, 419–420, 436, 452, 458, 460
Stable circular orbit(s), 117–119, 125, 127, 430, 435
Stable equilibrium, 193, 198
Standard candle(s), 280, 281
Static limit, 81, 438, 439
Static universe, 246, 315
Stationary weak-field approximation, 263
Stationary-source limit, 256
Stefan-Boltzmann relation, 119, 192, 197, 307, 333
Stellar cluster, 280, 286
Stress-energy
    dust, 232–235, 255
    gravitational field, 377
    electromagnetic, 241, 276
    perfect fluid, 235, 237, 238, 241, 256, 257, 261, 408
    physical meaning of components, 233, 234
    scalar, 296
    subsets, 235, 237
    tensor, 68, 72, 231–235, 237, 238, 241, 242, 244, 247, 251, 255, 256, 261, 263, 276, 295, 296, 298, 306, 307, 311, 350, 376, 377, 382, 388, 408, 418
    tensor components, 233, 408
String theory, 193, 198
Strong nuclear interaction, 342
Subscripted indices, 45, 66, 67, 69, 244
Subtle Is the Lord (Abraham Pais), 130, 136
Summation convention, 44
Supernovae, 118, 281, 283, 320, 321, 382
Superposition principle, 78
Superscripted indices, 32, 44–46, 56, 66–69, 202, 206, 207
Switching indices, 80
Symmetries of the Riemann tensor, 222, 225, 228, 230, 246, 249
Symmetry argument, 120, 395, 428
Symmetry of the Christoffel symbol, 201, 205, 209

T

Taylor, Edwin, 18, 21, 180, 400, 401
Taylor series, 37, 94, 138–140, 207, 213, 214, 319
Temperature
    accretion disk, 119
    black hole, 190, 192, 193, 197, 198, 461
    cosmic microwave background (CMB), 326
    neutrino decoupling, 330
    photon gas, 307, 336
Tensor
    equation(s), 65, 68, 69, 76, 78, 88, 111, 200, 203, 205, 214, 222, 225, 234, 235, 392
    field(s), 200, 352, 353, 377, 408
    formalism, 78
    generalization, 75, 79, 88, 232, 235, 244
    operations, 69, 73, 74

product, 69, 73, 75, 242
quantity, 68, 69, 74, 200, 210, 222, 227, 232, 235, 241, 244, 376
sum, 69, 251
tensor gradient (= absolute gradient), 200
transformation, 68, 73, 76, 88, 146, 353, 358, 376
Testing Newton's first law, 15
Thermal distribution(s), 329, 333
Thermodynamics
black-hole, 190, 192, 193, 197, 452
first law, 307, 338, 461
second law, 190, 193, 198, 460
Thin lens approximation, 157
Thorne, Kip, 4, 12, 254, 406, 453, 461
Threshold reaction, 40
Tidal effects (of gravitation), 7, 10, 11, 15, 212
Tidal forces, 230
Time evolution of the universe, 314, 316
Time of
inflation, 342
neutrino decoupling, 335, 338
photon decoupling, 331, 337, 338, 340
radiation-matter equality, 343, 347
recombination, 282, 331
Time-temperature relation, 328, 333, 334
Timelike
displacement, 181
geodesic(s), 90
separation, 90
worldline, 92, 175, 181, 185, 190, 191, 438
Topology, 297, 303
Torque, 410, 415
Trace of (matrix or tensor), 75, 356, 373, 386, 392
Trace-reverse(d) (metric perturbation), 352–354, 356, 360, 362, 365, 368, 377, 385, 386
Traceless (matrix or tensor), 369, 373, 386, 387, 392, 393
Transformation
basics (gauge), 353
coordinates, 56, 57, 69, 75, 96, 190, 207, 210, 223, 254, 263, 267, 353, 354, 440, 446
equation, 16, 24, 26, 34, 67, 353, 445
law, 16, 32, 34, 56, 57, 59, 62, 67, 68, 72, 73, 186, 200
Lorentz, 16, 18, 21–28, 30, 32, 34, 38, 42, 44–47, 51, 52, 57, 62, 69, 79, 88, 250, 360
matrix, 16, 45, 52
metric tensor, 56
partials, 70, 75, 76, 207, 353, 446
tensor, 56, 57, 68, 69, 73, 76, 88, 146, 200, 353, 358, 376
Transit of Venus, 285
Transition temperature, 343
Translational symmetry, 220
Transposed matrix, 445
Transverse-traceless (TT) gauge, 364–366, 369–373, 385, 386, 390, 391, 393, 395, 398, 399, 406
Trial metric, 266, 267, 271, 280, 294, 295, 306, 374, 419
Turner, D. G., 158, 290
Twin paradox, 128
Type Ia supernovae, 281, 283
Tyson, J. A., 158

U
Uniform acceleration, 9
Unit vector, 76, 82, 188, 213, 232, 386, 387, 391–393, 399, 410
Universe
critical density, 309, 316, 318, 330, 331, 337
De Sitter, 315
density, 282, 283
equilibrium stages, 329
evolution equation, 328, 332, 345
expansion, 283
flat, 251, 291, 296, 297, 309, 315, 316, 318, 320, 321, 325, 340–344, 347, 349, 350
gravitational constant, 2, 8–10, 75, 107
homogeneity, 282, 291, 294, 296, 303, 304, 306, 307, 315, 336
horizon problem, 341, 343, 349
initial conditions, 342
isotropy, 282, 291, 294–296, 304, 306, 315
Lemaître, 315
metric, 294–296, 303, 315, 325
Milne, 303, 304
model, 246, 282, 290, 303, 304, 315, 320, 321, 325, 328, 341
observed, 279
outer edge, 291
scale factor, 308, 315, 316, 328, 331, 332
spatial geometry, 309, 320, 331
stress-energy, 283, 311
visible, 318, 343

V
Vacuum solution (to Einstein equation), 418, 419
Vacuum stress-energy, 191, 242, 247, 251, 269, 276, 283, 295, 307–309, 315, 316, 325, 326, 328, 331, 342, 343, 347, 350
Vacuum-dominated (universe), 315, 316, 342
Variational principle, 90
Vector
bosons, 333, 342
calculus, 413, 414
field, 200–204, 210, 416
potential, 354
Velocity parameter, 25
Venus, 141, 165, 285
Vilenkin, A., 278
Virial theorem, 289, 292
Visible universe, 318, 343
Visualization, 14

W
Walsh, D., 164
Weak Energy Condition (WEC), 242
Weak interaction, 329, 342
Weak-field
approximation, 257, 263, 266, 352, 388
Einstein equation, 263, 353–355, 357, 361, 362, 364, 374, 384, 413
limit, 114, 254–260, 269, 308, 352–354, 358, 362, 364, 374, 377, 388, 408, 419, 425

Weak-field (*continued*)
   Ricci tensor, 361
   Riemann tensor, 260
Wedge angle, 133, 138
Weight, 5, 11
Weightless, 6
Weinberg, Steven, 130
Weisberg, Joel, 400, 401
Welch, Douglas, 159
Wheeler, John Archibald, 4, 8, 11, 12, 18, 21, 180, 254, 277, 376, 406, 426
White dwarf(s), 114, 157
Wien's law, 198
Wilkinson Microwave Anisotropy Probe (WMAP), 320, 341, 349
Wilson, Robert, 280, 282
Wobble function, 131, 133, 136, 141, 154, 160

Worldline(s)
   definition, 17
   of longest proper time, 90, 92, 93, 102, 175
   particle, 27, 93, 183
   photon, 23, 29, 35, 92, 128, 144, 182, 183

X

X-ray sources, 118, 119

Z

ZAMOs, 448
Zero-angular-momentum (particles or trajectories), 429, 448
Zero-angular-momentum observers (see ZAMOs)
Zero-gravity (path or trajectory), 154, 155
Zeroth-rank tensor (scalar), 68
Zeta Gemini (Cepheid), 290
Zwicky, Fritz, 292